STOCHASTIC PROCESSES IN CHEMICAL PHYSICS: THE MASTER EQUATION

STOCHASTIC PROCESSES IN CHEMICAL PHYSICS: THE MASTER EQUATION

Irwin Oppenheim

Kurt E. Shuler

George H. Weiss

The MIT Press
Cambridge, Massachusetts, and London, England

PUBLISHER'S NOTE

This format is intended to reduce the cost of publishing certain works in book form and to shorten the gap between editorial preparation and final publication. Detailed editing and composition have been avoided by photographing the text of this book directly from the author's typescript.

ACKNOWLEDGMENT

The authors would like to thank the following publishers for permission to reproduce material copyrighted by them:
American Institute of Physics for articles 2A, 2B, 2D, 3C, 3D, 5A, 5B, 5C, 5D, 6B, 6C, 6D, 7A, and 8A.
Institute of Mathematical Statistics for article 4A.
International Union of Pure and Applied Chemistry for article 8C.
John Wiley & Sons, Inc., for articles 1C, 4B, 6A, and 8B.
North-Holland Publishing Co. for articles 1A, 1B, and 1D.
Plenum Publishing Corp. for articles 2C and 7B.
Quarterly of Applied Mathematics for article 3A.

Printed and bound in the United States of America

Library of Congress Cataloging in Publication Data

Stochastic Processes in Chemical Physics: The Master Equation
Stochastic processes in chemical physics.

Includes bibliographies.
1. Stochastic processes--Addresses, essays, lectures.
2. Transport theory--Addresses, essays, lectures.
3. Mathematical physics--Addresses, essays, lectures.
4. Chemistry, Physical and theoretical--Addresses, essays, lectures. I. Oppenheim, Irwin. II. Shuler, Kurt Egon, 1922- III. Weiss, George Herbert, 1930-
QC20.7.S8S76 539'.01'5192 76-27843
ISBN 0-262-15017-4

CONTENTS

STOCHASTIC PROCESSES IN CHEMICAL PHYSICS: THE MASTER EQUATION

Chapter 1

INTRODUCTION

In recent years, there have been a large number of applications of stochastic techniques in various branches of physics and chemistry. Stochastic models have been used to gain insight into the equilibrium and time dependent properties of systems containing large numbers of atoms or molecules. While the ultimate justification of the use of stochastic techniques must lie in the dynamical laws governing the behavior of molecules, the initial applications of these techniques often precede their justification by many decades.

Our aim in writing this book is to introduce the student and the research worker in chemical physics to some of the stochastic tools which have proven to be of great use in treating physical and chemical systems. We have focused our attention on the *master* equation because many of the concepts of stochastic theory are included in its derivation and use and because it has been applied to many diversified physical problems. We restrict our attention to the master equation first discussed by Pauli because it is this form that has proved most useful. The Pauli equation is a linear first order differential-difference equation in discrete state space and a linear, first order differential-integral equation in continuous state space.

This book consists of two parts. The first part contains two chapters which treat the basic functions used in the description of stochastic processes and the probability theoretical (as contrasted to the "physical") derivation, properties, and solutions of the master equation. The second part consists of a collection of papers which have appeared in the chemical physics literature and which deal with the physical derivation, the properties and the applications of the master equation.

The papers that we have selected to reprint in this volume have been chosen to illustrate the application of the master equation to various areas in chemical physics. Our selection is arbitrary in the sense that it emphasizes the applications with which we are most familiar. Indeed, some important subjects such as the use of the master equation to describe phenomena in lasers and various biophysical applications are not included at all. The papers which are reprinted here are not necessarily the seminal ones, or the latest ones, nor the mathematically most profound contributions. They have been chosen primarily on the basis of providing useful and didactic examples.

The first group of papers discusses the derivation of the master equation from the fundamental dynamical equation of motion, the Liouville equation. The master equation describes the time dependence of the probability distribution of a set of dynamical variables of the system in an extremely simple form, in contrast to the Liouville equation. Thus, we would expect that there must be fairly stringent conditions on the validity of the master equation. These conditions are discussed and developed in this set of papers.

The second group of papers deals primarily with some useful properties of the master equation and in particular with the general theory of the relaxation to equilibrium. The third set of papers is concerned with the conditions under which the master equation can be approximated by the Fokker - Planck equation, which is a linear, second order partial differential equation. The succeeding group of papers discuss, in turn, the application of the master equation to first passage time problems, gas phase relaxation processes, chemical kinetics, spin relaxation processes and polymer dynamics.

There are, of course, many books available which treat probability theory and stochastic processes. The vast majority of these books, however, are written by mathematicians for mathematicians and make quite difficult reading for the chemical physicist. In addition, most of these books are not concerned with applications in the physical sciences. We have written the introductory chapters with the chemical physicist in mind and have tried to limit ourselves to simple concepts and techniques to arrive at our results. The same criterion has also governed the selections of the papers reprinted in this volume. For those interested in collateral reading we append a list of books which we have found useful in our work.

Two of us (I.O. and K.E.S.) gratefully acknowledge partial support by the National Science Foundation over a number of years for our work in stochastic processes.

ACKNOWLEDGEMENTS

We are grateful to Virginia Prescott, Nancy Crawford and Thelma Sparks for their excellent typing of this manuscript.

COLLATERAL READING

Bartlett, M. S., An Introduction to Stochastic Processes; Cambridge University Press, Cambridge (1960).

Bharucha-Reid, A. T., Elements of the Theory of Markov Processes and Their Applications; McGraw-Hill Book Co., Inc., New York (1960).

Blanc-Lapierre, A. and Fortet, R., Theory of Random Functions, Vols. I and II, Gordon and Breach Science Publishers, Inc., New York (1968).

Cox, D. R. and Miller, H. D., The Theory of Stochastic Processes; John Wiley & Sons, New York (1965).

Feller, W., Probability Theory and its Applications, Vol. I (1968) and Vol. II (1966); John Wiley & Sons, New York.

Gnedenko, B. V., The Theory of Probability; Chelsea Publishing Co., New York (1963).

Karlin, S., A First Course in Stochastic Processes; Academic Press, New York (1966).

Parzen, E., Modern Probability Theory and its Applications; John Wiley & Sons, New York (1960).

Stratonovich, R. L., Topics in the Theory of Random Noise, Vol. I; Gordon and Breach Science Publishers, Inc., New York (1963).

Wax, N., Editor, Selected Papers on Noise and Stochastic Processes; Dover Publications, New York (1954).

Chapter 2

BASIC FUNCTIONS OF STOCHASTIC PROCESSES

In this chapter we introduce the basic functions and concepts which enter into the theory of stochastic processes. In Section 1 we discuss the various probabilities, i.e., singlet, joint and conditional probabilities, and introduce the concepts of independent and Markov processes. In Section 2 we derive the Smoluchowski-Chapman-Kolmogoroff equation (SCK equation) which relates the conditional probabilities for Markov processes. In Section 3 we introduce some useful functionals of the probabilities, i.e., the moments and characteristic functions. A set of functions of great utility in the discussion of the correlation of stochastic variables is the cumulants which we consider in Section 4. In Sections 5 and 6 we treat Gaussian distributions and Gaussian Markov processes, respectively.

1) Joint and Conditional Probabilities

The definition of the various functions and functionals of random processes presented here will not be given in the most general or the most rigorous manner; rather our object is to emphasize techniques and applications which are particularly useful for physical problems. A definitive account of the measure-theoretic foundations of probability can be found in reference 2.1.1.

We are concerned with the properties of a random function $X(t)$ where the time t may be measured in discrete or continuous units. The properties of $X(t)$ can be described in terms of an infinite hierarchy of probabilities or probability densities, the former when $X(t)$ takes on a discrete set of values and the latter when $X(t)$ takes on a continuous set of values. For simplicity we assume that $X(t)$ is a scalar that takes on discrete values only. Then all of the relevant information about $X(t)$ is summarized in the hierarchy of probabilities

$\{P_r(n_1,t_1;n_2,t_2;\ldots;n_r,t_r)\}$ for $r = 1, 2, \ldots$, with

$$P_r(n_1,t_1;n_2,t_2;\ldots;n_r,t_r) = \text{prob}\ \{X(t_1) = n_1;\ X(t_2) = n_2;\ldots;X(t_r) = n_r\},\quad t_1<t_2<\ldots \tag{2.1.1}$$

where P_r is the joint probability that the stochastic variable $X(t)$ has the value n_1 at time t_1, the value of n_2 at time $t_2,\ldots$, and the value n_r at time t_r. The P_r, being probabilities, have the property $0 \le P_r \le 1$. Furthermore, the members of the hierarchy are related by

$$\sum_{n_j}\sum_{n_{j+1}}\cdots\sum_{n_r} P_r(n_1,t_1;\ldots;n_r,t_r) = P_{j-1}(n_1,t_1;\ldots;n_{j-1},t_{j-1})$$

$$\sum_{n_1} P(n_1,t_1) = 1 \qquad (2.1.2)$$

where $P(n_1,t_1) \equiv P_1(n_1,t_1)$ is the singlet probability and where the sums are over the entire set of values assumed by $X(t)$. Notice that the summations are over the "n" variables only and not over the time variables. A further elementary property of the joint probabilities P_r is

$$P_r(n_1,t;n_2,t;n_3,t_3;\ldots;n_r,t_r)$$
$$= P_{r-1}(n_1,t;n_3,t_3;\ldots;n_r t_r)\delta_{n_1 n_2} \qquad (2.1.3)$$

where $\delta_{nm} = 0$ if $n \neq m$ and $\delta_{nn} = 1$. This relation holds when two times in the argument are equal; it is a statement of the fact that the probability of the stochastic variable taking simultaneously both of the values n_1 and n_2, $n_1 \neq n_2$, at the time t is zero. There is no analogous relation when two values of the n variables are equal.

An important class of stochastic processes is formed by the so-called stationary processes. A stationary process is one for which

$$P_r(n_1,t_1;n_2,t_2;\ldots;n_r,t_r)$$
$$= P_r(n_1,t_1+\tau;n_2,t_2+\tau;\ldots;n_r,t_r+\tau) \qquad (2.1.4)$$

for all r, $\{n_j\}$, $\{t_j\}$ and τ. The symbol { } denotes a set of values. Stationarity implies that no origin in time is preferred or unique. These processes are of physical interest in the study of fluctuations and correlations in equilibrium and steady state systems.

It is convenient to introduce, in addition to the joint probabilities $\{P_r\}$, a set of conditional probabilities

$\{W_r(n_r,t_r|n_{r-1},t_{r-1};n_{r-2},t_{r-2};\ldots;n_1,t_1)\}$, where $W_r(n_r,t_r|n_{r-1},t_{r-1};\ldots;n_1t_1)$ is the probability that $X(t)$ has the value n_r at time t_r if it had the values n_1 at t_1, n_2 at $t_2,\ldots,n_{r-1}$ at t_{r-1} where $t_1<t_2\ldots<t_r$. The $\{W_r\}$ are probabilities and satisfy the conditions

$$0 \leq W_r \leq 1 \tag{2.1.5}$$

and

$$\sum_{n_r} W_r(n_r,t_r|n_{r-1},t_{r-1},\ldots,n_1,t_1) = 1.$$

The quantity $W_2(n_2,t_2|n_1,t_1)$ will be denoted by $W(n_2,t_2|n_1,t_1)$.

It follows from the above definitions that the joint probabilities and conditional probabilities are related by

$$P_r(n_1,t_1;\ldots;n_r,t_r)$$
$$= W_r(n_r,t_r|n_{r-1},t_{r-1};\ldots;n_1,t_1)$$
$$\times P_{r-1}(n_1,t_1;\ldots;n_{r-1},t_{r-1}). \tag{2.1.6}$$

Comparison of 2.1.6 with 2.1.3 yields

$$W_r(n_r,t_r|n_{r-1},t_r;\ldots;n_2,t_2) = \delta_{n_r,n_{r-1}}\,. \qquad (2.1.7)$$

For a stationary process

$$W_r(n_r,t_r|n_{r-1},t_{r-1};\ldots;n_1,t_1)$$

$$= W_r(n_r,t_r+\tau|n_{r-1},t_{r-1}+\tau;\ldots;n_1,t_1+\tau). \qquad (2.1.8)$$

All of the results given so far are easily generalized for vector $X(t)$ by a suitable proliferation of indices. They may also be extended to the situation where $X(t)$ can take on a continuum of values by using joint <u>probability densities</u> $p_r(x_1,t_1;x_2,t_2;\ldots;x_r,t_r)$ rather than probabilities. In this case the definition of 2.1.1 is replaced by

$$p_r(x_1,t_1;x_2,t_2;\ldots;x_r,t_r)dx_1dx_2\ldots dx_r$$

$$= \text{prob}\,\{x_1 \le X(t_1) < x_1+dx_1;x_2 \le X(t_2)$$

$$< x_2 + dx_2;\ldots;x_r \le X(t_r) < x_r + dx_r\} \qquad (2.1.9)$$

which is the joint probability that the stochastic variable $X(t)$ has a value between x_1 and x_1+dx_1 at time t_1, a value between x_2 and x_2+dx_2 at time $t_2,\ldots,$ and a value between x_r and x_r+dx_r at time t_r. It is important to notice that if the x_j have the dimensions of L the probability density p_r has the dimensions L^{-r}. The probabilities P_r, however, are dimensionless. The probability densities p_r have the properties

$$p_r(x_1,t_1;x_2,t_2;\ldots x_r,t_r) \geq 0$$

$$\int p(x_1,t_1)dx_1 = 1$$

$$\int \ldots \int p_r(x_1,t_1;\ldots;x_r,t_r)dx_j dx_{j+1}\ldots dx_r$$

$$= p_{j-1}(x_1,t_1;\ldots;x_{j-1}t_{j-1})$$

$$p_r(x_1,t;x_2,t;x_3,t_3;\ldots;x_r,t_r)$$

$$= p_{r-1}(x_1,t;x_3,t_3;\ldots;x_r,t_r)\delta(x_1-x_2) \qquad (2.1.10)$$

where $p(x_1,t_1) \equiv p_1(x_1,t_1)$. Equations 2.1.10 are analogous to Eqs. 2.1.2 and 2.1.3 with $\delta(x_1-x_2)$ being the Dirac delta function.

The <u>conditional probability density</u> $w_r(x_r,t_r|x_{r-1},t_{r-1};x_{r-2},t_{r-2};\ldots;x_1,t_1)$ is defined such that $w_r dx_r$ is the probability that the stochastic variable $X(t)$ has a value between x_r and x_r+dx_r at time t_r if it had the value x_{r-1} at $t_{r-1},\ldots,$ and the value x_1 at time t_1. The $\{w_r\}$ are probability densities and satisfy the conditions

$$w_r \geq 0 \qquad (2.1.11)$$

and

$$\int w_r(x_r,t_r|x_{r-1};t_{r-1};\ldots;x_1,t_1)dx_r = 1. \qquad (2.1.12)$$

The quantity $w_2(x_2,t_2|x_1,t_1)$ will be denoted by $w(x_2,t_2|x_1,t_1)$. If x_j has the dimensions L, then w_r has the dimensions L^{-1}. The conditional probability density w_r is related to the joint probability density p_r by

$$p_r(x_1,t_1;\ldots;x_r,t_r)$$

$$= w_r(x_r,t_r|x_{r-1},t_{r-1};\ldots;x_1,t_1)$$

$$\times\ p_{r-1}(x_1,t_1;\ldots;x_{r-1},t_{r-1}). \qquad (2.1.13)$$

For a stationary process

$$p_r(x_1,t_1;\ldots;x_r,t_r)$$

$$= p_r(x_1,t_1+\tau;x_2,t_2+\tau;\ldots;x_r,t_r+\tau) \qquad (2.1.14)$$

and

$$w_r(x_r,t_r|x_{r-1},t_{r-1};\ldots;x_1,t_1)$$

$$= w_r(x_r,t_r+\tau|x_{r-1},t_{r-1}+\tau;\ldots;x_1,t_1+\tau) \qquad (2.1.15)$$

for all r, $\{x_j\}$, $\{t_j\}$, and τ.

The quantities defined above are very general and pertain to all stochastic processes. Little progress can be made on this level of generality. To apply the theory of stochastic processes to physical phenomena it is necessary to place certain restrictions on the form of the joint and conditional probabilities. The first application of stochastic processes involved <u>independent processes</u>. These are processes in which the value of the stochastic variable X(t) at t_j does not depend on the previous history of the process. The precise mathematical definition of an independent process is

$$w_r(x_r,t_r|x_{r-1},t_{r-1};\ldots;x_1,t_1) = p(x_r,t_r) \qquad (2.1.16)$$

or, equivalently,

$$p_r(x_1,t_1;x_2,t_2;\ldots x_r,t_r) = \prod_{i=1}^{r} p(x_i,t_i) \qquad (2.1.17)$$

for all r. For nearly three hundred years probability theory was solely the study of independent processes. It is easy to construct examples of independent processes when the time variable can assume only discrete values. Some well known examples are successive rolls of a pair of dice which generate random variables barring imperfections of the dice; coin tossing; Buffon's needle problem, etc.(2.1.2) Physical phenomena, however, take place on a continuous time scale. In such a continuous time scale it is not possible to have a truly independent process in which the value of the stochastic variable $X(t)$ at time $t_1 + \tau$ is independent of its value at time t_1 for arbitrary values of $\tau \geq 0$. It is of course possible to consider, at least conceptually, physical processes on a discrete time scale in which the time intervals between observations are long compared to the interval between individual physical events. For instance, the direction of motion of a molecule with respect to a laboratory-fixed coordinate system in an equilibrium fluid is an independent random variable on a time scale in which the time interval τ between observations is long compared to the time between collisions.

It is clear that independent processes cannot provide a rich enough framework for the description of many aspects of physical reality. The idea of correlation (or "memory") must somehow be introduced particularly for continuous times. The decisive step in this direction was made by Markov in connection with a study of the alternation of vowels and consonants in Pushkin's Eugene Onegin.[2.1.3] Markov noted that the occurrence of a vowel or a consonant depended strongly on whether the immediately preceding letter was a vowel or a consonant and rather weakly on the character of earlier letters. One can idealize this situation by constructing a stochastic process which has the following properties: the value of $X(t_r)$ depends only on the value of $X(t_{r-1})$; previous values of $X(t)$ for $t < t_{r-1}$ do not affect $X(t_r)$. Dependent processes of this type are known as Markov processes. The precise mathematical definition of a Markov process is[2.1.4]

$$w_r(x_r,t_r|x_{r-1},t_{r-1};x_{r-2},t_{r-2};\ldots;x_1,t_1)$$

$$= w(x_r,t_r|x_{r-1},t_{r-1}) \qquad (2.1.18)$$

for all values of x and t. Equation 2.1.18 states that the conditional probability density w_r of $X(t)$ having a value between x_r and x_r+dx_r at time t_r depends only on the value of $X(t)$ at time t_{r-1}. For a stationary Markov process

$$w\,(x_r,t_r|x_{r-1},t_{r-1}) = w(x_r,t_r+\tau|x_{r-1},t_{r-1}+\tau) \tag{2.1.19}$$

where τ is arbitrary. The discrete state space analogs of Eqs. 2.1.18 and 2.1.19 are

$$W_r(n_r,t_r|n_{r-1},t_{r-1};\ldots;n_1,t_1) = W(n_r,t_r|n_{r-1},t_{r-1}) \tag{2.1.20}$$

and

$$W_r(n_r,t_r|n_{r-1},t_{r-1};\ldots;n_1,t_1) = W(n_r,t_r+\tau|n_{r-1},t_{r-1}+\tau). \tag{2.1.21}$$

For a Markov process, therefore, one needs to specify only a single conditional probability (or probability density) rather than a complete array $\{w_r\}$. Any of the joint probability densities $p_r(x_1,t_1;\ldots;x_r,t_r)$ with $t_1 \le t_2 \le \ldots \le t_r$ can then be expressed in terms of $p(x_1,t_1)$ and $w(x,t|x',t')$ as

$$p_r(x_1,t_1;x_2,t_2;\ldots;x_r,t_r) = p(x_1,t_1)w(x_2,t_2|x_1,t_1)w(x_3,t_3|x_2,t_2) \ldots w(x_r,t_r|x_{r-1},t_{r-1}). \tag{2.1.22}$$

through successive application of 2.1.13 and 2.1.18.

If the units of time are discrete multiples of a single unit τ, i.e., $t_s = s\tau$, and if there is at most a denumerable infinity of values which $X(t)$ can assume then the Markov process is commonly termed a

Markov chain. Any discrete stochastic process [in time and in the values of X(t)] with dependence on a fixed number of past events can be put in the form of a vector Markov chain. In the simplest case, a stochastic process which is completely determined by the conditional probabilities $W_3(n_3,s|n_2,s-1;n_1,s-2)$ is a simple Markov chain in the vectors $\underline{X}(s)$ with components $X(s) \equiv [n,s]$ and $X(s-1)$, i.e., $\underline{X}(s) = [X(s), X(s-1)]$ so that

$$W(n_3,s|n_2,s-1;n_1,s-2) = W(\underline{X}(s)|\underline{X}(s-1)) . \qquad (2.1.23)$$

This means that for discrete processes in discrete time the Markov process is the only intermediate between independence and a dependence on all previous events. When time is regarded as continuous this is not necessarily true.[(2.1.5)]

We pause here to give some illustrative examples of the different processes that have been defined. A simple example of an independent process is coin-tossing. At each trial the outcome is independent of all previous outcomes. An example of a Markov chain is the stochastic process which arises from the random drawing of a ball from an urn which contains an initially fixed finite number of black and white balls. At each trial (i.e., for each drawing of a ball) the outcome (i.e., the ratio of black to white balls after the drawings) depends upon the ratio of black to white balls at the time of the drawing, but is independent of the ratio at previous drawings.

Energy transfer in a dilute gas can be idealized as a Markov process in continuous time. The amount of energy transferred in a given collision depends upon the energy of the collision partners at the time of the collision and is independent of their energies at previous times. Finally, the case of complete dependence can be illustrated by the so-called excluded volume problem in polymer physics. A model for the configurations of a polymer chain can be constructed by considering a number of successively linked monomers in which each bond between monomers is oriented at random with respect to the preceding one. No two monomers can occupy the same region in space. If polymerization proceeds step by step, then any given monomer must avoid all previous parts of the polymer chain. Hence the outcome of any step in the polymerization depends on all of the previous steps.

2) The Smoluchowski - Chapman - Kolmorgorov Equation

In this section we derive the Smoluchowski-Chapman-Kolmogorov[2.2.1] equation (SCK equation) for both discrete and continuous stochastic processes. We shall demonstrate that the Markov conditions 2.1.18 and 2.1.20 on the conditional probabilities are sufficient, but not necessary, for the derivation of the SCK equation. There exist "Pseudo-Markovian" processes whose conditional probabilities obey the SCK equation but do not satisfy the Markov condition.[2.2.2]

For a stochastic process in a continuous time and discrete state space we can write

$$P_{r-1}(n_1,t_1;\ldots;n_{r-2},t_{r-2};n_r,t_r)$$

$$= \sum_{n_{r-1}} W_r(n_r,t_r|n_{r-1},t_{r-1};\ldots;n_1,t_1)$$

$$\times P_{r-1}(n_1,t_1;\ldots;n_{r-1},t_{r-1}) \qquad (2.2.1)$$

where we have used Eqs. 2.1.2 and 2.1.6. Using Eq. 2.1.6 we can also write

$$P_{r-1}(n_1,t_1;\ldots;n_{r-2},t_{r-2};n_r,t_r)$$

$$= W_{r-1}(n_r,t_r|n_{r-2},t_{r-2};\ldots;n_1,t_1)$$

$$\times P_{r-2}(n_1,t_1;\ldots;n_{r-2},t_{r-2}) \qquad (2.2.2)$$

and

$$P_{r-1}(n_1,t_1;\ldots;n_{r-1},t_{r-1})$$

$$= W_{r-1}(n_{r-1},t_{r-1}|n_{r-2},t_{r-2};\ldots;n_1,t_1)$$

$$\times P_{r-2}(n_1,t_1;\ldots;n_{r-2},t_{r-2}). \qquad (2.2.3)$$

Substitution of Eqs. 2.2.2 and 2.2.3 into Eq. 2.2.1 and cancellation of the common factor yields

$$W_{r-1}(n_r,t_r|n_{r-2},t_{r-2};\ldots;n_1,t_1)$$

$$= \sum_{n_{r-1}} W_r(n_r,t_r|n_{r-1},t_{r-1};\ldots;n_1,t_1) \times$$

$$\times W_{r-1}(n_{r-1},t_{r-1}|n_{r-2},t_{r-2};\ldots;n_1,t_1). \qquad (2.2.4)$$

Equation 2.2.4 is a recursion relation for the conditional probabilities valid for any stochastic processes in discrete state space. The analogue of Eq. 2.2.4 for stochastic processes in a continuous state space is

$$w_{r-1}(x_r,t_r|x_{r-2},t_{r-2};\ldots;x_1,t_1)$$
$$= \int dx_{r-1} w_r(x_r,t_r|x_{r-1},t_{r-1};\ldots;x_1,t_1)$$
$$\times w_{r-1}(x_{r-1},t_{r-1}|x_{r-2},t_{r-2};\ldots;x_1,t_1). \qquad (2.2.5)$$

In the remainder of this section we restrict our attention to continuous state space processes; the extension to the discrete state space is immediately obvious.

We note that Eq. 2.2.5 relates w_{r-1} to w_r and thus necessitates the solution of the complete hierarchy of equations to obtain the solution for w_{r-1}. It is of interest to investigate the conditions under which a closure of this hierarchy can be obtained. We consider Eq. 2.2.5 for the case $r = 3$ and write

$$w(x_3,t_3|x_1,t_1)$$
$$= \int dx_2 w_3(x_3,t_3|x_2,t_2;x_1,t_1)w(x_2,t_2|x_1,t_1). \quad (2.2.6)$$

We now write the conditional probability density w_3 in the form

$$w_3(x_3,t_3|x_2,t_2;x_1,t_1)$$
$$= w(x_3,t_3|x_2,t_2) + \Delta(x_3,t_3;x_2,t_2;x_1,t_1) \qquad (2.2.7)$$

where Δ is defined by Eq. 2.2.7. Substitution of Eq. 2.2.7 into Eq. 2.2.6 yields

$$w(x_3,t_3|x_1,t_1)$$
$$= \int dx_2\ [w(x_3,t_3|x_2,t_2)w(x_2,t_2|x_1,t_1)$$
$$+ \Delta(x_3,t_3;x_2,t_2;x_1,t_1)w(x_2,t_2|x_1,t_1)]. \qquad (2.2.8)$$

Equation 2.2.8 is still valid for all stochastic processes. Under the condition that

$$\int dx_2 \Delta(x_3,t_3;x_2,t_2;x_1,t_1)w(x_2,t_2|x_1,t_1) = 0 \qquad (2.2.9)$$

Eq. 2.2.8 reduces to

$$w(x_3,t_3|x_1,t_1) = \int dx_2 w(x_3,t_3|x_2,t_2)w(x_2,t_2|x_1,t_1). \qquad (2.2.10)$$

Equation 2.2.10 is the SCK equation for continuous state space stochastic processes. It is a closed equation for w. For a discrete state space the SCK equation reads

$$W(n_3,t_3|n_1,t_1) = \sum_{n_2} W(n_3,t_3|n_2,t_2)W(n_2,t_2|n_1,t_1). \qquad (2.2.11)$$

Equations 2.2.10 and 2.2.11 are trivially satisfied for a Markov process since from the definition,

Eq. 2.1.18, of a Markov process, $w_r = w$, it follows immediately that $\Delta = 0$. The SCK equations 2.2.10 and 2.2.11 thus provide a necessary condition for the conditional probabilities of a Markov process.

It is clear that Eqs. 2.2.10 and 2.2.11 cannot be used to obtain the form of the conditional probabilities w (or W) without additional information. For some simple physical systems, explicit expressions can be derived for the W_r and the validity of the SCK equations can be verified directly. This has been done for simple lattice dynamical models by Hemmer,[2.2.3] and Rubin,[2.2.4]. However, in most cases the applicability of a Markovian description can be tested only a posteriori by comparison between theory and experiment. The SCK equation can be used as a starting point in the derivation of the master equation and the Fokker-Planck equation.[2.2.5]

We shall not consider "pseudo-Markovian" processes in this book. Therefore the SCK equations 2.2.10 and 2.2.11 are equivalent to the conditions 2.1.18 and 2.1.20 on the conditional probabilities W_r. Thus, either the SCK equations 2.2.10 and 2.2.11, or the relations 2.1.18 and 2.1.20 can be used as definitions or criteria for a Markov process.

3) Moments and Characteristic Functions

We have seen that a set of joint and/or conditional probabilities can be used to specify various properties of the random variable X(t). It is usually easier to obtain certain functionals of the probability densities than the densities themselves. The most important of these functionals is the characteristic function which is also known as the moment generating function. In this section we define the moment and correlation functions of a stochastic process and then discuss the definition and properties of the characteristic function.

We define the state average $\langle F[X(t_1), X(t_2)\ldots, X(t_n)]\rangle$ of a function $F(x_1,x_2,\ldots,x_n)$ as

$$\langle F[X(t_1),X(t_2),\ldots,X(t_n)]\rangle$$
$$\equiv \int \ldots \int F(x_1,x_2,\ldots,x_n)p_n(x_1,t_1;\ldots;x_n,t_n)\, d^n\underset{\sim}{x} \tag{2.3.1}$$

where the integration is over all possible values of the state variable x. The multiple integral may be interpreted as a summation over $x_1,x_2,\ldots,x_n$ if the random variable $X(t)$ takes on discrete values only. The kth moment $\mu_k(t)$ of $p(x,t)$ is defined as $\int x^k p(x,t)dx$ which, by Eq. 2.3.1 is equal to the average value of x^k, i.e.,

$$\mu_k(t) \equiv \int x^k p(x,t)dx \equiv \langle X^k(t)\rangle \equiv \langle x^k\rangle_t . \tag{2.3.2}$$

The existence of the zeroth moment of the probability density $p(x,t)$ is assured by the general properties given in Eq. 2.1.10. However, it is not necessarily true that higher moments exist. As a simple example we consider the probability density function

$$p(x) = \varepsilon(1+x)^{-(1+\varepsilon)} \quad (2.3.3)$$

for $0 \leq x \leq \infty$ and $0 < \varepsilon \leq 1$. This probability density has a zeroth moment but has no higher integer moments.

The most frequently used moment functions are the first moment, $\mu_1(t)$, and the variance $\sigma^2(t)$ defined by

$$\sigma^2(t) \equiv \mu_2(t) - [\mu_1(t)]^2 = \langle [X(t) - \langle X(t)\rangle]^2 \rangle. \quad (2.3.4)$$

This function is a measure of the "dispersion" of the stochastic process since it is zero for the deterministic process in which $p(x,t)$ is a delta function, and is positive for all other $p(x,t)$.

If the stochastic process is stationary, $p(x,t)$ is independent of the time, i.e., $p(x,t) = p(x,0)$, and the moments are also independent of the time, i.e., $\langle X^k(t)\rangle = \langle X^k(0)\rangle$.

The <u>autocorrelation function</u> of order 2 is defined as

$$\rho_2(t_1,t_2) \equiv \rho\ (t_1,t_2) \equiv \langle X(t_1)X(t_2)\rangle. \quad (2.3.5)$$

For a stationary process this correlation function depends only on the time difference $\tau = t_2 - t_1$ or

$$\rho(t_1,t_2) = \rho(\tau) = \langle X(0)X(\tau)\rangle \quad (2.3.6)$$

and $\rho(\tau) = \rho(-\tau)$. It should be noted that for a stationary process

$$|\rho(\tau)| \leq \rho(0) = \mu_2. \quad (2.3.7)$$

This follows from the fact that

$$<[X(\tau) \pm X(0)]^2> = <X(\tau)^2> + <X(0)^2> \pm 2<X(\tau)X(0)>$$

$$= 2\mu_2 \pm 2\rho(\tau) \geq 0. \qquad (2.3.8)$$

The autocorrelation function of order n is

$$\rho_n(t_1,t_2,\ldots,t_n) = <X(t_1)X(t_2)\ldots X(t_n)>. \qquad (2.3.9)$$

For an independent process, 2.3.9 becomes

$$\rho_n(t_1,t_2,\ldots,t_n) = <X(t_1)> <X(t_2)> \ldots <X(t_n)>. \qquad (2.3.10)$$

It is clear from the definitions 2.3.1 and 2.3.2 that one can calculate all the moments of a distribution if the probability density p(x,t) is known. The question now arises whether it is possible to reconstruct the probability density from a knowledge of all the moments. The answer is: not necessarily. As an example, it is possible to show that the stationary probability density function

$$p(x) = x^{-\ell nx} [1 - \lambda \sin(4\pi \ell nx)]$$

$$0 \leq x \leq \infty$$

$$0 \leq \lambda \leq 1 \qquad (2.3.11)$$

has the same moments for all values of λ. Hence a knowledge of all the moments does not necessarily determine the probability density without some restrictions on the form of the moments.(2.3.1)

We now turn our attention to characteristic

functions. The <u>characteristic function</u> $\phi(s,t)$ for the scalar random variable $X(t)$ is defined by

$$\begin{aligned}\phi(s,t) &\equiv \langle e^{isX(t)}\rangle \\ &= \int_{-\infty}^{\infty} e^{isx} p(x,t)\,dx \end{aligned} \tag{2.3.12}$$

which is the Fourier transform of the probability density $p(x,t)$. Equation 2.3.12 can be inverted to yield

$$p(x,t) = \frac{1}{2\pi}\int_{-\infty}^{\infty} e^{-isx}\phi(s,t)\,ds. \tag{2.3.13}$$

It follows from the definition of $\phi(s,t)$ that $\phi(o,t) = 1$ and that

$$\begin{aligned}|\phi(s,t)| &\leq \int_{-\infty}^{\infty} |e^{ixs}|\,p(x,t)\,dx \\ &= \int_{-\infty}^{\infty} p(x,t)\,dx = 1, \end{aligned} \tag{2.3.14}$$

for all s and t.

The characteristic function $\phi(s,t)$ can be used to generate the moments. If we expand the exponential, then

$$\phi(s,t) = \left\langle \sum_{n=o}^{\infty} \frac{i^n}{n!} X^n(t) s^n \right\rangle = \sum_{n=o}^{\infty} \frac{i^n}{n!} \langle X^n(t)\rangle\, s^n. \tag{2.3.15}$$

In writing this equation we have assumed the existence of all the moments, and the possibility of interchanging summation and averaging. This assumption is usually valid in physical applications. Proceeding formally one finds

$$\mu_n(t) = \frac{1}{i^n} \frac{d^n}{ds^n} \phi(s,t) \Big|_{s = 0} \qquad (2.3.16)$$

so that if the characteristic function is known the moments can be calculated directly.

It is also useful to have an expression in terms of $\phi(s,t)$ for the probability that $X(t)$ is in some finite interval $(x, x + h)$ at time t. The cumulative probability $\underline{P}(x,t)$ is defined by

$$\underline{P}(x,t) = \int_{-\infty}^{x} p(y,t)dy. \qquad (2.3.17)$$

The probability that $X(t)$ is in $(x,x + h)$ at time t is $\underline{P}(x + h,t) - \underline{P}(x,t)$. From Eqs. 2.3.13 and 2.3.17 it then follows that[(2.1.1)]

$$\underline{P}(x + h,t) - \underline{P}(x,t) = \frac{1}{2\pi i} \int_{-\infty}^{\infty} e^{-isx} \frac{(1-e^{-ish})}{s} \phi(s,t)ds \qquad (2.3.18)$$

if $\underline{P}$ is continuous at x and $x + h$.

The preceding definition of the characteristic function is easily generalized to vector processes. For an n-vector $\underset{\sim}{X}(t) = (X_1(t), X_2(t), \ldots, X_n(t))$ the characteristic function $\phi_n(s_1,s_2,\ldots,s_n;t)$ is defined by

$$\phi_n(s_1,s_2,\ldots,s_n;t)$$

$$\equiv \left\langle e^{i(s_1X_1(t) + s_2X_2(t) + \ldots + s_nX_n(t))} \right\rangle$$

$$\equiv \langle e^{i\underline{s}\cdot\underline{X}(t)} \rangle \qquad (2.3.19)$$

where $\underline{s} = (s_1,s_2,\ldots,s_n)$. The n-variable probability density is then given by

$$p_n(x_1,x_2,\ldots,x_n;t)$$

$$= \frac{1}{(2\pi)^n} \int_{-\infty}^{\infty} \cdots \int_{-\infty}^{\infty} e^{-i\underline{s}\cdot\underline{x}} \phi(\underline{s},t)\, d^n\underline{s}\,. \qquad (2.3.20)$$

The moment $\mu_{k_1,k_2,\ldots,k_n}(t)$ is given by

$$\mu_{k_1,k_2,\ldots,k_n}(t) \equiv \left\langle X_1^{k_1}(t) X_2^{k_2}(t) \ldots X_n^{k_n}(t) \right\rangle$$

$$= \int x_1^{k_1} x_2^{k_2} \ldots x_n^{k_n}\, p_n(x_1,x_2,\ldots,x_n;t)\, dx_1 dx_2 \ldots dx_n$$

$$= \frac{1}{i^k} \left. \frac{\partial^k \phi(s_1,s_2,\ldots,s_n;t)}{\partial^{k_1} s_1 \partial^{k_2} s_2 \ldots \partial^{k_n} s_n} \right|_{s_1=s_2=\cdots=s_n=0} \qquad (2.3.21)$$

where $k=k_1+k_2+\ldots+k_n$.

The many-time characteristic function for a scalar stochastic process, $\phi_n(s_1,t_1;s_2,t_2;\ldots s_n,t_n)$, is defined by

$$\phi_n(s_1,t_1;s_2,t_2;\ldots;s_n,t_n)$$

$$\equiv \left\langle e^{i[s_1X(t_1) + s_2X(t_2) + \ldots s_nX(t_n)]} \right\rangle . \qquad (2.3.22)$$

The joint probability density is given by

$$p_n(x_1,t_1;x_2,t_2;\ldots;x_n,t_n)$$

$$= \frac{1}{(2\pi)^n} \int_{-\infty}^{\infty} \ldots \int_{-\infty}^{\infty} e^{-i[s_1x_1+s_2x_2+\ldots+s_nx_n]}$$

$$x\ \phi_n(s_1,t_1;s_2,t_2;\ldots;s_n,t_n)ds_1\ldots ds_n . \qquad (2.3.23)$$

The correlation function $\rho_{k_1,k_2,\ldots,k_n}(t_1,t_2,\ldots,t_n)$

is then given by

$$\rho_{k_1,k_2,\ldots,k_n}(t_1,t_2,\ldots,t_n)$$

$$\equiv \left\langle X^{k_1}(t_1)X^{k_2}(t_2)\ldots X^{k_n}(t_n) \right\rangle$$

$$= \frac{1}{i^k} \left. \frac{\partial^k \phi_n(s_1,t_1;s_2,t_2;\ldots;s_n,t_n)}{\partial^{k_1}s_1\, \partial^{k_2}s_2 \ldots \partial^{k_n}s_n} \right|_{s_1=s_2=\ldots=s_n=0} \qquad (2.3.24)$$

where $k = k_1+k_2+\ldots+k_n$. The autocorrelation function of order 2 then becomes

$$\rho(t_1,t_2) = - \left. \frac{\partial^2 \phi(s_1,t_1;s_2,t_2)}{\partial s_1 \partial s_2} \right|_{s_1=s_2=0} \qquad (2.3.25)$$

We note that the many time characteristic function for a scalar stochastic process is of the same form as the single time characteristic function of a vector stochastic process. In fact, Eqs. 2.3.22 through 2.3.25 are identical to the corresponding Eqs. 2.3.19 through 2.3.21 if we identify $X(t_1)$ with $X_1(t)$, $X(t_2)$ with $X_2(t)$, etc.

Other definitions of characteristic functions are also useful and may be more appropriate for certain applications. For example, for problems in which the random variable $X(t)$ can take on only positive values it is convenient to define a characteristic function (generating function) $g(s,t)$ by the Laplace transform

$$g(s,t) \equiv \langle e^{-sX(t)} \rangle = \int_0^\infty e^{-sx} p(x,t)\,dx. \tag{2.3.26}$$

4) Cumulants

Another set of functions of great utility in the analysis of stochastic processes is the cumulants or semi-invariants. These quantities which are simple algebraic functions of the moments are particularly useful in discussing correlations of stochastic variables. As will be seen below, they also arise naturally in relating the average of an exponential to the exponential of an average [see Eq. 2.4.5].

We shall define cumulants $K(\nu)$ of order ν, as

$$K(\nu) = i^{-\nu} \left. \frac{d^\nu \ln\phi(s)}{ds^\nu} \right|_{s=0} \tag{2.4.1}$$

This is equivalent to writing the characteristic function as

$$\phi(s) \equiv \langle e^{isX} \rangle = e^{L(s)} \tag{2.4.2}$$

where

$$L(s) = \sum_{\nu=1}^{\infty} \frac{i^{\nu}K(\nu)s^{\nu}}{\nu!} \ . \tag{2.4.3}$$

The above formulae can be easily generalized to vector random processes in which the stochastic variable $\underline{X} = (x_1, x_2, \ldots, x_n)$. Thus the cumulant $K^{(n)}(\nu_1, \nu_2, \ldots, \nu_n)$ of order ν is given by

$$K^{(n)}(\nu_1, \ldots, \nu_n) = i^{-\nu} \left. \frac{\partial^{\nu} \ln \phi^{(n)}(s_1, \ldots, s_n)}{\partial s_1^{\nu_1} \partial s_2^{\nu_2} \ldots \partial s_n^{\nu_n}} \right|_{s_1 = s_2 = \ldots s_n = 0} \tag{2.4.4}$$

where $\nu = \nu_1 + \nu_2 + \ldots + \nu_n$. The analogues of Eqs. 2.4.2 and 2.4.3 are

$$\phi^{(n)}(\underline{s}) \equiv \langle e^{i\underline{s}\cdot\underline{x}} \rangle = e^{L(\underline{s})} \tag{2.4.5}$$

and

$$L(\underline{s}) = \sum_{\nu_1=0}^{\infty} \ldots \sum_{\substack{\nu_n=0 \\ \nu_1+\ldots+\nu_n>0}}^{\infty} \frac{i^{\nu}K^{(n)}(\nu_1, \ldots, \nu_n) s_1^{\nu_1} s_2^{\nu_2} \ldots s_n^{\nu_n}}{\nu_1! \nu_2! \ldots \nu_n!} \tag{2.4.6}$$

Since the generating function can be written in terms of the moments, the cumulants can also be written as non-linear functionals of the moments. Let us consider a single random variable X. From Eqs. 2.4.2 and 2.4.3 one can derive the explicit relation

$$\frac{K(\nu)}{\nu!} = \sum_{r=1}^{\nu} \sum_{\{m\}} \frac{\mu_{m_1} \cdots \mu_{m_r}}{m_1! \cdots m_r!} \frac{(-1)^{r+1}}{r} \tag{2.4.7}$$

where

$m_1, \ldots, m_r > 0$, $\nu \geq 1$ and

$$m_1 + m_2 + \ldots + m_r = \nu \tag{2.4.8}$$

and where the moment μ_j is defined to be $\langle X^j \rangle$ [see Eq. 2.3.2]. The inverse relation between moments and cumulants is

$$\frac{\mu_\nu}{\nu!} = \sum_{r=1}^{\nu} \sum_{\{m\}} \left(\frac{1}{r!}\right) \frac{K(m_1) \ldots K(m_r)}{m_1! \ldots m_r!} \tag{2.4.9}$$

where $m_1, \ldots, m_r > 0$, $\nu \geq 1$ and

$$m_1 + m_2 \ldots + m_r = \nu \, . \tag{2.4.10}$$

The first few cumulants in terms of the moments μ_j are

$$K(1) = \mu_1$$

$$K(2) = \mu_2 - \mu_1^2 \equiv \sigma^2$$

$$K(3) = \mu_3 - 3\mu_2\mu_1 + 2\mu_1^3$$

$$K(4) = \mu_4 - 4\mu_3\mu_1 - 3\mu_2^2 + 12\mu_2\mu_1^2 - 6\mu_1^4$$

$$K(5) = \mu_5 - 5\mu_4\mu_1 - 10\mu_3\mu_2 + 20\mu_3\mu_1^2 + 30\mu_2^2\mu_1 - 60\mu_2\mu_1^3 + 24\mu_1^5. \tag{2.4.11}$$

Explicit expressions for the first ten cumulants in terms of the moments can be found in reference 2.4.1.

It should be noted that for a deterministic process in which the probability density $p(x,t)$ is a delta function all cumulants $K(\nu)$ are zero for $\nu > 1$. This follows readily from Eq. 2.4.7 since $\mu_j = \mu_1^j$ for deterministic processes or from Eqs. 2.4.1 and 2.4.2 since $\phi(s) = e^{is\langle X\rangle}$.

The generalization of Eqs. 2.4.11 to vector processes yields

$$K(1) = \langle X_1\rangle$$

$$K^{(2)}(1,1) = \langle X_1X_2\rangle - \langle X_1\rangle\langle X_2\rangle$$

$$K^{(3)}(1,1,1) = \langle X_1X_2X_3\rangle - \langle X_1\rangle\langle X_2X_3\rangle$$

$$- \langle X_2\rangle\langle X_1X_3\rangle - \langle X_3\rangle\langle X_1X_2\rangle$$

$$+ 2\langle X_1\rangle\langle X_2\rangle\langle X_3\rangle \tag{2.4.12}$$

and so forth where the moments $\langle X_iX_j\ldots\rangle$ are defined in Eq. 2.3.21. We use here the notation $K^{(n)}(\nu_1, o,\ldots,o) = K(\nu_1)$. Note that if the stochastic process X_1 is independent of the processes X_2 and X_3, the cumulants $K^{(2)}(1,1)$ and $K^{(3)}(1,1,1)$ are zero. In general, if X_1 is independent of all the other stochastic processes, $K^{(n)}(\nu_1,\nu_2,\ldots,\nu_n)$ is zero for $\nu_1 \geq 1$ and for $\nu - \nu_1 \geq 1$ where $\nu = \nu_1 + \ldots + \nu_n$. Thus cumulants can be used as a test of correlation among random variables.

Cumulants can also be written in terms of a "connected average" $\langle\ \rangle_c$,

$$K^{(n)}(\nu_1,\nu_2,\ldots,\nu_n) = \left\langle X_1^{\nu_1} X_2^{\nu_2} \ldots X_n^{\nu_n} \right\rangle_c . \qquad (2.4.13)$$

This expression which is due to Kubo[2.4.2] can be shown to be equivalent to our previous definition 2.4.4. A "connected average" is defined as one which vanishes when any of the stochastic variables, or any group of stochastic variables, is independent of the other stochastic variables. As examples of the calculation of connected averages we shall discuss $K^{(2)}(1,1)$ and $K^{(3)}(1,1,1)$. The connected average $\langle X_1X_2\rangle_c$ is obtained by starting with the average $\langle X_1X_2\rangle$ and subtracting from it the value of $\langle X_1X_2\rangle$ when X_1 and X_2 are independent. Thus

$$K^{(2)}(1,1) \equiv \langle X_1X_2\rangle_c = \langle X_1X_2\rangle - \langle X_1\rangle\langle X_2\rangle \qquad (2.4.14)$$

which is the same as 2.4.12. The calculation of the connected average $\langle X_1X_2X_3\rangle_c$ is begun by subtracting from $\langle X_1X_2X_3\rangle$ its value when X_1, X_2 and X_3 are initially independent, i.e., by forming $\langle X_1X_2X_3\rangle$ -

$- \langle X_1\rangle \langle X_2\rangle \langle X_3\rangle$. Next, we must subtract from this quantity its value when X_1 is independent of X_2 and X_3, i.e., $\langle X_1\rangle\{\langle X_2X_3\rangle - \langle X_2\rangle \langle X_3\rangle\}$; next we subtract from this combined quantity its value when X_2 is independent of X_1 and X_3, i.e., $\langle X_2\rangle\{\langle X_1X_3\rangle - \langle X_1\rangle\langle X_3\rangle\}$; and next subtract from this new quantity its value when X_3 is independent of X_1 and X_2, i.e., $\langle X_3\rangle\{\langle X_1X_2\rangle - \langle X_1\rangle \langle X_2\rangle\}$. The results of these manipulations lead to an expression for $K^{(3)}(1,1,1)$ which is identical to 2.4.12. Cumulants of higher order can be calculated in the same manner, but it is clear that the analytic technique of 2.4.7 is more convenient than the combinatorial method described here. The expression of cumulants in terms of "connected averages" makes explicit their cluster property; namely, the vanishing of the cumulants when any group of the stochastic variables is independent of the others.(2.4.2 - 2.4.5)

For future developments it is useful to introduce the <u>many time cumulant</u> $K_n(\nu_1,t_1;\nu_2,t_2,\ldots,\nu_n,t_n)$ for a scalar stochastic process. For simplicity we restrict ourselves to the case where $\nu_1=\nu_2=\ldots=\nu_n=1$. The many time cumulant is defined as [see 2.4.4]

$$K_n(t_1;t_2;\ldots;t_n)$$

$$\equiv i^{-n} \left. \frac{\partial^n \ell n\phi_n(s_1,t_1;s_2,t_2;\ldots;s_n,t_n)}{\partial s_1 \; \partial s_2 \; \ldots \; \partial s_n} \right|_{s_1=s_2=\ldots s_n=0} \tag{2.4.15}$$

or, in terms of connected averages, as

$$K_n(t_1;t_2;\ldots;t_n) = \langle X(t_1)X(t_2) \ldots X(t_n)\rangle_c. \qquad (2.4.16)$$

For a stationary process

$$K_n(t_1+\tau;\ t_2+\tau;\ldots;t_n+\tau) = K_n(t_1;t_2;\ldots;t_n). \qquad (2.4.17)$$

These many time cumulants describe the decay of correlations of the stochastic variables X(t) with time, i.e., they are a "measure" of the memory of the stochastic process. They become zero when the stochastic variable at any time t_i, $X(t_i)$, is independent of the variable at any other time.

The most frequently used many time cumulant is $K_2(t_1;t_2) \equiv K(t_1;t_2)$. From Eqs. 2.4.16 and 2.3.9 it follows that

$$K(t_1;t_2) = \langle X(t_1)X(t_2)\rangle - \langle X(t_1)\rangle\langle X(t_2)\rangle$$

$$= \rho(t_1,t_2) - \mu_1(t_1)\mu_1(t_2). \qquad (2.4.18)$$

The normalized two-time cumulant $\tilde{K}(t_1;t_2)$ is defined as

$$\tilde{K}(t_1;t_2) = \frac{K(t_1;t_2)}{K(t_1;t_1)} = \frac{\rho(t_1,t_2) - \mu_1(t_1)\mu_1(t_2)}{\sigma^2(t_1)}$$

(2.4.19)

where we have used Eq. 2.3.4 for $\sigma^2(t)$. This function has the property that $\tilde{K}(t_1;t_1) = 1$. If $t_2 - t_1$ is sufficiently large, $X(t_2)$ will no longer be correlated with $X(t_1)$ for most physical random processes.. In this case, $\tilde{K}(t_1;t_2)$ approaches zero as $t_2 - t_1$ approaches infinity. This can readily be seen from Eq. 2.4.19 since $\rho(t_1,t_2)$ approaches $\mu_1(t_1)\mu_1(t_2)$ as $t_2 - t_1$ approaches infinity.

It is possible to define a correlation time $\tau_c(t_1)$ as

$$\tau_c(t_1) = \int_{t_1}^{\infty} \tilde{K}(t_1;t_2)dt_2 \qquad (2.4.20)$$

which is a measure of the duration of the correlation of the stochastic process. For times t_2 such that $t_2 >> \tau_c(t_1)$ the value of $X(t_2)$ is effectively independent of $X(t_1)$. For a stationary process, $\tilde{K}(t_1;t_2) = K(\tau)$ where $\tau = t_2 - t_1$ and

$$\tau_c = \int_0^{\infty} \tilde{K}(\tau)d\tau. \qquad (2.4.21)$$

5) The Gaussian Distribution

Perhaps the most important distribution in the theory of probability and stochastic processes is the Gaussian (or normal) distribution. Gaussian processes are defined by the property that all their probability densities, $p(x,t)$, $p_2(x_1,x_2,t)$, $p_3(x_1,x_2,x_3,t)$,..., have the Gaussian form discussed below. They may be considered to be the simplest generalization of deterministic processes that can be introduced. This point of view is natural if we consider the cumulants. A deterministic process, $X(t)$, can be defined as one for which the first order cumulant may be non-zero, but higher order cumulants are equal to zero. The Gaussian distribution may have non-zero first and second order cumulants but all higher order cumulants are equal to zero. Often Gaussian processes are applied to physical problems simply on the basis of mathematical convenience; it is easy to perform many of the calculations in closed form when the underlying distributions are Gaussian. However, there is sometimes a deeper reason why such cavalier procedures do not lead to seriously incorrect results. The reason is to be found in the central limit theorem.[(2.5.1)]

The probability density $p(x,t)$ is <u>Gaussian</u> if it has the form

$$p(x,t) = \frac{1}{(2\pi)^{1/2}\,\sigma(t)} \exp\left\{ - \frac{[x - \mu_1(t)]^2}{2\sigma^2(t)} \right\} \quad (2.5.1)$$

where $\mu_1(t)$ is the first moment of the probability density as given by Eq. 2.3.2 and $\sigma^2(t)$ is the variance as given in Eq. 2.3.4. All the moments $\mu_j(t)$ of the Gaussian distribution for $j > 2$ can be expressed in terms of $\mu_1(t)$ and $\mu_2(t)$. The form of these higher moments $\mu_j(t)$ is such that all the cumulants $K(\nu)$ are zero for $\nu \geq 3$. As a matter of fact one can define a Gaussian distribution as one for which the cumulants of the distribution are equal to zero for $\nu \geq 3$.

The characteristic function $\phi(s,t)$ for the Gaussian probability density $p(x,t)$ of Eq. 2.5.1 is

$$\phi(s,t) = \exp\left[is\,\mu_1(t) - \frac{s^2\sigma^2(t)}{2} \right] \qquad (2.5.2)$$

a result which can readily be established from Eq. 2.3.12. Note the important result that the characteristic function also has a Gaussian form. Thus the Gaussian stochastic process can be characterized either by the Gaussian form of the probability density or, equivalently, by the Gaussian form of the characteristic function.

Considerable simplification of notation results when we consider stochastic variables $Y(t)$ which have a zero mean value and variance equal to one. The probability density $\tilde{p}(y,t)$ can readily be obtained from 2.5.1 if y is related to x by the equation

$$y = \frac{x - \mu_1(t)}{\sigma(t)} \qquad (2.5.3)$$

and if one uses the relation $\tilde{p}(y,t)dy = \dfrac{p(x,t)dx}{\sigma(t)}$.

We then find that

$$\tilde{p}(y,t) = \frac{1}{(2\pi)^{1/2}} e^{-\frac{y^2}{2}} . \qquad (2.5.4)$$

It follows from 2.5.4 that

$$\langle Y(t)\rangle \equiv \int_{-\infty}^{\infty} y\,\tilde{p}(y,t)dy = 0$$

$$\left\langle \left[Y(t) - \langle Y(t)\rangle \right]^2 \right\rangle = 1. \qquad (2.5.5)$$

The characteristic function $\phi(s,t)$ for the Gaussian probability density $\tilde{p}(y,t)$ of Eq. 2.5.4 becomes

$$\phi(s,t) = e^{-s^2/2}. \qquad (2.5.6)$$

Note that the probability density $\tilde{p}(y,t)$ and the characteristic function $\phi(s,t)$ do not depend explicitly on time. Therefore the process in the new variable $Y(t)$ is stationary as well as Gaussian.

The 2-variable probability density $p^{(2)}(x_1,x_2;t)$ is Gaussian if it has the form

$$p^{(2)}(x_1,x_2;t) = [2\pi\sigma_1(t)\sigma_2(t)(1 - \mathcal{K}^2)^{1/2}]^{-1}$$

$$\times \exp\left\{\frac{1}{2(\mathcal{K}^2 - 1)}\left[\left(\frac{x_1-\mu_{1,0}(t)}{\sigma_1(t)}\right)^2 + \left(\frac{x_2-\mu_{0,1}(t)}{\sigma_2(t)}\right)^2 - 2\mathcal{K}\left(\frac{x_1-\mu_{1,0}(t)}{\sigma_1(t)}\right)\left(\frac{x_2-\mu_{0,1}(t)}{\sigma_2(t)}\right)\right\} \qquad (2.5.7)$$

where $\sigma_1(t) \equiv \langle X_1^2\rangle - \langle X_1\rangle^2$, $\sigma_2(t) = \langle X_2^2\rangle - \langle X_2\rangle^2$ and where $\mathcal{K}$, commonly known as the <u>correlation coefficient</u>, is given by

$$\mathcal{K}(t) = \frac{\langle X_1X_2\rangle - \langle X_1\rangle \langle X_2\rangle}{\sigma_1(t)\sigma_2(t)} \quad . \tag{2.5.8}$$

In the space of the stochastic variable $\underline{Y}(t)$ with mean equal to zero and variance equal to one the correlation coefficient becomes

$$\tilde{\mathcal{K}}(t) = \langle Y_1Y_2\rangle \ . \tag{2.5.9}$$

The symbol $\tilde{\mathcal{K}}$ has been chosen here to emphasize the close relationship between the "correlation coefficient" and the normalized two-time cumulant $\tilde{K}(t_1;t_2)$ defined in Eq. 2.4.19. In the statistical literature, the expression $\langle X_1X_2\rangle - \langle X_1\rangle \langle X_2\rangle$ is known as the covariance. It can be shown from the Cauchy-Schwartz inequality(2.5.1) that $|\mathcal{K}(t)| \leq 1$. If the stochastic variables X_1 and X_2 are uncorrelated, $\mathcal{K}(t) = 0$ and Eq. 2.5.7 reduces to

$$p^{(2)}(x_1,x_2;t) = p(x_1,t)p(x_2,t). \tag{2.5.10}$$

An important property of Gaussian processes is that the joint probability densities can be factored into singlet probability densities even when the stochastic variables are correlated. Thus Eq. 2.5.7 can be rearranged to yield, using 2.5.4

$$\tilde{p}^{(2)}(y_1,y_2;t)$$

$$= [1 - \tilde{\mathcal{K}}^2(t)]^{-1/2}\tilde{p}(y_1)\ \tilde{p}\left[\frac{y_2 - \tilde{\mathcal{K}}(t)y_1}{[1 - \tilde{\mathcal{K}}^2(t)]^{1/2}}\right]. \tag{2.5.11}$$

If Y_1 and Y_2 are not correlated, Eq. 2.5.11 reduces to

$$\tilde{p}^{(2)}(y_1,y_2) = \tilde{p}(y_1)\tilde{p}(y_2) \ . \tag{2.5.12}$$

The characteristic function of the 2-variable probability density $p^{(2)}(x_1,x_2;t)$ is

$$\phi^{(2)}(s_1,s_2;t) = \exp\left[i\left(s_1\mu_{1,0}(t)+s_2\mu_{0,1}(t)\right) - \frac{1}{2}\left(s_1^2\sigma_1^2(t)+s_2^2\sigma_2^2(t)+2s_1s_2\sigma_1(t)\sigma_2(t)\mathcal{K}(t)\right)\right] \tag{2.5.13}$$

which is again a Gaussian form. For $\mathcal{K} = 0$, i.e., for uncorrelated stochastic variables,

$$\phi^{(2)}(s_1,s_2;t) = \phi(s_1,t)\phi(s_2,t) \tag{2.5.14}$$

where $\phi(s_j,t)$ is given by 2.5.2.

These results can be generalized to n-dimensional Gaussian processes.(2.5.2) The Gaussian form of the characteristic function implies the Gaussian form for the probability densities and that all cumulants $K(\nu)$ are zero for $\nu \geq 3$.

It is quite simple to calculate the moments and cumulants of a Gaussian process from the characteristic function. We shall carry out these calculations for a two-dimensional process in which the means of the two stochastic variables are zero. From the general relation [Eq. 2.3.21]

$$\langle X_1^{k_1} X_2^{k_2} \rangle = \frac{1}{i^k} \frac{\partial^k}{\partial s_1^{k_1} \partial s_2^{k_2}} \phi^{(2)}(\underline{s},t) \Big|_{\underline{s}=0} \qquad (2.5.15)$$

with $k = k_1 + k_2$ we find from Eq. 2.5.13 that

$$\langle X_1^{k_1} X_2^{k_2} \rangle = 0 \quad \text{for } k = \text{odd} \qquad (2.5.16)$$

and for k even

$$\langle X_1^{k_1} X_2^{k_2} \rangle = \frac{1}{2^{(k/2)}} \frac{k_1! k_2!}{(k/2)!}$$

$$\times \sum_{i_1=1}^{2} \cdots \sum_{i_k=1}^{2} (2 - \delta_{i_1 i_2})(2 - \delta_{i_3 i_4})$$

$$\cdots (2 - \delta_{i_{k-1} i_k}) \langle X_{i_1} X_{i_2} \rangle \langle X_{i_3} X_{i_4} \rangle \cdots$$

$$\cdots \langle X_{i_{k-1}} X_{i_k} \rangle \qquad (2.5.17)$$

where the summation is over all values of $(i_1, i_2, \ldots, i_k)$ such that the sum of exponents of X_1 is k_1, of X_2 is k_2. The factor $(2 - \delta_{ij})$ must be included

since the coefficient of $\langle X_i X_j \rangle$ is 2 if $i \neq j$ and 1 if $i = j$. Equations 2.5.16 and 2.5.17 show that all moments of the Gaussian distribution can be written in terms of the first and second moments.

As a simple example, let us compute $\langle X_1^2 X_2^2 \rangle$. In this example, $k_1 = 2$, $k_2 = 2$ and $k = 4$. Equation 2.5.17 now becomes

$$<X_1^2 X_2^2> = \frac{1}{2} \sum_{i_1} \cdots \sum_{i_4} (2 - \delta_{i_1 i_2})$$

$$\times (2 - \delta_{i_3 i_4}) <X_{i_1} X_{i_2}> <X_{i_3} X_{i_4}> \quad (2.5.18)$$

where $i_1,\ldots,i_4$ are either 1 or 2 subject to the condition that the sum of the exponents of X_1 is 2 and the sum of the exponents of X_2 is 2. This results in the following allowed combinations of i's:

i_1	1	2	1	1	2	2
i_2	1	2	2	2	1	1
i_3	2	1	1	2	1	2
i_4	2	1	2	1	2	1

These six combinations when used in Eq. 2.5.18 yield

$$<X_1^2 X_2^2> = <X_1^2> <X_2^2> + 2<X_1 X_2>^2. \quad (2.5.19)$$

The cumulants for a Gaussian distribution take on an especially simple form. From Eqs. 2.4.4 and 2.5.13 we obtain

$$K^{(2)}(\nu_1,\nu_2;t) = i^{-\nu} \frac{\partial^\nu}{\partial s_1^{\nu_1} \partial s_2^{\nu_2}}$$

$$[i(s_1\mu_{1,0}(t)+s_2\mu_{0,1}(t))$$

$$-\frac{1}{2}(s_1^2\sigma_1^2(t)+s_2^2\sigma_2^2(t)+2s_1s_2\sigma_1(t)\sigma_2(t)\mathcal{K}(t))] \quad (2.5.20)$$

evaluated at $s_1=s_2=0$ where $\nu = \nu_1+\nu_2$. Performing the indicated differentiations yields

$$K^{(2)}(1,0;t) = \mu_{1,0}(t)$$

$$K^{(2)}(0,1;t) = \mu_{0,1}(t)$$

$$K^{(2)}(1,1;t) = \sigma_1(t)\sigma_2(t)K(t) = \langle X_1X_2\rangle - \langle X_1\rangle\langle X_2\rangle$$

$$K^{(2)}(2,0;t) = \sigma_1^2(t) = \langle X_1^2\rangle - \langle X_1\rangle^2$$

$$K^{(2)}(0,2;t) = \sigma_2^2(t) = \langle X_2^2\rangle - \langle X_2\rangle^2$$

$$K^{(2)}(\nu_1,\nu_2;t) = 0 \qquad \text{for } \nu_1 + \nu_2 > 2. \qquad (2.5.21)$$

Thus the Gaussian distribution can be characterized by the requirement that all cumulants higher than second order are zero.

6) Gaussian Markov Processes

We shall now discuss some of the properties of stationary Gaussian Markov Processes which have an important bearing on a number of physical problems.

In the definition of the multivariate (i.e., n-dimensional) Gaussian distribution the random variables X_i may be considered components of a vector X at one time or as the values $X(t_i)$ of a scalar random variable $X(t)$ at time t_i. In the latter case, the moments $X(t_i)$ and the correlation function $\rho(t_i,t_j) = \langle X(t_i)X(t_j)\rangle - \langle X(t_i)\rangle\langle X(t_j)\rangle$ are functions of the time. For a stationary process the moments $\langle X(t_i)\rangle$ will be independent of time and the correlation function will be a function of the time difference $t_i - t_j$, i.e., $\rho(t_i,t_j) = \rho(|t_i-t_j|)$. Higher order correlation function can be defined in an analogous way, but since we are dealing here with Gaussian processes, the higher order correlation functions can all be written in terms of the first moment $\mu \equiv \langle X\rangle$ and the correlation function $\rho(t)$. Specifically, for a Gaussian process with zero mean, the correlation function $\rho_n(t_1,t_2,\ldots,t_n)$ can be written, for n even, as

$$\rho_n(t_1,t_2,\ldots\ldots,t_n) = \sum \rho(t_{i_1},t_{i_2})\rho(t_{i_3},t_{i_4}) \ldots\ldots \rho(t_{i_{n-1}},t_{i_n}). \tag{2.6.1}$$

The sum is taken over all partitions of the integers 1,2,...,n into pairs. For a stationary process, Eq. 2.6.1 becomes

$$\rho_n(t_1,t_2,\ldots\ldots,t_n) = \sum \rho(t_{i_1}-t_{i_2})\rho(t_{i_3}-t_{i_4})\ldots \rho(t_{i_{n-1}}-t_{i_n}) \tag{2.6.2}$$

In the subsequent development we restrict ourselves to stationary processes with mean zero.

From our previous definition 2.1.22, the joint probability density for a Markov process can be written as

$$p_r(x_1,t_1;x_2,t_2;\dots;x_r,t_r) = p(x_1,t_1)w(x_2,t_2|x_1,t_1) \times$$

$$w(x_3,t_3|x_2,t_2) \quad \cdots \quad w(x_r,t_r|x_{r-1},t_{r-1}). \qquad (2.6.3)$$

where $t_1 < t_2 < \dots < t_r$. A Gaussian Markov process is one which obeys Eq. 2.6.3 and whose joint probability density p_r has the Gaussian form. The conditional probability $w(x_2,t|x_1,0)$ can be written in the form

$$w(x_2,t|x_1,0) = \frac{p_2(x_1,0;x_2,t)}{p(x_1)} \qquad (2.6.4)$$

Use of Eqs. 2.5.11 and 2.5.4 for the probability densities p_2 and p then yields

$$w(x_2,t|x_1,0) = \left[2\pi\sigma^2\left(1 - \frac{\rho_1^2}{\sigma^4}\right)\right]^{-1/2} \exp\left[\frac{-\left(x_2 - \frac{\rho_1}{\sigma^2}x_1\right)^2}{2\sigma^2\left(1 - \frac{\rho_1^2}{\sigma^4}\right)}\right]$$

$$(2.6.5)$$

where $\rho \equiv \rho(t)$. Note that the conditional probability w also has the Gaussian form. The joint probability density p_r for a stationary Gaussian Markov process then becomes

$$p_r(x_1,t_1;x_2,t_2;\ldots;x_r,t_r) =$$

$$\frac{1}{(2\pi\sigma^2)^{r/2}\prod_{j=1}^{r=1}\left(1-\frac{\rho^2(j)}{\sigma^4}\right)^{1/2}} \exp\left[-\left\{\frac{1}{2\sigma^2}\,x_1^2 + \right.\right.$$

$$\left.\left.\sum_{j=1}^{r-1}\frac{\left(x_{j+1}-\frac{\rho(j)}{\sigma^2}\,x_j\right)^2}{\sigma^2\left(1-\frac{\rho^2(j)}{\sigma^4}\right)}\right\}\right] \tag{2.6.6}$$

where $\rho(j) \equiv \rho(t_{j+1}-t_j)$.

We shall now demonstrate that the correlation function $\rho(j)$ is determined by the Markovian character of the stochastic process. To do this we compute $\rho(t_3-t_1)$ from (2.6.6) by evaluating

$\iiint x_1x_3p_3(x_1,t_1;x_2,t_2;x_3,t_3)dx_1dx_2dx_3$ to find

$$\sigma^2\rho(t_3-t_1) = \rho(t_3-t_2)\rho(t_2-t_1) \tag{2.6.7}$$

Eq. 2.6.7 implies that $\sigma^2\rho(s+t) = \rho(s)\rho(t)$ for all s, $t>0$. Now, since $\rho(t)$ is an even function of t and $|\rho(t)|<\sigma^2$, either $\rho(t) \equiv 0$ (which would imply an independent process) or

$$\rho(t) \equiv \langle X(0)X(t)\rangle = \sigma^2 e^{-\beta t} \tag{2.6.8}$$

where $\beta > 0$. For a stationary Gaussian Markov process, the correlation function thus relaxes exponentially with time. The case $\beta = 0$ corresponds to $\rho(t) = \sigma^2$ in which case the conditional probability density $w(x_2,t|x_1,0)$ of Eq. 2.6.5 becomes $\delta(x_2-x_1)$ and the joint probability density p_r becomes

$$p_r(x_1,t_r;x_2,t_2;\ldots;x_r,t_r) = p(x_1)\delta(x_2-x_1)\delta(x_3-x_2)$$

$$\ldots\delta(x_r-x_{r-1}) \; . \qquad (2.6.9)$$

This corresponds to a stochastic process in which the stochastic variable X is not a function of time. The result in Eq. 2.6.8 for the exponential decay of the correlation function of a Gaussian Markov process has been discussed in detail by Doob (2.1.1). The result in Eq. 2.6.8 for the correlation function can be generalized to the multidimensional Gaussian Markov process. It can be shown (2.2.5) that if $\underset{\sim}{X}(t)$ is to be described by a Gaussian Markov process, then the matrix with element $\langle X_i(t)X_j(t+\tau)\rangle - \langle X_i(t)\rangle\langle X_j(t+\tau)\rangle$ must have a time dependendence of the form exp $(\underset{\sim}{Q}\tau)$ where $\underset{\sim}{Q}$ is a matrix independent of time.

References

2.1.1. J. L. Doob, Stochastic Processes, John Wiley and Sons, New York, 1953.

2.1.2. For a discussion of independent processes see, for instance, W. Feller, An Introduction to Probability Theory and its Applications, Vol. 1, John

Wiley and Sons, New York, 1957.

2.1.3. A. A. Markov, Izv. Akad. Nauk. SPG VI, Ser. 3, 61 (1907).

2.1.4. For a discussion of this definition see I. Oppenheim and K. E. Shuler, Phys. Rev. 138, B1007 (1965).

2.1.5. N. Levinson and H. P. McKean, Bull. Am. Math. Soc. 70, 128 (1964), H. P. McKean, Akad. Nauk. 8, 357(1963).

2.2.1. a) M. V. Smoluchowski, Ann. Phys. 21, 756 (1906); b) A. Kolmogoroff, Math. Ann. 104, 415 (1931); c) S. Chapman, Proc. Roy. Soc. A216, 279 (1916).

2.2.2. P. Levy, Compt. Rend. 228, 2004 (1949); C. J. Burke and M. Rosenblatt, Ann. Math. Stat. 29, 1112 (1958); W. Feller, Ann. Math. Stat. 30, 1252 (1959).

2.2.3. P. C. Hemmer, Dynamic and Stochastic Types of Motion in the Linear Chain, Thesis, University of Trondheim, 1959.

2.2.4. R. J. Rubin, J. Math. Phys. 2, 373 (1961).

2.2.5. See e.g., A. Kolmogoroff, ref. 2.2.1b; W. Feller, Trans. Amer. Math. Soc. 48, 488 (1940); M. C. Wang and G. E. Uhlenbeck, Rev. Mod. Phys. 17, 323 (1945).

2.3.1. J. A. Shohat and J. D. Tamarkin, The Problem of Moments, Amer. Math. Soc., New York, 1943.

2.4.1. M. G. Kendall and A. Stuart, The Advanced Theory of Statistics,, Vol. I, Stechert Hafner, New York, 1963.

2.4.2. R. Kubo, J. Phys. Soc. (Japan) 17, 1100 (1962).

2.4.3. H. D. Ursell, Proc. Camb. Phil. Soc. 23, 685 (1927).

2.4.4. E. G. D. Cohen, Physica 28, 1025 (1962) and subsequent papers.

2.4.5. B. Kahn (Thesis, Leiden University, 1938); reprinted in Studies in Statistical Mechanics, III; edited by J. de Boer and G. E. Uhlenbeck, Interscience Publishers, New York, 1965.

2.5.1. B. V. Gnedenko and A. N. Kolmogorov, Limit Distributions for Sums of Independent Random Variables, Addison-Wesley, Cambridge, Mass. 1954.

2.5.2. H. Cramer, Mathematical Methods of Statistics, Princeton University Press, Princeton, N. J., 1946, Ch. 24.

Chapter 3

THE MASTER EQUATION

In this chapter, we derive and discuss equations which describe the temporal development of joint and conditional probabilities of stochastic processes. Specifically, this chapter is devoted to a discussion of the master equation, which is, depending upon the nature of the state space, either a set of differential-difference or integro-differential equations for the probabilities. As is shown in the body of this chapter, the master equation describes the temporal behavior of the singlet and multivariate (one-time) probabilities for all stochastic processes. The master equation describes the temporal behavior of the conditional probabilities, however, only for Markov processes.

In Section 1, we present the derivation of the master equation and discuss the connection between master equations and Markov processes. Section 2 contains a discussion of the properties of the transition rates entering into the master equations. The formal solution of the master equation in terms of eigenfunction expansions is discussed in Section 3. Section 4 deals with the master equation in open systems and mean first passage times. In Section 5 we develop generating function techniques for the solution of master equations. In Section 6, we generalize the development of the previous section to multivariate processes.

1) Derivation of the Master Equation

In this section we derive equations which describe the temporal development of the lower order joint and conditional probabilities of arbitrary stochastic processes. We then specialize our results to Markov processes. In the physical literature the differential equations for the singlet probabilities are known as the master equations.(3.1.1, 3.1.2) For Markov processes, the differential equations for the conditional probabilities have a form analogous to the master equation; in the mathematical literature these are known as Kolmogorov's forward differential equations.(3.1.3) The master equation has been applied successfully to the description of a number of physical processes.

We start our derivation by rewriting Eq. 2.2.1 in the form

$$P_{r-1}(n_1,t_1;\ldots;n_{r-2},t_{r-2};n,t+\Delta)$$
$$= \sum_m W_r(n,t+\Delta|m,t;\ldots;n_1,t_1)$$
$$\times P_{r-1}(n_1,t_1;\ldots;m,t) \qquad (3.1.1)$$

where we have replaced n_r by n, t_r by $t+\Delta$, n_{r-1} by m, and t_{r-1} by t with $t+\Delta>t>t_{r-2}>\ldots>t_1$. Subtraction of the identity

$$P_{r-1}(n_1,t_1;\ldots;n_{r-2},t_{r-2};n,t) =$$

$$= \sum_m \delta_{n,m} P_{r-1}(n_1,t_1;\ldots;n_{r-2},t_{r-2};m,t) \qquad (3.1.2)$$

from Eq. 3.1.1 and division of the result by Δ leads to

$$\frac{1}{\Delta}\Big\{P_{r-1}(n_1,t_1;\ldots;n_{r-2},t_{r-2};n,t+\Delta) - P_{r-1}(n_1,t_1;\ldots;n_{r-2},t_{r-2};n,t)\Big\}$$

$$= \sum_m \frac{[W_r(n,t+\Delta|m,t;\ldots;n_1,t_1) - \delta_{n,m}]}{\Delta}$$

$$\times P_{r-1}(n_1,t_1;\ldots;n_{r-2},t_{r-2};m,t). \qquad (3.1.3)$$

We now wish to take the limit of Eq. 3.1.3 as $\Delta\to 0$. We already know that [Eq. 2.1.7]

$$\lim_{\Delta\to 0} W_r(n,t+\Delta|m,t;\ldots;n_1,t_1) = \delta_{n,m}. \qquad (3.1.4)$$

This quite reasonable relation is equivalent to the statement that the probability of the stochastic variables $X(t)$ taking simultaneously both of the values n and m, $n \neq m$, at time t is zero. To proceed further we postulate the existence of a function[(3.1.3)]

$$A_r(n,m,t;n_{r-2},t_{r-2};\ldots;n_1,t_1)$$

$$= \lim_{\Delta\to 0} \frac{W_r(n,t+\Delta|m,t;\ldots;n_1,t_1) - \delta_{n,m}}{\Delta} \qquad (3.1.5)$$

where the limit is assumed to exist for all values of n, m, $n_{r-2},\ldots,n_1$. Passage to the limit $\Delta\to 0$ in Eq.

3.1.3 now yields

$$\frac{\partial P_{r-1}(n_1,t_1;\ldots;n_{r-2},t_{r-2};n,t)}{\partial t}$$

$$= \sum_m A_r(n,m,t;n_{r-2},t_{r-2};\ldots;n_1,t_1)$$

$$\times P_{r-1}(n_1,t_1;\ldots;n_{r-2},t_{r-2};m,t). \quad (3.1.6)$$

Since, by definition,

$$P_{r-1}(n_1,t_1;\ldots;n_{r-2},t_{r-2},n,t)$$

$$= W_{r-1}(n,t|n_{r-2},t_{r-2};\ldots;n_1,t_1)$$

$$\times P_{r-2}(n_1,t_1;\ldots;n_{r-2},t_{r-2}), \quad (3.1.7)$$

Equation 3.1.6 can be rewritten as

$$\frac{\partial W_{r-1}(n,t|n_{r-2},t_{r-2};\ldots;n_1,t_1)}{\partial t}$$

$$= \sum_m A_r(n,m,t;n_{r-2},t_{r-2};\ldots;n_1,t_1)$$

$$\times W_{r-1}(m,t|n_{r-2},t_{r-2};\ldots;n_1,t_1). \quad (3.1.8)$$

Equations 3.1.6 and 3.1.8 are differential equations for the joint and conditional probabilities P_{r-1} and W_{r-1} respectively. Once the function A_r is known, Eqs. 3.1.6 and 3.1.8 can be solved, in principle, as initial value problems for P_{r-1} or W_{r-1}.

It is evident from the rate equations 3.1.6 and 3.1.8 that, for $m \neq n$, the A_r are the transition rates, i.e., $A_r dt$ is equal to the conditional probability for transitions from state m to n in the time interval (t,t+dt) under the condition that the stochastic variable had the value n_{r-2} at time $t_{r-2}, \ldots,$ and the value n_1 at time t_1.

For physical applications, we will be particularly interested in the singlet probabilities P(n,t) and the two-time conditional probabilities $W(n,t|n_1,t_1)$. Equation 3.1.6 for r = 2 becomes

$$\frac{\partial P(n,t)}{\partial t} = \sum_m A_{nm}(t)P(m,t) \qquad (3.1.9)$$

where we have denoted $A_2(n,m,t)$ by $A_{nm}(t)$. Equation 3.1.9 is valid for all stochastic processes for which A exists. To obtain the analogous equation for $W_2 \equiv W$, we set r = 3 in Eq. 3.1.8,

$$\frac{\partial W(n,t|n_1,t_1)}{\partial t} = \sum_m A_3(n,m,t;n_1,t_1)W(m,t|n_1,t_1). \qquad (3.1.10)$$

Equation 3.1.10 holds for all stochastic processes for which A_3 exists. For Markov processes,

$$A_r(n,m,t;n_{r-2},t_{r-2};\ldots;n_1,t_1) = A_{nm}(t) \qquad (3.1.11)$$

which follows from the definition of a Markov process in terms of the conditional probabilities [Eq. 2.1.18 and Eq. 3.1.5]. Substitution of Eq. 3.1.11 into Eq. 3.1.10 yields

$$\frac{\partial W(n,t|n_1,t_1)}{\partial t} = \sum_m A_{nm}(t)W(m,t|n_1,t_1). \qquad (3.1.12)$$

Equation 3.1.12 holds only for Markovian (and pseudo Markovian) processes. It is a closed equation for the conditional probabilities W since the transition rate A in turn is defined in terms of W. Thus, the conditional probability W is uniquely determined by Eq. 3.1.12, the initial condition 3.1.4 and the requirement that $0 \leq W \leq 1$ [Eq. 2.1.5].

We shall refer to Eq. 3.1.9 for the singlet probability P(n,t) as the master equation. This nomenclature is in accord with the usage in the literature of physics and chemistry.(3.1.2)

The continuous state space analogue of Eq. 3.1.9 is

$$\frac{\partial p(x,t)}{\partial t} = \int dy\ a(x,y,t)p(y,t) \qquad (3.1.13)$$

where the transition rate a(x,y,t) is given by

$$a(x,y,t) = \lim_{\Delta \to 0} \frac{w(x,t+\Delta|y,t) - \delta(x-y)}{\Delta}. \qquad (3.1.14)$$

The master equations 3.1.9 and 3.1.13 can readily be generalized to vector processes.

Since Eq. 3.1.12 for the conditional probabilities W is valid only for Markov processes it should properly be referred to as the Markov equation. However, since mathematically Eq. 3.1.12 has the same form as Eq. 3.1.9 we shall refer to both equations as "master equations" in discussing their properties and solutions. In those cases where our discussion refers specifically to the kinetic equation 3.1.12 for the conditional probability W, we shall denote it explicitly as the Markov equation.

The continuous state space analogue of Eq. 3.1.12 is

$$\frac{\partial w(x,t|x_1,t_1)}{\partial t} = \int dy\ a(x,y,t)w(y,t|x_1,t_1) \qquad (3.1.15)$$

with a(x,y,t) defined in Eq. 3.1.14.

It is of interest to inquire about what information is contained in the transition rate $A_{nm}(t)$ or, equivalently, its continuous state analogue a(x,y,t). We can make the following statements:(3.1.2)

1. Given $A_{nm}(t)$ for a stochastic process and no other information, it is not possible to determine whether the process is Markovian or non-Markovian. It is clear [compare Eqs. 3.1.10 and 3.1.12], that it is not possible to find the conditional probability $W(n,t|n_1,t_1)$. On the other hand, it follows from Eq. 3.1.9 that it is possible to determine the probability P(n,t) once the set of P(m,0) is given.

2. Given $A_{nm}(t)$ and the information that the stochastic process is non-Markovian, it is not possible to determine $W(n,t|n_1,t_1)$ since its time dependence is described by Eq. 3.1.10 which involves the transition rate A_3. Again, P(n,t) can be determined from Eq. 3.1.9 for a given initial set {P(n,0)}.

3. Given $A_{nm}(t)$ and the information that the stochastic process is Markovian, it is possible to determine the conditional probability $W(n,t|n_1,t_1)$ by making use of Eq. 3.1.12. Again $P(n,t)$ can be determined from Eq. 3.1.9 and the set $\{P(n,0)\}$.

For the special initial condition

$$P(n,0) = \delta_{n,n_1} \tag{3.1.16}$$

where n_1 is some fixed value of the stochastic variable, one obtains from Eq. 2.2.1

$$P(n,t) = W(n,t|n_1,0). \tag{3.1.17}$$

Substitution of Eq. 3.1.17 into Eq. 3.1.9 yields

$$\frac{\partial W(n,t|n_1,0)}{\partial t} = \sum_m A_{nm}(t)W(m,t|n_1,0). \tag{3.1.18}$$

Equation 3.1.18 is valid for all stochastic processes which have an initial condition given by Eq. 3.1.16. It can be solved for $W(n,t|m,0)$ once $A_{nm}(t)$ is given. While Eq. 3.1.18 is of the same form as the Markov equation 3.1.12, it should be noted that Eq. 3.1.18 is valid only for conditional probabilities of the form $W(n,t|m,0)$, i.e., the earlier time ($t = 0$) is that time for which the initial condition 3.1.16 applies.

The two-time conditional probability $W(n,t|n_1,t_1)$ or $w(x,t|x_1,t_1)$ is one of the most important functions in the theory of stochastic processes as regards its usefulness in physical application. The autocorrelation function $\rho(t,t_1)$, for instance, can be written in terms of the conditional probability W. The transport coefficients of non-equilibrium statistical mechanics, in turn, can be expressed in terms of the autocorrelation function through the Green-Kubo formalism.[3.2.1] Thus, the stochastic formulation of time-dependent processes in statistical mechanics can readily be based on the use of conditional probabilities.

2) Properties of the Transition Rates

We shall now discuss the significance and some of the properties of the transition rates. We limit our discussion to the case of stochastic processes in discrete state space. In this chapter the indices of the transition rates A_{ij} are always written such that the transition takes place from the right index (state j) to the left index (state i). The extension to processes in continuous state space can readily be carried out.

It might seem at first that in writing the master equation 3.1.12 we have chosen a notation which conceals its non-linearity since the transition rates $A_{nm}(t)$ after all depend upon the conditional probability W through Eq. 3.1.5. In physical applications one usually determines $W(n,t+\Delta|m,t)$ in the form

$$W(n,t+\Delta|m,t) = \delta_{n,m} + A_{nm}(t)\Delta + o(\Delta) \tag{3.2.1}$$

where $o(\Delta)$ has the property

$$\lim_{\Delta \to 0} \frac{o(\Delta)}{\Delta} = 0. \tag{3.2.2}$$

The form of the conditional probability shown in Eq. 3.2.1 arises naturally in perturbation calculations and phenomenological models of stochastic processes. It is clear from Eq. 3.2.1 that $A_{nm}(t)dt$, for $n \neq m$, is equal to the conditional probability for the transition $m \to n$ in the time interval $(t,t+dt)$. Once the conditional probability $W(n,t+dt|m,t)$ or, equivalently, $A_{nm}(t)$ has been determined, one can solve the master equation 3.1.12 for the desired conditional probability $W(n,t|n_1,t_1)$.

To demonstrate more clearly the concepts discussed above, we present a simple example. We consider a system composed of a very large number of radioactive atoms which emit α particles. The stochastic variable $X(t)$ is the number of α particles emitted in the time interval $(0,t)$. The quantity $P(n,t)$ is then defined as the probability that n particles have been emitted in the time interval $(0,t)$, and $W(n,t|n_1,t_1)$ is the conditional probability that if n_1 particles have been emitted in the time interval $(0,t_1)$, then n particles will have been emitted in the interval $(0,t)$ with $t > t_1$. Note that in this particular example, $W(n,t|n_1,t_1)$ is zero if $n < n_1$. We now assume that the probability of the emission of a single α particle during the time interval $(t,t+dt)$ is Λdt where Λ

is a constant. The probability that two or more particles are emitted during the time interval $(t,t+dt)$ is $o(dt)$ and thus can be taken to be zero. We may thus write

$$W(n,t+dt|n-1,t) = \Lambda dt$$

$$W(n,t+dt|m,t) = 0 \quad \text{for } m \neq n,\ n-1$$

$$W(n,t+dt|n,t) = 1 - \Lambda dt. \qquad (3.2.3)$$

Note that

$$\sum_n W(n,t+dt|m,t) = 1. \qquad (3.2.4)$$

It is clear from Eqs. 3.2.3 and 3.2.1 that the rate constant Λ is related to the transition rate A_{nm} by

$$A_{n,n-1}(t) = \Lambda \qquad n = 1,2,\ldots$$

$$A_{nn}(t) = -\Lambda \qquad n = 0,1,2,\ldots$$

$$A_{nm} = 0 \qquad \text{for } m \neq n,\ n-1. \qquad (3.2.5)$$

Since we have only the transition rate $A_{nm}(t)$ and no other information, it is not possible to determine whether the stochastic process of α-particle emission is Markovian or non-Markovian (see discussion of previous section). It is, however, reasonable to assume that this stochastic process is indeed Markovian. The conditional probability $W(n,t|n_1,t_1)$ can then be obtained from the master equation 3.1.12 in the form

$$\frac{\partial W(n,t|n_1,t_1)}{\partial t} = \sum_m A_{nm}(t)W(m,t|n_1,t_1) = \Lambda W(n-1,t|n_1,t_1) - \Lambda W(n,t|n_1,t_1) \quad (3.2.6)$$

under the condition $W(n,t|n_1,t_1) = 0$ for $n < n_1$. The probability $P(n,t)$ can be obtained in an analogous way from Eq. 3.1.9.

Several simple, but important, properties of the transition rates $A_{nm}(t)$ follow immediately from their definition, Eq. 3.1.5. If we sum Eq. 3.1.5 over all final states n and use Eq. 2.1.5 we find that

$$\sum_n A_r(n,m,t;n_{r-2},t_{r-2};\ldots;n_1,t_1) = 0 \quad (3.2.7)$$

which is an expression for the conservation of probability. For $r = 2$, this reduces to

$$\sum_n A_{nm}(t) = 0. \quad (3.2.8)$$

Furthermore, since the conditional probabilities are non-negative, the off-diagonal elements $A_{nm}(t)$, $n \neq m$, defined by

$$A_{nm}(t) = \lim_{\Delta\to 0} \frac{W(n,t+\Delta|m,t)}{\Delta} \quad (3.2.9)$$

are non-negative. On the other hand, the diagonal elements $A_{nn}(t)$ must be negative in order that Eq. 3.2.8 be satisfied, i.e.,

$$A_{nn}(t) = -\sum_m{}' A_{mn}(t) \quad (3.2.10)$$

where the prime on the sum indicates that the sum is over all values of m except m = n. The transition rates A_{nm} given in Eq. 3.2.5 clearly have the properties discussed above.

The master equations 3.1.9 and 3.1.12 can be re written, using Eq. 3.2.10, in the form

$$\frac{\partial P(n,t)}{\partial t} = \sum_m{}' A_{nm}(t)P(m,t) - P(n,t) \sum_m{}' A_{mn}(t), \quad (3.2.11)$$

or

$$\frac{\partial P(n,t)}{\partial t} = \sum_m [B_{nm}(t)P(m,t) - B_{mn}(t)P(n,t)] \qquad (3.2.12)$$

where

$$B_{nm}(t) = A_{nm}(t) \geq 0 \qquad \text{for } n \neq m \qquad (3.2.13)$$

and where $B_{nn}(t)$ is arbitrary since the term with m = n in Eq. 3.2.12 is zero. It follows from this that the sum over all m of $B_{mn}(t)$ need not be equal to zero. The master equation as written in Eq. 3.2.12 clearly exhibits the conservation of probability. The first sum of the r.h.s. represents transitions to state n from all other states and the second sum represents transitions from state n into all other states.

The preceding development can be extended in a straightforward manner to the case where the states form a continuum rather than discrete set. Thus, for instance, the master equation 3.2.12 now reads

$$\frac{\partial p(x,t)}{\partial t} = \int dy\ [b(x,y,t)p(y,t) - b(y,x,t)p(x,t)] \tag{3.2.14}$$

where $b(x,y,t) = a(x,y,t)$ for $x \neq y$ and where $b(y,y,t)$ is arbitrary.

In many physical applications, the transitions from one state of the system to another are due to interactions between the system and a reservoir whose properties remain constant in time. Under these conditions, although the quantity $A_{mn}P(n,t)$ varies with time, the transition rate itself is independent of time, i.e., $A_{mn}(t) = A_{mn}$. Such processes are called temporally homogeneous. For such processes, the conditional probability becomes a function of the time difference $(t - t_1)$ only and we can write the master equation 3.1.12 as

$$\frac{\partial W(n,\tau|n_1)}{\partial \tau} = \sum_m A_{nm}W(m,\tau|n_1) \tag{3.2.15}$$

where $\tau = t - t_1$. The quantity $W(n,\tau|n_1)$ is the conditional probability that if the stochastic variable had the value n_1, it will have the value n at a time interval τ later. It should be noted that while for a temporally homogeneous process the conditional probability $W(n,t|n_1,t_1)$ is a function of $t - t_1$ only, the joint probability $P_2(n_1,t_1;n,t)$ is a function of both t_1 and t and the singlet probability $P(n,t)$ is a function of t. This differs from the case of a stationary process (see Eqs. 2.1.4 and 2.1.8) where both the conditional probability and the joint probability are functions of the time interval $t - t_1$ only and where the singlet probability is independent of time.

3) Formal Solution of the Master Equation

In this section we discuss the formal solution of the master equation 3.1.9. We limit our consideration initially to temporally homogeneous processes for which the transition rates A_{nm} are constants, independent of time.

We shall consider first the formal solution of the master equation for a discrete state space. It is convenient for what follows to introduce the column vector $\underset{\sim}{P}(t) = (P(0,t), P(1,t), P(2,t)....)$ and the transition rate matrix $\underset{\sim}{A} = (A_{nm})$. In this notation we write the master equation 3.1.9 as

$$\frac{\partial \underset{\sim}{P}(t)}{\partial t} = \underset{\sim}{A}\,\underset{\sim}{P}(t)\,. \tag{3.3.1}$$

with the formal solution

$$\underset{\sim}{P}(t) = e^{\underset{\sim}{A}t}\,\underset{\sim}{P}(0). \tag{3.3.2}$$

It can be shown(3.3.1) that the conditions $A_{nn} \leq 0$, $A_{nm} \geq 0$, $(n \neq m)$, imply that the elements of $\exp(\underset{\sim}{A}t)$ are non-negative so that $\underset{\sim}{P}(t)$ is positive and bounded.

An alternative, and often quite useful, form of the

solution of the master equation can be written in terms of the eigenvalues and eigenfunctions of $\underset{\sim}{A}$. In the development below, we assume that the eigenvalues of $\underset{\sim}{A}$, denoted by $[\lambda_j]$, are discrete, and nondegenerate. In general $\underset{\sim}{A}$ has two sets of eigenfunctions, left and right eigenfunctions which we denote, respectively, by $\underset{\sim}{L}_j'$ and $\underset{\sim}{R}_j$ and which are solutions to

$$\underset{\sim}{L}_j' \underset{\sim}{A} = \lambda_j \underset{\sim}{L}_j'$$

$$j = 0, 1, 2, 3, \ldots$$

$$\underset{\sim}{A} \underset{\sim}{R}_j = \lambda_j \underset{\sim}{R}_j \tag{3.3.3}$$

where the prime denotes a row vector. The nondegeneracy of the λ_j implies that the $\underset{\sim}{L}_j'$ and $\underset{\sim}{R}_j$ are orthonormal,(3.3.1)

$$\underset{\sim}{L}_j' \underset{\sim}{R}_k = \delta_{jk} \tag{3.3.4}$$

where δ_{jk} is a Kronecker delta, since

$$\underset{\sim}{L}_j' \underset{\sim}{A} \underset{\sim}{R}_k = \lambda_j \underset{\sim}{L}_j' \underset{\sim}{R}_k = \lambda_k \underset{\sim}{L}_j' \underset{\sim}{R}_k . \tag{3.3.5}$$

Hence $\underset{\sim}{P}(t)$ can be expanded in terms of the complete set $\underset{\sim}{R}_j$ as follows:

$$\underset{\sim}{P}(t) = \sum_j c_j(t) \underset{\sim}{R}_j \tag{3.3.6}$$

where the $c_j(t)$ are given by

$$c_j(t) = \underset{\sim}{L}_j' \underset{\sim}{P}(t) = \underset{\sim}{L}_j' e^{\underset{\sim}{A}t} \underset{\sim}{P}(0) = \underset{\sim}{L}_j' \underset{\sim}{P}(0) e^{\lambda_j t} . \tag{3.3.7}$$

This last equality holds in consequence of the identities $\underset{\sim}{L}_j' \underset{\sim}{A}^n = \lambda_j^n \underset{\sim}{L}_j'$ and

$$\underset{\sim}{L}_j' e^{\underset{\sim}{A}t} = \sum_{n=0}^{\infty} \frac{t^n}{n!} \underset{\sim}{L}_j' \underset{\sim}{A}^n = \sum_{n=0}^{\infty} \frac{t^n}{n!} \lambda_j^n \underset{\sim}{L}_j' = \underset{\sim}{L}_j' e^{\lambda_j t}. \qquad (3.3.8)$$

The final expression for $\underset{\sim}{P}(t)$ now becomes

$$\underset{\sim}{P}(t) = \sum_{j=0}^{\infty} (\underset{\sim}{L}_j' \underset{\sim}{P}(0)) \, \underset{\sim}{R}_j e^{\lambda_j t}. \qquad (3.3.9)$$

Equation 3.3.9 is only a formal solution as it stands; in order for it to be useful it must be ascertained that the λ_n have negative or zero real parts, so that the probabilities remain bounded. If there is conservation of probability, so that the condition 3.2.8 applies, the matrix $\underset{\sim}{A}$ always has at least one zero eigenvalue, λ_o. If $\underset{\sim}{A}$ has a finite number of elements one can show that the real parts of the eigenvalues are non-positive by using Gershgorin's theorem. [(3.3.2)] When $\underset{\sim}{A}$ is infinite a more complicated proof is required, but the same result is obtained for all systems of physical interest. Hence the expansion of $\underset{\sim}{P}(t)$ in Eq. 3.3.9 can be written

$$\underset{\sim}{P}(t) = c_o \underset{\sim}{R}_o + \sum_{j=1}^{\infty} c_j \underset{\sim}{R}_j e^{\lambda_j t} \qquad (3.3.10)$$

where $c_j = \underset{\sim}{L}_j' \underset{\sim}{P}(0)$. The equilibrium state as $t \to \infty$, $\underset{\sim}{P}(\infty)$, is given by

$$\underset{\sim}{P}(\infty) = c_o \underset{\sim}{R}_o \qquad (3.3.11)$$

provided that the λ_j do not cluster around 0 and that $\lambda_o = 0$ is not a degenerate eigenvalue. The case where the eigenvalues have a limit point at zero will be discussed below.

No general necessary and sufficient conditions are known which insure that the λ_j do not have a limit point at zero but sufficient conditions are known (ref. 3.3.3) for matrices $\underset{\sim}{A}$ with elements of the form $A_{nm} = 0$ for $|n-m| > 1$. Such matrices which are known as Jacobian tridiagonal matrices apply to processes where transitions are allowed only between nearest neighbor states.

When there are degenerate eigenvalues, the solution $\underset{\sim}{P}(t)$ is no longer given by Eq. 3.3.10. When one of the eigenvalues λ_k, with $k \neq 0$, is r-fold degenerate, the solution $\underset{\sim}{P}(t)$ becomes

$$\underset{\sim}{P}(t) = c_o \underset{\sim}{R}_o + \sum_{j=1}^{\infty}{}' c_j \underset{\sim}{R}_j e^{\lambda_j t} + e^{\lambda_k t} \sum_{n=o}^{r-1} d_n \hat{\underset{\sim}{R}}_k(n) t^n \tag{3.3.12}$$

where the prime on the first sum indicates the absence of the j = k term in the summation. The symbol $\hat{\underset{\sim}{R}}_k(n)$ stands for a linear combination of eigenvectors belonging to the eigenvalue λ_k. The case where $\lambda_o = 0$ is a degenerate eigenvalue is considered below in connection with the discussion of the equilibrium solution $\underset{\sim}{P}(\infty)$.

In most physical applications of the master equation,

microscopic reversibility applies to the elementary dynamical events of the stochastic process. This implies that detailed balance obtains at equilibrium. The principle of detailed balance asserts that the elements of $\underset{\sim}{A}$ satisfy the equation

$$A_{mn} P(n,\infty) = A_{nm} P(m,\infty) \qquad (3.3.13)$$

for all n and m. The principle of detailed balance is a statement of the fact that each term in the sum over m in Eq. 3.2.11 is zero at equilibrium. For these important physical processes, one can prove the even stronger property that all of the non-zero eigenvalues of $\underset{\sim}{A}$ are real and negative.(3.3.4)

In order to prove reality of the eigenvalues λ_j we will show that the matrix $\underset{\sim}{A}$ can always be transformed to a symmetric matrix $\underset{\sim}{S}$ by a similarity transformation. Since the eigenvalues of a real symmetric matrix must necessarily be real,(3.3.1) it follows that those of $\underset{\sim}{A}$ are real. The matrix $\underset{\sim}{S}$ is constructed from $\underset{\sim}{A}$ by defining its elements to be

$$S_{nm} = [P(n,\infty)]^{-1/2} A_{nm} [P(m,\infty)]^{1/2} \qquad (3.3.14)$$

whence symmetry follows from Eq. 3.3.13. This is a similarity transformation $\underset{\sim}{S} = \underset{\sim}{U}^{-1} \underset{\sim}{A} \underset{\sim}{U}$ where $U_{ij} = [P(i,\infty)]^{1/2} \delta_{i,j}$. One can also verify that $\underset{\sim}{S}$ is the matrix corresponding to a negative semi-definite quadratic form, thereby confirming the fact that non-zero eigenvalues are negative. For this purpose we return to the expression for the A_{nm} in terms of the non-negative transition rates B_{nm} [see Eq. 3.2.13] and write

$$A_{nm} = (1 - \delta_{n,m})B_{nm} - \delta_{n,m} \sum_{r \neq n} B_{rn} \tag{3.3.15}$$

which is to be inserted into the expression for the quadratic form

$$\begin{aligned} S(y,y) &= \sum_{n,m} S_{nm} y_n y_m \\ &= \sum_{n,m} [P(n,\infty)]^{-1/2} [P(m,\infty)]^{1/2} y_n y_m \\ &\quad \times \{(1 - \delta_{n,m})B_{nm} - \delta_{nm} \sum_{r \neq n} B_{rn}\} \\ &= - \sum_{\substack{r,n \\ r \neq n}} B_{rn} y_n^2 + \sum_{\substack{n,m \\ n \neq m}} B_{nm} [P(m,\infty)]^{1/2} \\ &\quad \times [P(n,\infty)]^{-1/2} y_n y_m . \end{aligned} \tag{3.3.16}$$

If we now make use of the principle of detailed balance as expressed in Eq. 3.3.13, we find that $S(y,y)$ can be expressed as

$$\begin{aligned} S(y,y) &= - \frac{1}{2} \sum_{n,m} B_{nm} P(m,\infty) \\ &\quad \times \left[\frac{y_n}{P^{1/2}(n,\infty)} - \frac{y_m}{P^{1/2}(m,\infty)} \right]^2 \end{aligned} \tag{3.3.17}$$

which is manifestly negative semi-definite.

The time development of the probability $\underset{\sim}{P}(t)$ in Eq. 3.3.10 depends upon the eigenvalues λ_j, the eigenfunctions $\underset{\sim}{R}_j$ and $\underset{\sim}{L}_j'$ and the initial condition $\underset{\sim}{P}(0)$. While it is possible to derive, as shown above, some important general results on the properties of the eigenvalues λ_j from the general conditions, 3.2.8 and 3.2.9, on the transition rate matrix $\underset{\sim}{A}$, one cannot obtain similar general results for the eigenfunctions.

For the discrete state space discussed here, the eigenvalue spectrum may have several forms. When the number of states is finite, there will be a finite number of eigenvalues and the spectrum will be discrete. When the number of states is infinite, there will be an infinity of eigenvalues, and the spectrum may either be discrete or have limit points. While the time behavior of $\underset{\sim}{P}(t)$ may be quite complicated at intermediate times, its asymptotic time behavior as $t\to\infty$ is simply related to the eigenvalue spectrum. If neither the zero eigenvalue, λ_o, nor the first non-zero eigenvalue with smallest absolute value, λ_1, is a limit point, and the initial condition is such that $c_1 \neq 0$, then, asymptotically, $\underset{\sim}{P}(t)$ takes the form [see Eqs. 3.3.10 and 3.3.12]

$$\underset{\sim}{P}(t) \underset{t\to\infty}{\longrightarrow} \underset{\sim}{P}(\infty) + c_1 \underset{\sim}{R}_1 e^{\lambda_1 t}. \qquad (3.3.18)$$

For this case, the approach to equilibrium is exponential with a single relaxation time, $\tau = -\lambda_1^{-1}$, in the limit of long time. How long a time depends upon the structure of the eigenvalue spectrum. There are, however, many cases of physical interest for which the eigenvalue spectrum has a limit point and for which the mode of relaxation is not of the simple exponential form shown in Eq. 3.3.18. If the eigenvalue spectrum has a limit point at zero, the relaxation will not be exponential with time, and the equilibrium solution $\underset{\sim}{P}(\infty)$ may be zero.

We now consider briefly the formal solution of the master equation [3.1.13] in continuous state space. Since the method is quite analogous to that presented above for the discrete state space, we shall only sketch the development. No attempt is made to present a mathematically rigorous discussion. The right handed eigenfunctions $r_\lambda(y)$ satisfy the equation

$$\int a(x,y)r_\lambda(y)dy = \lambda r_\lambda(x) \tag{3.3.19}$$

which is the continuum analogue of a component of Eq. 3.3.3. The probability $p(y,t)$ can be written in terms of $r_\lambda(y)$ as

$$p(y,t) = \int_{-\infty}^{0} c_\lambda r_\lambda(y)e^{\lambda t}d\lambda \tag{3.3.20}$$

which is the continuum analogue of a component of Eq. 3.3.6. In writing Eq. 3.3.20 we have assumed that detailed balance applies, i.e.,

$$a(x,y)p(y,\infty) = a(y,x)p(x,\infty) \tag{3.3.21}$$

so that all the eigenvalues λ are real and non-positive. The coefficients c_λ are determined by the initial condition $p(y,0)$. It is clear that Eq. 3.3.20 is the solution of the master equation 3.1.13 if Eq. 3.3.19 is obeyed.

For the continuous state space, the eigenvalue spectrum may again have several forms.(3.3.5) The spectrum may be entirely discrete, it may be entirely continuous, it may be partly discrete and partly continuous and it may have limit points. If the two lowest eigenvalues $\lambda_o = 0$ and λ_1 are discrete, the asymptotic form of $p(x,t)$ is

$$p(x,t) \xrightarrow[t\to\infty]{} c_o r_o(x) + c_1 r_1(x) e^{\lambda_1 t} \tag{3.3.22}$$

where

$$p(x,\infty) = c_o r_o(x). \tag{3.3.23}$$

In this case, one again has a simple exponential asymptotic relaxation to equilibrium as $t\to\infty$. On the other hand, if the eigenvalues form a continuum starting with λ_o or λ_1, the asymptotic form of $p(x,t)$ as $t\to\infty$ will show an entirely different behavior. For example, if λ_o is discrete and $r_\lambda(x) = r(x)$ and $c_\lambda = c$ for all $|\lambda| \geq |\lambda_1|$, then the asymptotic form of $p(x,t)$ is

$$p(x,t) \xrightarrow[t\to\infty]{} p(x,\infty) + \frac{c\, r(x) e^{\lambda_1 t}}{t}. \tag{3.3.24}$$

Many of the stochastic processes of chemistry and physics show the simple asymptotic exponential relaxation of Eq. 3.3.22. However, there are processes described by master equations such as neutron thermalization[(3.3.5)] and certain problems in the theory of lattice dynamics[(3.3.6)] which lead to nonexponential asymptotic relaxation.

A question of great importance is the relation of the solution of the master equation as $t \to \infty$, $\underset{\sim}{P}(\infty)$, to the equilibrium solution, $\underset{\sim}{P}_{eq}$. In practice, two approaches to this problem have been used. In the first approach, which is more frequently used in physical applications, appropriate conditions, such as detailed balancing, are imposed on the transition rate matrix $\underset{\sim}{A}$, so that $\underset{\sim}{P}(\infty) \equiv \underset{\sim}{P}_{eq}$. The equilibrium distribution, $\underset{\sim}{P}_{eq}$, in such applications, is determined from equilibrium statistical mechanics independent of the master equation. In the second, and more fundamental approach, no extra conditions which would assure that $\underset{\sim}{P}(\infty) = \underset{\sim}{P}_{eq}$ are imposed on the transition rate matrix $\underset{\sim}{A}$ which has been derived from the dynamics of the physical process. It is now of interest to study the dependence of the form of $\underset{\sim}{P}(\infty)$ on the properties of the transition matrix $\underset{\sim}{A}$ and its eigenvalue spectrum. We again restrict our attention here initially to temporally homogeneous processes and discuss the situation for $\underset{\sim}{A} = \underset{\sim}{A}(t)$ briefly below. We also develop the arguments specifically for the case of a discrete state space, and then briefly discuss the implications of these results for continuous state space.

We have already shown above that the probabilities $\underset{\sim}{P}(t)$ remain bounded for all times t since all of the eigenvalues of $\underset{\sim}{A}$ have negative or zero real parts. The first question to be considered then is the condition on $\underset{\sim}{A}$ and its eigenvalue spectrum for $\underset{\sim}{P}(\infty)$ to be independent of the initial condition $\underset{\sim}{P}(0)$. It can be shown[(3.3.7)] that the necessary and sufficient condition for this to hold is that all of the states in the system communicate, i.e., that it is possible to go from any state to any other state in finite time. This requirement imposes certain conditions on the elements of the transition matrix $\underset{\sim}{A}$ and is equivalent to the statement that the eigenvalue $\lambda_o = 0$ of the matrix $\underset{\sim}{A}$ is not degenerate. It is easy to obtain an explicit expression for $\underset{\sim}{P}(\infty)$ from the eigenvalue expansion 3.3.10 for $\lambda_o = 0$ not degenerate and to show that a) $\underset{\sim}{P}(\infty)$ is independent of $\underset{\sim}{P}(0)$ and b) that $\underset{\sim}{P}(\infty) = \underset{\sim}{P}_{eq}$. It follows from Eqs. 3.3.3 and 3.2.8 and the non-degeneracy of $\lambda_o = 0$ that

$$\underset{\sim}{L}_o' = k(1,1...) \qquad (3.3.25)$$

where k is a constant to be determined. From the definition of c_o in Eq. 3.3.10, it follows that $c_o = k$. The normalization condition, Eq. 3.3.4, requires that

$$\underset{\sim}{L}_o{}'\underset{\sim}{R}_o = 1 = k \sum_i (\underset{\sim}{R}_o)_i \tag{3.3.26}$$

and thus

$$k = (\sum_i (\underset{\sim}{R}_o)_i)^{-1} \tag{3.3.27}$$

where $(\underset{\sim}{R}_o)_i$ is the i'th element of the column vector $\underset{\sim}{R}_o$. It now follows from Eq. 3.3.10 that

$$\underset{\sim}{P}(\infty) = c_o \underset{\sim}{R}_o = (\sum_i (\underset{\sim}{R}_o)_i)^{-1} \underset{\sim}{R}_o. \tag{3.3.28}$$

Since the eigenvector $\underset{\sim}{R}_o$ clearly depends only on the form of the matrix $\underset{\sim}{A}$, $\underset{\sim}{P}(\infty)$ is independent of the initial condition $\underset{\sim}{P}(0)$. Furthermore, it can readily be seen from Eqs. 3.3.3 and 3.3.28 that

$$\underset{\sim}{A}\,\underset{\sim}{P}(\infty) = 0. \tag{3.3.29}$$

The equilibrium solution P_{eq} is defined in terms of the master equation 3.3.1 by the requirement that

$$\frac{\partial \underset{\sim}{P}_{eq}}{\partial t} = \underset{\sim}{A}\,\underset{\sim}{P}_{eq} = 0. \tag{3.3.30}$$

If the eigenvalue $\lambda_o = 0$ is not degenerate, it then follows from Eqs. 3.3.29 and 3.3.30 that $\underset{\sim}{P}(\infty) = \underset{\sim}{P}_{eq}$.

It can also be shown(3.3.7) that the solution $\underset{\sim}{P}(\infty)$ will depend upon the initial state P(0) if and only if the eigenvalue $\lambda_o = 0$ is degenerate. A degenerate zero eigenvalue implies that not all states in the state space "communicate," i.e., it is not possible to go from any state to every other state.

Another question of interest is whether in an infinite state system $P(m,\infty)$, $m = 0, 1, \ldots, \infty$ is zero or greater than zero. This is easy to determine in physical applications from equilibrium considerations independent of the form of the master equation. On the other hand, if the only information available is the transition rate matrix $\underset{\sim}{A}$, the problem is much more difficult. It can be shown(3.3.7) that if all states "communicate," then either all of the $P(m,\infty)$ are zero or $P(m,\infty) > 0$ for all m. The statement $\underset{\sim}{P}(\infty) = 0$ is to be understood in the sense that even though each of its components $P(m,\infty)$ is vanishingly small for all m, the normalization condition $\sum_m P(m,\infty) = 1$ still applies. Which of the two possibilities will obtain depends on the behavior of the sum $\sum_i (\underset{\sim}{R}_o)_i$ of Eq. 3.3.26. If this sum diverges, it is clear from Eq. 3.3.28 that $\underset{\sim}{P}(\infty) = 0$. It is also clear from Eq. 3.3.28 that $\underset{\sim}{P}(\infty) = 0$ for $\underset{\sim}{R}_o = 0$.

In general, it is difficult to make any statement about the properties, i.e., convergence or divergence, of the function $\sum_i (\underset{\sim}{R}_o)_i$. However, for communicating processes in which transitions are obtained only between nearest neighbor states, the properties of $\sum_i (\underset{\sim}{R}_o)_i$ can readily be related to the values of the elements A_{nm} of the transition rate matrix $\underset{\sim}{A}$. For such processes, Eq. 3.3.29 can be written explicitly as the set of Eqs.

$$- A_{10}P(0,\infty) + A_{01}P(1,\infty) = 0$$

$$A_{n,n-1}P(n-1,\infty) - (A_{n-1,n} + A_{n+1,n})P(n,\infty)$$

$$+ A_{n,n+1}P(n+1,\infty) = 0; \quad n = 2,3,\ldots \qquad (3.3.31)$$

These equations can be solved recursively for $P(n,\infty)$ to yield

$$P(n,\infty) = \alpha_n P(0,\infty) \qquad (3.3.32)$$

where

$$\alpha_n = \frac{A_{10}A_{21}A_{32} \cdots A_{n,n-1}}{A_{01}A_{12}A_{23} \cdots A_{n-1,n}} \quad n = 1,2,\ldots \qquad (3.3.33)$$

is non-zero for a communicating set of states. A solution for $P(0,\infty)$ can be obtained from the normalization condition 2.1.2. This leads to the expression

$$P(0,\infty) = \left(1 + \sum_{n=1}^{\infty} \alpha_n\right)^{-1} \qquad (3.3.34)$$

so that

$$P(n,\infty) = \alpha_n \left(1 + \sum_{n=1}^{\infty} \alpha_n\right)^{-1}. \qquad (3.3.35)$$

It is now clear from Eq. 3.3.35 that when $\sum_{n=1}^{\infty} \alpha_n$ converges, $P(n,\infty)$ is non-zero for all n. Conversely, if the sum diverges, $P(n,\infty) = 0$ for all n. We can illustrate the two types of behavior by considering the special case $A_{n,n-1}/A_{n-1,n} = \alpha$, a constant for all n. For this case $\alpha_n = \alpha^n$ and Eq. 3.3.35 becomes

$$P(n,\infty) = \alpha^n(1-\alpha) \tag{3.3.36}$$

for $\alpha < 1$, and

$$P(0,\infty) = 0,$$

$$\sum_{n=0}^{\infty} P(n,\infty) = 1 \tag{3.3.37}$$

for $\alpha \geq 1$.

So far we have considered the solution to the master equation for temporally homogeneous processes only. There are, however, a number of processes for which the transition rates $A_{nm}(t)$ are functions of time. An example would be processes where the transitions from one state of a system to another are due to interaction with a reservoir whose properties change with time.

When the transition rates $A_{nm}(t)$ depend on time, neither the general solution given in Eq. 3.3.2 nor the eigenvalue expansion of Eq. 3.3.9 is valid. It is, however, possible to derive a formal series solution to the master equation with a time dependent rate matrix by converting it to an integral equation, i.e., by rewriting Eq. 3.3.1 as

$$\underset{\sim}{P}(t) = \underset{\sim}{P}(0) + \int_0^t \underset{\sim}{A}(\tau)\underset{\sim}{P}(\tau)d\tau. \tag{3.3.38}$$

Provided the series converges, the recursive solution to this integral equation is

$$\underset{\sim}{P}(t) = [\underset{\sim}{I} + \int_0^t \underset{\sim}{A}(\tau)d\tau + \int_0^t d\tau \int_0^\tau \underset{\sim}{A}(\tau)\underset{\sim}{A}(\tau_1)d\tau_1$$

$$+ \int_0^t d\tau \int^\tau d\tau_1 \int_0^{\tau_1} d\tau_2 \underset{\sim}{A}(\tau)\underset{\sim}{A}(\tau_1)\underset{\sim}{A}(\tau_2)$$

$$+ \ldots]\underset{\sim}{P} \tag{3.3.39}$$

where $\underset{\sim}{I}$ is the unit matrix. This series is equivalent to Dyson's solution to the Schrödinger equation in the interaction representation.[(3.3.8)] When $\underset{\sim}{A}(t)$ does not depend on time, the series of Eq. 3.3.39 reduces to the exponential form given in Eq. 3.3.2 as may readily be seen by performing the integrations. Furthermore when the matrices $\underset{\sim}{A}(t)$ and $\int_0^t \underset{\sim}{A}(\tau)d\tau$ commute, the series in Eq. 3.3.39 can again be summed to yield

$$\underset{\sim}{P}(t) = e^{\int_0^t \underset{\sim}{A}(\tau)d\tau} \underset{\sim}{P}(0). \tag{3.3.40}$$

Equation 3.3.39 is not the only possible expansion of the solution of the master equation with a time dependent transition matrix. The formal solution can be written as

$$\underset{\sim}{P}(t) = e^{\underset{\sim}{C}(t)} \underset{\sim}{P}(0) \tag{3.3.41}$$

where the matrix $\underset{\sim}{C}(t)$ is defined by Eq. 3.3.41. Magnus[(3.3.9)] has shown that the matrix $\underset{\sim}{C}(t)$ can be written as an infinite series $\underset{\sim}{C}(t) = \underset{\sim}{C}_1(t) + \underset{\sim}{C}_2(t) + \ldots$

where the first few $\underset{\sim}{C}$'s are

$$\underset{\sim}{C}_1(t) = \int_0^t \underset{\sim}{A}(\tau)d\tau$$

$$\underset{\sim}{C}_2(t) = -\frac{1}{2}\int_0^t d\tau \int_0^\tau d\tau_1[\underset{\sim}{A}(\tau_1), \underset{\sim}{A}(\tau)]$$

$$\underset{\sim}{C}_3(t) = \frac{1}{6}\int_0^t d\tau \int_0^\tau d\tau_1 \int_0^{\tau_1} d\tau_2 \left\{\left[\underset{\sim}{A}(\tau_2),[\underset{\sim}{A}(\tau_1),\underset{\sim}{A}(\tau)]\right] + \left[[\underset{\sim}{A}(\tau_2), \underset{\sim}{A}(\tau_1)], \underset{\sim}{A}(\tau)\right]\right\}. \tag{3.3.42}$$

This expansion is the analogue of the Baker-Hausdorff formula in group theory,[3.3.10, 3.3.11] and has been applied to some physical problems by Pechukas and Light[3.3.12] and Weiss and Maradudin[3.3.13].

The use of generating functions for the solution of master equations with time dependent transition rate matrices can sometimes lead to a differential equation which can be solved by standard techniques. This is the case, for instance, for two systems of oscillators which relax through mutual interactions.[3.3.14]

When $\underset{\sim}{A}(t)$ is of the form

$$\underset{\sim}{A}(t) = f(t)\underset{\sim}{A} \tag{3.3.43}$$

where $\underset{\sim}{A}$ is a constant, the master equation can be solved by introducing a new time variable

$$\tau(t) = \int_0^t f(s)ds. \tag{3.3.44}$$

With this transformation of the time variable, the master equation becomes

$$\frac{\partial \underset{\sim}{P}(\tau)}{\partial \tau} = \underset{\sim}{A}\,\underset{\sim}{P}(\tau) \tag{3.3.45}$$

with the solution

$$\underset{\sim}{P}(\tau) = e^{\underset{\sim}{A}\tau}\,\underset{\sim}{P}(0). \tag{3.3.46}$$

Little is known about the behavior of $\underset{\sim}{P}(t)$ as $t\to\infty$ when $\underset{\sim}{A}(t)$ is a function of time. However, one would expect that if $\lim_{t\to\infty} \underset{\sim}{A}(t) = \underset{\sim}{A}(\infty)$ exists, then the previously quoted results on the ergodicity or non-ergodicity of the stochastic process and the results on the form of $\underset{\sim}{P}(\infty)$, i.e., $\underset{\sim}{P}(\infty) = 0$ or $\underset{\sim}{P}(\infty) > 0$, would still be valid and will depend on the form of $\underset{\sim}{A}(\infty)$. It is clear that the properties of P_{eq} can again be determined from physical considerations if $\underset{\sim}{A}(\infty)$ exists.

The properties 3.2.8 and 3.2.9 of the transition rates A_{nm}, i.e., the non-negativeness of the off-diagonal elements and the fact that the columns sum to zero, in conjunction with the possibility of symmetrizing the transition rate matrix by invoking detailed balancing, have enabled us to determine some important general properties of the eigenvalues. These are that the eigenvalues are real, that there is at least one zero eigenvalue and that the rest of the eigenvalues are negative. Much more detailed information on the eigenvalue spectrum is required, however, before the formal solutions 3.3.10 and 3.3.20 of the master equation can be used to obtain explicit expressions for the probabilities $P(n,t)$ or $p(x,t)$ in physical applications. In addition, information is required on the form of the eigenfunctions. These are difficult problems, and it has not proven possible in general, to obtain useful solutions of master equations by the formal eigenvalue method outlined above. A different approach to the solution of master equations utilizes generating functions and will be discussed in some detail in Section 5 of this chapter.

So far, we have discussed the solution of the master equation for the probabilities $P(n,t)$ and $p(x,t)$. It is clear that the techniques developed above can be applied equally well to obtain formal solutions of the master equation for the conditional probability $W(n,\tau|n_1)$, Eq. 3.2.15, or its continuum analogue. In fact, $W(n,\tau|n_1)$ is identical with $P(n,\tau)$ subject to the initial condition $P(n,0) = \delta_{n,n_1}$.

4) Master Equations for Open Systems and First Passage Times

Our discussion so far has been for closed systems for which it is true for the set of states S that (for discrete random variables)

$$\sum_{n\varepsilon S} A_{nm}(t) = 0. \tag{3.4.1}$$

However, there is a class of problems, known mathematically as first passage time problems for which the state space can be partitioned into two (or more) sets of states in which accessibility to at least one set is not reversible. For example, if the set of states S is partitioned into two subsets $S = S_1 + S_2$, and if states of S_2 can be reached from states of S_1 but the reverse is not true, then the study of the statistics of the time to reach S_2 starting from some state in S_1 is known as a first-passage time problem. As an example considered in more detail by Montroll and Shuler,[3.4.1] the analysis of the dissociation of dilute systems of diatomic molecules can be looked upon as a first passage time problem. The two sets of states under consideration are the set of energy levels of the bound molecules, and the continuous energy state space of the dissociated molecules. In studying dissociation without subsequent recombination it is of interest to determine the statistical properties of the time for a bound molecule to become dissociated. The process is stochastic because the bound molecules can gain or lose energy, and may do so many times before the bound molecules gain enough energy to dissociate.

For systems partitioned into two sets of states, the rate matrix corresponding to open systems (i.e., systems with irreversibly accessible sets of states) can be written in partitioned form as

$$A = \begin{pmatrix} \underset{\sim}{a} & \underset{\sim}{0} \\ \underset{\sim}{b} & \underset{\sim}{c} \end{pmatrix} \begin{matrix} S_1 \\ S_2 \end{matrix} \qquad \text{(columns } S_1 \;\; S_2\text{)} \tag{3.4.2}$$

where $\underset{\sim}{a}$ contains rates for transitions within and out of S_1, $\underset{\sim}{b}$ contains rates for transitions from S_1 to S_2, and the $\underset{\sim}{0}$ matrix indicates that transition from S_2 to S_1 are not allowed. Very often one is interested only in the occupancy of S_1 states. Since there is a net "flow" from S_1 to S_2 we may expect that

$$\lim_{t\to\infty} \sum_{n\in S_1} P(n,t) = 0. \tag{3.4.3}$$

Let us consider the problem of calculating the statistics of residence times in S_1. For this purpose we partition the $\underset{\sim}{P}$ matrix as

$$\underset{\sim}{P} = \begin{pmatrix} \underset{\sim}{P}_1 \\ \underset{\sim}{P}_2 \end{pmatrix} \tag{3.4.4}$$

where $\underset{\sim}{P}_j$ contains the state probabilities for S_j. Thus the probability of being in S_1 at time t is just

$$F(t) = \sum_{n\in S_1} P(n,t). \tag{3.4.5}$$

The probability of making a transition from S_1 to S_2 in (t, t+dt) will be denoted by f(t)dt, and is given by

$$f(t)dt = F(t) - F(t+dt) \simeq -\frac{dF}{dt}\,dt\,. \tag{3.4.6}$$

The quantity f(t) is the first passage time density which can, in principle, be calculated by solving the equation

$$\dot{\underset{\sim}{P}}_1 = \underset{\sim}{a}\,\underset{\sim}{P}_1 \tag{3.4.7}$$

for the elements P(n,t), for n belonging to S_1. When it is not possible to carry this calculation out in detail, it may still be possible to calculate moments, μ_n, of the first passage time. These are defined to be

$$\mu_n \equiv \int_0^\infty t^n f(t)dt = n\int_0^\infty t^{n-1}\,F(t)dt \tag{3.4.8}$$

where the second integral is found from the first by an integration by parts. It is not difficult to show that if μ_n is finite, the term $\lim_{t\to\infty} t^n F(t)$ neglected in Eq. 3.4.8 is indeed 0. This follows from

$$\begin{aligned} t^n F(t) &= -t^n \int_t^\infty \frac{dF(\tau)}{d\tau}\,d\tau \\ &= t^n \int_t^\infty f(\tau)d\tau \le \int_t^\infty \tau^n f(\tau)d\tau \end{aligned}$$

and the fact that if $\mu_n < \infty$ the last integral tends to zero as $t\to\infty$.

For temporally homogeneous processes a formal expression for μ_n in terms of the matrix $\underset{\sim}{a}$ of Eq. 3.4.2

can be found by writing the solution to Eq. 3.4.7 as

$$\underset{\sim}{P}_1(t) = e^{\underset{\sim}{a}t}\, \underset{\sim}{P}_1(0). \tag{3.4.9}$$

The expression for F(t) in Eq. 3.4.5 can be rewritten as

$$F(t) = \underset{\sim}{U}\, \underset{\sim}{P}_1(t) \tag{3.4.10}$$

where $\underset{\sim}{U} = (1, 1, 1,\ldots,1)$. Thus, if we combine Eqs. 3.4.8, 3.4.9, and 3.4.10, we find that

$$\mu_n = n\, \underset{\sim}{U} \int_0^\infty e^{\underset{\sim}{a}t}\, t^{n-1}\, dt\, \underset{\sim}{P}_1(0). \tag{3.4.11}$$

It is easily shown that

$$\int_0^\infty t^{n-1}\, e^{\underset{\sim}{a}t}\, dt = (-1)^n\, (n-1)!\, \underset{\sim}{a}^{-n} \tag{3.4.12}$$

so that

$$\mu_n = (-1)^n\, n!\, \underset{\sim}{U}\, \underset{\sim}{a}^{-n} \underset{\sim}{P}_1(0) \tag{3.4.13}$$

provided that all the eigenvalues of $\underset{\sim}{a}$ have negative real parts. Although this property can be proved in detail mathematically, it is intuitively obvious. If any eigenvalue had a positive real part it would be impossible for the elements of $\underset{\sim}{P}_1(t)$ to remain bounded for all initial conditions. If any eigenvalue had a real part equal to zero, F(t) would not approach zero in contradiction to Eq. 3.4.3.

Somewhat more detailed results for the moments can be obtained for the case in which a takes the form of a tridiagonal matrix, that is, in which $a_{ij} = 0$ unless $j = i\pm1, i$. Let us suppose that the states of S_1 are labelled 0, 1, 2,..., N-1, and S_2 contains states whose index is N or greater. If we introduce the Laplace transforms

$$\rho_n(s) = \int_0^\infty P(n,t)\exp(-st)dt \tag{3.4.14}$$

then by a combination of Eqs. 3.4.5 and 3.4.8 we can find the moments in terms of the $\rho_n(s)$ and their derivatives at $s = 0+$ as

$$\mu_n = (-1)^{n-1}\, n \sum_{r=0}^{N-1} \frac{d^{n-1}}{ds^{n-1}} \rho_r(s) \Big|_{s=0} \tag{3.4.15}$$

In particular, the mean first passage time is

$$\mu_1 = \sum_{r=0}^{N-1} \rho_r(0). \tag{3.4.16}$$

We can calculate a more explicit expression for the μ_n for nearest neighbor transitions by solving recursion relations that can be obtained from the Laplace transform of the equation $\dot{\underset{\sim}{P}}_1 = \underset{\sim}{a}\,\underset{\sim}{P}_1$. This computation will be carried out in detail for the first moment under the assumption that the system is initially in state j, that is, $P(n,0) = \delta_{n,j}$. For simplicity we shall use the notation

$$a_{n,n+1} = c_{n+1},\ a_{n+1,n} = b_n$$

$$\theta_n = \frac{b_o b_1 \ldots b_{n-1}}{c_1 c_2 \ldots c_n},\ c_n\theta_n = b_{n-1}\,\theta_{n-1}. \tag{3.4.17}$$

The Laplace transform of the master equation is

$$s\rho_o(s) = c_1\rho_1(s) - b_o\rho_o(s)$$

$$\delta_{r,j} + s\rho_r(s) = c_{r+1}\rho_{r+1}(s) + b_{r-1}\rho_{r-1}(s)$$

$$- (c_r + b_r)\rho_r(s),\ r = 1,2,\ \ldots,\ N-2$$

$$s\rho_{N-1}(s) = b_{N-2}\rho_{N-2}(s) - (c_{N-1}+b_{N-1})\rho_{N-1}(s). \tag{3.4.18}$$

To calculate μ_1 we need only set $s = 0$ and solve the resulting equations for the $\rho_j(0)$. In this way we find that

$$\rho_r(0) = \theta_r\rho_o(0) - \Gamma_r \tag{3.4.19}$$

where

$$\Gamma_r = 0, \quad \text{for } r \leq j$$

and

$$\Gamma_r = \frac{1}{c_r}\left(1 + \frac{b_{r-1}}{c_{r-1}} + \frac{b_{r-1}b_{r-2}}{c_{r-1}c_{r-2}} + \ldots\right.$$

$$\ldots + \frac{b_{r-1}b_{r-2}\ldots b_{j+1}}{c_{r-1}c_{r-2}\ldots c_{j+1}}\Bigg) \tag{3.4.20}$$

for $r \geq j+1$. Alternatively the Γ_r can be calculated by a recurrence relation

$$\Gamma_j = 0,\ \Gamma_{j+1} = 1/c_{j+1}$$

$$c_{r+1}\,\Gamma_{r+1} = (c_r + b_r)\,\Gamma_r - b_{r-1}\,\Gamma_{r-1} \tag{3.4.21}$$

for $r \geq j+1$. Finally, $\rho_o(0)$ is calculated from the last line of Eq. 3.4.18 as

$$\rho_o(0) = \frac{1 + b_{N-1}\,\Gamma_{N-1}}{b_{N-1}\theta_{N-1}} = \frac{c_N\,\Gamma_N}{b_{N-1}\theta_{N-1}} = \frac{\Gamma_N}{\theta_N}. \tag{3.4.22}$$

Hence we find that the mean first passage time is

$$\mu_1 = \frac{\Gamma_N(1+\theta_1+\theta_2+\ldots+\theta_{N-1}) - \theta_N(\Gamma_{j+1}+\Gamma_{j+2}+\ldots+\Gamma_{N-1})}{\theta_N}. \tag{3.4.23}$$

As an example we consider the Montroll-Shuler model for the dissociation of a diatomic molecule,(3.4.1) in which the rate constants are

$$a_{n,n+1} = \kappa(n+1)$$

$$a_{n+1,n} = \kappa(n+1)e^{-\psi} \tag{3.4.24}$$

where κ and ψ are constants and $P(n,0) = \delta_{n,0}$. For these rate constants the parameters θ_n of Eq. 3.4.17, and Γ_n of Eq. 3.4.21 are easily seen to be:

$$\theta_n = e^{-n\psi}$$

$$\Gamma_n = \frac{1}{\kappa}\left(\frac{1}{n} + \frac{e^{-\psi}}{n-1} + \frac{e^{-2\psi}}{n-2} + \cdots + \frac{e^{-(n-1)\psi}}{1}\right). \quad (3.4.25)$$

Thus, the mean first passage time to go from state 0 to state N is, for this model,

$$\mu_1 = \frac{1}{\kappa(1-e^{-\psi})}\left\{(e^{N\psi}-1)\left[\frac{1}{N} + \frac{e^{-\psi}}{N-1} + \frac{e^{-2\psi}}{N-2} + \ldots + \frac{e^{-(N-1)\psi}}{1}\right] - (1-e^{-\psi})(\beta_{N-1} + e^{-\psi}\beta_{N-2} + e^{-2\psi}\beta_{N-3} + \ldots + e^{-(N-2)\psi}\beta_1)\right\} \quad (3.4.26)$$

in which β_r is defined to be

$$\beta_r = 1 + \frac{1}{2} + \frac{1}{3} + \ldots + \frac{1}{r}. \quad (3.4.27)$$

The expression in Eq. 3.4.26 can be reduced to the one originally given by Montroll and Shuler, and by Kim,(3.4.2) by replacing terms like $1/j$ by $\beta_j - \beta_{j-1}$ in the first square bracket. In this way μ_1 is found to be

$$\mu_1 = \frac{1}{\kappa(1-e^{-\psi})} \sum_{j=1}^{N} \frac{e^{j\psi}-1}{j}. \qquad (3.4.28)$$

Several comments are in order at this point. The first is that these calculations can be extended to derive explicit formulae for higher moments by differentiating Eq. 3.4.18 the appropriate number of times, setting s = 0, and solving the resulting equations recursively. Secondly, an equation analogous to 3.4.23 can be found when there are two absorbing barriers rather than one.[3.4.3] Finally, we emphasize that our derivation depended on two assumptions: that the process is one dimensional, and that transitions are allowed only between nearest neighbor states. In the absence of these conditions one generally uses the formula of Eq. 3.4.13.

5) Generating Function Techniques for Solution of Master Equations

We now return to the discussion of closed systems. As pointed out in Section 3, it is usually not possible to obtain a useful solution of the master equation by the eigenfunction expansion method. A more useful technique of obtaining a solution is provided by the introduction of generating functions. These generating functions are transforms of the distribution function and change the set of differential difference equations for the distribution function (the master equation) into either integro-differential or partial differential equations for the generating function.

We have discussed some of the fundamental properties of generating functions (characteristic functions) in 2.3. The particular form of the generating function one wants to use depends upon the range of the stochastic variable and upon the nature of the state space. Thus, if the stochastic variable $X(t)$ takes on discrete values between $-\infty$ and $+\infty$, one most frequently uses the generating function (see 2.3.12)

$$\Phi(s,t) = \sum_{n=-\infty}^{\infty} e^{ins} P(n,t). \tag{3.5.1}$$

If the stochastic variable takes on only positive discrete values between 0 and ∞, one generally uses the generating function (see 2.3.26)

$$g(s,t) = \sum_{n=0}^{\infty} e^{-ns} P(n,t) \tag{3.5.2}$$

or

$$G(z,t) = \sum_{n=0}^{\infty} z^n P(n,t). \tag{3.5.3}$$

For continuous state space, the sums in Eqs. 3.5.1 through 3.5.3 are replaced by integrals. For instance, Eq. 3.5.1 becomes

$$\phi(s,t) = \int_{-\infty}^{\infty} e^{isx} p(x,t)dx. \tag{3.5.4}$$

We shall carry out the general discussion of the use of generating functions for the forms given in Eqs. 3.5.3 and 3.5.4.

Other forms of the generating function can and have been used in a number of problems. An example is

$$G(z,t) = \sum_{n=0}^{\infty} \frac{z^n}{n!} P(n,t). \tag{3.5.5}$$

Generating functions involving sums over discrete times or integrals over continuous times are also frequently used.[3.5.1] The particular form of the generating function one wants to use to obtain a solution of the master equation depends upon the form of the master equation and the ingenuity of the investigator. No hard and fast rules can thus be given.

We now briefly discuss some of the properties of the generating functions which follow immediately from their definitions. We shall limit our discussion to the generating function $G(z,t)$ of Eq. 3.5.3; analogous results can be obtained for the other generating functions defined above. It follows immediately from Eqs. 3.5.3 and 2.1.2 that

$$G(1,t) = 1 \tag{3.5.6}$$

for all t. It is clear that

$$G(z,0) = \sum_{n=0}^{\infty} z^n P(n,0) \tag{3.5.7}$$

so that G(z,0) is determined from the initial condition for the probability distribution P(n,t). The kth derivative of G(z,t) with respect to the variable z, is given by

$$\frac{\partial^k G(z,t)}{\partial z^k} = \sum_{n=k}^{\infty} \frac{n!}{(n-k)!} z^{n-k} P(n,t) \tag{3.5.8}$$

so that

$$\left.\frac{\partial^k G(z,t)}{\partial z^k}\right|_{z=0} = k!\, P(k,t). \tag{3.5.9}$$

The moments $\mu_k(t)$ of the probability distribution are given by

$$\mu_k(t) \equiv \sum_{n=0}^{\infty} n^k P(n,t) = \left(z \frac{\partial}{\partial z}\right)^k G(z,t)\Bigg|_{z=1} . \tag{3.5.10}$$

The master equation in continuous state space is

$$\frac{\partial p(x,t)}{\partial t} = \int_{-\infty}^{\infty} a(x,y,t)\, p(y,t)\, dy. \tag{3.5.11}$$

Multiplication of both sides of this equation by e^{isx} and integration over x from $-\infty$ to ∞ yields

$$\frac{\partial \phi(s,t)}{\partial t} = \int_{-\infty}^{\infty} \int_{-\infty}^{\infty} a(x,y,t)\, e^{isx}\, p(y,t)\, dy\, dx. \tag{3.5.12}$$

We now use the relation (see Eq. 3.5.4)

$$p(y,t) = \frac{1}{2\pi} \int_{-\infty}^{\infty} \phi(s',t)\, e^{-is'y}\, ds' \tag{3.5.13}$$

to rewrite Eq. 3.5.12 as

$$\frac{\partial\phi(s,t)}{\partial t} = \int_{-\infty}^{\infty} \alpha(s,s',t)\, \phi(s',t)\, ds' \tag{3.5.14}$$

where

$$\alpha(s,s',t) = \frac{1}{2\pi} \int_{-\infty}^{\infty} \int_{-\infty}^{\infty} a(x,y,t)\, e^{isx}\, e^{-is'y}\, dy\, dx. \tag{3.5.15}$$

Equation 3.5.14, which is the Fourier transform of the master equation 3.5.11, is the desired integro-differential equation for the generating function $\phi(s,t)$. If Eq. 3.5.14 can be solved in closed form for $\phi(s,t)$, the distribution function $p(x,t)$ can be obtained as the Fourier transform of Eq. 3.5.4.

Little progress can be made on this level of generality. We shall thus consider some special forms of the transition rate elements $a(x,y,t)$. We shall find it convenient to rewrite Eq. 3.5.15 in the form

$$\alpha(s,s',t) = \frac{1}{2\pi} \int_{-\infty}^{\infty} \int_{-\infty}^{\infty} \bar{a}(u,y,t) e^{isu} e^{i(s-s')y} du\, dy \tag{3.5.16}$$

where $u = x-y$ and $\bar{a}(x-y, y,t) = a(x,y,t)$. We write the equation in this form because in many physical applications, a(x,y,t) depends strongly on the difference (x-y) and only weakly on y. For example, if the transition rate a(x,y,t) depends only on (x-y) and is independent of y, Eq. 3.5.16 becomes

$$\alpha(s,s',t) = \delta(s-s')$$

$$\times \int_{-\infty}^{\infty} \bar{a}(u,t)\, e^{isu} du = \delta(s-s')\, \bar{\alpha}(s,t) \qquad (3.5.17)$$

where $\bar{\alpha}(s,t)$ is defined by Eq. 3.5.17. Substitution of Eq. 3.5.17 into Eq. 3.5.14 then yields

$$\frac{\partial\phi(s,t)}{\partial t} = \bar{\alpha}(s,t)\,\phi(s,t). \qquad (3.5.18)$$

This can be solved immediately to give

$$\phi(s,t) = \phi(s,0)\exp\left[\int_0^t \bar{\alpha}(s,\tau)\, d\tau\right] \qquad (3.5.19)$$

where $\phi(s,0)$ is determined by the initial condition p(x,0) from Eq. 3.5.4. The distribution function p(x,t) can then be obtained, at least in principle, from the transform 3.5.13.

If the transition rate a(x,y,t) has the form

$$a(x,y,t) = y^n\, \bar{a}(u,t),$$

$$n = 0, 1, 2, \ldots \qquad (3.5.20)$$

then Eq. 3.5.16 becomes

$$\alpha(s,s',t) = \bar{\alpha}(s,t)\left(i\,\frac{\partial}{\partial s'}\right)^n \delta(s-s') \qquad (3.5.21)$$

and Eq. 3.5.14 now reads

$$\frac{\partial\phi(s,t)}{\partial t} = \bar{\alpha}(s,t)\left(-i\,\frac{\partial}{\partial s}\right)^n \phi(s,t). \qquad (3.5.22)$$

We have thus reduced the integro-differential equation 3.5.14 to a partial differential equation in the generating function. If a closed form solution can be obtained for $\phi(s,t)$ from Eq. 3.5.22 for a given set of boundary conditions, the distribution function $p(x,t)$ can again be evaluated, in principle, from the Fourier transform 3.5.13.

Note that for $n = 1$, Eq. 3.5.22 for the generating function becomes a first order partial differential equation for which a solution can always be obtained by the method of characteristics.(3.5.2) For transition rates linear in one of the state variables, a closed form expression for the generating function $\phi(s,t)$ can thus always be obtained. For $n > 1$, the partial differential equation 3.5.22 for the generating function becomes of an order higher than one, and it is not possible in general to obtain a closed form solution for $\phi(s,t)$.

We now discuss the application of the generating function $G(z,t)$ as given in Eq. 3.5.3 to the master equation in discrete state space,

$$\frac{\partial P(n,t)}{\partial t} = \sum_{m=0}^{\infty} A_{nm}(t)\, P(m,t). \tag{3.5.23}$$

Multiplication of 3.5.23 by z^n and summation from $n = 0$ to $n = \infty$ and use of the contour integral representation of $P(m,t)$ in terms of $G(y,t)$

$$P(m,t) = \frac{1}{2\pi i} \oint \frac{G(y,t)}{y^{m+1}}\, dy \tag{3.5.24}$$

yields

$$\frac{\partial G(z,t)}{\partial t} = \frac{1}{2\pi i} \oint \frac{G(y,t)\mathcal{A}(z,y,t)\, dy}{y} \tag{3.5.25}$$

where $\mathcal{A}(z,y,t)$ is defined by

$$\mathcal{A}(z,y,t) \equiv \sum_{m=0}^{\infty} \sum_{n=0}^{\infty} z^n\, y^{-m}\, A_{nm}(t). \tag{3.5.26}$$

The contour integral of Eq. 3.5.24 yields the coefficient of y^m in the expression of $G(y,t)$ in a power series in y. Eq. 3.5.25 is an integro-differential equation for the generating function $G(z,t)$. If Eq. 3.5.25 can be solved in closed form for the generating function $G(z,t)$ the distribution function $P(m,t)$ can be obtained from the contour integral of Eq. 3.5.24.

Again little progress can be made on this level of generality. We thus again consider some special simple forms of the transition rate elements $A_{nm}(t)$. We first rewrite Eq. 3.5.26 in the form

$$\mathcal{A}(z,y,t) = \sum_{m=0}^{\infty} \sum_{j=-m}^{\infty} (z/y)^m z^j \bar{A}(j,m,t) \quad (3.5.27)$$

where j = n-m and

$$\bar{A}(n-m, m,t) \equiv A_{nm}(t). \quad (3.5.28)$$

Interchanging the order of summation in Eq. 3.5.27 yields

$$\mathcal{A}(z,y,t) = \sum_{j=-\infty}^{-1} \sum_{m=-j}^{\infty} (z/y)^m z^j \bar{A}(j,m,t)$$

$$+ \sum_{j=0}^{\infty} \sum_{m=0}^{\infty} (z/y)^m z^j \bar{A}(j,m,t). \quad (3.5.29)$$

To give a specific example of this technique, we shall consider the case where the transition rate $\bar{A}(j,m,t)$ is of the form (with a = constant)

$$\bar{A}(j,m,t) = a(m+1)\delta_{j,1} + m\delta_{-1,j} - [a(m+1) + m]\delta_{j,0}$$

$$(3.5.30)$$

which implies that transitions are between nearest neighbors only with $j = n-m = \pm 1$ and that the transitions also depend upon the initial state m. The transition rate 3.5.30 is that for an harmonic oscillator in weak interaction with a heat bath. Substitution of Eq. 3.5.30 into Eq. 3.5.29 yields

$$\mathcal{A}(z,y,t) = \frac{1}{z} \sum_{m=1}^{\infty} m(z/y)^m +$$

$$+ \sum_{m=0}^{\infty} \left(\frac{z}{y}\right)^m [(z-1)\ a(m+1) - m] = - y\frac{\partial}{\partial y}\left[\frac{1}{y-z}\right]$$

$$+ a(z-1)\ \frac{y}{y-z} + [a(z-1) - 1]\left[- y\frac{\partial}{\partial y}\left(\frac{y}{y-z}\right)\right].$$

(3.5.31)

Substitution of Eq. 3.5.31 into Eq. 3.5.25 yields

$$\frac{\partial G(z,t)}{\partial t} = a(z-1)\left[(z-1/a)\ \frac{\partial G(z,t)}{\partial z} + G(z,t)\right]. \quad (3.5.32)$$

A closed form solution for the generating function G(z,t) can be obtained from Eq. 3.5.32 by the method of characteristics.[(3.5.2)] Substitution of this G(z,t) into Eq. 3.5.24 and performance of the indicated integration then yields the desired distribution function P(n,t).

For simple forms of the master equation in discrete state space one can obtain the differential equation for the generating function G(z,t) directly without going through the contour integration formalism described above. We shall present some simple examples to demonstrate this approach. The first example we consider is that of the emission of α-particles from nuclei. As was shown in Section 2, the master equation governing this process is

$$\frac{\partial P(n,t)}{\partial t} = \Lambda P(n-1,t) - \Lambda P(n,t),\ n = 1,\ 2....$$

$$\frac{\partial P(0,t)}{\partial t} = -\Lambda P(0,t) \tag{3.5.33}$$

with the initial condition $P(n,0) = \delta_{n,0}$. While these simple equations could readily be solved by recursion, we shall use here the generating function $G(z,t)$ of Eq. 3.5.3. Multiplying the equation for $\frac{\partial P(n,t)}{\partial t}$ by z^n and summing over all n, one finds

$$\frac{\partial G(z,t)}{\partial t} = -\Lambda(1-z)\ G(z,t) \tag{3.5.34}$$

with $G(z,0) = 1$. The solution of Eq. 3.5.34 is

$$G(z,t) = e^{-\Lambda(1-z)t} = e^{-\Lambda t} \sum_{n=0}^{\infty} \frac{(\Lambda t)^n}{n!} z^n. \tag{3.5.35}$$

To obtain the distribution function $P(n,t)$ we note that $P(n,t)$ is the coefficient of z^n in the expansion of $G(z,t)$ as given in Eq. 3.5.3. Thus,

$$P(n,t) = \frac{(\Lambda t)^n}{n!} e^{-\Lambda t}. \tag{3.5.36}$$

Using Eq. 3.5.10, we can obtain equations for the first moment $\mu_1(t)$ of the distribution, i.e., the average number of particles emitted in the time interval from 0 to t. Application of Eq. 3.5.10 to the expression for $G(z,t)$ of Eq. 3.5.35 yields

$$\mu_1(t) = \frac{\partial}{\partial z}\left[e^{-\Lambda(1-z)t}\right]_{z=1} = \Lambda t. \tag{3.5.37}$$

It is important to note that the moments of the distribution function $P(n,t)$ can be obtained <u>directly</u> from the generating function. This is very useful since in a number of cases of physical interest it is difficult to find the explicit form of $P(n,t)$, even though a closed form expression for the generating function $G(z,t)$ has been obtained.

As another simple example we consider the first order chemical reaction $B \xrightarrow{k} C$ where a substance B is converted to C with a rate coefficient k without back reaction. In the stochastic version of chemical kinetics one deals with the probabilities $P(n,t)$ that n molecules of reactant B are present at time t. The stochastic variable $X(t)$ is thus the number of molecules of species B at time t. It will be assumed that there are N molecules of species B at time t = 0 so that $P(n,0) = \delta_{n,N}$. The conditional probability that there will be j molecules of B at time t + dt if there were n molecules of B at time t is

$$W(j,t+dt|n,t) = kndt\delta_{j,n-1} + (1-kndt)\delta_{j,n}. \qquad (3.5.38)$$

That is to say, the probability that one molecule of species B will undergo reaction in the time interval (t,t+dt) is kndt if there are n molecules of B at time t. It is assumed that the probability that more than one molecule of B will undergo reaction in the time interval (t,t+dt) is zero. From Eqs. 3.2.1 and 3.5.38 it follows that the transition rates are

$$A_{n-1,n} = kn$$

$$A_{n,n} = -kn$$

$$A_{j,n} = 0, \quad j \neq n, n-1. \tag{3.5.39}$$

The master equation for this chemical reaction therefore reads

$$\frac{\partial P(n,t)}{\partial t} = k[(n+1)\, P(n+1,t) - nP(n,t)], \; n = 0,1,2,\ldots \tag{3.5.40}$$

To obtain the equation for the generating function $G(z,t)$ of Eq. 3.5.3 we multiply the master equation for $P(n,t)$ by z^n and sum over all values of n from zero to infinity

$$\begin{aligned}\frac{\partial G(z,t)}{\partial t} &= k \sum_{n=0}^{\infty} z^n \, [(n+1)\, P(n+1,t) - nP(n,t)] \\ &= k\left[\frac{\partial}{\partial z} \sum_{n=0}^{\infty} z^{n+1} P(n+1,t) - z \frac{\partial}{\partial z} \sum_{n=0}^{\infty} z^n \, P(n,t)\right] \\ &= k\left\{\frac{\partial}{\partial z} [G(z,t) - P(0,t)] - z \frac{\partial G(z,t)}{\partial z}\right\} \\ &= k(1-z) \frac{\partial G(z,t)}{\partial z}.\end{aligned} \tag{3.5.41}$$

Equation 3.5.41 is a first order partial differential equation which can be solved by the method of characteristics. We shall use here, however, a somewhat different and more direct method. We wish to rewrite

Eq. 3.5.41 in the form

$$\frac{\partial G(z,t)}{\partial \tau} = \frac{\partial G(z,t)}{\partial \zeta} \tag{3.5.42}$$

which can be done by choosing the new independent variables $\tau = kt$ and $\zeta = -\ln(1-z)$. This equation is satisfied by any differentiable function of the form

$$G(z,t) = f(\tau+\zeta). \tag{3.5.43}$$

Since the initial condition $P(n,0) = \delta_{n,N}$ implies that $G(z,0) = z^N$ we find

$$f(\zeta) \equiv f[-\ln(1-z)] = z^N. \tag{3.5.44}$$

The $f(\zeta)$ must be of the form

$$f(\zeta) = (1 - e^{-\zeta})^N \tag{3.5.45}$$

and, from Eq. 3.5.43 it follows that the desired solution is

$$G(z,t) = [1 - (1-z)\, e^{-kt}]^N. \tag{3.5.46}$$

Expanding the generating function $G(z,t)$ of Eq. 3.5.46 in a power series in z,

$$G(z,t) = \sum_{n=0}^{N} \binom{N}{n} (1-e^{-kt})^{N-n} (e^{-kt} z)^n \tag{3.5.47}$$

yields [see Eq. 3.5.3]

$$P(n,t) = \binom{N}{n} (1-e^{-kt})^{N-n}\, e^{-nkt} \tag{3.5.48}$$

for the probability P(n,t) of finding n molecules of species B at time t.

The first moment $\mu_1(t)$ of the distribution function corresponds to the average number of species B at time t, $\langle n \rangle_t$, in the reaction system. From Eqs. 3.5.10 and 3.5.46 it follows that

$$\mu_1(t) = \langle n \rangle_t = z \left.\frac{\partial G(z,t)}{\partial z}\right|_{z=1} = N e^{-kt}. \tag{3.5.49}$$

The discussion presented so far in this section may give the misleading impression that the generating function method is a panacea for the solution of master equations. This impression may be strengthened by our choice of the examples given above which were chosen precisely on the basis that a closed form solution for G(z,t) and P(n,t) could readily be obtained. It is important to point out that there is a large number of master equations of different forms for which an exact closed form solution for the probability distribution cannot be obtained, or, in some cases, has not yet been obtained, by the straightforward application of the generating function techniques discussed here. In some cases, solutions may, however, be obtained by some other means, such as, for instance, by recursion. We shall now give a short survey of the types of problems encountered. This discussion is limited to the master equation in discrete state space; analogous results are obtained for continuous state space.

a) The transition rates $\bar{A}(j,m)$ are nonlinear, i.e., are of higher degree than first in the variables m.

As an example, consider the master equation

$$\frac{\partial P(n,t)}{\partial t} = \Lambda[(n-1)^3 P(n-1,t) - n^3 P(n,t)]$$

$$n = 0,1,2, \ldots.. \qquad (3.5.50)$$

where $P(-1,t) = 0$. Introduction of the generating function 3.5.3 yields

$$\frac{\partial G(z,t)}{\partial t} = (z-1)\left(z\frac{\partial}{\partial z}\right)^3 G(z,t) \qquad (3.5.51)$$

The solution of Eq. 3.5.51 in terms of standard functions is not known and thus generating functions are not useful in obtaining a solution of Eq. 3.5.50. It is typical of master equations with transition rates nonlinear in the state variables that they lead to higher order (i.e., higher than first) partial differential equations for the generating function. Equations like Eq. 3.5.50 can, however, be solved by recursion.

b) The set of master equations cannot be written in terms of a general master equation valid for all values of the state variable.

As an example, let us consider the set of master equations

$$\frac{\partial P(0,t)}{\partial t} = \Lambda P(1,t)$$

$$\frac{\partial P(n,t)}{\partial t} = \Lambda[P(n+1,t) - P(n,t)], \; n = 1,2,\ldots \qquad (3.5.52)$$

Introduction of the generating function 3.5.3 leads to the equation

$$\frac{\partial G(z,t)}{\partial t} = \Lambda\left(\frac{1}{z} - 1\right)[G(z,t) - P(0,t)] \qquad (3.5.53)$$

where

$$P(0,t) = \frac{1}{2\pi i}\oint \frac{G(z',t)}{z'}\, dz' = G(0,t). \qquad (3.5.54)$$

Equation 3.5.53 is an integro-differential equation whose solution is difficult to obtain. In general, whenever the master equation cannot be written in terms of a general master equation valid for all values n of the state variable, one obtains an integro-differential equation for the generating function involving terms such as P(0,t), P(1,t), etc. These equations are, in general, difficult to solve in closed form(3.5.3) although the particular example in Eq. 3.5.52 can be solved recursively.

Approximate solutions for the generating function G(z,t) in equations such as 3.5.51 or 3.5.53 can be obtained by various methods. This is, however, of little use in obtaining solutions of the master equation or of the moments of the probability distribution for which the exact, analytic form of the generating function is required. When an exact solution of the partial differential equation or of the integro-differential equation for the generating function cannot be obtained, the approximation methods should be applied directly to the master equation.

6) Multivariate Processes

In the previous sections of this chapter we discussed the master equation and its solution for stochastic processes in a one variable state space. We shall now generalize the above results to a state space of r variables.

We begin by defining the r-variable probability distribution at time t, $P^{(r)}(n_1,n_2,\ldots,n_r;t)$ as the probability that the stochastic variable $\underset{\sim}{X}(t) = (X_1(t), X_2(t),\ldots,X_r(t))$ has the components $X_1(t) = n_1$, $X_2(t) = n_2,\ldots,X_r(t) = n_r$, all at time t. In vector notation, the r-variable probability distribution can be written as

$$P^{(r)}(n_1,n_2,\ldots,n_r;t) = P^{(r)}(\underset{\sim}{n};t). \tag{3.6.1}$$

Analogously we define the r-variable conditional probability $W^{(r)}(\underset{\sim}{n},t|\underset{\sim}{m},s)$ as the probability that if $X_1(t)$ had the value m_1 at time s, it will have the value n_1 at time t, if $X_2(t)$ had the value m_2 at time s, it will have the value n_2 at time t, etc. The multivariate probability distributions have the properties

$$\sum_{\underset{\sim}{n}} P^{(r)}(\underset{\sim}{n};t) = 1 \tag{3.6.2}$$

$$\sum_{n_j} P^{(r)}(\underset{\sim}{n};t) = P^{(r-1)}(n_1,n_2,\ldots,n_{j-1},n_{j+1},\ldots,n_r:t) \tag{3.6.3}$$

$$W^{(j)}(n_1,\ldots,n_j,t|m_1,\ldots,m_j,s)$$

$$= \sum_{n_{j+1},\ldots,n_r} \sum_{m_{j+1},\ldots,m_r}$$

$$\times \frac{W^{(r)}(\underset{\sim}{n},t|\underset{\sim}{m},s)\; P^{(r)}(\underset{\sim}{m},s)}{P^{(j)}(m_1,\ldots,m_j,s)}, \; j = 1,2,\ldots,r \tag{3.6.4}$$

which follows from [see Eq. 2.2.1]

$$P^{(j)}(n_1,\ldots,n_j;t) = \sum_{m_1,\ldots,m_j} \times W^{(j)}(n_1,\ldots,n_j,t|m_1,\ldots,m_j,s) \times P^{(j)}(m_1,\ldots,m_j,s) \tag{3.6.5}$$

$$\sum_{\underset{\sim}{n}} W^{(r)}(\underset{\sim}{n};t|\underset{\sim}{m},s) = 1 \tag{3.6.6}$$

$$W^{(r)}(\underset{\sim}{n},t|\underset{\sim}{m},t) = \prod_{j=1}^{r} \delta_{n_j,m_j} \equiv \delta_{\underset{\sim}{n},\underset{\sim}{m}}. \tag{3.6.7}$$

When a multivariate probability distribution

$P^{(r)}(\underset{\sim}{n};t)$ is not a product of single variable probability distributions, $P(n_1,t)$, then the stochastic variables $X_1(t)$, $X_2(t)$,..., are said to be correlated. Conversely, if the stochastic variables are completely uncorrelated, the multivariate probability distribution can be written as a product of the singlet probability distributions as

$$P^{(r)}(\underset{\sim}{n};t) = \prod_{i=1}^{r} P(n_i,t) \tag{3.6.8}$$

where $P^{(1)} \equiv P$. If the correlated subset $\alpha(\alpha \leq r)$ of the r stochastic variables is uncorrelated with the correlated subset $r-\alpha$ of the r stochastic variables then

$$P^{(r)}(\underset{\sim}{n},t) = P^{(\alpha)}(n_1,\ldots,n_\alpha;t)\ P^{(r-\alpha)}(n_{\alpha+1},\ldots,n_r;t). \tag{3.6.9}$$

The properties 3.6.8 and 3.6.9 are referred to as factorizations of the multivariate probability distributions $P^{(r)}(\underset{\sim}{n};t)$.

The transition rates $A^{(r)}(\underset{\sim}{n},\underset{\sim}{m},t)$ can be written in terms of the conditional probabilities $W^{(r)}(\underset{\sim}{n},t|\underset{\sim}{m},s)$ as

$$A^{(r)}(\underset{\sim}{n},\underset{\sim}{m},t) = \lim_{\Delta\to 0} \frac{W^{(r)}(\underset{\sim}{n},t+\Delta|\underset{\sim}{m},t) - \delta_{\underset{\sim}{n},\underset{\sim}{m}}}{\Delta}$$

$$= \left.\frac{\partial W^{(r)}(\underset{\sim}{n},t|\underset{\sim}{m},s)}{\partial t}\right|_{s=t} \tag{3.6.10}$$

The quantity $A^{(r)}(\underset{\sim}{n},\underset{\sim}{m},t)dt$, for $\underset{\sim}{n} \neq \underset{\sim}{m}$, is the probability for transitions from the states $\underset{\sim}{m}$ to states $\underset{\sim}{n}$ in the time interval $(t,t+dt)$. The master equation for the multivariate probability distribution $P^{(r)}(\underset{\sim}{n};t)$ is

$$\frac{\partial P^{(r)}(\underset{\sim}{n};t)}{\partial t} = \sum_{\underset{\sim}{m}} A^{(r)}(\underset{\sim}{n},\underset{\sim}{m},t)\, P^{(r)}(\underset{\sim}{m};t). \tag{3.6.11}$$

If the r-dimensional process is Markovian, then $W^{(r)}(\underset{\sim}{n},t|\underset{\sim}{m},s)$ obeys the equation [see Eq. 3.1.12]

$$\frac{\partial W^{(r)}(\underset{\sim}{n},t|\underset{\sim}{n}',s)}{\partial t} = \sum_{\underset{\sim}{m}} A^{(r)}(\underset{\sim}{n},\underset{\sim}{m},t)\, W^{(r)}(\underset{\sim}{m},t|\underset{\sim}{n}',s). \tag{3.6.12}$$

The equation for the j-variate probability distribution $P^{(j)}(n_1,\ldots,n_j;t)$ can be obtained from Eq. 3.6.11 by summing over $n_{j+1},\ldots,n_r$ which yields

$$\frac{\partial P^{(j)}(n_1,\ldots,n_j;t)}{\partial t} = \sum_{m_1,\ldots,m_j} \times\left\{\sum_{n_{j+1},\ldots,n_r}\ \sum_{m_{j+1},\ldots,m_r} \times \frac{A^{(r)}(\underset{\sim}{n},\underset{\sim}{m},t)\, P^{(r)}(\underset{\sim}{m};t)}{P^{(j)}(m_1,\ldots,m_j;t)}\right\}$$

$$\times \ P^{(j)}(m_1,\ldots,m_j;t) \tag{3.6.13}$$

where we have multiplied and divided each term on the r.h.s. by $P^{(j)}(m_1,\ldots,m_j;t)$. Comparison of Eq. 3.6.13 with the master equation 3.6.11 shows that

$$A^{(j)}(n_1,\ldots,n_j;m_1,\ldots,m_j;t) = \sum_{n_{j+1},\ldots,n_r} \sum_{m_{j+1},\ldots,m_r}$$

$$\times \frac{A^{(r)}(\underset{\sim}{n},\underset{\sim}{m},t)\ P^{(r)}\ (\underset{\sim}{m},t)}{P^{(j)}(m_1,\ldots,m_j;t)}. \tag{3.6.14}$$

Note that in general there is no simple recursion relation for the r-dimensional transition rates $A^{(r)}(\underset{\sim}{n},\underset{\sim}{m},t)$ as there is for the multivariate probability distributions $P^{(r)}(\underset{\sim}{n};t)$ as given in Eq. 3.6.3. In general, as can be seen from Eq. 3.6.14, the lower dimensional transition rates $A^{(j)}(\underset{\sim}{n},\underset{\sim}{m},t)$, for $j < r$, are functions of the r-dimensional probability distribution $P^{(r)}(\underset{\sim}{n};t)$. It is very difficult, in general, to obtain the solution $P^{(r)}(\underset{\sim}{n},t)$ from the r-dimensional master equation 3.6.11. Even if one had the solution $P^{(r)}(\underset{\sim}{n},t)$, it would still be very difficult to obtain the lower dimensional transition rates from $A^{(j)}(\underset{\sim}{n},\underset{\sim}{m},t)$ from equations like 3.6.14 and the lower dimensional probability distributions $P^{(j)}(\underset{\sim}{n};t)$ from equations like 3.6.13.

There are, however, a number of cases in which

significant simplifications can be made in the above scheme. If $P^{(r)}(\underset{\sim}{m};s)$ at time s is known to be of the delta function form

$$P^{(r)}(\underset{\sim}{m};s) = \delta_{\underset{\sim}{m},\underset{\sim}{m}'} \tag{3.6.15}$$

then Eqs. 3.6.4 and 3.6.14 reduce to

$$W^{(j)}(n_1,\ldots,n_j,t|m_1,\ldots,m_j,s)$$

$$= \sum_{n_{j+1},\ldots,n_r} W^{(r)}(\underset{\sim}{n},t|\underset{\sim}{m},s) \tag{3.6.16}$$

and

$$A^{(j)}(n_1,\ldots,n_j;m_1,\ldots,m_j;s)$$

$$= \sum_{n_{j+1},\ldots,n_r} A^{(r)}(\underset{\sim}{n},\underset{\sim}{m},s). \tag{3.6.17}$$

Another case in which a simplification is possible is that of stochastic processes with uncorrelated dynamics.[3.6.1] We define such processes by the factorization property of the conditional probabilities, i.e.,

$$W^{(r)}(\underset{\sim}{n},t|\underset{\sim}{m},s) = \prod_{i=1}^{r} W(n_i,t|m_i,s) \tag{3.6.18}$$

where $W = W^{(1)}$. Equation 3.6.18 simply states that the individual conditional probabilities $W(n_i,t|m_i,s)$

are independent. For such processes it follows from Eqs. 3.6.7 and 3.6.10 that

$$A^{(r)}(\underset{\sim}{n},\underset{\sim}{m},t) = \sum_{i=1}^{r} A(n_i,m_i,t) \prod_{\substack{k=1\\k\neq i}}^{r} \delta_{n_k,m_k} . \qquad (3.6.19)$$

It is clear that $W^{(j)}$ and $A^{(j)}$, $j = 1,\ldots,r$, are given by

$$W^{(j)}(\underset{\sim}{n},t|\underset{\sim}{m},s) = \prod_{i=1}^{j} W(n_i,t|m_i,s) \qquad (3.6.20)$$

and

$$A^{(j)}(n_1,\ldots,n_j;m_1,\ldots,m_j;t)$$

$$= \sum_{i=1}^{j} A(n_i,m_i,t) \prod_{\substack{k=1\\k\neq i}}^{j} \delta_{n_k,m_k} . \qquad (3.6.21)$$

The summation expression for the transition rates $A^{(r)}$ of Eq. 3.6.19 when substituted into the general form of the transition rate as given in Eq. 3.6.14 then yields

$$A^{(j)}(n_1,\ldots,n_j;m_1,\ldots,m_j;t) = \sum_{n_{j+1},\ldots,n_r} A^{(r)}(\underset{\sim}{n},\underset{\sim}{m},t). \qquad (3.6.22)$$

For the product form of the conditional probabilities $W^{(r)}(\underset{\sim}{n},t|\underset{\sim}{m},s)$ given in Eq. 3.6.18, the probability distribution $P^{(j)}(\underset{\sim}{n};t)$, $j = 1,2,\ldots,r$, can be obtained for a given initial condition $P^{(j)}(\underset{\sim}{n};s)$, from Eq. 3.6.5 and from $W(n_1,t|m_1,s)$ as determined from Eq. 3.1.10 if the process is non-Markovian, or Eq. 3.1.12 if the process is Markovian.

Multivariate stochastic processes are met with frequently in chemical reactions involving more than one species, transport processes in phase space where the relevant variables are, for example, position and momentum, processes which require for their complete specification both the energy and the position at time t as is the case, for instance, in absorption-desorption problems, random walks in r-dimensions, etc.

In order to effect a solution of multivariate master equations, one again makes use of the generating function technique. In analogy to our definition of the generating function G(z,t) of Eq. 3.5.3 for a discrete state space, we define the multivariate generating functions $G^{(r)}(\underset{\sim}{z},t) \equiv G^{(r)}(z_1,z_2,\ldots,z_r;t)$ by

$$G^{(r)}(\underset{\sim}{z};t) = \sum_{\underset{\sim}{n}} P^{(r)}(\underset{\sim}{n};t)\, z_1^{n_1} z_2^{n_2} \ldots z_r^{n_r}. \qquad (3.6.23)$$

Some of the important properties of the multivariate generating function $G^{(r)}(\underset{\sim}{z},t)$ are:

a) $G^{(r)}(z_1,z_2,\ldots,z_j,1,\ldots,1;t)$

$$= \sum_{n_1,\ldots,n_j} P^{(j)}(n_1,\ldots,n_j) z_1^{n_1} z_2^{n_2},\ldots,z_j^{n_j}$$

$$= G^{(j)}(z_1,\ldots,z_j;t), \quad j = 1,2,\ldots,r \tag{3.6.24}$$

where we have set $z_{j+1}, z_{j+2},\ldots,z_r = 1$. Equation 3.6.24 is the recursion relation for the multivariate generating function.

b) $$G^{(r)}(\underset{\sim}{1};t) = 1. \tag{3.6.25}$$

Equation 3.6.25 is the normalization condition for $G^{(r)}$.

c) $$\left(\prod_{i=1}^{r} \frac{\partial^{n_i}}{\partial z_i^{n_i}}\right) G^{(r)}(\underset{\sim}{z};t)\Bigg|_{\underset{\sim}{z}=\underset{\sim}{0}}$$

$$= P^{(r)}(\underset{\sim}{n};t) \prod_{i=1}^{r} (n_i!). \tag{3.6.26}$$

Equation 3.6.26 is the prescription for obtaining the desired multivariate probability distribution $P^{(r)}(\underset{\sim}{n};t)$ from $G^{(r)}(\underset{\sim}{z};t)$.

d) $$\mu_{\underset{\sim}{k}}(t) \equiv \sum_{\underset{\sim}{n}} P^{(r)}(\underset{\sim}{n};t)\, n_1^{k_1} n_2^{k_2},\ldots,n_r^{k_r}$$

$$= \prod_{i=1}^{r} \left(z_i \frac{\partial}{\partial z_i}\right)^{k_i} G^{(r)}(\underset{\sim}{z};t)\Bigg|_{\underset{\sim}{z}=1} \tag{3.6.27}$$

Equation 3.6.27 relates the multivariate moments $\mu_{\underset{\sim}{k}}(t)$ to $G^{(r)}(\underset{\sim}{z};t)$. The singlet moment μ_{k_1} is determined from Eq. 3.6.27 as

$$\mu_{k_1}(t) = \left(z_1 \frac{\partial}{\partial z_1}\right)^{k_1} G^{(r)}(\underset{\sim}{z};t)\Bigg|_{z_1=z_2=\ldots z_r=1}$$
$$= \left(z_1 \frac{\partial}{\partial z_1}\right)^{k_1} G(z_1,t)\Bigg|_{z_1=1} \tag{3.6.28}$$

where we have used Eq. 3.6.24 and have written $G^{(1)} \equiv G$.

e) $$G^{(r)}(\underset{\sim}{z};0) = \sum_{\underset{\sim}{n}} P^{(r)}(\underset{\sim}{n};0)z_1^{n_1}z_2^{n_2},\ldots,z_r^{n_r} \tag{3.6.29}$$

so that $G^{(r)}(\underset{\sim}{z};0)$ is determined from the initial condition on the multivariate probability distribution $P^{(r)}(\underset{\sim}{n};0)$. Equations 3.6.25 through 3.6.29 are completely analogous to the relations 3.5.6 through 3.5.10 for the singlet generating function $G(z;t)$.

For stochastic processes in continuous state space, the analogue to the singlet generating function $\phi(s,t)$ of Eq. 3.5.9 is [see Eq. 2.3.19]

$$\phi^{(r)}(\underset{\sim}{s},t) = \int_{-\infty}^{\infty} e^{i\underset{\sim}{s}\cdot\underset{\sim}{x}}\, p^{(r)}(\underset{\sim}{x};t)d\underset{\sim}{x} \tag{3.6.30}$$

where

$$\underset{\sim}{s}\cdot\underset{\sim}{x} = \sum_{i=1}^{r} s_i x_i \quad \text{and} \quad d\underset{\sim}{x} = \prod_{i=1}^{r} dx_i.$$

Relations for the generating function $\phi^{(r)}(\underset{\sim}{s};t)$ analogous to those of Eqs. 3.6.24 through 3.6.29 can readily be obtained. We shall, however, continue to restrict our attention here to multivariate stochastic processes in discrete space.

To obtain the partial differential equation for the generating function $G^{(r)}(\underset{\sim}{z};t)$ from the master equation 3.6.11, we multiply Eq. 3.6.11 by $\prod_{i=1}^{r} z_i^{n_i}$ and sum over all values of $\underset{\sim}{n}$,

$$\frac{\partial G^{(r)}(\underset{\sim}{z};t)}{\partial t} = \sum_{\underset{\sim}{n}} \sum_{\underset{\sim}{m}} A^{(r)}(\underset{\sim}{n},\underset{\sim}{m},t)P^{(r)}(\underset{\sim}{m};t)z_1^{n_1}z_2^{n_2} \ldots z_r^{n_r}. \tag{3.6.31}$$

Introducing the function [see Eq. 3.5.28]

$$\bar{A}^{(r)}(\underset{\sim}{n}-\underset{\sim}{m}, \underset{\sim}{m},t) = A^{(r)}(\underset{\sim}{n},\underset{\sim}{m},t) \tag{3.6.32}$$

we can write Eq. 3.6.31 as

$$\frac{\partial G^{(r)}(\underset{\sim}{z};t)}{\partial t} = \sum_{\underset{\sim}{m}} \Gamma^{(r)}(\underset{\sim}{m},\underset{\sim}{z},t)\ P^{(r)}(\underset{\sim}{m};t)z_1^{m_1}z_2^{m_2} \ldots z_r^{m_r} \tag{3.6.33}$$

where

$$\Gamma^{(r)}(\underset{\sim}{m},\underset{\sim}{z},t) \equiv \sum_{\underset{\sim}{n}-\underset{\sim}{m}} \bar{A}^{(r)}(\underset{\sim}{n}-\underset{\sim}{m}, \underset{\sim}{m},t) \prod_{i=1}^{r} z_i^{(n_i-m_i)}. \tag{3.6.34}$$

For stochastic processes with transition rates $A^{(r)}$ of a form such that $\Gamma^{(r)}(\underset{\sim}{m},\underset{\sim}{z},t)$ is an analytic function of $\underset{\sim}{z}$, Eq. 3.6.33 can be rewritten as[(3.6.2)]

$$\frac{\partial G^{(r)}(\underset{\sim}{z};t)}{\partial t} = \Gamma^{(r)}\left(\underset{\sim}{z}\frac{\partial}{\partial \underset{\sim}{z}}, \underset{\sim}{z};t\right) G^{(r)}(\underset{\sim}{z};t) \tag{3.6.35}$$

where $\Gamma^{(r)}$ is an operator which operates on the generating function $G^{(r)}$. Equation 3.6.35 is the desired partial differential equation for the generating function $G^{(r)}$. For $r = 1$, Eq. 3.6.35 is equivalent to the contour integral expression 3.5.25 for $G(z;t)$. Note that $\Gamma \equiv \Gamma^{(1)}$ is related to the quantity (z,y,t) of Eq. 3.5.26 by

$$(z,z,t) = \sum_m \Gamma(m,z,t). \tag{3.6.36}$$

For most multivariate stochastic processes, the partial differential equation 3.6.35 for the generating function is very complicated and it is unlikely that analytic solution for $G^{(r)}(\underset{\sim}{z};t)$ can be obtained. To illustrate the use of the generating function technique in the solution of a multivariate master equation, we present here a simple example for which a closed form solution for $G^{(2)}(z_1,z_2;t)$ has been obtained. We consider the set of chemical reactions

$$A + B \xrightarrow{\alpha} C + B \tag{3.6.37}$$

$$B \xrightarrow{\beta} D. \tag{3.6.38}$$

The stochastic variables in this reaction scheme are the number of molecules A at time t, $X_1(t)$, and the number of molecules B at time t, $X_2(t)$. We let $P^{(2)}(n_1,n_2,t)$ be the joint probability that there are

n_1 molecules of A and n_2 molecules of B at time t. The initial population of species A and B, respectively, are N_1 and N_2, so that

$$P^{(2)}(n_1,n_2,0) = \delta_{n_1,N_1}\,\delta_{n_2,N_2}. \tag{3.6.39}$$

We assume that the probability that one molecule of species A will undergo a reaction in the time interval (t,t+dt) is $\alpha n_1 n_2 dt$ and the probability that one molecule of species B will undergo a reaction in that time interval is $\beta n_2 dt$, where α and β are constants. It is furthermore assumed that the probability that more than one molecule of species A or B will undergo reaction in the time interval (t,t+dt) is zero. The non-zero transition rate elements for this process are

$$A^{(2)}(n_1-1,n_2;n_1,n_2) = \alpha n_1 n_2$$
$$A^{(2)}(n_1,n_2-1;n_1,n_2) = \beta n_2$$
$$A^{(2)}(n_1,n_2;n_1,n_2) = -[\alpha n_1 n_2 + \beta n_2]. \tag{3.6.40}$$

The master equation for this set of chemical reactions is

$$\frac{\partial P^{(2)}(n_1,n_2;t)}{\partial t} = \alpha(n_1+1)\,n_2 P^{(2)}(n_1+1,n_2;t) + \beta(n_2+1)\,P^{(2)}(n_1,n_2+1;t) - [\alpha n_1 n_2 + \beta n_2]\,P^{(2)}(n_1,n_2;t) \tag{3.6.41}$$

with $n_1 = 0,1,\ldots,N_1$ and $n_2 = 0,1,\ldots,N_2$. We set $P^{(2)}(n_1,n_2;t)$ equal to zero for $n_1 > N_1$ or $n_2 > N_2$ for all t.

The multivariate generating function $G^{(2)}(z_1,z_2;t)$ is [see Eq. 3.6.23]

$$G^{(2)}(z_1,z_2;t) = \sum_{n_1=0}^{N_1} \sum_{n_2=0}^{N_2} z_1^{n_1} z_2^{n_2} P^{(2)}(n_1,n_2;t). \tag{3.6.42}$$

Multiplication of the master equation 3.6.41 by $z_1^{n_1} z_2^{n_2}$ and performance of the summations indicated in Eq. 3.6.42 leads to

$$\frac{\partial G^{(2)}(z_1,z_2;t)}{\partial t} = \alpha z_2(1-z_1) \frac{\partial^2 G^{(2)}(z_1,z_2;t)}{\partial z_1 \partial z_2} + \beta(1-z_2) \times \frac{\partial G^{(2)}(z_1,z_2;t)}{\partial z_2} . \tag{3.6.43}$$

The solution to this partial differential equation with initial condition $G^{(2)}(z_1,z_2,0) = z_1^{N_1} z_2^{N_2}$ is[3.6.3]

$$G^{(2)}(z_1,z_2;t) = \sum_{j=0}^{N_1} \binom{N_1}{j} (z_1-1)^j \times \left[\frac{\beta+[\alpha j z_2+\beta(z_2-1)]\, e^{-(\beta+\alpha j)t}}{\beta+\alpha j}\right]^{N_2} . \tag{3.6.44}$$

The probability distribution $P^{(2)}(n_1,n_2;t)$ is obtained by expanding the r.h.s. of Eq. 3.6.44 as a power series in z_1 and z_2; as can be seen from Eq. 3.6.42, the distribution function $P^{(2)}(n_1,n_2;t)$ is the coefficient of $z_1^{n_1} z_2^{n_2}$ in this expansion. This yields

$$P^{(2)}(n_1;n_2,t) = \binom{N_1}{n_1}\binom{N_2}{n_2}\sum_{j=0}^{N_1-n_1}\binom{N_1-n_1}{j} \times$$

$$\times (-1)^j \left\{\left(\frac{\beta}{\beta+\alpha(n_1+j)}\right)\right.$$

$$\left.\times\left(1-e^{-[\beta+\alpha(n_1+j)]t}\right)\right\}^{N_2-n_2} \times e^{-n_2[\beta+\alpha(n_1+j)]t} .$$

(3.6.45)

We now use Eq. 3.6.28 to obtain the first moments $\mu_{1,0}(t) = \langle n_1\rangle_t$ and $\mu_{0,1}(t) = \langle n_2\rangle_t$ from the generating function $G^{(2)}(z_1,z_2;t)$ of Eq. 3.6.44. This yields

$$\langle n_1\rangle_t = N_1\left[\frac{\beta+\alpha e^{-(\alpha+\beta)t}}{\alpha+\beta}\right]^{N_2} \qquad (3.6.46)$$

$$\langle n_2\rangle_t = N_2 e^{-\beta t} . \qquad (3.6.47)$$

The moments $\langle n_1\rangle_t$ and $\langle n_2\rangle_t$ are, respectively, the average values of the number of molecules A and B in the reaction system at time t. It should be noted that $\langle n_1\rangle_t$ of Eq. 3.6.46 does not agree with the deterministic expression

$$n_1(t) = N_1 \exp\left[- \frac{\alpha}{\beta} N_2 (1-e^{-\beta t})\right] \qquad (3.6.48)$$

but will reduce to it in the limit as $N_1, N_2 \to \infty$[3.6.4].

References

3.1.1. The term "master equation" appears to have been used first by A. Nordsieck, W. E. Lamb, Jr., G. E. Uhlenbeck, Physica 7, 344 (1940) in connection with the theory of cosmic ray showers.

3.1.2. For a more detailed discussion see I. Oppenheim and K. E. Shuler, Phys. Rev. 138, B1007 (1965).

3.1.3. A. Kolmogoroff, Math. Ann. 104, 415 (1931); 108, 149 (1933).

3.1.4. W. Feller, Trans. Amer. Math. Soc. 48, 488 (1940).

3.2.1. See e.g. B. J. Berne and G. D. Harp, Adv. Chem. Phys. XVII, 63 (1970).

3.3.1. See e.g. R. Bellman, Introduction to Matrix Analysis, McGraw-Hill Book Co., New York, 1960.

3.3.2. S. Gershgorin, Izv. Akad. Nauk, S.S.S.R., 7, 749 (1931).

3.3.3. W. Ledermann and G. E. H. Reuter, Phil. Trans. Roy. Soc. A246, 321 (1953).

3.3.4. K. E. Shuler, Phys. Fluids 2, 442 (1959).

3.3.5. See e.g. M. R. Hoare, Adv. Chem. Phys. 20, 135 (1971).

3.3.6. See e.g. R. J. Rubin, J. Math. Phys. 1, 309 (1960).

3.3.7. See e.g. W. Feller, Probability Theory and Its

Applications, Vol. 1, John Wiley, Inc., New York, 1950, Ch. 15.

3.3.8. F. J. Dyson, Phys. Rev. 75, 486 (1949).

3.3.9. W. Magnus, Comm. Pure Appl. Math. 7, 649 (1954).

3.3.10. F. Hausdorff, Ber. Verhandl. Sachs. Akad. Wiss. Leipzig, 58, 19 (1905).

3.3.11. H. F. Baker, Proc. London Math. Soc. 34, 347 (1902); 35, 333 (1903).

3.3.12. P. Pechukas, J. C. Light, J. Chem. Phys. 44, 3897 (1966).

3.3.13. G. H. Weiss, A. A. Maradudin, J. Math. Phys. 3, 771 (1962).

3.3.14. K. E. Shuler, G. H. Weiss, J. Chem. Phys. 45, 1105 (1966).

3.4.1. E. W. Montroll and K. E. Shuler, Advances in Chemical Physics, ed. I. Prigogine, Interscience Press, New York, 1961, p. 361.

3.4.2. S. K. Kim, J. Chem. Phys. 28, 1057 (1957).

3.4.3. G. H. Weiss in Advances in Chemical Physics XIII, ed. I. Prigogine (John Wiley & Sons, New York, 1966), p. 1.

3.5.1. E. W. Montroll, Proc. Symp. Appl. Math. 16, 193 (1964); E. W. Montroll and G. H. Weiss, J. Math. Phys. 6, 167 (1965); E. W. Montroll, J. Phys. Soc. Japan Suppl. 26, 6 (1969); E. W. Montroll, J. Math. Phys. 10, 753 (1969).

3.5.2. G. B. Duff, Partial Differential Equations, University of Toronto Press, (1956).

3.5.3. This probelm is discussed in some detail in D. R. Cox and H. D. Miller, The Theory of Stochastic Processes, John Wiley and Sons, Inc., New York, 1965, pp. 193 et seq.

3.6.1. I. Oppenheim, K. E. Shuler, G. Weiss, J. Chem. Phys. 46, 4100 (1967).

3.6.2. See e.g. J. E. Moyal, J. Roy. Stat. Soc. B11, 150 (1949).

3.6.3. K. Dietz, J. Appl. Prob. 3, 375 (1966).

3.6.4. I. Oppenheim, K. E. Shuler and G. H. Weiss, J. Chem. Phys. 50, 460 (1969).

Chapter 4

REPRINTS

1) Derivation of the Master Equation

In order to assess the applicability of the master equation to the description of physical phenomena, it is essential to develop a procedure which obtains the master equation as a well-defined approximation to the exact equations of motion describing the system of interest. The papers in this section present derivations of the master equation from the quantum mechanical or classical exact Liouville equations. The master equation describes the temporal behavior of the reduced probability densities of the slowly changing variables of the system.

In the first paper in this section, van Hove derives the master equation for a quantum mechanical system in the weak coupling limit. The master equation describes the time development of the reduced probability density on the time scale $s = \lambda^2 t$ where λ is a measure of the rate of change of the slow variables and where the limit $\lambda \to 0$, $t \to \infty$, $\lambda^2 t$ fixed is taken. In the second paper, van Hove's pioneering developments are applied to a classical system by Prigogine and Resibois.

The third paper by Zwanzig utilizes powerful projection operator techniques which considerably simplify the diagrammatic analyses of the first two papers. Finally, in the fourth paper Zwanzig demonstrates that the generalized master equations derived by

Prigogine, Montroll, Nakajima and Zwanzig, which have somewhat different forms, are indeed identical.

THE APPROACH TO EQUILIBRIUM IN QUANTUM STATISTICS

A PERTURBATION TREATMENT TO GENERAL ORDER

by LÉON VAN HOVE

Instituut voor theoretische Fysica der Rijksuniversiteit, Utrecht, Nederland

Synopsis

The approach of a quantum system to statistical equilibrium under the influence of a perturbation is described by a well known transport equation, now often called master equation (see (1.1) hereunder). This equation holds only when the perturbation is taken into account to lowest non-vanishing order. It has been stressed in a recent paper that certain characteristic properties of the perturbation, easily seen to hold for actual systems (crystals, gases), play an essential role in determining the irreversible nature of the effects described by (1.1). On the basis of these properties it was possible to derive the lowest order master equation from the Schrödinger equation by making one assumption only, relative to the phases of the wave unction at the initial time. In contrast with the usual derivation which assumes the phases to be random at all times, the method just mentioned is capable of extension to higher orders in the perturbation. This extension is carried out in the present paper. The essential results are the establishment of a generalized master equation valid to arbitrary order in the perturbation, and the proof that the long time behaviour of its solution corresponds to establishment of microcanonical equilibrium (the latter being taken for the total hamiltonian, perturbation included). The generalized master equation exhibits with its lowest order version the essential difference that it corresponds to a non-markovian process. The transition from the exact master equation to its lowest order approximation is discussed in detail. It illustrates the existence of two time scales, a short one and a long one, for very slow irreversible processes, as well as their overlapping in the case of faster processes.

1. *Introduction.* We have studied in a previous paper [1]), to be referred to hereafter as A, the transport equation describing the approach to statistical equilibrium of a quantum system under the influence of a small perturbation. This equation, the form of which is well known

$$\mathrm{d}P_\alpha/\mathrm{d}t = \Sigma_\beta(W_{\alpha\beta}P_\beta - W_{\beta\alpha}P_\alpha), \tag{1.1}$$

involves the probability distribution P_α of the system over groups of eigenstates of the unperturbed hamiltonian H and expresses its irreversible time evolution under the action of the perturbation. It holds only to lowest order in the perturbation. Following recent practice we shall call it the *master equation*, thus avoiding possible confusion with transport equations of the Boltzmann type.

The first derivation of (1.1), given long ago by Pauli [2]), was based on the assumption that, when the total wave function is expanded in the eigenstates of H, the coefficients have *at all times* randomly distributed phases. As shown in A, this assumption, the unsatisfactory nature of which is obvious, can be avoided and replaced by an assumption on the wave function *at one initial time* (either randomness of phases or some other condition as stated in A) if one properly takes into account that the system has a very large number of degrees of freedom and that the perturbation satisfies certain special properties responsible for its irreversible effects and easily recognized on the systems met with in applications (gases, crystals, etc.). The formalism developed in A, while rather different from the one customary in quantum statistics, is better suited to the situation at hand because it gives a simple form to the relevant mathematical consequences of the facts just mentioned. With its help it was an elementary matter to derive the master equation from the Schrödinger equation under the conditions assumed for the initial value of the wave function.

While the assumption of random phases at all times can be used despite its unsatisfactory nature to derive the master equation to lowest order in the perturbation, it is strictly speaking incompatible with the Schrödinger equation itself, and would therefore be a completely erroneous starting point for an attempt at extending the master equation to higher order. The method used in A on the contrary, being rigorously correct, opens the possibility to solve this rather natural problem of deriving a generalized master equation valid to arbitrary order in the perturbation. Such a generalized master equation is obtained and analyzed in the present paper [3]).

A detailed description of the special properties of the hamiltonian, unperturbed part and perturbation, is given in the next section. Section 3 defines the quantity P, the time-dependence of which is to be expressed by the master equation, and discusses its physical significance for the time evolution of the quantum-mechanical system. The mathematical derivation of the master equation, eq. (4.31), is given in Section 4, to general order in perturbation. The following section shows how the general master equation reduces for small perturbations to the customary form (1.1). In Section 6 a detailed mathematical analysis is given for the asymptotic behaviour of the quantity P at long times. In Section 7, on the basis of proper assumptions on the symmetry of the transition rate matrix and the interconnection of states, it is established for finite but not too large perturbations that the long time limit of P is in accordance with the predictions of microcanonical ensemble theory, the microcanonical distribution being taken for the total hamiltonian, perturbation included.

The mathematical derivations carried out in Sections 4 to 7 are not simple.

Although we have relied more than once on earlier publications, the developments needed to derive the transport equation and to analyse its consequences are of considerable complication and length. In a number of places we have tried to alleviate their abstract character by inserting comments of a more physical nature. To help further the reader who likes to follow the general arguments by applying them to an example, we have considered briefly in the appendix, in its simplest form, a system composed of Bloch electrons and lattice vibrations (phonons) in a perfect crystal. Using the data of the appendix one can illustrate the various definitions and results of the general treatment.

2. *The hamiltonian and the basic representation.* We are interested in quantum-mechanical many-particle systems with a hamiltonian $H + \lambda V$ split into two parts. The first part, H, is assumed to give an essentially complete separation of variables and would therefore, taken alone, produce no approach to thermodynamical equilibrium. The perturbation λV (λ is a dimensionless parameter characterizing its size) mixes the many degrees of freedom left uncoupled by H and is entirely responsible for the irreversible behaviour. As examples we may quote non-conducting crystals (H contains then the harmonic part of the forces and λV is the potential of the anharmonic forces), the electron-phonon system in metals (H describes the free harmonic vibrations of the lattice and the conduction electrons in the periodic field of the ions at their equilibrium positions, the electron-phonon interaction is the perturbation), quantum gases (λV is here the potential of the intermolecular forces) etc.

In all such examples, considering the limiting case of a large system of given density (number of particles N and volume Ω tending to infinity with a constant ratio N/Ω), one finds a natural description of the unpertubed stationary states (eigenstates of H) in terms of elementary plane wave excitations. By the generic name excitation we understand phonons in crystals, phonons and electrons in metals, particles in gases, etc. Each of these excitations is characterized by a wave vector and possibly a polarization or spin index. The components of the wave vector vary by steps of order of magnitude $\Omega^{-\frac{1}{3}} \sim N^{-\frac{1}{3}}$ and behave consequently in the limit of a large system as continuous quantum numbers, while the unperturbed energy becomes a continuous function of them. Polarization or spin indices remain of course discrete.

This occurrence of continuous quantum numbers and the continuous nature of the unperturbed energies are the most important features through which the very large size of the system manifests itself in the properties of the unperturbed hamiltonian. We postulate them generally and simplify matters a little by leaving out of consideration all discrete quantum numbers like polarization or spin indices. The unperturbed hamiltonian is thus assumed to have a complete set of eigenstates $|\alpha\rangle$, each of which is characterized by a collection α of quantum numbers (the quantum numbers of all excitations present) behaving as continuous variables in the limit of a large system. α may contain an infinity of such quantum numbers since an infinity of excitations may be simultaneously present. The corresponding eigenvalue $\varepsilon(\alpha)$, defined by

$$H\,|\alpha\rangle = |\alpha\rangle\,\varepsilon(\alpha) \tag{2.1}$$

is assumed to be continuous in all the quantum numbers in α. We further normalize $|\alpha\rangle$ in such a manner that $\langle\alpha\,|\,\alpha'\rangle$ becomes, in the limit of a large system, a product of δ-functions for all variables involved. We write accordingly

$$\langle\alpha\,|\,\alpha'\rangle = \delta(\alpha - \alpha'). \tag{2.2}$$

How this normalization must be carried out in practice can be found in the appendix.

As for the perturbation λV, the fact that it always extends over the whole system implies in the limit $N \to \infty$ *) remarkable analytical properties of its matrix elements in the $|\alpha\rangle$-representation. These properties are best recognized in practical examples by expressing V for large but finite systems in terms of emission and absorption operators of free excitations and going then over to an infinite system. We describe these properties hereunder in general terms. It is useful to check them in simple cases, e.g. by means of the equations in the appendix. In certain cases the first property holds only after parts of the diagonal matrix elements $\lambda\,\langle\alpha|\,V\,|\alpha\rangle$ have been incorporated in the unperturbed hamiltonian H.

Property (i): Take the matrix element $\langle\alpha\,|\,V|\,\alpha'\rangle$ for those states α and α' for which it is not identically zero (i.e. for α and α' differing by the few excitations absorbed or emitted by V) and consider it for those states as function of all distinct quantum numbers contained in α and α'. In the limit of a large system this function exhibits a δ-singularity. This singularity expresses overall conservation of momentum (or wave vector). It does not imply a δ-singularity in the difference $\varepsilon(\alpha) - \varepsilon(\alpha')$.

Property (ii): Take a higher order matrix element of the type $\langle\alpha\,|VA_1V...A_nV|\,\alpha'\rangle$ where $A_1, \ldots A_n$ are diagonal operators in the $|\alpha\rangle$-representation

$$A_j\,|\alpha\rangle = |\alpha\rangle\,A_j(\alpha). \tag{2.3}$$

Assume each eigenvalue $A_j(\alpha)$ a smooth function of all quantum numbers involved in α. Consider this matrix element for those states α, α' for which it is not identically zero and regard it as function of all distinct quantum numbers in α and α'. In the limit of a large system, this function exhibits

*) By this limit we always understand $N \to \infty$, $\Omega \to \infty$ with finite density N/Ω.

singularities of δ-type originating from the singularities in $\langle\alpha''|V|\alpha'''\rangle$. In addition to the δ-factor expressing overall conversation of momentum (or wave vector) further δ-singularities may occur. They are caused by the fact that the number of intermediate states over which one has to sum when calculating (for a finite system) the expression

$$\langle\alpha|VA_1V\dots A_nV|\alpha'\rangle = \Sigma_{\alpha_1\dots\alpha_n}\langle\alpha|V|\alpha_1\rangle A(\alpha_1)\langle\alpha_1|V|\alpha_2\rangle\dots A(\alpha_n)\langle\alpha_n|V|\alpha'\rangle$$

may be larger by one or more factors N (or Ω) when α is in a special relation to α' than otherwise. The number of such factors always turns out to be one third of the number of relations between quantum numbers in α and α'. Since each quantum number (being a momentum component of an excitation) varies with steps of order $N^{-\frac{1}{3}}$, one gets precisely a δ-singularity when the limit $N\to\infty$ is taken. Higher singularities are never obtained. An important point is now the following. Among all δ-singularities which are thus possible, none except one implies the equality of the unperturbed energies of initial and final states, i.e. implies a δ-singularity in the difference $\varepsilon(\alpha)-\varepsilon(\alpha')$. The only exception is the δ-singularity obtained when the state α and the state α' are identical, i.e. when α and α' have the same number of excitations present and these excitations all have the same quantum numbers. In other words, it is a singularity in $\delta(\alpha-\alpha')$. That such a singularity involves a $\delta[\varepsilon(\alpha)-\varepsilon(\alpha')]$-factor is obvious. This particular singularity plays a central role in the dynamics of the system and we split the matrix element $\langle\alpha|VA_1V\dots A_nV|\alpha'\rangle$ in a term containing it and a rest term

$$\langle\alpha|VA_1V\dots A_nV|\alpha'\rangle = \delta(\alpha-\alpha')F_1(\alpha) + F_2(\alpha,\alpha'). \tag{2.4}$$

The term $\delta(\alpha-\alpha')\,F_1(\alpha)$ is called the *diagonal part* of the matrix element, and we call diagonal part of the operator $VA_1V\dots A_nV$ the operator $\{VA_1V\dots A_nV\}_d$ defined by

$$\{VA_1V\dots A_nV\}_d|\alpha\rangle = |\alpha\rangle F_1(\alpha). \tag{2.5}$$

The rest term $F_2(\alpha,\alpha')$ has no $\delta(\alpha-\alpha')$-singularity, nor has it any δ-singularity implying $\varepsilon(\alpha)=\varepsilon(\alpha')$.

Property (iii): All what has just been said on the matrix element $\langle\alpha|VA_1V\dots A_nV|\alpha'\rangle$ holds of course also for the partial matrix elements $\langle\alpha|VA_jVA_{j+1}V\dots A_kV|\alpha'\rangle$ with $1\leqslant j\leqslant k\leqslant n$. Since the latter are involved in the calculation of the former, one sees that the δ-singularities present in the latter will have to be taken into account when calculating the former by summation over intermediate states. In this summation (still for a finite system)

$$\langle\alpha|VA_1V\dots A_nV|\alpha'\rangle = \Sigma_{\alpha_1\dots\alpha_n}\langle\alpha|V|\alpha_1\rangle A(\alpha_1)\dots\langle\alpha_n|V|\alpha'\rangle, \tag{2.6}$$

the singularities of $\langle\alpha_{j-1}|VA_jV\dots A_kV|\alpha_{k+1}\rangle$ manifest themselves as follows (we assume $1\leqslant j\leqslant k\leqslant n$, $k-j<n-1$ and put $\alpha_0=\alpha$, $\alpha_{n+1}=\alpha'$):

when the state α_{j-1} is taken in some special relation to the state α_{k+1}, the number of intermediate states $\alpha_j, \alpha_{j+1} \dots \alpha_k$ becomes larger by a power of N compensating exactly the decrease in the number of possible choices of the pair $\alpha_{j-1}, \alpha_{k+1}$. We need not consider all possible situations of that kind. The only inportant ones for our purpose are those corresponding to the diagonal parts of partial matrices $\langle \alpha_{j-1} | VA_jV \dots A_kV | \alpha_{k+1} \rangle$, i.e. to their $\delta(\alpha_{j-1} - \alpha_{k+1})$-singularities. Their effect in the summation (2.6) is simply that the partial sum obtained by putting $\alpha_{j-1} = \alpha_{k+1}$ gives even when $N \to \infty$ a contribution of the same order of magnitude as the remaining part of the sum. More generally, non-negligible contributions are obtained by simultaneous consideration of diagonal parts of several submatrices, i.e. from partial summations in (2.6) where several pairs of intermediate states are kept equal

$$\alpha_{j_1-1} = \alpha_{k_1+1}, \alpha_{j_2-1} = \alpha_{k_2+1}, \dots (j_r \leqslant k_r, \quad r = 1, 2, \dots). \tag{2.7}$$

A very important point is now that such a pairing of intermediate states only produces (for $N \to \infty$) a non-negligible contribution when no two pairs are interlocked, i.e. when no relation of the form

$$j_r - 1 < j_s - 1 < k_r + 1 < k_s + 1 \qquad (r, s = 1,2 \dots) \tag{2.8}$$

holds. This is our property (iii). Just as the other properties it must be verified in all practical cases. The general reason for its validity can be stated as follows: the transitions contributing to the diagonal part of $\langle \alpha_{j-1} | VA_jV \dots$ $\dots A_kV | \alpha_{k+1} \rangle$ involve emission and reabsorption of excitations (or of holes in a sea of excitations) and one has to sum over a large number ($\sim N$) of states of these virtual excitations in order to get a non-vanishing contribution for $N \to \infty$. Therefore, in the case of interlocking diagonal parts as in (2.8), the state α_{k_r+1} cannot be kept identical to the state α_{j_r-1} because it involves the additional excitations (or holes) contributing to the diagonal part of $\langle \alpha_{j_s-1} | VA_{j_s}V \dots A_{k_s}V | \alpha_{k_s+1} \rangle$. In order for simultaneous diagonal parts (2.7) to contribute, their relative positions have to be given by one of the equations

$$j_r - 1 < k_r + 1 \leqslant j_s - 1 < k_s + 1 \tag{2.9}$$

or

$$j_r - 1 < j_s - 1 < k_s + 1 < k_r + 1 \,^{*)} \tag{2.10}$$

We have described the special properties exhibited by the perturbation in the limit of a large system, as they can be verified in the actual examples and as they are needed for the development of the general theory. They essentially amount to the occurrence of diagonal parts in operator products $VA_1V \dots A_nV$, as expressed in (2.4). The latter equation must be understood

*) The cases $j_r - 1 = j_s - 1 < k_s + 1 < k_r + 1$ and $j_r - 1 < j_s - 1 < k_s + 1 = k_r + 1$ reduce to the case (2.9) by a change of notation.

as expressing the relevant asymptotic properties of the matrix element $\langle \alpha | VA_1V \ldots A_nV | \alpha' \rangle$ for $N \to \infty$. The δ-function symbolizes the occurrence of a definite additional power of N in the value of the matrix element when α is identical to α'. One should always realize that both $F_1(\alpha)$ and $F_2(\alpha, \alpha')$ may still depend on N. In many applications these quantities actually have a complicated asymptotic behaviour for $N \to \infty$. As recently shown by Hugenholtz for systems near their groundstate, this behaviour is amenable to a thorough analysis leading to a neat separation of the effects of the perturbation in size-dependent and size-independent ones, in agreement with the physical expectations [4]). These questions, however, may be left out of consideration for our present purpose which is to establish to general order in λV a statistical equation for the approach to equilibrium. All we have to know concerning the asymptotic behaviour of matrix elements is expressed in (2.4).

As most readers will have observed, all what has been said until now on the hamiltonians of many-particle systems of quantum statistics applies as well to those of quantized fields in interaction. For this case diagonal parts of operators $VA_1V \ldots A_nV$ (V being the interaction between fields) correspond to the well known self-energy diagrams of the Feynman-Dyson theory, taken however not only for one-particle states but for general states (lower δ-singularities would correspond for example to vertex diagrams). The role played by the number of particles N or the volume Ω of the many-body system is taken over for fields by the volume of the large box in which one conventionally imagines them to the enclosed. We have shown in two previous papers [5]), to be referred to as I and II, how for fields the properties mentioned above imply the physical (self-energy and cloud) effects most characteristic of their interactions. While the occurrence of such effects has long been intuitively clear and the explicit calculation of self-energies is well known, most existing theoretical treatments circumvent the explicit consideration of cloud effects.

Despite the similarities between the hamiltonians, the many-particle systems of statistical mechanics have evidently a physical behaviour very different from interacting fields: the essential property we expect them to exhibit is a general tendency to approach thermodynamical equilibrium, i.e. an essentially dissipative type of motion, whereas nothing of that sort is to be found in the usual field-theoretical situations. The analogy between the hamiltonians is therefore not complete. One essential difference must be present, determining the dissipative or non-dissipative character of the motion under the influence of the perturbation. This difference has already been mentioned in I (section 4) and will be discussed at greater length hereafter. While I and II dealt with the non-dissipative case, the present paper is mainly concerned with the dissipative one.

A number of the results obtained in I, mainly those of Sections (I.2) and

(I. 3) *), will be used hereafter. They follow, as shown in I, from the properties of the hamiltonian listed above. Besides using the concept of diagonal part of a product $VA_1V \dots A_n V$ (with $A_1, \dots A_n$ diagonal in the $|\alpha\rangle$-representation), they rely on the concept of irreducible diagonal part. Its definition is the following. Consider the contribution to the matrix element (2.6) which is obtained when each submatrix $\langle\alpha_{j-1}|VA_jV \dots A_kV|\alpha_{k+1}\rangle$, $(1 \leqslant j \leqslant k \leqslant n,\ k - j < n - 1,\ \alpha_0 = \alpha,\ \alpha_{n+1} = \alpha')$ is taken with neglection of its diagonal part, i.e. is taken for $\alpha_{j-1} \neq \alpha_{k+1}$. For $N \to \infty$ this contribution to (2.6), considered as function of α and α', can be separated in a part involving a $\delta(\alpha - \alpha')$-singularity and a part involving at most weaker singularities. Let us write this separation in analogy to (2.4),

$$\delta(\alpha - \alpha')F_1'(\alpha) + F_2'(\alpha, \alpha').$$

We call $\delta(\alpha - \alpha')\, F_1'(\alpha)$ the *irreducible diagonal part* of the matrix element $\langle\alpha|VA_1V \dots A_nV|\alpha'\rangle$, and we define the irreducible diagonal part $\{VA_1V \dots A_nV\}_{id}$ of the operator $VA_1V \dots A_nV$ in analogy to (2.5) by

$$\{VA_1V \dots A_nV\}_{id}\,|\alpha\rangle = |\alpha\rangle\, F_1'(\alpha).$$

It is also useful to define the *non-diagonal part* $\{VA_1\, V \dots A_nV\}_{nd}$ of the operator $VA_1V \dots A_nV$. It is the operator, the matrix elements of which are obtained by keeping in (2.6) *all states* $\alpha, \alpha_1, \dots \alpha_n, \alpha'$ different from each other, the value of the matrix element for $\alpha = \alpha'$ being defined as the limit of its value for $\alpha \neq \alpha'$ when $\alpha' \to \alpha$.

The extension of these concepts to operators of the form $A_0VA_1V \dots$ $\dots A_nVA_{n+1}$ with all A's diagonal **), and to sums of such operators, is obvious and can be found in I, p. 907.

3. *The transition probabilities.* We shall define in the present section the quantity P, the time evolution of which will later be expressed by means of a master equation to general order in the perturbation. Let us consider the wave function φ_0 of the system at an initial time $t = 0$ and let us expand it in the $|\alpha\rangle$-representation

$$\varphi_0 = \int |\alpha\rangle\, d\alpha\, c(\alpha). \tag{3.1}$$

The expansion has been written for the limiting case $N \to \infty$, replacing summation over α by integration, since in this limit all quantum numbers are continuous. According to the orthonormalization equation (2.2) we have

$$\langle\varphi_0|\varphi_0\rangle = \int |c(\alpha)|^2\, d\alpha. \tag{3.2}$$

We assume this expression to have the value unity.

*) This notation for the sections of I, as well as the notations (I. 1.1.) or (A.1.1) for equations of I or A, are self-explanatory.

**) From now on the expression *diagonal operator* will be used for operators diagonal in the $|\alpha\rangle$-representation.

Choosing the units so that $\hbar = 1$, the wave function at time t is given by

$$\varphi_t = U_t \varphi_0, \tag{3.3}$$

$$U_t = \exp[-i(H + \lambda V)t]. \tag{3.4}$$

We are interested in the occupation probabilities at time t of groups of states $|\alpha\rangle$ which for finite but large N contain many states but are still narrow enough to give a very small variation of the quantum numbers over the group. Since for $N \to \infty$ all quantum numbers become continuous, such a group corresponds in this limit to an infinitesimal volume element $d\alpha$ in the space of the quantum numbers. The occupation probability of such a group can be written $p_t(\alpha)\, d\alpha$, and the probability density $p_t(\alpha)$ is such that, for a diagonal operator A with

$$A\,|\alpha\rangle = |\alpha\rangle\, A(\alpha), \tag{3.5}$$

one has

$$\langle \varphi_t |A| \varphi_t\rangle = \int A(\alpha) p_t(\alpha)\, d\alpha. \tag{3.6}$$

Inversely, the occupation probability density $p_t(\alpha)$ can be uniquely determined from a calculation of (3.6) for every diagonal A having as eigenvalue $A(\alpha)$ a smooth function of the quantum numbers in α. From (3.3) one has

$$\langle \varphi_t |A| \varphi_t\rangle = \langle \varphi_0 |U_{-t} A U_t| \varphi_0\rangle. \tag{3.7}$$

If one expands U_t and U_{-t} in powers of the perturbation, the operator $U_{-t}AU_t$ becomes a sum of a diagonal operator and an infinity of products $A_0 V A_1 \ldots V A_{n+1}$ with diagonal $A_0, \ldots A_{n+1}$. Consequently it has a diagonal part. We separate it out in the matrix element

$$\langle \alpha |U_{-t} A U_t| \alpha'\rangle = \delta(\alpha - \alpha')\, f_1(\alpha) + f_2(\alpha, \alpha'), \tag{3.8}$$

$f_2(\alpha, \alpha')$ having as usual only singularities weaker than $\delta(\alpha - \alpha')$. Obviously f_1 and f_2 depend linearly on the arbitrary diagonal operator A. These quantities are therefore linear functionals of the numerical function $A(\alpha'')$, in formulae

$$f_1(\alpha) = \int A(\alpha'')\, d\alpha''\, P(t \mid \alpha''\alpha), \tag{3.9}$$

$$f_2(\alpha, \alpha') = \int A(\alpha'')\, d\alpha''\, I(t \mid \alpha''\alpha\alpha'). \tag{3.10}$$

The functions $P(t \mid \alpha''\alpha)$ and $I(t \mid \alpha''\alpha\alpha')$ thus defined depend only on the system considered and on the variables indicated between brackets; they are independent of A. Inserting our equations in (3.7) we obtain

$$\begin{aligned}\langle \varphi_t |A| \varphi_t\rangle &= \int A(\alpha'')\, d\alpha'' \int P(t \mid \alpha''\alpha)\, d\alpha\, |c(\alpha)|^2 \\ &+ \int A(\alpha'')\, d\alpha'' \int I(t \mid \alpha''\alpha\alpha')\, d\alpha\, d\alpha'\, c^*(\alpha)\, c(\alpha')\end{aligned} \tag{3.11}$$

and for the occupation probability, by comparison with (3.6),

$$p_t(\alpha'') = \int P(t \mid \alpha''\alpha)\, d\alpha\, |c(a)|^2 + \int I(t \mid \alpha''\alpha\alpha')\, d\alpha\, d\alpha'\, c^*(\alpha)\, c(\alpha'). \tag{3.12}$$

The equations just obtained exhibit a very important feature, the separate occurrence in the righthand sides of terms depending only on the absolute squares of the initial amplitudes $c(\alpha)$ and of terms depending on the relative phases of $c(\alpha)$ for various α (i.e. on the α-variation of the phase). The former involve only the initial occupation probabilities $|c(\alpha)|^2$, whereas the latter are of the nature of interference terms (whence the notations P and I). The fact just mentioned, which could also be described as a separation of the density matrix, is a direct consequence of the occurrence of diagonal parts, i.e. of the special properties of the hamiltonian described in the previous section.

The separation which thus appears between phase-independent and phase-dependent terms leads directly to the distinction between initial states φ_0 with "random" or "incoherent phases" and those with "special phases". Loosely speaking, an initial state with random or incoherent phases will be such as to give a negligibly small value to the interference term

$$\int I(t \mid \alpha''\alpha\alpha')\, d\alpha\, d\alpha'\, c^*(\alpha)\, c(\alpha') \tag{3.13}$$

for every t in the time interval of interest, whereas a φ_0 with special phases would give it a non-vanishing value. It is difficult to find for this distinction a formulation at the same time complete and general. Of course, if instead of one initial state we are willing to consider an ensemble of such states with prescribed $|c(\alpha)|$ and random (more exactly uniformly distributed) phases, the average value of (3.13) will be zero. One must however expect much more to be true. The integral (3.13), when taken for a single initial state selected in some sense at random, must be vanishingly small most of the time, at least if one limits oneself to the values of t in a finite interval $|t| \leqslant T$, where T can be chosen large compared to the time needed for the approach to statistical equilibrium *). For example, if the phase of $c(\alpha)$ varies very rapidly with α as compared to the phase of $\exp[i\varepsilon(\alpha)T]$, the expression (3.13) will certainly be very small, because the phases of $I(t \mid \alpha''\alpha\alpha')$ can certainly not vary fast compared to the expression mentioned. Since it seems reasonable to expect the phases of an initial state, when chosen at random, to vary rapidly most of the time, we may expect that (3.13) will be negligible for an overwhelming majority of initial states.

Further support for this view can also be drawn from an analogy with the quantum theory of collision processes. Let us take the simplest example of collision, elastic scattering of a particle by a target at rest. The hamiltonian can be written in the form $H + \lambda V$, H representing the kinetic energy of the

*) The latter time is here finite because we treat the perturbation as finite. In A on the contrary, the limiting case $\lambda \to 0$ was considered, T tending to infinity as λ^{-2}. This is the reason why we could neglect in A all interference effects even for initial states with slowly varying phases: all these effects took place in a time short compared to T. Such is not the case here. Both here and in A the Poincaré cycles are infinite because we deal with the limiting case of an infinite system.

particle and λV the interaction between particle and static target. The states $|\alpha\rangle$ are plane waves characterized by continuous quantum numbers, just as was the case until now. The number of these continuous parameters is now however limited to three, and the perturbation V is such that none of the singularities described in Section 2 occur in matrix elements $\langle\alpha| VA_1V \ldots A_nV |\alpha'\rangle$. The quantity P defined by (3.9) consequently reduces to

$$P(t \mid \alpha'\alpha) = \delta(\alpha' - \alpha),$$

giving for the occupation probability density at time t

$$P_t(\alpha'') = | c(\alpha'')|^2 + \int I(t \mid \alpha''\alpha\alpha') \, d\alpha \, d\alpha' \, c^*(\alpha) \, c(\alpha').$$

The time-dependence of p_t, i.e. the very occurrence of scattering, is here entirely determined by the interference term. It is in this case quite clear in which sense can be asserted that the interference term is negligible for most initial states. Unless the $c(\alpha)$'s have special singularities, the initial state (3.1) represents a wave packet, and random choice of the phases of this wave packet implies random choice of its location in space. The occurrence of scattering on the contrary requires proper aiming of the incoming particle at the target, i.e. a proper correlation between initial position and direction of motion. Consequently the interference term (3.13) can in this case safely be said to vanish most of the time for random choice of the phases of the amplitudes $c(\alpha)$.

This rather lengthy discussion was presented to support the view that the interference term in (3.11) and (3.12) will only contribute for a minority of "special" initial states, the occupation probability reducing for all other states to the simple expression

$$p_t(\alpha') = \int P(t \mid \alpha'\alpha) \, d\alpha \, |c(\alpha)|^2. \tag{3.14}$$

The latter equation is equivalent to

$$\langle \varphi_t |A| \varphi_t\rangle = \int A(\alpha') \, d\alpha' \int P(t \mid \alpha'\alpha) \, d\alpha \mid c(\alpha)|^2. \tag{3.15}$$

Our further analysis will be devoted to a study of the quantity $P(t \mid \alpha'\alpha)$. This quantity is entirely defined in terms of the hamiltonian and of the $|\alpha\rangle$-representation, and its study is a purely quantum-mechanical problem from the treatment of which all extraneous arguments, statistical or others must and will be barred. From (3.14), for most initial states, the quantity $P(t \mid \alpha'\alpha)$ directly expresses the occupation probabilities at time t in terms of the occupation probabilities $p_0(\alpha) = | c(\alpha)|^2$ at time zero. Clearly $P(t | \alpha'\alpha)$ is the *transition probability density* from α to α' in the time interval from 0 to t, assuming randomness of phases at time 0. Although the present paper will be entirely concerned with this transition probability, we do not mean to imply that the interference function $I(t \mid \alpha''\alpha\alpha')$ is deprived of interest, nor that its study would present abnormally high difficulties. The physical importance of interference effects in dissipative systems can be illustrated by many examples. To quote only one, simultaneous excitation of two additional phonons in a crystal lattice in thermal equilibrium may give rise, for special configurations, to mutual scattering of the two phonons before they have covered their whole mean free parth and have vanished by dissipation. As for the quantum-mechanical study of $I(t \mid \alpha''\alpha\alpha')$, it can be tackled by the methods used hereafter and applied already in I and II where collision processes, i.e. interference effects, were extensively studied.

In A the quantum-mechanical calculation of $P(t \mid \alpha'\alpha)$ was performed to lowest order in λ, more exactly in the limiting case $\lambda \to 0$, $t \to \infty$, $\lambda^2 t$ finite. Written in our present notation (which is more convenient than the notation of A for a study to general order in λ), the main result of A was that the transition probability density verifies the differential equation (see (A.6.11))

$$dP(t \mid \alpha\alpha_0)/dt = 2\pi\lambda^2 \int \delta[\varepsilon(\alpha) - \varepsilon(\alpha')]\, W^{(0)}(\alpha\alpha')\, d\alpha'\, P(t \mid \alpha'\alpha_0) - 2\pi\lambda^2 \int d\alpha' \,.\, \delta[\varepsilon(\alpha') - \varepsilon(\alpha)]\, W^{(0)}(\alpha'\alpha) \,.\, P(t \mid \alpha\alpha_0). \quad (3.16)$$

This equation is in more explicit form and in our notation the master equation (1.1). The kernel $W^{(0)}(\alpha'\alpha)$ is defined by the identity

$$\{VAV\}_d \,|\alpha\rangle = |\alpha\rangle \int A(\alpha')\, d\alpha'\, W^{(0)}(\alpha'\alpha) \quad (3.17)$$

for arbitrary diagonal A with eigenvalues $A(\alpha')$. Equation (3.16), which holds for $t > 0$ but would apply to $t < 0$ after changing the sign of the righthand side, is characteristic of a stochastic process of Markov type, a feature which we will find not to hold to general order in the perturbation. The quantities $W^{(0)}$ play the role of transition rates, i.e. of transition probabilities per unit time. Eq. (3.16) must be supplemented by the initial condition

$$P(0 \mid \alpha\alpha_0) = \delta(\alpha - \alpha_0) \quad (3.18)$$

which follows immediately from the definition (3.8), (3.9) of P and is exact to all orders of the perturbation.

Two remarks will be made before we start in Section 4 the study of P. The first one concerns the dominant role played in all our considerations by the $|\alpha\rangle$-representation, despite the fact that it is of course not intrinsic to the many-particle system considered. In A, where the perturbation was assumed to be very small ($\lambda \to 0$), the use of the $|\alpha\rangle$-representation could be entirely motivated by the property that the $|\alpha\rangle$'s are the eigenstates of the hamiltonian in the limit $\lambda = 0$. Here however we assume the perturbation to be finite and the argument just mentioned consequently fails. The special significance of the $|\alpha\rangle$-representation for the many-particle systems of quantum statistics must be attributed to another fact, easily verified on all actual examples, to know: the simple relation of this representation to the physical quantities of greatest interest in irreversible processes. The latter quantities have usually a simple mathematical expression in terms of one-particle operators, and these operators themselves have in crystals as well as in gases simple matrix elements in the $|\alpha\rangle$-representation *). It has often been remarked that the irreversible behaviour of a many-particle system is not truly an intrinsic property of the system but is partly determined by which class of properties of the system the observer is looking at. In other words, it is not for every operator 0 that one can expect the expectation value $\langle\varphi_t|0|\varphi_t\rangle$ to tend in the course of time (starting from most initial states) toward the average value $\langle 0\rangle_{eq}$ calculated from equilibrium statistical mechanics. From the experimental evidence we only know this property to hold for special physical quantities 0, and it is an easy matter to construct other operators for which it does not hold. On the basis of experience the representation $|\alpha\rangle$ seems to provide a natural way to characterize operators 0 for which $\langle\varphi_t|0|\varphi_t\rangle$ can be expected to tend toward $\langle 0\rangle_{eq}$. To state it loosely, these operators will be such as to have simple matrix elements in the $|\alpha\rangle$-representation, even in the limiting case of a very large system. A sharper formulation of this statement is of course required, but no attempt to find one has been made until now. The only operators 0 to be considered in the present paper are diagonal operators A, for which the approach of $\langle\varphi_t|A|\varphi_t\rangle$ toward $\langle A\rangle_{eq}$ for most initial states will indeed be established on the basis of Eq. (3.15) (see section 7). As seen from (3.15), all we need for achieving this goal is the asymptotic value of $P(t\mid\alpha'\alpha)$ for large times. An extension to other operators 0 would of course be desirable. It could in principle be carried out by essentially the same modification of the techniques used here as would be needed to incorporate into the theory discrete quantum numbers for polarization and spin.

Our second remark concerns the vanishing of the interference term (3.13) for a particular type of initial states φ_0: wave packets of very narrow extension in α:

$$\varphi_0 = \int_{\Delta a} |\alpha\rangle\, d\alpha\, c(\alpha). \tag{3.19}$$

The integration extends over a very small domain $\Delta\alpha$ around a state α_0. Such a wave packet can be considered as an approximation to the unperturbed state $|\alpha_0\rangle$. Normalization requires.

$$\int_{\Delta a} |c(\alpha)|^2\, d\alpha = 1 \tag{3.20}$$

so that, from the Schwarz inequality,

$$\left|\int_{\Delta a} c(\alpha)\, d\alpha\right| \leqslant \Delta\alpha^{\frac{1}{2}}. \tag{3.21}$$

For very small $\Delta\alpha$ the interference term (3.13) is negligible. As a matter of fact it can be written

$$I(t\mid\alpha''\alpha_0\alpha_0)\left|\int_{\Delta a} c(\alpha)\, d\alpha\right|^2$$

*) This situation is radically different from what happens in field theory, where the unperturbed representation $|\alpha\rangle$ is formed by the unobservable bare particle states.

and from (3.21) it is seen to tend to zero for $\Delta\alpha \to 0$. The occupation probability (3.12) reduces in this limit to

$$p_t(\alpha) = P(t \mid \alpha\alpha_0). \tag{3.22}$$

This gives a new illustration of the physical significance of P as transition probability density.

4. *General properties of the system and master equation.* For the lowest order in the perturbation the transition probability $P(t \mid \alpha'\alpha)$ was calculated in A by taking the defining equations (3.8), (3.9) and carrying out a straight expansion of $U_{-t}AU_t$ in powers of λ, retaining the terms which do not vanish in the limit $\lambda \to 0$, $\lambda^2 t$ finite. The result obtained turned out to be the series expansion solution of the transport equation (3.16), i.e. the solution which would be derived from (3.16) and (3.18) by mere iteration. Besides its lack of elegance, such a method is very cumbersome to extend to higher orders in λ. The method to be followed presently is different and makes essential use of the resolvent

$$R_l = (H + \lambda V - l)^{-1}, \quad l \text{ complex number,} \tag{4.1}$$

an operator already applied in I and II to the study of non-dissipative systems.

The main properties of the resolvent for the type of hamiltonian here considered are the same for systems of statistical mechanics and for interacting fields; they have been derived in Section 3 of I. These properties concern the diagonal part D_l of the resolvent. Clearly, the study of the diagonal part of R_l on the basis of properties (i), (ii), (iii) of Section 2 presupposes that R_l can be expanded in powers of the perturbation, since this expansion is needed to bring R_l in the form of a sum of operators $A_0VA_1 \ldots$ $\ldots VA_{n+1}$ with diagonal A_j's. It is however only for l non-real that convergence of the expansion must be assumed. Furthermore, the formulation of the properties of D_l and the further development of the theory no longer use a complete expansion in powers of λ and thereby differ in an essential way from what ordinary perturbation calculus would give. One can say that all our equations differ from the corresponding results of ordinary perturbation theory through the fact that a number of partial summations have been carried out in closed form *). The possibility of such partial summations follows from the occurrence of diagonal parts, and our method seems to be well suited to take advantage of it. Still the summations explicitly carried out are only partial and we have to assume convergence for all remaining series.

Let us recall briefly the main properties of the resolvent for the limiting

*) Typical examples of these summations are found in I, section 3, Eq. (I.3.5) and (I.3.3).

case of a large system ($N \to \infty$, N/Ω finite). They have been derived in I under normal conditions of regularity of all functions involved, with due regard of course for the singularities implied by properties (ii), (iii) of section 2. The diagonal part D_l of R_l can be written in the form

$$D_l = (H - l - \lambda^2 G_l)^{-1} \tag{4.2}$$

where G_l is a diagonal operator satisfying the identity (see (I. 3.13))

$$G_l = \{VD_lV - \lambda VD_lVD_lV + \ldots\}_{id}. \tag{4.3}$$

This identity can be used to calculate G_l by successive approximations. The resolvent itself becomes, as stated in (I.3.15),

$$R_l = D_l - \lambda D_l \{V - \lambda VD_lV + \lambda^2 VD_lVD_lV - \ldots\}_{nd} D_l. \tag{4.4}$$

We denote by $G_l(\alpha)$ and $D_l(\alpha)$ the eigenvalues of G_l and D_l for the state $|\alpha\rangle$. From (4.2),

$$D_l(\alpha) = [\varepsilon(\alpha) - l - \lambda^2 G_l(\alpha)]. \tag{4.2 bis}$$

As a function of the complex variable l, $G_l(\alpha)$ is holomorphic in the whole complex plane except on a portion of the real axis *). It approaches zero as $|l|^{-1}$ when $l \to \infty$ **). It verifies

$$G_{l^*}(\alpha) = [G_l(\alpha)]^* \tag{4.5}$$

where the star denotes the complex conjugate. Furthermore

$$Im[G_l(\alpha)] > 0 \text{ for } Im(l) > 0, \text{ unless } G_l(\alpha) = 0 \quad \text{for all } l. \tag{4.5 bis}$$

For l approaching a point E of the real axis, $G_l(\alpha)$ approaches a finite limit

$$\lim_{0>\eta\to 0} G_{E\pm i\eta}(\alpha) = K_E(\alpha) \pm iJ_E(\alpha) \tag{4.6}$$

where $K_E(\alpha)$ is real and $J_E(\alpha)$ real non-negative. The latter quantity is positive on certain intervals of the E-axis and vanishes elsewhere; these intervals depend on α.

The above properties imply that $D_l(\alpha)$ is holomorphic in l in the whole complex plane except on a portion of the real axis and that it approaches zero as $|l|^{-1}$ for $l \to \infty$. Further, from (4.5) and (4.5 bis),

$$D_{l^*}(\alpha) = [D_l(\alpha)]^*, \; Im\,[D_l(\alpha)] > 0 \text{ for } Im(l) > 0. \tag{4.7}$$

In contrast to the case of $G_l(\alpha)$, the behaviour of $D_l(\alpha)$ for l approaching the real axis may be of two different types. Clearly, the limit of $D_l(\alpha)$ for $l \to E$, E real, is finite whenever one or both of the inequalities

$$J_E(\alpha) > 0, \; \varepsilon(\alpha) - E - \lambda^2 K_E(\alpha) \neq 0$$

*) $G_l(\alpha)$ and $D_l(\alpha)$ often have analytical continuations accross the real axis, from above and from below. They will however play no part in our considerations.

**) Except possibly when l remains close to the real axis. This provision will always have to be made when we talk about the behaviour of analytic functions at infinity.

hold, but it becomes infinite when one has simultaneously

$$J_E(\alpha) = 0,\ \varepsilon(\alpha) - E - \lambda^2 K_E(\alpha) = 0. \tag{4.8}$$

In the latter case $D_l(\alpha)$ has a pole at $l = E$ *). The absence or presence of poles in $D_l(\alpha)$ is therefore determined by the absence or presence of common roots for the equations (4.8).

The two possibilities thus encountered correspond to the distinction between the state $|\alpha\rangle$ having a dissipative or a non-dissipative behaviour under the effect of the perturbation. This statement deserves some comments in addition to what was said in I, Section 4. What we mean by it can be described most simply as follows. Select as in (3.19) an initial state φ_0 forming a wave packet of very narrow extension $\Delta\alpha$ around α_0. A time t later the state has become φ_t, given by (3.3) and (3.4). Let us calculate in the limit of small $\Delta\alpha$ the probability $q_t(\alpha_0)$ to find back the system in its initial state φ_0. It is

$$q_t(\alpha_0) = \lim_{\Delta\alpha\to 0} |\langle\varphi_0 \mid \varphi_t\rangle|^2. \tag{4.9}$$

A calculation similar to the derivation of (3.22) shows that $\langle\varphi_0 \mid \varphi_t\rangle$ is, in the limit of small $\Delta\alpha$, identical to the eigenvalue for α_0 of the operator

$$\begin{aligned}\{U_t\}_d &= (i/2\pi)\ \{\textstyle\int_\gamma \exp\,(-\ ilt)\ R_l\ \mathrm{d}l\}_d \\ &= (i/2\pi) \textstyle\int_\gamma \exp\,(-\ ilt)\ D_l\ \mathrm{d}l\end{aligned} \tag{4.10}$$

where γ is a contour in the complex plane encircling a sufficiently large portion of the real axis and is to be described counterclockwise. We have consequently

$$q_t(\alpha_0) = (2\pi)^{-2}\ |\textstyle\int_\gamma \exp\,(-\ ilt)\ D_l(\alpha_0)\ \mathrm{d}l|^2. \tag{4.11}$$

This formula directly implies that $q_t(\alpha_0)$ tends to zero for $t \to \infty$ when $D_l(\alpha_0)$ has no pole, and tends to a positive limit when $D_l(\alpha_0)$ has one or more poles **).

In the first case it is natural to describe the behaviour of the state $|\alpha_0\rangle$ as *dissipative*, because under the influence of the perturbation this state (or more exactly the properly normalized state φ_0 for very small $\Delta\alpha$) is so deeply modified that, eventually, it gets completely spread out over the other states $|\alpha\rangle$ and contains its initial value only as a negligibly small component. One should note that such a radical effect does not in any way require the perturbation to be strong. On the contrary dissipative behaviour can occur for arbitrarily small perturbations, the vanishing of the probability $q_t(\alpha_0)$ demanding then correspondingly long times (of order λ^{-2}).

---

*) As established in (I.5.8), (I.5.9) such a pole of $D_l(\alpha)$ is always of order one. $D_l(\alpha)$ has never non-real poles.

**) See also loc. cit.[3]). We may remark that $|\langle\varphi_0 \mid \varphi_t\rangle|^2$ always tends to zero for $t \to \infty$ if the limit of small $\Delta\alpha$ is not taken. This effect, also present when there is no perturbation, is just the ordinary spreading of a wave packet and is of no interest to us.

In the case where $D_l(\alpha_0)$ has one or more poles, the state $|\alpha_0\rangle$ (more properly φ_0 in the limit of small $\Delta\alpha$), although of course affected by the perturbation, always retains a non-vanishing component identical to its initial value. This situation, which we call *non-dissipative,* would occur in a convergent field theory, where the bare particle states are present with non-vanishing probability in the corresponding dressed (i.e. observable) particle states. We may also remark that in the quantum theory of collision processes, already quoted for comparison in Section 3, the function $D_l(\alpha_0)$ reduces to its unperturbed value $[\varepsilon(\alpha_0) - l]^{-1}$, so that $q_t(\alpha_0)$ is unity at all times. This means that a true plane wave state (corresponding to the limit of small $\Delta\alpha$) is perturbed by a collision process only in a negligibly small fraction of its totality, as is physically obvious since the collision can only take place in a limited region of configuration space whereas the extension of the plane wave is of course infinite.

In I and II we gave a systematic study of what may be called *non-dissipative systems,* i.e. systems for which all states $|\alpha\rangle$ are non-dissipative. We made the slightly more restrictive assumption that for each α the equations (4.8) have one single common root $E(\alpha)$, in the neighbourhood of which $J_E(\alpha)$ vanishes identically. This is typically the situation which would occur in a convergent field theory with positive masses for all fields (neither infra-red nor ultra-violet divergences). Formally a system can have part of its states dissipative, part of them non-dissipative. For the many-particle systems of quantum statistics one should expect the quasi-totality of unperturbed states to be dissipative *). It seems reasonable to call them *dissipative systems* and our main interest goes to them in the present paper. It is for them that consideration of the transition probabilities $P(t \mid \alpha'\alpha)$ and of a master equation is of actual importance. In the present section the properties of $P(t \mid \alpha'\alpha)$ and the master equation will however be established in full generality, irrespective of the dissipative or non-dissipative character of the $|\alpha\rangle$'s. The algebraic form of the master equation is completely independent of this distinction, although of course the nature of the solutions will be radically different in the two cases. This situation is already familiar in the limiting case of small perturbations, where the lowest order master equation (3.16) is of true interest only when

$$\delta[\varepsilon(\alpha') - \varepsilon(\alpha)]\, W^{(0)}\, (\alpha'\alpha) \neq 0. \tag{4.12}$$

The validity of this inequality for some states $|\alpha'\rangle$ can easily be shown to characterize the state $|\alpha\rangle$ as dissipative in the approximation considered. The master equation (3.16) remains nevertheless valid when the lefthand side of (4.12) vanishes. In the field-theoretical case for example, where all

*) One cannot expect all states to be dissipative. For instance the ground state should be non-dissipative. It might perhaps also be a fairly general feature that very low lying excited states behave non-dissipatively, as is suggested by the phenomena of superfluidity and superconductivity.

unperturbed states are non-dissipative, this quantity vanishes for all α and α', so that (3.16) reduces to

$$\mathrm{d}P(t \mid \alpha'\alpha)/\mathrm{d}t = 0,$$

or, on account of (3.18),

$$P(t \mid \alpha'\alpha) = \delta(\alpha' - \alpha) \text{ for all } t.$$

In the approximation considered ($\lambda \to 0$) this result is correct but of course of little interest. Its extension to general order will be given at the end of Section 6.

We now go over to the investigation of $P(t \mid \alpha'\alpha)$ to general order in λ. We use as in (4.10) the representation of U_t by a contour integral over the resolvent and introduce it in the definition (3.8), (3.9) of $P(t \mid \alpha'\alpha)$. Clearly

$$U_{-t}AU_t = -(2\pi)^{-2}\int_\gamma \mathrm{d}l \int_\gamma \mathrm{d}l' \exp[i(l-l')t]\, R_l A R_{l'}, \tag{4.13}$$

γ being the same contour as in (4.10). Let us define the function $X_{ll'}(\alpha'\alpha)$ by the identity

$$\{R_l A R_{l'}\}_d\, |\alpha\rangle = |\alpha\rangle \int A(\alpha')\, \mathrm{d}\alpha' X_{ll'}(\alpha'\alpha). \tag{4.14}$$

As usual A is an arbitrary diagonal operator of eigenvalues $A(\alpha')$. The quantity $X_{ll'}$ approaches zero as $|l|^{-1}$ for $l \to \infty$, as $|l'|^{-1}$ for $l' \to \infty$, as $|ll'|^{-1}$ when l and $l' \to \infty$. Introducing (4.14) into the diagonal part of (4.13) and comparing with (3.9) one finds

$$P(t \mid \alpha'\alpha) = -(2\pi)^{-2}\int_\gamma \mathrm{d}l \int_\gamma \mathrm{d}l' \exp[i(l-l')t]\, X_{ll'}(\alpha'\alpha). \tag{4.15}$$

We establish next a simple identity for $X_{ll'}(\alpha'\alpha)$. Define a new function $W_{ll'}(\alpha'\alpha)$ by the equation (A arbitrary diagonal operator)

$$\{(V-\lambda V D_l V+\ldots)\, A\, (V-\lambda V D_{l'} V+\ldots)\}_{id}\, |\alpha\rangle = |\alpha\rangle \int A(\alpha')\, \mathrm{d}\alpha'\, W_{ll'}(\alpha'\alpha). \tag{4.16}$$

Comparison with (3.17) shows that

$$\lim_{\lambda\to 0} W_{ll'}(\alpha'\alpha) = W^{(0)}(\alpha'\alpha). \tag{4.17}$$

Furthermore, one has also from the definition

$$\lim_{l,l'\to\infty} W_{ll'}(\alpha'\alpha) = W^{(0)}(\alpha'\alpha). \tag{4.18}$$

The diagonal part in (4.14), after insertion of (4.4), can be reduced to irreducible diagonal parts, whereby property (iii) of Section 2 plays a central role: all simultaneous diagonal parts occurring in this particular reduction have the relative configuration (2.10). The result is an expression of X in terms of W,

$$X_{ll'}(\alpha\alpha_0) = D_l(\alpha)D_{l'}(\alpha)\,\delta(\alpha-\alpha_0) + \lambda^2 D_l(\alpha)D_{l'}(\alpha)[W_{ll'}(\alpha\alpha_0) + \\ + \lambda^2 \int W_{ll'}(\alpha\alpha_1)D_l(\alpha_1)D_{l'}(\alpha_1)\,\mathrm{d}\alpha_1\, W_{ll'}(\alpha_1\alpha_0) + \ldots]\, D_l(\alpha_0)D_{l'}(\alpha_0). \tag{4.19}$$

On the other hand, the fundamental identity (4.3), taken for two values of l, gives by subtraction

$$G_l - G_{l'} = \{(V - \lambda V D_l V + \dots)(D_l - D_{l'})(V - \lambda V D_{l'} V + \dots)\}_{id}. \quad (4.20)$$

On account of (4.16) this is a simple identity for the function W:

$$G_l(\alpha) - G_{l'}(\alpha) = \int [D_l(\alpha') - D_{l'}(\alpha')] \, d\alpha' \, W_{ll'}(\alpha'\alpha). \quad (4.21)$$

It is convenient to introduce a new notation for the quantity under the integral

$$\tilde{W}_{ll'}(\alpha'\alpha) = i[D_l(\alpha') - D_{l'}(\alpha')] \, W_{ll'}(\alpha'\alpha). \quad (4.22)$$

Remark however that (4.2) gives

$$D_l - D_{l'} = [(l - l') + \lambda^2(G_l - G_{l'})] D_l D_{l'}. \quad (4.23)$$

Consequently the identity (4.21) can be written

$$(l - l') D_l(\alpha) D_{l'}(\alpha) = D_l(\alpha) - D_{l'}(\alpha) + i\lambda^2 \int d\alpha' \, \tilde{W}_{ll'}(\alpha'\alpha) D_l(\alpha) D_{l'}(\alpha). \quad (4.24)$$

It is then a simple matter to convert it into an identity for the function X. Multiplication of (4.19) by $l - l'$ and application of (4.24) gives

$$(l - l') \, X_{ll'}(\alpha\alpha_0) = [D_l(\alpha) - D_{l'}(\alpha)] \, \delta(\alpha - \alpha_0)$$
$$- i\lambda^2 \int \tilde{W}_{ll'}(\alpha\alpha') \, d\alpha' \, X_{ll'}(\alpha'\alpha_0) + i\lambda^2 \int d\alpha' \, \tilde{W}_{ll'}(\alpha'\alpha) \, X_{ll'}(\alpha\alpha_0). \quad (4.25)$$

The first part of the derivation is thereby ended. The only task left is to transform (4.25) into an equation for the time evolution of $P(t \mid \alpha\alpha_0)$. The form of the exponential in equation (4.15) shows that P depends only on an integral of $X_{ll'}$ over $l + l'$ for fixed difference $l - l'$. The identity (4.25) on the other hand refers to $X_{ll'}$ itself. As a consequence it was not found possible to derive an equation for the quantity P itself, but P could be written as an integral over another quantity $P_E(t \mid \alpha\alpha_0)$ for which (4.25) directly implies a master equation. The quantity $P_E(t \mid \alpha\alpha_0)$ is defined mathematically for $t \neq 0$ as

$$P_E(t \mid \alpha\alpha_0) = (2\pi^2)^{-1} \, s(t) \int_\gamma dl \, \exp(2ilt) \, X_{E+l,\, E-l}(\alpha\alpha_0), \quad (4.26)$$

E being real and γ being the same type of contour as before. The symbol $s(t)$, meaning sign of t, stands for $t^{-1}|t|$. We state that for $t \neq 0$ one has

$$P(t \mid \alpha\alpha_0) = \int_{-\infty}^{\infty} dE P_E(t \mid \alpha\alpha_0). \quad (4.27)$$

To establish this relation, replace in (4.26) the contour γ by two lines $l = E' \pm i\eta$ with $\eta > 0$ and very small. Only one of these lines (the one where $Re\,(ilt) > 0$) gives a non-vanishing contribution. After substitution of (4.26) into (4.27), transform the double integral obtained to the new variables $E \pm E'$. This is possible for $t \neq 0$ because the exponential factor involving t ensures convergence. The result is readily seen to be (4.15) with the contours taken as straight lines along the real axis.

On the other hand the time derivative of P_E contains according to (4.26) the combination $2l\, X_{E+l,\, E-l}$, which is the lefthand side of the identity (4.25) for the special suffices involved; we replace it by the righthand side. Let us now define

$$w_E(t \mid \alpha'\alpha) = (2\pi^2)^{-1} \int_\gamma dl \exp(2ilt)\, \tilde{W}_{E+l,\, E-l}(\alpha'\alpha). \tag{4.28}$$

One easily establishes

$$s(t) \int_\gamma dl \exp(2ilt)\, \tilde{W}_{E+l,\, E-l}\, (\alpha_3\alpha_2)\, X_{E+l,\, E-l}\, (\alpha_1\alpha_0) = \\ 4\pi^3 \int_0^t dt'\, w_E\, (t - t' \mid \alpha_3\alpha_2)\, P_E(t' \mid \alpha_1\alpha_0), \tag{4.29}$$

a result for which the vanishing of $\tilde{W}_{E+l,\, E-l}$ and $X_{E+l,\, E-l}$ for $l \to \infty$ is relevant. Introduce still the further definition

$$f_E(t \mid \alpha) = (2\pi^2)^{-1}\, is(t) \int_\gamma dl \exp(2ilt)\, [D_{E+l}(\alpha) - D_{E-l}(\alpha)]. \tag{4.30}$$

With the help of (4.29) and (4.30) the time derivative of P_E finally takes the form we were aiming at:

$$dP_E(t \mid \alpha\alpha_0)/dt = \delta(\alpha - \alpha_0)\, f_E(t \mid \alpha) + 2\pi\lambda^2 \int_0^t dt' \int w_E(t - t' \mid \alpha\alpha') \\ d\alpha' P_E(t' \mid \alpha'\alpha_0) - 2\pi\lambda^2 \int_0^t dt' \int d\alpha' w_E(t - t' \mid \alpha'\alpha)\, P_E(t' \mid \alpha\alpha_0). \tag{4.31}$$

The integro-differential equation (4.31) is the *master equation to general order in* λ. Its main structural differences with the lowest order equation (3.16) are apparent in the righthand side: they consist in the presence of an inhomogeneous term (the term in f_E) and in the time integration over the previous evolution of the system (non markovian nature of the process). Both features can be shown to be manifestations of the coherent phase relations present in the wave function φ_t at all times $t \neq 0$. We shall see in the next section how they become negligible in the limit of small perturbations.

The master equation (4.31) for P_E must be supplemented by an *initial condition* at $t = 0$. This condition is easily derived from the definition (4.26) of P_E. As we have seen, $X_{E-l,E+l}$ decreases as $|l|^{-2}$ for $l \to \infty$. Putting $t = 0$ in (4.26) and deforming the whole contour to infinity, one gets a vanishing result, so that

$$P_E(0 \mid \alpha\alpha_0) = 0. \tag{4.32}$$

With this initial condition the integro-differential equation (4.31) determines P_E uniquely for all times, positive or negative. One may wonder how (4.32) can be compatible with the fact, already noted in the previous section, Eq. (3.18), that the initial value of $P(t \mid \alpha\alpha_0)$ is $\delta(\alpha - \alpha_0)$. The reason is that (4.27) does not hold for $t = 0$. As stressed before, non-vanishing of t is necessary for P to be the integral of P_E over the energy E.

It must be noted that the quantities f_E, w_E and P_E are real for all values of their (real) arguments. For f_E the reality follows from (4.30) and the first equation (4.7). For w_E one notices that, from (4.16), (4.7) and the hermiticity of V,

$$[W_{ll'}(\alpha'\alpha)]^* = W_{l'^*l^*}(\alpha'\alpha). \tag{4.33}$$

Reality of (4.28) then easily results. The reasoning for P_E is similar, using

$$[X_{ll'}(\alpha'\alpha)]^* = X_{l'^*l^*}(\alpha'\alpha). \tag{4.34}$$

In view of the importance of P_E, – it is this quantity rather than P itself which satisfies the general master equation –, it is good to get some insight into its physical significance. One may say loosely that P_E expresses how much of the transition probability P is contributed by the total energy shell $H + \lambda V = E$. This interpretation can be justified by the following considerations, which show at the same time how it has to be understood in more accurate terms. Let A be a diagonal operator and $F(H')$ an arbitrary function of the total hamiltonian $H' = H + \lambda V$. Define the symmetrized product $[AF]_s$ by expansion of F in powers of H' and application of the special rule

$$[AH'^n]_s = 2^{-n} \textstyle\sum_{m=0}^{n} \binom{n}{m} H'^m A H'^{n-m} \tag{4.35}$$

which follows by induction from

$$[AH'^n]_s = \tfrac{1}{2}(H'[AH'^{n-1}]_s + [AH'^{n-1}]_s H').$$

Consider further, instead of (3.7), the expectation value $\langle\varphi_t\,|[AF]_s|\,\varphi_t\rangle$ for incoherent phases of the initial state φ_0. An easy calculation, similar to the derivation of (4.15) and (4.27), leads to

$$\langle\varphi_t\,|[AF]_s|\,\varphi_t\rangle = \int A(\alpha')\,\mathrm{d}\alpha'\,P'(t\,|\alpha'\alpha)\,\mathrm{d}\alpha\,|c(\alpha)|^2 \tag{4.36}$$

with

$$P'(t \mid \alpha'\alpha) = -\,(2\pi)^{-2}\int_\gamma \mathrm{d}l \int_\gamma \mathrm{d}l'\,F[\tfrac{1}{2}(l + l')]$$
$$\exp\,[i(l - l')t]X_{ll'}(\alpha'\alpha) = \int_{-\infty}^{\infty} \mathrm{d}E\,F(E)\,P_E(t \mid \alpha'\alpha). \tag{4.37}$$

Clearly, when F approximates a δ-function and thus picks out one energy shell, the quantity P' reduces to P_E. The relation of P_E to the total energy shell $H' = E$ will be confirmed later in the case of small perturbations (see eq. (5.9)) and, in the general case, by the long time expression of P_E for dissipative systems (see eq. (7.20) and the considerations thereafter).

5. *The limiting case of small perturbations.* The object of this section is to show how the general master equation (4.31) reduces to the familiar form (3.16) when the perturbation is very small. We note first that the time variation of the functions $w_E(t \mid \alpha'\alpha)$ and $f_E(t \mid \alpha')$, – more exactly of integrals over α' involving them –, takes place over time intervals, the order of magnitude T_0 of which does not change when $\lambda \to 0$. This can best be seen by studying the definitions (4.28) and (4.30) for small λ. Even in this limiting

case both functions converge to zero in the mean for t large compared to a finite time T_0. As a consequence a distinction must be made, in the righthand side of the master equation (4.31), between the inhomogeneous term containing f_E and the two homogeneous terms. The former is only present for times of the order of T_0 whereas the latter, in the case of dissipative systems, keep varying during all the time needed to reach statistical equilibrium. The order of magnitude T_1 of the latter time is $\lambda^{-2} w^{-1}$, where the energy w gives the order of magnitude of the quantity $W^{(0)}$.

In contrast with T_0, T_1 increases as λ^{-2} when $\lambda \to 0$. For very small perturbations we may therefore distinguish between two time scales, a short one of order T_0, essentially independent of λ, and a long one of order T_1, proportional to λ^{-2}. The inhomogeneous term of (4.31) is effective only when t is of the order of the short time scale T_0; it then completely dominates the homogeneous terms, the contribution of which is smaller by a factor λ^2. The equation can thus be written

$$\mathrm{d}P_E(t \mid \alpha\alpha_0)/\mathrm{d}t = \delta(\alpha - \alpha_0)\, f_E(t \mid \alpha) \qquad \text{for } |t| \sim T_0 \tag{5.1}$$

with neglection of terms of order λ^2. On the other hand, when $|t|$ exceeds the short time scale and becomes of order T_1, the situation is reversed. The inhomogeneous term becomes negligibly small, the whole time-dependence of P_E is determined by the homogeneous terms alone and it is consequently very slow:

$$\mathrm{d}P_E/\mathrm{d}t \sim P_E/T_1 \sim \lambda^2 w P_E.$$

This slow time variation implies a further simplification, because in the time integrals of the form

$$\textstyle\int_0^t \mathrm{d}t' w_E(t-t' \mid \alpha'\alpha)\, P_E(t' \mid \alpha_0'\alpha_0) = \int_0^t \mathrm{d}t_1 w_E(t_1 \mid \alpha'\alpha)\, P_E(t-t_1 \mid \alpha_0'\alpha_0) \tag{5.2}$$

the integration in t_1 extends over the short time scale, $t_1 \sim T_0$, i.e. over an interval in which P_E varies very little. We may therefore approximate (5.2) by

$$\begin{aligned}\textstyle\int_0^t \mathrm{d}t' w_E(t-t' \mid \alpha'\alpha)\, P_E(t' \mid \alpha_0'\alpha_0) &= P_E(t \mid \alpha_0'\alpha_0) \textstyle\int_0^\infty \mathrm{d}t_1 w_E(t_1 \mid \alpha'\alpha) = \\ &= (2\pi)^{-1} P_E(t \mid \alpha_0'\alpha_0)\, \tilde{W}_{E-i0,\,E+i0}(\alpha'\alpha).\end{aligned}$$

The last step, carried out under the assumption $t > 0$, follows from (4.28) through integration over t; the notation $E \pm i0$ is used for $E \pm i\eta$ with $\eta > 0$ very small. The master equation on the long time scale now becomes

$$\begin{aligned}\mathrm{d}P_E(t \mid \alpha\alpha_0)/\mathrm{d}t &= \lambda^2 \textstyle\int \tilde{W}_{E-i0,E+i0}(\alpha\alpha')\, \mathrm{d}\alpha'\, P_E(t \mid \alpha'\alpha_0) \\ &\quad - \lambda^2 \textstyle\int \mathrm{d}\alpha'\, \tilde{W}_{E-i0,E+i0}(\alpha'\alpha)\, P_E(t \mid \alpha\alpha_0) \qquad \text{for } t \sim T_1.\end{aligned} \tag{5.3}$$

It holds for positive t. The corresponding equation for negative times is obtained by interchanging the two indices of $\tilde{W}$.

The equations (5.1) and (5.3) have been derived under the assumption

$T_0 \ll T_1$ and they neglect corrections of relative order $T_0/T_1 \sim \lambda^2$. Consistency therefore requires that the functions f and $\tilde{W}$ which they contain should be taken to order λ only. To this order, as shown by (4.2), the operator D_l reduces to its unperturbed expression $(H - l)^{-1}$ and by substitution in (4.30) one finds

$$f_E(t \mid \alpha) = 2\pi^{-1} \cos [2(\varepsilon(\alpha) - E)t].$$

Integration of (5.1) for the short time scale is then elementary and gives

$$P_E(t \mid \alpha\alpha_0) = \pi^{-1}[\varepsilon(\alpha) - E]^{-1} \sin [2(\varepsilon(\alpha) - E)t]\ \delta(\alpha - \alpha_0) \text{ for } t \sim T_0. \quad (5.4)$$

At the upper end of the short time scale this expression approaches in the mean a limit which is obviously

$$P_E(t \mid \alpha\alpha_0) = \delta[\varepsilon(\alpha) - E]\ \delta(\alpha - \alpha_0) \quad \text{for} \quad T_0 \ll t \ll T_1. \quad (5.5)$$

The latter expression can now be used as initial value for the equation (5.3) relative to the long time scale. As we have seen, $\tilde{W}$ must be taken to first order in λ. Substituting $(H - l)^{-1}$ for D_l in (4.22) we find

$$\tilde{W}_{E-i0,E+i0}(\alpha'\alpha) = 2\pi\delta\ [\varepsilon(\alpha') - E]\ W^{(1)}_{E-i0,\ E+i0}\ (\alpha'\alpha)$$

where $W^{(1)}_{ll'}$ is the first order approximation of the function $W_{ll'}$ defined in (4.16). The master equation on the long time scale becomes

$$\mathrm{d}P_E(t \mid \alpha\alpha_0)/\mathrm{d}t = 2\pi\lambda^2 \int \delta[\varepsilon(\alpha) - E]\ W^{(1)}_{E-i0,E+i0}(\alpha\alpha')\ \mathrm{d}\alpha'$$

$$P_E(t \mid \alpha'\alpha_0) - 2\pi\lambda^2 \int \mathrm{d}\alpha'\ \delta[\varepsilon(\alpha') - E]\ W^{(1)}_{E-i0,E+i0}(\alpha'\alpha) P_E(t \mid \alpha\alpha_0). \quad (5.6)$$

It must be integrated with the initial condition (5.5). Clearly, from the occurrence of the δ-functions in the equation and the initial condition, the solution has the form

$$P_E(t \mid \alpha\alpha_0) = \delta[\varepsilon(\alpha) - E]\ \delta[\varepsilon(\alpha_0) - E]\ p(t \mid \alpha\alpha_0). \quad (5.7)$$

Substitution of this relation into (4.27) shows that

$$P(t \mid \alpha\alpha_0) = \delta[\varepsilon(\alpha) - \varepsilon(\alpha_0)]\ p(t \mid \alpha\alpha_0). \quad (5.8)$$

Consequently, to first order in λ, the transition probability $P(t \mid \alpha\alpha_0)$ contains a factor $\delta[\varepsilon(\alpha) - \varepsilon(\alpha_0)]$ implying conservation of the unperturbed energy. We note further from (5.7) and (5.8) that

$$P_E(t \mid \alpha\alpha_0) = \delta[\varepsilon(\alpha) - E]\ P(t \mid \alpha\alpha_0). \quad (5.9)$$

Substitution of this last expression into (5.6) gives, after eliminating a δ-factor,

$$\mathrm{d}P(t \mid \alpha\alpha_0)/\mathrm{d}t = 2\pi\lambda^2 \int \delta[\varepsilon(\alpha) - \varepsilon(\alpha')]\ \overline{W}^{(1)}(\alpha\alpha')\ \mathrm{d}\alpha'\ P(t \mid \alpha'\alpha_0)$$

$$- 2\pi\lambda^2 \int \mathrm{d}\alpha'\ \delta[\varepsilon(\alpha') - \varepsilon(\alpha)]\ \overline{W}^{(1)}(\alpha'\alpha)\ P(t \mid \alpha\alpha_0), \quad (5.10)$$

with the abbreviation

$$\overline{W}^{(1)}(\alpha\alpha') = W^{(1)}_{E-i0.E+i0}(\alpha\alpha') \quad \text{for} \quad E = \varepsilon(\alpha). \quad (5.11)$$

Equation (5.10) is the master equation on the long time scale, neglecting corrections of relative order λ^2. In contrast to the exact equation (4.31), it involves the transition probability P itself. It must be supplemented by the initial condition

$$P(t \mid \alpha\alpha_0) = \delta(\alpha - \alpha_0) \quad \text{for} \quad |t| \ll T_1 \tag{5.12}$$

directly obtained by comparison of (5.5) and (5.9). It should be noted that (5.12) holds for all times short compared to T_1, in particular for t of order T_0. Thus, to first order in λ, P is constant on the short time scale, whereas P_E has there a marked variation, given by (5.4).

Eq. (5.10) differs from the familiar lowest order equation (3.16) only by the fact that the transition rates $\overline{W}^{(1)}$ are correct to first order in λ, while (3.16) contains the zero order expression of these rates *). Our reduction of the general master equation to its well known lowest order form is thus complete. The interest of this reduction is of course not so much that it provides a new derivation for the familiar master equation. It lies rather in the clear and explicit picture obtained for the role of the two time scales, T_0 and T_1, and for the manner in which they remain completely separated if and only if all corrections of order $T_0/T_1 \sim \lambda^2$ are neglected. To be sure, the existence of a short time scale T_0 for all systems to which the familiar master equation applies has always been recognized. Still it seems to us quite instructive to have a more complete equation covering also the events on the short time scale and describing how the motions on the two time scales mix when T_0 and T_1 become of comparable order of magnitude.

6. *The long time behaviour of the transition probabilities.* Returning to the study of our system to general order in the perturbation, we want to investigate the asymptotic behaviour of the function $P_E(t \mid \alpha\alpha_0)$ and of the transition probability $P(t \mid \alpha\alpha_0)$ for very large times, $t \to \pm\infty$. From the identity

$$\int d\alpha \, P(t \mid \alpha\alpha_0) = 1 \tag{6.1}$$

(an obvious consequence of the definition of P), it follows that P and consequently P_E must have non-vanishing asymptotic expressions for $t \to \infty$. Our problem is to study them.

To find the asymptotic behaviour of P_E, which in turn implies through (4.27) the behaviour of P, one must according to (4.26) determine the singularities of $X_{E+l,E-l}$ as a function of the complex variable l. These singularities are located on the real axis. They can be either points where the function has a finite discontinuity for l crossing the real axis, or points in the neighbourhood of which the function becomes infinite. The former singular points fill continuous intervals of the real axis. Only the latter ones, which are

*) One should however say that in practice $\overline{W}^{(1)}$ is often identical with $W^{(0)}$, the difference $W_{ll'} - W^{(0)}$ being of order λ^2.

isolated and will be called hereafter *pseudopoles*, give any contribution to P_E in the limit of long times, as is readily seen from (4.26) by deforming the contour γ into lines $l = E' \pm i\eta$ with η very small. The name pseudopole is used because near such a singularity the function, while becoming infinite as in a pole, usually has in addition a finite discontinuity in every neighbouring point of the real axis. More precisely, the pseudopoles to be found for $X_{E+l,E-l}$ will be points E_0 such that the function

$$(l - E_0)\, X_{E+l,\,E-l}$$

has around E_0, as only singularities, finite discontinuities accross the real axis. Such pseudopoles may be called of degree one. Typical examples are given at $l = 0$ by the simple functions

$$l^{-1} \int_{-1}^{1} (x - l)^{-1}\, \mathrm{d}x,\ l^{-1} + \int_{-1}^{1} (x - l)^{-1}\, \mathrm{d}x.$$

We attempt to obtain information on the pseudopoles of $X_{E+l,E-l}$ from a discussion of the expansion (4.19). The function $W_{E+l,E-l}$, on account of its definition (4.16), has no pseudopoles. Its only singularities are finite discontinuities accross the real axis. Pseudopoles of $X_{E+l,E-l}$ may result in two ways from poles of the functions $D_{E\pm l}$ occurring in the expansion (4.19), i.e. from the presence of non-dissipative states in this expansion. Firstly pseudopoles may originate from the products

$$D_{E+l}(\alpha) D_{E-l}(\alpha), \quad D_{E+l}(\alpha_0) D_{E-l}(\alpha_0) \tag{6.2}$$

which occur as factors in each term of the expansion. For example, if the function $D_{l'}(\alpha)$ has a pole at $l' = E(\alpha)$ (the state $|\alpha\rangle$ being then non-dissipative), the first product (6.2) gives rise to two pseudopoles of degree one at

$$l = \pm\, [E(\alpha) - E]. \tag{6.3}$$

In the asymptotic expression of $P_E(t \mid \alpha\alpha_0)$ for large times such a pair of pseudopoles produces an oscillatory term involving

$$[E(\alpha) - E]^{-1} \sin\, [2(E(\alpha) - E)t].$$

This term would however disappear for large t in any integrated expression of the form

$$\int A(\alpha)\, \mathrm{d}\alpha\, P_E(t \mid \alpha\alpha_0) \tag{6.4}$$

under the condition that $E(\alpha)$ remains different from E in the integration. Since in practice integrations of the type (6.4) over α and similar integrations over α_0 (compare (3.15)) are usually taken, this sort of pseudopole is of little interest except in the case where the integral (6.4) includes points where $E(\alpha) = E$, i.e. when the poles (6.3) of the two factors in the product (6.2) become coincident, and in a similar case for an integration over α_0.

The latter circumstances are but examples of the second way in which poles of the functions $D_{E\pm l}$ give rise to pseudopoles of $X_{E+l,E-l}$. Generally any integral of the form

$$\int F(\alpha')D_{E+l}(\alpha')D_{E-l}(\alpha')\,d\alpha' \tag{6.5}$$

where $F(\alpha')$ is some function of α', will now be shown to have a pseudopole of degree one at $l = 0$ when the two functions $D_{E\pm l}$ have poles which become coincident in the domain of integration. Integrals as (6.5) occur in the expansion (4.19) of $X_{E+l,\,E-l}$ as a result of the integrations over intermediate states; they occur also when integrations like (6.4) are performed over the states α or α_0. To prove our assertion we let $E(\alpha')$ be a pole of $D_{l'}(\alpha')$. The quantity $J_E(\alpha')$ may be taken to vanish for E around $E(\alpha')$ and the function $G_{l'}(\alpha')$ is then holomorphic in this point. In its neighbourhood we may write

$$D_{l'}(\alpha') = N(\alpha')[E(\alpha') - l']^{-1} \tag{6.6}$$

with the notation, already introduced in (I. 5.8),

$$[N(\alpha')]^{-1} = 1 + \lambda^2\,[\partial G_l(\alpha')/\partial l]_{l=E(\alpha')}. \tag{6.7}$$

In those parts of the domain of integration in (6.5) where $E(\alpha') \neq E$ the two factors D have their poles at different l-values, the integrand has only poles of degree one, producing after integration finite discontinuities across the real axis. Consider now the case that the domain of integration contains a manifold where $E(\alpha') = E$. For α' in its immediate neighbourhood the poles of $D_{E\pm l}(\alpha')$ may be studied by means of the approximate expression (6.6), which gives in the integral

$$\int F(\alpha')[N(\alpha')]^2(E(\alpha') - E - l)^{-1}(E(\alpha') - E + l)^{-1}\,d\alpha' =$$
$$= (2l)^{-1}\int F(\alpha')[N(\alpha')]^2\,[(E(\alpha') - E - l)^{-1} - (E(\alpha') - E + l)^{-1}]\,d\alpha'. \tag{6.8}$$

In this form the occurrence of a pseudopole of degree one at $l = 0$ is obvious.

Since the expansion (4.19) contains terms with simultaneous integrations over several states and since, as we have just seen, each such integration may give rise to a pseudopole of degree one at $l = 0$, it might look as if more complicated singularities will result from superposition of pseudopoles of degree one. It will now be shown that such complications never occur. For this purpose we need considerations very similar to those of I, Section 4. We have there defined for each state $|\alpha\rangle$ the family x_α composed of the states $|\alpha'\rangle$ which play an effective part as intermediate state in the righthand side of the equation (see (4.3))

$$|\alpha\rangle\, G_l(\alpha) = \{VD_lV - \lambda VD_lVD_lV + \ldots\}_{id}\,|\alpha\rangle.$$

We can here formulate this definition more simply in terms of the function $W_{ll'}(\alpha'\alpha)$ introduced by (4.16). The family x_α is composed of the states $|\alpha'\rangle$ for which the function $W_{ll'}(\alpha'\alpha)$ of the complex variables l, l' is not identically zero. This family can in practice always be pictured as a set of continuous manifolds in the space of the quantum numbers characterizing the states $|\alpha'\rangle$. In Section 4 of I another set of states was also defined, the family y_α obtained by taking together all states of x_α, all states belonging to the families $x_{\alpha'}$ associated with the latter states, and so on. On account of (4.19) an alternative definition of y_α is the following: y_α is composed of the states $|\alpha'\rangle$ for which the function $X_{ll'}(\alpha'\alpha)$ of the variables l, l' does not vanish identically. The physical interpretation of the family y_α was given in I for the special case of non-dissipative systems. This interpretation is valid for a state $|\alpha\rangle$ as soon as it is itself non-dissipative, irrespectively of the properties of the other states. Another interpretation holds more generally, also when $|\alpha\rangle$ is dissipative; y_α is simply the set of all states $|\alpha'\rangle$ which get involved in the time evolution of a wave packet initially concentrated very narrowly around $|\alpha\rangle$. The family x_α has no special physical significance; in practice it is always found to coincide with y_α.

The property of the family y_α which we have to use presently is the following. Whenever $D_l(\alpha)$ has a pole at $l = E(\alpha)$, the functions $D_l(\alpha')$ belonging to all states $|\alpha'\rangle$ of y_α are regular around $l = E(\alpha)$ and those belonging to states $|\alpha'\rangle$ such that $|\alpha\rangle$ is in $y_{\alpha'}$ have a finite discontinuity for l crossing the real axis at $l = E(\alpha)$. As was shown in Section (I. 4), this property is a direct consequence of the identity (I. 3.13) or (4.3). Consider now the general term of the expansion (4.19) of $X_{E+l,E-l}$. It reads

$$\lambda^{2(n+1)} D_{E+l}(\alpha) D_{E-l}(\alpha) \int W_{E+l,E-l}(\alpha\alpha_n) D_{E+l}(\alpha_n) D_{E-l}(\alpha_n)\, d\alpha_n \ldots$$
$$\ldots d\alpha_1\, W_{E+l,E-l}(\alpha_1\alpha_0) D_{E+l}(\alpha_0) D_{E-l}(\alpha_0). \tag{6.9}$$

Take any sequence of states α, α_n, ... α_1, α_0 for which the integrand does not vanish. The family y belonging to any element of the sequence contains all elements of the sequence situated more to the left. From the property just mentioned it follows then immediately that if in (6.9) the two factors of a product

$$D_{E+l}(\alpha_j) D_{E-l}(\alpha_j) \tag{6.10}$$

have coincident poles at $l = 0$ no other such product appearing in (6.9) can have the same singularity.

The singularities of $X_{E+l,E-l}(\alpha\alpha_0)$ which are produced by poles of D_l-functions are thereby completely described: except for finite discontinuities accross the real axis and uninteresting pseudopoles which disappear when integrations are carried out over α and α_0, the only possible singularity is a pseudopole of degree one at $l = 0$. It is remarkable that the latter type of singularity also occurs when none of the $D_{l'}$-functions involved in the expansion of X has a pole at $l' = E$ in the domain of integration. This is best seen on the basis of the simple equation

$$D_{E+l}(\alpha_0) - D_{E-l}(\alpha_0) = 2l \int d\alpha\, X_{E+l,\, E-l}(\alpha\alpha_0) \tag{6.11}$$

which is easily derived by taking the diagonal part of the well known identity

$$R_l - R_{l'} = (l - l')\, R_l R_{l'}$$

and applying the definition (4.14) of X. Obviously, whenever $D_{l'}(\alpha_0)$ has a finite discontinuity for l' crossing the real axis at the point E (i.e. whenever $J_E(\alpha_0) \neq 0$) the integral in the righthand side of (6.11) has a pseudopole of degree one at $l = 0$. Consequently the function $X_{E+l,\, E-l}\,(\alpha\alpha_0)$ must then have the same singularity, at least for some range of values of α. This pseudopole must occur irrespectively of the presence or absence of poles in the D-functions involved in X. We also note that it cannot reduce to a true pole, because for non-vanishing $J_E(\alpha_0)$ the limit of the whole expression (6.11) for $l \to 0$ is different depending on the sign of $Im(l)$.

When, in the expansion (4.19) taken for $X_{E+l,E-l}$, none of the $D_{l'}$-functions involved has a pole at $l' = E$, the individual terms of the expansion have no singularities other than finite discontinuities across the real axis. The occurrence, for $J_E(\alpha_0) \neq 0$, of a pseudopole in the total expression $X_{E+l,E-l}$ $(\alpha\alpha_0)$ must then be attributed to the fact that the expansion becomes divergent when l vanishes. This lack of convergence can be verified by considering for the case under discussion the convergence properties of the expansion (4.19) in the limit of small λ. In this limit we may put everywhere, according to (4.17),

$$W_{ll'}(\alpha'\alpha) = W^{(0)}(\alpha'\alpha)$$

and we may approximate all functions D by

$$D_{l'}(\alpha) = \varepsilon(\alpha) - l' - \lambda^2 K_{\epsilon(\alpha)}(\alpha) \mp i\lambda^2 J_{\epsilon(\alpha)}(\alpha) \tag{6.12}$$

where the upper (lower) sign must be taken for positive (negative) value of $Im(l')$. The terms in λ^2 are important only when l' is near $\varepsilon(\alpha')$. Using the approximation (6.12) we find

$$D_{E+l}(\alpha) D_{E-l}(\alpha) = \{[\varepsilon(\alpha) - E - \lambda^2 K_{\epsilon(\alpha)}(\alpha)]^2 + [\lambda^2 J_{\epsilon(\alpha)}(\alpha) \mp il]^2\}^{-1}. \tag{6.13}$$

We are interested here in the case where $D_{l'}$ has no pole at $l' = E$. In the approximation (6.12) this condition is

$$J_{\epsilon(\alpha)}(\alpha) \neq 0. \tag{6.14}$$

When it is verified (6.13) remains bounded in the neighbourhood of $l = 0$ and in this point an integral of the form (6.5) becomes approximately

$$\int F(\alpha') D_{E+l}(\alpha')\, D_{E-l}(\alpha')\, d\alpha' = \int F(\alpha') \{[\varepsilon(\alpha') - E - \lambda^2 K_{\epsilon(\alpha')}(\alpha')]^2 + \\ + \lambda^4 [J_{\epsilon(\alpha')}(\alpha')]^2\}^{-1}\, d\alpha' = \pi\lambda^{-2} \int [J_{\epsilon(\alpha')}(\alpha')]^{-1} F(\alpha')\, \delta[\varepsilon(\alpha') - E]\, d\alpha'. \tag{6.15}$$

It is consequently very large, of order λ^{-2}. Returning now to the expansion (4.19) for $X_{E+l,\, E-l}$, still in the case of small λ, we observe that when going from an arbitrary term of the expansion to the next, one gets an additional factor λ^2 and an additional integration of the type (6.15). The coefficient λ^{-2} occurring in the latter integral for $l = 0$ compensates the factor λ^2, thus explaining why convergence of the expansion cannot be achieved by taking λ small enough. It should be noted that this compensation occurs only at the point $l = 0$. For all other values of l (even infinitely close to the real axis) the integral (6.15) remains finite when $\lambda \to 0$ and nothing prevents the increasing powers of λ in front of the successive terms of the expansion to ensure its convergence. No other singular points than $l = 0$ can therefore originate from lack of convergence of the series (4.19) for $X_{E+l,\, E-l}$.

We have encountered two different origins for the pseudopoles of the expression $X_{E+l,\, E-l}$, the first one being a pole of $D_{l'}(\alpha')$ at $l' = E$ for states $|\alpha'\rangle$ included in the domains of integration (or appearing as initial or final state), and the second one the divergence of the expansion of $X_{E+l,\, E-l}$ at $l = 0$. Just as we have seen that poles of $D_{l'}$ never can have a cumulative effect and produce higher order singularities by superposition of pseudopoles of degree one, no such cumulative effect can originate from the simultaneous occurrence of poles of D-functions and of a divergence in the expansion. To verify this point, we consider in the expansion of $X_{E+l,\, E-l}$ all terms (6.9) of order $n \geqslant m \geqslant 1$, leave out the integrations over the intermediate states $\alpha_1, \ldots \alpha_m$, but perform those over all other intermediate states. The result is

$$\lambda^{2m} X_{E+l,\, E-l}(\alpha\alpha_m)\, W_{E+l,\, E-l}(\alpha_m\alpha_{m-1}) D_{E+l}(\alpha_{m-1}) D_{E-l}(\alpha_{m-1}) \ldots$$
$$\ldots D_{E+l}(\alpha_1) D_{E-l}(\alpha_1)\, W_{E+l,\, E-l}(\alpha_1\alpha_0) D_{E+l}(\alpha_0) D_{E-l}(\alpha_0). \tag{6.16}$$

Assume now that α_{m-1} would be such that $D_{l'}(\alpha_{m-1})$ has a pole at $l' = E$. It is then just as impossible for $X_{E+l,E-l}(\alpha\alpha_m)$ to have a pseudopole due to lack of convergence as to have one caused by poles in other D-functions. Indeed, from our assumption and from the property of the family y already used before, $D_l(\alpha')$ is regular at $l' = E$ for all states α' in the family $y_{\alpha_{m-1}}$, thus in particular for the state α_m. This implies

$$\lim_{l\to 0} \left[D_{E+l}(\alpha_m) - D_{E-l}(\alpha_m)\right] = 0.$$

Comparing with (6.11) we must indeed conclude that no pseudopole occurs in $X_{E+l,E-l}(\alpha\alpha_m)$.

This ends our determination of the singularities of $X_{E+l,E-l}(\alpha\alpha_0)$. Leaving out of consideration, when one of the states α, α_0 is non-dissipative, pseudopoles which disappear when an integration over these states is carried out, the only singularities we have found are finite discontinuities across the real axis and a pseudopole of degree one at $l = 0$. It is this last singularity which determines the asymptotic behaviour of $P_E(t \mid \alpha\alpha_0)$ for long times. Indeed it follows from Eq. (4.26) that

$$\lim_{t\to\pm\infty} P_E(t \mid \alpha\alpha_0) = \pi^{-1} \lim_{0<\eta\to 0} \eta X_{E\mp i\eta,\, E\pm i\eta}(\alpha\alpha_0) \tag{6.17}$$

where upper (lower) signs must be taken together. When the state α or

α_0 is non-dissipative this relation holds after averaging over the time or integrating over α or α_0. Otherwise it is strictly exact. From (6.17) the asymptotic value of the transition probability follows

$$\lim_{t\to\pm\infty} P(t \mid \alpha\alpha_0) = \pi^{-1} \int_{-\infty}^{\infty} dE \lim_{0<\eta\to 0} \eta X_{E\mp i\eta, E\pm i\eta}(\alpha\alpha_0). \qquad (6.18)$$

We note that the quantity (6.17) is non-negative; this follows from the definition (4.14), which implies more generally that X_{ll^*} is never negative.

The last task is the determination of the limit in the righthand side of (6.17). On physical grounds the result is expected to depend very strongly on the nature of the system and on the dissipative or non-dissipative character of the states involved. For this reason a unified treatment applicable to all cases is not possible. An exhaustive analysis of the various possible situations seems at present very difficult. Here we shall restrict ourselves to the extension to general order of the case of dissipative behaviour which is usually discussed in lowest order on the basis of the master equation (3.16). However, before studying this case in the next section, we want to mention for the sake of completeness the value taken by (6.17) for the non-dissipative systems studied in I and II. The calculation of the limit (6.17) for such systems is very simple because the pseudopole of $X_{E+l, E-l}$ is here entirely determined by the poles of the D-functions. Assuming as in I and II that each function $D_l(\alpha)$ has only one pole $E(\alpha)$, one finds

$$\lim_{t\to+\infty} P_E(t \mid \alpha\alpha_0) = [N(\alpha)]^2\, \delta[E(\alpha) - E]\, \delta(\alpha - \alpha_0)$$
$$+ \lambda^2 [N(\alpha)]^2\, \delta[E(\alpha) - E] \int W_{E-i0, E+i0}(\alpha\alpha_1)\, d\alpha_1 X_{E-i0, E+i0}(\alpha_1\alpha_0)$$
$$+ \lambda^2 \int X_{E-i0, E+i0}(\alpha\alpha_1)\, d\alpha_1\, W_{E-i0, E+i0}(\alpha_1\alpha_0) [N(\alpha_0)]^2\, \delta[E(\alpha_0) - E]$$
$$+ \lambda^4 \int X_{E-i0, E+i0}(\alpha\alpha_3)\, d\alpha_3\, W_{E-i0, E+i0}(\alpha_3\alpha_2) [N(\alpha_2)]^2$$
$$\delta[E(\alpha_2) - E]\, d\alpha_2\, W_{E-i0, E+i0}(\alpha_2\alpha_1)\, d\alpha_1 X_{E-i0, E+i0}(\alpha_1\alpha_0).$$

The limit for $t \to -\infty$ is found by interchanging the indices $E \pm i0$. The symbol $N(\alpha)$ was defined in (6.7). This result is susceptible of a simple interpretation in terms of the perturbed stationary states of the system, know from II, but we shall not describe it in the present paper.

7. *The approach to statistical equilibrium.* It has often been shown how the lowest order master equation (3.16) implies that the system studied approaches microcanonical equilibrium under the influence of the perturbation [6]). The assumptions on which the derivation is most customarily based are the symmetry of the transition rate matrix, reading in our notation

$$W^{(0)}(\alpha'\alpha) = W^{(0)}(\alpha\alpha'), \qquad (7.1)$$

and the interconnection of all states of equal unperturbed energy, by which is meant that for any two states α, α' verifying

$$\varepsilon(\alpha) = \varepsilon(\alpha')$$

there exists a sequence of states connecting them

$$\alpha = \alpha_0, \alpha_1, \dots \alpha_n, \alpha_{n+1} = \alpha' \tag{7.2}$$

with the properties

$$\varepsilon(\alpha_j) = \varepsilon(\alpha_{j+1}), W^{(0)}(\alpha_{j+1}\, \alpha_j) \neq 0, (j = 0, \dots n). \tag{7.3}$$

Under these assumptions the long time limit of P is found to be in our notation

$$\lim_{t\to\pm\infty} P(t \mid \alpha\alpha_0) = \{\int \delta[\varepsilon(\alpha') - \varepsilon(\alpha)]\, d\alpha'\}^{-1} \,.\, \delta[\varepsilon(\alpha) - \varepsilon(\alpha_0)]. \tag{7.4}$$

It corresponds to the microcanonical equilibrium distribution in absence of the perturbation, as is consistent with the lowest order nature of (3.16) *). It should be noted that the assumption of interconnection of states automatically implies that all states α are dissipative; indeed it implies that each state α is contained in its own family of states y_α, a situation incompatible with the occurrence of poles in $D_l(\alpha)$ (see I, section 4 or section 6 of the present paper).

The symmetry property (7.1), often referred to as principle of microscopic reversibility or principle of detailed balance, is usually claimed to be an immediate consequence of the hermiticity of the perturbation V, on the ground that $W^{(0)}(\alpha'\alpha)$ is essentially the absolute square of the matrix element $\langle\alpha'|V|\alpha\rangle$. Although of course partly valid, this argument is oversimplified. Taken in the framework of our analysis, it disregards the fact that the diagonal part must be taken in the definition (3.17) of $W^{(0)}$. Let us consider for example (see appendix) the system composed of one Bloch electron in interaction with the lattice vibrations of a crystal. We choose for α one electron states with the lattice in its unperturbed ground state and for α' states where, in addition to the electron, a phonon is present. The quantity $W^{(0)}(\alpha\alpha')$ then vanishes identically, whereas $W^{(0)}(\alpha'\alpha)$ has non-vanishing values, even on the energy shell $\varepsilon(\alpha) = \varepsilon(\alpha')$. Clearly (7.1) does not hold. Still the master equation (3.16) is fully applicable and describes in the limit of small phonon-electron interaction the dissipation of a state α with one electron and no phonon present into states where one, two, etc. phonons have been emitted by the electron. The vanishing of $W^{(0)}(\alpha\alpha')$ corresponds to the physically obvious fact that a state with one electron and one phonon will never dissipate, i.e. never go over for arbitrary phases, into a state without phonon; in other words phonon absorption by the electron is a phase dependent, transient process.

It is only for states with a very large number of elementary excitations present (of the order of the size of the system), in the foregoing example

*) The customary discussion of these matters treats the states α as discrete. In our presentation they are considered as continuous, but evidently we could also, by groupings of states, go over to a formulation in terms of discrete indices.

states with many phonons, that the symmetry property (7.1) can be expected to hold. For such states, – and most states are of this type for a large system not too close to its ground state –, one expects that if transitions in the neighbourhood of $\alpha_1 \rightarrow \alpha_2$ contribute to the value of $W^{(0)}(\alpha'\alpha)$ at $\alpha = \alpha_1$, $\alpha' = \alpha_2$ (see (3.17)), the transitions in the neighbourhood of $\alpha_2 \rightarrow \alpha_1$ will contribute the same amount to $W^{(0)}(\alpha_1\alpha_2)$, thus giving rise to the symmetry (7.1). Similarly, for such states, if a succession of transitions $\alpha_1 \rightarrow \alpha_2 \rightarrow \ldots \rightarrow \alpha_n$ and its neighbouring ones contribute to the value at $\alpha = \alpha_1$, $\alpha' = \alpha_n$ of the function $W_{ll'}(\alpha'\alpha)$ defined in (4.16), the transitions around $\alpha_n \rightarrow \ldots \ldots \alpha_2 \rightarrow \alpha_1$ are expected to contribute to $W_{ll'}(\alpha_1\alpha_n)$ and the hermiticity of V now gives the generalized symmetry relation

$$W_{ll'}(\alpha'\alpha) = W_{l'l}(\alpha\alpha'). \tag{7.5}$$

For a detailed verification of this conclusion in an actual case, e.g. for the electron-phonon system of the appendix (considering only states where the number of phonons is of order N or Ω), the labelling of the states α by the wave vectors of the elementary excitations, as described in Section 2, is not very convenient and it would be replaced with profit by a different one, namely a labelling by the numbers of excitations of each sort per small cell of wave vector space. The coordinates of the cells where these numbers are changed in the transitions caused by the perturbation would then play the role of the continuous variables present in our integrals.

For our discussion of the approach toward equilibrium to general order in the perturbation λV, we adopt the property (7.5) as the basic assumption generalizing (7.1). As for the assumption of interconnection of states, we keep it in the same form as above: for every pair α, α' of states of equal unperturbed energy there exists a sequence (7.2) verifying (7.3). We note again that it implies the dissipative property for all states α, and thus the boundedness of the function $D_{E\pm i0}(\alpha)$. Under these assumptions, we shall establish that for λ not too large one has

$$\lim_{t\rightarrow\pm\infty} P_E(t \mid \alpha\alpha_0) = [\int \Delta_E(\alpha')\, d\alpha']^{-1} \,.\, \Delta_E(\alpha)\Delta_E(\alpha_0), \tag{7.6}$$

with the abbreviation

$$\Delta_E(\alpha) = (2\pi i)^{-1}[D_{E+i0}(\alpha) - D_{E-i0}(\alpha)]. \tag{7.7}$$

If one remembers the relation (4.27) between P_E and P, and notices that (7.7) reduces to $\delta[\varepsilon(\alpha) - E]$ in the limit of small λ, one readily sees that in this limit (7.6) gives the lowest order value (7.4) for the long time expression of P. We now proceed to prove (7.6). We shall show afterwards that this result corresponds to establishment of the microcanonical equilibrium distribution in presence of the perturbation.

Introducing the notation

$$\lim_{t\rightarrow\pm\infty} P_E(t \mid \alpha\alpha_0) = q_E^{\pm}(\alpha\alpha_0), \tag{7.8}$$

we recall the main result of Section 6, equation (6.17),

$$q_E^{\pm}(\alpha\alpha_0) = \pi^{-1} \lim \eta X_{E\mp i\eta,\, E\pm i\eta}(\alpha\alpha_0) \tag{7.9}$$

where η is a positive number tending to zero. We note incidentally that in view of the relation (4.19) between X and W, the symmetry relation (7.5) implies a similar property for X:

$$X_{ll'}(\alpha'\alpha) = X_{l'l}(\alpha\alpha') \tag{7.10}$$

and consequently, from (7.9),

$$q_E^{+}(\alpha\alpha_0) = q_E^{-}(\alpha_0\alpha). \tag{7.11}$$

We have mentioned earlier that the assumption of interconnection of states requires the function $D_{E\pm i0}(\alpha)$ to be bounded. As we have seen in detail in Section 6, the limit (7.9) originates then entirely from divergence of the series (4.19) when l and l' approach from opposite sides the point E of the real axis. From the boundedness of $D_{E\pm i0}(\alpha)$, the first term in the righthand side of (4.19) gives a vanishing contribution to the limit (7.9). The latter consequently verifies the two following equations:

$$q_E^{\pm}(\alpha\alpha_0) = \lambda^2 D_{E+i0}(\alpha) D_{E-i0}(\alpha) \int W_{E\mp i0,\, E\pm i0}(\alpha\alpha')\, d\alpha'\, q_E^{\pm}(\alpha'\alpha_0), \tag{7.12}$$

$$q_E^{\pm}(\alpha\alpha_0) = \lambda^2 \int q_E^{\pm}(\alpha\alpha')\, d\alpha'\, W_{E\mp i0,\, E\pm i0}(\alpha'\alpha_0)\, D_{E+i0}(\alpha_0) D_{E-i0}(\alpha_0). \tag{7.13}$$

Our determination of q_E will be based on the first; one could however use the second one as well *). Equation (7.12) shows that q_E is eigenfunction of an eigenvalue problem. It belongs to the eigenvalue 1.

Consider this eigenvalue problem in the limit of small λ. Since we are dealing with the dissipative case, equation (6.14) holds and we may use equation (6.15), which states that one has approximately

$$\lambda^2 D_{E+i0}(\alpha) D_{E-i0}(\alpha) = \pi [J_{\epsilon(\alpha)}(\alpha)]^{-1}\, \delta[\varepsilon(\alpha) - E].$$

The quantity J must be taken in the limit $\lambda \to 0$; its value is then

$$J_{\epsilon(\alpha)}(\alpha) = \pi \int d\alpha'\, \delta[\varepsilon(\alpha') - \varepsilon(\alpha)]\, W^{(0)}(\alpha'\alpha). \tag{7.14}$$

The limiting form of the eigenvalue equation is

$$q_E^{\pm}(\alpha\alpha_0) = \pi [J_{\epsilon(\alpha)}(\alpha)]^{-1}\, \delta[\varepsilon(\alpha) - E] \int W^{(0)}(\alpha\alpha')\, d\alpha'\, q_E^{\pm}(\alpha'\alpha_0). \tag{7.15}$$

The solution is necessarily of the form

$$q_E^{\pm}(\alpha\alpha_0) = \delta[\varepsilon(\alpha) - E] f(\alpha)$$

with

$$f(\alpha) = \pi [J_{\epsilon(\alpha)}(\alpha)]^{-1} \int W^{(0)}(\alpha\alpha')\, \delta[\varepsilon(\alpha) - \varepsilon(\alpha')]\, d\alpha' f(\alpha').$$

The latter equation is to be taken on the energy shell $\varepsilon(\alpha) = E$ only. The

*) An equation closely connected to (7.12) would be obtained by inserting (7.8) in the master equation (4.31). The form (7.12) is however more practical for the following considerations.

function $f(\alpha) =$ constant is a solution, as results immediately from (7.14) and the symmetry relation (7.5) in its lowest order form (7.1). Using the non-negative nature of $W^{(0)}(\alpha\alpha')$ and the assumed interconnection of all states on the energy shell $\varepsilon(\alpha) = E$, one furthermore concludes by a well known argument [6]) that the solution mentioned is the only one. Finally the transposed eigenvalue equation of (7.15), which is

$$q'(\alpha') = \pi \int \mathrm{d}\alpha\, q'(\alpha)[J_{\varepsilon(\alpha)}(\alpha)]^{-1}\, \delta[\varepsilon(\alpha) - E]\, W^{(0)}(\alpha\alpha')$$

is easily seen from (7.14) to have the solution

$$q'(\alpha) = \pi \int \mathrm{d}\alpha'\, \delta[\varepsilon(\alpha') - E]\, W^{(0)}(\alpha'\alpha).$$

This solution is not orthogonal to the unique solution we have found for (7.15). As a consequence of known theorems [7]) we can therefore conclude that the eigenvalue 1 is a simple root of the characteristic equation of the eigenvalue problem.

The latter property, just established in the limit of small λ, implies by continuity that for $|\lambda|$ smaller than a positive quantity λ_c the characteristic equation of the exact eigenvalue problem (7.12) has one and only one root in the neighbourhood of 1, and that this root is simple. The corresponding eigenfunction is consequently unique (except for an arbitrary multiplicative constant). We show now that the root in question is 1 and that the corresponding eigenfunction is $\Delta_E(\alpha)$, as defined by (7.7). For this purpose we insert $\Delta_E(\alpha')$ instead of q_E in the righthand side of (7.12). Applying the symmetry relation (7.5), we obtain for the integral

$$(2\pi i)^{-1} \int [D_{E+i0}(\alpha') - D_{E-i0}(\alpha')]\, \mathrm{d}\alpha' W_{E\pm i0\, E\mp i0}(\alpha'\alpha).$$

In view of the identity (4.21) this expression is simply, for both values of the double signs,

$$(2\pi i)^{-1}\, [G_{E+i0}(\alpha) - G_{E-i0}(\alpha)].$$

The righthand side of (7.12) is thus equal to

$$(2\pi i)^{-1}\, \lambda^2\, [G_{E+i0}(\alpha) - G_{E-i0}(\alpha)] D_{E+i0}(\alpha) D_{E-i0}(\alpha). \tag{7.16}$$

We make use of the identity (4.23) for $l = E + i0$, $l' = E - i0$. In view of the boundedness of $D_{E\pm i0}$, the term in $l - l'$ gives no contribution and (7.16) reduces to

$$(2\pi i)^{-1}\, [D_{E+i0}(\alpha) - D_{E-i0}(\alpha)] = \Delta_E(\alpha).$$

Our statement is thereby established.

On the basis of the foregoing, the determination of $q_E^{\pm}$ is easily completed. From the unicity of the eigenfuction $\Delta_E(\alpha)$, it is clear that q_E must have the form

$$q_E^{\pm}(\alpha\alpha_0) = \Delta_E(\alpha)\, f_E^{\pm}(\alpha_0). \tag{7.17}$$

Remembering the identity

$$D_{E+l}(\alpha_0) - D_{E-l}(\alpha_0) = 2l \int d\alpha\, X_{E+l,\, E-l}(\alpha\alpha_0)$$

already mentioned in (6.11) we find from (7.9) and (7.7)

$$\int d\alpha\, q_E^{\pm}(\alpha\alpha_0) = \Delta_E(\alpha_0).$$

Insertion of (7.17) gives

$$f_E^{\pm}(\alpha_0) = [\int \Delta_E(\alpha')\, d\alpha']^{-1}\, \Delta_E\,(\alpha_0)$$

independently of the double sign. We have thereby established, for $|\lambda| < \lambda_c$, the announced expression (7.6) for the long time limit of P_E.

From (7.6), the long time limit of P is

$$\lim_{t\to\pm\infty} P(t \mid \alpha\alpha_0) = \int_{-\infty}^{\infty} dE\, [\int \Delta_E\,(\alpha')\, d\alpha']^{-1}\, \Delta_E(\alpha)\, \Delta_E(\alpha_0). \tag{7.18}$$

We shall denote by T a time such that this limit is practically attained for $|t| \gtrsim T$. This is the time T already considered in Section 3. Let our system be at time $t = 0$ in the quantum state

$$\varphi_0 = \int |\alpha\rangle\, d\alpha\, c(\alpha) \tag{7.19}$$

and let us assume this initial state to have phases sufficiently incoherent for the interference term (3.13) to remain negligible over the time interval $|t| \lesssim T$. The expectation value at time t of a diagonal operator A is then given for $|t| \lesssim T$ by the formula

$$\langle A\rangle_t = \int A(\alpha)\, d\alpha\, P(t \mid \alpha\alpha_0)\, d\alpha_0 \mid c(\alpha_0)|^2.$$

In view of (7.18) it approaches for $t \to \pm T$ the limit

$$\lim \langle A\rangle_t = \int_{-\infty}^{\infty} dE\, \langle A\rangle_E p_E, \tag{7.20}$$

where we have put

$$\langle A\rangle_E = [\int \Delta_E(\alpha')\, d\alpha']^{-1} \int A(\alpha)\, \Delta_E(\alpha)\, d\alpha, \tag{7.21}$$

$$p_E = \int |c(\alpha)|^2\, \Delta_E(\alpha)\, d\alpha. \tag{7.22}$$

As will now be established, this limit agrees with the value which would be calculated for A from the microcanonical distribution taken for the complete hamiltonian $H + \lambda V$. We introduce to this end the projection operator Q_E on the energy shell $H + \lambda V = E$, given by

$$Q_E = (2\pi i)^{-1} \lim_{\eta\to 0}(R_{E+i\eta} - R_{E-i\eta}), \quad \eta > 0,$$

and already used in II, Section 2. Note that the diagonal part $\{Q_E\}_d$ of this operator is simply the diagonal operator Δ_E, the eigenvalues of which are given by (7.7). The microcanonical average of a diagonal operator A on the energy shell $H + \lambda V = E$ can be written as (Sp denotes the trace)

$$Sp(AQ_E)/Sp(Q_E).$$

Calculating the trace in the $|\alpha\rangle$-representation, one can replace Q_E by its diagonal part Δ_E, and, in the limit of an infinite system, one finds the expression (7.21). Consequently the quantity $\langle A\rangle_E$ is the microcanonical average of A on the shell of total energy E. On the other hand, when the system is in its initial state (7.19) the probability for the total energy $H+\lambda V$ to have a value between E and $E+\mathrm{d}E$ is

$$\langle\varphi_0\,|Q_E|\,\varphi_0\rangle\,\mathrm{d}E. \tag{7.23}$$

In view of the phase incoherence of the amplitudes $c(\alpha)$, the only contribution to (7.23) comes from the diagonal part of Q_E. The expectation value $\langle\varphi_0\,|Q_E|\,\varphi_0\rangle$ thus reduces to (7.22) and we conclude that p_E gives the probability distribution of the total energy in the initial state. The announced result is thereby reached: the righthand side of (7.20) is the microcanonical average of A corresponding to the statistical distribution of total energy in the initial state φ_0.

We end with a few comments on the results of this section. It is only for diagonal operators A and for initial states φ_0 with rapidly varying phases that we have established the approach to microcanonical equilibrium values. The first restriction has already been discussed in Section 3. One would expect the approach to equilibrium values to hold true for a wider class of operators, to know the operators which have simple matrix elements in the $|\alpha\rangle$-representation, even in the limit of a large system. Such an extension of our results would however require a proper generalization of our mathematical treatment. As for the second restriction, its only purpose is to make the contribution of the interference term (3.13) negligible for a time interval as long as T *). The more rapidly the phases of the initial amplitudes vary, the longer will be the time interval over which no interference effects occur. More cannot be said in general, however, because one can imagine initial states such that $\langle A\rangle_t$, having reached the equilibrium value (7.20) at a time $t\sim T$, retains it for a time much longer than T whereupon a new deviation from the equilibrium value sets in because of a sudden appearance of contributions from the interference term.

It should be clear that the assumption of rapidly varying phases for the initial state φ_0 has little in common and is even in contradiction with the conventional assumption of incoherent phases at all times. The former assumption singles out the initial state and implies for all $t\neq 0$ coherent phase relations between the amplitudes $c_t(\alpha)$ of

$$\varphi_t=\exp\left[-\,i(H+\lambda V)\,t\right]\varphi_0=\int|\alpha\rangle\,\mathrm{d}\alpha\;c_t(\alpha).$$

These phase relations are all-important when the effects of the perturbation are taken into account to general order. It is only to lowest order that their

*) We note that for $\Delta\alpha$ small enough initial states of the type considered in (3.19), (3.20) satisfy this property. They form another class of states for which (7.20) holds.

influence is weakened: it may then for example be neglected in the derivation of the master equation but remains essential when time inversion is applied to the system [1]). To general order in the perturbation, the initial time $t = 0$ is completely singled out by its incoherence of phases. Starting from it, the system behaves dissipatively both toward the future and toward the past. This is why the general master equation (4.31) imparts a special role to the instant $t = 0$. This special role is no longer explicitly visible in the lowest order master equation (3.16), but it is still implied by it, because, as is well known, a non-stationary solution of (3.16) cannot be continued indefinitely toward the past without taking negative values. In connection with the coherence of phases of the amplitudes $c_t(\alpha)$ for all non-vanishing times, we may also mention the close relation of this property with the non-markovian nature of the general master equation (4.31). This non-markovian nature can be understood as resulting from interference effects between the various waves produced by the perturbation. Such interference effects are a manifestation of definite phase relationships. As was discussed in detail in A, they become negligible for small perturbations, and this circumstance is responsible for the markovian character of the lowest order equation (3.16).

Our discussion of the approach to statistical equilibrium has been carried out under the assumptions of symmetry (7.5) and of interconnection of states. It is for systems with a large number of excitations present (of the order of the number of particles in the system) that we may expect these assumptions to hold. For such systems, the master equation (4.31) can be used to follow in detail the time evolution of the system toward equilibrium. There are however also other situations to which the master equation is directly applicable, namely all situations where initially a very few excitations are present in the system, so that the dissipative process consists in their decay into an ever increasing number of other modes of motion. An example of this sort, the system composed of one Bloch electron in interaction with a lattice initially in its ground state, has already been quoted and can in principle be studied completely starting from the equations in the appendix. In such cases of shower-like processes it is no longer strictly possible to describe the long time behaviour of the system as an approach to equilibrium, because the total excitation energy is too low. One must rather think in terms of a shower phenomenon which, governed by the master equation, continues as long as dissipative states are involved and can only be interrupted if and when low-lying non-dissipative states are reached.

Appendix. We illustrate in this appendix a few aspects of the general formalism used throughout the paper by considering a special example. We assume a perfect non-conducting crystal, add a few electrons in the lowest unoccupied Bloch band and consider their interaction with longitudinal lattice vibrations (to be called phonons). The hamiltonian of this system is composed of a part H for the non-interacting electrons and phonons, and a part λV describing the interaction. In the simplest approximation one has

$$H = \sum_{\mathbf{k}} \varepsilon(\mathbf{k})\, \alpha^*_{\mathbf{k}} \alpha_{\mathbf{k}} + \sum_{\mathbf{q}} \omega(\mathbf{q}) a^*_{\mathbf{q}} a_{\mathbf{q}}, \tag{A.1}$$

$$\lambda V = i\lambda\, (8\pi^3/\Omega)^{\frac{1}{2}}\, \gamma \sum_{\mathbf{k},\mathbf{q}} [\omega(\mathbf{q})]^{\frac{1}{2}} (a^*_{\mathbf{q}} \alpha^*_{\mathbf{k}-\mathbf{q}} \alpha_{\mathbf{k}} - \alpha^*_{\mathbf{k}} \alpha_{\mathbf{k}-\mathbf{q}} a_{\mathbf{q}}). \tag{A.2}$$

We have put $\hbar = 1$. The wave vector and energy of an electron are represented by $\mathbf{k}$, $\varepsilon(\mathbf{k})$, the corresponding quantities for a phonon by $\mathbf{q}$, $\omega(\mathbf{q})$. One often adopts for the energies the simple expressions

$$\varepsilon(\mathbf{k}) = |\mathbf{k}|^2/2m, \; \omega(\mathbf{q}) = s|\mathbf{q}|. \tag{A.3}$$

A very important feature is that the electron velocity $|\mathbf{k}|/m$ is usually much larger than the phonon velocity s. This fact is essential for the dissipative nature of the electron states. The operators a, α and a^*, α^* are the annihilation and creation operators for phonons and electrons. We have

$$a_{\mathbf{q}} a^*_{\mathbf{q}} - a^*_{\mathbf{q}} a_{\mathbf{q}} = 1, \quad \alpha_{\mathbf{k}} \alpha^*_{\mathbf{k}} + \alpha^*_{\mathbf{k}} \alpha_{\mathbf{k}} = 1. \tag{A.4}$$

All other commutators of a, a^* and all other anticommutators of α, α^* vanish. Every a or a^* further commutes with every α or α^*. The volume of the crystal is Ω. We assume it to have a cubic shape and quantize the wave vectors $\mathbf{q}$, $\mathbf{k}$ so as to verify periodic boundary conditions. The constant γ can be written

$$\gamma = (d/m)^{\frac{1}{2}} \tag{A.5}$$

where d is the lattice constant. The dimensionless constant λ is then not far from having the order of magnitude one in realistic cases. We remark that electron-electron and phonon-phonon interactions as well as the electron spin are completely neglected in the hamiltonian (A.1), (A.2).

The unperturbed representation $|\alpha\rangle$, as introduced in Section 2, is composed of the following states

$$|\mathbf{k}_1, \ldots \mathbf{k}_n, \mathbf{q}_1, \ldots \mathbf{q}_r\rangle = (\Omega/8\pi^3)^{(n+r)/2}\, \alpha^*_{\mathbf{k}_1} \ldots \alpha^*_{\mathbf{k}_n} a^*_{\mathbf{q}_1} \ldots a^*_{\mathbf{q}_r} |0\rangle. \tag{A.6}$$

$|0\rangle$ denotes the no-electron, no-phonon state. It corresponds to the case where the Bloch band is empty and the lattice in its ground state. We assume for simplicity all $\mathbf{k}$-vectors to be different, and we assume them to be written in succession according to a prescribed order. The same assumption is made for the $\mathbf{q}$-vectors. We avoid in this way formal complications which are known from the practice of quantum field theory to play no role in actual calculations. Under our assumptions we have, in the limit $\Omega \to \infty$,

$$\langle \mathbf{k}_1, \ldots \mathbf{k}_n, \mathbf{q}_1, \ldots \mathbf{q}_r \; \mathbf{k}'_1, \ldots \mathbf{k}'_{n'}, \mathbf{q}_1, \ldots \mathbf{q}'_{r'}\rangle =$$
$$\delta_{nn'}\delta_{rr'}\, \delta(\mathbf{k}_1 - \mathbf{k}'_1) \ldots \delta(\mathbf{k}_n - \mathbf{k}'_n)\, \delta(\mathbf{q}_1 - \mathbf{q}'_1) \ldots \delta(\mathbf{q}_r - \mathbf{q}'_r) \tag{A.7}$$

where the δ-functions originate from Kronecker symbols through

$$\lim_{\Omega\to\infty} (\Omega/8\pi^3)\, \delta_{kk'} = \delta(\mathbf{k} - \mathbf{k}') \tag{A.8}$$

and similarly for $\mathbf{q}$-vectors. (A.7) corresponds to eq. (2.2) of the text. In the representation (A.6) the perturbation V has matrix elements with a simple limiting expression for $\Omega \to \infty$. One finds for example by application of (A.8)

$$\langle \mathbf{k}, \mathbf{q}\, |V|\, \mathbf{k}' \rangle = i\gamma\, [\omega(\mathbf{q})]^{\frac{1}{2}}\, \delta(\mathbf{k} + \mathbf{q} - \mathbf{k}'), \tag{A.9}$$

$$\langle \mathbf{k}, \mathbf{q}_1, \mathbf{q}_2\, |V|\, \mathbf{k}', \mathbf{q}' \rangle = i\gamma[\omega(\mathbf{q}_1)]^{\frac{1}{2}}\delta(\mathbf{k} + \mathbf{q}_1 - \mathbf{k}') \times \\ \times\, \delta(\mathbf{q}_2 - \mathbf{q}') + i\gamma\, [\omega(\mathbf{q}_2)]^{\frac{1}{2}}\, \delta(\mathbf{k} + \mathbf{q}_2 - \mathbf{k}')\, \delta(\mathbf{q}_1 - \mathbf{q}'). \tag{A.10}$$

The calculation of matrix elements $\langle\alpha|\, VA_1\, V \ldots A_nV\, |\alpha'\rangle$ with diagonal $A_1, \ldots A_n$ can be performed either by application of equations of the type (A.9), (A.10), or by direct use of (A.2) for finite Ω, the limit $\Omega \to \infty$ being taken at the end of the calculation. We illustrate on the case $n = 1$ the occurrence of diagonal parts. In the evaluation of $\langle \mathbf{k}, \mathbf{q}\, |VAV|\, \mathbf{k}', \mathbf{q}'\rangle$ the possible transitions are

$$\mathbf{k}, \mathbf{q} \leftarrow \mathbf{k}'' \leftarrow \mathbf{k}', \mathbf{q}',$$
$$\mathbf{k}, \mathbf{q} \leftarrow \mathbf{k}'', \mathbf{q}, \mathbf{q}' \leftarrow \mathbf{k}', \mathbf{q}',$$
$$\mathbf{k}, \mathbf{q} \leftarrow \mathbf{k}'', \mathbf{q}', \mathbf{q}'' \neq \mathbf{q} \leftarrow \mathbf{k}', \mathbf{q}'.$$

In the last transition scheme the same phonon $\mathbf{q}''$ is emitted and reabsorbed, whereas in the last but one $\mathbf{q}$ is emitted and $\mathbf{q}'$ absorbed. The calculation gives

$$\langle \mathbf{k}, \mathbf{q}\, |VAV|\, \mathbf{k}', \mathbf{q}'\rangle = \gamma^2\, \delta(\mathbf{k} - \mathbf{k}')\, \delta(\mathbf{q} - \mathbf{q}') \int A(\mathbf{k} - \mathbf{q}'', \mathbf{q}, \mathbf{q}'')\, \omega(\mathbf{q}'')\, d\mathbf{q}'' \\ + \gamma^2\, \delta(\mathbf{k} + \mathbf{q} - \mathbf{k}' - \mathbf{q}')\, [A(\mathbf{k} + \mathbf{q}) + A(\mathbf{k}' - \mathbf{q}, \mathbf{q}, \mathbf{q}')]\, [\omega(\mathbf{q})\, \omega(\mathbf{q}')]^{\frac{1}{2}}. \tag{A.11}$$

The first term in the righthand side, which originates from the third transition scheme, is the diagonal part of the matrix element. The second term stems from the two first transition schemes.

We take next the case $n = 3$ to illustrate the concept of irreducible diagonal part. The transition schemes contributing to the diagonal part of $\langle \mathbf{k}, \mathbf{q}\, |VA_1VA_2VA_3V|\, \mathbf{k}', \mathbf{q}'\rangle$ are easily seen to be

$$\mathbf{k}, \mathbf{q} \leftarrow \mathbf{k}_3, \mathbf{q}, \mathbf{q}_2 \leftarrow \mathbf{k}_2, \mathbf{q}, \mathbf{q}_1, \mathbf{q}_2 \leftarrow \mathbf{k}_1, \mathbf{q}, \mathbf{q}_1 \leftarrow \mathbf{k}, \mathbf{q}$$
$$\mathbf{k}, \mathbf{q} \leftarrow \mathbf{k}_1, \mathbf{q}, \mathbf{q}_1 \leftarrow \mathbf{k}_2, \mathbf{q}, \mathbf{q}_1, \mathbf{q}_2 \leftarrow \mathbf{k}_1, \mathbf{q}, \mathbf{q}_1 \leftarrow \mathbf{k}, \mathbf{q}$$
$$\mathbf{k}, \mathbf{q} \leftarrow \mathbf{k}_2, \mathbf{q}, \mathbf{q}_2 \leftarrow \mathbf{k}, \mathbf{q} \leftarrow \mathbf{k}_1, \mathbf{q}, \mathbf{q}_1 \leftarrow \mathbf{k}, \mathbf{q}$$

Only the first scheme contributes to the irreducible diagonal part. The second one involves the diagonal part of the subproduct VA_2V, while the third scheme involves the diagonal parts of VA_1V and VA_3V. Similar examples can be worked out for states involving more than one electron. The formal analogy of such calculations with the quantum theory of fields is of course striking. One can in particular introduce a graphical representation of the transition schemes by means of Feynman diagrams, a method which has already been applied to a Fermi gas with interactions by Goldstone [8]) and Hugenholtz [4]).

Before closing we still calculate to lowest order in λ the functions $G_l(\alpha)$ and $W_{ll'}(\alpha'\alpha)$. One finds directly from (A.11)

$$G_l(\mathbf{k}, \mathbf{q}) = \gamma^2 \int [\varepsilon(\mathbf{k} - \mathbf{q}') + \omega(\mathbf{q}) + \omega(\mathbf{q}') - l]^{-1} \omega(\mathbf{q}')\, d\mathbf{q}',$$

$$W_{ll'}(\mathbf{k}', \mathbf{q}_1', \mathbf{q}_2'; \mathbf{k}, \mathbf{q}) = \gamma^2 \omega(\mathbf{q}_1')\, \delta[\mathbf{q}_2' - \mathbf{q}]\, \delta[\mathbf{k}' + \mathbf{q}_1' - \mathbf{k}].$$

The phonon $\mathbf{q}$ is seen to play no role in these functions, except for a shift in l, l'. One can easily calculate such functions to higher order and determine in this way the quantities f_E and w_E which enter the master equation (4.31) for the case at hand.

Received 16-3-57.

REFERENCES

1) Van Hove, L., Physica **21** (1955) 517, here quoted as A.
2) Pauli, W., *Festschrift zum 60. Geburtstage A. Sommerfelds*, Hirzel, Leipzig (1928) p. 30.
3) Part of our results have been presented briefly at the "Colloque International sur les Phénomènes de Transport en Mécanique Statistique" (Bruxelles, sept. 1956). See the Proceedings of this meeting, to appear shortly.
4) Hugenholtz, N. M., Physica **23** (1957), 481 (this issue).
5) Van Hove, L., Physica **21** (1955) 901 and **22** (1956) 343, here quoted as I and II. The first of these papers contains an unfortunate misprint in the statement of what we call here property (iii) of the perturbation: the inequalities on p. 907, line 12 from below, should read as (2.8) in the present paper.
6) Jordan, P., *Statistische Mechanik auf quantentheoretischen Grundlagen*, Vieweg, Braunschweig (1933) p. 25 and foll. For a discussion of various possible choices of assumptions, see Thomsen, J. S., Phys. Rev. **91** (1953) 1263.
7) For the case of finite matrices, one has to use the theory of elementary divisors and the reduction of the matrix to the so-called normal form. For an extension to continuously varying indices. i.e. to integral equations, see for example Goursat, E., *Cours d'Analyse Mathématique*, Gauthier-Villars, t. III, esp. the reduction of an integral equation to canonical form (p. 408 to 412 of the 4th edition).
8) Goldstone, J., Proc. Roy. Soc. **A 239** (1957) 267.

ON THE KINETICS OF THE APPROACH TO EQUILIBRIUM

I. PRIGOGINE and P. RÉSIBOIS *)

Faculté des Sciences de l'Université Libre de Bruxelles, Bruxelles, Belgium.

Résumé

Nous appliquons la technique de diagrammes développée par l'un de nous (I. P.) et R. Balescu, combinée au formalisme de la résolvante, pour établir les équations générales d'évolution d'un gaz classique à partir de l'équation de Liouville.

Nous obtenous des équations valables à tous les ordres dans le paramètre de couplage et dans la concentration, et pour tous les temps. Pour des temps courts, ces équations sont *non-Markoviennes* et tiennent compte de la durée finie d'une collision; dans la limite des temps longs ($t \to \infty$), elles se ramènent aux expressions asymptotiques Markoviennes obtenues précédemment dans l'étude de l'approche vers l'équilibre [2])

Nous discutons rapidement le lien entre cette théorie et le schéma proposé par Bogolioubov. Enfin, nous montrons que, pour le calcul des coefficients de transport le caractère non Markovien de la cinétique de l'approche vers l'équilibre ne joue pas de rôle: ces coefficients sont entièrement déterminés par les sections efficaces asymptotiques.

1. *Introduction.* Recently F. Henin, F. Andrews and the authors[1)2)3)] have established a general H theorem, valid both for classical and quantum mechanics. This theorem permits to show that all reduced properties depending on a finite number of degrees of freedom tend for sufficiently long times to their equilibrium value, as calculated with the canonical distribution:

$$\rho \sim \exp -\beta H \quad (\beta = 1/kT) \tag{1-1}$$

where H is the Hamiltonian of the system including the interactions.

The assumptions which have to be formulated about the phase distribution function at the initial time $t = 0$, in order to insure the validity of this H-theorem will not be discussed here (see specially [7)]). We want however to mention that we are always working in the limit of a large system (volume $\Omega \to \infty$, $N \to \infty$, $N/\Omega = C =$ finite) and that we suppose at time $t = 0$ the two following assumptions to be satisfied:

a) the correlation between two particles vanishes when the distance between these two particles becomes infinite.

---

*) Aspirant au Fonds National de la Recherche Scientifique de Belgique.

b) all intensive properties are well defined quantities in the limit $N \to \infty$, $\Omega \to \infty$, $N/\Omega = C$ finite.

We may notice that the equations of evolution are such that these conditions, if valid at $t = 0$, will be satisfied at any time $t > 0$.

In order to obtain the generalized H-theorem, it is obviously sufficient to study the long time behaviour of the system. Here, on the contrary, we want to derive the form of the kinetic equations describing the evolution of the system for arbitrary short or long times.

An essential feature is then the following: in the case of a strongly coupled system there is no sharp separation between the collision time τ_c and the relaxation time τ_{rel}. For instance if we consider a gas of particles of radius a, mean velocity $\bar{v}$ and concentration C, the relaxation time will be, for purely dimensional reasons, of the form:

$$\tau_{rel}^{-1} = a^2 C \bar{v} [1 + (a^3 C) + (a^3 C)^2 + \dots] \tag{1-2}$$

while the collision time is of the form:

$$\tau_c^{-1} = \frac{\bar{v}}{a}$$

The ratio of these two times is therefore:

$$\tau_c / \tau_{\text{rel}} = a^3 C [1 + (a^3 C) + \dots] \tag{1-3}$$

When we want to retain corrections of order (a^3C) *in the relaxation time* (1-2), *we can no more consider the collisions as instantaneous.* It is precisely for this reason that the analysis published in the papers [1][2][3] applies to the long time behaviour but does *not* describe the kinetics of the approach to equilibrium for short times.

This latter point will be our essential problem here. Our main tools will be the diagram technique we have introduced previously (see [4]) as well as the resolvant formalism (see [2][3]). We restrict ourselves to the case of a classical system but the extension to a quantum system is obvious (see [3]).

In paragraph 2 we analyze the most important topological features of the diagrams which describe the dynamics of the system and we summarize briefly the analytical properties of the contributions associated with the main types of these diagrams (this question is studied in greater detail in [7]).

In § 3 we derive the equations of evolution for the Fourier coefficients of the phase distribution function valid for arbitrary times. These equations are valid to all orders in the coupling constant or the concentration. From the mathematical point of view these equations appear as identities valid for a large system for which the two conditions mentionned at the beginning of this paragraph (no correlation of infinite range, existence of reduced properties) are satisfied at the initial time. No approximations are involved.

A remarkable feature of these equations is their non-markovian character.

The change of the distribution function at a given time depends on its value at earlier times. This "memory" is due to the existence of characteristic times (or lengths) like the duration of a collision τ_c or the range of correlations.

We want to mention that, using the random phase approximation, Van Hove [8]) had derived previously a transport equation of a non-markovian type. It is however not easy to compare Van Hove's equation with ours. We postpone therefore a detailed discussion but we want to make the following two remarks:

a) Van Hove's equation and ours apply to different initial states; from this point of view we want to note that, as Philippot [10]) has shown, the random phase approximation is generally inconsistent with the specification of the energy.

b) our equations apply to *all* Fourier coefficients (or in other words to both diagonal and non diagonal elements of the von Neumann matrix) while Van Hove's equation applies to a rather complicated transform of the diagonal elements of the density matrix.

In § 4, we show that for times long in respect to all characteristic times involved in the diagrams, our general equations of evolution reduce to the Markovian expressions we had derived previously [2]), the only minor difference being that here we take explicitly into account the finite duration of the collisions. We can therefore refer to 2) for the proof of the general H-theorem. An inportant feature of these evolution equations is that for such times the velocity distribution function (or the one particle d.f. for inhomogeneous systems) satisfies a separate equation, independent of the evolution of the correlations.

In § 5, we compare our results with Bogolioubov's theory; we show that after times of the order of τ_c, all quantities which depend on the value of the distribution functions for distances of the order of the intermolecular forces may be expressed in terms of functionals of the one particle distribution function. This as well as the existence of a separate equation for this one particle d.f. are precisely the basic assumptions of Bogolioubov's theory.

Finally we discuss briefly the problem of the macroscopic transport coefficients under time independent conditions (for example the *static* electrical conductivity). We only study a simple example because this question is studied by Balescu [6]) in a separate paper.

We show in this example that such transport coefficients are determined by the asymptotic value of the cross-section and that quantities related to the finite duration of the collision play no role. For the same reason the non Markovian character of the kinetic equations is here irrelevant. However it is no more so when the external perturbation is time dependent.

2. *Formal solution of the Liouville equation*: Let us recall rapidly the basic

formulas we have proved in reference 2 (appendix). Considering a system of N particles enclosed in a box of volume Ω, we write its Hamiltonian:

$$H = H_0 + \lambda V = \sum_i \frac{v_i^2}{2} + \sum_{i>j} V_{ij}(|\boldsymbol{r}_i - \boldsymbol{r}_j|) \qquad (m = 1) \tag{2-1}$$

The Liouville equation for the N particle distribution function is:

$$i\partial_t \rho = i[H, \rho] = L\rho \tag{2-2}$$

where we have introduced the Liouville operator L:

$$L = L_0 + \lambda \delta L = i \sum_j \boldsymbol{v}_j \frac{\partial}{\partial \boldsymbol{r}_j} - i\lambda \sum_j \frac{\partial V}{\partial \boldsymbol{r}_j} \frac{\partial}{\partial \boldsymbol{v}_j} \tag{2-3}$$

The formal solution of equation (2-2) can be written (see 2), formula (A-2)):

$$\rho(t) = \frac{-1}{2\pi i} \oint_c \mathrm{d}z \, \frac{\exp -izt}{L - z} \rho(0) \tag{2-4}$$

where c is the contour indicated in fig. 1 and where:

$$(L - z)^{-1} = \sum_{n=0}^{\infty} (L_0 - z)^{-1}[-\lambda \delta L \, (L_0 - z)^{-1}]^n \tag{2-5}$$

is the resolvant operator of the problem.

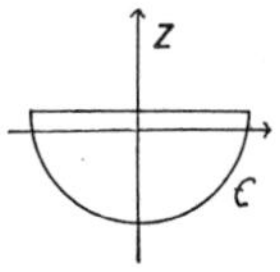

Fig. 1.

We express (2-4) and (2-5) in the Fourier representation $|\{\boldsymbol{k}\}\rangle$ and we get (see ref. 2), formulas (A-6), (A-7) and (2-8)):

$$\rho_{\{\boldsymbol{k}\}}(\{\boldsymbol{v}\}, t) = \frac{-1}{2\pi i} \oint_c \exp -izt \, \mathrm{d}z \times$$

$$\times \sum_{n=0}^{\infty} \sum_{\{\boldsymbol{k}'\}} \langle \{\boldsymbol{k}\} |(L_0 - z)^{-1} [-\lambda \delta L (L_0 - z)^{-1}]^n | \{\boldsymbol{k}'\}\rangle \, \rho_{\{\boldsymbol{k}'\}}(\{\boldsymbol{v}\}, 0) \tag{2-6}$$

where we have:

$$\rho_{\{\boldsymbol{k}\}}(\{\boldsymbol{v}\}, t) = \int \rho(\{\boldsymbol{r}\}, \{\boldsymbol{v}\}, t) \exp -\mathrm{i} \sum_j \boldsymbol{k}_j \boldsymbol{r}_j \, \mathrm{d}^N \boldsymbol{r} \tag{2-7}$$

$$\langle \{\boldsymbol{k}\} |(L_0 - z)^{-1}| \{\boldsymbol{k}'\}\rangle = (\sum_j \boldsymbol{k}_j \boldsymbol{v}_j - z)^{-1} \, \delta^{Kr}_{\{\boldsymbol{k}\}\{\boldsymbol{k}'\}} \tag{2-8}$$

$$\langle \{\boldsymbol{k}\} |\delta L \,|\{\boldsymbol{k}'\}\rangle = \sum_{m>n} \langle \{\boldsymbol{k}\} |\delta L_{nm} | \{\boldsymbol{k}'\}\rangle \tag{2.9}$$

with:

$$\langle\{\boldsymbol{k}\}\,|\delta L_{nm}|\,\{\boldsymbol{k}'\}\rangle = \Omega^{-1} V_{|\boldsymbol{k}_n - \boldsymbol{k}_n'|}(\boldsymbol{k}_n - \boldsymbol{k}_n')\left(\frac{\partial}{\partial \boldsymbol{v}_n} - \frac{\partial}{\partial \boldsymbol{v}_m}\right) \times$$
$$\times\, \delta^{Kr}(\boldsymbol{k}_n + \boldsymbol{k}_m - \boldsymbol{k}_n' - \boldsymbol{k}_m') \prod_{l \neq n,m} \delta^{Kr}(\boldsymbol{k}_l - \boldsymbol{k}_l') \qquad (2\text{-}10)$$

In order to avoid a too heavy notation we have suppressed all volume factor in the definition of the Fourier coefficients $\rho_{\{\boldsymbol{k}\}}$ (see for ex. 2) formula (2-4)). Moreover we shall limit ourselves to homogeneous systems ($\sum_i \boldsymbol{k}_i = 0$)

Using the diagram technique developed by one of us (I.P.) and R. Balescu [4]), we represent each elementary transition $\langle\{\boldsymbol{k}\}\,|\delta L_{nm}|\,\{\boldsymbol{k}'\}\rangle$ by one of the six vertices of fig. 2.

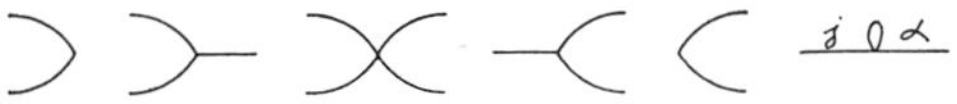

Fig. 2. The six basic vertices.

Combining these different vertices, we associate in the usual way a diagram to each contribution of the series (2-6) and we classify these diagrams according to their topological structure.

Let us introduce what we shall call a *"destruction vertex"* D. We shall give this name to:

a) any vertex involving the interaction of two initially excited lines (fig. 3*a*),

b) the last vertex in a set of transitions where an initially excited particle transfers its total wave number to another particle (fig. 3*b*).

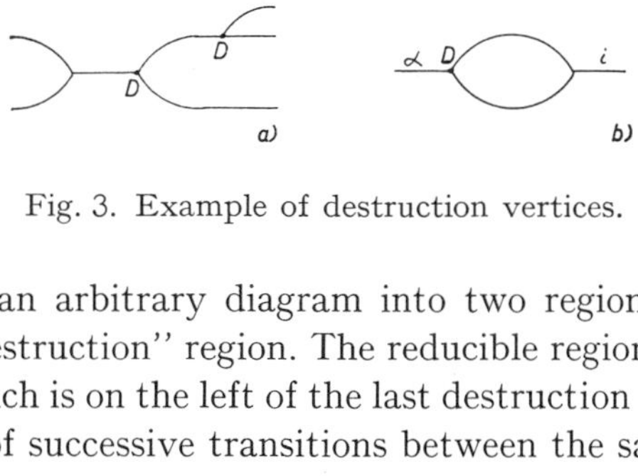

Fig. 3. Example of destruction vertices.

We split then an arbitrary diagram into two regions, the "reducible" region and the "destruction" region. The reducible region (R.R.) is the part of the diagram which is on the left of the last destruction vertex (see fig. 4*a*). However if a set of successive transitions between the same lines occurs at the same time as this last destruction vertex (see fig. 4*b*) it is convenient to absorb it in the destruction region.

The main interest in this splitting is the following: we shall always assume that *the correlations at the initial time are of finite extension in configurational space*; thus, after a finite time, initially correlated particles tend to separate and can thus no more interact with each other. But, by its very definition, the destruction region corresponds to the part of the diagram where these initially correlated particles interact; this region describes thus events which take place during a finite time (a collision time if we assume correlations over molecular distances only) even when we analyze the behaviour of the whole diagram in the limit $t \to \infty$. Only the reducible region can give an asymptotically growing contribution in the limit $t \to \infty$.

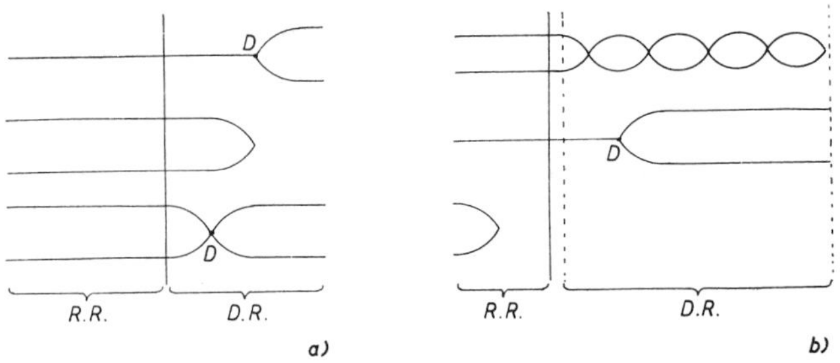

Fig. 4. Examples of reducible and destruction transitions.

Let us analyse the reducible region with some more details; starting from a state $|\{k'\}\rangle$ we shall in general reach the final state $|\{k\}\rangle$ by the following scheme: we have first an arbitrary number of *diagonal transitions* bringing the system to the same intermediate state $|\{k'\}\rangle$ and then if $|\{k'\}\rangle$ is different from $|\{k\}\rangle$ we shall finally have some *creation transition* $|\{k'\}\rangle \to |\{k\}\rangle$. The reason why this last transition is necessarily a creation transition, bringing the system to a state of *higher* correlation $|\{k\}\rangle$, is that we have included all possible destructions in the destruction region. More precisely, if we define

a) *a diagonal fragment* (D.F.): as any set of transitions $|\{k'\}\rangle \to |\{k'\}\rangle$ with no intermediate state identical to $|\{k'\}\rangle$.

b) *a creation fragment* (C.F.): as any set of transition $|\{k'\})\rangle \to |\{k\}\rangle$ with no intermediate state identical to $|\{k'\}\rangle$,

the most general reducible region will be decomposed into:

a) either $m(m \geq 0)$ diagonal fragments if $|\{k'\}\rangle \equiv |\{k\}\rangle$

b) or $m(m \geq 0)$ diagonal framents $|\{k'\}\rangle \to |\{k'\}\rangle$ followed by one creation fragment $|\{k'\}\rangle \to |\{k\}\rangle$ $(|\{k'\}\rangle \neq |\{k\}\rangle)$.

Examples of these situations will be found in *fig.* 5.

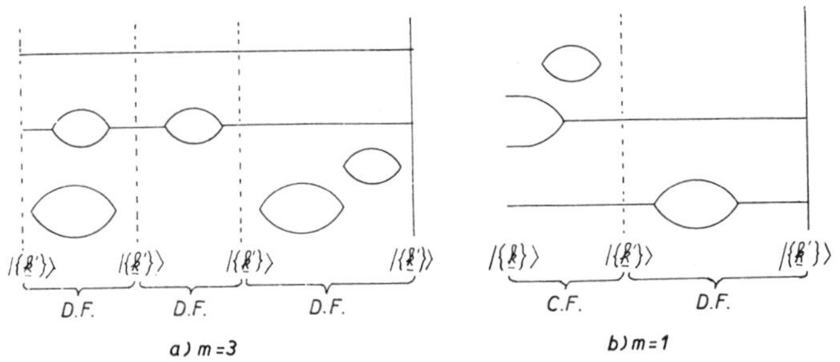

Fig. 5. Examples of reducible regions.

With these definitions, we can now establish very easily the equations of evolution for the distribution function ρ valid for an arbitrary time.

3. *The equations of evolution.* Let us use the following notations:

$$\psi_{\{\boldsymbol{k}\}\{\boldsymbol{k}\}}(z) = \sum_{n=2}^{\infty} (-\lambda)^n \langle\{\boldsymbol{k}\}\,|\delta L \left[\frac{1}{L_0 - z}\,\delta L\right]^n |\,\{\boldsymbol{k}\}\rangle \tag{3-1}$$

$$= \text{sum of all possible diagonal fragments } |\{\boldsymbol{k}\}\rangle \to |\{\boldsymbol{k}\}\rangle \tag{3-1'}$$

$$\tilde{C}_{\{\boldsymbol{k}\}\{\boldsymbol{k}'\}}(z) = \sum_{n=1}^{\infty} (-\lambda)^n \langle\{\boldsymbol{k}\} \left| \left[\frac{1}{L_0 - z}\,\delta L\right]^n \right| \{\boldsymbol{k}'\}\rangle_{C.F.} \tag{3-2}$$

$$= \text{sum of all possible creation fragments } |\{\boldsymbol{k}'\}\rangle \to |\{\boldsymbol{k}\}\rangle \tag{3.2'}$$

$$\mathscr{D}_{\{\boldsymbol{k}'\}\{\boldsymbol{k}''\}}(z) = \sum_{n=1}^{\infty} (-\lambda)^n \langle\{\boldsymbol{k}'\} \left| \left[\delta L\,\frac{1}{L_0 - z}\right]^n \right| \{\boldsymbol{k}''\}\rangle_{\text{D.R.}} \tag{3-3}$$

$$= \text{sum of all possible destruction regions} \tag{3-3'}$$

The few first terms of particular operators ψ, $\tilde{C}$ and $\mathscr{D}$ are indicated diagramatically in fig. 6, 7 and 8.

Fig. 6. The first terms of $\psi_{\{0\}\{0\}}(z)$.

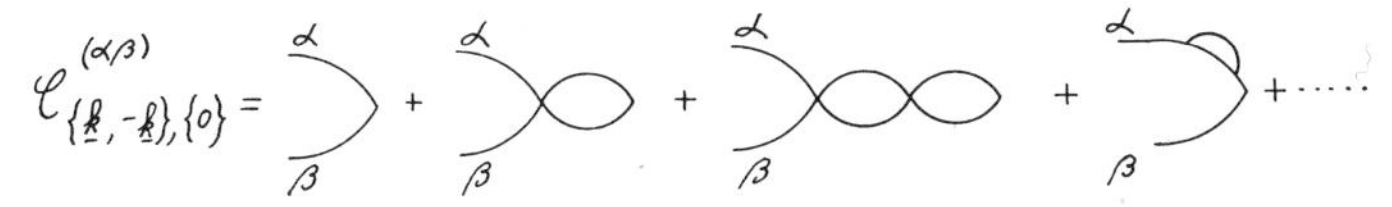

Fig. 7. The first terms of $\tilde{C}^{(\alpha\beta)}_{\{\boldsymbol{k}-\boldsymbol{k}\},\{0\}}(z)$

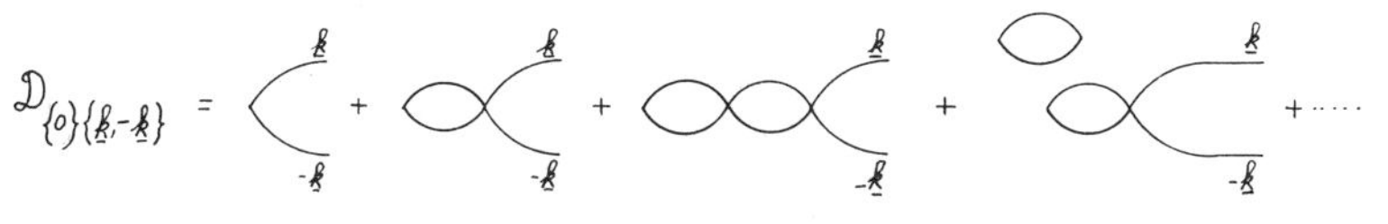

Fig. 8. The first terms of $\mathscr{D}_{\{0\},\{\boldsymbol{k}-\boldsymbol{k}\}}(z)$.

The basic properties we shall use for these quantities are the following:

1) $\psi_{\{\boldsymbol{k}\}\{\boldsymbol{k}\}}(z)$, $\tilde{C}_{\{\boldsymbol{k}\}\{\boldsymbol{k}'\}}(z)$ and $\sum_{\{\boldsymbol{k}''\}} \mathscr{D}_{\{\boldsymbol{k}'\}\{\boldsymbol{k}''\}}(z)\, \rho_{\{\boldsymbol{k}''\}}(\{\boldsymbol{v}\}, 0)$

are analytical functions of z in the whole complex plane, except for a finite discontinuity on the real axis. More exactly we may define two functions of the type $\psi^{+}_{\{\boldsymbol{k}\}\{\boldsymbol{k}\}}(z)$ and $\psi^{-}_{\{\boldsymbol{k}\}\{\boldsymbol{k}\}}(z)$, according to Im $z > 0$ or Im $z < 0$. The first is analytical in the upper half plane and the second in the lower half.

Similarly we may introduce

$$\tilde{C}^{\pm}_{\{k\}\{k'\}}(z) \text{ and } \sum_{\{k''\}} \mathscr{D}^{\pm}_{\{k'\}\{k''\}}(z)\, \rho_{\{k''\}}(\{v\}, 0).$$

2) The functions

$$\tilde{C}^{+}_{\{k\}\{k\}}(z),\ \varphi^{+}_{\{k\}\{k'\}}(z) \text{ and } \Sigma_{\{k\}''}\ \mathscr{D}^{+}_{\{k'\}\{k''\}}(z)\, \rho_{\{k''\}}(\{v\}, 0),$$

analytical for Im $z > 0$, have an analytical continuation in the lower half-plane; moreover we shall always assume that these continuations are regular except for poles at a finite distance from the origin.

The first assertion is an exact mathematical consequence of the definitions (for more details see ref. [2][7]); the second part of the theorem has to be considered as a *sufficient condition* for the validity of the equations of evolution we shall establish later; how this condition is effectively realized in some particular cases has been discussed previously (see ref. 2, app. A; ref. 3, app. A5); however, this condition is not necessary; for instance, R. Balescu [11] has discussed the case where ψ^+ has logarithmic branch-points on the real or on the negative imaginary axis. In general, the analytical properties of ψ^+, $\tilde{C}^+$ and $\mathscr{D}^+$ will depend on the law of interaction $V(\boldsymbol{r})$ and it is very difficult to state their most general behaviour. We shall thus limit ourselves here to the simplest case where condition 2) is fulfilled.

Let us now write the formal solution for the velocity distribution function $\rho_0(\{v\}, t)$: as there is no possible creation fragment ending with $\{k\} = 0$, the most general contribution to $\rho_0(\{v\}, t)$ is due to a destruction region followed by m diagonal fragments (see an example on fig. 9).

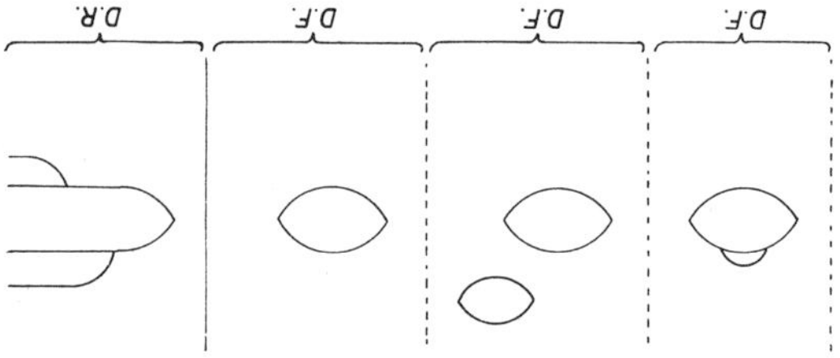

Fig. 9. A particular contributions to $\rho_0(\{v\}, t)$.

We have thus:

$$\rho_0(\{v\},t) = \frac{-1}{2\pi i} \oint_c dz \exp - izt \sum_{n=0}^{\infty} \frac{1}{-z} \left[\psi^{+}_{00}(z) \frac{1}{-z}\right]^n [\rho_0(\{v\}, 0) +$$

$$+ \Sigma_{\{k''\}\neq 0}\ \mathscr{D}^{+}_{0\{k''\}}(z)\, \rho_{\{k''\}}(\{v\}, 0)] \qquad (3\text{-}4)$$

$$= \frac{-1}{2\pi i} \oint_c dz \exp -izt\, \tilde{\rho}_0(z) \qquad (3\text{-}5)$$

where the last formula is a definition for the Laplace transform of $\rho_0(t)$.

Let us also define an operator:

$$D^+_{00}(z) = \sum_{n=0}^{\infty} \frac{1}{-z} \left[\psi^+_{00}(z) \frac{1}{-z} \right]^n \tag{3-6}$$

which obviously obeys the equation:

$$D^+_{00}(z) = \frac{1}{-z} + \frac{1}{-z} \psi^+_{00}(z) D^+_{00}(z) \tag{3-7}$$

Putting (3-6) and (3-7) in (3-4) and differentiating with respect to t, we get:

$$\partial_t \rho_0(\{\boldsymbol{v}\}, t) = \frac{-1}{2\pi} \oint_c \mathrm{d}z \exp -izt \sum_{\{\boldsymbol{k}''\} \neq 0} \mathscr{D}^+_{0,\{\boldsymbol{k}''\}}(z) \, \rho_{\{\boldsymbol{k}''\}}(\{\boldsymbol{v}\}, 0) +$$

$$+ \frac{-1}{2\pi} \oint_c \mathrm{d}z \exp -izt \, \psi^+_{00}(z) D^+_{00}(z) [\rho_0(\{\boldsymbol{v}\}, 0) + \sum_{\{\boldsymbol{k}''\} \neq 0} \mathscr{D}^+_{0,\{\boldsymbol{k}''\}} \rho_{\{\boldsymbol{k}''\}}(\{\boldsymbol{v}\}, 0)] \tag{3-8}$$

We define a time dependent functional of the initial conditions given by the following Laplace transform:

$$\mathscr{D}_0(t, \rho_{\{k''\}}(\{\boldsymbol{v}\}, 0)) = \frac{-1}{2\pi} \oint_c \exp -izt \sum_{\{\boldsymbol{k}''\} \neq 0} \mathscr{D}^+_{0\{\boldsymbol{k}''\}}(z) \, \rho_{\{\boldsymbol{k}''\}}(\{\boldsymbol{v}\}, 0) \tag{3-9}$$

and also the Laplace transform of the diagonal fragment:

$$G_{00}(\tau) = \frac{+1}{2\pi i} \oint_c \exp -iz\tau \, \psi^+_{00}(z) \, \mathrm{d}z. \tag{3-10}$$

Applying the convolution theorem of Laplace transform to the second term of the r.h.s. of (3-8), we then obtain

$$\partial_t \rho_0(\{\boldsymbol{v}\}, t) = \mathscr{D}_0(t, \rho_{\{\boldsymbol{k}''\}}(\{\boldsymbol{v}\}, 0)) + \int_0^t G_{00}(t - t') \, \rho_0(\{\boldsymbol{v}\}, t') \, \mathrm{d}t' \tag{3-11}$$

This generalized "master equation" describes the *exact behaviour* of the velocity distribution function *for any time*; its physical interpretation is the following:

the inhomogeneous term gives the contribution, at time t, of the initially excited Fourier components which interact with each other in order to give a state without correlation; the second term, which is *non-Markovian*, expresses the fact that a collision process is in general a non instantaneous event and connects thus the distribution function at a given time t to the same function at previous times.

This very general result is valid whatever the initial correlations are; it allows to study exactly strongly coupled systems, where the "random-phase approximation" (i.e. the assumption $\rho_{\{\boldsymbol{k}''\}} = 0$ $\{\boldsymbol{k}''\} \neq 0$) is certainly not generally valid (see Philippot [10])).

The non-diagonal Fourier components $\rho_{\{k\}}(\{v\}, t)$ can be studied similarly; however we shall have to retain the possibility of creation fragments; for this reason, we split $\rho_{\{k\}}(\{v\}, t)$ into two parts:

$$\rho_{\{k\}}(\{v\}, t) = \rho'_{\{k\}}(\{v\}, t) + \rho''_{\{k\}}(\{v\}, t) \tag{3-12}$$

where:

$$\rho'_{\{k\}}(\{v\}, t) = \frac{-1}{2\pi i} \oint_c dz \exp -izt \sum_{n=0}^{\infty} \frac{1}{\sum_j k_j v_j - z} \left[\psi^+_{\{k\}\{k\}}(z) \frac{1}{\sum_j k_j v_j - z} \right]^n \times$$
$$\times \left[\rho_{\{k\}}(\{v\}, 0) + \sum_{\{k''\}} \mathscr{D}^+_{\{k\}\{k''\}}(z)\, \rho_{\{k''\}}(\{v\}, 0)\right] \tag{3-13}$$
$$\equiv \frac{-1}{2\pi i} \oint_c dz \exp -izt\, \tilde{\rho}'_{\{k'\}}(z) \tag{3-14}$$

contains all diagrams with no creation fragment (see fig. 10*a*), whereas:

$$\rho''_{\{k\}}(\{v\}, t) = \sum_{\{k'\}} \frac{-1}{2\pi i} \oint_c dz \cdot \exp -izt\, \tilde{C}_{\{k\}\{k'\}}(z) \sum_{n=0}^{\infty} \frac{1}{\sum_j k'_j v_j - z} \times$$
$$\times \left[\psi^+_{\{k'\}\{k'\}}(z) \frac{1}{\sum_j k'_j v_j - z} \right]^n \left[\rho_{\{k'\}}(\{v\}, 0) + \sum_{\{k''\}} \mathscr{D}^+_{\{k'\}\{k''\}}(z) \rho_{\{k''\}}(\{v\}, 0)\right] \tag{3-15}$$

includes the remaining contributions, with a creation fragment at the left (see fig. 10*b*)

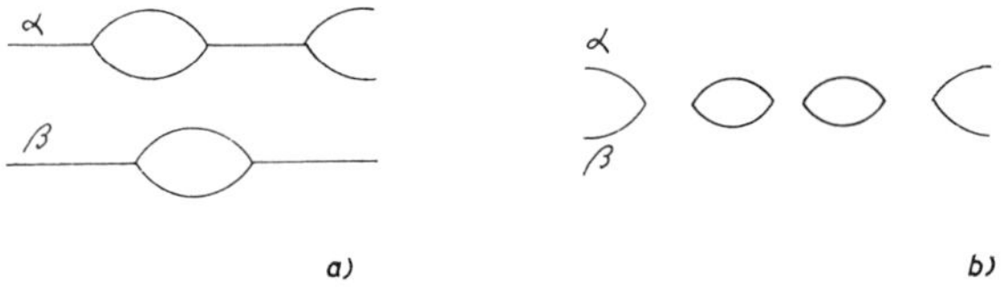

Fig. 10. Examples of contribution to ρ'_{k-k} and ρ''_{k-k}.

The first part, given by (3-13), can be treated as $\rho_0(\{v\}, t)$; we obtain finally:

$$\partial_t \rho'_{\{k\}}\{v\}, t) + i \sum_j k_j v_j \rho'_{\{k\}}(\{v\}, t) = \mathscr{D}^+_{\{k\}}(t, \rho_{\{k''\}}(\{v\}, 0)) +$$
$$+ \int_0^t G^+_{\{k\}\{k\}}(t - t')\, \rho'_{\{k\}}(t')\, dt'. \tag{3-16}$$

where the operators $\mathscr{D}^+_{\{k\}}$ and $G^+_{\{k\}\{k\}}$ are defined by expressions similar to (3-9) and (3-10) with $\{k\}$ replacing $\mathbf{0}$. Moreover, comparing (3-15) with (3-13) and (3-14), we see that:

$$\rho''_{\{k\}}(\{v\}, t) = \frac{-1}{2\pi i} \sum_{\{k'\}} \oint_c dz \exp -izt\, \tilde{C}_{\{k\}\{k'\}}(z)\, \tilde{\rho}'_{\{k'\}}(z) \tag{3-17}$$

Defining a time dependant creation operator:

$$C_{\{k\}\{k'\}}(\tau) = \frac{-1}{2\pi} \oint_c dz \cdot \exp -iz\tau \, \tilde{C}_{\{k\}\{k'\}}(z) \tag{3-18}$$

we easily prove that:

$$\rho''_{\{k\}}(\{v\}, t) = \sum_{\{k'\}} \int_0^t C_{\{k\}\{k'\}}(t - t') \, \rho'_{\{k'\}}(t') \, dt'. \tag{3-19}$$

Equations (3-16) and (3-19) give the exact time evolution of the non diagonal elements of the distribution function. The first one (3-16) can be interpreted exactly as (3-11): it describes the dissipation of the initial correlations by non instantaneous collisions with initially uncorrelated particles; the second one (3-19) corresponds to the continuous creation of "fresh" correlations by direct mechanical interaction of less excited states; moreover, these new correlations tend also to dissipate, which is expressed mathematically by an immediate consequence of our basic theorem:

$$C_{\{k\}\{k'\}}(\tau) \underset{\tau \to \infty}{\to} 0 \tag{3-20}$$

In the same way we have also:

$$G_{\{k\}\{k\}}(\tau) \underset{\tau \to \infty}{\to} 0 \tag{3-21}$$

$$\mathscr{D}_{\{k\}}(\tau) \underset{\tau \to \infty}{\to} 0 \tag{3-22}$$

Formula (3-21) expresses the fact that the duration of a collision is always finite while (3-22) has the following physical meaning: with our choice of initial conditions, where the correlations are of finite extension in configurational space, initially excited states can only interact during a finite time, as we explained in paragraph 2).

4. *Long time equations.* The equations of evolution we have studied in the last paragraph are valid for any time t: however they simplify considerably in the limit where $t \to \infty$. It would be possible to analyze this long time behaviour directly on the formulas (3-11) (3-15) and (3-19); however, it is much simpler to study the starting equations (3-4) (3-13) and (3-15). As the method is completely similar to the analysis given previously for the proof of an H-theorem [1][2][3], we shall be very brief and limit ourselves to the evolution of the velocity distribution function $\rho_0(\{v\}, t)$.

As is well-known, the long time behaviour of $\rho_0(\{v\}, t)$ is entirely determined by the behaviour of its Laplace transform $\tilde{\rho}(z)$ near the real axis. Using our theorem concerning the analytical properties of $\psi^+_{00}(z)$ and $\sum_{\{k\}} \mathscr{D}^+_{0\{k\}}(z) \, \rho_{\{k\}}(\{v\})$ as functions of z, we infer immediately that the only relevant contributions in a term of (3-4) containing n diagonal fragments come from the $(n+1)$-th order pole at $z = 0$. Applying the residue theorem, we get:

$$\rho_0(\{\boldsymbol{v}\}, t) = -\sum_{n=0}^{\infty} \frac{(+it)^n}{n!} \sum_{u=0}^{\infty} \frac{1}{u!} \cdot$$
$$\cdot \left\{ \frac{\partial}{\partial z^u} (\psi_{00}^+(z))^{n+u} [\rho_0(\{\boldsymbol{v}\}, 0) + \sum_{\{k''\}} \mathscr{D}_{0\{k''\}}(z)\, \rho_{\{k''\}} [(\{\boldsymbol{v}\}, 0)] \right\}_{z=+i0}$$

Differentiating with respect to t:

$$\partial_t \rho_0(\{\boldsymbol{v}\}, t) \underset{t\to\infty}{=} +\frac{1}{i} \sum_{n=0}^{\infty} \frac{(+it)^n}{n!} \sum_{k=0}^{\infty} \frac{1}{u!} \cdot$$
$$\cdot \left\{ \frac{\partial}{\partial z^u} [(\psi_{00}^+(z))^{n+u+1} [\rho_0(\{\boldsymbol{v}\}, 0) + \sum_{\{k''\}} \mathscr{D}_{0\{k''\}}(z)\, \rho_{\{k''\}}(\{\boldsymbol{v}\}, 0)] \right\}_{z=+i0}. \tag{4-2}$$

We use the following identity:

$$\frac{1}{u!} \left\{ \frac{\partial}{\partial z^u} (\psi_{00}^+(z))^{n+u+1} [\rho_0(\{\boldsymbol{v}\}, 0) + \sum_{\{k''\}} \mathscr{D}_{0\{k''\}}^+(z)\, \rho_{\{k''\}}(\{\boldsymbol{v}\}, 0)] \right\}_{z=+i0} =$$
$$= \sum_{\alpha=0}^{u} {}^{(0)}\Omega_{u-\alpha}(+i0)\, \psi_{00}^+(+i0) \frac{1}{\alpha!} \left\{ \frac{\partial}{\partial z^\alpha} [(\psi_{00}^+(z))^{n+\alpha} [\rho_0(\{\boldsymbol{v}\}, 0) + \right.$$
$$\left. + \sum_{\{k''\}} \mathscr{D}_{0\{k''\}}^+(z)\, \rho_{\{k''\}}(\{\boldsymbol{v}\}, 0)]] \right\}_{+i0} \tag{4-3}$$

where ${}^{(0)}\Omega_j(+i0)$ are operators defined by the following recurrence relations:

$${}^{(0)}\Omega_0 = 1 \tag{4-4}$$

$${}^{(0)}\Omega_j(z) = \frac{1}{j} \sum_{i=0}^{j-1} \left\{ \frac{\partial}{\partial z} {}^{(0)}\Omega_{j-1-i}(z)\, \psi_{00}^+(z) \right\} {}^{(0)}\Omega_i(z) \tag{4-5}$$

We shall not prove these formulas here [7]; the calculations can be done by recurrence and involve a few algebraic manipulations.

We put (4-3) into (4-2) and rearrange the dummy variables u and α; it is then very easy to obtain the following differential (Markovian) equation:

$$\partial_t \rho_0(\{\boldsymbol{v}\}, t) \underset{t\to\infty}{=} {}^{(0)}\Omega(+i0)\, \psi_{00}^+(+i0)\, \rho_0(\{\boldsymbol{v}\}, t) \tag{4-6}$$

with

$${}^{(0)}\Omega(z) = \sum_{l=0}^{\infty} {}^{(0)}\Omega_l(z) \tag{4-7}$$

This result is the analog of formula (8-1) of reference 2, written down in a more compact way; the only minor difference is the introduction of the operator ${}^{(0)}\Omega$, which takes explicity into account the finite duration of a collision. However, the proof of the H-theorem is unaffected by this operator because it has been shown in ref. 2 that:

$$\psi_{00}(+i0) \exp -\beta H_0 = 0 \tag{4-8}$$

Then, a fortiori:

$$^{(0)}\Omega(+i0)\ \psi_{00}(+i0) \exp -\beta H_0 = 0 \tag{4-9}$$

We could extend without difficulty the result we have obtained here to the non diagonal elements of the distribution function $\rho(\{\boldsymbol{r}\}, \{\boldsymbol{v}\}, t)$; as no new characteristic feature would appear, we shall omit these calculations here (see 7)).

5. *Discussion of Bogolioubov's theory*: In his well-known work [5]) Bogolioubov has derived a general theory of irreversible processes in gases, using the following two assumptions:

1°) After a time of the order of τ_c, the velocity distribution function obeys a closed equation, which we may write symbolically:

$$\partial_t \rho_0(\{\boldsymbol{v}\}, t) = O_{00}(\rho_0(\{\boldsymbol{v}\}, t)) \tag{5-1}$$

2°) After times of the same order, all the correlation functions are functionals of ρ_0:

$$\rho_{\{k\}}(\{\boldsymbol{v}\}, t) = O_{\{k\}0}(\rho_0(\{\boldsymbol{v}\}, t)) \tag{5-2}$$

For non homogeneous systems, the velocity distribution would be replaced in (5-1) (5-2) by the one particle distribution function (see 7)).

We shall show here that, under special conditions, the general equations (3-11) (3-12) (3-19) are indeed of the form (5-1) (5-2).

Let us first assume that *all initial correlations are of molecular origin.* This condition, which is much more restrictive than the assumption that the correlations have a finite extension in configurational space, has for consequence that equation (3-22) is satisfied for times larger than a collision time τ_c. Indeed, if the particles are only correlated on a molecular scale, after a time τ_c they tend to separate off and thus can not interact any more with each other (see fig. 11).

We have then (see (3-11)):

$$\partial_t \rho_0(\{\boldsymbol{v}\}, t) = \int_0^t G_{00}(t - t')\ \rho_0(t')\ \mathrm{d}t' \qquad t > \tau_c \tag{5-3}$$

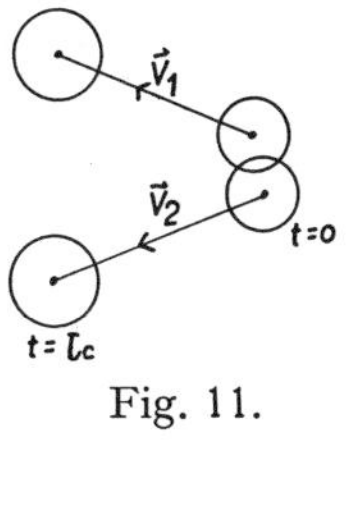

Fig. 11.

Moreover (see (3-12) (3-19):

$$\partial_t \rho'_{\{k\}}(\{\boldsymbol{v}\}, t) + i \sum_j \boldsymbol{k}_j \boldsymbol{v}_j \, \rho'_{\{k\}}(\{\boldsymbol{v}\}, t) = \int_0^t G_{\{k\}\{k\}}(t-t')\, \rho'_{\{k\}}(t')\, \mathrm{d}t' \qquad (5\text{-}4)$$

$$t > \tau_c$$

$$\rho''_{\{k\}}(\{\boldsymbol{v}\}, t) = \sum_{\{k'\}} \int_0^t C_{\{k\}\{k'\}}(t-t')\, \rho'_{\{k'\}}(t')\, \mathrm{d}t'. \qquad (5\text{-}5)$$

Equation (5-3) is already of the general form (5-1).

In order to express the correlation functions as a functional of ρ_0, we have to make one supplementary assumption: let us suppose that we are only interested in *"thermodynamical quantities"*, i.e. averages of quantities decaying rapidly to zero when the molecular distance increases.

To be more specific, let us consider the case of the average potential energy per particle:

$$\bar{V}(t) = \int \mathrm{d}\boldsymbol{r}_{12} V(|r_{12}|)\, f(|\boldsymbol{r}_{12}|, t) \qquad (5\text{-}6)$$

where $f(|\boldsymbol{r}_{12}|, t)$ is the 2 particle distribution function.

Expressing (5-6) in terms of Fourier transforms we obtain:

$$\bar{V}(t) = \int \mathrm{d}^{3N}\boldsymbol{v} \int \mathrm{d}^3\boldsymbol{k} V_{-\boldsymbol{k}}[\rho'^{(12)}_{\boldsymbol{k}-\boldsymbol{k}}(\{\boldsymbol{v}\}, t) + \rho''^{(12)}_{\boldsymbol{k}-\boldsymbol{k}}(\{\boldsymbol{v}\}, t)] \qquad (5\text{-}7)$$

However, it is very easy to verify on (5-7) that the term in ρ' cannot give any contribution for $t > \tau_c$. Indeed we may write:

$$\rho'^{(12)}_{\boldsymbol{k}-\boldsymbol{k}}(\{\boldsymbol{v}\}, t) = \exp -i\boldsymbol{k}\boldsymbol{v}_{12}t\; \bar{\rho}'^{(12)}_{\boldsymbol{k}-\boldsymbol{k}}(\{\boldsymbol{v}\}, t) \qquad (5\text{-}8)$$

i.e. we express the Fourier coefficient ρ' in interaction representation, where $\bar{\rho}'$ is a slowly varying function of time which in only modified by collisions.

We have thus, for the first term of (5-7):

$$\bar{V}'(t) = \int \mathrm{d}^{3N}\boldsymbol{v} \int \mathrm{d}^3\boldsymbol{k}\; V_{-\boldsymbol{k}}\, \bar{\rho}'^{(12)}_{\boldsymbol{k}-\boldsymbol{k}}(\{\boldsymbol{v}\}, t) \exp -i\boldsymbol{k}\boldsymbol{v}_{12}t \qquad (5\text{-}9)$$

As the factor $V_{-\boldsymbol{k}}\, \bar{\rho}^{(12)}_{\boldsymbol{k}-\boldsymbol{k}}$ is maximum when $|k|^{-1} \simeq \kappa^{-1} = a$, characteristic length of the order of the range of the forces, this integral will be vanishingly small for times

$$t > \kappa v = \tau_c \qquad (5\text{-}10)$$

This result may be understood as a special case of the relations between the spectral extension of a wavepacket and its life time (here $\nu = kv_{12}$)

$$\Delta\nu \Delta t \simeq 1$$

Thus, due simply to the free propagation of the correlations, we see that $\rho'_{\{k\}}(t)$ does not contribute to the computation of "thermodynamical quantities" for $t > \tau_c$.

For the same reason, it would be easy to verify that the only part of

$\rho''_{\{k\}}(t)$ which gives a non negligible contribution in the limit $t > \tau_c$ is the one corresponding to creation of correlations from the state $\{\boldsymbol{k}\} = \boldsymbol{0}$.

In other words, in the "thermodynamical case" we may replace the set of equations (5-3) (5-4) (5-5) by the simpler system:

$$\partial_t \rho_0(\{\boldsymbol{v}\}, t) = \int_0^t G_{00}(t - t')\, \rho_0(\{\boldsymbol{v}\}, t')\, \mathrm{d}t' \qquad t > \tau_c \tag{5-11}$$

$$\rho'_{\{k\}}(\{\boldsymbol{v}\}, t) = 0 \qquad t > \tau_c \tag{5-12}$$

$$\rho''_{\{k\}}(\{\boldsymbol{v}\}, t) = \int_0^t C_{\{k\},0}\,(t - t')\, \rho_0(\{\boldsymbol{v}\}, t')\, \mathrm{d}t'. \qquad t > \tau_c \tag{5-13}$$

This system is precisely of the general form (5-1) (5-2) postulated by Bogolioubov.

6. *Transport coefficients*: Let us consider a very simple model in which the cycle (λ^2 order) is the only allowed transition; moreover, we do *not* make the usual assumptions $\tau_c \ll \tau_{\text{rel.}}$, so that the general equations of evolution remain non-Markovian. We have (see (3-11)): *)

$$\partial_t \rho_0(\{\boldsymbol{v}\}, t)|_{\text{coll.}} = \lambda^2 \int_0^t G_{00}(t - t')\, \rho_0(t')\, \mathrm{d}t' \tag{6-1}$$

where we have explicitely indicated that the change in time is due to collisions only (no external forces).

To order λ^2, we have:

$$G_{00}(\tau) = \frac{+1}{2\pi i} \oint_c \exp -iz\tau\, \psi_{00}^{+(2)}(z)\, \mathrm{d}z \tag{6-2}$$

$$\psi_{00}^{+(2)}(z) = \langle \boldsymbol{0}\, | \delta L \frac{1}{L_0 - z} \delta L |\, \boldsymbol{0} \rangle \tag{6-3}$$

Let us analyze the long time solution of (6-1); when t goes to infinity, we may write (6-1) as (see (3-21)):

$$\partial_t \rho_0(\{\boldsymbol{v}\}, t)|_{\text{coll.}} = \lambda^2 \int_0^\infty G_{00}(\tau)\, \rho_0(t - \tau)\, \mathrm{d}\tau \qquad t \to \infty \tag{6-4}$$

We expand $\rho_0(t - \tau)$:

$$\rho_0(t - \tau) = \rho_0(t) - \tau \left. \frac{\partial \rho_0}{\partial t} \right|_{\text{coll.}} + \frac{\tau^2}{2!} \left. \frac{\partial^2 \rho_0}{\partial t^2} \right|_{\text{coll.}} + \ldots \tag{6-5}$$

*) This equation, valid up to order λ^2, has been first derived by Dr. J. Philippot (private communication).

and we get thus:

$$\partial_t \rho_0(\{\boldsymbol{v}\}, t)|_{\text{coll.}} = \lambda^2 \int_0^\infty G_{00}(\tau)\, \mathrm{d}\tau\, \rho_0(t) -$$

$$-\lambda^2 \int_0^\infty G_{00}(\tau)\, \tau\, \mathrm{d}\tau \frac{\partial \rho_0(t)}{\partial t}\bigg|_{\text{coll.}} + \lambda^2 \int_0^\infty G_{00}(\tau) \frac{\tau^2}{2!}\, \mathrm{d}\tau \frac{\partial^2 \rho_0}{\partial t^2}\bigg|_{\text{coll.}} + \ldots \tag{6-6}$$

To the first order in λ^2, we obtain:

$$\begin{aligned} \partial_t \rho_0(\{\boldsymbol{v}\}, t) &= \lambda^2 \int_0^\infty G_{00}(\tau)\, \mathrm{d}\tau\, \rho_0(t) \\ &= i\lambda^2 \psi_{00}^{+(2)}(+i0)\, \rho_0(t) \end{aligned} \tag{6-7}$$

as is readily verified by performing the time integral in (6-2).

To order λ^4, we have to take into account the first correction term:

$$\partial_t \rho_0(\{\boldsymbol{v}\}, t) = i\lambda^2 [\psi_{00}^{+(2)}(+i0) + \lambda^2 \int_0^\infty G_{00}(\tau)\, \tau\, \mathrm{d}\tau\, \psi_{00}^{+(2)}(+i0)]\, \rho_0(\{\boldsymbol{v}\}, t) \tag{6-8}$$

But we have

$$\int_0^\infty G_{00}(\tau)\, \tau\, \mathrm{d}\tau = \frac{\partial}{\partial z} \psi_{00}^{+}(z)|_{z=+i0} \tag{6-9}$$

More generally one can prove by successive approximations that:

$$\begin{aligned} \partial_t \rho_0(\{\boldsymbol{v}\}, t) &= i\left[1 + \frac{\partial}{\partial z} \psi_{00}^{+(2)}(z)|_{z=+i0} + \ldots\right] \psi_{00}^{+(2)}(+i0)\, \rho_0(\{\boldsymbol{v}\}, t) \\ &= i^{(0)}\Omega^{(2)}\, \psi_{00}^{+(2)}(+i0)\, \rho_0(\{\boldsymbol{v}\}, t) \end{aligned} \tag{6-10}$$

which is a special case of (4-6) when we retain only the second order cycle. This method of derivation shows clearly that *this operator* ${}^{(0)}\Omega$ *only appears as a consequence of the non-stationary character of the distribution function.*

We expect therefore that in a transport problem, where we look for a stationary solution of the distribution function in an time-independent externed field, these corrections will disappear. It is easy to verify that this is verified in the case of electrical conductivity; indeed, supposing charged particles in an external field $\boldsymbol{E}$, the distribution function is affected by a flow term:

$$\frac{\partial \rho_0}{\partial t}\bigg|_{\text{flow}} = e\boldsymbol{E} \sum_{j=1}^{N} \frac{\partial \rho_0}{\partial \boldsymbol{v}_j} \tag{6-11}$$

A *stationary* state will be characterized by:

$$\frac{\partial \rho_0}{\partial t}\bigg|_{\text{flow}} = \frac{\partial \rho_0}{\partial t}\bigg|_{\text{coll.}}$$

or

$$eE \sum_j \frac{\partial \rho_0(\{\boldsymbol{v}\})}{\partial \boldsymbol{v}_j} = \int_0^\infty G_{00}(\tau)\,\mathrm{d}\tau\,\rho_0(\{\boldsymbol{v}\}) = i\psi_{00}^{+(2)}(+i0)\,\rho_0(\{\boldsymbol{v}\}) \tag{6-13}$$

Linearizing this equation:

$$\rho_0 = \rho_0^{\mathrm{eq}} + \rho_0^{(1)} \qquad \left(\rho_0^{\mathrm{eq}} = \frac{\exp -\beta H_0}{\int \exp -\beta H_0\,\mathrm{d}^N\boldsymbol{v}}\right) \tag{6-14}$$

we get, to the first order in the field:

$$eE \sum_j \frac{\partial \rho_0^{\mathrm{eq}}(\{\boldsymbol{v}\})}{\partial \boldsymbol{v}_j} = i\psi_{00}^{+(2)}\,(+i0)\,\rho_0^{(1)}(\{\boldsymbol{v}\}) \tag{6-15}$$

The solution $\rho_0^{(1)}$ of this equation allows to compute the stationary current and thus also the electrical conductivity. We see that it is only the collision operator ψ_{00}^{+} which plays a role and that we have no corrections Ω due to "refinements of time integration", even though we have nowhere made the assumption $\tau_{\mathrm{rel}} \gg \tau_c$; in a stationary problem, the finite duration of the collision plays no role.

This rather remarkable result has also been obtained by R. Balescu for more general situations and is discussed in a separate paper 6).

7. *Conclusions*: We see as the consequence of our discussion that the form of the kinetic equations depends essentially on the time scale:

1) In the general case the kinetic equations are of the non-Markovian type described § 3;

2) For times long in respect to all characteristic times involved in the diagrams, they reduce to the Markovian equations of § 4.

3) For quasistationary situations as they are realized for example in a static external field, only the asymptotic form of the cross sections is relevant, all terms depending on the finite duration of the collision disappear from the transport equations.

These conclusions show in a striking way the direct link which exists between the time scale and the "precision" of the molecular description which is required. The long time description involves only a solution of the mechanical aspects equivalent to an asymptotic S matrix formalism. On the contrary the short time description involves a complete solution and therefore the specification of the resolvant in the whole complex plane.

Acknowledgements. We want to thank Prof. L. Van Hove for a stimulating discussion.

Received 13-2-61.

REFERENCES

1) Prigogine, I. and Henin, F., J. math. Phys. **1** (1960) 349.
2) Henin, F., Résibois, P., and Andrews, F., J. math. Phys. **2** (1961) 68.
3) Résibois, P., Phys. Rev. Letters **5** (1960) 411; Physica (to appear 1961).
4) Prigogine, I. and Balescu, R., Physica **25** (1959) 281, 302.
5) Bogolioubov, N., *Problems of a Dynamical Theory in Statistical Physics* (in russian) Moscow (1947).
6) Balescu, R., Physica, **27** (1961) 693.
7) Prigogine, I., *Non Equilibrium Statistical Mechanics* (Interscience; to appear 1961).
8) Van Hove, L., Physica **23** (1957) 441.
9) Prigogine, I. and Balescu, R., Physica **26** (1960) 145.
10) Philippot, J., Physica **27** (to appear 1961).
11) Balescu, R., *Statistical Mechanics of Charged Particles.* (Interscience, New York; to appear (1961).

STATISTICAL MECHANICS OF IRREVERSIBILITY*

Robert W. Zwanzig
National Bureau of Standards
Washington, D. C.

Introduction

These lectures have two objectives. One is to explain in an elementary way some recent advances in the theory of irreversibility. The other is to describe several mathematical ideas and devices that are useful in this subject.

The recent advances to which we refer are due mainly to Van Hove[1,2] and to Prigogine and his co-workers.[3,4] Van Hove and Prigogine have developed an elaborate mathematical technique for following the temporal evolution of initial states of many-body systems. They use an infinite order perturbation theory, along with various prescriptions, based on physical arguments, for keeping specific terms in the perturbation expansion, and leaving out the rest.

The idea of starting out with a precisely, (although not necessarily completely), defined initial state is useful and attractive. It distinguishes the work of Van Hove and the Prigogine school from most earlier work in this field. (Kubo's theory of linear transport processes,[5] discussed in this volume by Montroll (p. 221), is in the same spirit.)

However, the infinite order, many-body perturbation theory often seems to be more complicated than is really needed. One has to pay close attention to the details of tedious combinatorial and topological (or "diagrammatic") arguments; this can obscure fundamental questions of a more physical kind.

We shall show in these lectures that much of the combinatorial topology can be avoided entirely by analytic methods that are at least as clear, and considerably more compact. These methods are presented in Sections II, III and IV.

A significant difference between the work of Prigogine and that of Van Hove is that Prigogine studied mainly the classical ensemble distribution function and the classical Liouville equation, while Van Hove was concerned with the quantum mechanical wave function and the Schroedinger equation. Van Hove had to deal with the additional formal complication, not occurring in classical statistical mechanics, that average values of observables are bilinear in the wave function.

In these lectures we show how this extra complication can be

* Presented at the THEORETICAL PHYSICS INSTITUTE, University of Colorado, Summer 1960

avoided by using the density matrix. We thereby display a striking formal similarity between weak-coupling problems in classical and quantum statistical mechanics.

Section VI contains two examples of our general method. The first is a new derivation of the Pauli equation, and the second is a derivation of a classical analogue, due originally to Prigogine and Brout.

Some new material on the evaluation of time correlation functions is presented in Section VII.

From the point of view of fundamental theory of irreversibility, Section V is perhaps the most important in these lectures. This section has to do with the time dependence of physical quantities, and with some mathematical methods that are pertinent. We work out a special prototype example in great detail, with emphasis on asymptotic time dependence, and on the effect of the finite size of a real physical system.

I. The Dynamical Problem

In both classical and quantum mechanics, dynamical problems can be formulated in terms of operators in Hilbert space. Because this is a powerful method, and because it has universal applicability, we shall discuss it first.

The Hilbert space formulation of quantum mechanics has been common knowledge for a long time; the corresponding formulation of classical mechanics has existed almost as long, but is not nearly so widely known. In this section I shall mention also the relatively unfamiliar formulation of quantum mechanics in terms of the Hilbert space of operators (rather than of states); this should be quite useful in quantum statistical mechanics.

In all cases we shall find that equations of motion can be written in the simple form

$$i \frac{dx}{dt} = Lx, \tag{I.1}$$

where x is a vector in Hilbert space and L is a Hermitian operator.

In classical mechanics the equation of motion of any dynamical quantity $\alpha(R, p)$, depending on coordinates R and momenta p, but not explicitly on time, can be written in Poisson bracket form:

$$\frac{d\alpha}{dt} = - \{H, \alpha\}_{P.B.} \quad . \tag{I.2}$$

Here H is the Hamiltonian function and $\{ \ \}_{P.B.}$ denotes the Poisson bracket. Let us define the Liouville operator L by

$$i \frac{d\alpha}{dt} = - L\alpha = - i\{H, \alpha\}_{P.B.} \quad . \tag{I.3}$$

More specifically, L is

$$L = i \frac{\partial H}{\partial R} \frac{\partial}{\partial p} - i \frac{\partial H}{\partial p} \frac{\partial}{\partial R} . \tag{I.4}$$

The operator L was first introduced by Koopman[6]. Note that L is Hermitian in the Hilbert space of phase functions.

Liouville's equation for the time dependence of the phase space ensemble density f(R, p; t) is

$$i \frac{\partial f}{\partial t} = Lf. \tag{I.5}$$

The change in sign from Eq. (3) to Eq. (5) is important.

The classical equations of motion possess formal solutions as initial value problems. These solutions are given by exponential operators,

$$\alpha(t) = \exp(itL) \cdot \alpha(0), \tag{I.6}$$

$$f(t) = \exp(-itL) \cdot f(0). \tag{I.7}$$

The operator exp(± itL) is unitary in Hilbert space. It is often referred to as a Green's function or a propagator. Its Laplace transform with respect to time,

$$\int_0^\infty dt\, e^{-pt}\, e^{itL} = \frac{1}{p - iL} , \tag{I.8}$$

is also called a Green's function, propagator and, more often, a resolvent operator.

Equations (6· and (7) illustrate a kind of Heisenberg picture in classical mechanics. The expectation value of a variable $\alpha(R, p)$, weighted according to the ensemble density f(R, p; t), is

$$< \alpha;\ f(t) > = \int\int dR\, dp\ \alpha(R, p)\, f(R, p; t). \tag{I.9}$$

The integration covers all of phase space. But this average is also

$$\begin{aligned} < \alpha;\ f(t) > &= < \alpha;\ e^{-itL} f(0) > \\ &= < e^{+itL} \alpha;\ f(0) > \\ &= < \alpha(t);\ f(0) > . \end{aligned} \tag{I.10}$$

The expectation at t can be calculated in two ways: by following the evolution of either the ensemble density or the dynamical variable. It is a matter of convenience which is preferred.

In quantum mechanics, an observable α is represented by a matrix or an operator in Hilbert space. The evolution of α is determined by Heisenberg's equation of motion,

$$i \frac{d\alpha}{dt} = -\frac{1}{\hbar}[H, \alpha] = -\frac{1}{\hbar}(H\alpha - \alpha H). \tag{I.11}$$

We can define a linear Hermitian operator L by

$$L\alpha = \frac{1}{\hbar}[H, \alpha]; \tag{I.12}$$

then Eq. (11) becomes

$$i \frac{d\alpha}{dt} = -L\alpha. \tag{I.13}$$

This L is an operator that works in the Hilbert space of operators rather than the space of states.

The quantum mechanical L has been mentioned by Kubo,[7] and Fano[8] has shown how to calculate it explicitly in, for example, spin problems.

Liouville's equation for the density matrix $\rho(t)$ is

$$i \frac{d\rho}{dt} = \frac{1}{\hbar}[H, \rho] = L\rho . \tag{I.14}$$

Again note the change in sign from Eq. (13), which reflects the Heisenberg-Schroedinger duality.

The quantum mechanical equations of motion also have formal operator solutions, which can be written in the familiar Hamiltonian or the unfamiliar Liouville form,

$$\alpha(t) = e^{iHt/\hbar}\alpha(0)\, e^{-iHt/\hbar} = e^{itL}\alpha(0), \tag{I.15}$$

$$\rho(t) = e^{-iHt/\hbar}\rho(0)\, e^{iHt/\hbar} = e^{-itL}\rho(0). \tag{I.16}$$

The expectation of α, weighted according to $\rho(t)$, is

$$< \alpha;\ \rho(t) > = \text{Trace}\{\alpha\, \rho(t)\} . \tag{I.17}$$

It is easy to verify the quantum form of Eq. (10).

This shows that all equations of motion can be written in the general form

$$i \frac{dx}{dt} = Lx ,$$

where L is Hermitian. We have introduced the formal, abstract operator solution

$$\chi(t) = G(t)\,\chi(0), \tag{I.18}$$

$$G(t) = \exp(-itL). \tag{I.19}$$

Much of the current work in statistical mechanics consists of finding useful approximations for G(t).*

II. The Diagonal Part of G(t)

On many occasions one is not interested in full knowledge of G(t), but only in its diagonal matrix elements. We shall give an example later; for the present let us just assume that this is so.

There is an obvious, though hard, way of calculating G(t). This is to solve the Liouville equation, or, the same thing, to evaluate exp(-itL). But there is another more devious way of proceeding, and this is the subject of the present section. The idea is to replace Eq. (I.1), which involves all matrix elements of G, by a new equation containing only the diagonal part of G.

The procedure followed here is completely equivalent to what has been done in the past by diagrams and perturbation theory. However, the present method is more direct and does not involve expansions.

We introduce an operator D, which selects the diagonal part of the entire matrix on which it operates, and which discards the off-diagonal part. Then

$$(DA)_{jk} = A_{jj}\,\delta_{jk}\,. \qquad \text{(II.1)}$$

Note that when D operates on a product it sees both factors, not only the nearest one:

$$(DAB)_{jk} = \sum_{\ell} A_{j\ell}\,B_{\ell j}\,\delta_{jk}\,. \qquad \text{(II.2)}$$

Furthermore, D is a projection operator:

$$D^2 = D, \qquad D(1-D) = 0. \qquad \text{(II.3)}$$

It is possible to write D as a tetradic operator. The (ij) component of DA is

$$(DA)_{ij} = \sum_k \sum_{\ell} D_{ijk\ell}\,A_{k\ell}\,. \qquad \text{(II.4)}$$

In order to recover Eq. (1), we require that

$$D_{ijk\ell} = \delta_{ij}\,\delta_{ik}\,\delta_{j\ell}\,. \qquad \text{(II.5)}$$

Since the combination (1-D) occurs also, we need the unit tetradic; this is just

$$1_{ijk\ell} = \delta_{ik}\,\delta_{j\ell}\,.$$

* As long as the density matrix is represented as a two-index matrix, the quantum mechanical L cannot be represented in the same way. In fact, L is a tetradic operator, with four indices. Because of this, classical and quantum mechanical treatments in matrix notation must be made independently.

Using D, the Green's function can be divided into diagonal and non-diagonal parts:

$$G_d = \mathbf{D} \cdot G, \qquad G_{o.d.} = (1-D) \cdot G, \qquad (II.6)$$

and

$$G = G_d + G_{o.d.} \ . \qquad (II.7)$$

In the same way the kinetic equation for G,

$$i \frac{dG}{dt} = LG, \qquad (II.8)$$

can be divided into two equations. First we multiply by D:

$$D\, i \frac{dG}{dt} = i \frac{dG_d}{dt} = DLG = DLG_d + DLG_{o.d.} \ . \qquad (II.9)$$

Then we do the same with (1-D):

$$(1-D)\, i \frac{dG}{dt} = i \frac{dG_{o.d.}}{dt} = (1-D)LG = (1-D)LG_d + (1-D)LG_{o.d.} \qquad (II.10)$$

Now we formally solve the second of these, Eq. (10), for $G_{o.d.}$ In doing this a modified Green's function appears:

$$\mathcal{G}(t) \equiv \exp[-it(1-D)L] \ . \qquad (II.11)$$

This function can be defined either by its series expansion or by its use in solving the equation for matrix A,

$$i \frac{dA}{dt} = (1-D)LA, \qquad (II.12)$$

with the initial condition A = A(0). The solution is then

$$A(t) = e^{-it(1-D)L} A(0) \equiv \mathcal{G}(t) \cdot A(0). \qquad (II.13)$$

The operator (1-D) in the exponent of $\mathcal{G}(t)$ sees not only the L that follows it directly, but everything to its right, including the matrix on which $\mathcal{G}$ operates.

Returning to the solution of Eq. (10), one can verify by substitution that

$$G_{o.d.}(t) = -i \int_0^t ds \ \mathcal{G}(t-s)(1-D)L\, G_d(s)$$

$$= -i \int_0^t ds \ \mathcal{G}(s)\,(1-D)L\, G_d(t-s). \qquad (II.14)$$

The initial condition $G_{o.d.}(0) = 0$ was used.

When Eq. (14) is substituted in Eq. (9), we obtain the desired result,

$$i \frac{dG_d}{dt} = DLG_d(t) - i\int_0^t ds\ DL\mathcal{G}(s)(1-D)LG_d(t-s). \qquad \text{(II.15)}$$

After this equation has been solved for $G_d(t)$, Eq. (14) may be used to calculate $G_{o.d.}$

The same equations can be written out in matrix form. (The following statements are not correct when L is the quantum mechanical Liouville operator; then, as we shall see, the commutator definition of L leads to more complicated results.) A diagonal element of G is denoted by G_{kk} and its kinetic equation is

$$i \frac{dG_{kk}(t)}{dt} = L_{kk}G_{kk}(t) - i\int_0^t ds[L\mathcal{G}(s)(1-D)L]_{kk}G_{kk}(t-s). \qquad \text{(II.16)}$$

The various diagonal elements of G are not coupled. Similarly, the off-diagonal elements ($j \neq k$) are

$$G_{jk}(t) = -i\int_0^t ds\ [\mathcal{G}(s)\ (1-D)L]_{jk}G_{kk}(t-s). \qquad \text{(II.17)}$$

In a later section the perturbation expansion of Eq. (15) will be presented. It will then become clear that the results of this section are entirely equivalent to those found by "diagram" methods.

III. Generalization

In the last section we considered only the diagonal part of a Green's function. While this is useful and important, one may be interested in other more complicated situations. We shall consider several variations on the "diagonal part" theme in this section.

It is instructive to begin by checking over the preceding derivation to see just where the diagonal property of D was used. In fact, it appeared explicitly only in the assertion that G is initially diagonal, or $DG(0) = G(0)$. Otherwise D could have been quite arbitrary.

Suppose that we want to find the diagonal part of the density matrix ρ, which satisfies the equation

$$i \frac{d\rho}{dt} = L\rho\ . \qquad \text{(III.1)}$$

The procedure used before cannot work here exactly, because the initial density matrix may have off-diagonal elements. This means that the initial value $\rho_{o.d.}(0)$ must appear. We shall just give the general result; the method of derivation will be obvious. We find

$$i\,\frac{d\rho_d(t)}{dt} = D\,L\,\rho_d(t) - i\int_0^t ds\; D\,L\,\mathcal{G}(s)\,(1-D)L\,\rho_d(t-s) + D\,L\,\mathcal{G}(t)\,\rho_{o.d.}(0) \qquad \text{(III. 2)}$$

and

$$\rho_{o.d.}(t) = -i\int_0^t ds\;\; \mathcal{G}(s)\,(1-D)L\,\rho_d(t-s) + \mathcal{G}(t)\rho_{o.d.}(0). \qquad \text{(III. 3)}$$

These equations may be used as the starting point for the derivation of the Pauli equation (weakly-coupled master equation) of quantum statistical mechanics. We shall go through the derivation later.

If ρ is initially diagonal, Eqs. (2) and (3) reduce to the same formal structure as the equations we have already found for the diagonal part of a Green's function. But the difference in interpretation is important. In the case of the density matrix, we are not solving for a Green's function in the same sense as in the preceding section. The Green's function that goes with the density matrix has the property

$$\rho_{jk}(t) = \sum_{\ell}\sum_{m} G_{jk,\,\ell m}(t)\;\rho_{\ell m}(0), \qquad \text{(III. 4)}$$

and its diagonal part, $G_{jk,\,jk}(t)$, connects an element of ρ at time t with the initial value of the same element:

$$\rho_{jk}(t) = G_{jk,\,jk}(t)\;\rho_{jk}(0) \;+\; \text{off-diagonal terms}, \qquad \text{(III.5)}$$

while Eq. (2) refers to the quantities $\rho_{jj}(t)$ only.

The use that one makes of the diagonalization device depends on what kind of information is desired, and what kind of information is available initially.

Finally we describe one more generalization. Let us consider the abstract problem

$$i\,\frac{dx}{dt} = L\,x\,, \qquad \text{(III. 6)}$$

where x and L are, respectively, a vector and an operator in some Hilbert space. Let P denote a projection operator in this space; it defines a subspace of Hilbert space, and Px is the part of the vector x that lies in the subspace. One can easily verify that Eqs. (2) and (3) remain correct if one makes the following changes:

$$\begin{aligned} D &\to P, \\ \rho_d &\to Px, \\ \rho_{o.d.} &\to (1-P)x, \\ \mathcal{G}(s) &\to \exp[-is(1-P)L] . \end{aligned} \qquad \text{(III. 7)}$$

This is the most general result of its kind; it includes the D-operator as a special case.

IV. Perturbation Theory

A. Weak Perturbations.

When L can be represented conveniently as a diagonal matrix, there are no dynamical problems, because the equation of motion can be integrated immediately. If L has a small off-diagonal part, then a perturbation theory is useful. Many problems can be reduced to this situation. Later we shall discuss in particular the master equations for classical and quantum mechanical weakly-coupled systems--the Prigogine-Brout and the Pauli equations. But in this section we consider for simplicity only a special case, the perturbation theory of the diagonal part of a matrix Green's function.

Suppose that L can be separated into two parts,

$$L = L^o + \lambda L' , \qquad \text{(IV. 1)}$$

where L^o is diagonal (and hence commutes with D: $L^oD = DL^o$), and L' has no diagonal elements. The parameter λ measures the strength of the perturbation L'.

We want to see the form taken by the exact equation (II. 15),

$$i\frac{dG_d(t)}{dt} = DLG_d(t) - i\int_0^t ds\; DL\,\mathcal{G}(s)(1-D)LG_d(t-s), \qquad \text{(IV. 2)}$$

in the limit of weak perturbation or small λ. The first term is easy:

$$DLG_d = L^o\,G_d, \qquad \text{(IV.3)}$$

because we take the diagonal part of L times a diagonal matrix G_d :

$$(DLG_d)_{jj} = (LG_d)_{jj} = \sum_k L_{jk}(G_d)_{kj} = L_{jj}(G_d)_{jj} = (L^oG_d)_{jj}. \qquad \text{(IV. 4)}$$

The contribution from the integral in Eq. (2) is a bit harder to work out. The (1-D) on the right operates on L times a diagonal matrix G_d, so the L^o part drops out:

$$D L \mathcal{G}(1-D) L G_d = D L \mathcal{G}(1-D) \lambda L' G_d . \qquad \text{(IV. 5)}$$

Next we observe that $\mathcal{G}(1-D)\lambda L' G_d$ has no diagonal elements. This is most easily seen by the pedestrian method of expanding; for any arbitrary matrix M,

$$\begin{aligned} \mathcal{G}(1-D)M &= \exp[-it(1-D)L](1-D)M \\ &= [1 - it(1-D)L + (1/2)(it)^2(1-D)L(1-D)L - \ldots](1-D)M \\ &= (1-D)M - it(1-D)L(1-D)M \qquad \text{(IV. 6)} \\ &\quad + (1/2)(it)^2(1-D)L(1-D)L(1-D)M + \ldots \end{aligned}$$

Each term in the expansion begins with a (1-D). This is in fact the same as the expansion

$$\mathcal{G}(1-D)M = (1-D)\exp[-itL(1-D)]\, M \qquad \text{(IV. 7)}$$

and the whole quantity has only off-diagonal elements. Now the first D on the left of the integral in Eq. (2) operates on L times a non-diagonal matrix of the form of Eq. (7), so that L can be replaced by $\lambda L'$.

In this way Eq. (2) becomes, exactly,

$$i\frac{dG_d(t)}{dt} = L^o G_d(t) - i\lambda^2 \int_0^t ds\, D L' \mathcal{G}(s)(1-D)L' G_d(t-s). \qquad \text{(IV. 8)}$$

A factor λ^2 appears explicitly, and $\mathcal{G}(s)$ depends on λ:

$$\mathcal{G}(s) = \exp[-is(1-D)(L^o + \lambda L')] . \qquad \text{(IV. 9)}$$

(It should be noted that one may not simplify $\mathcal{G}(s)$ in the following apparently obvious way:

$$\mathcal{G}(s) \to \exp[-is(1-D)\lambda L'] , \qquad \text{(IV. 10)}$$

even though L^o is diagonal and $(1-D)L^o$ appears to vanish. The reason is that (1-D) operates on everything to its right. See the expansion in Eq. (6) and the equivalent Eq. (7).)

If we want only the weak coupling limit, we can replace the general $\mathcal{G}(s;\lambda)$ by its limit $\mathcal{G}(s;0)$, given by

$$\mathcal{G}(s;0) = \exp[-is(1-D)L^o] ; \qquad \text{(IV. 11)}$$

and, for example by expansion, one can see that

$$\mathcal{G}(s;0)(1-D) = \exp[-is(1-D)L^o](1-D) = e^{-isL^o}(1-D), \qquad \text{(IV. 12)}$$

so only the unperturbed Green's function $\exp(-isL^o)$ appears.

Thus Eq. (8) becomes

$$i\frac{dG_d(t)}{dt} = L^o G_d(t) - i\lambda^2 \int_0^t ds\ DL'\, e^{-isL^o}(1-D)L'G_d(t-s) + O(\lambda^3 G_d). \qquad \text{(IV.13)}$$

If one writes this out with subscripts, the result is

$$i\frac{dG_{kk}(t)}{dt} = L^o_{kk}\, G_{kk}(t) - i\lambda^2 \int_0^t ds \sum_{j\neq k} L'_{kj}\, e^{-isL^o_{jj}}\, L'_{jk} G_{kk}(t-s) + O(\lambda^3 G). \qquad \text{(IV. 14)}$$

The assumptions that went into the derivation of Eqs. (13) and (14) are purely mathematical and have nothing to do with physical ideas. They are:

1) G is initially diagonal,

2) L^o is diagonal, L' is non-diagonal,

3) The lowest order in λ only is kept.

B. Perturbation Theory to Higher Order.

In order to demonstrate the relation of our results to other more familiar ones, and to provide a convenient way of going to higher orders in λ, we introduce Laplace transforms.

Let

$$\varphi(p) = \int_0^\infty dt\ e^{-pt}\, G_d(t) \qquad \text{(IV.15)}$$

be the Laplace transform of $G_d(t)$. The transform of $\mathcal{G}$ is

$$\int_0^\infty dt\ e^{-pt}\ \mathcal{G}(t) = \frac{1}{p+i(1-D)L} \qquad \text{(IV.16)}$$

and the transform of the (exact) Eq. (8) is

$$ip\varphi(p) - iG_d(0) = L^o\varphi(p) - i\lambda^2 DL'\frac{1}{p+i(1-D)L} \times (1-D)L'\varphi(p). \qquad \text{(IV.17)}$$

We have used the property that the transform of a convolution is the product of the transforms.

Solving Eq. (17) for $\varphi(p)$, we find

$$\varphi(p) = \left[p + iL^{o} + \lambda^{2} DL' \left(\frac{1}{p+i(1-D)L} \right) (1-D)L' \right]^{-1} G_{d}(0). \quad \text{(IV.18)}$$

The standard formula for inverting a Laplace transform is

$$G_{d}(t) = \frac{1}{2\pi i} \int_{\epsilon - i\infty}^{\epsilon + i\infty} dp\, e^{pt} \varphi(p) \; ; \quad \text{(IV.19)}$$

the real constant ϵ must be chosen so that (for positive t) the path of integration lies to the right of all singularities of $\varphi(p)$.

The perturbation theory now comes in through the expansion in powers of λ of $\kappa(p;\lambda)$,

$$\kappa(p;\lambda) = DL' \frac{1}{p+i(1-D)(L^{o}+\lambda L')} (1-D)L'. \quad \text{(IV.20)}$$

First we expand the middle factor:

$$\frac{1}{p+i(1-D)(L^{o}+\lambda L')} = \frac{1}{p+i(1-D)L^{o}} - \frac{1}{p+i(1-D)L^{o}} i(1-D)\lambda L' \times \frac{1}{p+i(1-D)L^{o}}$$
$$+ \frac{1}{p+i(1-D)L^{o}} i(1-D)\lambda L' \frac{1}{p+i(1-D)L^{o}} \times i(1-D)\lambda L' \frac{1}{p+i(1-D)L^{o}} + \ldots \quad \text{(IV.21)}$$

Taking the Laplace transform of Eq. (12), we find that

$$\frac{1}{p+i(1-D)L^{o}} (1-D) = \frac{1}{p+iL^{o}} (1-D). \quad \text{(IV.22)}$$

Let

$$\gamma^{o} = \frac{1}{p+iL^{o}} \quad \text{(IV.23)}$$

be the transform of the unperturbed Green's function. Then

$$\begin{aligned}\kappa(p;\lambda) &= DL'\gamma^{o}(1-D)L' \\ &\quad - i\lambda DL'\gamma^{o}(1-D)L'\gamma^{o}(1-D)L' \\ &\quad + (i\lambda)^{2} DL'\gamma^{o}(1-D)L'\gamma^{o}(1-D)L'\gamma^{o}(1-D)L' \\ &\quad - \ldots \; . \end{aligned} \quad \text{(IV.24)}$$

Since κ is to operate on the identity $G_d(0) = 1$, and since each term begins with D, we need only the diagonal part of κ itself. Each factor γ^o is diagonal, and the only non-diagonal contributions come from L'. If we write the general term of κ as a matrix product, each L' forming a separate factor, the effect of the (1-D) is that no matrix subscript in the interior of the product can equal the initial or final subscript on the product. In the language of the diagram method, the matrix products are connected.

Incidentally, this provides a method of calculating partition functions in quantum mechanics. If we want

$$Q = \text{Trace } e^{-\beta H} = \sum_j (e^{-\beta H})_{jj}, \tag{IV.25}$$

where H is the Hamiltonian operator and $\beta = 1/kT$, then we want the sum of the diagonal elements of the matrix

$$G = e^{-\beta H}. \tag{IV. 26}$$

But G satisfies the Bloch equation,

$$\frac{\partial G}{\partial \beta} = -HG, \tag{IV. 27}$$

with the initial condition $G(0) = 1$. This is a natural setup for the preceding perturbation theory. We replace (iL) by H, and write, as with L,

$$H = H^o + \lambda H'. \tag{IV. 28}$$

We replace t by β, take the Laplace transform of $Q(\beta)$, and find that (as in Eq. (18)),

$$\begin{aligned} \int_0^\infty d\beta\, e^{-\beta z} Q(\beta) &= \text{Trace } \frac{1}{z+H} \\ &= \sum_j \left(z + H^o_{jj} - \left[H \frac{1}{z+(1-D)H} (1-D) H \right]_{jj} \right)^{-1} \\ &= \sum_j \left(z + H^o_{jj} - \lambda^2 \left[H' \frac{1}{z+(1-D)H} H' \right]_{jj} \right)^{-1}. \end{aligned} \tag{IV.29}$$

All matrices are expressed in the representation where H^o is diagonal. Equation (29) is a suitable starting point for the investigation of various kinds of approximation schemes in the calculation of partition functions.

Perturbation expansions are only natural when λ is actually small and when all pertinent quantities exist (mathematically speaking). At this point the advantage of our general method becomes evident. Most of our results are independent of whether or not a perturbation expansion exists or converges, and other methods of evaluating the memory operator $\kappa(p; \lambda)$ may be sought in case perturbation theory does not work.

V. Time Dependence

The ultimate objective of this work is to find the time dependence of various quantities of interest in statistical mechanics. In order to avoid excessive detail, we shall consider only a very special prototype example.

We shall look for the solution of

$$\frac{dG(t)}{dt} = -\lambda^2 \int_0^t ds\, K(s; \lambda)\, G(t-s) \tag{V.1}$$

with the initial condition $G(0)=1$. Here $G(t)$ and $K(s; \lambda)$ are just functions, not operators or matrices, and we assume that $K(s; \lambda)$ is regular for small λ. In particular we are interested in the solution for small λ.

Equation (1) has the form of Eq. (IV.8) without the L^o.

In the weak coupling limit (small λ) one can proceed in several ways. Perhaps the most elementary argument is that when λ is small, then dG/dt is small and $G(t-s)$ can be replaced by the term $G(t) + O(\lambda^2)$. In this way Eq. (1) becomes

$$\frac{dG(t)}{dt} = -\lambda^2 \int_0^t ds\, K(s; \lambda)\, G(t) \quad + \quad O(\lambda^4). \tag{V.2}$$

This is now a simple differential equation rather than an integral equation. Its solution is

$$G(t) = \exp\left[-\lambda^2 \int_0^t ds \int_0^s ds'\, K(s'; \lambda) + O(\lambda^4) \right]. \tag{V.3}$$

We assume that K is integrable from zero to infinity, so that there is some time τ_m beyond which

$$\int_0^t ds\, K(s; \lambda) \simeq \int_0^{\tau_m} ds\, K(s; \lambda) \tag{V.4}$$

to sufficient accuracy. Then for t much larger than τ_m,

$$\int_0^t ds \int_0^s ds' K(s';\lambda) \simeq t \int_0^{\tau_m} ds\, K(s;\lambda) , \qquad (V.5)$$

and for such large times, G decays exponentially:

$$G(t) \cong \exp\left[-\lambda^2 t \int_0^{\tau_m} ds\, K(s;\lambda) + O(\lambda^4)\right]. \qquad (V.6)$$

Note that small λ is required in this argument.

The preceding "derivation" is not rigorous, and the result is not quite correct. In this paragraph we give a more rigorous but less informative derivation. This is based on the substitutions

$$x = \lambda^2 t, \qquad (V.7)$$

$$g(x) = G(t) . \qquad (V.8)$$

Now Eq. (1) can be written

$$\frac{dg(x)}{dx} = - \int_0^{x/\lambda^2} ds\, K(s;\lambda)\, g(x-\lambda^2 s). \qquad (V.9)$$

We take the limit

$$\lambda \to 0, \qquad t \to \infty , \qquad x \text{ fixed}. \qquad (V.10)$$

Then in the limit,

$$\frac{dg(x)}{dx} = - \int_0^{\infty} ds\, K(s;0)\, g(x), \qquad (V.11)$$

or

$$g(x) = \exp\left[-x \int_0^{\infty} ds\, K(s;0)\right], \qquad (V.12)$$

which is equivalent to

$$G(t) = \exp\left[-\lambda^2 t \int_0^{\infty} ds\, K(s;0)\right] . \qquad (V.13)$$

Again we find an exponential decay, with the same sort of rate constant as before, but only in the double limit $\lambda \to 0$, $t \to \infty$, $\lambda^2 t$ fixed. This limiting process, introduced by Van Hove, is elegant and easy to apply, but tends to obscure an important feature of the time dependence when λ does <u>not</u> go to zero, but remains finite.

To see that all may not be well, let us assume that as $t \to \infty$, G decays exponentially:

$$G(t) \sim \exp(-\alpha t). \qquad (V.14)$$

When this is put into Eq. (1) we get

$$\alpha e^{-\alpha t} \sim \lambda^2 \int_0^t ds\, K(s;\lambda)\, e^{-\alpha(t-s)} \qquad (V.15)$$

as $t \to \infty$. The factor $\exp(-\alpha t)$ can be cancelled out, leaving

$$\alpha = \lambda^2 \int_0^\infty ds\, K(s;\lambda)\, e^{+\alpha s} \qquad (V.16)$$

in the limit $t \to \infty$. This equation determines the rate constant for the assumed decay.

If there is actually a decay, then α is positive and $\exp(\alpha s)$ increases exponentially in s. For α to exist, the memory $K(s;\lambda)$ must <u>decrease</u> exponentially at least as fast as $\exp(-\alpha s)$. The Laplace transform

$$\kappa(p;\lambda) = \int_0^\infty ds\, e^{-ps}\, K(s;\lambda), \qquad (V.17)$$

in particular, must be analytic on the imaginary p-axis.

But this cannot be so, in the weak coupling limit anyhow. There the form of $K(s;\lambda)$ is known; it is typically, as in Eq. (IV.14),

$$K(s;\lambda) = \sum_{k\neq j} L'_{kj}\, e^{-isL^o_{jj}}\, L'_{jk} \qquad (V.18)$$

and the Laplace transform is

$$\kappa(p;\lambda) = \kappa(p) = \sum_{k\neq j} L'_{kj}\, \frac{1}{p+iL^o_{jj}}\, L'_{jk}. \qquad (V.19)$$

Since L^o is Hermitian, $\kappa(p)$ has poles on the imaginary axis. We are forced to conclude that G(t) will <u>not</u> decrease exponentially as $t \to \infty$

What is the trouble here? It is that $K(s;\lambda)$, according to Eq. (18), is a periodic or almost periodic function and is not integrable. This is connected with the problem of recurrences. Again, by taking a special example we can see what is going on in general.

Let us assume that

$$K(s;\lambda) = \sum_{\substack{j=-N \\ j\neq 0}}^{+N} \frac{1}{2N} \exp(-isj/N)$$

$$= \frac{1}{N} \sum_{1}^{N} \cos(sj/N). \tag{V. 20}$$

This corresponds to Eq. (18) with

$$j = 0, \qquad L'_{k0} = L'_{0k} = 1/\sqrt{2N}, \quad \text{if } k \neq 0,$$
$$L^{o}_{kk} = k/N, \qquad \text{for } -N \leq k \leq +N, \tag{V. 21}$$

where N is supposed to be very large. In the limit $N \to \infty$, we can replace the sum by an integral, and we find

$$K(s;\lambda) \to \frac{\sin s}{s}, \tag{V. 22}$$

which is certainly not periodic or even almost periodic.

But why should we take the limit $N \to \infty$? We may be dealing with a finite system, so N is not infinite. It is important to see what limitations are imposed by finite N. In the case of Eq. (22), for example, the approximation $(\sin s)/s$ is valid only for certain s. We note first that by its definition $K(s;\lambda)$ is periodic:

$$K(s + 2\pi N;\lambda) = K(s;\lambda), \tag{V. 23}$$

although the period is very large. Next we observe that the sum in Eq. (20) can be calculated exactly, since it is a geometric series. We find

$$K(s;\lambda) = \frac{1}{2N} (\cos s - 1) + \frac{s}{2N} \cot(s/2N) \frac{\sin s}{s}, \tag{V. 24}$$

which shows what is left out in Eq. (22). Only when s is much smaller than N, and only if terms of order $1/N$ are neglected, will Eq. (22) be valid (in the first period anyhow).

Now we observe that G(t) is determined only by values of K(s) for $s < t$; this follows from Eq. (1). Use of the approximation in Eq. (22) is valid only for $s << N$, and only if microscopic "noise" in K, of order $1/N$, is of no interest. The G(t) calculated using Eq. (22) is valid only under the same restrictions. In particular, the behavior of G(t), as $t \to \infty$ but N remains finite, is not correctly calculated in this way.

If we want only the "early behavior, $s << N$, then we can find it by the following procedure.

We shall solve Eq. (1) by Laplace transforms. Let

$$\varphi(p) = \int_0^\infty dt\, e^{-pt}\, G(t). \tag{V.25}$$

Then Eq. (1) becomes

$$p\,\varphi(p) - 1 = -\lambda^2 \kappa(p)\,\varphi(p), \tag{V.26}$$

where κ (p) is the transform of K,

$$\kappa(p) = \frac{1}{2i} \log \frac{p+i}{p-i}\,. \tag{V.27}$$

Note that $\kappa(p)$ has branch points on the imaginary axis at $p = \pm\, i$. So even with the approximate K, G does not decay exponentially for long t.

Solving for $\varphi(p)$ we find

$$\varphi(p) = \frac{1}{p + \lambda^2 \kappa(p)}\,. \tag{V.28}$$

The standard formula for the inverse Laplace transform gives

$$G(t) = \frac{1}{2\pi i} \int_{\epsilon - i\infty}^{\epsilon + i\infty} dp\, \frac{e^{pt}}{p + \lambda^2 \kappa(p)}\,, \tag{V.29}$$

where ϵ is any positive real number. The calculation of G(t) has thus been reduced to quadratures.

The Laplace transform $\kappa(p)$ was defined only for $\mathrm{Re}\,(p) > 0$. If we want to use the methods of theory of complex variables, deforming contours, evaluating residues, etc., we have to find the analytic continuation of κ (p) to the negative half plane, $\mathrm{Re}\,(p) < 0$. Because the singularities of κ(p) are branch points, we must cut the plane, and this may be done in any way that is convenient. The choice that seems easiest to use is indicated by the heavy line in Fig. 1, p. 124.

With this choice of analytic continuation of κ (p), it turns out that κ is analytic near the origin:

$$\kappa(p) = \frac{\pi}{2} + O(p), \tag{V.30}$$

so that $\varphi(p)$ has a pole at the root of

$$p + \lambda^2 \kappa(p) = 0, \tag{V.31}$$

namely,

$$p = -\lambda^2 \frac{\pi}{2} + O(\lambda^4) . \qquad (V.32)$$

The presence of the pole at this point is due entirely to the particular choice of cut. If the plane were cut instead from p = -i to p = +i, there would be no pole in the cut plane.

Now the contour in Eq. (29) may be deformed as shown in Fig. 1. Then G(t) falls into two parts, a term from the residue at

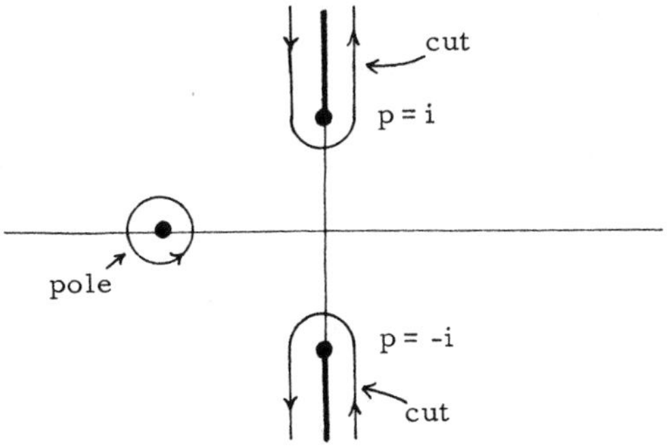

Fig. 1

the pole and the integrals around the cuts. We shall not evaluate the latter integrals here; they are tedious, but easy to find when λ is small. We determine that

$$G(t, \text{ from cuts}) = -\lambda^2 \int_1^\infty d\omega \frac{\cos \omega t}{\omega^2} + O(\lambda^4) \qquad (V.33)$$

in the limit of small λ. The pole gives a contribution that decays exponentially:

$$G(t, \text{from pole}) = [1 + \lambda^2 + O(\lambda^4)] \exp[-(\lambda^2\pi/2)t + O(\lambda^4 t)] . \quad (V.34)$$

The sum has the structure

$$G(t) = \exp[-(\lambda^2 \pi/2)t + O(\lambda^4 t)] + \lambda^2 \times \text{(a function bounded in t, but not exponentially decaying)} . \qquad (V.35)$$

The most instructive feature of this result is that the exponential decay represented by the first term is ultimately submerged in the non-exponential decay of the second term. In the Van Hove limit $\lambda \to 0$, $t \to \infty$, $\lambda^2 t$ fixed, the second term drops out and one gets simple exponential behavior:

$$G(t) \xrightarrow{\text{V.H.}} \exp(-\lambda^2 \kappa t/2). \qquad (V.36)$$

But for finite λ and t, the exponential decay persists only for times

$$t << O\left(\frac{1}{\lambda^2}\log(1/\lambda^2)\right) \qquad (V.37)$$

If we had cut the p-plane from p = -i to p = +i instead, we would not have been able to analytically continue κ (p) through the origin, and the contour integral would not see a pole. The function G(t) must be the same no matter how it is evaluated; but in this way the separation into an exponential decay and a small remainder would not have been so obvious.

If the parameter λ is not small, it is no longer clear that G will show exponential behavior for any significant time interval. Much more study of solutions of Eq. (1), with various λ and K(s), is needed.

We have considered in this section only a rather special example of the general problem of finding time dependence. However, several general conclusions are suggested. First, exponential decay is not universal, and if it appears, it may ultimately be hidden in some other kind of time dependence. Second, when one takes advantage of the approximations that can be made on a system with a large number of degrees of freedom, the resulting G(t) may be valid only for a limited time interval, and only if small scale "noise" is ignored.

VI. Examples of Weak Perturbation Theory

In this section we shall study two specific examples of the preceding formalism. The first example is the Pauli equation; the second is a classical analogue.

It must be emphasized that entirely satisfactory derivations of both equations are known. In particular, Van Hove's derivation of the Pauli equation is responsible, to a large extent, for much of the work going on these days in the theory of irreversibility, especially that in which diagram and perturbation techniques are used. We shall discuss new derivations here just to illustrate the convenience and directness of the preceding formalism.

A. The Pauli Equation.

Let us consider the Pauli equation first. The system is described by a Hamiltonian

$$H = H^{o} + \lambda H' ; \qquad (VI.1)$$

we write all operators as matrices in the representation where H^o is diagonal. We require further that H' has no diagonal elements. Now suppose that we are interested only in averages of operators that commute with H^o, or are diagonal in the unperturbed representation. Since the average of any observable α is given by

$$< \alpha > = \mathrm{Trace}\,\{\alpha\, \rho(t)\}\,, \qquad \text{(VI. 2)}$$

where $\rho(t)$ is the density matrix, it is evident that we need to know only the diagonal elements of $\rho(t)$. Because of this limitation the theory we have just discussed is immediately applicable.

We make one further restriction. Instead of the most general initial condition, we shall use only ensembles whose density matrices are initially diagonal. This corresponds to initially making the phases random.

So we want to find the small-λ approximation of the following exact equation (see Eq. (III. 2)) for the diagonal part ρ_d of the density matrix (assuming $\rho_{o.d.}(0) = 0$):

$$i\,\frac{d\rho_d(t)}{dt} = DL\rho_d(t) - i\int_0^t ds\; DL\mathcal{G}(s)(1-D)L\rho_d(t-s), \qquad \text{(VI. 3)}$$

where

$$L = L^o + \lambda L' \qquad \text{(VI. 4)}$$

and

$$\mathcal{G}(s) = \exp[-is(1-D)L]\;. \qquad \text{(VI. 5)}$$

Taking the (kk) element of ρ_d, we have

$$i\,\frac{d\rho_{kk}(t)}{dt} = [L\rho_d(t)]_{kk} - i\int_0^t ds[L\mathcal{G}(s)(1-D)L\rho_d(t-s)]_{kk}. \qquad \text{(VI. 6)}$$

Consider the first term on the right; this is

$$(L\rho_d)_{kk} = \frac{1}{\hbar}[H, \rho_d]_{kk} = \frac{1}{\hbar}\,(H\rho_d - \rho_d H)_{kk}. \qquad \text{(VI. 7)}$$

But because ρ_d is diagonal, the diagonal elements of $(H\rho_d - \rho_d H)$ vanish, and we find

$$DL\rho_d = 0. \qquad \text{(VI. 8)}$$

Now consider the " memory" term; this begins with

$$[L \cdot \mathcal{G} \cdot (1-D) \cdot L\rho_d]_{kk} = \frac{1}{\hbar}[H, \mathcal{G} \cdot (1-D) \cdot L\rho_d]_{kk}, \quad \text{(VI. 9)}$$

and because $H = H^o + \lambda H'$ with H^o diagonal, we obtain

$$\frac{1}{\hbar}[\lambda H', \mathcal{G} \cdot (1-D) \cdot L\rho_d]_{kk} \,. \quad \text{(VI.10)}$$

Next we note that

$$\mathcal{G}(s) \cdot (1-D) = [e^{-isL^o} + O(\lambda)] \; (1-D), \quad \text{(VI.11)}$$

just as in our earlier discussions of perturbation theory. The term of order λ in Eq. (11) gives rise to a term of order λ^3 in the result, so we neglect it here. According to Eq. (I. 16),

$$e^{-isL^o}(1-D)L\rho_d = e^{-iH^o s/\hbar}[(1-D)L\rho_d]\, e^{iH^o s/\hbar}, \quad \text{(VI.12)}$$

so the memory term becomes

$$\frac{1}{\hbar}[\lambda H', e^{-iH^o s/\hbar}\big((1-D)L\rho_d(t-s)\big)\, e^{iH^o s/\hbar}]_{kk}. \quad \text{(VI. 13)}$$

Next we put in

$$(1-D)L\rho_d = \frac{1}{\hbar}(1-D)[H, \rho_d] = \frac{1}{\hbar}[\lambda H', \rho_d] \; ; \quad \text{(VI.14)}$$

the H^o term drops out. In terms of conventional matrices, the memory is

$$\frac{\lambda^2}{\hbar^2}\Big[H', \; e^{-iH^o s/\hbar}[H', \rho_d(t-s)]\, e^{iH^o s/\hbar}\Big]_{kk}. \quad \text{(VI.15)}$$

Incidentally this is just the same as

$$\lambda^2 D L' e^{-isL^o} L' \rho_d \,, \quad \text{(VI.16)}$$

which illustrates clearly the formal similarity of the quantum mechanical theory to the simpler result given in Eqs. (IV. 13) and (IV. 14).

In order to write out the result completely in matrix form, we need the eigenvalues E_k of the unperturbed Hamiltonian H^o. Then it is easy to work out the commutators in Eq. (15); we get

$$\frac{\lambda^2}{\hbar^2}\sum_{\ell \neq k} |H'_{k\ell}|^2 \cdot 2\cos\Big(\frac{E_k - E_\ell}{\hbar}s\Big)\{\rho_{kk}(t-s) - \rho_{\ell\ell}(t-s)\}. \text{(VI.17)}$$

Let us define

$$\omega_{k\ell}(s) = \omega_{\ell k}(s) = \frac{2}{\hbar^2} |H'_{k\ell}|^2 \cos\left(\frac{E_k - E_\ell}{\hbar} s\right) \quad \text{(VI.18)}$$

Then the equation for $\rho_{kk}(t)$ is

$$\frac{d\rho_{kk}(t)}{dt} = \lambda^2 \sum_{\ell \neq k} \int_0^t ds\, \omega_{k\ell}(s)\,[\rho_{\ell\ell}(t-s) - \rho_{kk}(t-s)] + O(\lambda^3 \rho). \quad \text{(VI.19)}$$

This is almost the Pauli equation. We still have to take care of the time convolution, and we have not yet introduced the simplifications that are possible with a large system.

We follow the first solution method described in Section V. Although it is not as neat as the Laplace transform method, it is much easier to use. In this scheme $\rho(t-s)$ is replaced by $\rho(t)$, terms of order $(\lambda^2\rho)$ are neglected, and we get

$$\frac{d\rho_{kk}(t)}{dt} = \lambda^2 \sum_{\ell \neq k} \int_0^t ds\, \omega_{k\ell}(s)\,[\rho_{\ell\ell}(t) - \rho_{kk}(t)] + O(\lambda^3 \rho). \quad \text{(VI.20)}$$

The integration of ω is trivial:

$$\int_0^t ds\, \omega_{k\ell}(s) = \frac{2}{\hbar^2} |H'_{k\ell}|^2 \frac{\hbar}{E_k - E_\ell} \sin\left(\frac{E_k - E_\ell}{\hbar} t\right). \quad \text{(VI. 21)}$$

In the limit of large t, the final factor goes over into a delta function:

$$\frac{\hbar}{E_k - E_\ell} \sin\left(\frac{E_k - E_\ell}{\hbar} t\right) \rightarrow \pi \hbar\, \delta(E_k - E_\ell). \quad \text{(VI. 22)}$$

But we shall not yet take this limit.

So far the unperturbed states have been labelled with single subscripts. However, in a many-body system each unperturbed state can generally be labelled with many quantum numbers. We choose the total energy E and we let α denote the eigenvalues of all operators that commute with H^0. Then each state is marked by (E, α). The matrix elements of the perturbation are

$$H'_{k\ell} \rightarrow H(E\alpha;\, E'\alpha') \quad \text{(VI. 23)}$$

and the diagonal elements of the density matrix will be called

$$\rho_{kk}(t) \rightarrow \rho(E\alpha\,; t). \quad \text{(VI. 24)}$$

In this notation Eq. (20) is

$$\frac{d\rho(E\alpha;t)}{dt} = \lambda^2 \sum_{E'}\sum_{\alpha'} \frac{2}{\hbar^2} | H(E\alpha; E'\alpha') |^2 \frac{\hbar}{E-E'} \sin\left(\frac{E-E'}{\hbar} t \right)$$

$$\times \{\rho(E'\alpha';t) - \rho(E\alpha;t)\} . \qquad \text{(VI. 25)}$$

In order to take advantage of the δ-function limit in Eq. (22), we should like to be able to replace the sum over E' by an integral. Let us investigate the conditions under which this is allowed.

When t is large, the sum over E' is limited, because of the sine factor, to those states for which

$$| E - E' | = O(\hbar /t). \qquad \text{(VI. 26)}$$

When the system is large, the energy levels will be closely spaced, and we can introduce a density function N(E) such that $N(E)\Delta E$ is the number of states whose energies lie between E and $E + \Delta E$. The number of states that are included in the sum in Eq. (25) is roughly

$$N(E) \times O(\hbar/t). \qquad \text{(VI. 27)}$$

In order to be able to convert the sum to an integral, this number of states must be considerably larger than one:

$$N(E) \times O(\hbar /t) >> 1, \qquad \text{(VI.28)}$$

or

$$\frac{t}{N(E)\hbar} << O(1). \qquad \text{(VI. 29)}$$

But the density of states in a large system is proportional to the number of degrees of freedom. This means that the time interval for which the Pauli equation is valid is limited by the size of the system. (This is the same kind of limitation that we have discussed in Section V.)

We should also be prepared to find that for any finite λ the Pauli equation will only be valid for a finite time interval. This is connected with the possibility of branch points, as we have remarked in Section V.

In the limits used by Van Hove: (1) the size of the system goes to infinity first, (2) $\lambda \to 0$, and (3) $t \to \infty$ with $\lambda^2 t$ fixed. Then the Pauli equation is valid. The various corrections to be made, when these limits are not taken, must be based on a much more careful analysis of Eq. (19).

In the limiting case we can replace the sum by an integral:

$$\sum_{E'} \to \int dE' N(E'), \tag{VI. 30}$$

and we can use Eq. (22). Then Eq. (25) becomes

$$\frac{d\rho(E\alpha;t)}{dt} = \lambda^2 \sum_{\alpha'} \int dE' N(E') \frac{2\pi}{\hbar} \delta(E-E') \left| H(E\alpha; E'\alpha') \right|^2$$

$$\times \{\rho(E'\alpha';t) - \rho(E\alpha;t)\} . \tag{VI. 31}$$

Performing the integration, and using the abbreviation

$$W(\alpha',\alpha) = \frac{2\pi}{\hbar} \left| H(E\alpha'; E\alpha) \right|^2 N(E) = W(\alpha,\alpha'), \tag{VI. 32}$$

we find that Eq. (31) is

$$\frac{d\rho(E\alpha;t)}{dt} = \lambda^2 \sum_{\alpha'} \{W(\alpha,\alpha')\, \rho(E\alpha';t) - W(\alpha',\alpha)\, \rho(E\alpha;t)\} . \tag{VI.33}$$

This is the Pauli equation

We have had to make various assumptions to get the desired result:

1) the density matrix is initially diagonal,
2) λ is small,
3) the system has many degrees of freedom,
4) t is large (of order $1/\lambda^2$), and
5) matrix elements are "smooth" functions of E.

The fifth assumption was made tacitly. The first and second assumptions are needed to arrive at Eq. (19); the others are needed to go the rest of the way to Eq. (33). If the third through fifth are not satisfied, one can always use Eq. (19), which is a kind of extension of the Pauli equation.

B. The Prigogine-Brout Equation.

As our second example of perturbation theory, we shall consider a system in classical mechanics where the unperturbed motion is that of free particles, and the perturbation is due to (weak) forces between the particles. We shall derive a "master" equation which was obtained first by Prigogine and Brout. Another derivation, based on diagram techniques, is given in this volume by Balescu (see p. 382).

The Hamiltonian for our system is

$$H = \sum_{j=1}^{N} \frac{1}{2m} p_j^2 + \frac{1}{2} \lambda \sum_{\substack{j \neq k \\ =1}}^{N} \sum V(|r_j - r_k|). \tag{VI.34}$$

In this equation, p_j is the momentum of the j-th particle, m is its mass and $V(|r_j - r_k|)$ is the spherically symmetric potential energy of interaction of the particles situated at r_j and r_k. The Liouville operator is (as in Eq. (I.14))

$$L = L^o + \lambda L , \tag{VI.35}$$

$$L^o = -i \sum_{j=1}^{N} \frac{1}{m} p_j \frac{\partial}{\partial r_j} , \tag{VI.36}$$

$$L = +i \frac{1}{2} \sum_{j \neq k}^{N} \sum \frac{\partial V(|r_j - r_k|)}{\partial r_j} \left(\frac{\partial}{\partial p_j} - \frac{\partial}{\partial p_k} \right). \tag{VI.37}$$

The "vector" in Hilbert space is the phase space distribution function f(p, r; t). We use p and r to denote collectively all momenta and positions whenever convenient.

We can approach the derivation of the desired kinetic equation in two different but equivalent ways. The method used by Balescu, for example, is to expand f as a Fourier series in positions, keeping the momenta in the Fourier coefficients. Then the Liouville equation becomes a hybrid matrix (in Fourier components of position) and differential operator (in momentum space) equation. The Prigogine-Brout master equation is connected with a particular diagonal element of the Green's function exp(-itL) in this representation. It is instructive to carry through the derivation in this way, but we shall not do so here.

Instead we use the idea of a projection operator. Suppose that we are concerned with calculating averages of functions of momenta only. For this purpose we do not have to know all of f(p, r; t); we need only its projection on the subspace of functions of momenta. Let P be the operator that removes positions by integrating over the volume Ω of the system:

$$P = \frac{1}{\Omega^N} \int_\Omega dr_1 \ldots \int_\Omega dr_N . \tag{VI.38}$$

When P operates on a function that is already independent of positions, it gives back the same function, since the volume integrals then cancel out the factor Ω^{-N}. This means that $P^2 = P$, or P is a projection operator.

The projected distribution function

$$f_1(p;t) = P \cdot f(p, r; t) \qquad \text{(VI.39)}$$

is all we need to know for calculating averages of functions of momenta. We shall now find the kinetic equation for f_1.

The starting point is Eq. (III. 2), translated, according to the prescription at the end of Section III, into

$$i \frac{\partial f_1(t)}{\partial t} = P L f_1(t) - i \int_0^t ds\, P L e^{-is(1-P)L} (1-P)L f_1(t-s)$$

$$+ P L e^{-it(1-P)L}(1-P) f(p, r; 0). \qquad \text{(VI.40)}$$

The initial value of the position-dependent part of f, namely

$$(1-P) f(p, r; 0) = f(p, r; 0) - f_1(p; 0),$$

appears in this equation.

Once a suitable solution of Eq. (40) has been found, the remaining part, $(1-P)f(p, r; t)$, can be calculated by the translation of Eq. (III. 3).

Consider first the term

$$PLf_1 = PL^o f_1 + \lambda PL' f_1. \qquad \text{(VI. 41)}$$

Because f_1 does not depend on position (by definition) and because L^o contains derivatives with respect to position, the term $PL^o f_1$ vanishes. The other part, $PL'f_1$, vanishes also, but for a different reason. The only position dependence of $L'f_1$ is contained in factors like $\partial V(r_{jk}) / \partial r_{jk}$. When the integrations required by P are performed, they see only such factors. The integral over all space of the force vanishes; so

$$P L' f_1 = 0.$$

Now let us consider the memory term. We note first that whenever PL^o operates on something it gives a vanishing result. This is because L^o differentiates with respect to position and P integrates these derivatives over all space. (We assume that the resulting surface integrals all vanish.) So the memory term begins with $\lambda PL'$. The final $Lf_1(t-s)$ in the memory term can be replaced by $\lambda L'f_1(t-s)$ because L^o differentiates with respect to position and f_1 has no position dependence. The memory term is thus

$$PLe^{-is(1-P)L}(1-P)Lf_1(t-s) = \lambda^2 PL'e^{-is(1-P)L}(1-P)L'f_1(t-s). \qquad \text{(VI. 42)}$$

Just as in our previous examples of weak perturbation theory, we can now make the approximation

$$e^{-is(1-P)L}(1-P) = e^{-isL^o}(1-P) + O(\lambda), \tag{VI.43}$$

leading to

$$P L e^{-is(1-P)L}(1-P)L f_1(t-s) = \lambda^2 PL' e^{-isL^o}(1-P)L' f_1(t-s) + O(\lambda^3) . \tag{VI.44}$$

We have already shown that $PL'f_1 = 0$, so that

$$(1-P)L'f_1 = L'f_1 ;$$

we can therefore drop the (1-P) from the right-hand side of Eq. (44).

To the same order of approximation, the contribution from the initial (1-P)f is

$$\lambda P L' e^{-isL^o}(1-P) f(p, r; 0) + O(\lambda^2 (1-P)f) . \tag{VI.45}$$

Let us assume that f(0) has only a macroscopic variation in space, or that it is essentially constant over the range of intermolecular forces. Then the integration required by the P on the left sees essentially only the position dependence of the force in L', and the integral is negligibly small.

Furthermore, let us accept the argument of Prigogine and Balescu that the quantity (1-P) f(0) is to be regarded formally as being itself of order λ. Their reasoning is based on the idea that the spatial variation of a distribution function will generally be of the same order of magnitude as if the system were in thermal equilibrium. Since the spatial dependence of the equilibrium distribution is proportional to

$$\exp\left(-\lambda \sum_{j<k}\sum V(|r_j - r_k|)/kT \right) ,$$

in lowest order of λ the equilibrium distribution looks like

$$\begin{aligned} f(\text{equilibrium}) = \; & (\text{function of momenta}) \\ & + \lambda \times (\text{function of momenta and positions}) \\ & + O(\lambda^2) . \end{aligned} \tag{VI. 46}$$

Then (1-P) f(equilibrium) is of order λ. This argument appears to be sensible as long as the nonequilibrium f does not deviate much from its equilibrium limit; otherwise it does not appear to be valid.

Anyhow, if we accept their argument, the extra term

$$O(\lambda^2(1-P)f)$$

in Eq. (45) is formally of order λ^3 and can be omitted.

The result of the preceding discussion is

$$\frac{\partial f_1(p;t)}{\partial t} = -\lambda^2 \int_0^t ds\, P L' e^{-isL^o} L' f_1(p;t-s) + O(\lambda^3). \qquad \text{(VI.47)}$$

We shall conclude the derivation by showing that in the limit of large time and small λ (with $\lambda^2 t$ held constant), Eq. (47) reduces to the Prigogine-Brout equation.

It will be clear, just as in our previous discussions, that the $\lambda^2 t$ limit gives

$$\frac{\partial f_1(p;t)}{\partial t} = -\lambda^2 \int_0^\infty ds\, P L' e^{-isL^o} L' f_1(p;t) . \qquad \text{(VI. 48)}$$

Putting in the definition of L', we obtain

$$P L' e^{-isL^o} L' = \frac{1}{2} \sum_{j \neq k}\sum \frac{1}{2} \sum_{\mu \neq \nu}\sum \left(\frac{\partial}{\partial p_j} - \frac{\partial}{\partial p_k} \right)$$

$$\times \frac{1}{\Omega^N} \int_\Omega dr_1 \ldots \int_\Omega dr_N \frac{\partial V(|r_j - r_k|)}{\partial r_j} e^{-isL^o}$$

$$\times \frac{\partial V(|r_\mu - r_\nu|)}{\partial r_\mu} \left(\frac{\partial}{\partial p_\mu} - \frac{\partial}{\partial p_\nu} \right). \qquad \text{(VI. 49)}$$

From here on we will use Balescu's notation. The Fourier expansion of V is

$$V(r) = \frac{1}{\Omega} \sum_\ell V_{\vec{\ell}}\, e^{i\vec{\ell}\cdot\vec{r}} \qquad \text{(VI. 50)}$$

The volume integral in Eq. (49) becomes

$$\frac{1}{\Omega^2} \sum_\ell \sum_{\ell'} (i\vec{\ell} V_{\vec{\ell}})(i\vec{\ell}' V_{\vec{\ell}'}) \frac{1}{\Omega^N} \int_\Omega dr_1 \ldots \int_\Omega dr_N\, e^{i\ell(r_j - r_k)}$$

$$\times e^{-isL^o} e^{i\ell'(r_\mu - r_\nu)} . \qquad \text{(VI. 51)}$$

According to the definition of L^o,

$$\exp(-isL^o) = \exp\left(-s\sum_j \frac{1}{m} p_j \frac{\partial}{\partial r_j}\right), \tag{VI. 52}$$

so that

$$e^{-isL^o} e^{i\ell'(r_\mu - r_\nu)} = e^{i\ell'(r_\mu - r_\nu)} e^{-is\ell'(p_\mu - p_\nu)/m}. \tag{VI. 53}$$

It is now easy to do the volume integrals; the only nonvanishing contributions come from ($j = \mu$, $k = \nu$) or ($j = \nu$, $k = \mu$), and Kronecker deltas in ℓ and ℓ' appear. We get, after some rearrangement,

$$PL' e^{-isL^o} L' = -\frac{1}{2}\sum_{j \neq k}\sum\left(\frac{\partial}{\partial p_j} - \frac{\partial}{\partial p_k}\right)\frac{1}{\Omega}\sum_\ell \vec{\ell}\vec{\ell}\, |v_\ell|^2$$

$$\times \exp[-is\ell\cdot(p_j - p_k)/m]\left(\frac{\partial}{\partial p_j} - \frac{\partial}{\partial p_k}\right). \tag{VI.54}$$

The time integral in Eq. (48) sees only the factor

$$\exp[-is\ell(p_j - p_k)/m],$$

and

$$\int_0^\infty ds\, e^{-is\ell(p_j - p_k)/m} = \pi\delta_-\left(\ell\cdot\frac{p_j - p_k}{m}\right) \tag{VI. 55}$$

in terms of the δ_- function.

The final equation is

$$\frac{\partial f_1(p;t)}{\partial t} = \frac{\lambda^2}{2}\sum_{j \neq k}\sum\left(\frac{\partial}{\partial p_j} - \frac{\partial}{\partial p_k}\right)\cdot\frac{\pi}{\Omega}\sum_\ell \vec{\ell}\vec{\ell}\, |v_\ell|^2\, \delta_-\left(\ell\cdot\frac{p_j - p_k}{m}\right)$$

$$\times\left(\frac{\partial}{\partial p_j} - \frac{\partial}{\partial p_k}\right) f_1(p;t). \tag{VI. 56}$$

This is the result of our derivation, in a form and notation that agree entirely with Balescu's.

VII. Time-Correlation Functions

In a lot of recent work in the theory of irreversibility (see, for example, Montroll's lectures, this volume, p. 221) the evaluation of time-correlation functions is of decisive importance. In this section we shall show that such functions (at least in classical statistical mechanics) obey Volterra equations of the sort that were discussed in Section V.

Consider some dynamical quantity $U(p, q)$, and some equilibrium ensemble density $f(p, q)$ such that

$$
\begin{aligned}
< U > &= \int\int dp\, dq\, U(p, q)\, f(p, q) = 0 , \\
< U^2 > &= \int\int dp\, dq\, U^2(p, q)\, f(p, q) = 1.
\end{aligned}
\qquad \text{(VII. 1)}
$$

There is no loss of generality in restricting $< U >$ and $< U^2 >$ to the values zero and one; this can always be arranged.

Let $p(t)$, $q(t)$ be the momenta and positions at time t in terms of initial momenta and positions, and write

$$U(t) = U\Big(p(t),\ q(t) \Big) .$$

We want to find the time dependence of

$$A(t) = < U(0)\, U(t) > . \qquad \text{(VII. 2)}$$

Note that $A(0) = 1$ because of the normalization of U.

We shall demonstrate that

$$\frac{dA(t)}{dt} = - \int_0^t ds\, K(s)\, A(t-s) \qquad \text{(VII. 3)}$$

with $K(s)$ suitably defined. We use the method of projection operators.

The time-correlation function will be rewritten in the suggestive way:

$$A(t) = \int\int dp\, dq\, U(p, q)\, \mathcal{F}(p, q; t) , \qquad \text{(VII. 4)}$$

where the quasi-distribution function $\mathcal{F}$ is

$$\mathcal{F}(p, q; t) = U\Big(p(t),\ q(t) \Big) f(p, q) \qquad \text{(VII. 5)}$$

$$= [e^{itL}\, U(p, q)]\ f(p,q) \qquad \text{(VII. 6)}$$

$$= e^{itL}\, [U(p, q)\, f(p, qP] . \qquad \text{(VII. 7)}$$

In going from Eq. (5) to (6) we have used Eq. (I. 6). To get from (6) to (7) we took advantage of the assumption that f is an equilibrium distribution, and therefore independent of time. Also, it is clear that

$$\mathfrak{F}(p, q; t) = e^{itL} \mathfrak{F}(p, q; 0)$$
$$\mathfrak{F}(p, q; 0) = U(p, q) f(p, q) \quad ; \qquad \text{(VII. 8)}$$

or

$$i \frac{\partial \mathfrak{F}}{\partial t} = - L\mathfrak{F} \qquad \text{(VII. 9)}$$

with the initial condition given by Eq. (8).

The basic point of the rest of this calculation is that we do not want the general properties of the Green's function exp(itL). We need only its effect on a special "initial state" U f , and we want only as much of $\mathfrak{F}(p, q; t)$ as is needed for the averaging of U (Eq. (4)). One way of putting this is that we want the diagonal element of exp(itL) between "unit vectors" $U f^{1/2}$ in Hilbert space. However, we can also more conveniently use the projection operator formalism.

Let us introduce a projection operator P, and its associated subspace in Hilbert space. The subspace is supposed to include functions of the form U f, and it should allow the following :

$$\iint dp\, dq\, U\mathfrak{F} = \iint dp\, dq\, U\, P\, \mathfrak{F} \ . \qquad \text{(VII.10)}$$

The simplest projection that does this is the following one, defined by its effect on some arbitrary function G:

$$PG = U(p, q)\, f(p, q) \iint dp'\, dq'\, U(p', q')\, G(p', q'). \qquad \text{(VII. 11)}$$

This P selects only the component of G in the direction of U. Because

$$P^2G = P(PG) = PG$$

(as a result of $< U^2 > = 1$), P is clearly a projection. This is the reason why we originally normalized U to unity.

With this P, the requirements that we set up may be shown to be satisfied. For example,

$$PUf = Uf \iint dp'\, dq'\, U \cdot Uf = Uf , \qquad \text{(VII. 12)}$$

and

$$\iint dp\, dq\, UP\mathfrak{F} = \iint dp\, dq\, U \cdot Uf \iint dp' dq' U\, \mathfrak{F}$$
$$= \iint dp'\, dq'\, U\, \mathfrak{F} \qquad \text{(VII. 13)}$$

Since we want only $P\mathcal{F}$, since $P\mathcal{F}(p, q; 0) = \mathcal{F}(p, q; 0)$ from Eq. (12), and since $\mathcal{F}$ satisfies Liouville's equation, we have exactly the situation discussed in Section III. The methods of that section can be used directly to find a kinetic equation for $P\mathcal{F}$:

$$i \frac{\partial P\mathcal{F}(t)}{\partial t} = - PLP\mathcal{F}(t) - i \int_0^t ds\ PL\ e^{is(1-P)L} (1-P)LP\mathcal{F}(t-s). \quad \text{(VII.14)}$$

(There is a change in sign from the equations of Section III. This is because Eq. (9) has a (-L) instead of a (+L).)

Now we multiply through by U and integrate over phase space. The left-hand side becomes

$$\iint dp\, dq\ U\ i \frac{\partial P\mathcal{F}(t)}{\partial t} = i \frac{\partial A(t)}{\partial t} . \quad \text{(VII.15)}$$

On the right-hand side we note that

$$\iint dp\, dq\ UPLP\mathcal{F} = \iint dp\, dq\ ULP\mathcal{F} , \quad \text{(VII.16)}$$

because of the definition of P. Next we put in

$$P\mathcal{F} = Uf \int U \mathcal{F}\, dp'\, dq' , \quad \text{(VII. 17)}$$

so that

$$\iint dp\, dq\ UPLP\mathcal{F} = \iint dp\, dq\ ULUf \iint dp'\, dq'\ U\mathcal{F} . \quad \text{(VII.18)}$$

But in general

$$\iint dp\, dq\ ULUf = < ULU > = 0. \quad \text{(VII. 19)}$$

This is because LU is, aside from factors, just $\dot{U}$. Then ULU is proportional to $U\dot{U}$, or proportional to dU^2 / dt. The equilibrium average of a time derivative vanishes; q. e. d. The first term on the right of Eq. (14) does not contribute.

In exactly the same way one can easily show that the second term on the right is

$$\begin{aligned} PL\,e^{is(1-P)L} & (1-P)LP\mathcal{F}(t-s) \\ & = \left\{ \iint dp\, dq\ ULe^{is(1-P)L}\ LUf \right\} \\ & \qquad \times \iint dp'\, dq'\ U(p', q')\ \mathcal{F}(p', q'; t) \\ & = K(s)\ A(t-s) , \end{aligned} \quad \text{(VII.20)}$$

where the memory function K(s) is

$$K(s) = \int\int dp\, dq\, U L e^{is(1-P)L} L U f$$
$$= \langle U L e^{is(1-P)L} L U \rangle . \qquad \text{(VII.21)}$$

Consequently, Eq. (14) is equivalent to

$$\frac{dA(t)}{dt} = - \int_0^t ds\, K(s)\, A(t-s), \qquad \text{(VII.22)}$$

which is what we set out to show.

The memory can be expressed in a more suggestive way by using Eq. (I. 3), or

$$L U = - i \dot{U} , \qquad \text{(VII.23)}$$

and by taking advantage of the Hermitian character of L. Then we get

$$K(s) = \langle \dot{U} e^{is(1-P)L} \dot{U} \rangle . \qquad \text{(VII.24)}$$

Thus K(s) is the time correlation of $\dot{U}$, but evaluated with a modified propagator.

We conclude this section with a few comments on the evaluation of transport coefficients with Eq. (22). As many authors have shown (see Montroll's lectures, this volume, for examples), transport coefficients can often be expressed in the form

$$\sigma = \lim_{\epsilon \to 0^+} \int_0^\infty dt\, e^{-\epsilon t} A(t). \qquad \text{(VII.25)}$$

Let us define $\sigma(\epsilon)$ as the Laplace transform

$$\sigma(\epsilon) = \int_0^\infty dt\, e^{-\epsilon t} A(t). \qquad \text{(VII. 26)}$$

Equation (22) may be solved by Laplace transforms: see Section V for the method in some detail. The solution is

$$\sigma(\epsilon) = \left[\epsilon + \int_0^\infty dt\, e^{-\epsilon t} K(t) \right]^{-1} . \qquad \text{(VII.27)}$$

If there are no peculiar divergence troubles, the limit σ is

$$\sigma = \lim_{\epsilon \to 0^+} \sigma(\epsilon) = \left[\lim_{\epsilon \to 0^+} \int_0^\infty dt\, e^{-\epsilon t} K(t) \right]^{-1} . \qquad \text{(VII.28)}$$

All one needs is the limiting behavior of the transform of the memory function.

The preceding result has a close relationship to what one gets from the Markoffian approximation

$$\frac{dA(t)}{dt} - \int_0^\infty K(s)\, ds \cdot A(t)\ . \tag{VII. 29}$$

This equation can be solved easily to give

$$A(t) = \exp\left[-t\int_0^\infty ds\ K(s)\right], \tag{VII. 30}$$

and σ is given by

$$\sigma = \int_0^\infty dt\ A(t) = \left[\int_0^\infty ds\ K(s)\right]^{-1} \tag{VII. 31}$$

It is interesting to see that if integrals and limits exist, etc., the Markoffian approximation actually gives correct results for transport coefficients.

References

This list is not meant to be exhaustive.

1. L. Van Hove, Physica 21, 517 (1955). This is the first application of infinite order, many-body perturbation theory to the theory of irreversibility. The Pauli equation is derived here. Van Hove makes use of the $\lambda^2 t$ limit.
2. L. Van Hove, Physica 23, 441 (1957). The perturbation method of the preceding article is extended here to general order. The resolvent technique is used. Van Hove obtains a kinetic equation of the non-Markoffian, Volterra type that we discuss in these lectures.
3. R. Brout and I. Prigogine, Physica 22, 621 (1956).
4. R. Balescu and I. Prigogine, Physica 25, 281, 302 (1959). These and many intervening articles are concerned with perturbation theory in classical mechanics. The lectures by Balescu in this volume, p. 382, should be consulted for further material.
5. R. Kubo, Lectures at the Summer Institute of Theoretical Physics, Vol. I, (University of Colorado, 1958). These lectures give an excellent summary of Kubo's important work. The lectures by Montroll in this book also contain valuable material about Kubo's work.
6. B. O. Koopman, Proc. Natl. Acad. Sci. U.S. 17, 315 (1931). The original suggestion of the Hilbert space formulation of classical mechanics is contained here.
7. R. Kubo, J. Phys. Soc. Japan 12, 570 (1957). The quantum mechanical Liouville operator is referred to briefly here. Although I have not traced it any further back than this, it very likely appears in Von Neumann's work on rings of operators in Hilbert space.

8. U. Fano, Revs. Modern Phys. 29, 74 (1957). This valuable article contains an explicit construction of the quantum mechanical Liouville operator in the Hilbert space of operators. Another way of combining classical and quantum mechanical methods, which makes use of the Wigner function, is also discussed here.
9. R. J. Rubin, J. Math. Phys. 1, 309 (1960). Although there is no specific reference to this article in the lectures, it is nevertheless worth looking up. Rubin gives a thoroughly worked out example of how a simple classical many-body system can appear to be irreversible. The treatment is elementary, and makes no use of diagrammatic or field theoretic techniques. The discussion of time dependence in Section V of these lectures is greatly influenced by Rubin's work.

ON THE IDENTITY OF THREE GENERALIZED MASTER EQUATIONS

by ROBERT ZWANZIG

National Bureau of Standards, Washington, D.C.

Synopsis

Three apparently different quantum mechanical master equations, derived by Prigogine and Résibois, by Montroll, and independently by Nakajima and Zwanzig, are shown to be identical. The derivation by Zwanzig, based on projection operator and Liouville operator techniques, is repeated in greater detail than in previous articles. The results of Prigogine and Résibois, and of Montroll, are found by making changes in notation.

Introduction. Considerable attention has been given in recent years to the rigorous derivation of quantum mechanical master equations, valid to all orders in a perturbation. This article contains a demonstration of the identity of three apparently different results.

Four exact master equations have been found (to the best of my knowledge). In chronological order, they are as follows:

The first was obtained by Van Hove [1]) in 1957. Although his derivation was based on the limit of an infinite system, and in particular, on his diagonal singularity condition, the same equation has been obtained by Swenson [2]) without making any assumptions at all about the nature of the system. This equation is referred to as VHS in the following.

The second master equation was obtained by Nakajima [3]) in 1958, and independently and in greater detail by myself [4]) [5]) in 1960. Both derivations used projection operator and Liouville operator methods. The resulting master equation is referred to here as NZ.

The third master equation was derived by Résibois [6]) in 1961. His method of derivation was modified in a subsequent article by Prigogine and Résibois [7]), also in 1961. The final result is given explicitly in a 1963 article by Résibois [8]), where he refers to it as a master equation derived by Prigogine and Résibois; so we refer to it as PR.

The fourth master equation was derived by Montroll [9]) in 1961. An earlier article [10]) in 1960 gave the derivation to lowest order in the perturbation, and suggested the general approach.

In a recent article [8]) Résibois discussed the relation between VHS and

PR. He concluded that both equations were exact, and that they both could be used to calculate the same quantities, but that they were not identical. Résibois did not mention, however, the other master equations NZ and M.

The main purpose of the present article is to show that PR and M, while apparently different in structure from NZ, are actually identical with NZ aside from differences in notation. Specifically, M is obtained from NZ by using determinants and minors to calculate (formally) the inverse of a certain tetradic operator. Also, PR is obtained from NZ by replacing matrix subscripts (m, n) by their difference and their arithmetic mean. Thus, the results of this article, taken together with those of ref. 8, can be summarized in the equation NZ $=$ PR $=$ M $\neq$ VHS.

A secondary purpose of the present article is to call attention once more to the methods used to derive NZ, and to go into greater detail concerning the derivation than seemed desirable in my previous articles. These methods were devised in the first place to be fast and economical, so that one would not have to go through complicated and tedious arguments of the sort found in refs. 6 and 7. While concise and direct, these methods involve a certain amount of unfamiliar and abstract operator technique. It is hoped that the following detailed exposition will serve to make them more accessible and easily used.

Liouville operators and von Neumann's equation. The master equations with which we are concerned are kinetic equations for the diagonal elements of a density matrix. The density operator is $\hat{\rho}$; its matrix elements in some particular representation are ρ_{mn}.

The time dependence of the density matrix $\rho_{mn}(t)$ is determined by von Neumann's equation. In operator form, this is (with $\hbar = 1$)

$$\frac{\partial \hat{\rho}}{\partial t} = -i\,(\hat{H}\hat{\rho} - \hat{\rho}\hat{H}), \tag{1}$$

where $\hat{H}$ is the Hamiltonian operator. In the chosen representation, von Neumann's equation is

$$\frac{\partial \rho_{mn}}{\partial t} = -i \sum_{l} (H_{ml}\,\rho_{ln} - \rho_{ml}\,H_{ln}). \tag{2}$$

We wish to extract from this an equation, determining the time dependence of the diagonal elements of $\hat{\rho}$, which does not contain the nondiagonal elements of $\hat{\rho}$.

We solve von Neumann's equation by means of Liouville operators. These operators are defined as follows. Given an arbitrary operator $\hat{A}$, another operator $\hat{C}$ can be constructed by the rule

$$\hat{C} = (\hat{H}\hat{A} - \hat{A}\hat{H}). \tag{3}$$

In Heisenberg quantum mechanics this is tantamount to turning an operator into its time derivative. The operation of going from $\hat{A}$ to $\hat{C}$ will be denoted by L, and L will be called the Liouville operator,

$$\hat{C} = L\hat{A}. \tag{4}$$

It is clearly a linear operation. In the chosen representation it turns a matrix with two subscripts into another matrix with two subscripts; therefore, it can be represented by a *tetradic* with four subscripts,

$$C_{mn} = \sum_{m'} \sum_{n'} L_{mnm'n'} A_{m'n'}. \tag{5}$$

Because of the definition of L as the commutator with $\hat{H}$, the explicit form of the tetradic L is

$$L_{mnm'n'} = H_{mm'} \delta_{nn'} - \delta_{mm'} H_{n'n}. \tag{6}$$

The multiplication rule for tetradics

$$(L_1L_2)_{mnm'n'} = \sum_{a} \sum_{b} (L_1)_{mnab} (L_2)_{abm'n'} \tag{7}$$

can be found by evaluating repeated commutators. The identity tetradic is evidently

$$(1)_{mnm'n'} = \delta_{mm'} \delta_{nn'}. \tag{8}$$

Tetradics behave very much like ordinary matrices. In fact, their algebra can be reduced to that of matrices by the following trick. In representing an operator by a matrix, we may pick some arbitrary way of ordering pairs of subscripts, so that the pair (m, n) is denoted by a single integral subscript (α). In this way the matrix A_{mn} is replaced by the linear array or vector $A_{(\alpha)}$. Equation (5) can now be written as

$$C_{(\alpha)} = \sum_{(\beta)} L_{(\alpha)(\beta)} A_{(\beta)}. \tag{9}$$

We see that the tetradic L has been replaced by a matrix with two subscripts.

This means that the algebra of tetradics is isomorphic to the algebra of matrices. For finite tetradics, e.g. for interacting spin systems, the isomorphism is perfect. For infinite tetradics, convergence difficulties can arise as a result of the completely arbitrary ordering of subscript pairs. On this account we assume, as is customary in theoretical physics, that all infinite series converge unless there is good reason to believe otherwise; and we do not go into this question further.

We show later that the representation used by Résibois [6]) has an especially attractive property: the tetradic L is diagonal in a pair of bsuscripts. Because of this circumstance, he is able to use more conventional methods involving matrices with two subscripts.

In the Liouville operator notation, von Neumann's equation is

$$\frac{\partial \hat{\rho}}{\partial t} = -iL\hat{\rho}. \tag{10}$$

It has the formal solution, as an initial value problem,

$$\hat{\rho}(t) = \mathrm{e}^{-itL}\,\hat{\rho}(0). \tag{11}$$

The tetradic operator $\exp(-itL)$ is just as well defined as the more familiar $\exp(-it\hat{H})$. It can be calculated by expansion of the exponential function, followed by use of tetradic multiplication to get the powers of L. The useful identity

$$(\mathrm{e}^{-itL})_{mnm'n'} = (\mathrm{e}^{-it\hat{H}})_{mm'}\,(\mathrm{e}^{it\hat{H}})_{nn'} \tag{12}$$

can be verified by differentiation with respect to time, and application of eq. (6). On applying this identity to eq. (11) we get the familiar Heisenberg solution

$$\hat{\rho}(t) = \mathrm{e}^{-it\hat{H}}\,\hat{\rho}(0)\,\mathrm{e}^{it\hat{H}}. \tag{13}$$

The Liouville operator solution has two advantages over the usual Heisenberg operator solution. One advantage is the extra compactness in performing perturbation expansions. We divide the Hamiltonian $\hat{H}$ into an unperturbed part $\hat{H}_0$ and a perturbation $\hat{H}_1$,

$$\hat{H} = \hat{H}_0 + \hat{H}_1. \tag{14}$$

The perturbation expansion of $\exp(-it\hat{H})$ is well known,

$$\mathrm{e}^{-it\hat{H}} = \mathrm{e}^{-it\hat{H}_0} - i\int_0^t \mathrm{d}t_1\,\mathrm{e}^{-i(t-t_1)\hat{H}_0}\,\hat{H}_1\,\mathrm{e}^{-it_1\hat{H}_0} + \\ + (-i)^2\int_0^t \mathrm{d}t_1\int_0^{t_1}\mathrm{d}t_2\,\mathrm{e}^{-i(t-t_1)\hat{H}_0}\,\hat{H}_1\,\mathrm{e}^{-i(t_1-t_2)\hat{H}_0}\,\hat{H}_1\,\mathrm{e}^{-it_2\hat{H}_0} \\ + \ldots\ldots \tag{15}$$

When this is applied to eq. (13), the perturbation appears on both sides of $\hat{\rho}(0)$. To collect terms of the n-th order in the perturbation, one must combine contributions of various orders from the two sides.

But we may divide L into an unperturbed part L_0 and a perturbation L_1,

$$L = L_0 + L_1. \tag{16}$$

The perturbation expansion of $\exp(-itL)$ has exactly the same structure as that of $\exp(-it\mathscr{H})$,

$$\begin{aligned} e^{-itL} &= e^{-itL_0} - i\int_0^t dt_1\, e^{-i(t-t_1)L_0}\, L_1{}^{-it_1L_0} + \\ &+ (-i)^2 \int_0^t dt_1 \int_0^{t_1} dt_2\, e^{-i(t-t_1)L_0}\, L_1\, e^{-i(t_1-t_2)L_0}\, L_1\, e^{-it_2L_0} + \\ &+ \ldots . \end{aligned} \tag{17}$$

When this expansion is applied to eq. (13), the perturbation always appears on the left of $\hat{\rho}(0)$, and it is not necessary to collect and combine contributions from two sides.

But the real advantage of the Liouville operator solution lies in the *resolvent* form of solution. Let us solve von Neumann's equation by means of Laplace transforms. The transform of the density operator $\hat{\rho}$ is denoted by $\hat{g}(p)$,

$$\hat{g}(p) = \int_0^\infty dt\, e^{-pt}\, \hat{\rho}(t). \tag{18}$$

On transforming eq. (10), we get

$$p\hat{g}(p) - \hat{\rho}(0) = -\, iL\, \hat{g}(p). \tag{19}$$

The formal operator solution is

$$\hat{g}(p) = \frac{1}{p + iL}\, \hat{\rho}(0). \tag{20}$$

This is very much simpler than the Laplace transform solution using Hamiltonians. If we define the resolvent of the Hamiltonian by

$$R(z) = \frac{1}{\hat{H} - z}, \tag{21}$$

then the solution $\hat{g}(p)$ is given by

$$\hat{g}(p) = \frac{1}{2\pi} \oint dz\, R(z - ip)\, \hat{\rho}(0)\, R(z). \tag{22}$$

The contour of integration in the z plane encloses that part of the real axis occupied by the exact eigenvalues of $\hat{H}$.

Equation (22) is equivalent to the starting point of Van Hove's derivation; the only difference is the trivial one that we have taken the Laplace transform where he took the Fourier transform.

Separation into relevant and irrelevant parts. The derivation given in refs. (3), (4), and (5) was based on the separation of the density operator into relevant and irrelevant parts by means of a projection operator. The relevant part is clearly the diagonal part of $\hat{\rho}$; and the irrelevant part is the nondiagonal part of $\hat{\rho}$.

To select the diagonal part, we use the projection operator D. In tetradic form, D is

$$(D\hat{\rho})_{mn} = \rho_{mm}\, \delta_{mn},$$
$$D_{mnm'n'} = \delta_{mn}\, \delta_{mm'}\, \delta_{nn'}. \tag{23}$$

It is easy to verify that this has the desired property, and that it obeys the fundamental requirement of a projection, $D^2 = D$. The irrelevant or nondiagonal part is selected by the projection operator $1-D$.

Thus the density operator separates into the diagonal part $\hat{\rho}_1$ and the nondiagonal part $\hat{\rho}_2$,

$$\hat{\rho} = \hat{\rho}_1 + \hat{\rho}_2,$$
$$\hat{\rho}_1 = D\hat{\rho}, \qquad \hat{\rho}_2 = (1 - D)\hat{\rho}. \tag{24}$$

In just the same way, the Laplace transform $\hat{g}(p)$ separates into diagonal and nondiagonal parts,

$$\hat{g}(p) = \hat{g}_1(p) + \hat{g}_2(p),$$
$$\hat{g}_1(p) = D\hat{g}(p), \qquad \hat{g}_2(p) = (1-D)\, \hat{g}(p). \tag{25}$$

Next, we use D and $1-D$ to separate von Neumann's equation into two parts,

$$Dp\hat{g}(p) - D\hat{\rho}(0) = -iDL\, \hat{g}(p),$$
$$(1-D)\, p\hat{g}(p) - (1-D)\, \hat{\rho}(0) = -i(1-D)\, L\hat{g}(p), \tag{26}$$

or

$$p\hat{g}_1(p) - \hat{\rho}_1(0) = -iDL\, \hat{g}_1(p) - iDL\hat{g}_2(p),$$
$$p\hat{g}_2(p) - \hat{\rho}_2(0) = -i(1-D)L\hat{g}_1(p) - i(1-D)L\hat{g}_2(p). \tag{27}$$

We solve the second equation for $\hat{g}_2(p)$,

$$[p + i(1-D)L]\, \hat{g}_2(p) = \hat{\rho}_2(0) - i(1-D)L\, \hat{g}_1(p);$$
$$\hat{g}_2(p) = \frac{1}{p + i(1-D)L}\, \hat{\rho}_2(0) - \frac{1}{p + i(1-D)L}\, i(1-D)L\, \hat{g}_1(p); \tag{28}$$

and we put the solution back into the first equation,

$$p\hat{g}_1(p) - \hat{\rho}_1(0) = -iDL\, \hat{g}_1(p) - iDL \frac{1}{p + i(1-D)L}\, \hat{\rho}_2(0) -$$
$$- DL\, \frac{1}{p + i(1-D)L}\, (1-D)L\, \hat{g}_1(S). \tag{29}$$

The result of the preceding derivation is the Laplace transforms of our generalized master equation. Let us invert the transform, using the well known theorem that the inverse of a product is a convolution. The result is

$$\frac{\mathrm{d}\hat{\rho}_1(t)}{\mathrm{d}t} = -\,iDL\,\hat{\rho}_1(t) - iDL\,\mathrm{e}^{-i(1-D)Lt}\,\hat{\rho}_2(0) - \\ -\int_0^t \mathrm{d}t_1\,DL\,\mathrm{e}^{-it_1(1-D)L}(1-D)L\,\hat{\rho}_1(t-t_1). \tag{30}$$

In most applications of the master equation, the density operator is diagonal initially, or

$$\hat{\rho}_2(0) = 0. \tag{31}$$

This initial condition is commonly referred to as the assumption of initial random phases. When it applies, eq. (30) makes no reference to the non-diagonal elements of $\hat{\rho}$. Thus the master equation has the desired property of containing only the diagonal elements of $\hat{\rho}$.

When a more general initial condition applies, then all of eq. (30) must be used. Even so, the nondiagonal elements enter only through the initial value $\hat{\rho}_2(0)$. This structure is characteristic of the other master equations, M, PR, and VHS.

If the time dependence of the nondiagonal elements is also of interest, one needs the Laplace inversion of eq. (28). Th.s connects $\hat{\rho}_2(t)$ to $\hat{\rho}_1(t)$.

For the rest of this article we discuss only the diagonal elements, and use only the initial condition $\hat{\rho}_2(0) = 0$.

Explicit form of the master equation. Equation (30) is expressed in an abstract operator notation. Here we change to the more explicit subscript notation. We use the Liouville operator given in eq. (6), and the projection operator given in eq. (23).

At this point we require that the representation be chosen so that the unperturbed energy is diagonal. (This is true of the other master equations too). The unperturbed energy eigenvalues will be denoted by E_n, so that

$$(H_0)_{mn} = E_m\,\delta_{mn}. \tag{32}$$

Then the unperturbed Liouville operator is

$$(L_0)_{mnm'n'} = (E_m - E_n)\,\delta_{mm'}\,\delta_{nn'}. \tag{33}$$

It is a trivial calculation to verify that

$$DL_0 = L_0D = 0. \tag{34}$$

Also, the product DLD vanishes for any Hamiltonian. This is seen from the definitions, which lead to

$$(DLD)_{mnm'n'} = \delta_{mn}\,L_{mmm'm'}\,\delta_{m'n'}. \tag{35}$$

But because of eq. (6), L_{mmnn} vanishes. Therefore we have

$$DLD = 0. \tag{36}$$

Equation (36) serves to eliminate the first term on the right of eq. (30). The reason is

$$DL\,\hat{\rho}_1 = DLD\,\hat{\rho}. \tag{37}$$

The rest of eq. (30), without the initial value $\hat{\rho}_2(0)$, has the explicit form

$$\frac{\mathrm{d}\hat{\rho}_{mm}(t)}{\mathrm{d}t} = -\int_0^t \mathrm{d}t_1\,[K(t_1)\,\hat{\rho}_1(t-t_1)]_{mm}. \tag{38}$$

We have abbreviated

$$K(t) = DL\,\mathrm{e}^{-it(1-D)L}\,(1-D)L. \tag{39}$$

Because $\hat{\rho}_1$ is diagonal, eq. (38) is

$$\frac{\mathrm{d}\rho_{mm}(t)}{\mathrm{d}t} = -\int_0^t \mathrm{d}t_1 \sum_n K_{mmnn}(t_1)\,\rho_{nn}(t-t_1). \tag{40}$$

As a result of the identities in eq. (34), the memory kernel $K_{mmnn}(t)$ can be simplified slightly to

$$K_{mmnn}(t) = [L_1\,\mathrm{e}^{-it(1-D)L}(1-D)L_1]_{mmnn}. \tag{41}$$

Note that this is formally of the second order in the perturbation; it contains two explicit factors L_1. Higher order dependence on the perturbation comes from the exponential operator in K.

Next we prove the sum rule

$$\sum_n K_{mmnn}(t) \equiv 0. \tag{42}$$

The demonstration starts with

$$\begin{aligned}\sum_n K_{mmnn} &= \sum_n \sum_a \sum_b [L_1\,\mathrm{e}^{-it(1-D)L}(1-D)]_{mmab}\,(L_1)_{abnn} = \\ &= \{\sum_a \sum_b [L_1\,\mathrm{e}^{-it(1-D)L}(1-D)]_{mmab}\} \times \sum_n (L_1)_{abnn}.\end{aligned} \tag{43}$$

But according to the definition of the Liouville operator,

$$\sum_n (L_1)_{abnn} = \sum_n (H_1)_{an}\,\delta_{bn} - \sum_n \delta_{an}\,(H_1)_{nb} \tag{44}$$

which proves our assertion .

The sum rule can be used as follows,

$$\sum_{n\neq m} K_{mmnn} = -K_{nn}, \tag{45}$$

so that our master equation takes the familiar gain-loss form

$$\frac{\mathrm{d}\rho_{mm}(t)}{\mathrm{d}t} = -\int_0^t \mathrm{d}t_1 \sum_{n\neq m} K_{mmnn}(t_1)\,[\rho_{nn}(t-t_1) - \rho_{mm}(t-t_1)] \tag{46}$$

with the kernel K given explicitly by eq. (41).

Montroll's master equation. We demonstrate here the identity of the generalized master equation derived by Montroll [9]) and our own master equation.

Several changes in notation are required to conform with Montroll's work. We replace t_1 by τ, m and n by j and k, and $\hat{H}_1$ by λU.

Montroll's equation has the general structure of eq. (46). To prove the identity of the two equations, we must obtain his memory kernel from ours. We start with the definition of K,

$$K_{jjkk}(\tau) = [L_1\,\mathrm{e}^{-i\tau(1-D)L}\,(1-D)L_1]_{jjkk}. \tag{47}$$

By tetradic multiplication this is

$$= \sum_a \sum_b \sum_c \sum_d (L_1)_{jjab}\,[\mathrm{e}^{-i\tau(1-D)L}]_{abcd}\,(L_1)_{cdkk}. \tag{48}$$

Note that the terminal $(1-D)$ in eq. (47) is redundant, because L_{cckk} always vanishes. For convenience we abbreviate

$$\mathrm{e}^{-i\tau(1-D)L} = \mathscr{G}(\tau). \tag{49}$$

On using the definition of the tetradic L, we obtain

$$K_{jjkk}(\tau) = \lambda^2 \sum_a \sum_b \sum_c \sum_d U_{ba}\,U_{cd}\,\mathscr{G}_{abcd}(\tau) \times$$
$$\times\,(\delta_{bj} - \delta_{aj})\,(\delta_{dk} - \delta_{ck}). \tag{50}$$

Montroll's calculation is equivalent to evaluating the elements of the tetradic G by means of Laplace transforms. Let us define

$$\mathscr{D}(p) = \int_0^\infty \mathrm{d}t\,\mathrm{e}^{-pt}\,\mathscr{G}(t). \tag{51}$$

By direct integration this is also

$$\mathscr{D}(p) = \frac{1}{p + i(1-D)L} \tag{52}$$

We observed earlier in this article that tetradics can be manipulated as if they were matrices, by the device of associating subscript pairs with single integers. Thus the inverse tetradic $\mathscr{D}(p)$ may be calculated in terms of the determinant and minors of the tetradic in the denominator,

$$M = p + i(1-D)L \tag{53}$$

or

$$M_{abcd} = p\,\delta_{ac}\,\delta_{bd}\,\delta_{ab} + i\,\omega_{ab}\,\delta_{ac}\,\delta_{bd} + \\ + i(1-\delta_{ab})\,\{\lambda U_{ac}\,\delta_{bd} - \lambda\,\delta_{ac}\,U_{db}\}. \qquad (54)$$

(In writing eq. (54) we used the abbreviation $E_a - E_b = \omega_{ab}$). The determinant of M is $D(p)$. The minor associated with the $[(ab), (cd)]$ element of M is $D(ab/cd; p)$. Thus the inverse is

$$\mathscr{D}_{abcd}(p) = \mathscr{D}(ab/cd; p) = \frac{D(ab/cd; p)}{D(p)}. \qquad (55)$$

These quantities are all identical with those introduced by Montroll, and we have used his notation.

Next, we use the operator identity

$$\frac{1}{p+i(1-D)L} = \frac{1}{p+i(1-D)L_0} - \frac{1}{p+i(1-D)L_0}\,i(1-D)L_1\,\frac{1}{p+i(1-D)L}. \qquad (56)$$

Equation (52) introduces $\mathscr{D}$, and eq. (34) eliminates the quantity DL_0. In this way we obtain

$$\mathscr{D}(p) = \frac{1}{p+iL_0} - \frac{1}{p+iL_0}\,i(1-D)L_1\,\mathscr{D}(p). \qquad (57)$$

The unperturbed resolvent, in subscript notation, is

$$\left(\frac{1}{p+iL_0}\right)_{abcd} = \frac{1}{p+i\omega_{ab}}\,\delta_{ac}\,\delta_{bd}, \qquad (58)$$

so that

$$\mathscr{D}_{abcd}(p) = \frac{1}{p+i\omega_{ab}}\,\delta_{ac}\,\delta_{bd} - \\ - \frac{i\lambda}{p+i\omega_{ab}}(1-\delta_{ab})\sum_m \{U_{am}\,\mathscr{D}_{mbcd}(p) - U_{mb}\,\mathscr{D}_{amcd}(p)\}. \qquad (59)$$

The inverse Laplace transform of eq. (59) is

$$\mathscr{G}_{abcd}(\tau) = \mathrm{e}^{-i\omega_{ab}\tau}\,\delta_{ac}\,\delta_{bd} - \\ - \frac{\lambda}{2\pi}\int_{C-i\infty}^{C+i\infty} \mathrm{d}p\;\mathrm{e}^{p\tau}\,\frac{1-\delta_{ab}}{p+i\omega_{ab}}\sum_m \{U_{am}\mathscr{D}_{mbcd}(p) - U_{mb}\mathscr{D}_{amcd}(p)\}. \qquad (60)$$

When this is substituted in eq. (50) to give the kernel $K_{jjkk}(\tau)$, and the kernel is put into eq. (46), we obtain Montroll's form of the master equation.

Prigogine and Résibois' master equation. We demonstrate here the identity of the generalized master equation derived by Prigogine and Résibois and our own master equation.

They use a second-quantized Hamiltonian, in the occupation number representation. This is quite unnecessary, however, and their results are (in a formal sense) considerably more general.

The states of the unperturbed Hamiltonian $\hat{H}_0$ are labelled by the quantum number n. (In the occupation number representation, these are the numbers of elementary excitations in each unperturbed state). We do not have to give any physical interpretation to these quantum numbers. The matrix elements of the unperturbed Hamiltonian are

$$(\hat{H}_0)_{nn'} = E_n \, \delta_{nn'}, \tag{61}$$

and the perturbation $\hat{H}_1 = \lambda V$ has matrix elements $V_{nn'}$.

Résibois uses a special notation for matrix elements. For an arbitrary operator $\hat{A}$, let

$$A_{nn'} = A_{n-n'}\left(\frac{n+n'}{2}\right), \tag{62}$$

and use the abbreviations

$$\nu = n - n'; \; N = \frac{n+n'}{2}. \tag{63}$$

Thus the average of $\hat{A}$ is

$$\begin{aligned} \langle \hat{A} \rangle &= \sum_n \sum_{n'} A_{nn'} \, \rho_{n'n} = \\ &= \sum_\nu \sum_N A_\nu(N) \, \rho_{-\nu}(N). \end{aligned} \tag{64}$$

When $\hat{A}$ is diagonal in the unperturbed representation, we have

$$\begin{aligned} A_{nn'} &= A_{nn} \, \delta_{nn'}; \\ A_\nu(N) &= A_0(N) \, \delta_{\nu 0}; \end{aligned} \tag{65}$$

so that the average of $\hat{A}$ is

$$\langle \hat{A} \rangle = \sum_N A_0(N) \, \rho_0(N). \tag{66}$$

Thus our master equation is concerned with the N and t dependence of the particular matrix elements $\rho_0(N, t)$.

The Liouville operator is expressed in this notation as follows. We start with

$$L_{mnm'n'} = H_{mm'} \, \delta_{nn'} - \delta_{mm'} \, H_{n'n} \tag{67}$$

and use the substitutions

$$m = N + \frac{\nu}{2}, \quad n = N - \frac{\nu}{2},$$

$$m' = N' + \frac{\nu'}{2}, \quad n' = N' - \frac{\nu'}{2}. \tag{68}$$

Then the Liouville operator is

$$L_{\nu\nu'}(N, N') = H_{\nu-\nu'}(N + \tfrac{1}{2}\nu')\, \delta\left(N' - N - \frac{\nu' - \nu}{2}\right) -$$

$$- H_{\nu-\nu'}(N - \tfrac{1}{2}\nu')\, \delta\left(N' - N + \frac{\nu' - \nu}{2}\right). \tag{69}$$

Following Résibois, we introduce the shift operator $\eta^{\pm\nu}$ which replaces a function of N by the same function of $N \pm \frac{1}{2}\nu$,

$$\eta^{\pm\nu} f(N) = f(N \pm \tfrac{1}{2}\nu). \tag{70}$$

Then the Liouville operator is

$$L_{\nu\nu'}(N, N') = \eta^{\nu'} H_{\nu-\nu'}(N)\, \eta^{-\nu}\, \delta(N - N') -$$

$$- \eta^{-\nu'} H_{\nu-\nu'}(N)\, \eta^{\nu}\, \delta(N - N'). \tag{71}$$

On introducing Résibois' matrix element (still an operator with respect to N),

$$\langle \nu | \mathscr{H}(N) | \nu' \rangle = \eta^{\nu'} H_{\nu-\nu'}(N)\, \eta^{-\nu} - \eta^{-\nu'} H_{\nu-\nu'}(N)\, \eta^{\nu}, \tag{72}$$

we obtain

$$L_{\nu\nu'}(N, N') = \langle \nu | \mathscr{H}(N) | \nu' \rangle\, \delta(N - N'). \tag{73}$$

This representation has the remarkable property of being "diagonal" in N, for the perturbed system as well as the unperturbed one.

The diagonality referred to here is exactly the same as one sees in the coordinate representation of matrix mechanics. For example, the matrix of the momentum operator in the coordinate representation is

$$p(\boldsymbol{r} - \boldsymbol{r}') = - i\nabla_{\boldsymbol{r}}\, \delta(\boldsymbol{r} - \boldsymbol{r}'). \tag{74}$$

This matrix is "diagonal" in the sense that it contains the delta function in positions. As is well known, one can drop the delta function and the matrix notation, and deal exclusively with Schrödinger operators. Résibois does this with respect to the quantum numbers N. Therefore, his operator $\mathscr{H}(N)$ is just another way of writing the Liouville operator L; and his matrix-operator $\langle \nu | \mathscr{H}(N) | \nu' \rangle$ is a matrix with respect to one set of quantum numbers and an operator with respect to the other.

In this notation, von Neumann's equation becomes

$$\frac{\partial \rho_\nu(N,t)}{\partial t} = \sum_{\nu'} \sum_{N'} L_{\nu\nu'}(N,N')\, \rho_{\nu'}(N',t) = $$
$$= \sum_{\nu'} \langle \nu | \mathscr{H}(N) | \nu' \rangle\, \rho_{\nu'}(N,t). \tag{75}$$

For the moment we keep the full tetradic notation.

The projection operator defined by eq. (23), when rewritten by means of the substitutions in eq. (68), becomes

$$D_{\nu\nu'}(N,N') = \delta(\nu)\,\delta(\nu')\,\delta(N-N'). \tag{76}$$

Note that the projection operator also is diagonal in N.

Now we are ready to go from eq. (38) to the PR master equation. In present notation, eq. (38) is

$$\frac{\mathrm{d}\rho_0(N,t)}{\mathrm{d}t} = -\int_0^t \mathrm{d}t_1 \sum_{N'} K_{00}\,(N,N';t_1)\,\rho_0(N',t-t_1), \tag{77}$$

and the kernel is

$$K_{00}(N,N';t) = [L_1\,\mathrm{e}^{-it(1-D)L}\,(1-D)\,L_1]_{00,NN'}. \tag{78}$$

For simplicity of notation we have written the (N,N') dependence as subscripts.

Each tetradic in the kernel contains a delta function in N; see eqs. (73) and (76). Now we drop these delta functions and deal exclusively with *operators* in N space. The tetradic $L_{\nu\nu'}(N,N)$ is replaced by the matrix-operator $\langle \nu|\mathscr{H}(N)|\nu'\rangle$, and the projection operator tetradic $D_{\nu\nu'}(N,N')$ is replaced by the matrix

$$\langle \nu|\mathscr{D}|\nu'\rangle = \delta(\nu)\,\delta(\nu'). \tag{79}$$

The operator corresponding to the matrix $K_{00}(N,N';t)$ will be denoted by $-G(N,t)$. Evidently its explicit formula is

$$-G(N,t) = \langle 0|\,\mathscr{H}_1\,\mathrm{e}^{-it(1-\mathscr{D})\mathscr{H}}\,(1-\mathscr{D})\mathscr{H}_1\,|0\rangle, \tag{80}$$

where $\langle 0| - |0\rangle$ means the matrix element between $\nu = 0$ and $\nu = 0$. The N dependence of $\mathscr{H}$ was left implicit to save space.

In this operator notation, eq. (77) is

$$\frac{\mathrm{d}\rho_0(N,t)}{\mathrm{d}t} = \int_0^t \mathrm{d}t_1\, G(N,t_1)\,\rho_0(N,t-t_1). \tag{81}$$

This has precisely the form of PR. All that remains is to show the identity of G, as defined in eq. (80), with the quantity G_{00} defined by Résibois.

By means of residue theory, we may express G in terms of a new quantity $\psi(Z)$,

$$G(N, t) = \frac{1}{2\pi i} \oint \mathrm{d}z\, \mathrm{e}^{-itz} \psi(Z) \tag{82}$$

where $\psi(Z)$ must have the value

$$\psi(Z) = -\left\langle 0 \middle| \mathscr{H}_1 \frac{1}{Z-(1-\mathscr{D})\mathscr{H}} (1-\mathscr{D})\, \mathscr{H}_1 \middle| 0 \right\rangle . \tag{83}$$

The Hamiltonian is separated into two parts, the unperturbed $\hat{H}_0$ and the perturbation $\hat{H}_1 = \lambda V$; the Liouville operator $\mathscr{H}$ is separated in just the same way into $\mathscr{H}_0$ and $\mathscr{H}_1 = \lambda \mathscr{V}$,

$$\begin{aligned} \hat{H} &= \hat{H}_0 + \lambda V, \\ \mathscr{H} &= \mathscr{H}_0 + \lambda \mathscr{V}. \end{aligned} \tag{84}$$

Now we make the familiar power series expansion

$$\frac{1}{Z-(1-\mathscr{D})\mathscr{H}} = \frac{1}{Z-(1-\mathscr{D})\mathscr{H}_0} + \frac{1}{Z-(1-\mathscr{D})\mathscr{H}_0} (1-\mathscr{D})\lambda\mathscr{V} \cdot$$
$$\cdot \frac{1}{Z-(1-\mathscr{D})\mathscr{H}_0} + \frac{1}{Z-(1-\mathscr{D})\mathscr{H}_0} (1-\mathscr{D})\lambda\mathscr{V} \frac{1}{Z-(1-\mathscr{D})\mathscr{H}_0} (1-\mathscr{D})\lambda\mathscr{V} \cdot$$
$$\cdot \frac{1}{Z-(1-\mathscr{D})\mathscr{H}_0} + \ldots . \tag{85}$$

Because DL_0 vanishes, as in eq. (34), it is obvious that $\mathscr{D}\mathscr{H}_0$ also vanishes and that

$$Z - (1-\mathscr{D})\mathscr{H}_0 = Z - \mathscr{H}_0 . \tag{86}$$

Then the expansion of $\psi(Z)$ is

$$\psi(Z) = -\left\langle 0 \middle| \lambda\mathscr{V} \left\{ \frac{1}{Z-\mathscr{H}_0} + \frac{1}{Z-\mathscr{H}_0} (1-\mathscr{D})\, \lambda\mathscr{V} \frac{1}{Z-\mathscr{H}_0} + \right.\right.$$
$$\left.\left. + \frac{1}{Z-\mathscr{H}_0} (1-\mathscr{D})\, \lambda\mathscr{V} \frac{1}{Z-\mathscr{H}_0} (1-\mathscr{D})\, \lambda\mathscr{V} \frac{1}{Z-\mathscr{H}_0} + \ldots \right\} (1-\mathscr{D})\, \lambda\mathscr{V} \middle| 0 \right\rangle . \tag{87}$$

The projection operators $1 - \mathscr{D}$ contained in each term have the effect that in calculating matrix products, no intermediate states with $\nu = 0$ may occur. This is the same as the irreducibility criterion of Résibois. We drop the $1 - \mathscr{D}$ and instead put a subscript $\nu \neq 0$ on the matrix elements,

$$\psi(Z) = -\left\langle 0 \middle| \lambda\mathscr{V} \left\{ \frac{1}{Z-\mathscr{H}_0} + \frac{1}{Z-\mathscr{H}_0} \lambda\mathscr{V} \frac{1}{Z-\mathscr{H}_0} + \ldots \right\} \lambda\mathscr{V} \middle| 0 \right\rangle_{\nu \neq 0} . \tag{88}$$

But this is precisely the formula given by Résibois [8]) for his operator $\psi^{+}_{00}(Z)$. We conclude that our $G(N, t)$ is identical with his $G_{00}(N, t)$. Thus the master equations PR and NZ are identical.

Received 22-11-63

REFERENCES

1) Van Hove, L., Physica **23** (1957) 441.
2) Swenson, R. J., J. math. Phys. **3** (1962) 1017.
3) Nakajima, S., Prog. theor. Phys. **20** (1958) 948.
4) Zwanzig, R., J. chem. Phys. **33** (1960) 1338.
5) Zwanzig, R., Lectures in theoretical Physics (Boulder) **3** (1960) 106.
6) Résibois, P., Physica **27** (1961) 541.
7) Prigogine, I. and Résibois, P., *ibid.* (1961) 629.
8) Résibois, P., Physica **29** (1963) 721.
9) Montroll, E. W., Fundamental Problems in Statistical Mechanics, compiled by E. G. D. Cohen, pp. 230–249. North-Holland Publishing Co., (1962). This volume is the proceedings of a Summer Course held in August 1961.
10) Montroll, E. W., Lectures in theoretical Physics (Boulder) **3** (1960) 221.

2) Some General Properties of the Master Equation

Some general properties of master equations and their solutions have been presented in Chapter 3 of the text. In the papers in this section, properties of master equation solutions for special forms of the transition rate matrix are discussed. The first two papers in this section discuss the conditions under which an initial multinomial or canonical form of the probabilities (or probability densities) remains invariant with time. In the first paper systems are considered in which a multinomial form of the probability is preserved in time. In the second paper, the necessary and sufficient conditions on the transition rate matrix for the exact preservation of a canonical distribution are derived. The problem of the approximate preservation of an initial Gaussian distribution has been considered recently by Kubo et al[4.2.1]. In the third paper, the necessary and sufficient conditions on the transition rate matrix are derived under which reduced probabilities obey master equations and under which Ursell functions decay to their equilibrium values faster than the probability distributions. The Ursell functions are measures of correlations and their temporal behavior is of great importance.

In the fourth paper the relationship between the solutions of random walks and master equations is discussed. This paper together with the paper by Kenkre, Montroll and Schlesinger[4.2.2] describes the conditions under which random walk problems can be treated by master equations.

References

4.2.1. R. Kubo, K. Matsuo and K. Kitahara, J. Stat. Phys. 9, 51, (1973).

4.2.2. V. M. Kenkre, E. W. Montroll and M. F. Schlesinger, J. Stat. Phys. 9, 45 (1973).

Stochastic Equations for Nonequilibrium Processes*†

P. M. MATHEWS,‡ I. I. SHAPIRO, AND D. L. FALKOFF§
Department of Physics, Brandeis University, Waltham, Massachusetts
(Received August 3, 1959)

A system of weakly interacting particles is described by a time-dependent joint probability distribution in the occupation numbers of the individual particle states. The "master" equation for the distribution is obtained by considering the time evolution of the system as a Markoff process with transition probabilities per unit time given by first-order quantum mechanical perturbation theory. This is done for particles obeying classical and quantum statistics. The resulting equations include the usual rate equations for the average occupation numbers as special cases; but they also yield all higher moments and correlations in the occupation numbers. The general solution and its properties are discussed for the case in which a relaxing subsystem interacts via binary collisions with a larger system having a fixed but not necessarily thermal distribution. The explicit solution for the joint distribution in occupation numbers for all time is constructed for the case of identical harmonic oscillators which have an arbitrary initial distribution. These interact via binary collisions with a reservoir of similar oscillators, the coupling being linear in each oscillator coordinate. This model is also generalized and solved for a case in which the number of interacting particles is not conserved.

I. INTRODUCTION

A BASIC problem of theoretical physics is that of describing the approach to equilibrium, from an arbitrary initial state, of a system composed of a large number of interacting particles. Ever since the classic work of Boltzmann, it has been recognized[1] that statistical, as well as dynamical, considerations must play an important role in describing the change in time of such systems. In this paper we shall characterize the state of the whole system at any time by a joint probability distribution (JPD), $P(n_1,\cdots,n_i,\cdots;t)$, for the occupation numbers, n_i, of the individual particle states, i. Further, assuming that the change of state of the system may be described as a Markoff stochastic process, we shall obtain the "master equation"[2] appropriate to such a process for the systems under consideration. The transition probabilities per unit time are taken from first-order quantum mechanical perturbation theory.

This procedure leaves open the question of how far such probabilistic equations are derivable from the exact quantum dynamical equations of motion. Implicit in our approach is the assumption (which has been used, for example, by Pauli[3] in his derivation of the H-theorem) that the occupation numbers remain good quantum numbers for all time. Since the occupation numbers correspond to diagonal elements of a density matrix, it is not obvious why, during the course of time, nondiagonal matrix elements should not become equally important. Pauli eliminated them by invoking random phase averages at all times. Considerable progress has been made recently in clarifying this situation, mainly through the work of Van Hove,[4] Brout,[5] and Prigogine[6] which provides both a critique of and a justification for the use of a master equation. In particular, Van Hove has succeeded in deriving the

* A preliminary account of this work was given in Bull. Am. Phys. Soc. 4, 15 (1959).
† This research was supported by the Office of Naval Research, the Air Force Cambridge Research Center, and the National Science Foundation.
‡ Present address: Department of Physics, University of Madras, Madras 25, India.
§ Now on leave at CERN, Geneva, Switzerland.
[1] P. Ehrenfest and T. Ehrenfest, Encyclopaedie der Math. Wissenschaften, Vol. IV, pt. 32, 1911.
[2] This terminology is due to G. E. Uhlenbeck. It refers to the differential equation governing the time dependence of the JPD of the occupation numbers for a Markoff process. See, for example, G. E. Uhlenbeck, Higgins Lectures, Princeton University, 1954 (unpublished).
[3] W. Pauli, *Festschrift zum 60 Gebürtstag A. Sommerfelds* (S. Hirzel, Leipzig, 1928), p. 30.
[4] L. Van Hove, Physica **21**, 517 (1955); **23**, 441 (1957).
[5] R. Brout, Physica **22**, 509 (1956).
[6] I. Prigogine and co-workers in numerous papers in Physica **23**, **24**, and **25** (1957–1959).

master equation, under certain conditions, as a valid first approximation from the quantum mechanical equations of motion for a system of a large number of weakly interacting particles.

The understanding of the approach to equilibrium described by using the master equation has been greatly aided by the rapid development of the theory of stochastic processes.[7–11] On the one hand, general theorems have been proved which establish such relevant results as the existence, for a wide class of processes, of a unique stationary distribution to which a system will relax irrespective of its initial distribution. On the other hand, exact treatments of various special cases, i.e., models, have been given which have provided greater physical insight into the statistical behavior of systems approaching equilibrium. Indeed, although the instructiveness of such models has long been appreciated,[12,13] it is only in recent years that even the simpler schematic models (such as the Ehrenfests' "dog-flea"[12] and "wind-tree" models) have been treated adequately.[14,15]

This paper is a contribution in the latter direction, namely, that of delving more deeply into the consequences of the probabilistic formulation of relaxation phenomena. In Sec. II, we set up master equations for the JPD (joint probability distribution) in the occupation numbers for the cases in which the interacting particles obey classical or quantum statistics.[16] Then in Sec. III, we exhibit some general consequences of these equations when the interacting particles may be considered to consist of two sets, one of which (the relaxing set) is free to change its distribution, while the other (the reservoir set) has its distribution maintained constant in time by some external agency.

In Sec. IV, a complete solution is given for a special physical situation: the relaxation of an arbitrarily excited gas of harmonic oscillators interacting via binary collisions with a (not necessarily thermal) reservoir of similar oscillators. Although this process is of considerable interest in the study of collisional and radiative relaxation of vibrationally excited gases, and has been treated in this connection by Shuler,[17] Montroll,[18] and Rubin,[19] we shall here be concerned with it solely as a model for which one can exhibit in detail the occupation number JPD which describes the approach to equilibrium.

Finally, in Sec. V, a generalization of this model is solved in which the number of relaxing oscillators is no longer kept constant, thus allowing for the possibility of formation and dissociation of diatomic molecules.

The essential point in our development which makes possible a more complete statistical description of relaxation phenomena than that given by the conventional rate equations[20] is this: the latter are equations for the average occupation numbers, $\bar{n}_i$. However, the n_i are random variables and we determine their joint distribution $P(n_1,n_2,\cdots;t)$ as a function of time. In obtaining the master equation governing this JPD, the same quantum mechanical transition probabilities are used as in obtaining the usual rate equations. These latter, of course, follow from our master equation by suitable averaging. But from our JPD, $P(n_1,n_2,\cdots;t)$, we can obtain in addition all higher moments, e.g., correlations, $\langle n_i n_j\rangle_{av}$,[21] and fluctuations, $\langle(n_i-\bar{n}_i)^2\rangle_{av}$, in the occupation numbers for all time.

II. MASTER EQUATION FOR THE PROBABILITY DISTRIBUTION AND ITS GENERATING FUNCTION

A. The Master Equation

In this section we derive the master equation for both classical (M. B.) and quantum (B. E. and F. D.) statistics and deduce from it the associated equation for the generating function of the JPD.

This joint distribution, $P(\langle n|;t)\equiv P(n_1,n_2,\cdots;t)$, satisfies

$$P(\langle n|;t+dt) = \sum_{\langle m|} P(\langle m|;t)P(\langle m|;t\rightarrow\langle n|;t+dt), \quad (2.1)$$

where $P(\langle m|;t\rightarrow\langle n|;t+dt)$ is the conditional probability of finding the set of occupation numbers $\langle n|$ at

[7] M. C. Wang and G. E. Uhlenbeck, Revs. Modern Phys. **17**, 323 (1945).

[8] S. Chandresekhar, Revs. Modern Phys. **15**, 1 (1943).

[9] M. Kac, *Probability and Related Topics in Physical Sciences* (Interscience Publishers, New York, 1959).

[10] J. E. Moyal, J. Roy. Stat. Soc. **B11**, 150 (1949).

[11] References 6–10 are concerned mainly with physical applications. More mathematical treatments are given in the books: W. Feller, *Introduction to Probability Theory* (John Wiley & Sons, Inc., New York, 1957); M. S. Bartlett, *Introduction to Stochastic Processes* (Cambridge University Press, New York, 1955); J. S. Doob, *Stochastic Processes* (John Wiley & Sons, Inc., New York, 1953).

[12] P. Ehrenfest and T. Ehrenfest, Physik. Z. **8**, 311 (1907).

[13] M. v. Smoluchowski, Physik. Z. **17**, 557 (1916).

[14] M. Kac, Am. Math. Monthly **54**, 369 (1947).

[15] A. J. F. Siegert, Phys. Rev. **76**, 1708 (1949).

[16] This formulation of the relaxation problem is due to Siegert and Moyal (references 15 and 10), who, however, were primarily concerned with the method rather than with obtaining explicit solutions for specific models. Their work is noteworthy in giving the Boltzmann Stosszahlansatz a proper probabilistic formulation for both classical and quantum statistics; our work is a direct continuation of theirs.

[17] K. E. Shuler, J. Phys. Chem. **61**, 849 (1957).

[18] E. W. Montroll and K. E. Shuler, J. Chem. Phys. **26**, 454 (1957).

[19] R. J. Rubin and K. E. Shuler, J. Chem. Phys. **26**, 137 (1957).

[20] See references 17–19 for the rate equations describing interacting harmonic oscillators. The rate equations for the average occupation numbers are used in most kinetic treatments of entropy or the *H*-theorem. See, for example, R. Tolman, *Principles of Statistical Mechanics* (Clarendon Press, Oxford, 1938); D. Ter Haar, *Elements of Statistical Mechanics* (Rinehart and Company, New York, 1954), Appendix I; or W. Pauli, reference 3. Such equations are also standard in the theory of paramagnetic relaxation, e.g., N. Bloembergen, R. V. Pound, and E. M. Purcell, Phys. Rev. **73**, 679 (1948).

[21] Because of typesetting limitations, the symbol "$\langle\ \rangle_{av}$" is used in place of a bar to indicate averages of products of random variables.

time $t+dt$, given the set $\langle m|$ at t. The summation extends over all possible sets $\langle m|$ of the occupation numbers. We restrict ourselves to processes with stationary transition probabilities so that

$$P(\langle m|;t\rightarrow\langle n|;t+dt)$$

depends only on the time difference dt and not on t.

Assuming that the limit exists, we define

$$Q(\langle m|\rightarrow\langle n|) = \lim_{dt\rightarrow 0}\left[\frac{P(\langle m|;t\rightarrow\langle n|;t+dt)-\prod_i\delta(n_i,m_i)}{dt}\right], \quad (2.2)$$

where $\delta(n_i,m_i)$ is unity if $n_i=m_i$ and is zero otherwise. We note that Q satisfies

$$Q(\langle m|\rightarrow\langle n|)\geqslant 0 \quad \text{for} \quad \langle m|\neq\langle n|, \quad (2.3a)$$

and

$$Q(\langle m|\rightarrow\langle m|)=-\sum_{\langle n|\neq\langle m|}Q(\langle m|\rightarrow\langle n|). \quad (2.3b)$$

These relations follow from

$$P(\langle m|\rightarrow\langle n|;t)\equiv P(\langle m|;t_0\rightarrow\langle n|;t_0+t)\geqslant 0, \quad (2.4a)$$

and

$$\sum_{\langle n|}P(\langle m|\rightarrow\langle n|;t)=1. \quad (2.4b)$$

If $\langle m|\neq\langle n|$, then $Q(\langle m|\rightarrow\langle n|)$ is just the probability per unit time that the system undergoes in a transition from a state with the set of occupation numbers $\langle m|$ to another with the set of occupation numbers $\langle n|$.

The matrix elements $Q(\langle m|\rightarrow\langle n|)$ differ from zero only when the conservation laws are satisfied. From (2.1) and (2.2) one gets the master equation

$$\frac{\partial P(\langle n|;t)}{\partial t}=\sum_{\langle m|}P(\langle m|;t)Q(\langle m|\rightarrow\langle n|), \quad (2.5)$$

or, in more succinct notation,

$$\partial\langle P(t)|/\partial t=\langle P(t)|Q, \quad (2.5a)$$

where $\langle P(t)|$ is a row vector with elements $P(\langle n|;t)$ and Q is the matrix whose rows and columns are indexed by the sets of occupation numbers $\langle m|$ and $\langle n|$, respectively. The formal solution to (2.5a) is

$$\langle P(t)|=\langle P(0)|e^{Qt}, \quad (2.6)$$

where $\langle P(0)|$ is the row vector of the arbitrary initial distribution, $P(\langle n|;0)$. This deceptively simple-looking formal solution is of little use in practice since it can entail exponentiating an infinite dimensional matrix if the number of states is infinite. Indeed, the master equation (2.5) then represents an infinite set of ordinary differential equations, one for each possible set of occupation numbers. The solution of this set of equations for the JPD, with the integral-valued arguments n_i, is often facilitated by transforming from P to a probability generating function (PGF), Φ, defined by the equation

$$\Phi(z_1,z_2,\cdots;t)\equiv\Phi(|z\rangle;t) =\sum_{\langle n|}z_1^{n_1}z_2^{n_2}\cdots P(\langle n|;t). \quad (2.7)$$

Equation (2.7) determines the one-to-one correspondence between Φ and P. A formal advantage of working with Φ is that it is an analytic function in each of its continuous-valued arguments z_i; it is thus amenable to the methods of analysis rather than to those of algebra. In fact it will be shown below that $\Phi(|z\rangle;t)$ satisfies a single partial differential equation. One also readily obtains from Φ all of the moments of $P(\langle n|;t)$ as well as any desired marginal distributions by noting that the summing of P over all values of n_i, for any i, is equivalent to setting $z_i=1$ in Φ. Thus, using the definition of probability and Eq. (2.7), we find that

$$\Phi(|z\rangle;t)_{z_1=z_2=\cdots=1}=1, \quad (2.8)$$

and that the moments of the occupation numbers are given by

$$\langle n_in_j\cdots n_k\rangle_{\mathrm{av}} =\left[z_i\frac{\partial}{\partial z_i}z_j\frac{\partial}{\partial z_j}\cdots z_k\frac{\partial}{\partial z_k}\Phi\right]_{z_1=z_2=\cdots=1}. \quad (2.9)$$

A marginal distribution such as, say, $P(n_1,n_2;t)$ is just the set of coefficients of $z_1^{n_1}z_2^{n_2}$ in the series expansion of $\Phi(z_1,z_2;t)$, the marginal PGF (probability generating function), which one gets from (2.7) by setting all $z_i=1$, except z_1 and z_2.

The equation satisfied by Φ follows from (2.5):

$$\frac{\partial\Phi}{\partial t}=\sum_{\langle n|}\sum_{\langle m|}P(\langle m|;t)Q(\langle m|\rightarrow\langle n|)\prod_i z_i^{n_i}. \quad (2.10)$$

We define

$$\Gamma(\langle m|;|z\rangle)\equiv\sum_{\langle n|}Q(\langle m|\rightarrow\langle n|)\prod_i z_i^{n_i-m_i} =\sum_{\langle n-m|}Q'(\langle m|;\langle n-m|)\prod_i z_i^{n_i-m_i}, \quad (2.11)$$

where

$$Q'(\langle m|;\langle n-m|)\equiv Q(\langle m|\rightarrow\langle n|).$$

On using (2.3b), this can also be written as

$$\Gamma(\langle m|;|z\rangle)=\sum_{\langle m-n|\neq\langle 0|}Q'(\langle m|;\langle n-m|)\times\left[\prod_i z_i^{n_i-m_i}-1\right]. \quad (2.12)$$

The right-hand side of (2.10) now becomes

$$\sum_{\langle m|} \Gamma(\langle m|;|z\rangle)(\prod_i z_i^{m_i})P(\langle m|;t)$$
$$=\Gamma(\langle z\partial/\partial z|;|z\rangle)\Phi(|z\rangle;t), \quad (2.13)$$

where the derivatives act only on $\Phi(|z\rangle;t)$. (We use this notation to eliminate the explicit dependence on the integral-valued occupation numbers.) Hence (2.10) may be written as the symbolic operator equation

$$\frac{\partial\Phi(|z\rangle;t)}{\partial t}=\Gamma(\langle z\partial/\partial z|;|z\rangle)\Phi(|z\rangle;t). \quad (2.14)$$

This is the fundamental equation which we shall employ in the following sections.[22]

Before proceeding further, we must specify in more detail the transition probability matrix $Q(\langle m| \rightarrow \langle n|)$. We shall assume that the changes in the state of the whole system take place through events in which the occupation numbers of several of the states i change simultaneously. For an event in which p particles disappear from states $i_1, i_2, \cdots, i_p$ and q particles appear in states $k_1, k_2, \cdots, k_q$ (the occupation numbers of all other states remaining unchanged), we take the transition probability per unit time to be given by

$$Q(n(i_1), \cdots, n(i_p); n(k_1), \cdots, n(k_q) \rightarrow n(i_1)-1, \cdots, n(i_p)-1; n(k_1)+1, \cdots, n(k_q)+1)$$
$$=A(i_1,\cdots,i_p;k_1,\cdots,k_q)n(i_1)\cdots n(i_p) \times[1+\theta n(k_1)]\cdots[1+\theta n(k_q)]. \quad (2.15)$$

Here $n(k_q)$ is the occupation number of the state k_q; and $\theta=0$, $+1$, and -1 apply, respectively, to particles obeying classical (M. B.), B. E., and F. D. statistics. In the latter two cases the expression (2.15) follows from first order quantum mechanical perturbation theory[23]; the coefficients $A(i_1,\cdots,i_p;k_1,\cdots,k_q)$ are the squares of the absolute values of matrix elements of the interaction term in the Hamiltonian between initial and final states of the system. For systems, like a gas at low density, that are adequately described by classical statistics, (2.15) represents the familiar Boltzmann Stosszahlansatz. In this case $q=p$, and the factor $A(i_1,\cdots,i_p;k_1,\cdots,k_p)$ can be interpreted as the probability per collision for the transition $(i_1,\cdots,i_p\rightarrow k_1,\cdots,k_p)$ while the product $n(i_1)\cdots n(i_p)$ is proportional to the probability per unit time for a collision involving p particles in the initial states $i_1, i_2, \cdots, i_p$.

Equation (2.15) is applicable only to transitions in which the states involved change their occupation numbers by unity. However, in general, there can be transitions in which the occupation numbers change by arbitrary amounts. The corresponding transition probability per unit time is

$$Q(n(i_1), \cdots, n(i_p); n(k_1), \cdots, n(k_q) \rightarrow n(i_1)-r(i_1), \cdots, n(i_p)-r(i_p); n(k_1)+r(k_1), \cdots, n(k_q)+r(k_q))$$
$$=A(i_1,\cdots,i_p;k_1,\cdots,k_q|r(i_1),\cdots,r(i_p);r(k_1),\cdots,r(k_q))(\prod_{j=1}^{p}\{n(i_j)[n(i_j)-1]\cdots[n(i_j)-r(i_j)+1]\})$$
$$\times\prod_{j=1}^{q}\{[1+\theta n(k_j)][1+\theta(n(k_j)+1)]\cdots[1+\theta(n(k_j)+r(k_j)-1)]\}. \quad (2.16)$$

In particular,

$$A(i_1,\cdots,i_p;k_1,\cdots,k_q|1,\cdots,1;1,\cdots,1)=A(i_1,\cdots,i_p;k_1,\cdots,k_q),$$

and in this case (2.16) reduces to (2.15). From the definition of Γ it now follows that

$$\Gamma(\langle n|;|z\rangle)=\sum_{p,q}\sum_{\substack{r(i_1),\cdots,r(i_p),\\ r(k_1),\cdots,r(k_q)}}\sum_{\substack{i_1,\cdots,i_p,\\ k_1,\cdots,k_q}}{}' A(i_1,\cdots,i_p;k_1,\cdots,k_q|r(i_1),\cdots,r(i_p);r(k_1),\cdots,r(k_q))$$
$$\times(\prod_{j=1}^{p}[z(i_j)]^{-r(i_j)})(\prod_{j=1}^{q}[z(k_j)]^{r(k_j)}-1)(\prod_{j=1}^{p}\{n(i_j)[n(i_j)-1]\cdots[n(i_j)-r(i_j)+1]\})$$
$$\times\prod_{j=1}^{q}\{[1+\theta n(k_j)][1+\theta(n(k_j)+1)]\cdots[1+\theta(n(k_j)+r(k_j)-1)]\}, \quad (2.17)$$

where the prime on the summation sign indicates that in summing over all single particle states no two of the indices $i_1, \cdots, i_p, k_1, \cdots, k_q$ should be taken equal in the same term, and no event should be counted more than once. To be consistent with lowest order pertur-

[22] Equations of this form were first introduced in this context by Moyal and Bartlett; see references 10 and 11.

[23] See Tolman's book, reference 20, Sec. 100, or D. I. Blochinzev, *Osnovi Kvantovoi Mechaniki* (State Publishing House, Moscow, 1949), Sec. 115 (in Russian).

bation theory, one should retain only the terms of lowest order in the expansion (2.17) of Γ.

It may be noted that the sum over all sets of final state occupation numbers appearing in the definition (2.11) of Γ is equivalent to the sum over all transitions of a given type [characterized by fixed values of p, q, $r(i_1)$, $r(i_2)$, $\cdots$, $r(i_p)$, $r(k_1)$, $r(k_2)$, $\cdots$, $r(k_q)$]—i.e., the sum over states in (2.17)—followed by a sum over all types of transitions.

Substitution of (2.17) into (2.14) now yields the partial differential equation for Φ equivalent to the master equation (2.5) for P.

In the sequel we will be concerned with the simple case in which interactions between particles occur exclusively through binary collisions. Then only four types of terms survive in (2.17), namely those for which

$$\left.\begin{matrix}[p=2, r(i_1)=r(i_2)=1] \\ \text{or } [p=1, r(i_1)=2]\end{matrix}\right\} \text{ and } \left\{\begin{matrix}[q=2, r(k_1)=r(k_2)=1] \\ \text{or } [q=1, r(k)=2].\end{matrix}\right.$$

The resulting equation for Φ is

$$\frac{\partial\Phi}{\partial t}=\tfrac{1}{4}\sum_{i,j,k,l} A_{ij}{}^{kl}(z_kz_l-z_iz_j)[\partial_i\partial_j(1+\theta z_k\partial_k)(1+\theta z_l\partial_l)]\Phi+\tfrac{1}{2}\sum_{i,k,l} A_{ii}{}^{kl}(z_kz_l-z_i^2)[\partial_i^2(1+\theta z_k\partial_k)(1+\theta z_l\partial_l)]\Phi$$
$$+\tfrac{1}{2}\sum_{i,j,k} A_{ij}{}^{kk}(z_k^2-z_iz_j)[\partial_i\partial_j(1+\theta z_k\partial_k)(1+\theta(z_k\partial_k+1))]\Phi$$
$$+\sum_{i,k} A_{ii}{}^{kk}(z_k^2-z_i^2)[\partial_i^2(1+\theta z_k\partial_k)(1+\theta(z_k\partial_k+1))]\Phi\equiv\Gamma_2(\langle z\partial/\partial z|;|z\rangle)\Phi, \quad (2.18)$$

where $\partial_i\equiv\partial/\partial z_i$ and $A_{ij}{}^{kl}\equiv A(i,j;k,l)$. The factors $\frac{1}{4}$ and $\frac{1}{2}$ compensate for multiple counting of identical events. (In this equation and in the sequel, if any pair of summation indices are equal in a given term, then this term is excluded.)

In the case of a gas, if the individual particle states of definite momentum (and energy) are labelled by i, j, $\cdots$, then the coefficients $A_{ij}{}^{kl}$ vanish whenever $i=j$ or $k=l$, by energy-momentum conservation, so that the last three terms in (2.18) drop out, reducing the equation to

$$\frac{\partial\Phi}{\partial t}=\tfrac{1}{4}\sum_{i,j,k,l} A_{ij}{}^{kl}(z_kz_l-z_iz_j)\times[\partial_i\partial_j(1+\theta z_k\partial_k)(1+\theta z_l\partial_l)]\Phi. \quad (2.19)$$

This is the form considered by Moyal[10] and by Siegert[15] (with $\theta=0$). However there are systems for which this simplification is not applicable. For example, in a gas of diatomic molecules whose vibrational states are labelled by i, j, $\cdots$, it is possible for two molecules which are initially in the same vibrational state to transfer energy during a collision and go to different final states so that (2.18) rather than (2.19) is required.

No general solution is known for Eq. (2.18) or (2.19). Siegert[15] has obtained a solution of (2.19) for a 2-state system for all time. Moyal[10] has obtained a time-independent solution to (2.19) by assuming statistical independence of the occupation numbers of the various states. His solution corresponds to the grand canonical ensemble equilibrium distribution with the average occupation numbers $\bar{n}_i$ given by

$$\bar{n}_i/(1+\theta\bar{n}_i)=Ce^{-\beta\epsilon_i}, \quad (2.20)$$

where ϵ_i is the energy of a particle in state i, and C is a normalization constant. [As usual, $\beta=(kT)^{-1}$.]

In Sec. III we shall specialize Eq. (2.19) still further so as to be able to exhibit a solution for all time for the harmonic oscillator problem of references 17–19.

B. Rate Equations

By taking suitable derivatives of the PGF equation (2.18) or (2.19), and then setting all the z's equal to unity according to (2.9), one can obtain rate equations for the various moments of the occupation numbers. In particular, the equation one gets for the rate of change of the averages, $\bar{n}_i$, is not the rate equation commonly used in transport theory[24] or in discussions of the H-theorem.[3] To exhibit the essential difference, and the additional assumptions necessary for the conventional equations to be valid, we write down first the rate equation one gets from (2.19) for the M. B. case $(\theta=0)$[25]:

$$\frac{d\bar{n}_k}{dt}=\tfrac{1}{2}\sum_{i,j,l}[A_{ij}{}^{kl}\langle n_in_j\rangle_{\text{av}}-A_{kl}{}^{ij}\langle n_kn_l\rangle_{\text{av}}]. \quad (2.21)$$

This equation relates the time variation of $\bar{n}_k$ to the time-dependent correlations $\langle n_in_j\rangle_{\text{av}}$. Similarly the rate of change of the latter depends on $\langle n_in_jn_k\rangle_{\text{av}}$, etc. Hence, we obtain a hierarchy of interconnected equations. Only with the additional assumption

$$\langle n_in_j\rangle_{\text{av}}=\bar{n}_i\bar{n}_j \quad (2.22)$$

does Eq. (2.21) reduce to the form of the (nonlinear) Boltzmann equation appropriate for the case of a spatially homogeneous gas with no external forces.

[24] E. A. Uehling and G. E. Uhlenbeck, Phys. Rev. 43, 552 (1933).

[25] See also Siegert, reference 15, Appendix II.

The corresponding equation to (2.21) for F. D. or B. E. particles is readily obtained from (2.19):

$$\frac{d\bar{n}_k}{dt}=\tfrac{1}{2}\sum_{i,j,l}\left[A_{ij}{}^{kl}\langle n_in_j(1+\theta n_k)(1+\theta n_l)\rangle_{\rm av}-A_{kl}{}^{ij}\langle n_kn_l(1+\theta n_i)(1+\theta n_j)\rangle_{\rm av}\right]. \quad (2.23)$$

This equation will also not reduce to the Boltzmann form with the modified Stosszahlansatz (used in the transport theory of degenerate quantum gases[24]) unless one makes the additional assumption[26]:

$$\langle n_in_j(1+\theta n_k)(1+\theta n_l)\rangle_{\rm av}=\bar{n}_i\bar{n}_j(1+\theta\bar{n}_k)(1+\theta\bar{n}_l). \quad (2.24)$$

III. SYSTEMS IN INTERACTION WITH A RESERVOIR

A. General Properties

Let us suppose that the interacting particles of a system can be divided into two sets (subsystems): a relaxing set and a reservoir set. Assume that within each set the particles are all of the same kind, although the two sets do not necessarily consist of the same kind of particles or have the same statistics. However, we require that the energy level spacings of the particles in the two subsystems be compatible so that energy exchange is possible.

The number of particles in the reservoir set is, by definition, very large compared to that in the relaxing system. Consequently, in studying the change in the occupation number distribution of the relaxing set, only the interaction between particles of this set and those of the reservoir need be considered, the effect of interactions within the relaxing set being negligible in comparison. On the other hand, the occupation number distribution of the reservoir set is assumed to be unaffected by its interaction with the relaxing set, and its time dependence may be prescribed arbitrarily.

Such a decomposition is appropriate for many physical systems. Some examples are:

(i) a "Rayleigh Gas" [27] consisting of a dilute mixture of low mass atoms (relaxing set) with an arbitrary initial velocity distribution and a homogeneous spatial distribution within a vessel of much heavier atoms (reservoir) which are in thermal equilibrium;

(ii) a gas of identical diatomic molecules (reservoir), a small fraction of which (relaxing set) has been excited to various vibrational states by a transient source of radiation[17–19];

(iii) a paramagnetic salt, with excited spin state populations (relaxing set) which are coupled to the lattice vibrations or phonons (reservoir).

It should be noted that the reservoir need not be thermal, i.e., it need not have an occupation number distribution corresponding to some definite temperature. In fact, the reservoir need not even be stationary. For example, in (iii) the spin states with energy difference $h\nu$ are coupled to phonons of frequency ν whose population could be made anomalous (greater than the average appropriate for the temperature of the lattice bath) by pumping with an ultrasonic generator at the resonance frequency ν. The phonon distribution would then be neither thermal nor stationary.

1. Master Equation for the Relaxing System

We shall now derive the master equation for the relaxing system. We first note that there are now two sets of occupation numbers, one for the relaxing system and the other for the reservoir. We denote these by $\langle n|$ and $\langle n'|$, respectively, and we denote the PGF of their JPD by $\Phi(|z\rangle, |z'\rangle; t)$. Assuming binary collisions between particles, we find that Φ satisfies

$$\frac{\partial\Phi}{\partial t}=\Gamma_2(\langle z\partial/\partial z|\,;\,|z\rangle)\Phi+\Gamma_2(\langle z'\partial/\partial z'|\,;\,|z'\rangle)\Phi+\sum_{i,j,k,l}(z_kz_l'-z_iz_j')\times A_{ij}{}^{kl}[\partial_i\partial_j'(1+\theta z_k\partial_k)(1+\theta z_l'\partial_l')]\Phi. \quad (3.1)$$

(The right side may also contain terms representing the effect of any external agency which acts on the reservoir.) The operator Γ_2 is defined in (2.18) and is appropriate when only binary collisions occur. The three terms in (3.1) correspond, respectively, to collisions within the relaxing system, collisions within the reservoir, and collisions between particles of the relaxing system and those of the reservoir. The first of these can be ignored in virtue of the discussion at the beginning of this section. We also note that the assumption of statistical independence between the reservoir and the relaxing system allows us to write

$$\Phi(|z\rangle,|z'\rangle;t)=\Phi_S(|z\rangle;t)\Phi_R(|z'\rangle;t), \quad (3.2)$$

where Φ_S and Φ_R are the separate PGF's for the relaxing system and for the reservoir, respectively. Setting $|z'\rangle=|1\rangle$ in (3.1) and using (2.18), we find that the second term vanishes identically, yielding the following equation for Φ_S:

$$\frac{\partial\Phi_S(|z\rangle;t)}{\partial t}=\sum_{i,j,k,l}A_{ij}{}^{kl}[\bar{n}_j'(t)+\theta'\langle n_j'n_l'(t)\rangle_{\rm av}]\times(z_k-z_i)[\partial_i(1+\theta z_k\partial_k)]\Phi_S(|z\rangle;t). \quad (3.3)$$

Hence,

$$\frac{\partial\Phi(|z\rangle;t)}{\partial t}=\sum_{i,k}a_i{}^k(z_k-z_i)\partial_i(1+\theta z_k\partial_k)\Phi(|z\rangle;t), \quad (3.4)$$

where

$$a_i{}^k=\sum_{j,l}A_{ij}{}^{kl}[\bar{n}_j'(t)+\theta'\langle n_j'n_l'(t)\rangle_{\rm av}]. \quad (3.5)$$

[26] That this additional assumption is often implicitly made in applying (2.15) has also been pointed out by Blochinzev, reference 23, p. 474.

[27] Lord Rayleigh, *Scientific Papers* (Cambridge University Press, New York, 1903), Vol. III, p. 473.

(We have dropped the subscript S on Φ since all future PGF's will refer to the relaxing system.) Equation (3.5) shows that only the means and correlations of the reservoir occupation numbers affect the relaxing system when the interaction is via binary collisions.

The form of the master equation for the PGF of the relaxing system as given by (3.4) is quite general (subject, however, to the statistical independence assumption): This form applies whatever the mechanism of interaction between particles, as long as only one relaxing particle is involved in a collision. The specific form of interaction, e.g., binary collisions, affects only the explicit structure of the a_i^k.

In a similar way one could obtain the master equation for the change with time of the reservoir PGF by setting $|z\rangle=|1\rangle$ in (3.1). However, we shall assume that the reservoir distribution is stationary. This can be realized physically to a very good approximation whenever (a) the number of reservoir particles is very much greater than those of the relaxing set, and (b) in the case of a stationary, but not thermal equilibrium distribution, the coupling of reservoir particles to the external agency is much stronger than their interaction with each other.

One can give a simple proof of, and thereby gain insight into, a general property of the a_i^k when the reservoir is thermal. *If the reservoir is in thermal equilibrium corresponding to a temperature $T=1/(\beta k)$ and if its particles are*

(i) M. B. *with* $\bar{n}_i'=Ce^{-\beta\epsilon_i}$, *or*

(ii) F. D. *or* B. E. *with uncorrelated occupation numbers*:

$$\bar{n}_i'/(1+\theta'\bar{n}_i')=Ce^{-\beta\epsilon_i},$$
$$\langle n_i'n_j'\rangle_{\mathrm{av}}=\bar{n}_i'\bar{n}_j',$$

then

$$a_i^k=a_k^i\exp[\beta(\epsilon_i-\epsilon_k)]. \qquad (3.6)$$

Proof: Since the proof for (i) is a special case of that for (ii), consider case (ii). We can write (3.5) as

$$a_i^k=\sum_{j,l}A_{ij}^{kl}[\bar{n}_j'(1+\theta'\bar{n}_l')]$$
$$=\sum_{j,l}A_{ij}^{kl}\{[\bar{n}_j'/(1+\theta'\bar{n}_j')][(1+\theta'\bar{n}_l')/\bar{n}_l']\}\times\bar{n}_l'(1+\theta'\bar{n}_j').$$

In virtue of (ii), the factor in curly brackets is $\exp\beta(\epsilon_l-\epsilon_j)$ and, hence, depends only on the energy transfer and not on the individual energies of the colliding particles. But from the conservation of energy in the transition $(i,j)\rightarrow(k,l)$, it follows that $\epsilon_l-\epsilon_j=\epsilon_i-\epsilon_k$; i.e., the quantity in square brackets is independent of j and l. It may therefore be factored outside the sum so that

$$a_i^k=[\exp\beta(\epsilon_i-\epsilon_k)]\sum_{j,l}A_{ij}^{kl}\bar{n}_l'(1+\theta'\bar{n}_j')$$
$$=[\exp\beta(\epsilon_i-\epsilon_k)]\sum_{j,l}A_{kl}^{ij}\bar{n}_l'(1+\theta'\bar{n}_j')$$
$$=a_k^i\exp\beta(\epsilon_i-\epsilon_k).$$

In the second step microscopic reversibility, $A_{ij}^{kl}=A_{kl}^{ij}$, has been assumed. The basic property (3.6) of the a_i^k is quite general. The validity of the proof is not restricted by the assumption of binary collisions; and, in fact, a completely analogous proof can be given which includes all higher order collisions as well as the possibility of emission and absorption.[28]

2. *Rate Equations*

Rate equations for the relaxing system are easily obtained from (3.4) by using (2.9). Thus for the mean values of the occupation numbers one gets

$$d\bar{n}_k/dt=\sum_i\{a_i^k[\bar{n}_i+\theta\langle n_in_k\rangle_{\mathrm{av}}]-a_k^i[\bar{n}_k+\theta\langle n_in_k\rangle_{\mathrm{av}}]\}. \qquad (3.7)$$

This set of equations is only a partial characterization of the relaxation; additional equations are required to determine $\langle n_in_k(t)\rangle_{\mathrm{av}}$. If the occupation numbers are uncorrelated the set of equations (3.7) is closed but nonlinear. However, if the relaxing system consists of M. B. particles, (3.7) reduces to

$$d\bar{n}_k/dt=\sum_i(a_i^k\bar{n}_i-a_k^i\bar{n}_k)=\sum_i\Lambda_{ik}\bar{n}_i, \qquad (3.8)$$

where

$$\Lambda_{ik}=a_i^k-\delta_i^k\sum_j a_i^j. \qquad (3.9)$$

(δ_i^k is the Kronecker delta.)

This is a system of linear differential equations with the formal solution

$$\langle\bar{n}(t)|=\langle\bar{n}(0)|T(t), \qquad (3.10)$$

where $\bar{n}_k(t)$, $k=0, 1, 2, \cdots$, are the components of the row vector $\langle\bar{n}_k(t)|$, and the square matrix $T(t)$ is that solution of the matrix equation

$$dT/dt=T\Lambda, \qquad (3.11)$$

which satisfies the initial condition

$$T_{ij}(0)=\delta_i^j. \qquad (3.12)$$

When Λ is finite dimensional there exists a unique matrix $T(t)$ which satisfies (3.11) and (3.12), namely

$$T(t)=e^{\Lambda t}. \qquad (3.13)$$

When Λ is an infinite dimensional matrix, there are cases[29] in which there can be more than one solution $T(t)$ satisfying (3.11) and (3.12). We shall assume that for any Λ of physical interest the matrix $T(t)$ defined

[28] Although the property (3.6) is well known, the usual proofs entail more restrictive assumptions. See, for example, M. J. Klein and P. H. E. Meijer, Phys. Rev. **96**, 250 (1954); C. Kittel, *Elementary Statistical Physics* (John Wiley & Sons, Inc., New York, 1958), Sec. 39; J. H. Van Vleck, Suppl. Nuovo cimento **6**, 1081 (1957). Our proof is a generalization of one given by R. T. Cox, Revs. Modern Phys. **22**, 238 (1950). Cox adopts the point of view of Gibbs in that he takes the index i to refer to the state of a whole macroscopic system rather than to an individual particle state as we do (following Boltzmann). The relation of the form (3.6) which he obtains applies to two states i and j of subsystems of a canonical ensemble.

[29] See Appendix I and the references cited there.

by (3.11) and (3.12) is unique and has a unique spectral representation

$$T(t)=\sum_k e^{\lambda_k t}|x(k)\rangle\langle y(k)|, \tag{3.14}$$

where λ_k, $k=0, 1, 2, \cdots$, are eigenvalues of Λ, and $|x(k)\rangle$ and $\langle y(k)|$ are the corresponding column and row eigenvectors which satisfy the completeness and orthogonality relations

$$\sum_k|x(k)\rangle\langle y(k)|=I, \tag{3.15}$$

and

$$\langle y(k)|x(l)\rangle=\delta_k{}^l, \tag{3.16}$$

where I is the unit matrix.

Explicit determination of the eigenvalues λ_k and the eigenvectors $|x(k)\rangle$ and $\langle y(k)|$ is usually not simple. However, for the case of interacting harmonic oscillators, considered in the next section, Montroll and Shuler[18] obtained the solution to (3.8) by a generating function method. We shall extract from this solution the eigenvalues and eigenvectors appropriate to that problem. (See Appendix II.)

B. General Solution for M.B. Particles

We now turn to the problem of exhibiting an explicit solution to Eq. (3.4). When $\theta=\mp 1$, corresponding to F.D. and B.E. particles, Eq. (3.4) is a second order partial differential equation which we have been unable to solve. However, for $\theta=0$, the case of classical M.B. particles, the equation reduces to

$$\frac{\partial\Phi}{\partial t}=\sum_{i,k} a_i{}^k(z_k-z_i)\frac{\partial\Phi}{\partial z_i}=\sum_{i,k}\Lambda_{ik}z_k\frac{\partial\Phi}{\partial z_i}, \tag{3.17}$$

which, being a first order partial differential equation, can be solved by the method of characteristics.[30] The differential equations for the characteristics are

$$dt=-dz_i/\sum_j\Lambda_{ij}z_j, \quad i=0, 1, 2, \cdots, \tag{3.18}$$

or, in matrix form,

$$d|z\rangle/dt=-\Lambda|z\rangle. \tag{3.19}$$

The solution of this equation is

$$T(t)|z\rangle=\text{constant vector}. \tag{3.20}$$

Changing the independent variables in (3.17) to t and $|\eta\rangle$ where

$$|\eta\rangle=T(t)|z\rangle, \tag{3.21}$$

we may write the solution for $\Phi(|z\rangle;t)$ of (3.17) as

$$\Phi(|z\rangle;t)=f(|\eta\rangle),$$

where f is an arbitrary function of the $|\eta\rangle$ which, however, is uniquely determinable once the initial conditions are specified. These conditions may, for example, be the number of particles $n_i{}^0$ in each state i at $t=0$. More generally we shall suppose that $P(\langle n|;0)$ at $t=0$ is given.[31] Thus let

$$\Phi(|z\rangle;0)\equiv\Phi_0(|z\rangle) \tag{3.22}$$

be the initial PGF. Then from (3.20) and (3.21), we have

$$\Phi_0(|z\rangle)=f(|z\rangle),$$

and hence,

$$\Phi(|z\rangle;t)=f(|\eta\rangle)=\Phi_0(|\eta\rangle)$$
$$=\Phi_0(T(t)|z\rangle). \tag{3.23}$$

Equations (3.23) and (3.10) show that the time dependent behavior of both the JPD of the occupation numbers and the mean values of the occupation numbers is determined by the same matrix $T(t)=e^{\Lambda t}$. In particular, it follows that if one knows the mean values of the occupation numbers $\bar{n}_i(t)$ for all time and for an arbitrary set of initial values, then this is sufficient to determine $P(\langle n|;t)$ and, hence, all higher moments $\langle n_i^2(t)\rangle_{\text{av}}$, $\langle n_in_j(t)\rangle_{\text{av}}$, etc., for all time.

1. The Stationary Solution

We now consider the stationary, i.e., time independent, solutions of (3.8) and (3.17). In particular, we show how to construct the solution of one from that of the other. First we list several known properties[32] of the stochastic matrix $T(t)$:

(i) $T_{ij}(t)\geqslant 0$;
(ii) $\sum_j T_{ij}(t)=1$;
(iii) The limit, $T(\infty)\equiv\lim_{t\to\infty}T(t)$ exists, or equivalently, there is at least one λ_k equal to zero in the spectral representation (3.14). All the nonzero λ_k have negative real parts. [That Λ has in fact at least one zero eigenvalue, say λ_0, follows from $\sum_j\Lambda_{ij}=0$ which implies that the column vector $|x(0)\rangle$ with all components equal:

$$x_i(0)=1, \quad i=1, 2, \cdots, \tag{3.24}$$

is a right eigenvector belonging to $\lambda_0=0$.]

We shall assume, in addition:

(iv) The relaxing system is ergodic, i.e., that for every pair of states i, k there exists a sequence of states $j_1, j_2, \cdots, j_r$ such that $\Lambda_{ij_1}\Lambda_{j_1j_2}\cdots\Lambda_{j_rk}$ is nonzero. (Physically, ergodicity means that every state is accessible from every other.) The limit $T_{ij}(\infty)$ is then independent of i. This is equivalent to the nondegeneracy of the zero eigenvalue in the spectral representation (3.14). Indeed, using this fact and (iii), one gets

$$\lim_{t\to\infty}T(t)=|x(0)\rangle\langle y(0)|, \tag{3.25}$$

[30] See, for example, R. Courant and D. Hilbert, *Methoden der Mathematischen Physik* (Verlag Julius Springer, Berlin, 1927), Vol. II. Also, M. S. Bartlett, J. Roy. Stat. Soc. **B11**, 211 (1949).

[31] Specifying the number of particles in each state at $t=0$ corresponds to choosing the particular initial distribution $P(\langle n|;0)=\Pi_i\delta(n_i,n_i{}^0)$.

[32] M. Fréchet, *Recherches Théorique Modernes sur le Calcul des Probabilités* (Gauthier-Villars, Paris, 1952).

so that by (3.24)

$$T_{ij}(\infty)=y_j(0). \tag{3.26}$$

Finally, we shall use the fact:

(v) $T(\infty)$ is a stationary solution of (3.11). In fact, if $F(t)$ is a solution of a differential equation which is of the first order in time, such as (3.8), (3.17), or (2.5), then if $F(\infty)\equiv\lim_{t\to\infty}F(t)$ exists it is a stationary solution F_s, [i.e., $F_s(t)=F_s(t+\tau)$ for all t], and conversely. Hence, we may obtain the unique stationary solution as the $\lim_{t\to\infty}F(t)$ when this latter exists: $F_s=F(\infty)$.

We can now completely characterize the stationary solutions of (3.8) and (3.17): *The stationary solutions* $\langle\bar{n}_s|$ *and* $P_s(\langle n|)$ [*or* $\Phi_s(|z\rangle)$] *are unique, independent of initial conditions, and depend only on the row eigenvector* $\langle y(0)|$ *of* Λ. *If* N *is the total number of particles in the relaxing system, and* $\langle\bar{n}(t)|$ *and* $P(\langle n|;t)$ *are any solutions for the mean values of the occupation numbers and for the* JPD *respectively, then*

$$\langle\bar{n}_s|=\lim_{t\to\infty}\langle\bar{n}(t)|=N\langle y(0)|, \tag{3.27}$$

where $\langle y(0)|$ *is the row eigenvector of* Λ *belonging to* $\lambda_0=0$; *and*

$$P_s(\langle n|)=\lim_{t\to\infty}P(\langle n|;t)$$

$$=\frac{N!}{\prod_i n_i!}\prod_i[y_j(0)]^{n_j}$$

$$=\frac{N!}{\prod_i n_i!}\prod_i\left[\frac{\bar{n}_i(\infty)}{N}\right]^{n_i},\quad \sum_i n_i=N. \tag{3.28}$$

The associated PGF *is given by*

$$\Phi_S(|z\rangle)=\langle y(0)|z\rangle^N. \tag{3.29}$$

These equations follow directly on using (v), (3.24), and (3.25). Thus,

$$\begin{aligned}\langle\bar{n}(\infty)|&=\lim_{t\to\infty}\langle\bar{n}(0)|T(t)\\&=\langle\bar{n}(0)|x(0)\rangle\langle y(0)|\\&=N\langle y(0)|,\end{aligned} \tag{3.30}$$

and

$$\begin{aligned}\Phi(|z\rangle;\infty)&=\lim_{t\to\infty}\Phi_0(T(t)|z\rangle)\\&=\Phi_0(|x(0)\rangle\langle y(0)|z\rangle)\\&=\sum_{\langle n|}P(\langle n|;0)\prod_i\langle y(0)|z\rangle^{n_i}\\&=\langle y(0)|z\rangle^N\sum_{\langle n|}P(\langle n|;0)\\&=\langle y(0)|z\rangle^N.\end{aligned} \tag{3.31}$$

Equation (3.28) for $P_s(\langle n|)$ then follows from (3.31) by using the multinomial expansion.

From Eq. (3.31) one can readily obtain all desired moments and marginal distributions of the stationary JPD (3.28). For example, $\Phi(z_i;\infty)$, the marginal PGF, is obtained by setting all z's, except z_i, equal to unity in (3.31). Thus, in virtue of (3.16) and (3.24), we find

$$\Phi(z_i;\infty)=[1-(1-z_i)y_i(0)]^N. \tag{3.32}$$

The corresponding stationary marginal probability distribution of n_i is then readily seen to be

$$P(n_i;\infty)=\binom{N}{n_i}\left(\frac{\bar{n}_i}{N}\right)^{n_i}\left(1-\frac{\bar{n}_i}{N}\right)^{N-n_i}, \tag{3.33}$$

which is a binomial distribution with mean $\bar{n}_i$ and relative variance

$$\frac{\langle n_i^2(\infty)\rangle_{\rm av}-\bar{n}_i^2(\infty)}{\bar{n}_i^2(\infty)}=\left(\frac{1}{\bar{n}_i}-\frac{1}{N}\right). \tag{3.34}$$

Similarly, the marginal PGF for the joint occupation number distribution for any two states i and j is

$$\Phi(z_i,z_j;\infty)=[1-(1-z_i)y_i(0)-(1-z_j)y_j(0)]^N, \tag{3.35}$$

from which it follows that

$$P(n_i,n_j;\infty)=\frac{N!}{n_i!n_j!n_k!}\left(\frac{\bar{n}_i}{N}\right)^{n_i}\left(\frac{\bar{n}_j}{N}\right)^{n_j}\left(\frac{\bar{n}_k}{N}\right)^{n_k}, \tag{3.36}$$

where

$$n_k=N-(n_i+n_j).$$

The correlation coefficient of these occupation numbers is given by

$$\frac{\langle n_i(\infty)n_j(\infty)\rangle_{\rm av}-\bar{n}_i(\infty)\bar{n}_j(\infty)}{\bar{n}_i(\infty)\bar{n}_j(\infty)}=-\frac{1}{N},\quad i\neq j. \tag{3.37}$$

Generally, for the average of the product of the occupation numbers for r states $i_1, i_2, \cdots, i_r$, one gets

$$\langle n_{i1}(\infty)\cdots n_{i_r}(\infty)\rangle_{\rm av}=\frac{N!}{(N-r)!}\prod_{j=1}^r\left(\frac{\bar{n}_{ij}(\infty)}{N}\right), \tag{3.38}$$

where all the i_j are distinct.

We now state certain obvious consequences of the preceding development which are of interest inasmuch as they extend the familiar results of statistical mechanics for systems in thermal equilibrium to the stationary distribution of an ergodic system interacting with any stationary (*not* necessarily thermal) reservoir. (For the case of a thermal reservoir, see below.)

(i) The stationary JPD given by Eq. (3.28) is *not* a product of independent distributions for each state. This is due to the constraint $\sum n_i=N$ which induces a weak but nonvanishing negative correlation in the occupation numbers. [See Eq. (3.37).]

(ii) The stationary JPD is expressible wholly in terms of the average occupation numbers $\bar{n}_i(\infty)$.

(iii) For N sufficiently large and $\bar{n}_i \ll N$, we find that $P(n_i;\infty)$, given by Eq. (3.33), approaches a Poisson distribution

$$P(n_i;\infty)=e^{-\bar{n}_i}\bar{n}_i^{n_i}/n_i!, \quad (3.39)$$

where $\bar{n}_i \equiv \bar{n}_i(\infty)$.

(iv) The formulas of this section all reduce to those for a canonical ensemble at the temperature $T=1/(k\beta)$ when the $\bar{n}_i(\infty)$ are given by the Maxwell-Boltzmann values

$$\bar{n}_i = Ne^{-\beta\epsilon_i}/\sum_j e^{-\beta\epsilon_j}. \quad (3.40)$$

(v) The mean occupation numbers $\bar{n}_i(\infty)$ for the general stationary solution of the rate equation (3.8) do *not* satisfy the detailed balance condition

$$a_i^k \bar{n}_i(\infty)=a_k^i \bar{n}_k(\infty), \quad \text{all } i, k. \quad (3.41)$$

For this condition to be satisfied the a_i^k must satisfy the consistency relations

$$a_i^k a_k^l a_l^i = a_i^l a_l^k a_k^i, \quad i\neq l\neq k\neq i. \quad (3.42)$$

(vi) The form of the stationary JPD (3.28) is a multinomial distribution in the occupation numbers. This form has the remarkable property that *if the initial* JPD *has this form, then the* JPD *will preserve this form for all time* with the mean occupation numbers, $\bar{n}_i(t)$, being given by the solution (3.10) of the rate equation (3.8) for the given initial means $\bar{n}_i(0)$. This conclusion follows readily from (3.29) and (3.23):

$$\Phi(|z\rangle;t)=\Phi_0(T(t)|z\rangle)=(1/N^N)\langle\bar{n}(0)|T(t)|z\rangle^N$$
$$=(1/N^N)\langle\bar{n}(t)|z\rangle^N.$$

2. *Stationary Solution with Thermal Reservoir*

Let us now assume that the relaxing set interacts with a thermal reservoir at the temperature T. Since the form of the JPD is given by (3.28), it remains only to determine the unique set of $\bar{n}_i(\infty)$ appropriate to this case. We do this by noting that in virtue of (3.6), the a_i^k for the case of a thermal reservoir do satisfy the consistency relations (3.42). Hence, a possible solution of the rate equation (3.8) is that for which the $\bar{n}_i(\infty)$ satisfy the detailed balance relation (3.41). From this and (3.6) one gets the M.B. average occupation numbers (3.40). From the discussion of the preceding subsection, it follows that this solution is unique and corresponds to the canonical ensemble.[33]

It was, of course, to be expected that a system interacting with a thermal reservoir would approach a thermal distribution at the reservoir temperature. However, in Sec. IV we give an example showing that it is possible for the relaxing system to approach thermal equilibrium even when the reservoir is not thermal.

[33] For a different proof, see Siegert, reference 15.

3. *Time Dependent Correlations and Fluctuations*

We now return to the general time dependent solution $\Phi(|z\rangle;t)$ of (3.17) and extract from it the time dependent correlations and fluctuations in the occupation numbers. To obtain these in terms of their initial values at $t=0$, we apply (2.9) to the solution (3.23) for the PGF in the form $\Phi_0(|\eta\rangle)$. Then Φ_0 incorporates the initial conditions and $|\eta\rangle \equiv T(t)|z\rangle$ incorporates the time dependence. The following relations are now useful:

$$d\eta_i/dz_j = T_{ij}(t), \quad (3.43)$$

and

$$|\eta\rangle = |x(0)\rangle \quad \text{when all } z_j=1, \quad (3.44)$$

i.e., $\eta_i=1$ for all i when all $z_j=1$. The first relation follows immediately from (3.21) while the latter follows from the additional fact that $\sum_j T_{ij}(t)=1$. Using these relations, one readily finds the correlations of the occupation numbers:

$$\langle n_i(t)n_j(t)\rangle_{\text{av}}=\sum_k T_{ki}[(\delta_i^j - T_{kj})\bar{n}_k(0)$$
$$+\sum_l T_{lj}\langle n_k n_l(0)\rangle_{\text{av}}]. \quad (3.45)$$

In a similar manner one can obtain the time dependence of all other higher moments, correlations, and fluctuations in the occupation numbers.

In the limit $t\to\infty$, we find[34]

$$T_{ki}\to x_k(0)y_i(0)=y_i(0)=\bar{n}_i(\infty)/N, \quad (3.46)$$

and we easily recover from (3.45) the results for the variance and covariance obtained earlier for the stationary distribution.

It is noteworthy that, unlike the stationary solutions, the JPD at any finite time in general cannot be specified solely in terms of the mean occupation numbers at that time.

IV. RELAXING SET OF HARMONIC OSCILLATORS

We now consider in detail an example of a relaxing system in interaction with a reservoir. We suppose that both the system and the reservoir consist of identical harmonic oscillators (e.g., the vibrational modes of identical diatomic molecules). The major portion of these (the reservoir) is prepared and maintained with a stationary occupation number distribution, while the remaining N relaxing oscillators may have an arbitrary distribution.

Relaxation takes place through collision with the reservoir oscillators. It is assumed, following Landau and Teller,[35] that the interaction energy of two colliding oscillators depends linearly on the vibrational coordinate of each of them. This interaction is effective during the "collision time" and causes simultaneous changes in the states of the two oscillators. The transi-

[34] Note that the argument of $|x\rangle$ or $\langle y|$ is the eigenvalue to which the eigenvector belongs, whereas time is the argument of the occupation numbers n_i.

[35] L. Landau and E. Teller, Physik. Z. Sowjetunion **10**, 34 (1936).

tion probabilities per collision, $(i,j) \to (k,l)$, as obtained from first order quantum mechanical perturbation theory, are[36]

$$A_{ij}^{kl}=0 \quad \text{unless} \quad k=i\pm 1 \quad \text{and} \quad l=j\mp 1,$$
$$A_{ij}^{i+1,j-1}=C(i+1)j, \tag{4.1}$$
$$A_{ij}^{i-1,j+1}=Ci(j+1),$$

where the positive constant C is a measure of the coupling strength between the oscillators. (For a discussion of the application of this model to chemical kinetics and the relaxation of vibrationally excited gases, see references 17–19.)

It follows from (3.5) (with $\theta=0$) and (4.1) that

$$a_i^k=0 \quad \text{unless} \quad k=i\pm 1,$$
$$a_{i+1}^i=C(i+1)\sum_{j=0}^{\infty}(j+1)\bar{n}_j', \tag{4.2}$$
$$a_i^{i+1}=C(i+1)\sum_{j=0}^{\infty}j\bar{n}_j',$$

where the $\bar{n}_j'$ are the mean occupation numbers for the (not necessarily thermal) reservoir.

For oscillators with frequency ν, the energy levels are $\epsilon_i=(i+\frac{1}{2})h\nu$, and the average energy of a reservoir oscillator is (in units of $h\nu$)

$$\bar{\epsilon}=\textstyle\sum_j (j+\frac{1}{2})\bar{n}_j'/\sum_j \bar{n}_j'. \tag{4.3}$$

Defining

$$\epsilon_{\pm}=\bar{\epsilon}\pm\tfrac{1}{2}, \tag{4.3}$$

we may write

$$\frac{a_i^{i+1}}{a_{i+1}^i}=\frac{(\bar{\epsilon}-\frac{1}{2})}{(\bar{\epsilon}+\frac{1}{2})}=\frac{\epsilon_-}{\epsilon_+}. \tag{4.4}$$

This ratio depends only on the average energy per particle of the reservoir. In particular, it is independent of the state i. If the reservoir is thermal (4.4) reduces to (3.6). Note also that

$$a_{i+1}^i-a_i^{i+1}=C(i+1)\textstyle\sum_j \bar{n}_j'=c(i+1), \tag{4.5}$$

where $c=C\sum \bar{n}_j'$.

Thus, for this model the a_i^j, and hence the entire probabilistic description of the relaxation process, depend only on the two physical parameters c and $\bar{\epsilon}$.

[36] This expression for the transition probability per collision is correct only so long as $Cij \ll 1$, since its derivation tacitly assumes that, during the effective interaction time, each of the oscillators participating in a collision undergoes only one transition (or none at all). The expression is no longer valid for collisions between oscillators in very high energy states (i, j large) because there is then a finite probability for the occurrence of several transitions during the course of one collision. This would leave the oscillators in final states whose quantum numbers, in general, would differ from those of the initial states by more than one, contrary to our assumption. We shall avoid this difficulty by confining our attention to systems in which the populations of such high energy states are negligibly small. Hence, while continuing to use the same expression for the transition probability, we shall refrain from interpreting Cij as a transition probability per collision when ij is large.

In terms of these the Λ matrix can be written as

$$\Lambda=c\begin{bmatrix} -\epsilon_- & \epsilon_- & 0 & 0 & \cdots \\ \epsilon_+ & -(\epsilon_+ +2\epsilon_-) & 2\epsilon_- & 0 & \cdots \\ 0 & 2\epsilon_+ & -(2\epsilon_+ +3\epsilon_-) & 3\epsilon_- & \cdots \\ \vdots & \vdots & \vdots & \vdots & \ddots \end{bmatrix}. \tag{4.6}$$

The problem now is to use this Λ matrix to determine $\langle \bar{n}(t)|$, the solution to the rate equation (3.8). Then the solution (3.23) for $\Phi(|z\rangle;t)$ will also be determined. Montroll and Shuler[18] obtained $\langle \bar{n}(t)|$ by solving for the generating function $G(u;t)$ defined by Eq. (A II.5). They were not concerned with, and therefore did not obtain, the matrix elements $T_{ij}(t)$, or the eigenvalues and eigenvectors of the matrix Λ which are needed to determine the time dependent correlations in the occupation numbers. [See Eqs. (3.45) and (3.14).] In Appendix II, we shall obtain the eigenvalues and eigenvectors for a slightly more general matrix than (4.6). On specializing that result to the present case, we find that the double generating function

$$G(u,v;t)=\sum_{i,j=0}^{\infty}T_{ij}(t)v^iu^j \tag{4.7}$$

of $T_{ij}(t)$ is given by

$$G(u,v;t)=\{[(\epsilon_+-u\epsilon_-)-\epsilon_-(1-u)e^{-ct}] \\ -v[(\epsilon_+-u\epsilon_-)-\epsilon_+(1-u)e^{-ct}]\}^{-1}. \tag{4.8}$$

Expansion of (4.8) leads to the two alternative expressions

$$T_{ij}(t)=[\epsilon_+(1-e^{-ct})]^{i-j}[\epsilon_--\epsilon_+e^{-ct}]^j \\ \times[\epsilon_+-\epsilon_-e^{-ct}]^{-i-1}\,{}_2F_1(-j,i+1;1;s), \tag{4.9a}$$

or

$$T_{ij}(t)=[\epsilon_-(1-e^{-ct})]^{j-i}[\epsilon_--\epsilon_+e^{-ct}]^i \\ \times[\epsilon_+-\epsilon_-e^{-ct}]^{-j-1}\,{}_2F_1(-i,j+1;1;s), \tag{4.9b}$$

where

$$s=e^{-ct}[(\epsilon_+-\epsilon_-e^{-ct})(\epsilon_+e^{-ct}-\epsilon_-)]^{-1}, \tag{4.10}$$

and the ${}_2F_1$ are hypergeometric functions simply related to the Jacobi polynomials. The identity[37]

$$(1-w)^{a-1}(1-w+sw)^{-a} \\ =\sum_{n=0}^{\infty}w^n\,{}_2F_1(-n,a;1;s); \\ |w|<1, \quad |w(1-s)|<1, \tag{4.11}$$

has been used in obtaining the above expressions.

In the present case, the eigenvalues λ_k in the spectral representation (3.14) of $T(t)$ are (see Appendix II):

$$\lambda_k=-kc, \quad k=0,1,2,\cdots. \tag{4.12}$$

[37] See, for example, A. Erdelyi *et al.*, *Higher Transcendental Functions* (McGraw-Hill Book Company, Inc., New York, 1953), Vol. I, p. 82.

The corresponding row and column eigenvectors have the respective generating functions

$$G_R(u,k) \equiv \sum_i y_i(k) u^i = \epsilon_-^{\,k}(1-u)^k(\epsilon_+ - \epsilon_- u)^{-k-1}, \quad (4.13)$$

and

$$G_C(v,k) \equiv \sum_i x_i(k) v^i = \epsilon_-^{\,-k}(1-v)^{-k-1}(\epsilon_- - \epsilon_+ v)^k. \quad (4.14)$$

These equations, in conjunction with (3.14) and (3.23), give the time dependent JPD in the occupation numbers explicitly for the system under consideration. We now examine several properties of our solution which are specific to this model.

Consider first the stationary JPD $P(\langle n|;\infty)$ given by (3.28). To determine this JPD, one needs the $\bar{n}_i(\infty)=Ny_i(0)$. From (4.13), one gets

$$y_i(0)=\epsilon_+^{-1}(\epsilon_-/\epsilon_+)^i = \frac{(\epsilon_-/\epsilon_+)^{i+\frac{1}{2}}}{\sum_{j=0}^{\infty}(\epsilon_-/\epsilon_+)^{j+\frac{1}{2}}}. \quad (4.15)$$

Let us now introduce a parameter T defined by the relation[38]

$$T=(1/k)\ln(\epsilon_+/\epsilon_-). \quad (4.16)$$

Then the average occupation numbers for the stationary distribution of the relaxing oscillators are given by

$$\bar{n}_i(\infty)=N\frac{\exp[-(i+\frac{1}{2})/kT]}{\sum_j \exp[-(j+\frac{1}{2})/kT]}. \quad (4.17)$$

These are precisely the Maxwell-Boltzmann values of the $\bar{n}_i(\infty)$ [see (3.40)] for a system of harmonic oscillators of frequency ν in thermal equilibrium at the temperature T.[39] Thus, the stationary JPD of the occupation numbers of the relaxing set of harmonic oscillators is the same as that for a canonical ensemble corresponding to thermal equilibrium at a temperature T defined by (4.16) *irrespective of the reservoir distribution*. The equilibrium temperature T of the relaxing system is determined only by the average energy per reservoir particle. The reservoir itself need not be in thermal equilibrium!

One can give another formal argument why the stationary values of the average occupation numbers should correspond to a thermal distribution for the case of interacting oscillators considered here. Namely, in Sec. III.B.2, it was seen that if the $a_i^{\,j}$ satisfy (3.6) the $\bar{n}_i(\infty)$ for the relaxing set will be Maxwellian. But in virtue of (4.2), (4.4), and the definition (4.16), the relation (3.6) is satisfied for this system irrespective of the specification of the average occupation numbers $\bar{n}_i'$ of the reservoir. One may readily verify that this striking result would no longer be true if the interaction energy between harmonic oscillators were taken as proportional to the cube of each oscillator coordinate rather than linear in it, or if the energy levels of the interacting system were discrete but had at least one spacing incommensurable with the others.

For most initial conditions the explicit expression for $\Phi(|z\rangle;t)$ is not particularly perspicuous. However, there is one important exception which we now consider. Let the initial distribution of the relaxing system be thermal at a temperature $T_0 \neq T$, where T is defined by (4.16). Then the initial PGF is

$$\Phi_0(|z\rangle)=[\sum_j e^{-j\beta_0}(1-e^{-\beta_0})z_j]^N. \quad (4.18)$$

A simple computation based on (3.10) and (3.23) then shows that at any later time

$$\Phi(|z\rangle;t)=[\sum_j e^{-j\beta(t)}(1-e^{-\beta(t)})z_j]^N, \quad (4.19)$$

where

$$\beta(t)=\ln\left(\frac{e^{-ct}(1-e^{\beta_\infty-\beta_0})-e^{\beta_\infty}(1-e^{-\beta_0})}{e^{-ct}(1-e^{\beta_\infty-\beta_0})-(1-e^{-\beta_0})}\right), \quad (4.20)$$

$\beta_0=1/(kT_0)$, and $\beta_\infty=1/(kT)$. [By differentiating (4.20), one can easily show that $\beta(t)$ is a monotonic function of t.] This establishes that *if the* JPD *of the occupation numbers is initially thermal it remains thermal with a temperature $T(t)$ which varies monotonically with time from the given initial T_0 to a final value T at $t=\infty$ determined by the reservoir* [*see* (4.16)]. This is not surprising in view of the corresponding result for the mean values of the occupation numbers obtained by Montroll and Shuler[18] for this model, and the general property, mentioned in remark (vi) of Sec. III.B.1, that the multinomial JPD preserves its form for all time.

Another interesting aspect of the approach to equilibrium, which was noted by Montroll and Shuler,[18] is that the mean energy content $\bar{E}(t)$ of the relaxing system approaches its equilibrium value exponentially. This result follows directly on using the eigenvalues and eigenvectors of Λ. Thus, using (3.10), we see that

$$\bar{E}(t)=\sum_j(j+\tfrac{1}{2})\bar{n}_j(t)=\sum_{i,j}(j+\tfrac{1}{2})T_{ij}\bar{n}_i(0), \quad (4.21)$$

and noting that $\sum_j jT_{ij}(t)$ is the coefficient of v^i in the expansion of $[\partial G(u,v;t)/\partial u]_{u=1}$ we obtain

$$\bar{E}(t)=[\bar{E}(0)-N\bar{\epsilon}]e^{-ct}+N\bar{\epsilon}. \quad (4.22)$$

Hence,

$$\bar{E}(\infty)=N\bar{\epsilon}, \quad (4.23)$$

and

$$\bar{E}(t)-\bar{E}(\infty)=[\bar{E}(0)-\bar{E}(\infty)]e^{-ct}, \quad (4.24)$$

which shows that the average energy of the relaxing system at any time depends only on the average energy at $t=0$ and not on how it was distributed initially among the various states. We can, using our more

[38] In particular, if the reservoir is thermal at a temperature T_R, then (4.3) yields the familiar average energy per oscillator $\bar{\epsilon}=\frac{1}{2}\coth(1/2kT_R)$, and (4.16) yields $T=T_R$.

[39] Note that all energies are expressed in units of $h\nu$.

general method, obtain a similar result for the variance of the energy. Thus,

$$\sigma_E^2(t)\equiv\langle E^2(t)\rangle_{\text{av}}-\bar{E}^2(t)$$
$$=\sum_{i,j}(i+\tfrac{1}{2})(j+\tfrac{1}{2})\sum_k\{\bar{n}_k(0)[T_{ki}\delta_i{}^j-T_{ki}T_{kj}]$$
$$+\sum_l[\langle n_kn_l(0)\rangle_{\text{av}}-\bar{n}_k(0)\bar{n}_l(0)]T_{li}T_{kj}\}. \quad (4.25)$$

Using the generating function $G(u,v;t)$ as before, we obtain, after some calculation

$$\frac{[\sigma_E^2(t)-\sigma_E^2(\infty)]-\epsilon[\bar{E}(t)-\bar{E}(\infty)]}{[\sigma_E^2(0)-\sigma_E^2(\infty)]-\epsilon[\bar{E}(0)-\bar{E}(\infty)]}=e^{-2ct}. \quad (4.26)$$

V. RELAXATION INCLUDING NONCONSERVATION OF PARTICLES

In the example treated in Sec. IV, the number of harmonic oscillators remained constant throughout the relaxation process.

In this section we consider a generalization of this model, allowing for the possibilities that during the relaxation process

(1) Harmonic oscillators appear in the various states i at a rate which is a function only of i (and not of n_i), and

(2) Harmonic oscillators disappear from the various states i at a rate which is a function of i and n_i.

The first possibility arises when a chemical reaction is taking place which produces harmonic oscillators in various states at varying rates. The second occurs in practice due to the dissociation of the harmonic oscillators (or, rather, of the diatomic molecules) which are in states with energy greater than the dissociation energy.[40]

The determination of the JPD in these cases, as in the simpler case treated in the last section, is intimately related to the solution of the equations for the average occupation numbers $\bar{n}_i$. It has not been possible to solve these equations explicitly with a completely realistic energy dependence for the dissociation rate, but the explicit solution can be obtained if this rate increases linearly with energy. We present this derivation below, since it gives a further illustration of the method outlined in the last section; it may also be expected that the solution exhibits the main features of the actual problem with dissociation.

We make the following definitions:

(1) ν_i=probability per unit time that in the relaxing subsystem a new oscillator is created in the state i, and

(2) $n_i\mu_i$=probability per unit time that an oscillator disappears (through "dissociation") from state i, given that the occupation number of this state is n_i.

The equation for the JPD of the occupation numbers now follows from (2.5) when the appropriate Q is introduced. We readily find

$$\frac{\partial P(\langle n|;t)}{\partial t}=\sum_{\langle m|}P(\langle m|;t)Q_I(\langle m|\to\langle n|)$$
$$+\sum_i[-(\nu_i+n_i\mu_i)P(\langle n|;t)$$
$$+\nu_iP(n_1,n_2,\cdots,n_i-1,\cdots;t)$$
$$+\mu_iP(n_1,n_2,\cdots,n_i+1,\cdots;t)]. \quad (5.1)$$

Here Q_I is that part of Q which refers to collisions between particles of the relaxing system and those of the reservoir. In the equation for the generating function Φ, it leads to a term identical with the right side of (3.4) (with $\theta=0$). We thus have

$$\frac{\partial\Phi(|z\rangle;t)}{\partial t}=\sum_i\left[-\sum_j a_i{}^j(z_i-z_j)\frac{\partial}{\partial z_i}+(z_i-1)\left(\nu_i-\mu_i\frac{\partial}{\partial z_i}\right)\right]\Phi. \quad (5.2)$$

It must be noted that the operator acting on the right side of (5.2) is not of the form obtained by replacing n by $z(\partial/\partial z)$ in (2.17). The reason is that the Q of the present case has a part that depends on ν and hence represents the production of particles by an external agency (independent of the occupation numbers). This part is not of the form (2.16) on which (2.17) is based.

If we now set $z_i-1=w_i$, Eq. (5.2) becomes

$$\frac{\partial\Phi}{\partial t}=\sum_i\left[\sum_j L_{ij}w_j\frac{\partial\Phi}{\partial w_i}+\nu_iw_i\Phi\right], \quad (5.3)$$

where

$$L_{ij}=a_i{}^j-\delta_i{}^j(\textstyle\sum_k a_i{}^k+\mu_i). \quad (5.4)$$

(The symbol Φ is retained for convenience.)

Note that the mean values of the occupation numbers obey the equation

$$(\partial/\partial t)\langle\bar{n}(t)|=\langle\bar{n}|L+\langle\nu|, \quad (5.5)$$

with the formal solution

$$\langle\bar{n}(t)|=\langle\bar{n}(0)|e^{Lt}-\langle\nu|L^{-1}(1-e^{Lt}). \quad (5.6)$$

Here L is the matrix with elements L_{ij} given by (5.4), and L^{-1} is its inverse. If L is singular, $L^{-1}(1-e^{Lt})$ is to be understood in the sense of a limit.

Passing now to the solution of (5.3), we observe that the method of characteristics is applicable; the characteristics are

$$e^{Lt}|w\rangle=|\phi\rangle, \quad (5.7)$$

where $|\phi\rangle$ is a constant vector. On replacing the w's by new variables ϕ defined by (5.7), Eq. (5.3) reduces to

$$\partial\Phi/\partial t=\textstyle\sum_i\nu_iw_i(|\phi\rangle;t)\Phi\equiv\langle\nu|e^{-Lt}|\phi\rangle\Phi. \quad (5.8)$$

[40] Related problems have been considered by E. W. Montroll and K. Shuler, *Advances in Chemical Physics* (Interscience Publishers, Inc., New York, 1958), Vol. I; and F. Buff and D. Wilson, J. Chem. Phys. 32, 677 (1960).

Integration of this equation is trivial; reverting to the old variables w, and applying the initial condition

$$\Phi(|w\rangle;0)=\Phi_0(|w\rangle), \tag{5.9}$$

we finally obtain

$$\Phi(|w\rangle;t)=\exp[-\langle\nu|L^{-1}(1-e^{Lt})|w\rangle] \times\Phi_0(e^{Lt}|w\rangle). \tag{5.10}$$

This is the general solution to our problem. Both (5.10) and (5.6) depend on the same matrix operator e^{Lt}. An explicit determination of this operator can be made in the case of harmonic oscillators, with the "dissociation" rate $\mu_i=i\mu$, interacting with a reservoir characterized as in the last section by the constants $\bar{\epsilon}$ and c. This is done in Appendix II. The result is that the double generating function

$$G(u,v;t)=\sum_{i,j}(e^{Lt})_{ij}v^iu^j \tag{5.11}$$

of the matrix elements of e^{Lt} is given by

$$G(u,v;t)=\frac{(\gamma-\alpha)e^{-c''t}}{[(\gamma-u)-(\alpha-u)e^{-c't}]-v[\alpha(\gamma-u)-\gamma(\alpha-u)e^{-c't}]}, \tag{5.12}$$

where the constants c', c'', α, γ $(0<\alpha\leqslant 1<\gamma)$ are defined in Appendix II. The eigenvalues of L appearing in the spectral expansion of e^{Lt} are

$$l_k=-[kc'+c'']; \quad k=0,1,\cdots. \tag{5.13}$$

The corresponding column and row eigenvectors, $|p(k)\rangle$ and $\langle q(k)|$, given in (A II.25) and (A II.26), respectively, form a complete set in terms of which (5.10) can be expressed. Thus

$$\Phi(|w\rangle;t)=\exp\left(-\sum_k\langle\nu|p(k)\rangle\frac{1-e^{l_kt}}{l_k}\langle q(k)|w\rangle\right) \times\Phi_0\Big(\sum_k e^{l_kt}|p(k)\rangle\langle q(k)|w\rangle\Big). \tag{5.14}$$

Neither (5.14) nor the expressions for the correlations, etc. that follow from it simplify appreciably when we use the specific eigenvectors and eigenvalues for our problem. Therefore, in the following we shall restrict ourselves to a consideration of the qualitative differences in asymptotic behavior between (5.14) with $\mu\neq0$, $\nu\neq0$ and the special case for which $\mu=\nu=0$ (which was treated in Sec. IV).

When $\mu\neq0$, all the l_k are negative so that $\exp(l_kt)\to0$ as $t\to\infty$, and

$$\Phi(|w\rangle;\infty)=\exp[-\textstyle\sum_k\langle\nu|p(k)\rangle l_k^{-1}\langle q(k)|w\rangle]. \tag{5.15}$$

Here we have made use of the fact that

$$[\Phi_0(|w\rangle)]_{w_1=w_2=\cdots=0}\equiv[\Phi_0(|z\rangle)]_{z_1=z_2=\cdots=1}=1.$$

Evidently (5.15) can be separated into factors, each of which contains only one w_i. Recalling that $w_i=z_i-1$, we can rewrite (5.15) as

$$\Phi(|z\rangle;\infty)=\prod_{i=0}^{\infty}\Phi_i(z_i;\infty), \tag{5.16}$$

where

$$\Phi_i(z_i;\infty)=\exp[-\textstyle\sum_k\langle\nu|p(k)\rangle l_k^{-1}q_i(k)(z_i-1)]. \tag{5.17}$$

Correspondingly,

$$P(\langle n|;\infty)=\prod_{i=0}^{\infty}P_i(n_i;\infty), \tag{5.18}$$

where

$$P_i(n_i;\infty)=e^{-\bar{n}_i}\bar{n}_i^{n_i}/n_i!, \tag{5.19}$$

and

$$\bar{n}_i\equiv\bar{n}_i(\infty)=-\textstyle\sum_k\langle\nu|p(k)\rangle l_k^{-1}q_i(k); \tag{5.20}$$

that is, the occupation numbers n_i have *independent* Poisson distributions with means $\bar{n}_i$ given by (5.20).

This may be contrasted with the case $\mu=\nu=0$, for which (5.14) goes over into the form (3.31) when $t=\infty$.[41] In this case the n_i are correlated since the total number of relaxing particles is fixed, though the correlation tends to zero as the total number of particles becomes very large.

ACKNOWLEDGMENT

One of us (D. L. F.) would like to express his appreciation of the hospitality extended to him by the Department of Physics, Harvard University, during part of his tenure of a National Science Foundation Senior Postdoctoral Fellowship.

APPENDIX I

In the infinite dimensional case, the necessary and sufficient conditions for the uniqueness of the solution $T(t)$ to Eqs. (3.11) and (3.12) have been established only for a special class of Λ matrices (occurring in the theory of birth and death processes), and with the requirement that the elements $T_{ij}(t)$ remain non-negative and bounded.[42] But in most physical problems one may safely assume that the Λ matrix is such as to lead to a unique solution $T(t)$ satisfying these constraints. In our case, the constraints arise from the physical significance of the $\bar{n}_i$: These are average occupation numbers and hence are non-negative and cannot exceed the total number of particles.

It is not clear that similar reasons exist to constrain the elements of the matrix which appears in the solution (3.20) of (3.19), because the z's are auxiliary variables with no physical significance. This causes no difficulty, however, because this matrix must be identical with that in (3.10)—as we have already indicated by the

[41] Note that for $\mu=\nu=0$, we have $l_0\to\lambda_0=0$; $l_k\to\lambda_k\langle0$ for $k=1,2,\cdots$; $|p(k)\rangle\to|x(k)\rangle$; and $\langle q(k)|\to\langle y(k)|$.

[42] S. Karlin and J. McGregor, Proc. Natl. Acad. Sci. 41, 387 (1955); also, W. Feller, Ann. Math. 65, 527 (1957).

use of the same notation $T(t)$ in both cases—on account of the relation (2.9) between Φ and $\bar{n}_i$.

The spectral representation (3.14) of $T(t)$ is valid when Λ is finite dimensional and has a complete set of eigenvectors. When Λ is of infinite dimensionality, a spectral representation for $T(t)$ as a sum involving a denumerable set of eigenvalues λ_k cannot always be found; in fact, the representation may be an integral with respect to a continuous variable λ over a finite or infinite region. Examples of simple matrices Λ with both discrete and continuous spectra are given by Ledermann *et al.*[43] We shall, however, assume that all Λ matrices of interest to us have discrete spectra.

It may be mentioned here that the particular matrix Λ considered in detail in Sec. IV does satisfy the conditions for uniqueness of $T(t)$, given by Karlin and McGregor.[42] It also has a discrete spectrum and hence (3.14) is valid. The matrix L, Eq. (A II.2), is of a more general type than the matrices considered in the literature, but it too satisfies all our assumptions.

APPENDIX II

To obtain the spectral decomposition (for the case $\mu_i = i\mu$) of the operator e^{Lt} appearing in (5.10), we shall consider the solution of the matrix equation

$$(d/dt)\langle\bar{n}(t)| = \langle\bar{n}(t)|L, \qquad \text{(A II.1)}$$

where

$$L = c\begin{pmatrix} -\epsilon_- & \epsilon_- & 0 & \cdots \\ \epsilon_+ & -(\epsilon_+ + 2\epsilon_- + \mu/c) & 2\epsilon_- & \cdots \\ 0 & 2\epsilon_+ & -(2\epsilon_+ + 3\epsilon_- + 2\mu/c) & \cdots \\ \vdots & \vdots & \vdots & \ddots \end{pmatrix}. \qquad \text{(A II.2)}$$

The results required in Sec. IV will then be obtained as a special case by setting $\mu = 0$ [in which case L reduces to the matrix of (4.6)].

We observe by comparison with (5.5) that (A II.1) describes the changes that occur in the mean occupation numbers $\bar{n}_i$ of harmonic oscillators in states i through "dissociation" and through collisions with reservoir oscillators. Note that

$$\sum_i \bar{n}_i(t) \leqslant \sum_i \bar{n}_i(0) = N_0, \qquad \text{(A II.3)}$$

where N_0 is the finite number of initial particles in the relaxing system. The equality sign in (A II.3) holds when $\mu = 0$ (no "dissociation").

Consider now the formal solution of (A II.1) in the form

$$\langle\bar{n}(t)| = \sum_k \langle\bar{n}(0)|p(k)\rangle e^{l_k t}\langle q(k)|, \qquad \text{(A II.4)}$$

where the l_k are eigenvalues of L, and $|p(k)\rangle$ and $\langle q(k)|$ are the corresponding column and row eigenvectors. We want to determine these. This is most easily done in terms of generating functions. Define the generating function

$$G(u;t) = \sum_{i=0}^{\infty} \bar{n}_i(t)u^i \equiv \langle\bar{n}(t)|u\rangle, \qquad \text{(A II.5)}$$

where $|u\rangle$ is a column vector with elements $u_i = u^i$. In view of (A II.3), the generating function G is certainly analytic for $u \leqslant 1$. From (A II.4) we then have

$$G(u;t) = \sum_k \langle\bar{n}(0)|p(k)\rangle e^{l_k t} G_R(u,k), \qquad \text{(A II.6)}$$

where

$$G_R(u,k) = \langle q(k)|u\rangle \qquad \text{(A II.7)}$$

is the generating function of the elements of the row eigenvector $\langle q(k)|$ of L. If we denote by $G(u,v;t)$ the special case of (A II.6) corresponding to the particular initial conditions

$$\bar{n}_i(0) = v^i; \quad v < 1, \qquad \text{(A II.8)}$$

then

$$G(u,v;t) = \sum_k G_C(v,k) e^{l_k t} G_R(u,k), \qquad \text{(A II.9)}$$

where

$$G_C(v,k) \equiv \langle v|p(k)\rangle \equiv \sum_i v^i p_i(k) \qquad \text{(A II.10)}$$

is the generating function of the elements of the column eigenvector $|p(k)\rangle$.

To obtain the eigenvalues l_k, and the corresponding eigenfunctions $|p(k)\rangle$ and $\langle q(k)|$ (via their respective generating functions G_C and G_R), we need only use the method of generating functions to solve (A II.1) with the special initial conditions (A II.8), and then expand the function $G(u,v;t)$ thus obtained, in the form (A II.9). We proceed to do this below.

It can easily be shown from (A II.1) and (A II.2) that the equation satisfied by $G(u;t)$ is

$$\partial G/\partial t = c\epsilon_-(u-1) + [c\epsilon_- u^2 - (2c\bar{\epsilon} + \mu)u + c\epsilon_+]\partial G/\partial u. \qquad \text{(A II.11)}$$

The characteristic equation for this first order partial differential equation is

$$dt = -\frac{du}{c\epsilon_- u^2 - (2c\bar{\epsilon} + \mu)u + c\epsilon_+} = -\frac{du}{c\epsilon_-(u-\alpha)(u-\gamma)}, \qquad \text{(A II.12)}$$

where

$$\alpha = \frac{1}{2c\epsilon_-}[2c\bar{\epsilon} + \mu - (c^2 + \mu^2 + 4c\bar{\epsilon}\mu)^{\frac{1}{2}}]$$
$$\beta = \frac{1}{2c\epsilon_-}[2c\bar{\epsilon} + \mu + (c^2 + \mu^2 + 4c\bar{\epsilon}\mu)^{\frac{1}{2}}]. \qquad \text{(A II.13)}$$

---

[43] W. Ledermann and G. E. H. Reuter, Phil. Trans. Roy. Soc. (London) **A246**, 321 (1953–54).

[44] It is clear from inspection of (A II.2) that any number whatever is an eigenvalue of L since, for any l, one can construct the components of the associated eigenvector by solving the resultant system of linear equations recursively. However, since the elements of allowable eigenfunctions are restricted by (A II.3), we obtain the discrete spectrum (A II.22).

The solution of (A II.12) is

$$\left(\frac{u-\alpha}{u-\gamma}\right)e^{-c\epsilon_-(\gamma-\alpha)t}=\psi=\text{constant}. \quad \text{(A II.14)}$$

On using ψ as a new variable instead of u, Eq. (A II.11) reduces to

$$\left(\frac{\partial G}{\partial t}\right)_\psi=c\epsilon_-\left(\frac{\gamma\psi-\alpha e^{-c't}}{\psi-e^{-c't}}-1\right)G, \quad \text{(A II.15)}$$

where $c'=c\epsilon_-(\gamma-\alpha)$. The solution to this equation is

$$G=e^{c\epsilon_-(\gamma-1)t}(\psi-e^{-c't})h(\psi), \quad \text{(A II.16)}$$

where $h(\psi)$ is an arbitrary function of ψ, to be determined by the initial condition. If the latter is

$$G(u;0)=\sum_i \bar{n}_i(0)u^i=G_0(u), \quad \text{(A II.17)}$$

we finally obtain

$$G(u;t)=\frac{(\gamma-\alpha)e^{-c''t}}{(\gamma-u)-(\alpha-u)e^{-c't}} \times G_0\left(\frac{\alpha(\gamma-u)-\gamma(\alpha-u)e^{-c't}}{(\gamma-u)-(\alpha-u)e^{-c't}}\right), \quad \text{(A II.18)}$$

where $c''=c\epsilon_-(1-\alpha)$.

This gives the generating function of the $\bar{n}_i(t)$ for arbitrary initial conditions. To determine the eigenvalues and eigenvectors we now consider the special condition (A II.8) for which

$$G_0(u)=1/(1-uv). \quad \text{(A II.19)}$$

Hence,

$$G(u,v;t)=\frac{(\gamma-\alpha)e^{-c''t}}{[(\gamma-u)-(\alpha-u)e^{-c't}]-v[\alpha(\gamma-u)-\gamma(\alpha-u)e^{-c't}]} \quad \text{(A II.20)}$$

$$=\sum_{k=0}^{\infty}(\gamma-\alpha)\frac{(\alpha-u)^k}{(\gamma-u)^{k+1}}\frac{(1-v\gamma)^k}{(1-v\alpha)^{k+1}} \times\exp\{-[kc'+c'']t\}. \quad \text{(A II.21)}$$

Comparing this with (A II.9), we deduce that

$$l_k=-[kc'+c''],\quad k=0,1,\cdots, \quad \text{(A II.22)}$$

$$G_C(v,k)=(1-v\gamma)^k(1-v\alpha)^{-k-1}, \quad \text{(A II.23)}$$

and

$$G_R(u,k)=(\gamma-\alpha)(\alpha-u)^k(\gamma-u)^{-k-1}. \quad \text{(A II.24)}$$

Using the identity (4.11), we expand the last two equations in powers of v and u, respectively, and obtain

$$p_i^k=\gamma^i\,{}_2F_1(-i,k+1;1;1-\alpha/\gamma), \quad \text{(A II.25)}$$

and

$$q_j^k=(1-\alpha/\gamma)(\alpha/\gamma)^k \times\alpha^{-j}\,{}_2F_1(-j,k+1;1;1-\alpha/\gamma). \quad \text{(A II.26)}$$

Since

$$G(u,v;t)=\langle v|e^{Lt}|u\rangle, \quad \text{(A II.27)}$$

a direct expansion of (A II.20) in powers of u and v [rather than in powers of the exponential as in (A II.21)] yields the following two alternative expressions for $(e^{Lt})_{ij}$:

$$(e^{Lt})_{ij}=(\gamma-\alpha)e^{-c''t}[\alpha\gamma(1-e^{-c't})]^{i-j}, \quad \text{(A II.28a)}$$

or

$$(e^{Lt})_{ij}=(\gamma-\alpha)e^{-c''t}(1-e^{-c't})^{j-i}(\alpha-\gamma e^{-c't})^i \times(\gamma-\alpha e^{-c't})^{-j-1}\,{}_2F_1(-i,j+1;1;s), \quad \text{(A II.28b)}$$

where

$$s=\frac{(\gamma-\alpha)^2e^{-c't}}{(\gamma-\alpha e^{-c't})(\gamma e^{-c't}-\alpha)}. \quad \text{(A II.29)}$$

To obtain the results used in Sec. IV, we set $\mu=\nu=0$ in the above formulas, so that

$$\alpha=1,\quad \gamma=\epsilon_+/\epsilon_-,\quad c'=c,\quad c''=0, \quad \text{(A II.30)}$$

and the expressions (A II.22), (A II.23), and (A II.24) for l_k, G_C, and G_R go over into (4.12), (4.14), and (4.13), respectively. The expressions (4.9a, b) for $T_{ij}\equiv(e^{\Lambda t})_{ij}$ are obtained from (A II.28a, b).

Exact Conditions for the Preservation of a Canonical Distribution in Markovian Relaxation Processes

H. C. ANDERSEN* AND I. OPPENHEIM†
Department of Chemistry, Massachusetts Institute of Technology, Cambridge, Massachusetts
AND
KURT E. SHULER AND GEORGE H. WEISS‡
National Bureau of Standards, Washington, D. C.
(Received 9 November 1963)

Necessary and sufficient conditions have been determined for the exact preservation of a canonical distribution characterized by a time-dependent temperature (canonical invariance) in Markovian relaxation processes governed by a master equation. These conditions, while physically realizable, are quite restrictive so that canonical invariance is the exception rather than the rule. For processes with a continuous energy variable, canonical invariance requires that the integral master equation is exactly equivalent to a Fokker–Planck equation with linear transition moments of a special form. For processes with a discrete energy variable, canonical invariance requires, in addition to a special form of the level degeneracy, equal spacing of the energy levels and transitions between nearest-neighbor levels only. Physically, these conditions imply that canonical invariance is maintained only for weak interactions of a special type between the relaxing subsystem and the reservoir. It is also shown that canonical invariance is a sufficient condition for the exponential relaxation of the mean energy. A number of systems (hard-sphere Rayleigh gas, Brownian motion, harmonic oscillators, nuclear spins) are discussed in the framework of the above theory. Conditions for *approximate* canonical invariance valid up to a certain order in the energy are also developed and then applied to nuclear spins in a magnetic field.

I. INTRODUCTION

THERE exists a wide class of nonequilibrium systems whose relaxation can be described, at least to a very good approximation, by a *master equation* characteristic of a Markovian stochastic process. One class of such systems, with which this paper is concerned, is that of a dilute subsystem (with number density n_s) dispersed in a heat bath (with number density n_h) with $n_s \ll n_h$. It is then assumed that the subsystem, which has been prepared in an initial nonequilibrium distribution, relaxes to the (time-invariant) equilibrium distribution characteristic of the heat bath solely through interactions with the heat bath, the inequality $n_s \ll n_h$ being taken sufficiently strong that interactions between the subsystem particles can be neglected compared to the subsystem–heat-bath interactions.

In previous studies on the relaxation of such subsystem–heat-bath ensembles, it has been shown, exactly and analytically, that for *some* such ensembles an initial canonical distribution of the subsystem (i.e, a Maxwell–Boltzmann distribution in energy or velocity) will relax to the final equilibrium canonical distribution corresponding to that of the heat bath via a continuous (in time) sequence of canonical distributions with a time-dependent "temperature". We shall refer to this property as *canonical invariance*. Examples of such exact canonical invariance are: the vibrational relaxation of a set of harmonic oscillators subject to Landau–Teller transition probabilities in contact with a heat bath,[1,2] the (translational) relaxation of a hard-sphere Rayleigh gas,[3] and the (translational) relaxation of a Lorentz gas with a Maxwellian (r^{-5}) force law.[4] For many other systems, however, it has been shown that an initial canonical distribution is not preserved during the relaxation process.[5] Some examples of these are: the relaxation of a subsystem of harmonic oscillators with exponentially varying transition probabilities,[6] the relaxation of a subsystem of anharmonic oscillators,[7] the relaxa-

* Summer Student, National Bureau of Standards.
† Consultant, National Bureau of Standards.
‡ Consultant, National Bureau of Standards. Present Address: Rockefeller Institute, New York, N. Y.

[1] R. J. Rubin and K. E. Shuler, J. Chem. Phys. **25**, 59 (1956).
[2] E. W. Montroll and K. E. Shuler, J. Chem. Phys. **26**, 454 (1957).
[3] K. Andersen and K. E. Shuler, J. Chem. Phys. **40**, 633 (1964).
[4] O. I. Osipov, Bull. Moscow Univ. Ser. III **1**, 13 (1961) and Ref. 3.
[5] We are discussing here exact analytical results and are not concerned, at this time, with approximate theoretical treatments or with the approximate fitting of experimental data.
[6] R. J. Rubin and K. E. Shuler, J. Chem. Phys. **25**, 68 (1956).
[7] N. W. Bazley, E. W. Montroll, R J. Rubin, and K. E. Shuler, J. Chem. Phys. **28**, 700 (1958)

tion of a subsystem of rigid rotators,[8] the relaxation of a hard-sphere Lorentz gas,[3] and the relaxation of a subsystem of spins in contact with a lattice.[9]

The only variables entering into the relaxation equation (the master equation) which can effect the preservation or nonpreservation of the initial distribution during relaxation are the transition probabilities per unit time between the quantum states (or the energy states), the degeneracies of the energy states, and, in discrete quantum systems, the spacing of the energy levels and their total number, i.e., finite or infinite. It is of interest to determine the necessary and sufficient conditions on these variables for canonical invariance for Markovian relaxation processes. This program is carried through in the subsequent sections of this paper.

Sections II and III are devoted primarily to the formal mathematical development. We begin our treatment in Sec. II by deriving, from the master equation in energy space, the necessary and sufficient conditions for canonical invariance for systems with continuous energy-level spectra. We then apply these results to three specific examples: Brownian Motion (Ornstein–Uhlenbeck processes), the Rayleigh gas, and classical harmonic oscillators. In Sec. III we establish the connection between the quantum-state master equation and the master equation in energy space, and then derive the necessary and sufficient conditions for systems with discrete energy levels, i.e., quantum systems, for both finite- and infinite-level systems. The quantal results are then applied to the specific examples of the relaxation of (Landau–Teller) harmonic oscillators and to spin-lattice relaxation systems. In Sec. IV we derive the conditions for the *approximate* preservation of a canonical distribution for a finite discrete energy-level system with a level spacing small compared to kT. This case is of interest in connection with certain spin-relaxation problems. In Sec. V we then consider the physical implications of our analysis, and the relation of our results to some previous work on relaxation processes.

Our findings may be summarized broadly as follows. The conditions for exact canonical invariance, while physically realizable, are quite restrictive. Canonical invariance in a relaxation process is thus the "exception" rather than the rule. Whenever canonical invariance obtains, the mean energy of the subsystem undergoes a simple exponential relaxation of the type discussed previously by Shuler, Weiss, and Andersen.[10] Canonical invariance is, however, only a sufficient and not a necessary condition for such an exponential relaxation of the mean energy. The analysis presented here permits the ready determination of canonical invariance from the form of the relevant parameters (transition-probability kernel, degeneracy, level spacing, etc.), and obviates the need for an explicit solution of the relaxation equation. Left open in the present study is the important question as to approximate canonical invariance, i.e., the extent of the deviation from canonical invariance, under a weakening of the conditions for exact invariance. We plan to consider this problem in a subsequent paper.

II. SYSTEMS WITH A CONTINUOUS ENERGY SPECTRUM

Necessary and Sufficient Conditions

We first consider subsystems, such as a classical gas, in which the individual particles can have any energy greater than zero. Let us define $P(\epsilon, t)d\epsilon$ as the probability that a subsystem particle has an energy between ϵ and $\epsilon + d\epsilon$ at time t. We assume that $P(\epsilon, t)$ satisfies the master equation

$$\frac{\partial P(\epsilon, t)}{\partial t} = \int_0^\infty [B(\epsilon \mid \epsilon')P(\epsilon', t) - B(\epsilon' \mid \epsilon)P(\epsilon, t)]\, d\epsilon'. \tag{2.1}$$

The transition probabilities per unit time, for transitions from state ϵ' to ϵ, $B(\epsilon \mid \epsilon')$, are taken to be independent of time, i.e., we consider a stationary process. We also use detailed balancing,

$$B(\epsilon \mid \epsilon')P(\epsilon', \infty) = B(\epsilon' \mid \epsilon)P(\epsilon, \infty). \tag{2.2}$$

As $t \to \infty$, the function $P(\epsilon, t)$ will approach the Maxwell–Boltzmann distribution

$$P(\epsilon, \infty) = g(\epsilon)e^{-\beta(\infty)\epsilon}/Q[\beta(\infty)], \tag{2.3}$$

where $\beta(\infty) \equiv [kT(\infty)]^{-1}$ with $T(\infty)$ being the heat-bath temperature, the partition function $Q[\beta] = \int_0^\infty g(\epsilon)e^{-\beta\epsilon}\, d\epsilon$, and $g(\epsilon)$ is the degeneracy, i.e., the density of states with energy ϵ.

We wish to investigate the conditions under which a canonical distribution is preserved during the relaxation process, i.e., the conditions under which Eq. (2.1) has a solution of the form

[8] R. Herman and K. E. Shuler, J. Chem. Phys. 29, 366 (1958).

[9] Except for the trivial case of spin ½, which involves only two energy levels. The relaxation of spin systems will be discussed in more detail in the body of the paper.

[10] K. E. Shuler, G. H. Weiss, and K. Andersen, J. Math. Phys. 3, 550 (1962).

$$P(\epsilon, t) = g(\epsilon)e^{-\beta(t)\epsilon}/Q[\beta(t)]. \tag{2.4}$$

Such a solution would enable us to define a time-dependent temperature $T(t) = 1/k\beta(t)$ for all times $0 \le t \le \infty$ for the relaxation process. To find the necessary conditions we assume that a solution of the form (2.4) exists. Then

$$\frac{\partial P(\epsilon, t)}{\partial t} = -\left\{\dot\beta(t)\epsilon + \frac{\dot Q[\beta(t)]}{Q[\beta(t)]}\right\}P(\epsilon, t), \tag{2.5}$$

where the dot denotes differentiation with respect to t. Since

$$\frac{\dot Q[\beta(t)]}{Q[\beta(t)]} = -\frac{\dot\beta(t)\int_0^\infty \epsilon g(\epsilon)e^{-\beta(t)\epsilon}\,d\epsilon}{\int_0^\infty g(\epsilon)e^{-\beta(t)\epsilon}\,d\epsilon} = -\dot\beta(t)\bar\epsilon(t), \tag{2.6}$$

where $\bar\epsilon(t)$ is the average energy of the particles at time t, Eq. (2.5) reduces to

$$\partial P(\epsilon, t)/\partial t = \dot\beta(t)[\bar\epsilon(t) - \epsilon]P(\epsilon, t). \tag{2.7}$$

The detailed balance condition (2.2) can be used to rewrite Eq. (2.1) as

$$\frac{\partial P(\epsilon, t)}{\partial t} = P(\epsilon, t)\int_0^\infty B(\epsilon' \mid \epsilon) \times [\exp\{-[\beta(t) - \beta(\infty)](\epsilon' - \epsilon)\} - 1]\,d\epsilon'. \tag{2.8}$$

With the definition

$$\alpha(t) \equiv \beta(t) - \beta(\infty), \tag{2.9}$$

the master equation (2.1) can finally be written as

$$\dot\alpha(t)[\bar\epsilon(t) - \epsilon] = \int_0^\infty B(\epsilon' \mid \epsilon) \times \{\exp[-\alpha(t)(\epsilon' - \epsilon)] - 1\}\,d\epsilon', \tag{2.10}$$

where we have made use of Eqs. (2.7) and (2.8) above. Equation (2.10) will serve as our starting point for the determination of the conditions for canonical invariance.

We assume that the right side of Eq. (2.10) can be expressed as

$$\int_0^\infty B(\epsilon' \mid \epsilon)\{\exp[-\alpha(t)(\epsilon' - \epsilon)] - 1\}\,d\epsilon' = \sum_{m=1}^\infty \frac{(-1)^m}{m!}\,b_m(\epsilon)\alpha^m, \tag{2.11}$$

where

$$b_m(\epsilon) = \int_0^\infty B(\epsilon' \mid \epsilon)(\epsilon' - \epsilon)^m\,d\epsilon'. \tag{2.12}$$

We now determine the coefficients $b_m(\epsilon)$ by deriving an equation for $\bar\epsilon$ as a function of time and then relating $\bar\epsilon(t)$ to $\alpha(t)$ for a particular class of degeneracies $g(\epsilon)$. This will yield an explicit expression in terms of $\alpha(t)$ for the left-hand side (lhs) of Eq. (2.10). The use of a Taylor expansion of the lhs of Eq. (2.10) in powers of α will then permit us to determine the $b_m(\epsilon)$.

Combining (2.10) and (2.11) we obtain

$$\dot\alpha(t)[\bar\epsilon(t) - \epsilon] = \sum_{m=1}^\infty \frac{(-1)^m}{m!}\,b_m(\epsilon)\alpha^m. \tag{2.13}$$

We first note that, for Eq. (2.13) to hold for all ϵ, the $b_m(\epsilon)$ must be of the form

$$b_m(\epsilon) = b_m^0 + \epsilon b_m^1; \qquad m = 1, 2, \cdots, \tag{2.14}$$

where b_m^0 and b_m^1 are constants independent of ϵ. This follows from the fact that the lhs of Eq. (2.13) is linear in ϵ. We can now find the equation satisfied by $\bar\epsilon(t)$. From Eq. (2.12) it follows that, for $m = 1$,

$$\int_0^\infty B(\epsilon' \mid \epsilon)(\epsilon' - \epsilon)\,d\epsilon' = b_1^0 + \epsilon b_1^1. \tag{2.15}$$

It has been shown in Ref. (10) that a transition moment $b_1(\epsilon)$ of the form displayed in Eq. (2.15) leads directly to the relation

$$d\bar\epsilon(t)/dt = b_1^0 + b_1^1\bar\epsilon(t) \tag{2.16}$$

for the exponential relaxation of the mean energy.

To find the relationship between $\alpha(t)$ and $\bar\epsilon(t)$ we must introduce an ansatz about the density of states $g(\epsilon)$. We shall use the form[11]

$$g(\epsilon) = \epsilon^n e^{p\epsilon}, \tag{2.17}$$

where $n > -1$ (but not necessarily integer) to ensure the normalizability of the canonical distribution. The density of states of a classical free particle is of this form with $p = 0$ and $n = -\frac{1}{2}$, 0, or $\frac{1}{2}$, depending upon whether the motion takes place in one, two, or three dimensions.

For the density of states given by (2.17), the mean energy $\bar\epsilon(t)$ is

$$\bar\epsilon(t) = \int_0^\infty \epsilon g(\epsilon)e^{-\beta(t)\epsilon}\,d\epsilon \Big/ \int_0^\infty g(\epsilon)e^{-\beta(t)\epsilon}\,d\epsilon = \frac{n+1}{\alpha(t) + \beta(\infty) - p}. \tag{2.18}$$

An expression for $\alpha(t)$ can be obtained by differentiating (2.18) and using Eq. (2.16),

$$\dot\alpha(t) = -[\alpha(t) + \beta(\infty) - p] \times \left\{\frac{b_1^0[\alpha(t) + \beta(\infty) - p]}{n+1} + b_1^1\right\}. \tag{2.19}$$

[11] The form of $g(\epsilon)$ in Eq. (2.17) can be shown to be the continuum analog of the discrete degeneracy g_i whose necessary and sufficient form for canonical invariance can be determined unequivocally (see Sec. III and Appendix III).

When $\alpha = 0$, $\beta(t) = \beta(\infty)$ and the ensemble will be in its equilibrium distribution with $\dot{\alpha}(t) \equiv \dot{\beta}(t) = 0$. From this it follows that

$$b_1^1 = -b_1^0[\beta(\infty) - p]/(n+1), \qquad (2.20)$$

and

$$\dot{\alpha}(t) = [-b_1^0\alpha(t)/(n+1)][\alpha(t) + \beta(\infty) - p]. \quad (2.21)$$

Combining these results with Eq. (2.18), we finally obtain

$$\dot{\alpha}(t)\bar{\epsilon}(t) = -b_1^0\alpha(t) \qquad (2.22)$$

as the desired relation between $\dot{\alpha}$ and $\bar{\epsilon}$. We can now use Eqs. (2.20)–(2.22) to reexpress the lhs of Eq. (2.10) and equate it to the expansion (2.11). This yields

$$b_1^0\left\{\left[\frac{(\beta(\infty) - p)\epsilon}{n+1} - 1\right]\alpha + \frac{\epsilon}{n+1}\alpha^2\right\} = \sum_{m=1}^{\infty}\frac{(-1)^m}{m!}b_m(\epsilon)\alpha^m. \qquad (2.23)$$

Comparing coefficients of α we then find

$$\begin{aligned} b_1(\epsilon) &= b_1^0\{1 - \epsilon[\beta(\infty) - p]/(n+1)\} = b_1^0(1 - A\epsilon), \\ b_2(\epsilon) &= 2b_1^0\epsilon/(n+1) = b_1^0(A'\epsilon), \qquad (2.24) \\ b_m(\epsilon) &= 0; \qquad m \geq 3. \end{aligned}$$

Clearly these relations could not be realized if $B(\epsilon' \mid \epsilon)$ were a function in the ordinary sense, since $B(\epsilon' \mid \epsilon) \geq 0$ and $(\epsilon' - \epsilon)^m > 0$ for $\epsilon' \neq \epsilon$ for even m. Hence $B(\epsilon' \mid \epsilon)$ must be expressed in terms of distributions, i.e.,

$$B(\epsilon' \mid \epsilon) = b_0(\epsilon)\delta(\epsilon' - \epsilon) - b_1(\epsilon)\delta^{(1)}(\epsilon' - \epsilon) + \tfrac{1}{2}b_2(\epsilon)\delta^{(2)}(\epsilon' - \epsilon), \qquad (2.25)$$

where $\delta^{(m)}(\epsilon)$ is the mth derivative of the Dirac delta function. When the above expression is substituted into the master equation (2.1) and the resulting equation integrated by parts, one obtains

$$\frac{\partial P(\epsilon, t)}{\partial t} = -\frac{\partial}{\partial\epsilon}[b_1(\epsilon)P(\epsilon, t)] + \tfrac{1}{2}\frac{\partial^2}{\partial\epsilon^2}[b_2(\epsilon)P(\epsilon, t)], \qquad (2.26)$$

with $b_1(\epsilon)$ and $b_2(\epsilon)$ given by Eq. (2.24). We have thus obtained the interesting result that the transition probability $B(\epsilon' \mid \epsilon)$ which gives rise to canonical invariance is of the form which leads to the exact equivalence between the integral master equation (2.1) and the Fokker–Planck equation (2.26). Put in other words, it is only for relaxation processes described by the Fokker–Planck equation (2.26) with the $b_n(\epsilon)$ given by (2.24) that canonical invariance obtains for the distribution function. It can readily be verified by direct substitution that the canonical distribution (2.4) is a solution of the relaxation equation (2.26).

To summarize then, we have shown here that the necessary and sufficient conditions for canonical invariance for a Markovian relaxation process with a continuous energy variable are:

(a) *The transition probability $B(\epsilon' \mid \epsilon)$ is a sum of Dirac delta functions and their derivatives as shown in Eq. (2.25).*

(b) *The moments $b_m(\epsilon)$ are given by Eq. (2.24).*

Condition (a) implies that the master equation is exactly equivalent to the Fokker–Planck equation (2.26). The above results pertain to degeneracies $g(\epsilon)$ as given by Eq. (2.17).

Relaxation of the Temperature and Mean Energy

An explicit expression for the relaxation of the temperature $T(t) \equiv 1/k\beta(t)$ can readily be obtained from Eqs. (2.20) and (2.21). The integration of (2.21) leads to

$$\left[\frac{T(t) - T(\infty)}{T(0) - T(\infty)}\right]\left[\frac{T(0) - T_p}{T(t) - T_p}\right] = e^{b_1{}^1 t}, \qquad (2.27)$$

with $T_p \equiv 1/kp$. When $p = 0$, so that $g(\epsilon) = \epsilon^n$, one obtains the simple exponential temperature relaxation

$$[T(t) - T(\infty)]/[T(0) - T(\infty)] = e^{b_1{}^1 t}. \qquad (2.28)$$

It can readily be verified from Eq. (2.28) that $b_1^1 < 0$.

The differential equation for the relaxation of the mean energy has already been given [Eq. (2.16)]. Its solution is

$$[\bar{\epsilon}(t) - \bar{\epsilon}(\infty)]/[\bar{\epsilon}(0) - \bar{\epsilon}(\infty)] = e^{b_1{}^1 t}. \qquad (2.29)$$

This result is independent of the value of p in Eq. (2.17) for the density of states. Since $\bar{\epsilon}$ is proportional to kT for $g(\epsilon) = \epsilon^n$ with a canonical $P(\epsilon, t)$, the exponential temperature relaxation in (2.28) is equivalent to the exponential energy relaxation of Eq. (2.29).

We have thus demonstrated that canonical invariance is a sufficient condition for the exponential relaxation of the mean energy.[12]

[12] Canonical invariance is, however, not a necessary condition. This can readily be seen from the fact that it is necessary to specify *both* $b_1(\epsilon)$ and $b_2(\epsilon)$ for canonical invariance [Eq. (2.24)], whereas only $b_1(\epsilon)$ needs to be specified explicitly for the exponential relaxation of the mean energy [see Ref. (10)].

The Conditional Probability

The Fokker–Planck equation (2.26) can be used to obtain the conditional probability $W(\epsilon, t; \epsilon', 0)$ which corresponds to the transition probabilities $B(\epsilon' \mid \epsilon)$ given in Eq. (2.25). The conditional probability $W(\epsilon, t; \epsilon', 0)$, which is defined so that $W(\epsilon, t; \epsilon', 0)d\epsilon$ is the probability that a molecule has energy between ϵ and $\epsilon + d\epsilon$ at time t given that it has energy ϵ' at time $t = 0$, is the solution of Eq. (2.26) for the initial condition

$$P(\epsilon, 0) = \delta(\epsilon' - \epsilon). \tag{2.30}$$

To simplify the following expressions, we transform to dimensionless units of energy x and time τ, by letting

$$x = \epsilon[\beta(\infty) - p], \qquad \tau = -b_1^1 t. \tag{2.31}$$

Equation (2.26) then becomes

$$\frac{\partial P(x, \tau)}{\partial \tau} = P(x, \tau) + (x - n + 1)\frac{\partial P(x, \tau)}{\partial x} + x\frac{\partial^2 P(x, \tau)}{\partial x^2}. \tag{2.32}$$

Before evaluating $W(\epsilon, t; \epsilon', 0)$ we shall first develop the general solution of (2.32). With the transformation

$$P(x, \tau) = x^n e^{-x}\psi(x, \tau), \tag{2.33}$$

Eq. (2.32) becomes

$$\frac{\partial \psi(x, \tau)}{\partial \tau} = (n + 1 - x)\frac{\partial \psi(x, \tau)}{\partial x} + x\frac{\partial^2 \psi(x, \tau)}{\partial x^2}. \tag{2.34}$$

The rhs of Eq. (2.34) defines the linear operator

$$\mathfrak{D} \equiv (n + 1 - x)(\partial/\partial x) + x(\partial^2/\partial x^2),$$

so that (2.34) can be written compactly as

$$\partial\psi(x, \tau)/\partial\tau = \mathfrak{D}\psi(x, \tau). \tag{2.35}$$

The eigenfunctions of the operator $\mathfrak{D}$ are the generalized Laguerre polynomials L_m^n, i.e.,

$$\mathfrak{D}L_m^n(x) = -mL_m^n(x). \tag{2.36}$$

Since the L_m^n form a complete set of functions, we can write

$$\psi(x, \tau) = \sum_{m=0}^{\infty} c_m L_m^n(x)\Omega_m(\tau). \tag{2.37}$$

It can easily be shown that

$$\Omega_m(\tau) = e^{-m\tau}. \tag{2.38}$$

The general eigenfunction solution of (2.32) is thus

$$P(x, \tau) = x^n e^{-x} \sum_{m=0}^{\infty} c_m L_m^n(x)e^{-m\tau}. \tag{2.39}$$

The Laguerre polynomials have the orthogonality relation

$$\int_0^\infty x^n e^{-x} L_m^n(x)L_r^n(x)\,dx = \frac{\delta_{mr}\Gamma(m + n + 1)}{m!}, \tag{2.40}$$

so that

$$c_m = \frac{m!}{\Gamma(n + m + 1)}\int_0^\infty L_m^n(x)P(x, 0)\,dx. \tag{2.41}$$

For the initial condition $P(x, 0) = \delta(x - x_0)$, we obtain

$$c_m = m!\,L_m^n(x_0)/\Gamma(n + m + 1), \tag{2.42}$$

which yields

$$W(x, \tau; x_0, 0) = x^n e^{-x}\sum_{m=0}^{\infty}\frac{m!}{\Gamma(n + m + 1)} \times L_m^n(x_0)L_m^n(x)e^{-m\tau} = \left(\frac{x}{x_0}\right)^{\frac{1}{2}n}(1 - e^{-\tau})^{-1}e^{\frac{1}{2}n\tau} \times \exp\left[\frac{(x + x_0e^{-\tau})}{1 - e^{-\tau}}\right]I_n\left[\frac{2(xx_0e^{-\tau})^{\frac{1}{2}}}{1 - e^{-\tau}}\right], \tag{2.43}$$

where I_n is the nth-order modified Bessel function of the first kind.[13] This is the unique form of the conditional probability which preserves the canonical distribution. For $n = \frac{1}{2}$, Eq. (2.43) becomes[3]

$$W(x, \tau; x_0, 0) = \frac{e^{\frac{1}{4}\tau}}{2[\pi x_0(1 - e^{-\tau})]^{\frac{1}{2}}}\left\{\exp\left[-\frac{[x^{\frac{1}{2}} - (x_0e^{-\tau})^{\frac{1}{2}}]^2}{1 - e^{-\tau}}\right] - \exp\left[-\frac{[x^{\frac{1}{2}} + (x_0e^{-\tau})^{\frac{1}{2}}]^2}{1 - e^{-\tau}}\right]\right\}, \tag{2.44}$$

where the conditional probability can be seen to be the difference between two Gaussian terms. It is of course well known that Gaussian conditional probabilities (in velocity space) lead exactly to Fokker–Planck equations. The development presented here shows that while a Gaussian conditional probability is sufficient for the equivalence of the master equation and the Fokker–Planck equation, it is by no means a necessary form of $W(x, \tau; x_0, 0)$.

[13] Bateman Manuscript Project, *Higher Transcendental Functions*, edited by A. Erdélyi (McGraw-Hill Book Company, Inc., New York, 1953), Vol. 2, p. 189.

Examples

We shall now apply the above results to some specific examples.

A. Harmonic Oscillator

Rubin and Shuler[1] have derived a partial differential equation to describe the collisional relaxation of an ensemble of harmonic oscillators in a heat bath for the case that $h\nu/kT(\infty) \ll 1$. (The frequency of the oscillator is ν and h is Planck's constant.) Physically, this corresponds to replacing the discrete quantum oscillator by a quasiclassical one whose transitions are still governed by the quantal transition probabilities. Their Fokker–Planck equation is

$$\frac{1}{k_{10}}\frac{\partial u(y,t)}{\partial t} = \theta u + (\theta y + 1)\frac{\partial u}{\theta y} + y\frac{\partial^2 u}{\partial y^2}, \qquad (2.45)$$

where $u(y, t)$ is defined so that for integral y it is equal to the probability that an oscillator is in the yth quantum state at time t, θ is $h\nu/kT(\infty)$, and the unit of energy is chosen such that $h\nu = 1$. We can compare their result with the Fokker–Planck equation (2.26) derived here. When the indicated differentiation is performed, Eq. (2.26) becomes

$$\frac{\partial P(\epsilon, t)}{\partial t} = \frac{b_1^0}{n+1}\left\{\epsilon\frac{\partial^2 P}{\partial \epsilon^2} + [\epsilon[\beta(\infty) - p] + 1 - n]\frac{\partial P}{\partial \epsilon} + [\beta(\infty) - p]P\right\}. \qquad (2.46)$$

With $n = 0$, $p = 0$ for the harmonic oscillator $[g(\epsilon) = 1]$, it can readily be seen that Eqs. (2.45) and (2.46) are identical except for notation. The canonical invariance and exponential relaxation of the mean energy found by Rubin and Shuler for their quasiclassical oscillators is thus seen to be consistent with the general formulation developed here.

B. Rayleigh Gas

Andersen and Shuler[3] derived the following Fokker–Planck equation for the relaxation of a hard-sphere Rayleigh gas:

$$\frac{\partial P(x, \tau)}{\partial \tau} = k_R\frac{\partial}{\partial x}\left\{(x - \tfrac{3}{2})P(x, \tau) + \frac{\partial}{\partial x}[xP(x, \tau)]\right\}. \qquad (2.47)$$

Here x is the reduced dimensionless energy, $x = \epsilon/kT_2$, where T_2 is the temperature of the heat bath and ϵ is the kinetic energy of the subsystem particles. We can transform our Eq. (2.26) to one analogous to (2.47) by transforming to the reduced energy variable $x = \epsilon[\beta(\infty) - p]$. Equation (2.26) then becomes

$$\frac{\partial P(x, \tau)}{\partial \tau} = \frac{b_1^0[\beta(\infty) - p]}{n+1}\frac{\partial}{\partial x} \times \left\{(x - n - 1)P(x, \tau) + \frac{\partial}{\partial x}[xP(x, \tau)]\right\}. \qquad (2.48)$$

For the three-dimensional Rayleigh gas, $n = \frac{1}{2}$ and $p = 0$, and (2.48) is thus identical with (2.47) with $2b_1^0/3kT_2 = k_R$. The translation relaxation of a hard-sphere Rayleigh gas in energy space is thus shown to be, as already demonstrated explicitly in Ref. 3, another specific example of canonical invariance.

C. Ornstein–Uhlenbeck Process

Bowen and Meijer[14] have shown that for a one-dimensional Ornstein–Uhlenbeck relaxation process (i.e., Brownian motion in the absence of external forces), a velocity distribution which is initially Gaussian will remain Gaussian in form. This is true for two and three dimensions as well. The Fokker–Planck equation for the *velocity* probability density function for three-dimensional Brownian motion is[15]

$$\partial W(\mathbf{u}, t)/\partial t = \gamma\nabla_u(W\mathbf{u}) + [\gamma/\beta(\infty)m]\nabla_u^2 W, \qquad (2.49)$$

where $\mathbf{u}$ is the velocity of the particle, m is its mass, and γ is the friction coefficient (Chandrasekhar's β). When this equation is converted to an equation for the *energy* probability density function, the result is of the form of Eq. (2.26) with

$$b_1(\epsilon) = [3\gamma/\beta(\infty)][1 - \tfrac{2}{3}\epsilon\beta(\infty)], \qquad b_2(\epsilon) = [3\gamma/\beta(\infty)](\tfrac{4}{3}\epsilon). \qquad (2.50)$$

This is in agreement with Eqs. (2.24), since $p = 0$ and $n = \frac{1}{2}$ for three-dimensional Brownian motion. The "Gaussian invariance" of the Ornstein–Uhlenbeck process is thus shown to be another specific example of canonical invariance.

III. SYSTEMS WITH A DISCRETE ENERGY SPECTRUM

Transition from the Quantum State to the Energy-Level Master Equation

We now consider systems which have a discrete energy spectrum such that

[14] J. J. Bowen and P. H. E. Meijer, Physica **26**, 485 (1960).
[15] S. Chandrasekhar, Rev. Mod. Phys. **15**, 1 (1943).

$$0 = \epsilon_0 < \epsilon_1 < \epsilon_2 < \cdots,$$

where every energy-level ϵ_i can be reached from every level ϵ_k in a finite number of steps, i.e., the energy-level system is irreducible in the sense of Markov chains. Each level has a degeneracy, g_i, and there may be a finite or infinite number of levels. In the finite case let there be $N + 1$ levels, so that the highest energy is ϵ_N.

We use two quantum numbers, i and σ_i, to designate a quantum state. The first denotes the energy level and the second denotes the quantum state within that level. The value of σ_i is an integer between 1 and g_i. We define $P_i^{\sigma_i}(t)$ to be the probability that a system is in state i, σ_i at time t, and assume that it satisfies a master equation:

$$\frac{dP_i^{\sigma_i}(t)}{dt} = \sum_{j=0}^{N} \sum_{\sigma_j=1}^{g_j} [B_{ij}^{\sigma_i\sigma_j}P_j^{\sigma_j} - B_{ji}^{\sigma_j\sigma_i}P_i^{\sigma_i}],$$
$$i = 0, 1, \cdots N; \quad \sigma_i = 1, 2, \cdots g_i. \tag{3.1}$$

The transition probabilities $B_{ij}^{\sigma_i\sigma_j}$, which denote the probability per unit time for a transition from state j, σ_j to state i, σ_i, are again taken to be independent of time. We define $B_{ii}^{\sigma_i\sigma_i} = 0$, for $i = 0, 1, \cdots N$, and $\sigma_i = 1, 2, \cdots g_i$. The equilibrium distribution is given by

$$P_i^{\sigma_i}(\infty) = e^{-\beta(\infty)\epsilon_i}/Q[\beta(\infty)], \tag{3.2}$$

where $Q[\beta] = \sum_{j=0}^{N} \sum_{\sigma_j=1}^{g_j} e^{-\beta\epsilon_j} = \sum_{j=0}^{N} g_j e^{-\beta\epsilon_j}$. We require that the degeneracies g_j be such that $Q[\beta]$ remains finite for the case $N = \infty$. Detailed balancing holds in the form

$$B_{ij}^{\sigma_i\sigma_j}P_j^{\sigma_j}(\infty) = B_{ji}^{\sigma_j\sigma_i}P_i^{\sigma_i}(\infty). \tag{3.3}$$

We now wish to examine the conditions under which the *quantum-state master equation* preserves the form of the canonical ensemble, i.e., the conditions under which Eq. (3.1) has a solution of the form

$$P_i^{\sigma_i}(t) = e^{-\beta(t)\epsilon_i}/Q[\beta(t)]. \tag{3.4}$$

In Appendix I it is shown that this is possible if and only if:

(a) the corresponding energy-level master equation

$$\frac{dP_i(t)}{dt} = \sum_{j=0}^{N} [B_{ij}P_j - B_{ji}P_i] \tag{3.5}$$

has a canonically invariant solution of the form

$$P_i(t) = g_i e^{-\beta(t)\epsilon_i}/Q[\beta(t)], \tag{3.6}$$

and (b)

$$\sum_{\sigma_j=1}^{g_j} B_{ij}^{\sigma_i\sigma_j} \quad \text{is independent of} \quad \sigma_i. \tag{3.7}$$

In Eq. (3.5), the quantity $P_i(t)$ is defined as the probability that a system is in the energy level ϵ_i at time t, and B_{ij} is the transition probability per unit time for a transition from energy level ϵ_j to ϵ_i. These "energy level" variables are related to the corresponding "quantum state" variables by

$$P_i(t) = \sum_{\sigma_i=1}^{g_i} P_i^{\sigma_i}(t), \tag{3.8}$$

$$B_{ij} = \frac{1}{g_j} \sum_{\sigma_i=1}^{g_i} \sum_{\sigma_j=1}^{g_j} B_{ij}^{\sigma_i\sigma_j}, \tag{3.9}$$

with $B_{ii} = 0$. As is shown in Appendix I, the B_{ij} satisfy detailed balancing with

$$B_{ij}P_j(\infty) = B_{ji}P_i(\infty). \tag{3.10}$$

The condition (3.7) on the quantum-state transition probabilities $B_{ij}^{\sigma_i\sigma_j}$ has the following physical interpretation. If the population of a state is dependent only upon its energy (as is the case in a Boltzmann distribution) then the rate at which transitions are made from level j to the states in level i must be the same for all states in level i. This condition ensures that, if at one time $P_i^{\sigma_i}$ is independent of σ_i, then this independence holds for all subsequent times.

Necessary and Sufficient Conditions

We now wish to find the necessary and sufficient conditions for Eq. (3.5) to have a solution of the form of (3.6). That is, we wish to determine the conditions imposed, if any, on the degeneracies g_i, the transition probabilities per unit time B_{ij}, and the level spacing $\epsilon_i - \epsilon_j$ for $P_i(t)$ to be canonically invariant. As in the previous section [see Eq. (2.10)] we find that the existence of such a solution, together with the detailed balance condition, implies that

$$\alpha(t)[\bar{\epsilon}(t) - \epsilon_i] = \sum_{j=0}^{N} B_{ji}[e^{-\alpha(t)(\epsilon_j - \epsilon_i)} - 1], \tag{3.11}$$

where $\alpha(t) = \beta(t) - \beta(\infty)$. We define the function $B(\epsilon; i)$ of the continuous variable ϵ and the discrete variable i by

$$B(\epsilon; i) = \sum_{j=0}^{N} B_{ji}\delta(\epsilon - \epsilon_j). \tag{3.12}$$

The Laplace transform $\mathcal{B}(\alpha; i)$ of this function is $\mathcal{B}(\alpha; i) \equiv \mathcal{L}\{B(\epsilon; i)\}$

$$\equiv \int_0^\infty e^{-\alpha\epsilon}B(\epsilon; i)\, d\epsilon = \sum_{j=0}^{N} B_{ji}e^{-\alpha\epsilon_j}. \tag{3.13}$$

Equation (3.11) can now be written as

$$\dot{\alpha}[\bar{\epsilon} - \epsilon_i] = e^{\alpha\epsilon_i}\mathcal{B}(\alpha; i) - \mathcal{B}(0; i). \tag{3.14}$$

From this we find

$$\mathcal{B}(\alpha; i) = e^{-\alpha\epsilon_i}[\mathcal{B}(0; i) + \dot{\alpha}\bar{\epsilon} - \epsilon_i\dot{\alpha}] \tag{3.15}$$

and

$$B(\epsilon; i) = \mathcal{L}^{-1}\{e^{-\alpha\epsilon_i}[\mathcal{B}(0; i) + \dot{\alpha}\bar{\epsilon} - \epsilon_i\dot{\alpha}]\}$$
$$= \mathcal{B}(0; i)\delta(\epsilon - \epsilon_i) + \mathcal{L}^{-1}\{e^{-\alpha\epsilon_i}[\dot{\alpha}\bar{\epsilon} - \epsilon_i\dot{\alpha}]\}, \tag{3.16}$$

where $\mathcal{L}^{-1}$ denotes the inverse Laplace-transform operator. By successively setting $i = 0$ and $i = 1$ we can rewrite the last term of (3.16) to obtain (see Appendix II)

$$B(\epsilon; i) = \left[\mathcal{B}(0; i) - \frac{\epsilon_i}{\epsilon_1}\mathcal{B}(0; 1) + \left(\frac{\epsilon_i}{\epsilon_1} - 1\right)\mathcal{B}(0; 0)\right]\delta(\epsilon - \epsilon_i)$$
$$- \left(\frac{\epsilon_i}{\epsilon_1} - 1\right)\sum_{j=0}^{N} B_{j0}\delta(\epsilon - \epsilon_i - \epsilon_j) + \frac{\epsilon_i}{\epsilon_1}\sum_{j=0}^{N} B_{j1}\delta(\epsilon - \epsilon_i - \epsilon_j + \epsilon_1). \tag{3.17}$$

A comparison of Eq. (3.17) with Eq. (3.12) then yields

$$B_{ki} = \frac{\epsilon_i}{\epsilon_1}\sum_{j=0}^{N} B_{j1}\Delta(\epsilon_i + \epsilon_j - \epsilon_1, \epsilon_k) - \left(\frac{\epsilon_i}{\epsilon_1} - 1\right)\sum_{j=0}^{N} B_{j0}\Delta(\epsilon_i + \epsilon_j, \epsilon_k), \tag{3.18}$$

where $\Delta(a, b)$ is the Kronecker delta

$$\Delta(a, b) = 1, \qquad a = b,$$
$$= 0, \qquad a \neq b.$$

We shall now show that, if the system has more than two energy levels, Eq. (3.18) implies that $\epsilon_i = i\epsilon_1$ for $i = 0, 1, 2, \cdots$. For let us consider the expression for B_{ki} for $i \geq 2$ and $i > k$, where B_{ki} is to be greater than zero. Then the second summation can never contribute since $\epsilon_i + \epsilon_j > \epsilon_k$ for all j. The first sum contributes a term only for values of j such that

$$\epsilon_i - \epsilon_1 + \epsilon_j = \epsilon_k. \tag{3.19}$$

This can only occur (for $i \geq 2$) for $j = 0$. For $j = 1$, $B_{ki} = 0$ since $B_{11} = 0$; for $j \geq 2$, Eq. (3.19) cannot hold with $i > k$. Hence, Eq. (3.19) reduces to

$$\epsilon_i - \epsilon_1 = \epsilon_k. \tag{3.20}$$

The argument for $i < k \geq 2$ follows from detailed balance. Since we are concerned here only with sets of levels which are irreducible (in the sense of Markov chains), it follows that

$$\epsilon_i = i\epsilon_1, \tag{3.21}$$

as asserted. The energy levels are thus uniformly spaced.

Next we will show that only a small number of the B_{j0} and B_{j1} can differ from zero, and that they can be expressed in terms of B_{01}. Consider first the expression for B_{1i} for $i \geq 2$. According to Eqs. (3.18) and (3.21) and the argument of the last paragraph we must have

$$B_{12} = 2B_{01},$$
$$B_{1i} = 0, \qquad i > 2. \tag{3.22}$$

From detailed balance we conclude that

$$B_{i1} = 0, \qquad i > 2,$$
$$B_{21} = (2g_2/g_1)B_{01}e^{-\beta(\infty)\epsilon_1}. \tag{3.23}$$

Furthermore, it is easily verified that $B_{0i} = 0$ for $i \geq 2$ since

$$B_{0i} = i\sum_j B_{j1}\Delta(i - 1 + j, 0)$$
$$- (i - 1)\sum_j B_{j0}\Delta(i + j, 0), \tag{3.24}$$

and both Kronecker deltas are equal to zero when $i > 1$. The transition probability B_{10} can be written in terms of B_{01} as

$$B_{10} = (g_1/g_0)B_{01}e^{-\beta(\infty)\epsilon_1}. \tag{3.25}$$

Collecting now the results of Eqs. (3.18), (3.21), (3.23), and (3.25), we can finally write the general expression

$$B_{ki} = B_{01}\left\{\frac{\epsilon_i}{\epsilon_1}\Delta(i - 1, k) + e^{-\beta(\infty)\epsilon_i}\right.$$
$$\left.\times\left[2\frac{\epsilon_i}{\epsilon_1}\frac{g_2}{g_1} - \left(\frac{\epsilon_i}{\epsilon_1} - 1\right)\frac{g_1}{g_0}\right]\Delta(i + 1, k)\right\}, \tag{3.26}$$

which is valid both for $i > k$ and $i < k$. It is immediately evident that B_{ki} is nonzero only when $k = i \pm 1$. We have thus obtained the interesting result that, for canonical invariance, transitions can occur *only* between nearest-neighbor levels.

The above results are in accord with the results in Sec. II for continuous energy-level systems where we have shown that canonical invariance obtains only when the kinetic equation is a Fokker–Planck equation as displayed in (2.26). The kinetic equation for discrete energy-level systems with equally spaced levels and with transitions only between adjacent levels is equivalent, in the limit of vanishing level spacing, to a Fokker–Planck equation.

We shall now derive the necessary and sufficient conditions on the degeneracy g_k for canonical invariance. We rewrite Eq. (3.26) as

$$B_{ki} = B_{01}[i\Delta(i-1,k) + e^{-\beta(\infty)\epsilon_1}(ai+b)\Delta(i+1,k)], \quad (3.27)$$

where we have used the result $\epsilon_i = i\epsilon_1$ of Eq. (3.21) and where we have set

$$a = \frac{2g_2}{g_1} - \frac{g_1}{g_0} \quad \text{and} \quad b = \frac{g_1}{g_0}. \quad (3.28)$$

From detailed balancing we find

$$\frac{g_{k+1}}{g_k} = \frac{B_{k+1,k}e^{-\beta(\infty)\epsilon_1}}{B_{k,k+1}} = \frac{a(k+f)}{k+1}, \quad (3.29)$$

where $f = b/a$. From Eq. (3.29) it then follows that

$$g_k = \binom{k+f-1}{k} a^k g_0. \quad (3.30)$$

Equation (3.29) is the general condition on the degeneracies for canonical invariance. It is valid for both an infinite-level system and a finite-level system. If the number of levels is infinite, we can readily show that $a \geq 1$. It is easily seen with the aid of Eq. (3.29) that a cannot be zero or negative. Also, if $0 \leq a < 1$, then Eq. (3.29) leads to $\lim_{k\to\infty} g_{k+1}/g_k = a < 1$; hence from (3.30), it follows that $\lim_{k\to\infty} g_k = 0$, which is impossible since all the g_k are positive integers. Therefore $a \geq 1$. In this case f must be positive, and Eq. (3.30) is the general form of the degeneracy for canonical invariance of an *infinite*-level system. The requirement that $Q(\beta)$ be finite when $N = \infty$ implies that

$$1 < ae^{-\epsilon_1\beta(t)}, \quad (3.31)$$

and therefore that $\beta(t) > 0$.

A further condition applies when N, the index of the highest energy level is finite. Since transitions still take place only between nearest neighbors, it is sufficient to consider an infinite set of levels and require that there be no transitions between levels N and $N+1$, i.e., that $B_{N+1,N} = 0$. According to Eq. (3.27), this requires that $aN + b = 0$ or

$$f = -N. \quad (3.32)$$

This result is also necessary. While this value of f cannot be substituted directly into Eq. (3.30) since we would then have gamma functions of negative integer arguments, we can substitute it into Eq. (3.29) and again use recurrence relations. In this way we obtain the recurrence formula

$$\frac{g_{k+1}}{g_k} = \frac{b(1-k/n)}{k+1}, \quad (3.33)$$

and the form of g_k for finite N is then found to be

$$g_k = \binom{N}{k}\left(\frac{g_1}{Ng_0}\right)^k g_0. \quad (3.34)$$

It can now readily be verified by direct substitution that the canonical distribution (3.6), with the g_i, ϵ_i, and B_{ij} as given above, is a solution of the relaxation equation (3.5), so that the above necessary conditions are also sufficient.

To summarize then, we have shown that the necessary and sufficient conditions for canonical invariance for a Markovian relaxation process with a discrete energy variable are:

(a) *The sum* $\sum_{\sigma_j} B_{ij}^{\sigma_i\sigma_j}$ *is independent of* σ_i *[Eq. (3.7)].* This means that, if the population of a state is dependent only on its energy (as it is in the Boltzmann distribution), then the rate at which transitions are made from level j to the states in level i must be the same for all states in level i.

(b) *The energy levels are uniformly spaced with* $\epsilon_i = i\epsilon_1$ *[Eq. (3.21)].*

(c) *Transitions take place only between nearest-neighbor levels [Eq. (3.26)].*

(d) *The transition probability per unit time* B_{ki} *is a sum of Kronecker deltas as shown in Eq. (3.27).*

(e) *The degeneracy* g_k *is as given in Eq. (3.30) for* $N = \infty$ *and in Eq. (3.34) for* N *finite.*

Relaxation of the Temperature and Mean Energy

We now investigate the relaxation of the mean energy $\bar{\epsilon}(t)$ and the temperature $T(t) \equiv [k\beta(t)]^{-1}$ for the canonically invariant distribution function $P_i(t)$ as given by Eq. (3.6). Substitution of Eq. (3.27) into Eq. (3.11), together with the use of the definition

$$\bar{\epsilon}(t) = \sum_k g_k\epsilon_k e^{-\beta(t)\epsilon_k} \Big/ \sum_k g_k e^{-\beta(t)\epsilon_k} \quad (3.35)$$

and the relation $\epsilon_k = k\epsilon_1$ [Eq. (3.21)] with g_k as given by Eq. (3.30), leads to the differential equation

$$\dot{\beta}(t) = -(B_{01}/\epsilon_1) \times \{[e^{\beta(t)\epsilon_1} - a][e^{-\beta(\infty)\epsilon_1} - e^{-\beta(t)\epsilon_1}]\} \quad (3.36)$$

for the time-dependent temperature function $\beta(t)$. The solution of this equation is

$$\beta(t) = \frac{1}{\epsilon_1} \ln \left[\frac{e^{\beta(\infty)\epsilon_1} + aDe^{-b_{01}t}}{1 + De^{-b_{01}t}} \right], \quad (3.37)$$

where

$$D = [e^{\beta(0)\epsilon_1} - e^{\beta(\infty)\epsilon_1}]/[a - e^{\beta(0)\epsilon_1}] \quad (3.38)$$

and

$$b_{01} = B_{01}[1 - ae^{-\beta(\infty)\epsilon_1}]. \quad (3.39)$$

The inequality in Eq. (3.31) assures that $b_{01} > 0$. This is also consistent with Eq. (3.37) as $t \to \infty$. Equation (3.37) is the desired result for the relaxation of the "temperature" $\beta(t) \equiv [kT(t)]^{-1}$ for a canonically invariant distribution function. It will be noted that, in the discrete energy-level system (quantum case), the temperature does not undergo a simple exponential relaxation.

We now wish to investigate the relaxation of the mean energy $\bar{\epsilon}(t)$. It can readily be verified from Eq. (3.27) that

$$\sum_{k=0}^{\infty} B_{ki}(\epsilon_k - \epsilon_i) = -b_{01}\epsilon_i + b\epsilon_1 B_{01} e^{-\beta(\infty)\epsilon_1} = -b_{01}\epsilon_i + d, \quad (3.40)$$

with d being a constant independent of ϵ_i. Equation (3.40) is the discrete analogue of Eq. (2.15) with the quantity $\sum_k B_{ki}(\epsilon_k - \epsilon_i)$ being the discrete analog of the transition moment $b_1(\epsilon)$. As is shown in some detail in Ref. 10, Eq. (3.40) leads to the relaxation equation

$$d\bar{\epsilon}(t)/dt = -b_{01}\bar{\epsilon} + d \quad (3.41)$$

for the mean energy $\bar{\epsilon}(t)$, with the solution

$$[\bar{\epsilon}(t) - \bar{\epsilon}(\infty)]/[\bar{\epsilon}(0) - \bar{\epsilon}(\infty)] = e^{-b_{01}t} \quad (3.42)$$

We have thus shown that canonical invariance implies the exponential relaxation of the mean energy.

It should be noted that, for a finite number of levels, nothing we have done restricts $\beta(t) \equiv [kT(t)]^{-1}$ to positive values. All the above necessary and sufficient conditions for canonical invariance and the results obtained for the relaxation of the temperature and mean energy will hold and are still valid for subsystems with an initial canonical distribution with *negative* temperatures. For an ensemble consisting of a subsystem with an initial negative temperature $T_s(0) < 0$ and a heat bath with a positive temperature $T_h > 0$, the temperature relaxation of a canonically invariant subsystem distribution would proceed from $T_s(0) < 0$ to $T_s(\infty) = T_h$ via an "infinite temperature" $T_s(t) = \infty$ at some time t.

Finally we return to mention an apparent discrepancy between the analyses in the continuous and the discrete case. We have shown, in the discrete case, that the equations themselves determine the allowable form of the degeneracies [cf. Eq. (3.29) *et seq.*]. On the other hand, we chose, apparently somewhat arbitrarily, a particular form for $g(\epsilon)$ in the continuous case which conveniently allowed us to evaluate several integrals. It can be shown however (Appendix III), by starting from the results for a discrete set of levels and passing properly to the limit where the spacing approaches zero, that the resulting form of the degeneracy $g(\epsilon)$ is the one which we used in Eq. (2.17).

Examples

There are two physically important systems which satisfy all the conditions for the exact preservation of the form of the canonical ensemble. The first is a nuclear spin of $\frac{1}{2}$ (or a group of identical, noninteracting spins of $\frac{1}{2}$) in a magnetic field interacting with a lattice (heat bath). This, however, is a somewhat trivial example since a two-level system can always be characterized by a temperature.

The other example is that of a system of harmonic oscillators in weak interaction with a heat bath. Montroll and Shuler[3] derived an equation for the collisional relaxation of an ensemble of harmonic oscillators of which a small fraction is excited to an initial vibrational nonequilibrium distribution, while a large excess of unexcited oscillators serves as a heat bath. This equation is

$$\frac{dx_n(t)}{dt} = \frac{k_{10}}{1 - e^{-\theta}} \{ne^{-\theta}x_{n-1} - [n + (n+1)e^{-\theta}]x_n + (n+1)x_{n+1}\} \quad (3.43)$$

$$n = 0, 1, \cdots,$$

where $x_n(t)$ is the fraction of excited oscillators in level n at time t, and $\theta = h\nu/kT(\infty)$. They found that an initial Boltzmann distribution relaxes to the final Boltzmann distribution at the heat-bath temperature via a continuous sequence of Boltzmann distributions. The harmonic-oscillator energy levels are nondegenerate, so that our Eq. (3.27) reduces to

$$B_{ki} = B_{01}[i\Delta(i-1, k) + e^{-\beta(\infty)\epsilon_1}(i+1)\Delta(i+1, k)]. \quad (3.44)$$

This is the form of the transition probability used by Montroll and Shuler based on the Landau-Teller prescription; its use in the master equation (3.5) yields

$$\frac{dP_i(t)}{dt} = B_{01}\{ie^{-\beta(\infty)\epsilon_1}P_{i-1}(t) - [i + (i+1)e^{-\beta(\infty)\epsilon_1}] \times P_i(t) + (i+1)P_{i+1}(t)\}, \quad i = 0, 1, \cdots, \tag{3.45}$$

which is identical with Eq. (3.43).

It should be pointed out that spin systems in general do not obey the conditions derived here for canonical invariance. While the levels will be equally spaced in a magnetic field, and while transitions may be only between adjacent levels, the degeneracies are not of the required form (3.34). The relaxation of spin systems (except for the trivial case of spin $\frac{1}{2}$) can therefore not be described by a "spin temperature" in a rigorous sense. However, as will be discussed in the next section, there can be under certain physical conditions an approximate preservation of the canonical distribution which would permit one to ascribe an approximate "spin temperature" to the relaxation of such systems.

It should be noted that the analysis of this section clearly shows that the relaxation of a system of quantal rotators cannot be canonically invariant since the energy levels are not uniformly spaced ($\epsilon_i \neq i\epsilon_1$) for such systems. This result is in agreement with the calculations of Herman and Shuler[8] on the relaxation of a system of rigid rotators.

IV. APPROXIMATE CANONICAL INVARIANCE

Up to this point we have only considered the conditions under which the form of the canonical ensemble is preserved *exactly*. These conditions are very restrictive and rule out many systems of physical interest. For example, a system of nuclear spins with $I = 1$ in a magnetic field cannot exhibit exact canonical invariance since the degeneracies do not satisfy Eq. (3.34).

It may, however, be a useful procedure under certain conditions to consider the *approximate* preservation of the canonical distribution during a Markovian relaxation process. In this section we shall study the special case in which there are a finite number of energy levels with a spacing $\Delta\epsilon$ small compared with kT, i.e., $\Delta\epsilon/kT \ll 1$. If the form of the canonical ensemble were preserved exactly, then the order of magnitude of both sides of the master equation (3.1) would be $B\beta\epsilon$, where B is a typical value of the $B_{ij}^{\sigma_i\sigma_j}$, β is the larger of $\beta(0)$, and $\beta(\infty)$, and ϵ is approximately $\epsilon_N - \epsilon_0$. We shall investigate the conditions under which canonical invariance is maintained when quantities of the order of $B\beta^2\epsilon^2$ are neglected. Only an outline of the proofs will be given since the methods are similar to those already described in the previous sections.

The quantal time-dependent canonical distribution function is

$$P_i^{\sigma_i}(t) = \frac{e^{-\beta(t)\epsilon_i}}{\sum_{i=0}^{N} g_i e^{-\beta(t)\epsilon_i}} = \frac{1 - \beta(t)\epsilon_i + O(\beta^2\epsilon^2)}{R - \beta(t)S + O(\beta^2\epsilon^2)}$$

$$= \frac{1}{R}\left[1 + \beta(t)\left(\frac{S}{R} - \epsilon_i\right)\right] + O(\beta^2\epsilon^2), \quad i = 0, 1, \cdots N, \tag{4.1}$$

where $R = \sum_{i=0}^{N} g_i$ and $S = \sum_{i=0}^{N} g_i\epsilon_i$. We now assume that these $P_i^{\sigma_i}$ will satisfy Eq. (3.1) when quantities of order $B\beta^2\epsilon^2$ are neglected. By the same methods used in Appendix I for the exact invariance it can be shown that this is the case if and only if:

(a) the equations

$$\frac{dP_i(t)}{dt} = \sum_{j=0}^{N} [B_{ij}P_j - B_{ji}P_i], \quad i = 0, 1, \cdots N \tag{4.2}$$

possess a solution of the form

$$P_i(t) = \frac{g_i}{R}\left[1 + \beta(t)\left(\frac{S}{R} - \epsilon_i\right)\right] + O(\beta^2\epsilon^2), \tag{4.3}$$

and (b)

$$\sum_{j=0}^{N}\left\{g_j\left[B_{ij} - \frac{g_i}{g_j}\sum_{\sigma_j=1}^{g_j} B_{ij}^{\sigma_i\sigma_j}\right] - g_i\left[B_{ji} - \sum_{\sigma_j=1}^{g_j} B_{ji}^{\sigma_j\sigma_i}\right]\right\} + O(B\beta^2\epsilon^2) = 0, \tag{4.4}$$

$$\sum_{j=0}^{N}\left\{g_j\left(\frac{S}{R} - \epsilon_j\right)\left[B_{ij} - \frac{g_i}{g_j}\sum_{\sigma_j=1}^{g_j} B_{ij}^{\sigma_i\sigma_j}\right] - g_i\left(\frac{S}{R} - \epsilon_i\right)\left[B_{ji} - \sum_{\sigma_j=1}^{g_j} B_{ji}^{\sigma_j\sigma_i}\right]\right\} + O(B\beta\epsilon^2) = 0,$$

$$\sigma_i = 1, 2, \cdots g_i. \tag{4.5}$$

The B_{ij} are defined in Eq. (3.9). As might be expected, these necessary and sufficient conditions are less stringent than those derived for the *exact* preservation of the form of the canonical ensemble. A simple example for which condition (b) is satisfied is that of nondegenerate energy levels where all the g_k are unity.

As in the exact case [see Eq. (3.11)], we find that condition (a) implies that

$$\dot{\beta}(t)\left(\frac{S}{R} - \epsilon_i\right) = -[\beta(t) - \beta(\infty)] \times \sum_{j=0}^{N} B_{ji}(\epsilon_j - \epsilon_i) + O(B\beta^2\epsilon^2), \quad (4.6)$$

which can be rearranged to give

$$\frac{\dot{\beta}(t)}{\beta(t) - \beta(\infty)} = \frac{-\sum_{j=0}^{N} B_{ji}(\epsilon_j - \epsilon_i)}{S/R - \epsilon_i} + O(B\beta\epsilon). \quad (4.7)$$

Equation (4.7) can be fulfilled only if

(i) $\dfrac{\sum_{j=0}^{N} B_{ji}(\epsilon_j - \epsilon_i)}{S/R - \epsilon_i}$ is independent of i

for $\epsilon_i \neq S/R$,

when quantities of order $B\beta\epsilon$ are neglected, or

(ii) $\sum_{j=0}^{N} B_{ji}(\epsilon_j - \epsilon_i)$ is of $O(B\beta\epsilon^2)$

when $\epsilon_i = S/R$.

If we now set the lhs of Eq. (4.7) equal to a constant b (independent of i) we obtain

$$[\beta(t) - \beta(\infty)]/[\beta(0) - \beta(\infty)] = e^{-bt} \quad (4.8)$$

for the relaxation of the "temperature" $\beta(t) \equiv 1/kT(t)$ with $1/b$ equal to the "relaxation time".

To summarize then, in order for a quantum-state master equation to preserve the form of the canonical distribution to within first order in $\beta\epsilon$, it is necessary and sufficient that

(a) $\dfrac{\sum_{j=0}^{N} B_{ji}(\epsilon_j - \epsilon_i)}{\dfrac{S}{R} - \epsilon_i}$

be independent of i to within quantities of order B. (4.9)

$$\text{(b)} \quad \sum_{j=0}^{N} \left\{ g_i \left[B_{ij} - \frac{g_i}{g_j} \sum_{\sigma_j=1}^{g_j} B_{ij}^{\sigma_i\sigma_j} \right] - g_i \left[B_{ji} - \sum_{\sigma_j=1}^{g_i} B_{ji}^{\sigma_j\sigma_i} \right] \right\} + O(B\beta^2\epsilon^2) = 0. \quad (4.10)$$

$$\text{(c)} \quad \sum_{j=0}^{N} \left\{ g_i \left(\frac{S}{R} - \epsilon_j\right) \left[B_{ij} - \frac{g_i}{g_j} \sum_{\sigma_j=1}^{g_i} B_{ij}^{\sigma_i\sigma_j} \right] - g_i \left(\frac{S}{R} - \epsilon_i\right) \left[B_{ji} - \sum_{\sigma_j=1}^{g_i} B_{ji}^{\sigma_j\sigma_i} \right] \right\} + O(B\beta\epsilon^2) = 0. \quad (4.11)$$

There are $(N + 1)^2$ quantities in the B_{ij} matrix. All the diagonal elements, B_{ii}, are taken to be zero, so there can be as many as $N(N + 1)$ nonzero B_{ij}'s. The detailed balance conditions impose $N[\frac{1}{2}(N + 1)]$ independent linear relationships among the B_{ij}, and the condition that b be independent of i imposes at most N additional linear relationships. The number of matrix elements B_{ij} (with $i \neq j$) minus the number of relationships is at least $N[\frac{1}{2}(N - 1)]$. Therefore, it is always possible in the case considered here to construct a set of B_{ij} (which may or may not represent a valid description of a physical system) which lead to canonical invariance to within order $\beta\epsilon$, independent of the spacing of the energy levels (as long as $\Delta\epsilon/kT \ll 1$) and independent of the form of the degeneracies. This conclusion follows directly from the conditions (4.9)–(4.11) above for $N \geq 2$. When $N = 1$, there are only two energy levels in the system and the canonical distribution can always be preserved exactly.

Example

The ortho-hydrogen molecule in a magnetic field forms an example of a system which exhibits the approximate preservation of the canonical distribution of the type discussed above.[16] There are three nondegenerate nuclear states with energies 0, ϵ_1, and $2\epsilon_1$. The condition that b be independent of i leads to the following two conditions:

$$B_{10} + 2B_{20} = 2B_{02} + B_{12} + O(B\beta\epsilon), \qquad -B_{01} + B_{21} = O(B\beta\epsilon). \quad (4.12)$$

The detailed balance conditions are

$$B_{01} = B_{10} + O(B\beta\epsilon), \quad B_{02} = B_{20} + O(B\beta\epsilon), \quad B_{12} = B_{21} + O(B\beta\epsilon). \quad (4.13)$$

The solution of Eqs. (4.12) and (4.13) is

$$B_{12} = B_{21} + O(B\beta\epsilon) = B_{01} + O(B\beta\epsilon) = B_{10} + O(B\beta\epsilon), \quad B_{02} = B_{20} + O(B\beta\epsilon). \quad (4.14)$$

The relaxation time b is then given by

$$b = B_{10} + 2B_{20}. \quad (4.15)$$

Needler and Opechowski[16] found that the transition probabilities applicable to their model satisfied these equations.

V. DISCUSSION

The conditions for the preservation of an exact canonical distribution in Markovian relaxation processes are extremely severe. They involve restrictions

[16] G. T. Needler and W. Opechowski, Can. J. Phys. **39**, 870 (1961).

upon the interaction of the system of interest with the reservoir and upon the energy-level spectrum of the system itself. In this section, we discuss briefly the physical import of some of the imposed conditions, the relationship of the process described here to Brownian motion in an arbitrary field of force, and the connection between our results and those of previous investigators.

The forms of the transition probabilities given by Eq. (2.25) in the continuum case and by Eq. (3.27) in the discrete case, imply that each interaction between the system and the reservoir must be weak. However, while weak interactions are a necessary condition for canonical invariance, they are not a sufficient condition. This is evident since the forms of the coefficients $b_1(\epsilon)$ and $b_2(\epsilon)$ must be specified by Eq. (2.24) in the continuum case, and the forms of the transition probabilities $B_{i,i+1}$ and $B_{i,i-1}$ must be specified by Eq. (3.27) in the discrete case. The deeper *physical* significance, if any, of these particular forms of $b_m(\epsilon)$ and $B_{i,j}$ are not clear to us at this time.

The forms of the $b_m(\epsilon)$ and $B_{i,j}$ for canonical invariance also yield, as is to be expected, the correct expressions for the energy fluctuation of a canonical distribution at the temperature $T(t)$. In the continuum case, for instance, it can readily be shown that the mean value of the jth power of the energy is related to the jth power of the mean energy by

$$\overline{\epsilon^j}(t) = [(j+n)!/n!\,(n+1)^j][\overline{\epsilon(t)}]^j, \qquad (5.1)$$

when $g(\epsilon)$ is given by (2.17) with $p = 0$. From Eq. (5.1) one can readily obtain the following expression for the fluctuation (mean-square deviation) of the energy:

$$\overline{\epsilon^2}(t) - [\bar{\epsilon}(t)]^2 = (n+1)^{-1}[\bar{\epsilon}(t)]^2. \qquad (5.2)$$

The fluctuation of the energy for a canonical ensemble is usually written as

$$\overline{\epsilon^2} - (\bar{\epsilon})^2 = kT^2C_v, \qquad (5.3)$$

where C_v is the heat capacity at constant volume. For $\bar{\epsilon}(t) = (n+1)kT(t)$, as will be the case for $g(\epsilon) = \epsilon^n$ and with $C_v(t) = d\bar{\epsilon}(t)/dT(t)$, Eq. (5.2) is equivalent to (5.3). Thus, the form of the $b_m(\epsilon)$ assures that the fluctuation of the energy at all time t are those of a canonical ensemble at the temperature $T(t)$. An analogous result can be obtained for the discrete energy-level spectrum. The forms of the b's and B's are necessary and sufficient for this canonical fluctuation.

We shall now investigate whether Brownian motion in an arbitrary field of force preserves the canonical form of the probability density $P(\mathbf{R}, \mathbf{u}, t)$. The quantity $P(\mathbf{R}, \mathbf{u}, t)\, d\mathbf{R}\, d\mathbf{u}$ is the probability that the Brownian particle has a postion between $\mathbf{R}$ and $\mathbf{R} + d\mathbf{R}$ and a velocity between $\mathbf{u}$ and $\mathbf{u} + d\mathbf{u}$ at time t. The Fokker–Planck equation for Brownian motion in a field of force $\mathbf{F}(\mathbf{R})$ can be derived from the Langevin equation and is given by

$$\frac{\partial P}{\partial t} + \mathbf{u}\cdot\nabla_R P + \frac{\mathbf{F}(\mathbf{R})}{m} - \nabla_u P = \gamma\nabla_u\cdot(P\mathbf{u}) + \frac{\gamma}{m\beta(\infty)}\nabla_u^2 P, \qquad (5.4)$$

where m is the mass of the Brownian particle, γ is the friction coefficient, and $\beta(\infty) = [kT(\infty)]^{-1}$ where $T(\infty)$ is the temperature of the heat bath and the final equilibrium temperature of the system. Substitution of the canonical form

$$P(\mathbf{R}, \mathbf{u}, t) = e^{-\beta(t)H(\mathbf{r},\mathbf{u})} \Big/ \int e^{-\beta(t)H(\mathbf{r},\mathbf{u})}\, d\mathbf{R}\, d\mathbf{u}, \qquad (5.5)$$

where the Hamiltonian $H(\mathbf{R}, \mathbf{u})$ is given by

$$H(\mathbf{R}, \mathbf{u}) = \tfrac{1}{2}mu^2 + V(\mathbf{R}), \qquad (5.6)$$

and the potential $V(\mathbf{R})$ is related to the force $\mathbf{F}(\mathbf{R})$ by

$$\mathbf{F}(\mathbf{R}) = -\nabla_R V(\mathbf{R}), \qquad (5.7)$$

into (5.4) yields

$$\dot{\beta}(t)[\bar{\epsilon}(t) - H(\mathbf{R}, \mathbf{u})] = \gamma[\beta(t)mu^2 - 3][\beta(t)/\beta(\infty) - 1], \qquad (5.8)$$

with a mean energy $\bar{\epsilon}(t)$ given by

$$\bar{\epsilon}(t) = \int H(\mathbf{R}, \mathbf{u})P(\mathbf{R}, \mathbf{u}, t)\, d\mathbf{R}\, d\mathbf{u}. \qquad (5.9)$$

When $H(\mathbf{R}, \mathbf{u})$ is a function of $\mathbf{R}$, Eq. (5.8) can be satisfied only when $\dot{\beta}(t) = 0$ and $\beta(t) = \beta(\infty)$. Thus, a canonical distribution cannot be preserved for all times t for Brownian motion *in a field of force*.

Other investigators have studied solutions of equations similar to Eq. (3.5) whose forms remain invariant in time. Mathews, Shapiro, and Falkoff[17] have shown that the joint probability distribution $P(n_1, \cdots n_i, \cdots; t)$ which is the probability that there are n_1 particles in state 1, n_2 particles in state 2, $\cdots$, at time t will preserve the multinomial form

$$P(n_1, \cdots n_i, \cdots; t) = (N!/\prod_i n_i!) \prod_i [P_i(t)]^{n_i} \qquad (5.10)$$

[17] P. M. Mathews, I. I. Shapiro, and D. L. Falkoff, Phys. Rev. **120**, 1 (1960).

during a Markovian relaxation process for an initial multinomial distribution $P(n_1, \cdots n_i, \cdots ; 0)$. In Eq. (5.10), $P_i(t) \equiv \bar{n}_i(t)/N$ is the probability that a particle is in state i at time t and $N = \sum n_i$ is the total number of particles in the system. The multinomial invariance of the joint probability distribution $P(n_1, \cdots n_i, \cdots ; t)$ as shown in (5.10) requires none of the restrictive assumptions which had to be made in the body of this paper to assure the canonical invariance of the distribution function $P_i(t)$. This is not too surprising since the requirement of a specific time-invariant form for the distribution function $P_i(t)$ itself is much more stringent than the more general multinomial form of the joint probability distribution. Sher and Primakoff[18] have considered spin systems in which there are interactions between the particles (or degrees of freedom) of the subsystem as well as between the subsystem and the reservoir. This more general question of possible exact or approximate canonical invariance, when there are both subsystem–subsystem interactions and subsystem–reservoir interactions, is an important extension of the present study. We plan to investigate this problem in subsequent publications.

APPENDIX I. CONDITIONS ON THE $B_{ij}^{\sigma_j\sigma_i}$

We assume that Eq. (3.1) has a solution of the form of Eq. (3.4). If we define

$$P_i(t) \equiv \sum_{\sigma_i=1}^{g_i} P_i^{\sigma_i} = g_i P_i^{\sigma_i} = g_i e^{-\beta(t)\epsilon_i}/Q[\beta(t)], \qquad i = 0, 1, \cdots N, \qquad \text{(I1)}$$

we have from Eq. (3.1)

$$\frac{dP_i}{dt} = \sum_{\sigma_i=1}^{g_i}\sum_{j=0}^{N}\sum_{\sigma_j=1}^{g_j}[B_{ij}^{\sigma_i\sigma_j}P_j^{\sigma_j} - B_{ji}^{\sigma_j\sigma_i}P_i^{\sigma_i}]$$

$$= \sum_{j=0}^{N}[B_{ij}P_j - B_{ji}P_i], \quad i = 0, 1, \cdots N, \qquad \text{(I2)}$$

where the B_{ij} are defined by Eq. (3.9), i.e., $B_{ij} = 1/g_j \sum_{\sigma_i}\sum_{\sigma_j} B_{ij}^{\sigma_i\sigma_j}$. [We are at liberty to choose $B_{ii} = 0$, $i = 0, 1, \cdots N$, since their values do not enter into Eq. (I.2)]. We can obtain another equation for dP_i/dt from Eq. (3.1),

$$\frac{dP_i}{dt} = g_i\frac{dP_i^{\sigma_i}}{dt} = \sum_{j=0}^{N}\left[P_j\left(\sum_{\sigma_j=1}^{g_j}\frac{g_i}{g_j}B_{ij}^{\sigma_i\sigma_j}\right) - P_i\left(\sum_{\sigma_j=1}^{g_j}B_{ji}^{\sigma_j\sigma_i}\right)\right], \qquad \sigma_i = 1, 2, \cdots g_i. \qquad \text{(I3)}$$

Comparing Eqs. (I2) and (I3) and using Eq. (I1) we find that

$$0 = \sum_{j=0}^{N}\left\{g_je^{-\beta\epsilon_j}\left[B_{ij} - \frac{g_i}{g_j}\sum_{\sigma_j=1}^{g_j}B_{ij}^{\sigma_i\sigma_j}\right] - g_ie^{-\beta\epsilon_i}\left[B_{ji} - \sum_{\sigma_j=1}^{g_j}B_{ji}^{\sigma_j\sigma_i}\right]\right\}. \qquad \text{(I4)}$$

Equation (I4) must hold for all β between $\beta(0)$ and $\beta(\infty)$. This is possible only if the coefficient of each $e^{-\beta\epsilon_j}$ is identically zero. This condition yields

$$B_{ij} - \frac{g_i}{g_j}\sum_{\sigma_j=1}^{g_j}B_{ij}^{\sigma_i\sigma_j} = 0. \qquad \text{(I5)}$$

This in turn implies that the sum $\sum_{\sigma_j=1}^{g_j} B_{ij}^{\sigma_i\sigma_j}$ is independent of σ_i.

By reversing the argument, it is possible to show that conditions (I5) and (3.6) are sufficient, as well as necessary, to ensure that the quantum-state master equation preserves the canonical ensemble.

If we use Eqs. (3.3) and (I1) we find

$$B_{ij}P_j(\infty) = \left(\frac{1}{g_j}\sum_{\sigma_j=1}^{g_j}\sum_{\sigma_i=1}^{g_i}B_{ij}^{\sigma_i\sigma_j}\right)[g_jP_j^{\sigma_j'}(\infty)]$$

$$= \left[\frac{1}{g_i}\sum_{\sigma_j=1}^{g_j}\sum_{\sigma_i=1}^{g_i}B_{ji}^{\sigma_j\sigma_i}\frac{P_i^{\sigma_i}(\infty)}{P_j^{\sigma_j}(\infty)}\right][g_jP_j^{\sigma_j'}(\infty)]$$

$$= \left(\frac{1}{g_i}\sum_{\sigma_j=1}^{g_j}\sum_{\sigma_i=1}^{g_i}B_{ji}^{\sigma_j\sigma_i}\right)[g_iP_i^{\sigma_i'}(\infty)]$$

$$= B_{ji}P_i(\infty). \qquad \text{(I6)}$$

for the detailed balance relation for the B_{ij}.

APPENDIX II. EVALUATION OF $\mathcal{L}^{-1}\{e^{-\alpha\epsilon_i}[\dot{\alpha\epsilon} - \epsilon_i\dot{\alpha}]\}$

We can solve for $\mathcal{L}^{-1}\{\dot{\alpha\epsilon}\}$ by setting $i = 0$ in Eq. (3.16). This yields

$$\mathcal{L}^{-1}\{\dot{\alpha\epsilon}\} = B(\epsilon; 0) - \mathcal{B}(0; 0)\delta(\epsilon). \qquad \text{(II1)}$$

Using a familiar result from the theory of Laplace transforms we note that

$$\mathcal{L}^{-1}\{e^{-\alpha\epsilon_i}\dot{\alpha\epsilon}\} = B(\epsilon - \epsilon_i; 0) - \mathcal{B}(0; 0)\delta(\epsilon - \epsilon_i). \qquad \text{(II2)}$$

This result can now be substituted into Eq. (3.16) to yield

$$B(\epsilon; i) = [\mathcal{B}(0; i) - \mathcal{B}(0; 0)]\delta(\epsilon - \epsilon_i) + B(\epsilon - \epsilon_i; 0) - \epsilon_i\mathcal{L}^{-1}\{e^{-\alpha\epsilon_i}\dot{\alpha}\}. \qquad \text{(II3)}$$

An expression for the last term in Eq. (II3) can be obtained by setting $i = 1$. This leads to the final result

[18] A. Sher and H. Primakoff, Phys. Rev. **119**, 178 (1960); **130**, 1267 (1963).

$$B(\epsilon; i) = \left[\mathcal{B}(0; i) - \frac{\epsilon_i}{\epsilon_1}\mathcal{B}(0; 1) + \left(\frac{\epsilon_i}{\epsilon_1} - 1\right)\mathcal{B}(0; 0)\right]\delta(\epsilon - \epsilon_i) - \left(\frac{\epsilon}{\epsilon_1} - 1\right)B(\epsilon - \epsilon_i; 0) + \frac{\epsilon_i}{\epsilon_1}B(\epsilon - \epsilon_i + \epsilon_1; 1)$$

$$= \left[\mathcal{B}(0; i) - \frac{\epsilon_i}{\epsilon_1}\mathcal{B}(0; 1) + \left(\frac{\epsilon_i}{\epsilon_1} - 1\right)\mathcal{B}(0; 0)\right]\delta(\epsilon - \epsilon_i)$$

$$- \left(\frac{\epsilon_i}{\epsilon_1} - 1\right)\sum_{j=0}^{N} B_{j0}\delta(\epsilon - \epsilon_i - \epsilon_j) + \frac{\epsilon_i}{\epsilon_1}\sum_{j=0}^{N} B_{j1}\delta(\epsilon - \epsilon_i + \epsilon_1 - \epsilon_j). \quad \text{(II4)}$$

APPENDIX III. PASSAGE FROM g_i TO $g(\epsilon)$

Let us assume that ϵ_1 is so small that $\beta\epsilon_1 \ll 1$, where β is the larger of $\beta(0)$ and $\beta(\infty)$. In this case, the discrete spectrum can be approximately represented by a continuous one. Let us define a density of states, $g(\epsilon)$, which is a continuous function of ϵ. In order for $g(\epsilon)$ to represent the discrete spectrum, it must be true that

$$g(i\epsilon_1) \approx g_i/\epsilon_1 \quad \text{(III1)}$$

if g_i is a slowly varying function of i. With the aid of Eq. (3.29) we then find that

$$\frac{g[(i+1)\epsilon_1]}{g(i\epsilon_1)} = \frac{a(i+f)}{i+1}, \quad \text{(III2)}$$

or

$$\frac{g(\epsilon+\epsilon_1)}{g(\epsilon)} = \frac{a(\epsilon/\epsilon_1+f)}{\epsilon/\epsilon_1+1}, \quad \text{(III3)}$$

with $f=b/a$ [see Eq. (3.29)]. Since $g(\epsilon+\epsilon_1) \approx g(\epsilon) + \epsilon_1[dg(\epsilon)/d\epsilon] + O(\epsilon_1^2)$, one can write, using (III3),

$$\frac{dg(\epsilon)}{d\epsilon} = g(\epsilon)\left[\frac{(\epsilon/\epsilon_1)(a-1)+af-1}{\epsilon+\epsilon_1}\right]. \quad \text{(III4)}$$

For $\epsilon \gg \epsilon_1$, the solution of this equation is

$$g(\epsilon) = \epsilon^{b-1}e^{[(a-1)/\epsilon_1]\epsilon}. \quad \text{(III5)}$$

If we now let $p = (a-1)/\epsilon_1$ and $n = b - 1$, we obtain Eq. (2.17).

Decay of Correlations. IV. Necessary and Sufficient Conditions for a Rapid Decay of Correlations

Dick Bedeaux,[1] Kurt E. Shuler,[1] and Irwin Oppenheim[2]

Received February 26, 1971

We consider N-particle systems whose probability distributions obey the master equation. For these systems, we derive the necessary and sufficient conditions under which the reduced n-particle ($n < N$) probabilities also obey master equations and under which the Ursell functions decay to their equilibrium values faster than the probability distributions. These conditions impose restrictions on the form of the transition rate matrix and thus on the form of its eigenfunctions. We first consider systems in which the eigenfunctions of the N-particle transition rate matrix are completely factorized and demonstrate that for such systems, the reduced probabilities obey master equations and the Ursell functions decay rapidly if certain additional conditions are imposed. As an example of such a system, we discuss a random walk of N pairwise interacting walkers. We then demonstrate that for systems whose N-particle transition matrix can be written as a sum of one-particle, two-particle, etc. contributions, and for which the reduced probabilities obey master equations, the reduced master equations become, in the thermodynamic limit, those for independent particles, which have been discussed by us previously. As an example of such N-particle systems, we discuss the relaxation of a gas of interacting harmonic oscillators.

KEY WORDS: Decay of correlations; master equations, stochastic processes; Ursell functions; reduced distribution functions; reduced master equations.

Supported in part (grants to D.B. and K.E.S.) by the Advanced Research Projects Agency of the Department of Defense as monitored by the U.S. Office of Naval Research under Contract N00014-69-A-0200-6018, and in part (grant to I.O.) by the National Science Foundation.

[1] Department of Chemistry, University of California—San Diego, La Jolla, California.

[2] Department of Chemistry, Massachusetts Institute of Technology, Cambridge, Massachusetts.

1. INTRODUCTION

In our previous papers in this series[(1–3)] (hereafter referred to as I, II, III, respectively), on the decay of correlations in various model systems, we have consistently found the relations

$$P_n(\alpha^n; t) - P_n^{(0)}(\alpha^n) \sim P_1(\alpha; t) - P_1^{(0)}(\alpha); \qquad n \geqslant 1 \tag{1}$$

$$R_n(\alpha^n; t) - R_n^{(0)}(\alpha^n) \sim [P_1(\alpha; t) - P_1^{(0)}(\alpha)]^n \tag{2}$$

where $P_n(\alpha^n; t)$ stands for the n-particle distribution function at time t with dynamical variables $\alpha^n \equiv (\alpha_1, \alpha_2, ..., \alpha_n)$; $R_n(\alpha^n; t)$ stands for the n-particle *relaxants*[3] exemplified by the Ursell function U_n in papers I and II and the C-function in paper III; P_1 denotes the one-particle distribution function; and the superscript zero denotes the equilibrium value. Equation (1) describes the asymptotic relaxation of the n-particle distribution functions to their equilibrium value $P_n^{(0)}$ and Eq. (2) describes the asymptotic relaxation of the n-particle relaxant to its equilibrium value $R_n^{(0)}$. Equation (2) implies the important result that $P_n(t)$, $n > 1$, relaxes to a *functional* of lower-order distribution functions $[P_{n-1}(t), P_{n-2}(t), ..., P_1(t)]$ as $[P_1(\alpha; t) - P_1^{(0)}(\alpha)]^n$. These results hold for such diverse systems as noninteracting, initially correlated particles in contact with a heat bath,[(1)] an infinite chain of coupled harmonic oscillators,[(2)] and the relaxation of spins in Glauber's[(4)] one-dimensional Ising model.[(3)] It is evident that there must be an underlying physical basis for these identical results on the decay of correlations.

In this paper, we study the general conditions under which the n-particle distribution functions and the Ursell functions will be of the form shown in Eqs. (1) and (2). We limit our consideration here to many-body systems whose time development is governed by a master equation as exemplified in papers I and III.

In Section 2, we develop eigenfunction expansions for the total probability $\mathbf{P}_N(t)$ and for the reduced probabilities $\mathbf{P}_n(t)$, $n = 1, ..., N - 1$. We derive the necessary and sufficient conditions on the form of the eigenfunctions and the master operator $\mathbf{A}_N$ under which the reduced probabilities obey master equations and under which the Ursell functions relax asymptotically to their equilibrium values faster than the reduced probability distributions relax to their equilibrium values.

In Section 3, we consider systems for which the eigenfunctions of $\mathbf{A}_N$ are completely factorized into single-particle functions. We show that for such systems, the reduced probabilities obey master equations and the Ursell functions decay more rapidly than the reduced probabilities.

In Section 4, we consider a model of N interacting random walkers as an example of a system for which the eigenfunctions of $\mathbf{A}_N$ are completely factorized. For the particular random walk chosen, in which the only transitions are those in which one particle takes a step to the right while another particle takes a step to the left, $\mathbf{U}_{N-n}$ decays asymptotically at the same rate as $\mathbf{U}_n$. In the thermodynamic limit, the reduced

[3] The term "relaxant" has been suggested to us by Prof. Michael E. Fisher as a generic term for functions which are useful in studying the decay of correlations.

master equations reduce to those for independent particles, and the asymptotic relaxation of $\mathbf{U}_n$ and $\mathbf{P}_n$ is again that found in Paper I, and given by Eqs. (1) and (2).

In Section 5, we generalize the results of Section 4 and demonstrate that for systems for which $\mathbf{A}_N$ can be written as a sum of one-particle, two-particle, etc. contributions, and for which the reduced probabilities obey master equations, the reduced master equations become, in thermodynamic limit, those for independent particles. As an example, we consider in some detail a system of N interacting harmonic oscillators which undergo transitions as a result of two-body resonant collisions.

2. EIGENFUNCTION EXPANSIONS

We consider a system containing N particles and with a time-dependent multivariate probability $P_N(\alpha^N; t)$, where $\alpha^N \equiv (\alpha_1, \alpha_2, \ldots, \alpha_N)$, with α_i some property or set of properties of the ith particle. We assume that the time dependence of P_N is governed by the master equation

$$\partial P_N(\alpha^N; t)/\partial t = \sum_{\gamma^N} A_N(\alpha^N, \gamma^N) P_N(\gamma^N; t) \tag{3}$$

where the transition rate $A_N(\alpha^N, \gamma^N)$ is related to the N-particle "gain and loss" transition rates $B_N(\alpha^N, \gamma^N)$ by

$$A_N(\alpha^N, \gamma^N) = B_N(\alpha^N, \gamma^N) - \delta_{\alpha^N, \gamma^N} \sum_{\mu^N} B_N(\mu^N, \alpha^N) \tag{4}$$

Here and in the equations to follow, $\delta_{\alpha^N, \gamma^N}$ is the Kronecker delta for discrete variables and the Dirac delta function for continuum variables. We shall write the solution to Eq. (3) in terms of eigenfunction expansions and investigate under what conditions the reduced probabilities $P_n(\alpha^n; t)$, $n = 1, \ldots, N-1$, obey master equations of the form

$$\partial P_n(\alpha^n; t)/\partial t = \sum_{\gamma^n} A_n(\alpha^n, \gamma^n) P_n(\gamma^n; t) \tag{5}$$

We shall also investigate the conditions under which the Ursell functions[1] $U_n(\alpha^n; t)$ decay to their equilibrium value faster than the $P_n(\alpha^n; t)$ decay to their equilibrium values.

Equation (3) can be written in operator notation as

$$(\partial/\partial t)\, \mathbf{P}_N(t) = \mathbf{A}_N \mathbf{P}_N(t) \tag{6}$$

with the formal solution

$$\mathbf{P}_N(t) = (\exp \mathbf{A}_N t) \cdot \mathbf{P}_N(0) \tag{7}$$

The right and left eigenfunctions $\mathbf{R}_{N,\lambda_N}$ and $\mathbf{L}_{N,\lambda_N}$ respectively of $\mathbf{A}_N$ obey the equations

$$\mathbf{A}_N \cdot \mathbf{R}_{N,\lambda_N} = \lambda_N \mathbf{R}_{N,\lambda_N} \tag{8}$$

$$\mathbf{L}_{N,\lambda_N} \cdot \mathbf{A}_N = \lambda_N \mathbf{L}_{N,\lambda_N} \tag{9}$$

$$\mathbf{L}_{N,\lambda_N} \cdot \mathbf{R}_{N,\lambda_N'} = \delta_{\lambda_N, \lambda_N'} \tag{10}$$

where the λ_N are the eigenvalues, whose spectrum may be discrete and/or continuous. We consider here only systems for which the equilibrium distribution is unique, i.e., $\lambda_N = 0$ is a nondegenerate eigenvalue. The other eigenvalues, however, may be degenerate and the eigenfunctions should have a label denoting the state as well as the eigenvalue. We shall suppress this label for ease of notation. Equation (10) is understood to imply that the eigenfunctions for different states are orthogonal. The probability $\mathbf{P}_N(t)$ can now be written

$$\mathbf{P}_N(t) = \sum_{\lambda_N} e^{\lambda_N t}[\mathbf{L}_{N,\lambda_N} \cdot \mathbf{P}_N(0)]\, \mathbf{R}_{N,\lambda_N} \tag{11}$$

where the symbol $\sum$ denotes the sum over the discrete spectrum and the integral over the continuous spectrum of eigenvalues.

From detailed balance,

$$A_N(\alpha^N, \gamma^N)\, P_N^{(0)}(\gamma^N) = A_N(\gamma^N, \alpha^N)\, P_N^{(0)}(\alpha^N) \tag{12}$$

where

$$P_N^{(0)}(\alpha^N) \equiv P_N(\alpha^N; \infty) \tag{13}$$

is the equilibrium probability distribution, and from the fact that $\mathbf{A}_N$ is a stochastic matrix, it follows that the eigenvalues γ_N are real and nonpositive[5] and that

$$L_{N,\lambda_N}(\alpha^N) = [P_N^{(0)}(\alpha^N)]^{-1}\, R_{N,\lambda_N}(\alpha^N) \tag{14}$$

If $\lambda_N = 0$ is part of the discrete spectrum, then

$$\mathbf{P}_N^{(0)} = \mathbf{R}_{N,0} \tag{15}$$

If $\lambda_N = 0$ is part of the continuous spectrum, then

$$\mathbf{P}_N^{(0)} = 0 \tag{16}$$

and in Eqs. (12) and (14), $\mathbf{R}_{N,0}$ should be used instead of $\mathbf{P}_N^{(0)}$.

The reduced probabilities $P_n(\alpha^n; t)$, $n = 1,\ldots, N-1$, for the n particles, $i_1,\ldots, i_n$, are defined by

$$P_n(\alpha^n; t) = \sum_{\alpha^{N-n}} P_N(\alpha^N; t) \tag{17}$$

where the sum is over the variables of the $N - n$ other particles. It follows from Eq. (11) that

$$P_n(\alpha^n; t) = \sum_{\lambda_N} e^{\lambda_N t}[\mathbf{L}_{N,\lambda_N} \cdot \mathbf{P}_N(0)] \sum_{\alpha^{N-n}} R_{N,\lambda_N}(\alpha^N) \tag{18}$$

We now investigate the conditions under which $P_n(\alpha^n; t)$ is a solution to Eq. (5), i.e., when $\mathbf{P}_n(t)$ is given by

$$\mathbf{P}_n(t) = (\exp \mathbf{A}_n t) \cdot \mathbf{P}_n(0) \tag{19}$$

or

$$P_n(\alpha^n; t) = \sum_{\lambda_n} e^{\lambda_n t} [\mathbf{L}_{n,\lambda_n} \cdot \mathbf{P}_n(0)] \, R_{n,\lambda_n}(\alpha^n) \tag{20}$$

where $\mathbf{R}_{n,\lambda_n}$ and $\mathbf{L}_{n,\lambda_n}$ are the right and left eigenfunctions, respectively, and λ_n the eigenvalues of $\mathbf{A}_n$. It is obvious that the λ_n must form a subset of the λ_N.

Clearly, Eq. (18) reduces to Eq. (20) if and only if, for each λ_N, either

$$\text{(i)} \qquad \sum_{\alpha^{N-n}} R_{N,\lambda_N}(\alpha^N) \equiv R_{n,\lambda_N}(\alpha^n) = 0 \tag{21}$$

or

$$\text{(ii)} \qquad [\mathbf{L}_{N,\lambda_N} \cdot \mathbf{P}_N(0)] = [\mathbf{L}_{n,\lambda_N} \cdot \mathbf{P}_n(0)] \tag{22}$$

Let us consider a λ_N for which Eq. (21) does not apply; for this λ_N, Eq. (22) must apply and we can write

$$L_{N,\lambda_N}(\alpha^N) = L_{n,\lambda_N}(\alpha^n) \tag{23}$$

since Eq. (22) must be true for all $\mathbf{P}_N(0)$. This subset of λ_N for which Eq. (23) applies is identical to the set of λ_n. It then follows from Eqs. (14), (15), and (23) that

$$R_{N,\lambda_N}(\alpha^N) = [R_{N,0}(\alpha^N)/R_{n,0}(\alpha^n)] \cdot R_{n,\lambda_N}(\alpha^n) \tag{24}$$

The physical significance of the fact that the reduced distribution functions obey the master equation (5) is more readily understood from the condition that must be imposed on $\mathbf{A}_N$. This condition, which is of course equivalent to Eq. (23), is

$$A_n(\alpha^n, \gamma^n) = \sum_{\alpha^{N-n}} A_N(\alpha^N, \gamma^N) \tag{25}$$

i.e., *the transition rates for the n-particle subsystem must be independent of the initial states of the other $N - n$ particles.* Equation (25) imposes severe restrictions on the properties of $\mathbf{A}_N$ but these conditions are not severe enough to determine uniquely its functional form. We discuss some sufficient conditions for Eq. (25) to hold in Section 3.

We now investigate the conditions on the eigenfunctions of the master operator under which the Ursell functions relax to their equilibrium form faster than the probability distributions relax to their equilibrium form. We start with the discussion of $\mathbf{U}_2$ and then extend our considerations to $\mathbf{U}_n$. The two-particle Ursell function is defined by

$$U_2(\alpha_1, \alpha_2; t) = P_2(\alpha_1, \alpha_2; t) - P_1(\alpha_1; t) \, P_1(\alpha_2; t) \tag{26}$$

with the property

$$\sum_{\alpha_1} U_2(\alpha_1, \alpha_2; t) = \sum_{\alpha_2} U_2(\alpha_1, \alpha_2; t) = 0 \tag{27}$$

The time dependence of U_2 is given explicitly by

$$U_2(\alpha_1, \alpha_2; t) = \sum_{\lambda_N} (\exp \lambda_N t)[\mathbf{L}_{N,\lambda_N} \cdot \mathbf{P}_N(0)]\, R_{2,\lambda_N}(\alpha_1, \alpha_2)$$
$$- \sum_{\lambda_N, \lambda_N'} \{\exp[(\lambda_N + \lambda_N')\, t]\}[\mathbf{L}_{N,\lambda_N} \cdot \mathbf{P}_N(0)][\mathbf{L}_{N,\lambda_N'} \cdot \mathbf{P}_N(0)]$$
$$\times R_{1,\lambda_N}(\alpha_1)\, R_{1,\lambda_N'}(\alpha_2) \tag{28}$$

where we have used Eq. (18) and

$$R_{n,\lambda_N}(\alpha^n) \equiv \sum_{\alpha^{N-n}} R_{N,\lambda_N}(\alpha^N) \tag{29}$$

Equation (28) has the properties of Eq. (21) since

$$\sum_{\alpha^N} R_{N,\lambda_N}(\alpha^N) = \delta_{\lambda_N,0} \tag{30}$$

which follows from Eqs. (10) and (14). If no restrictions are imposed on the eigenfunctions $\mathbf{R}_{N,\lambda_N}$, the asymptotic decay of $\mathbf{U}_2$ is the same as the asymptotic decay of $\mathbf{P}_2$, which is governed by $\lambda_N{}^\circ$, which is the highest nonzero eigenvalue for which $\mathbf{R}_{2,\lambda_N}$ is nonzero. The asymptotic decay of $\mathbf{P}_1$ will be the same as the asymptotic decay of $\mathbf{P}_2$ if $\mathbf{R}_{1,\lambda_N}$ is nonzero for the highest nonzero eigenvalue; if $\mathbf{R}_{1,\lambda_N}$ is zero for this eigenvalue, $\mathbf{P}_1$ will decay faster than $\mathbf{P}_2$.

In order for $\mathbf{U}_2$ to decay faster than $\mathbf{P}_2$, it is necessary that the coefficient of the term in Eq. (28) containing the highest non zero eigenvalue be zero for all initial conditions. This will be the case if and only if

$$R_{2,\lambda_N{}^\circ}(\alpha_1, \alpha_2) = R_{1,0}(\alpha_1)\, R_{1,\lambda_N{}^\circ}(\alpha_2) + R_{1,\lambda_N{}^\circ}(\alpha_1)\, R_{1,0}(\alpha_2) \tag{31}$$

for each eigenstate of $\lambda_N{}^\circ$.

The general expression for $\mathbf{U}_n$ is given by

$$U_n(\alpha^n; t) \equiv \sum_{\xi} (-1)^k (k-1)!\, P_{m_1}(\alpha_{i_1}, \ldots, \alpha_{i_{m_1}}; t) \cdots P_{m_k}(\alpha_{i_{m-m_k+1}}, \ldots, \alpha_{i_m}; t)$$
$$= \sum_{\xi} (-1)^k (k-1)! \sum_{\lambda_N^{(1)}, \ldots, \lambda_N^{(k)}} \{\exp[(\lambda_N^{(1)} + \cdots + \lambda_N^{(k)})\, t]\}$$
$$\times \prod_{j=1}^{k} [\mathbf{L}_{N,\lambda_N^{(j)}} \cdot \mathbf{P}_N(0)][R_{m_j,\lambda_N^{(j)}}(\alpha_{i_1}, \ldots, \alpha_{i_{m_j}})] \tag{32}$$

where the sum is over all partitions ξ of n particles in subgroups, k is the number of subgroups, and m_i is the number of particles in the ith subgroup. The n-particle Ursell function has the important property

$$\sum_{\alpha_i} U_n(\alpha^n; t) = 0 \tag{33}$$

where the subscript i denotes any of the n particles. An extension of the argument of the last paragraph yields the result that the $\mathbf{U}_m$, $m = 2,..., n$, decay asymptotically faster than $\exp(\lambda_N{}^\circ t)$ if and only if

$$R_{n,\lambda_N{}^\circ}(\alpha^n) = \sum_{i=1}^{n} R_{n-1,0}(\alpha^{n-1})\, R_{1,\lambda_N{}^\circ}(\alpha_i) \tag{34}$$

Note that in this case, the time dependence of U_n is independent of whether or not the equilibrium probability distribution $\mathbf{P}_n^{(0)}$ factorizes into single-particle functions.

If $\mathbf{P}_n(t)$, $n = 1,..., N - 1$, is a solution of the reduced master equation (5), so that Eqs. (21)–(24) apply, the Ursell functions will not decay any faster than the probabilities unless additional assumptions are made. Sufficient additional assumptions are:

(i) The equilibrium distribution is factorized into single-particle functions, i.e.

$$P_n^{(0)}(\alpha^n) = \prod_{i=1}^{n} P_1^{(0)}(\alpha_i) \tag{35}$$

(ii) The eigenfunction $R_{1,\lambda_N{}^\circ}(\alpha_1) \neq 0$ for each of the eigenstates of $\lambda_N{}^\circ$ for one of the particles of the set n.

Under these conditions, it follows from Eq. (24) that for each eigenstate of $\lambda_N{}^\circ$,

$$R_{1,\lambda_N{}^\circ}(\alpha_j) = 0, \qquad j \neq i \tag{36}$$

and thus Eq. (24) is identical with Eq. (34). Therefore, $\mathbf{U}_m$, $m = 2,..., n$, decays asymptotically faster than $\mathbf{P}_m$, $m = 1,..., n$.

3. COMPLETELY FACTORIZED EIGENFUNCTIONS

As we have shown in the last section, the specification of the *necessary* conditions for the fast decay of the Ursell functions is quite complicated. In this section, we shall consider some fairly stringent *sufficient* conditions for the fast decay of the Ursell functions and for the validity of the master equation for the reduced distribution functions, Eq. (5).

We assume that all of the eigenfunctions of the master operator $\mathbf{A}_N$ are completely factorized into single-particle functions, i.e.,

$$R_{N,\lambda_N}(\alpha^N) = \prod_{i=1}^{N} R_{\nu_i}(\alpha_i) \tag{37}$$

where we use the notation

$$\lambda_N = \lambda_N(\nu_1, ..., \nu_N) \equiv \lambda_N(\nu^N) \tag{38}$$

with $\lambda_N(0,..., 0) = 0$. Under these conditions, Eq. (11) can be written

$$P_N(\alpha^N; t) = \sum_{\nu^N} \{\exp[\lambda_N(\nu^N)\, t]\}[\mathbf{L}_{N,\lambda_N} \cdot \mathbf{P}_N(0)] \prod_{i=1}^{N} R_{\nu_i}(\alpha_i) \tag{39}$$

where

$$L_{N,\lambda_N}(\alpha^N) = \prod_{i=1}^{N} [R_{\nu_i}(\alpha_i)/R_0(\alpha_i)] \equiv \prod_{i=1}^{N} L_{\nu_i}(\alpha_i) \tag{40}$$

and $R_0(\alpha_i) = p_1^{(0)}(\alpha_i)$ is the one-particle equilibrium distribution. Equation (40) follows from Eq. (14). The orthonormality condition (10) is equivalent to

$$\sum_{\alpha_i} L_{\nu_i}(\alpha_i)\, R_{\nu_i'}(\alpha_i) = \delta_{\nu_i,\nu_i'} \tag{41}$$

If we take $\nu_1 = 0$, this yields

$$\sum_{\alpha_i} R_{\nu_i'}(\alpha_i) = \delta_{0,\nu_i'} \tag{42}$$

Equation (18) for the reduced probabilities $\mathbf{P}_n$ now becomes

$$P_n(\alpha^n; t) = \sum_{\nu^N} \{\exp[\lambda_N(\nu^N)\, t]\}[\mathbf{L}_{N,\lambda_N} \cdot \mathbf{P}_N(0)] \prod_{i=1}^{n} R_{\nu_i}(\alpha_i) \prod_{i=n+1}^{N} \delta_{0,\nu_i} \tag{43}$$

where we have used Eq. (42). The sum over ν_{n+1} ,..., ν_N in Eq. (43) can be immediately performed to yield

$$P_n(\alpha^n; t) = \sum_{\nu^n} \{\exp[\lambda_n(\nu^n)\, t]\}[\mathbf{L}_{n,\lambda_n} \cdot \mathbf{P}_n(0)] \prod_{i=1}^{n} R_{\nu_i}(\alpha_i) \tag{44}$$

where

$$\lambda_n(\nu^n) \equiv \lambda_N(\nu^n, 0,..., 0) \tag{45}$$

and

$$L_{n,\lambda_n}(\alpha^n) \equiv \prod_{i=1}^{n} L_{\nu_i}(\alpha_i) \tag{46}$$

Equation (44) is obviously of the form of Eq. (20), and thus the reduced probability $\mathbf{P}_n$ obeys the master equation (5). *Therefore, the factorization of the eigenfunctions* $\mathbf{R}_{N,\lambda_N}$ *as given in Eq.* (37) *is a sufficient condition for the validity of Eq.* (5).

We shall now investigate the properties of the Ursell functions when all the eigenfunctions factorize. It follows from (44) and (46) that the terms in Eq. (44) for which $\nu_j = 0$, $j = 1,..., n$, are of the form $P_{n-1}(\alpha^{n-1}; t)\, P_1^{(0)}(\alpha_j)$. Thus, Eq. (44) can be rewritten

$$P_n(\alpha^n; t) = \sum_{\eta} P_{n-m}(\alpha^{n-m}; t)\, P_m^{(0)}(\alpha^m)$$
$$+ \sum_{\nu^n}{}' \{\exp[\lambda_n(\nu^n)\, t]\}[\mathbf{L}_{n,\lambda_n} \cdot \mathbf{P}_n(0)] \prod_{i=1}^{n} R_{\nu_i}(\alpha_i) \tag{47}$$

where the sum over η is over all partitions of n particles into two subgroups where m goes from 1 to n. The notation $\sum'_{\nu^n}$ implies the sum over all possible nonzero values of $\nu_1, ..., \nu_n$. Substitution of Eq. (47) into Eq. (32) for $\mathbf{U}_n(t)$ yields

$$U_n(\alpha^n; t) = \sum_{\xi} (-1)^k (k-1)! \sum_{\nu^n}{}' \exp\{[\lambda_{m_1}(\nu^{m_1}) + \cdots + \lambda_{m_k}(\nu^{m_k})]\, t\}$$
$$\times \prod_{i=1}^{k} [L_{m_i, \lambda_{m_i}} \cdot P_{m_i}(0)] \prod_{i=1}^{n} R_{\nu_i}(\alpha_i) \tag{48}$$

where we have used the fact that factorized probabilities do not contribute to the Ursell functions. Note that because of the restriction on the sum over ν^n, all the λ_{m_i} that appear in Eq. (48) are nonzero. The asymptotic decay of $\mathbf{U}_n$ is determined by the maximum value of $\lambda_{m_1}(\nu^{m_1}) + \cdots + \lambda_{m_k}(\nu^{m_k})$, $\nu_1, ..., \nu_n \neq 0$.

The stringent conditions on the eigenfunctions in Sections 2 and 3 for the rapid decay of the Ursell function $\mathbf{U}_N(t)$ do not provide much physical insight. It would be preferable if these conditions could be stated in terms of properties of the operators $\mathbf{A}_N$ and $\mathbf{A}_n$ of the N-particle and reduced n-particle master equations. We have been unable to do that in general. In Sections 4 and 5, we present and discuss some *sufficient* conditions on the $\mathbf{A}_N$ and $\mathbf{A}_n$ for the rapid decay of correlations. These conditions do yield physical insight into the types of interactions that lead to a rapid decay of correlations in stochastic processes governed by a master equation.

4. RANDOM WALK WITH INTERACTIONS

In this section, we consider N pairwise interacting random walkers on a finite one-dimensional lattice with periodic boundary conditions. The sites on the lattice might represent, for example, the internal states of gas particles. The steps of the random walkers on the lattice mirror the results of the binary collisions of the gas particles. The features of the problem that are of particular interest to us are the asymptotic decay of the Ursell functions and the structure of the transition rate matrix of the master equation in the thermodynamic limit.

We consider N random walkers with binary interactions on a finite one-dimensional lattice with l sites and periodic boundary conditions. The position of particle i on the lattice is denotes by $m_i = 1, ..., l$, with $i = 1, ..., N$. In our model, any site m can be occupied by more than one particle, subject to the restriction that the total number of particles equals N. We write the master equation of the N-particle system as

$$\partial P_N(m^N; t)/\partial t = (\alpha/N) \sum_{\substack{i \neq j \\ 1}}^{N} [P_N(m_1, ..., m_i - 1, ..., m_j + 1, ..., m_N; t) - P_N(m^N; t)] \tag{49}$$

where α is a constant transition rate for the transitions $m_i \to m_i - 1$; $m_j \to m_j + 1$. This equation describes a system with two-particle interactions such that one particle moves one step to the right while the other particle moves one step to the left while all other particles remain in the same state. In this model, $\sum_{i=1}^{N} m_i$ is conserved in the sense that for all states that occur in Eq. (49),

$$\sum_{i=1}^{N} m_i = C + kl \tag{50}$$

where C is a constant and k is an integer.

The eigenfunctions of the master operator are

$$\begin{aligned} R_{N,\nu^N}(m^N) &= l^{-N} \exp[(2\pi i/l)\, \nu^N m^N] \\ &= \prod_{j=1}^{N} R_{\nu_j}(m_j), \qquad \nu_j = 0, 1, \dots, l-1 \end{aligned} \tag{51}$$

where

$$R_{\nu_j}(m_j) = l^{-1} \exp[(2\pi i/l)\, \nu_j m_j] \tag{52}$$

The eigenvalues are

$$\lambda_N(\nu^N) = (2\alpha/N) \sum_{\substack{i<j \\ 1}}^{N} [\cos(2\pi/l)(\nu_i - \nu_j) - 1] \tag{53}$$

It is clear that

$$\lambda_N(\nu^N) = 0$$

if and only if

$$\nu_k = \nu, \qquad k = 1, \dots, N \tag{54}$$

for all $\nu_k = 0, \dots, l-1$. The zero eigenvalue λ_N is l-fold degenerate, which is a consequence of the conservation law expressed in Eq. (50). The highest nonzero eigenvalue $\lambda_N{}^\circ$ for $\mathbf{P}_N$ and $\mathbf{U}_N$ is

$$\lambda_N{}^\circ \equiv \lambda_N(\nu + 1, \nu, \nu, \dots, \nu) = (2\alpha/N)(N-1)[\cos(2\pi/l) - 1] \equiv -\Lambda \tag{55}$$

The N-particle eigenfunctions in this model are completely factorized into one-particle functions and thus we are dealing with a special case of the systems discussed in Section 3. The reduced probabilities $\mathbf{P}_n$ obey master equations of the form of Eq. (5). The reduced eigenfunctions are given by

$$R_{n,\nu^n}(m^n) = \prod_{j=1}^{n} R_{\nu_j}(m_j) \tag{56}$$

and the reduced eigenvalues are

$$\begin{aligned} \lambda_n(\nu^n) &= \lambda_N(\nu^n, 0, \dots, 0) \\ &= \frac{2\alpha(N-n)}{N} \sum_{j=1}^{n} \left[\cos\left(\frac{2\pi}{l}\nu_j\right) - 1\right] + \frac{2\alpha}{N} \sum_{\substack{i<j \\ 1}}^{n} \left[\cos\left(\frac{2\pi}{l}\right)(\nu_i - \nu_j) - 1\right] \end{aligned} \tag{57}$$

It should be noted that

$$\lambda_n(\nu^n) = 0, \qquad n = 1,\ldots, N-1 \tag{58}$$

if and only if

$$\nu_k = 0, \qquad k = 1, 2,\ldots, n \tag{59}$$

Thus, the zero eigenvalue for all *reduced* probabilities is nondegenerate.

The highest nonzero eigenvalue $\lambda_n{}^\circ$ for $P_n(m^n)$ is

$$\lambda_n{}^\circ \equiv \lambda_n(1, 0,\ldots, 0) = (2\alpha/N)(N-1)[\cos(2\pi/l) - 1] = -\Lambda \tag{60}$$

and thus all the reduced probabilities decay asymptotically to their equilibrium values as

$$P_n(m^n; t) - P_n^{(0)}(m^n) \sim c_n e^{-\Lambda t} \tag{61}$$

where c_n is a constant which depends on the initial conditions and where

$$P_n^{(0)}(m^n) = l^{-n} \tag{62}$$

The expression for $U_n(m^n; t)$ is given in Eq. (48). The highest nonzero eigenvalue appearing in this expression is

$$\begin{aligned}\lambda_n(1, 1,\ldots, 1) &= [2\alpha n(N-n)/N][\cos(2\pi/l) - 1] \\ &\equiv -[(N-n)/(N-1)]n\Lambda\end{aligned} \tag{63}$$

for $n = 1,\ldots, N-1$. Thus, $U_n(m^n; t)$, $n = 2,\ldots, N-1$, decays asymptotically to its equilibrium value $\mathbf{U}_n^{(0)} = 0$ as

$$U_n(m^n; t) \sim c_n{}' \exp\{-[n(N-n)/(N-1)]\Lambda t\} \tag{64}$$

where $c_n{}'$ is a constant which depends on the initial condition. Note that $U_1(m; t) \equiv P_1(m; t)$.

From the form of $\lambda_N{}^\circ$ in Eq. (55) it follows that the N-particle functions $\mathbf{P}_N(t)$ and $\mathbf{U}_N(t)$ decay asymptotically to their equilibrium forms as

$$\mathbf{P}_N(t) - \mathbf{P}_N^{(0)} \sim c_N e^{-\Lambda t} \tag{65}$$

and

$$\mathbf{U}_N(t) - \mathbf{U}_N^{(0)} \sim c_N{}' e^{-\Lambda t} \tag{66}$$

where

$$\mathbf{U}_N(\infty) = \mathbf{P}_N^{(0)} - l^{-N} \tag{67}$$

and $\mathbf{P}_N^{(0)}$ depends on the initial conditions because of the degenerate zero eigenvalue as given in Eq. (54).

We now discuss the behavior of the Ursell functions $\mathbf{U}_n(t)$ for finite N and in the thermodynamic limit as $N \rightarrow \infty$. For N finite, $\mathbf{U}_{N-n}(t)$ decays asymptotically at the same rate as $\mathbf{U}_n(t)$, which follows immediately from Eq. (64). The fastest-decaying Ursell functions are those for which $n \simeq N/2$. The slowest-decaying Ursell functions are those with $n \ll N$ and $n \simeq N$. The Ursell functions for $n \simeq N$ decay slowly owing to the conservation law given in Eq. (50). In the thermodynamic limit, the asymptotic decay of the reduced probabilities $\mathbf{P}_n(t)$ is still given by Eq. (61), while the decay of the Ursell functions $\mathbf{U}_n(t)$ is given by

$$\mathbf{U}_n(t) \sim c_n' e^{-nAt} \tag{68}$$

which follows from Eq. (64). In the thermodynamic limit, our model system thus has the identical asymptotic behavior for $\mathbf{P}_n(t)$ and $\mathbf{U}_n(t)$ as previously found in I and II.

The following question immediately arrises: Why does this model system *with interactions* behave, in the thermodynamic limit, in the same way as the independent-particle systems considered in I? The operator $\mathbf{A}_N$ in the master equation (49) can be written as

$$A_N(m^N, q^n) = (\alpha/N) \sum_{\substack{i\neq j\\1}}^{N} \left[\left(\delta_{m_i-1,q_i}\delta_{m_j+1,q_j} \prod_{\substack{k\neq i,j\\1}}^{N} \delta_{m_k,q_k}\right) - \delta_{m^N,q^N}\right] \tag{69}$$

The operator $\mathbf{A}_n$ for the reduced master equation is

$$A_n(m^n, q^n) = (\alpha/N) \sum_{\substack{i\neq j\\1}}^{n} \left[\left(\delta_{m_i-1,q_i}\delta_{m_j+1,q_j} \prod_{\substack{k\neq i,j\\1}}^{n} \delta_{m_k,q_k}\right) - \delta_{m^n,q^n}\right]$$

$$+ [\alpha(N-n)/N] \sum_{i=1}^{n} \left\{\left[(\delta_{m_k-1,q_i} + \delta_{m_i+1,q_i}) \prod_{\substack{k\neq i\\1}}^{n} \delta_{m_k,q_k}\right] - 2\delta_{m^n,q^n}\right\} \tag{70}$$

In the thermodynamic limit, $\mathbf{A}_n$ becomes

$$A_n(m^n, q^n) = \alpha \sum_{i=1}^{n} [\delta_{m_i-1,q_i} + \delta_{m_i+1,q_i} - 2\delta_{m_i,q_i}] \prod_{\substack{k\neq i\\1}}^{n} \delta_{m_k,q_k} \tag{71}$$

Thus, in the thermodynamic limit, the n-particle transition rate $\mathbf{A}_n$ is the sum of single-particle transition rates and is identical in form to Eq. (2.15) of paper I for independent-particle dynamics.

5. MASTER EQUATION MODELS FOR INTERACTING GASES

In this section, we consider systems containing N identical interacting particles in which the operator $\mathbf{A}_N$ for an N-particle master equation can be written as a sum of one-particle, two-particle, etc. contributions. We show that in the thermodynamic limit as $N \to \infty$, the reduced master equation operators $\mathbf{A}_n$ can be written as a sum of one-particle terms as long as Eq. (25) applies. We illustrate our discussion by treating a system of N interacting harmonic oscillators.

We consider an N-particle transition rate matrix of the form

$$A_N(\alpha^N, \gamma^N) = \sum_{i=1}^{N} A_1^{(1)}(\alpha_i\,;\gamma_i) \prod_{\substack{k\neq i\\1}}^{N} \delta_{\alpha_k,\gamma_k}$$
$$+ (1/2N) \sum_{\substack{i\neq j\\1}}^{N} A_2^{(2)}(\alpha_i\,,\alpha_j\,;\gamma_i\,,\gamma_j) \prod_{\substack{k\neq i,j\\1}}^{N} \delta_{\alpha_k,\gamma_k}$$
$$+ (1/6N^2) \sum_{\substack{i\neq j\neq l\\1}}^{N} A_3^{(3)}(\alpha_i\,,\alpha_j\,,\alpha_l\,;\gamma_i\,,\gamma_j\,,\gamma_l) \prod_{\substack{k\neq i,j,l\\1}}^{N} \delta_{\alpha_k,\gamma_k} + \cdots \tag{72}$$

where $A_2^{(2)}$, $A_3^{(3)}$, etc. are symmetric when the particle indices are permuted. Here, $\mathbf{A}_1^{(1)}$ is one-particle transition rate matrix which has the property

$$\sum_{\alpha_i} A_1^{(1)}(\alpha_i\,;\gamma_i) = 0 \tag{73}$$

$\mathbf{A}_2^{(2)}$ is a two-particle transition rate matrix with the property

$$\sum_{\alpha_i,\alpha_j} A_2^{(2)}(\alpha_i\,,\alpha_j\,;\gamma_i\,,\gamma_j) = 0 \tag{74}$$

$\mathbf{A}_3^{(3)}$ is a three-particle transition rate matrix with the property

$$\sum_{\alpha_i,\alpha_j,\alpha_l} A_3^{(3)}(\alpha_i\,,\alpha_j\,,\alpha_l\,;\gamma_i\,,\gamma_j\,,\gamma_l) = 0 \tag{75}$$

etc. The form of $\mathbf{A}_N$ in Eq. (72) implies that the systems under consideration can be described in terms of one-particle, two-particle, three-particle, etc. interactions as is the case, for instance, for weakly interacting systems with short-range forces. The factors $N^{-(i-1)}$ in front of the sums take account of the frequencies of i-particle collisions. Thus, the $\mathbf{A}_i$ are independent of N. We assume that for the operator $\mathbf{A}_N$ given in Eq. (72), the condition (25) holds in the thermodynamic limit as $N \to \infty$. In this case, $\mathbf{A}_n$ will be given by

$$A_n(\alpha^n; \gamma^n) = \sum_{i=1}^{n} A_1(\alpha_i\,;\gamma_i) \prod_{\substack{k\neq i\\1}}^{n} \delta_{\alpha_k,\gamma_k} \tag{76}$$

where

$$\mathbf{A}_1 = \mathbf{A}^{(1)} + \mathbf{A}^{(2)} + \mathbf{A}^{(3)} + \cdots \tag{77}$$

and where

$$A^{(1)}(\alpha_i\,;\gamma_i) = A_1^{(1)}(\alpha_i\,;\gamma_i) \tag{78}$$

$$A^{(2)}(\alpha_i\,,\gamma_i) = \lim_{N\to\infty}(1/N) \sum_{\substack{j=1\\j\neq i}}^{N} \sum_{\alpha_j} A_2^{(2)}(\alpha_i\,,\alpha_j\,;\gamma_i\,,\gamma_j) \tag{79}$$

$$A^{(3)}(\alpha_i\,,\gamma_i) = \lim_{N\to\infty}(1/N^2) \sum_{\substack{j,l=1\\i\neq j\neq l}}^{N} \sum_{\alpha_j}\sum_{\alpha_l} A_3^{(3)}(\alpha_i\,,\alpha_j\,,\alpha_l\,;\gamma_i\,,\gamma_j\,,\gamma_l) \tag{80}$$

etc. In writing Eqs. (79) and (80), we have assumed that the two-particle, three-particle, etc. contributions to $\mathbf{A}_N$ obey the condition (25) in the thermodynamic limit. In many systems, $\sum_{\alpha_j} A_2^{(2)}(\alpha_i, \alpha_j; \gamma_i, \gamma_j)$ will be independent of γ_j; in other systems, as we shall see below, it is necessary to take the sum over j to obtain a quantity independent of γ_j in thermodynamic limit.

Note that if $\mathbf{A}_n$ has the form given in Eq. (76), the results given in I for independent-particle systems will again apply.

We now consider the example of a gas of interacting harmonic oscillators which exchange only vibrational energy. The N-particle master equation has the form[6]

$$\begin{aligned}\partial P_N(m^N; t)/\partial t &= \sum_{q^N} A_N(m^N, q^N)\, P_N(q^N; t) \\ &= (a/N) \sum_{\substack{i \neq j \\ 1}}^{N} [m_i(m_j+1)\, P_N(m_1, ..., m_i - 1, ..., m_j + 1, ..., m_N; t) \\ &\quad - (m_i+1)\, m_j P_N(m^N; t)] \qquad (81)\end{aligned}$$

where a is a rate coefficient and where $m_i = 0, 1, ...,$ with $i = 1, ..., N$, denotes the states of the ith oscillator with energy $E(m_i) = m_i h\nu$. The dimensionless mean vibrational energy per particle ϵ is given by

$$\epsilon = (1/Nh\nu) \sum_{m_i} E(m_i) = (1/N) \sum_{m_i=1}^{N} m_i \qquad (82)$$

Clearly, this mean energy is a constant which is conserved during the relaxation process. In order for the relaxation process to take place on the constant-energy shell with energy, $N\epsilon$, it is necessary to choose initial conditions such that $P_N(m^N; 0)$ is nonzero only when $\sum_{m_i} m_i = N\epsilon$. The N-particle master operator has the form

$$A_N(m^N; q^N) = (1/2N) \sum_{\substack{i \neq j \\ 1}}^{N} A_2^{(2)}(m_i, m_j; q_i, q_j) \prod_{\substack{k=1 \\ k \neq i,j}}^{N} \delta_{m_k, q_k} \qquad (83)$$

where

$$\begin{aligned}A_2^{(2)}(m_i, m_j; q_i, q_j) &= a\{m_i(m_j+1)\, \delta_{m_i-1, q_i}\delta_{m_j+1, q_j} + (m_i+1)\, m_j \delta_{m_i+1, q_i}\delta_{m_j-1, q_j} \\ &\quad - [(m_i+1)\, m_j + m_i(m_j+1)]\, \delta_{m_i, q_i}\delta_{m_j, q_j}\} \qquad (84)\end{aligned}$$

Equation (83) is clearly a special case of Eq. (72). The two-particle transition rate matrix $\mathbf{A}_2^{(2)}$ has the properties

$$\begin{aligned}C(m_i; q_i, q_j) &\equiv \sum_{m_j} A_2^{(2)}(m_i, m_j; q_i, q_j) \\ &= aq_j[m_i\delta_{m_i-1, q_i} + (m_i+1)\, \delta_{m_i+1, q_i} - (2m_i+1)\, \delta_{m_i, q_i}] \\ &\quad + a[(m_i+1)\, \delta_{m_i+1, q_i} - m_i\delta_{m_i, q_i}] \qquad (85)\end{aligned}$$

and

$$A^{(2)}(m_i\,;q_i) = \lim_{N\to\infty}(1/N)\sum_{j=1;j\neq 1}^{N} C(m_i\,;q_i\,,q_j)$$
$$= a[\epsilon m_i \delta_{m_i-1,q_i} + (\epsilon + 1)(m_i + 1)\,\delta_{m_i+1,q_i} - (2\epsilon m_i + \epsilon + m_i)\,\delta_{m_i,q_i}] \tag{86}$$

where the limit $N \to \infty$ is taken keeping ϵ fixed. Equation (86) leads to an $A^{(2)}$ of the form given in Eq. (79). From the relation given in Eq. (77), it then follows that $\mathbf{A}_n$ is of the form of Eq. (76) with the n-particle master operator given as a sum of one-particle transition rates. The results of paper I are thus directly applicable to the system of interacting oscillators. The eigenfunctions and eigenvalues of this one-particle transition rate matrix $A_1(\alpha_1, \gamma_1)$ can be found in the paper by Montroll and Shuler.[7]

The results of this section can be briefly summarized as follows. If the transition rate matrix $\mathbf{A}_N$ for the N-particle master equation can be written in the form of Eq. (72), which is the case, for instance, for systems with weak interactions and short-range forces, and if the operator $\mathbf{A}_N$ is of the form given in Eq. (25), i.e., if the N-particle master equation reduces to an n-particle master equation by summation over the other $N - n$ particles, then in the thermodynamic limit, $\mathbf{A}_N$ reduces to a sum of one-particle transition rates $\mathbf{A}_1$. This implies that for all such systems,

$$\mathbf{U}_n(t) - \mathbf{U}_n^{(0)} \sim [P_1(t) - P_1^{(0)}]^n \tag{87}$$

$$\mathbf{P}_n(t) - \mathbf{P}_n^{(0)} \sim [P_1(t) - P_1^{(0)}] \tag{88}$$

These results are identical with those obtained by us in I for systems described by independent-particle dynamics which are also characterized by one-particle transition rates $\mathbf{A}_1$.

REFERENCES

1. I. Oppenheim, K. E. Shuler, and G. H. Weiss, *J. Chem. Phys.* **46**:4100 (1967).
2. I. Oppenheim, K. E. Shuler, and G. H. Weiss, *J. Chem. Phys.* **50**:3662 (1969).
3. D. Bedeaux, K. E. Shuler, and I. Oppenheim, *J. Stat. Phys.* **2**:1 (1970).
4. R. Glauber, *J. Math. Phys.* **4**:294 (1963).
5. K. E. Shuler, *Phys. Fluids* **2**:442 (1959).
6. K. E. Shuler, *J. Chem. Phys.* **32**:1692 (1960).
7. E. W. Montroll and K. E. Shuler, *J. Chem. Phys.* **26**:454 (1957).

On the Relation between Master Equations and Random Walks and Their Solutions*

Dick Bedeaux, Katja Lakatos-Lindenberg, and Kurt E. Shuler

Department of Chemistry, University of California San Diego, La Jolla, California 92037

(Received 18 January 1971)

It is shown that there is a simple relation between master equation and random walk solutions. We assume that the random walker takes steps at random times, with the time between steps governed by a probability density $\psi(\Delta t)$. Then, if the random walk transition probability matrix **M** and the master equation transition rate matrix **A** are related by $\mathbf{A} = (\mathbf{M} - 1)/\tau_1$, where τ_1 is the first moment of $\psi(t)$ and thus the average time between steps, the solutions of the random walk and the master equation approach each other at long times and are essentially equal for times much larger than the maximum of $(\tau_n/n!)^{1/n}$, where τ_n is the nth moment of $\psi(t)$. For a Poisson probability density $\psi(t)$, the solutions are shown to be identical at all times. For the case where $\mathbf{A} \neq (\mathbf{M} - 1)/\tau_1$, the solutions of the master equation and the random walk approach each other at long times and are approximately equal for times much larger than the maximum of $(\tau_n/n!)^{1/n}$ if the eigenvalues and eigenfunctions of **A** and $(\mathbf{M} - 1)/\tau_1$ are approximately equal for eigenvalues close to zero.

1. INTRODUCTION

There exists an extensive literature on master equations and random walks and their solutions.[1] We show in this paper that there is a close relation between random walks and master equations and their solutions. We consider random walks[2] in which the walker takes his steps at random times $t_1, t_2, \cdots$ and where the random variables $T_1 = t_1, T_2 = t_2 - t_1, \cdots, T_n = t_n - t_{n-1}$ have a common probability density $\psi(T)$. A random walk with constant time intervals $T_1 = T_2 = \cdots \equiv \tau$ between steps is the special case with $\psi(t) = \delta(t - \tau)$.

In Sec. 2 the random walk and the master equation are formally solved in terms of Green's functions. It is shown that a simple relation between the Green's functions exists if the master operator **A** and the random walk transition matrix **M** are related by $\mathbf{A} = (\mathbf{M} - 1)/\tau_1$, where τ_1 is the first moment of $\psi(t)$, i.e., the average time between steps.

In Sec. 3 it is shown that for a Poisson process, characterized by $\psi(t) = (1/\tau_1)e^{-t/\tau_1}$, the solutions for a random walk and the corresponding master equation with $\mathbf{A} = (\mathbf{M} - 1)/\tau_1$ are equal *for all times*. This is the only time step distribution for which this is the case. It is also shown that the random walk equation and the corresponding master equation are identical for a Poisson process. It should be stressed that this equivalence is valid independently of the value of the average time between steps, τ_1, and that it is not necessary to go to the limit $\tau_1 \longrightarrow 0$.

In Sec. 4 processes with arbitrary $\psi(t)$ are investigated. We discuss there the implications of the main result of this paper which can be stated here loosely as the following: If $\mathbf{A} = (\mathbf{M} - 1)/\tau_1$, then the random walk and master equation solutions approach each other at long times, and are approximately equal for times much greater than the maximum of $(\tau_n/n!)^{1/n}$, where τ_n is the nth moment of the distribution $\psi(t)$. For $\mathbf{A} \neq (\mathbf{M} - 1)/\tau_1$ some additional conditions must be imposed. These results are stated more precisely in Sec. 4 and are proved in the Appendix. We also present in Sec. 4 a mathematically precise formulation and a rigorous proof of the often stated equivalence of the random walk and master equations in the limit as the time interval between steps tends to zero.

2. FORMAL SOLUTION OF THE RANDOM WALK AND MASTER EQUATIONS

The general equation for a random walk is

$$P(\alpha; n+1) = \sum_{\alpha'} M_{\alpha\alpha'} P(\alpha'; n) = \mathbf{M}\, P(\alpha; n), \tag{2.1}$$

where $P(\alpha; n)$ is the probability that the walker is in state α after the nth step, $M_{\alpha\alpha'}$ is the probability that the walker goes from α to α' in one step, and **M** is the transition probability matrix. We impose no restrictions on the number of states between α and α', i.e., random walks with nonnearest neighbor transitions are included in our subsequent analysis. If there is a continuum of states, the sum over α is understood to be an integral over the continuous part and a sum over the discrete part of state space.

To calculate the probability $P(\alpha; t)$ that the random walker is in state α at time t, we must specify the probability that the random walker makes a step in a given time interval. We shall assume[2] that jumps are made at random times $t_1, t_2, t_3, \cdots$, where the random variables $T_i \equiv (t_i - t_{i-1})$, $i = 1, 2, \cdots$, with $t_0 = 0$, have a common probability density $\psi(T)$.

The general form of the master equation is

$$\frac{\partial}{\partial t} Q(\alpha; t) = \sum_{\alpha'} A_{\alpha\alpha'} Q(\alpha'; t) = \mathbf{A}\, Q(\alpha, t), \tag{2.2}$$

where $Q(\alpha; t)$ is the probability that the system is in state α at time t, $A_{\alpha\alpha'}$ is the transition rate from state α' to α, and **A** is the transition rate matrix. The transition rate $A_{\alpha\alpha'}$ is related to the more usually employed gain and loss rates $B_{\alpha\alpha'}$ by

$$A_{\alpha\alpha'} = B_{\alpha\alpha'} - \delta_{\alpha\alpha'} \sum_{\alpha''} B_{\alpha''\alpha}, \tag{2.3}$$

in which $\delta_{\alpha,\alpha'}$ is the Kronecker δ for discrete states and the Dirac δ function for continuous states.

We consider only processes which are temporally homogeneous, i.e., for which **M** and $\psi(T)$ are independent of n and t, and A is independent of t.

The formal solution of the random walk equation, Eq. (2.1), is

$$P(\alpha; n) = \mathbf{M}^n P(\alpha; 0). \tag{2.4}$$

The generating function[3] for the random walk is defined by

$$P(\alpha; z) \equiv \sum_{n=0}^{\infty} z^n P(\alpha; n) = (1 - z\mathbf{M})^{-1} P(\alpha; 0), \tag{2.5}$$

and the corresponding Green's function is

$$G_{\alpha\alpha'}(z) = (1 - z\mathbf{M})^{-1} \delta_{\alpha,\alpha'}. \tag{2.6}$$

The formal solution of the random walk problem in continuous time is given by[2]

$$P(\alpha; t) = \sum_{n=0}^{\infty} \Phi_t(n) P(\alpha; n), \tag{2.7}$$

where $\Phi_t(n)$ is the probability that the walker has made exactly n steps at time t. This probability is related to the probability density $\psi_n(t)$ that the walker makes his nth step at time t, $t = t_n$, by

$$\Phi_t(n) = \int_0^t \psi_n(\tau) \int_{t-\tau}^{\infty} \psi(\tau')d\tau' d\tau, \tag{2.8}$$

where

$$\psi_n(t) = \int_0^t \psi_{n-1}(\tau)\psi(t-\tau)\, d\tau, \qquad n > 1,$$

$$\psi_1(t) = \psi(t), \qquad \psi_0(t) = \delta(t). \tag{2.9}$$

The generating function for the random walk as a function of time is defined by the Laplace transform:

$$P(\alpha; s) \equiv \int_0^{\infty} e^{-st} P(\alpha; t)dt. \tag{2.10}$$

Substitution of Eqs. (2.7)–(2.9) into Eq. (2.10) yields

$$P(\alpha; s) = \left(s - \frac{s\tilde{\psi}(s)}{1 - \tilde{\psi}(s)}(\mathbf{M} - 1)\right)^{-1} P(\alpha; 0), \tag{2.11}$$

where

$$\psi(s) \equiv \int_0^{\infty} e^{-st} \psi(t)dt \tag{2.12}$$

and $P(\alpha; 0) = P(\alpha; n = 0) = P(\alpha; t = 0)$. The corresponding Green's function is

$$G_{\alpha\alpha'}(s) = \left(s - \frac{s\tilde{\psi}(s)}{1 - \tilde{\psi}(s)}(\mathbf{M} - 1)\right)^{-1} \delta_{\alpha,\alpha'}. \tag{2.13}$$

This Green's function is related to the one in Eq. (2.6) by

$$G_{\alpha\alpha'}(s) = \frac{[1 - \tilde{\psi}(s)]}{s} G_{\alpha\alpha'}[z = \tilde{\psi}(s)]. \tag{2.14}$$

The formal solution of the master equation, Eq. (2.2), is

$$Q(\alpha; t) = e^{\mathbf{A}t} Q(\alpha; 0). \tag{2.15}$$

The generating function for the master equation is defined by the Laplace transform:

$$Q(\alpha; s) \equiv \int_0^{\infty} e^{-st} Q(\alpha; t)dt = (s - \mathbf{A})^{-1} Q(\alpha; \mathrm{o}). \tag{2.16}$$

The corresponding Green's function is

$$F_{\alpha\alpha'}(s) = (s - \mathbf{A})^{-1} \delta_{\alpha,\alpha'}. \tag{2.17}$$

It will be shown in the following sections that the solutions of the random walk problem and the master equation are closely related if we make the identification

$$\mathbf{A} = (\mathbf{M} - 1)/\tau_1, \tag{2.18}$$

where

$$\tau_1 \equiv \int_0^{\infty} t\psi(t)dt \tag{2.19}$$

is the average time between steps. $F_{\alpha\alpha'}(s)$ and $G_{\alpha\alpha'}(s)$ are then related by

$$G_{\alpha\alpha'}(s) = \frac{[1 - \tilde{\psi}(s)]}{\tau_1 s\tilde{\psi}(s)} F_{\alpha\alpha'}\left(\frac{1 - \tilde{\psi}(s)}{\tau_1\tilde{\psi}(s)}\right), \tag{2.20}$$

while $F_{\alpha\alpha'}(s)$ and $G_{\alpha\alpha'}(z)$ are related by

$$G_{\alpha\alpha'}(z) = \tau_1 z\, F_{\alpha\alpha'}\left(\frac{1 - z}{\tau_1 z}\right). \tag{2.21}$$

It is thus clear that the solution of any one of the three problems in terms of Green's functions immediately gives the solutions of the other two problems in terms of Green's functions. All three problems can in principle be solved by the diagonalization of the same operator.

In terms of the eigenfunctions and eigenvalues of the operator **A**, the solutions of the random walk and the master equation can be written as

$$P(\alpha; n) = \sum_a (1 + \tau_1 a)^n f_a(\alpha)\, P(a; 0), \tag{2.22}$$

$$P(\alpha; t) = \sum_a \left[\frac{1}{2\pi i}\int_{c-i\infty}^{c+i\infty} ds e^{st}\left(s - \frac{s\tilde{\psi}(s)\tau_1 a}{1 - \tilde{\psi}(s)}\right)^{-1}\right] \times f_a(\alpha)\, P(a; 0), \tag{2.23}$$

$$Q(\alpha; t) = \sum_a e^{at} f_a(\alpha)\, Q(a; 0), \tag{2.24}$$

where

$$P(a; 0) = \sum_a g_a(\alpha)\, P(\alpha; 0), \tag{2.25}$$

$$Q(a; 0) = \sum_a g_a(\alpha)\, Q(\alpha; 0), \tag{2.26}$$

and $a, g_a(\alpha)$, and $f_a(\alpha)$ are the eigenvalues, left eigenfunctions, and right eigenfunctions of **A**, respectively. If the spectrum of **A** has continuous parts, the sum over a is understood to be an integral over these parts. The quantity c in Eq. (2.23) is any positive constant.

In order that $P(\alpha;n)$ be a well-defined probability, $0 \leqslant P(\alpha;n) \leqslant 1$, it is clear that the Euclidean norm of the operator M must be bounded:

$$\|M\| \equiv \max\|Mh\| \leq 1, \qquad \|h\| = 1. \tag{2.27}$$

Because of the relation between $\mathbf{A}$ and $\mathbf{M}$ in Eq. (2.18), the bound on $\mathbf{M}$ gives upper and lower bounds to the spectrum of $\mathbf{A}$:

$$-2/\tau_1 \leq a \leq 0. \tag{2.28}$$

Since we have made no restrictive statements on conservation of probability, the results of this paper are also valid for "open" systems, i.e., systems where probability is not preserved in that some or all of the walkers are removed in time (trapped, absorbed, evaporated, etc.).

The existence of a lower bound on the eigenvalues of the operator $\mathbf{A}$ is crucial to our technique of relating the solutions of the random walk and the master equations via their Green's function, Eqs. (2.14), (2.20), and (2.21). Since, however, eigenvalues with a large absolute value do not contribute appreciably to the long-time behavior of the solutions, it is possible to relate the long-time behavior of the solutions of a master equation with those of a random walk equation if the eigenvalues of $\mathbf{A}$ do not obey Eq. (2.28). In that case it is, however, impossible to relate $\mathbf{A}$ and $\mathbf{M}$ by Eq. (2.18); hence it is also impossible to give a simple relation between the various Green's functions. This will be explored further in Sec. 4.

3. POISSON PROCESSES

The following question immediately arises: Does there exist a probability density $\psi(t)$ such that the solutions of the master equation (2.2) and of the random walk equation in continuous time (2.7) are identical at all times for identical initial conditions? It follows from the Green's functions for both problems (2.13) and (2.17) that this will be the case if and only if $\mathbf{A}$ and $\mathbf{M}$ are related by Eq. (2.18) and if

$$\frac{s\tilde{\psi}(s)}{1-\tilde{\psi}(s)} = \frac{1}{\tau_1}. \tag{3.1}$$

Equation (3.1) has the solution

$$\tilde{\psi}(s) = (\tau_1 s + 1)^{-1}, \tag{3.2}$$

which then yields upon inversion

$$\psi(t) = \tau_1^{-1}\, e^{-t/\tau_1}. \tag{3.3}$$

This is the probability density for a Poisson process.[4] For such a process, the probability that the walker has made exactly n steps at time t is the Poisson distribution

$$\Phi_t(n) = \frac{1}{n!}\left(\frac{t}{\tau_1}\right)^n e^{-t/\tau_1}. \tag{3.4}$$

It is possible to show directly that the difference equation for a random walk with a Poisson density $\psi(t)$ is equivalent to a master equation at all times t. From the formal solution [Eq. (2.4)] of the random walk equation and the relation (2.7) one obtains

$$P(\alpha;t) = \sum_{n=0} \Phi_t(n)\, \mathbf{M}^n P(\alpha;0), \tag{3.5}$$

where $\sum_{n=0}^{\infty} \Phi_t(n)\mathbf{M}^n$ can be considered as an operator which translates the initial distribution to the distribution at time t.

Equation (3.5) yields:

$$\begin{aligned}\frac{P(\alpha;t+h)-P(\alpha;t)}{h} &= h^{-1}\left[\exp\left(\frac{(t+h)}{\tau_1}(\mathbf{M}-1)\right) - \exp\left(\frac{t}{\tau_1}(\mathbf{M}-1)\right)\right] P(\alpha;0)\\ &= h^{-1}\left[\exp\left(\frac{h}{\tau_1}(\mathbf{M}-1)\right) - 1\right] \exp\left(\frac{t}{\tau_1}(\mathbf{M}-1)\right) P(\alpha;0)\\ &= h^{-1}\left[\exp\left(\frac{h}{\tau_1}(\mathbf{M}-1)\right) - 1\right] P(\alpha;t).\end{aligned} \tag{3.6}$$

It is clear from what has been said above that Eq. (3.6) is valid for all $h > 0$. In the limit as $h \to 0$, Eq. (3.6) becomes

$$\frac{\partial P(\alpha;t)}{\partial t} = \frac{(\mathbf{M}-1)}{\tau_1} P(\alpha;t) = \mathbf{A}\,P(\alpha;t), \tag{3.7}$$

which is the master equation as given in Eq. (2.2).

The Poisson density $\psi(t)$ of Eq. (3.3) of time intervals between distinct events is characteristic of a large class of uncorrelated random processes developing in time. For such stochastic processes, where the random walk formulation with discrete steps is completely equivalent to the master equation formulation in continuous time for all times t, it is evidently only a matter of personal choice which equation one wants to employ in the solution of the problem at hand.

4. PROCESSES WITH GENERAL DENSITIES $\psi(t)$

In this section we discuss the conditions under which the solutions of the random walk in continuous time and the random walk as a function of step number approach and are approximately equal to the solution of the master equation at long times. This analysis is subject to identical initial conditions for all processes, i.e.,

$$P(\alpha; n = 0) = P(\alpha; t = 0) = Q(\alpha; 0). \qquad (4.1)$$

We will first analyze this problem for $\mathbf{A} = (\mathbf{M} - 1)\tau_1$ using the solutions of the random walk in continuous time and of the master equation, Eqs. (2.23) and (2.24):

$$P(\alpha; t) = \sum_a \theta_a(t) f_a(\alpha) P(a; 0), \qquad (4.2)$$

$$Q(\alpha; t) = \sum_a e^{at} f_a(\alpha) P(a; 0), \qquad (4.3)$$

where

$$\theta_a(t) = \frac{1}{2\pi i} \int_{c-i\infty}^{c+i\infty} ds e^{st} \left(s - \frac{s\tilde{\psi}(s)\tau_1 a}{1 - \tilde{\psi}(s)} \right)^{-1}. \qquad (4.4)$$

Let τ_n by the nth moment of the probability density $\psi(t)$:

$$\tau_n \equiv \int_0^\infty t^n \psi(t) dt \qquad (4.5)$$

and let us define

$$\gamma \equiv \sup(\tau_n / n!)^{1/n}. \qquad (4.6)$$

The following theorem, which is the main result of this paper, is proved in the Appendix:

$$P(\alpha, t) - Q(\alpha, t) = Q(\alpha, t)\, O(\gamma/t) + O(e^{-t/\gamma}) \qquad (4.7)$$

for $t \gg \gamma, \tau_1$ of order γ, and $\mathbf{A} = (\mathbf{M} - 1)/\tau_1$.

The order symbol O denotes that if $f(x) = O(x)$, then $f(x)/x$ remains bounded for all x. This theorem states that the solutions of the random walk in continuous time and the master equation are essentially equal for times much larger than γ. Our subsequent discussion in this section explores the implications of this theorem.

In order for theorem (4.7) to hold for all densities $\psi(t)$, it is necessary to note the following caveat. For a certain restricted class of sharply peaked densities $\psi(t)$, which are precisely defined in Eq. (A19) of the Appendix [an example would be $\psi(t) = \delta(t - T)$], it is necessary to exclude random walks with oscillatory solutions which persist at long times.[5] This corresponds to the exclusions of master equations with an operator $\mathbf{A}$ which has eigenvalues a in the range $0 \leq a + 2/\tau_1 \ll 1/\gamma$.

If the spectrum of $\mathbf{A}$ contains eigenvalues which satisfy the condition

$$0 < -a \ll 1/\gamma, \qquad (4.8)$$

then the theorem implies that the two solutions are essentially equal *before* equilibrium is reached.

If any moment of $\psi(t)$ is infinite, it is clear from theorem (4.7) that the two solutions will become equal only *at* equilibrium. For probability densities $\psi(t)$ that decay at least exponentially at long times, all moments τ_n are finite. Typical examples of probability densities $\psi(t)$ with infinite γ are those that decay with negative powers of t for large times.

One quite frequently sees the statement that in the limit as the time interval between steps goes to zero, the random walk equation becomes equivalent to a master equation. We now give a mathematically precise formulation of this statement. We consider a sequence of random walks characterized by a sequence of transitions matrices $\{\mathbf{M}_i\}$ and a sequence of densities $\{\psi_i(t)\}$ with the property

$$\lim_{i\to\infty} \gamma_i = 0. \qquad (4.9)$$

The meaning of condition (4.9) is that our sequence of random walks is so constructed that as the sequence index i increases, the moments of the probability density $\psi_i(t)$ all go to zero as specified. If we now define

$$P^{(\infty)}(\alpha; t) \equiv \lim_{i\to\infty} P^{(i)}(\alpha; t), \qquad (4.10)$$

then $P^{(\infty)}(\alpha; t)$ is a solution of the master equation

$$\frac{\partial}{\partial t} P^{(\infty)}(\alpha; t) = \mathbf{A}^{(\infty)} P^{(\infty)}(\alpha; t) \qquad (4.11)$$

if and only if the limit [see Eq. (2.18)]

$$\mathbf{A}^{(\infty)} \equiv \lim_{i\to\infty} (\mathbf{M}_i - 1)/\tau_{1,i} \qquad (4.12)$$

exists, where as before, $\tau_{1,i}$ is of order γ_i. Therefore, the random walk equation indeed becomes equivalent to a master equation "in the limit as the time interval between steps goes to zero." Equation (4.12) is an analog of the well-known Kolmogoroff condition.[6]

It should be pointed out that if one considers the random walk in the continuous time and continuous space limits, which in general yields a diffusion-like equation, it is still necessary that Eq. (4.9) be satisfied. The conditions on the density $\psi(t)$ for passage to a diffusion equation are therefore the same as for passage to a master equation.

It is also possible to relate the solution of a random walk as a function of step number to the solution of the corresponding master equation. This is easily seen with the choice $\psi(t) = \delta(t - \tau_1)$, in which case $P(\alpha; n) = P(\alpha; t = n\tau_1)$. The use of theorem (4.7) yields

$$P(\alpha;n) - Q(\alpha;n\tau_1) = Q(\alpha;n\tau_1)\, O(1/n) + O(e^{-n}) \tag{4.13}$$

for $n \gg 1$. Since the δ function is an example of a sharply peaked distribution as defined in Eq. (A19) of the Appendix, we must exclude transition operators **M** with eigenvalues m such that $0 \le m + 1 \ll 1$.

We will now consider cases for which $A \neq (\mathbf{M}-1)/\tau_1$ As will be proved in the Appendix, our main theorem, Eq. (4.7), and its consequences as discussed above, are still true, subject to a condition on the eigenvalues and eigenfunctions of **A** and **M**. If we define

$$\mathbf{B} \equiv (\mathbf{M}-1)/\tau_1, \tag{4.14}$$

then this condition can be stated as follows: For eigenvalues a, b of the operators **A**, **B** which lie in the range

$$0 < -a \ll 1/\gamma, \qquad 0 < -b \ll 1/\gamma, \tag{4.15}$$

there must be a one-to-one correspondence between the right eigenfunctions f_a, h_b of **A** and B such that

$$|a-b| \ll -a, \tag{4.16}$$

$$f_a(\alpha) - h_b(\alpha) = f_a(\alpha)\, O(\gamma a). \tag{4.17}$$

Theorem (4.7) thus still holds if the eigenvalues and eigenfunctions of A and $(\mathbf{M}-1)/\tau_1$ are approximately equal for eigenvalues close to zero. This extended version of theorem (4.7) specifies the class of random walk problems which have the same long-time behavior as that of a given master equation or vice versa.

The relation between **A** and **M** is determined by the physics of the problem. Consider, as an example, an open system in which the total probability is not conserved, due to irreversible trapping or evaporation. The transition rate A can now be written as the sum of two terms, $\mathbf{A} = \mathbf{A}_1 - \mathbf{A}_2$, where $\mathbf{A}_1$ conserves probability and $\mathbf{A}_2$ describes the irreversible loss process. Then the physically appropriate choice of **M** is[7]

$$\mathbf{M} = (1+\tau_1\mathbf{A}_1)(1+\tau_1\mathbf{A}_2)^{-1}. \tag{4.18}$$

The matrix **M** of Eq. (4.18) describes a situation in which the random walk within the system and the loss process therefrom are statistically independent and hence enter multiplicatively. That this choice of **M** is physically more reasonable for open systems than Eq. (2.18) is easily seen if one takes the case where $\mathbf{A}_2$ is a constant, $\mathbf{A}_2 = k$. Then $(1+\tau_1\mathbf{A}_2)^{-1} = (1+\tau_1 k)^{-1}$ is the probability per step that the random walker remains in the system. This probability ranges between 0 and 1 as the rate k in the master equation ranges between ∞ and 0. If $k \ll 1/\gamma$, then the conditions of Eqs. (4.15)–(4.17) are satisfied and the random walk solution and the master equation solution approach each other at long times.

A number of other examples could be discussed involving various physically plausible relations between **A** and **M** different from that in Eq. (2.18) for simple closed systems. In all such cases, conditions (4.14)–(4.17) determine whether and how rapidly the solution of the random walk and the master equation approach each other.

ACKNOWLEDGMENTS

One of us (K. L.-L.) would like to thank Dr. R. M. Pearlstein and Dr. R. P. Hemenger for many helpful and stimulating discussions.

APPENDIX:

We will now prove the theorem stated in Eq. (4.7).

We begin by considering densities $\psi(t)$ for which γ defined in Eq. (4.6) is finite. If this is the case, then $\tilde{\psi}(s)$ is analytic at $s = 0$ and can be expanded in a power series:

$$\tilde{\psi}(s) = 1 - \tau_1 s + \frac{\tau_2 s^2}{2!} - \frac{\tau_3 s^3}{3!} + \cdots, \tag{A1}$$

where the first term in the series is unity because $\psi(t)$ is normalized. The τ_n are defined in Eq. (4.5). Since $\psi(t)$ is positive-definite, all its moments τ_n are positive. Therefore the singularity of $\tilde{\psi}(s)$ closest to the origin will be on the negative real axis. The distance R of this pole from the origin is then the radius of convergence of the power series in Eq. (A1). This radius of convergence can be related to the moments of $\psi(t)$ by Hadamard's formula[8]

$$R = \left[\overline{\lim_{n\to\infty}}\, (\tau_n/n!)^{1/n}\right]^{-1} \ge \gamma - 1, \tag{A2}$$

where $\overline{\lim}$ indicates the limes superior, which is the greatest limit point of the sequence. This limit exists, since the sequence has the upper bound γ. Since $\tilde{\psi}(s)$ is analytic on the real axis to the right of $-R$, $\tilde{\psi}(s)$ is analytic for all s for which $\mathrm{Re}\, s > -R$, and for such s it is given by[9]

$$\tilde{\psi}(s) = \int_0^\infty e^{-st}\,\psi(t)dt, \quad \mathrm{Re}\, s > -R. \tag{A3}$$

The function $\theta_a(t)$ of Eq. (4.4) can be expressed in terms of the singularities of the integrand. Only the singularities to the right of $-R$ are important for the times we are interested in. For $\mathrm{Re}\, s > -R$, the only singularities are simple poles, which, for $a \neq 0$, are the zeros of the function $\tilde{\psi}(s) - (\tau_1 a + 1)^{-1}$ in that region:

$$\tilde{\psi}(s) - 1/(\tau_1 a + 1) = 0, \quad \mathrm{Re}\, s > -R. \tag{A4}$$

If a zero s_a is of nth order, then

$$\left.\frac{d^k\tilde{\psi}(s)}{ds^k}\right|_{s=s_a} = 0, \quad k = 1, 2, \cdots, n-1, \tag{A5}$$

and the pole in the integrand of Eq. (4.4) is also of nth order. For $a = 0$, it is easy to see that $\theta_0(t) = 1$.

Since $\tilde{\psi}(s)$ is real for s real and $s > -R$, it follows from the reflection principle[10] that

$$\tilde{\psi}(s^*) = \tilde{\psi}^*(s) \tag{A6}$$

for all s. This implies that the solutions of Eq. (A4) which do not lie on the real axis occur in complex conjugate pairs. From Eq. (A3) it follows that $\tilde{\psi}(s)$ is a monotonic decreasing function of s for s real and $s > -R$. Since $\tilde{\psi}(0) = 1$, Eq. (A4) has exactly one solution on the real axis for $-R < s \le 0$ if

$$0 \ge a > \tau_1^{-1}\,[(\tilde{\psi}^+)^{-1} - 1] \ge -\tau_1^{-1}, \tag{A7}$$

where

$$\tilde{\psi}^+ \equiv \lim_{\epsilon \to 0^+} \tilde{\psi}(-R + \epsilon),$$

with ϵ positive and real. This solution is of first order. From Eq. (A3) it follows that

$$\mathrm{Re}\tilde{\psi}(r + i\Gamma) \le \mathrm{Re}\tilde{\psi}(r) = \tilde{\psi}(r), \qquad r > -R, \tag{A8}$$

where r and Γ are real. From this it follows that the complex solutions of Eq. (A4) do not lie to the right of the real solution discussed above. If Eq. (A7) is not satisfied, Eq. (A4) will have neither real nor complex solutions for $-1/\tau_1 \le a \le 0$. For $-2/\tau_1 < a < -1/\tau_1$, $(1 + \tau_1 a)^{-1}$ is negative. In this case Eq. (A4) can only have solutions on the real axis left of $-R$, but may very well have complex solutions to the right of $-R$. If we denote the solution on the real axis by s_a and the complex solutions in the upper half-plane by $s_{a,j} = r_{a,j} + i\,\Gamma_{a,j}$, with $r_{a,j}$ and $\Gamma_{a,j}$ real, then

$$-R < r_{a,j} \le s_a \quad \text{for} \quad -1/\tau_1 \le a \le 0. \tag{A9}$$

The complex solutions in the lower half-plane are than given by $s_{a,j} = r_{a,j} - i\,\Gamma_{a,j}$, which follows from Eq. (A6).

For $t \gg \gamma$, the only contributions to $\theta_a(t)$ which are not of $O(e^{-t/\gamma})$ come from poles for which

$$\begin{aligned} &0 \le -s_a \le -r_{a,j} \ll 1/\gamma \quad \text{for} \quad -1/\tau_1 \le a \le 0, \\ &0 \le -r_{a,j} \ll 1/\gamma \quad \text{for} \quad -2/\tau_1 < a < -1/\tau_1. \end{aligned} \tag{A10}$$

The function $\tilde{\psi}(s)$ can be expanded in a power series about any of the solutions $s_{a,j}$:

$$\tilde{\psi}(s) = \frac{1}{\tau_1 a + 1} - \tau_1^j (s - s_{a,j}) + \frac{\tau_2^j}{2!} \times (s - s_{a,j})^2 - \cdots, \tag{A11}$$

where

$$\tau_n^j \equiv \int_0^\infty t^n e^{-s_{a,j}t}\psi(t)dt. \tag{A12}$$

It is clear that

$$\begin{aligned} |\tau_n^j| &\le \int_0^\infty t^n e^{-r_{a,j}t}\psi(t)dt = \sum_{m=0}^{\infty} \frac{(-r_{a,j})^m}{m!}\,\tau_{n+m} \\ &\le \tau_n + \gamma^n \sum_{m=1}^{\infty} \frac{(n+m)!}{m!}(-r_{a,j}\gamma)^m \\ &= \tau_n + \gamma^n \frac{d^n}{dx^n}\frac{x^{n+1}}{1-x}\Big|_{x=-r_{a,j}\gamma}. \end{aligned} \tag{A13}$$

But

$$\frac{d^n}{dx^n}\frac{x^{n+1}}{(1-x)}\Big|_{x=-r_{a,j}\gamma} \ll (n+1)! \tag{A14}$$

This yields

$$|\tau_n^j| \le \tau_n + \Delta, \qquad \Delta \ll (n+1)!\gamma^n. \tag{A15}$$

From Eqs. (A15) and (A8) and the fact that $|(1 + \tau_1 a)^{-1}| \ge 1$, it follows that poles with real parts that fulfill Eq. (A10) can occur only for values of a in the ranges $0 \le -a \ll 1/\gamma$ and $0 \le a + 2/\tau_1 \ll 1/\gamma$. In addition, we will show that the complex poles with real parts $r_{a,j}$ satisfying Eq. (A10) occur only in the extreme case where $\psi(t)$ is a superposition of very sharp peaks. For such $s_{a,j}$, Eqs. (A8), (A11), and (A15) show for the real and imaginary parts of $\tilde{\psi}(i\Gamma_{a,j})$ that

$$0 \le 1 - \left|\int_0^\infty \cos\Gamma_{a,j}t\,\psi(t)dt\right| \ll 1 \tag{A16}$$

and

$$\left|\int_0^\infty \sin\Gamma_{a,j}t\,\psi(t)dt\right| \ll 1. \tag{A17}$$

Since $\psi(t)$ is normalized to unity, Eqs. (A16) and (A17) can hold for $0 \le -a \ll 1/\gamma$ only if $\psi(t)$ is appreciably different from zero only for

$$\begin{aligned} &|t - 2\pi n/\Gamma_{a,j}| \ll 2\pi/\Gamma_{a,j}, \\ &n = 0 \text{ or } n = 1 \text{ or } n = 2 \text{ or } \cdots, \end{aligned} \tag{A18}$$

with $\Gamma_{a,j}$ of order $1/\gamma$ or greater. For $0 \le a + 2/\tau_1 \ll 1$, Eqs. (A16) and (A17) can hold only if $\psi(t)$ is appreciably different from zero only for

$$\begin{aligned} &|t - 2\pi(n + 1/2)/\Gamma_{a,j}| \ll 2\pi/\Gamma_{a,j}, \\ &n = 0 \text{ or } n = 1 \text{ or } n = 2 \text{ or } \cdots, \end{aligned} \tag{A19}$$

with $\Gamma_{a,j}$ of order $1/\gamma$ or greater. It is possible that a density $\psi(t)$ belongs to both of the cases described above. An example of such a density is $\psi(t) = \delta(t - \Delta t)$. The values of $\psi(t)$ outside of these peaks must be so small that their contribution to $\int_0^\infty \psi(t)dt$ is $\ll 1$. This proves that complex poles satisfying Eq. (A10) can occur only if $\psi(t)$ is sharply peaked in the manner described above.

We now proceed to show that the poles corresponding to $0 \le -a \ll 1/\gamma$, which lie close to the imaginary s axis, Eq. (A10), are of first order and that the distances between them are of the order $2\pi/\gamma$. We first consider the real pole s_a. For $|s| \ll 1/\gamma$, Eq. (A1) becomes

$$\tilde{\psi}(s) = 1 - \tau_1 s + O(s^2\gamma^2). \tag{A20}$$

For τ_1 of order γ, the third term in Eq. (A20) is much smaller than the second term. Equation (A4) then yields

$$s_a = a[1 + O(\gamma a)], \tag{A21}$$

$$|s_a - a| \ll -a. \tag{A22}$$

From the way in which this solution was constructed, with τ_1 of order γ, it is clear that there are no other solutions of Eq. (A4) fulfilling Eq. (A22).

The poles due to a sharply peaked density $\psi(t)$ of the type described above which lie within a distance $\ll 1/\gamma$ of the imaginary s axis lie within a distance $\ll 1/\gamma$ of a finite subset of the poles of an appropriately chosen superposition of δ functions. To analyze the behavior of such sharply peaked densities, it is therefore sufficient to study distributions which are superpositions of δ functions. Consider, therefore,

$$\psi(t) = \sum_{n=0}^{\infty} c_n \, \delta(t - n\Delta t), \qquad c_n \geq 0, \tag{A23}$$

with Δt of order γ or less. Then

$$\tilde{\psi}(s) = \sum_{n=0}^{\infty} c_n \, e^{-ns\Delta t}. \tag{A24}$$

This is a periodic function:

$$\tilde{\psi}(s + 2\pi i/\Delta t) = \tilde{\psi}(s). \tag{A25}$$

Therefore, if s_a is the solution of Eq. (A4) on the real axis, then

$$s_{a,j} = s_a + 2\pi i j/\Delta t \tag{A26}$$

are also solutions of Eq. (A4). Since s_a is a first-order pole, so are the $s_{a,j}$. The distance between the poles is of the order $2\pi/\gamma$.

The results up to this point can be summarized as follows. It is possible to divide the probability densities $\psi(t)$ with finite γ and τ_1 of order γ into three classes:

(i) Densities which are superpositions of sharp peaks as specified in Eq. (A19). For such distributions we will consider only master operators **A** which have no eigenvalues in the range $0 \leq a + 2/\tau_1 \ll 1/\gamma$.

(ii) Densities which are superpositions of sharp peaks as specified in Eq. (A18) excluding those which are of the first class.

(iii) All other densities.

Densities in the third class produce only a single pole in the integrand of $\theta_a(t)$ which contributes for long times. This pole is real and of first order. Densities in the first and second classes produce, in addition, first order complex poles which lie sufficiently close to the imaginary s axis to contribute to $\theta_a(t)$ at long times for $0 \leq -a \ll 1/\gamma$. Densities in the first class produce additional poles in the range $0 \leq a + 2/\tau_1 \ll 1/\gamma$. Since in this class of densities we only consider master operators **A** which have no eignvalues in this range, these poles do not contribute to $P(\alpha, t)$. This corresponds to the exclusion of transition matrices **M** of the random walk with eigenvalues which cause oscillations in the random walk solution that persist at long times. These eigenvalues cause no persistent oscillations for distributions of the second and third classes and therefore need not be excluded for these cases.

We will first consider densities which are in the third class. In this case Eq. (4.4) can be written as

$$\theta_a(t) = -\tau_1 a(1 + \tau_1 a)^{-2} s_a^{-1} e^{s_a t} \times \left(\frac{d}{ds} \tilde{\psi}(s)\Big|_{s=s_a}\right)^{-1} + O(e^{-t/\gamma}) \tag{A27}$$

Via Eqs. (A20) and (A21), $\theta_a(t)$ becomes

$$\theta_a(t) = e^{at}[1 + O(\gamma a)] + O(e^{-t/\gamma}). \tag{A28}$$

The only values of a for which the first term is not of the same order as the second term are those for which at is of order 1 or less. Therefore,

$$\theta_a(t) = e^{at}[1 + O(\gamma/t)] + O(e^{-t/\gamma}). \tag{A29}$$

We will now show that this result is also valid for densities in the first and second classes. For these cases Eq. (4.4) can be written as

$$\begin{aligned}\theta_a(t) = &-\tau_1 a(1 + \tau_1 a)^{-2} s_a^{-1} e^{s_a t} \left(\frac{d}{ds} \tilde{\psi}(s)\Big|_{s=s_a}\right)^{-1} \\ &- 2\tau_1 a(1 + \tau_1 a)^{-2} \sum_j e^{r_{a,j} t} \\ &\times \left| s_{a,j} \frac{d}{ds} \tilde{\psi}(s)\Big|_{s=s_{a,j}}\right|^{-1} \cos[\Gamma_{a,j} t + \eta(s_{a,j})] \\ &+ O(e^{-t/\gamma}),\end{aligned} \tag{A30}$$

where the phase shifts are

$$\exp[i\eta(s_{a,j})] \equiv \left| s_{a,j} \frac{d\tilde{\psi}(s)}{ds}\Big|_{s=s_{a,j}} \right/ s_{a,j} \frac{d\tilde{\psi}(s)}{ds}\Big|_{s=s_{a,j}} \tag{A31}$$

Equation (A30) differs from Eq. (A27) only in the second term of Eq. (A30):

$$\xi_a(t) = -2\tau_1 a(1 + \tau_1 a)^{-2} \sum_j e^{r_{a,j} t} \times \left| s_{a,j} \frac{d}{ds} \tilde{\psi}(s)\Big|_{s=s_{a,j}}\right|^{-1} \cos[\Gamma_{a,j} t + \eta(s_{a,j})]. \tag{A32}$$

For a density $\psi(t)$ which is a superposition of δ functions, we obtain

$$\xi_a(t) = 2\tau_1 a[1 + O(\gamma a)]e^{at}\sum_{j=1}^{\infty}\frac{\Delta t}{2\pi j\tau_1}\sin\frac{2\Delta j t}{\Delta t}$$

$$= a\Delta t[1 + O(\gamma a)]e^{at}\left(\tfrac{1}{2} - \frac{t}{\Delta t} + \left[\frac{t}{\Delta t}\right]\right), \qquad \text{(A33)}$$

where $[t/\Delta t]$ is the integer part of $t/\Delta t$. Since Δt is of order γ or less,

$$\xi_a(t) = O(\gamma a)e^{at}. \qquad \text{(A34)}$$

If $\psi(t)$ is a superposition of peaks of nonzero width as given in Eq. (A18) or Eq. (A19), then the poles $s_{a,j}$ which lie within a distance $\ll 1/\gamma$ of the imaginary s axis lie within a distance $\ll 1/\gamma$ of a finite subset of the poles of the corresponding superposition of δ functions. Then

$$\xi_a(t) = 2\tau_1 a[1 + O(\gamma a)]e^{at}\sum_j e^{(r_{a,j}-s_a)t} \times \frac{\Delta t}{2\pi n_j\tau_1}\sin\frac{2\pi n_j t}{\Delta t}, \qquad \text{(A35)}$$

where the sum over j is over a finite number of poles and where n_j is the number of the pole of the corresponding superposition of δ functions. Since $r_{a,j} \leq s_a$ and the sum is a finite sum, this case also yields Eq. (A34). Substitution of Eq. (A34) into (A30) again yields (A28) for $\theta_a(t)$. Therefore, for both classes of distributions, Eq. (A29) holds.

Substitution of Eq. (A29) into (4.2) yields

$$P(\alpha, t) = Q(\alpha, t)\,[1 + O(\gamma/t)] + O(e^{-t/\gamma}) \qquad \text{(A36)}$$

for τ_1 of order γ, which proves theorem (4.7) for distributions $\psi(t)$ with finite γ.

For densities with infinite γ, the theorem is trivially true and is essentially empty. If γ is infinite, either $\tilde{\psi}(s)$ has an essential singularity at $s = 0$, or $s = 0$ is an accumulation point of singularities of $\tilde{\psi}(s)$. In both cases, the integrand of $\theta_a(t)$ will contribute in every open neighborhood of $s = 0$. Therefore, $\theta_a(t)$ can never approach an exponential at any time. For such distributions, $P(\alpha, t)$ and $Q(\alpha, t)$ will never approach each other even after long times, and will only be equal at equilibrium.

We will now prove the extended version of theorem (4.7). For this case the solution of the random walk equation can be written as

$$P(\alpha; t) = \sum_b \theta_b(t)\, h_b(\alpha)\, P(b; 0). \qquad \text{(A37)}$$

A consequence of the conditions (4.15)–(4.17) is that Eq. (A37) can be written as

$$P(\alpha; t) = \sum_a{}' \theta_b(t)\, h_b(\alpha)\, P(b; 0) + O(e^{-t/\gamma}), \qquad \text{(A38)}$$

where the prime on the sum indicates that the sum is taken only over those values of a that obey condition (4.15), and a and b are related by the one-to-one correspondence between the eigenfunctions f_a and h_b, Eqs. (4.16) and (4.17). A consequence of conditions (4.16) and (4.17) is that

$$\theta_b(t)\, h_b(\alpha)\, P(b; 0) = \theta_a(t) f_a(\alpha)\, P(a; 0)\,[1 + O(\gamma a)], \qquad \text{(A39)}$$

which immediately yields the desired result.

Note added in proof: A quantity that is often of physical interest is the mean time for a walker to reach state α for the first time. Let $R(\alpha, n \mid \alpha', 0)$ be the probability that a walker starting at state α' reaches state α for the first time on the nth step. Then

$$P(\alpha, n \mid \alpha', 0) = \sum_{n'=1}^{n} R(\alpha, n' \mid \alpha', 0) P(\alpha, n - n' \mid \alpha, 0).$$

The generating function

$$R_{\alpha\alpha'}(z) \equiv \sum_{n=1}^{\infty} z^n R(\alpha, n \mid \alpha', 0)$$

is then related to $G_{\alpha\alpha'}(z)$ by $R_{\alpha\alpha'}(z) = (G_{\alpha\alpha'}(z) - \delta_{\alpha\alpha'})/G_{\alpha\alpha}(z)$. The mean first passage time to state α is

$$\tau_{\alpha\alpha'} = \int_0^{\infty} dt \sum_{n=1}^{\infty} t\psi_n(t) R(\alpha, n \mid \alpha', 0)$$

$$= \tau_1 \frac{d}{dz} R_{\alpha\alpha'}(z)\Big|_{z=1} = \tau_1 \langle n \rangle_{\alpha\alpha'}$$

where $\langle n \rangle_{\alpha\alpha'} = (d/dz) R_{\alpha\alpha'}(z)|_{z=1}$ is the mean number of steps to reach state α for the first time. The mean first passage time is thus the same for all random walks for which τ_1 is the same, including the master equation. Higher moments of the first passage time can be similarly calculated. The nth moment will, in general, be a function of $\tau_1, \tau_2, \ldots, \tau_m$.

* Supported in part by the National Science Foundation, under Grant No. GP10536 and by the Advanced Research Projects Agency of the Department of Defense, monitored by the U.S. Office of Naval Research, under Contract No. N00014-69-A-0200-6018.

1 E.g., W. Feller, *An Introduction to Probability Theory and Its Applications* (Wiley, New York, 1966, 1968), 3rd ed., Vol. I, and 1st ed., Vol. II; F. Spitzer, in *Principles of Random Walks* (Van Nostrand, Princeton, N.J., 1964); *Selected Papers on Noise and Stochastic Processes*, edited by N. Wax (Dover, New York, 1954); I. Oppenheim, K. E. Shuler, and G. H. Weiss, Advan. Mol. Relaxation Process. **1**, 13 (1967).
2 E. W. Montroll and G. H. Weiss, J. Math. Phys. **6**, 167 (1965).
3 E. W. Montroll, Proc. Symp. Appl. Math. **16**, 193 (1964).
4 W. Feller, Ref. 1., Vol. I, p. 447, and Vol. II, p. 58.
5 For a discussion of such solutions and the associated eigenvalue spectrum see, e.g., J. G. Kemeny and J. L. Snell, *Finite Markov Chains* (Van Nostrand, Princeton, N.J., 1960) Chap. 5; E. W. Montroll, *Energetics in Metallurgical Phenomena* (Gordon and Breach, New York, 1967), Vol. 3, p. 139 et seq.
6 A. Kolmogoroff, Math. Ann. **104**, 451 (1931); **108**, 149 (1933).
7 K. Lakatos-Lindenberg, R. P. Hemenger, and R. M. Pearlstein (unpublished).
8 L. V. Ahlfors, *Complex Analysis* (McGraw-Hill New York, 1953), p. 141.
9 D. V. Widder, *The Laplace Transform* (Princeton U.P., Princeton, N.J., 1946), p. 36.
10 R. V. Churchill, *Complex Variables and Applications* (McGraw-Hill, New York, 1960), p. 265.

3) The Relation Between the Fokker-Planck and Master Equation

The master equation represents one linear limiting form of the nonlinear SCK equation. Even though the master equation is linear, its solution can still present formidable mathematical difficulties. A second type of linear approximation to the SCK equation is the Fokker-Planck equation which is either derived directly from the SCK equation, or can be obtained as an approximation to the master equation. An important property of the solution to the Fokker-Planck equation is that it is non-negative definite, i.e., is never negative. Paper 1 by Keilson and Storer discusses a problem for which one can formulate a master equation and solve it exactly. The authors then derive a Fokker-Planck equation for the same problem and examine its adequacy in light of the exact solution. It is shown that there are delta function solutions of the master equation that cannot be obtained from the Fokker-Planck equation, implying the possibility of qualitative differences between the results of these two approaches. The second paper by van Kampen deals with a systematic perturbation approach to the derivation of the Fokker-Planck and higher approximations. The results are expressed as the solution to partial differential equations of higher order than the second order Fokker-Planck equation. Kubo *et al*[4.2.1] have presented some additional heuristic arguments for the validity of van Kampen's procedure. The third paper, by Pawula, gives the mathematically rigorous result that it is impossible to find a partial differential equation other than the Fokker-Planck equation, that approximates to the master equation and leads to a non-negative definite solution. This conclusion may not invalidate practical results derived by a perturbation method such as van Kampen's because the region of negativity of the solution so derived may be physically insignificant. In the fourth paper, the conditions under which master equations can be adequately approximated by a set of partial differential equations (Fokker-Planck equations) with boundary conditions are described.

On Brownian Motion, Boltzmann's Equation, and the Fokker-Planck Equation

by

Julian Keilson and James E. Storer

Cruft Laboratory, Harvard University

Cambridge, Massachusetts

Abstract

In order to describe Brownian motion rigorously, Boltzmann's integral equation must be used. The Fokker-Planck type of equation is only an approximation to the Boltzmann equation and its domain of validity is worth examining.

A treatment of the Brownian motion in velocity space of a particle with known initial velocity based on Boltzmann's integral equation is given. The integral equation, which employs a suitable scattering kernel, is solved and its solution compared with that of the corresponding Fokker-Planck equation. It is seen that when M/m, the mass ratio of the particles involved, is sufficiently high and the dispersion of the velocity distribution sufficiently great, the Fokker-Planck equation is an excellent description. Even when the dispersion is small, the first and second moments of the Fokker-Planck solution are reliable. The higher moments, however, are then in considerable error--an error which becomes negligible as the dispersion increases.

- - - - - - -

I

In the treatment of Brownian motion, it is customary to assume a Langevin equation and simple dynamical statistics of the individual collisions and then to deduce a Fokker-Planck equation describing the random motion of the heavy particle. The Fokker-Planck equation obtained is a second-order partial differential equation and the absence of higher-order differential terms is inferred directly from the above assumptions. As will be seen, the solution of this Fokker-Planck equation does not provide a completely satisfactory physical description. Consequently, the assumptions underlying the equation cannot be correct[1,2,3,4] and the extent of their approximate validity comes under question.

That the solution of the Fokker-Planck equation is not a wholly satisfactory representation of Brownian motion may be seen in the following way. Consider a heavy particle known to have the velocity v_o at $t = 0$. For all subsequent time, there is a finite probability that the particle will have undergone no collision. It must, therefore, be expected that the probability density $w(v,t)$* describing the stochastic motion in velocity space will always have a singular component of the form $f(t)\delta(v-v_o)$, where $\delta(v-v_o)$ is the Dirac delta-function. If one were to try to describe the motion by the Fokker-Planck equation

$$\frac{\partial w(v,t)}{\partial t} = \frac{1}{2} D\nabla^2 w(v,t) + \eta\nabla\cdot\left\{vw(v,t)\right\} \quad , \tag{1}$$

subject to the initial condition

$$w(v,0) = \delta(v-v_o) \quad , \tag{2}$$

no such singular component would be available in the solution. The immediate disappearance of an initial singularity is, indeed, characteristic of all partial differential equations of finite order. Only by means of an integral equation can such a singularity be maintained.

All of this, of course, is in keeping with the fact that fundamental to the description of Brownian motion is Boltzmann's equation, an integral equation of the type desired.[4] If $A(v',v)dv$ is the probability per unit time that a particle with velocity v' will undergo a transition to a volume dv about v, the Boltzmann equation describing the motion is

- - - - - - -

* The function $w(v,0)$ obeying (2) is often represented in the literature by $P_2(v_o/v;t)$, the probability density for velocity v, t seconds after there is a known velocity v_o.

$$\frac{\partial w(v,t)}{\partial t} = \int w(v',t)A(v',v)dv' - w(v,t)\int A(v,v')dv' \qquad (3)$$

This is simply an expression of the fact that the rate of change of the population of a cell in velocity space is the difference between the rate of departures from the cell and the rate of arrivals.

From this Boltzmann equation a corresponding Fokker-Planck equation may be derived. If Eq. (3) is multiplied by an arbitrary, but suitably behaved function R(v), and integrated over v,

$$\begin{aligned}\int R(v)\frac{\partial w(v,t)}{\partial t}dv &= \iint R(v)w(v',t)A(v',v)dv'dv \\ &\quad - \iint R(v)w(v,t)A(v,v')dvdv' \\ &= \iint\left\{\sum_{0}^{\infty}\frac{(v-v')^{(n)}}{n!}\circ\nabla'^{(n)}R(v')\right\}w(v',t)A(v',v)dvdv' \\ &\quad - \iint R(v)w(v,t)A(v,v')dvdv' \\ &= \iint\left\{\sum_{1}^{\infty}\frac{(v-v')^{(n)}}{n!}\circ\nabla'^{(n)}R(v')\right\}w(v',t)A(v',v)dv'dv.\end{aligned} \qquad (4)$$

Here $(v-v')^{(n)}\circ\nabla'^{(n)}$ is to be understood as the dot product of two n^{th}-rank tensors. Integrating by parts, one has

$$\int R(v')\frac{\partial w(v',t)}{\partial t}dv'$$

$$= \iint R(v')\sum_{0}^{\infty}\frac{1}{n!}\nabla'^{(n)}\circ\left\{(v'-v)^{(n)}A(v',v)W(v',t)\right\}dv'dv. \qquad (5)$$

Since R(v') is an arbitrary function, the associated coefficients may be equated to yield

$$\frac{\partial w(v,t)}{\partial t} = \sum_{1}^{\infty} \frac{1}{n!} \nabla^{(n)} \circ \left\{ A_n(v) w(v,t) \right\} , \qquad (6)$$

where $A_n(v)$ is the tensor

$$A_n(v) = \int (v-v')^{(n)} A(v,v') dv' . \qquad (7)$$

Equations (3) and (6) are equivalent and provide an exact description of Brownian motion.

This treatment may be readily generalized to include Brownian motion in coordinate and velocity space.

II

In those treatments of Brownian motion based on Langevin's equation, moments higher than the second are found to vanish, and the Fokker-Planck equation (3) is obtained. As already noted, such an equation is certainly unsatisfactory when the dispersion is small. It would, therefore, be desirable to try to treat the Boltzmann equation directly. Plainly an exact kernel $A(v,v')$ is unavailable and its use is almost certainly not feasible. However, it is possible to introduce a kernel which provides a reasonably accurate description of the microscopic scattering process and which is, at the same time, amenable to treatment. Such a kernel is of the form $A(v,v') = \mathcal{A}(v'-\gamma v)$, where γ is a dynamical damping parameter close in value to, but less than, one. Some justification for this form may be found along the following lines:

Let $B(v,v')dv'$ be the probability per unit time of a particle with initial velocity v making a transition to a volume element dv' about v', when all the particles with which the heavy particle collides are stationary. If the lighter particles have an equilibrium distribution $w(v'')$, then

$$A(v,v') = \int B(v-v'',v'-v'')w(v'')dv'' . \tag{8}$$

Since the particle under observation is very much heavier than the particles with which it collides, B(v,v') is a highly localized function of v', centered roughly about γv where again γ is very close to but less than unity. If B(v,v') is assumed to have the form $B(v'-\gamma v)$, then

$$A(v,v') = \int B[v'-v''-\gamma(v-v'')]w(v'')dv'' = \int B[v'-\gamma v-(1-\gamma)v'']w(v'')dv'' ,$$

so that this will have the form of $\mathcal{A}(v'-\gamma v)$.

Note that the form of $\mathcal{A}(v'-\gamma v)$ implies that the mean free time τ of a heavy particle is independent of its velocity, since

$$\frac{1}{\tau(v)} = \int A(v,v')dv' = \int \mathcal{A}(v'-\gamma v)dv' = \int \mathcal{A}(v'')dv'' , \tag{9}$$

a constant. This behavior is proper to Brownian motion where the heavy particle moves so slowly compared to the lighter particles that the mean relative velocity of the heavy particle does not vary significantly.

It would appear offhand that the functional form of $\mathcal{A}(v)$ could be chosen arbitrarily. However, this is not the case since $\mathcal{A}(v,v')$ must satisfy the equilibrium condition:

$$\underline{w}(v')A(v',v) = \underline{w}(v)A(v,v') \tag{10}$$

where $\underline{w}(v)$ is the equilibrium distribution of the heavy particles which the particle will ultimately assume. If it is also demanded that $\underline{w}(v)$ depend only on $|v|$, the two restrictions imply that $\mathcal{A}(v)$ must have the form $\mathcal{A}_0 e^{-\beta v^2}$ and that $\underline{w}(v)$ must have the corresponding form $\underline{w_0} e^{-\beta(1-\gamma^2)v^2}$, where $\mathcal{A}_0$ and $\underline{w_0}$ are constants (see Appendix 1). That the Gaussian character of the equilibrium distribution follows from the form of $\mathcal{A}(v-\gamma v')$ is reassuring.

Thus, the Boltzmann equation to be solved is

$$\frac{\partial w(v,t)}{\partial t} = a_0 \int w(v',t)e^{-\beta(v-\gamma v')^2}dv' - \frac{1}{\tau} w(v,t) , \qquad (11)$$

where

$$\frac{1}{\tau} = \int a(v'-\gamma v)dv' = a_0 \int e^{-\beta(v'-\gamma v)^2}dv' = a_0 \int e^{-\beta v''^2}dv'' = a_0(\frac{\pi}{\beta})^{3/2} . \qquad (12)$$

Before discussing the solution of this equation, it is worth while to put down the corresponding Fokker-Planck equation. The first moment will be given by

$$A_1 = a_0 \int (v-v')e^{-\beta(v'-\gamma v)^2}dv' = a_0 \int [(\gamma v-v')+(1-\gamma)v]e^{-\beta(v'-\gamma v)^2}dv'$$

$$= a_0(1-\gamma)v \int e^{-\beta(v'-\gamma v)^2}dv'$$

$$= a_0(\frac{\pi}{\beta})^{3/2}(1-\gamma)v = (\frac{1-\gamma}{\tau})v . \qquad (13)$$

Similarly for the second moment,

$$A_2 = a_0 \int (v-v')(v-v')e^{-\beta(v'-\gamma v)^2}dv'$$

$$= a_0 \int [(v'-\gamma v)(v'-\gamma v)+(1-\gamma)^2 vv]e^{-\beta(v'-\gamma v)^2}dv'$$

$$= \frac{a_0}{2} \pi^{3/2} \beta^{-5/2} \varepsilon + (1-\gamma)^2 vv \; a_0(\frac{\pi}{\beta})^{3/2}$$

$$= \frac{1}{2\beta\tau} \varepsilon + \frac{(1-\gamma)^2 vv}{\beta\tau} , \qquad (14)$$

where ε is the unit tensor.

If the latter part of A_2 is ignored since $(1-\gamma)^2$ is small and if the higher moments (whose effect will be small for $t \gg \tau$) are ignored, Eq. (1) is regained where now

$$D = \frac{1}{2\beta\tau} \quad \text{and} \quad \eta = \frac{1-\gamma}{\tau} \quad . \tag{15}$$

As is seen in Appendix 2, the solution of the Boltzmann equation (11) subject to condition (2) is given by

$$w_B(v,t) = e^{-t/\tau}\left[\delta(v-v_o)+\sum_{1}^{\infty}\frac{1}{n!}\left(\frac{t}{\tau}\right)^n\left(\frac{\beta}{\pi\Delta_n}\right)^{3/2} e^{-(\beta/\Delta_n)(v-\gamma^n v_o)^2}\right], \tag{16}$$

where

$$\Delta_n = \frac{1-\gamma^{2n}}{1-\gamma^2} \quad . \tag{17}$$

The solution of the Fokker-Planck equation (1) subject to condition (2) may be taken directly from Wang and Uhlenbeck[1] and is given by

$$w_{FP}(v,t) = \left[\frac{\eta}{\pi D(1-e^{-2\eta t})}\right]^{3/2} e^{\frac{-\eta(v-v_o e^{-\eta t})^2}{D(1-e^{-2\eta t})}} \quad . \tag{18}$$

Note that the singularity $\delta(v-v_o)$ is preserved in the solution of the integral equation but is not in the solution of the Fokker-Planck equation. For $t \gg \tau$, however, the delta-function ceases to play a prominent role.

From Eqs. (16), (17) one finds that the equilibrium distribution for the solution of the Boltzmann equation is given by

$$\underline{w}_B(v) = \lim_{t\to\infty} w_B(v,t) = \left[\frac{\beta(1-\gamma^2)}{\pi}\right]^{3/2} e^{-\beta(1-\gamma^2)v^2} \quad . \tag{19}$$

For the Fokker-Planck equation,

$$\underline{w}_{FP}(v) = \left(\frac{\eta}{\pi D}\right)^{3/2} e^{-(\eta/D)v^2} \quad . \tag{20}$$

Inserting the values of η, D from (15), this becomes:

$$\underline{w}_{FP}(v) = \left(\frac{2(1-\gamma)\beta}{\pi}\right)^{3/2} e^{-2\beta(1-\gamma)v^2} \quad . \tag{21}$$

Plainly if γ is sufficiently close to one, then

$$(1-\gamma^2) = (1-\gamma)(1+\gamma) \simeq 2(1-\gamma)$$

and the two equilibrium distributions are identical.

It is also of interest to compare the manner in which the average velocity and the variance vary in time. These quantities are defined by

$$\bar{v}(t) = \int v w(v,t) dv$$

and

$$\sigma^2(t) = \int (v^2 - \bar{v}^2) w(v,t) dv.$$

$\bar{v}_B(t)$, $\bar{v}_{FP}(t)$, $\sigma_B^2(t)$, and $\sigma_{FP}^2(t)$ may be computed directly from their respective equations of motion. Thus, if Eq. (1) is multiplied on both sides by v and integrated over v, then

$$\frac{d\bar{v}}{dt} = -\eta\bar{v}, \text{ so that } \bar{v}_{FP}(t) = v_o e^{-\eta t} = v_o e^{-(\frac{1-\gamma}{\tau})t} \quad . \tag{22}$$

Similarly, multiplying by v^2 and integrating, it is found that

$$\frac{d\overline{v^2}}{dt} = 3D - 2\eta\overline{v^2} \quad , \tag{23}$$

which yields in turn

$$\frac{d\sigma^2}{dt} = \frac{d}{dt}(\overline{v^2} - \bar{v}^2) = 3D - 2\eta\sigma^2 \quad ; \tag{24}$$

and so

$$\sigma_{FP}^2(t) = \frac{3D}{2\eta}(1-e^{-2\eta t}) = \frac{3}{4\beta(1-\gamma)}\left(1-e^{-\frac{2}{\tau}(1-\gamma)t}\right). \tag{25}$$

The same procedure may be applied to the integral equation (11) to give

$$\bar{v}_B(t) = v_o e^{-(\frac{1-\gamma}{\tau})t} \qquad (26)$$

and

$$\overline{v_B^2}(t) = \frac{3}{2\beta(1-\gamma^2)}\left[1-e^{-(\frac{1-\gamma^2}{\tau})t}\right] + v_o^2 e^{-(\frac{1-\gamma^2}{\tau})t} . \qquad (27)$$

Correspondingly, one finds that

$$\sigma_B^2(t) = \frac{3}{2\beta(1-\gamma^2)}\left[1-e^{-\frac{(1-\gamma^2)t}{\tau}}\right] + v_o^2\left[e^{-(\frac{1-\gamma^2}{\tau})t} - e^{-2(\frac{1-\gamma}{\tau})t}\right] . \qquad (28)$$

These same results could also have been obtained from the solutions (16) and (18), but the computations are more tedious.

It is seen that $\bar{v}_B(t)$ and $\bar{v}_{FP}(t)$ are identical and that $\sigma_B^2(t)$ and $\sigma_{FP}^2(t)$ are nearly identical. Indeed, if the smaller term in A_2 had not been ignored in obtaining the corresponding Fokker-Planck approximation, $\sigma_B^2(t)$ and $\sigma_{FP}^2(t)$ would have been precisely the same. For consider the Boltzmann equation in its differential form:

$$\frac{\partial w}{\partial t} = \sum_{1}^{\infty} \nabla^{(n)} \circ (A_n w) .$$

The above procedure yields

$$\frac{d\overline{v^2}}{dt} = \int v^2\, \nabla^{(1)} \circ (A_1 w)dv + \int v^2\, \nabla^{(2)} \circ (A_2 w)dv ,$$

since all integrals involving higher moments vanish when integration by parts is carried out. Moreover, from the choice of $A(v,v')$, the two integrals are simple functions of $\overline{v^2}$ and the above differential equation does determine $\overline{v^2}(t)$. The same procedure applied to the Fokker-Planck equation can only yield the same result, because all contributing terms are present.

It is seen then that the validity of the Fokker-Planck approximation is excellent when γ is sufficiently close to one. For the ordinary domain of Brownian motion this will certainly be the case. For the elastic collision of hard spheres, for example, it is easily found that

$$\overline{\delta v} = \frac{-4}{3}\frac{m}{M+m}v \quad ,$$

where $\overline{\delta v}$ is the mean change in velocity suffered by a particle of mass M and velocity v in a single collision with particles of mass m. Then

$$\frac{d\bar{v}}{dt} \simeq \frac{\overline{\delta v}}{\tau} = \frac{-4}{3}\frac{m}{(M+m)\tau}v \quad ,$$

so that, from Eqs.(15) and (22)

$$\eta = \frac{(1-\gamma)}{\tau} \cong \frac{4}{3}\frac{m}{(M+m)\tau} \quad .$$

$1-\gamma$ then is given by $\frac{4}{3}\frac{m}{M+m}$ and for typical Brownian motion will be extremely small.

If it were possible to treat the exact kernel $A(v,v')$, one would still expect to find excellent agreement between the Fokker-Planck and Boltzmann solutions for $t \gg \tau$. Even when $t \sim \tau$, the first and second moments of the Fokker-Planck equation should be reliable. But for $t \sim \tau$, higher-order moments would be in considerable error. However, for $t \gg \tau$, these errors will become entirely negligible.

References

1. M. C. Wang and G. E. Uhlenbeck,"On the Theory of the Brownian Motion II,"Rev. Mod. Phys. 17, 323 (1945).
2. S. Chandrasekhar, Rev. Mod. Phys. 15, 1 (1943).
3. Lawson and Uhlenbeck, Threshold Signals (MIT) Radiation Laboratory Series, Vol. 24, McGraw-Hill (1950), Chap. III.
4. J. Keilson, The Statistical Nature of Inverse Brownian Motion in Velocity Space, Technical Report No. 127, Cruft Laboratory, Harvard University, May 10, 1951.

Appendix 1

The Restrictions on $\mathcal{Q}(v)$ and $\underline{w}(v)$ Imposed by Equilibrium

Denoting the rectangular components of v by v_i, $i = 1,2,3$, and letting

$$\psi(v_1,v_2,v_3) = \ln \mathcal{Q}(v) ,$$

the equilibrium relation (10) can be written in the form

$$\ln \underline{w}(v')+\psi(v_1-\gamma v_1',v_2-\gamma v_2',v_3-\gamma v_3') = \ln \underline{w}(v)+\psi(v_1'-\gamma v_1,v_2'-\gamma v_2,v_3'-\gamma v_3). \tag{1-1}$$

Taking the partial derivative of Eq. (1-1) with respect to v_i, v_j', and noting that

$$\frac{\partial^2}{\partial v_i \partial v_j'} \ln \underline{w}(v') = \frac{\partial^2}{\partial v_i \partial v_j'} \ln \underline{w}(v) = 0 ,$$

it is seen that

$$-\gamma\psi_{ij}(v_1-\gamma v_1',v_2-\gamma v_2',v_3-\gamma v_3') = -\gamma\psi_{ij}(v_1'-\gamma v_1,v_2'-\gamma v_2,v_3'-\gamma v_3), \tag{1-2}$$

where

$$\psi_{ij} = \frac{\partial^2}{\partial v_i \partial v_j} \psi(v_1,v_2,v_3) .$$

Setting $v_i' = \gamma v_i$, $\bar{v}_i = (1-\gamma^2)v_i$ in (1-2), it is further observed that

$$\psi_{ij}(\bar{v}_1,\bar{v}_2,\bar{v}_3) = \psi_{ij}(0,0,0) = -\beta_{ij} ,$$

where β_{ij} is a constant. Hence $\psi(v_1,v_2,v_3)$ must be of the form

$$\psi(v_1,v_2,v_3) = -\sum_{i=1}^{3} \sum_{j=1}^{3} \beta_{ij}v_iv_j + \sum_{1}^{3} \alpha_i v_i + \Delta ,$$

where α_i and Δ are constants. Inserting this result into (1-1), one finds that

$$\ell n\ \underline{w}(v) = -\sum_{j=1}^{3}\sum_{j=1}^{3}\beta_{ij}(1-\gamma^2)v_iv_j + \sum_{1}^{3}\alpha_i(1+\gamma)v_i + \Delta'. \tag{1-3}$$

But, since the distribution of small particles is assumed to be isotropic, one has

$$\underline{w}(v) = \underline{w}(|v|) = \underline{w}\left(\sqrt{v_1^2+v_2^2+v_3^2}\right). \tag{1-4}$$

The only possible way (1-3) can satisfy this condition is for

$$\beta_{ij} = \beta\delta_{ij}, \qquad \alpha_i = 0\ . \tag{1-5}$$

Hence

$$\mathcal{Q}(v) = \mathcal{Q}_0 e^{-\beta v^2}, \qquad \underline{w}(v) = \underline{w}_0 e^{-\beta(1-\gamma^2)v^2}\ ,$$

where $\mathcal{Q}_0$ and $\underline{w}_0$ are constants.

Appendix 2

Solution of the Boltzmann Equation

It is desired to solve Eq. (11) subject to the condition (2). Two methods will be given.

One procedure is to introduce the Fourier transform of $w(v,t)$, i.e.,

$$T_w(k,t) = \int e^{ik\cdot v}\, w(v,t)dv$$

with

$$T_w(k,0) = \int e^{ik\cdot v}\, \delta(v-v_0)dv = e^{ik\cdot v_0} .$$

Taking the Fourier transform of Eq. (11) one obtains the following equation for $T_w(k,t)$:

$$\frac{\partial}{\partial t} T_w(k,t) = A(k)\, T_w(\gamma k,t) - \frac{1}{\tau} T_w(k,t) , \qquad (2\text{-}1)$$

where

$$A(k) = \int e^{ik\cdot v}\, \mathcal{a}(v)dv = \mathcal{a}_0 \int e^{ik\cdot v-\beta v^2}\, dv$$

$$= \mathcal{a}_0 \left(\frac{\pi}{\beta}\right)^{3/2} e^{-k^2/4\beta} = \frac{1}{\tau}\, e^{-k^2/4\beta} .$$

It is now convenient to introduce the Laplace transform

$$L_w(k,s) = \int_0^\infty e^{-st}\, T_w(k,t)dt ;$$

then

$$\int_0^\infty e^{-st} \frac{\partial}{\partial t} T_w(k,t)dt = -T_w(k,0) + s\, L_w(k,s) = -e^{ik\cdot v_0} + s\, L_w(k,s).$$

Thus, taking the Laplace transform of (2-1), one obtains the equation for $L_W(k,s)$,

$$-e^{ik\cdot v_o} + s\, L_W(k,s) = \frac{1}{\tau}\, e^{-k^2/4\beta}\, L_W(\gamma k,s) - \frac{1}{\tau}\, L_W(k,s) .$$

This may be rearranged to give

$$L_W(k,s) = \frac{1}{s+\frac{1}{\tau}}\, e^{ik\cdot v_o} + \frac{1}{\tau}\,\frac{1}{s+\frac{1}{\tau}}\, e^{-k^2/4\beta}\, L_W(\gamma k,s) . \quad (2\text{-}2)$$

The finite difference equation (2-2) may be solved by the following procedure: Replace k by γk. This yields

$$L_W(\gamma k,s) = + \frac{1}{s+\frac{1}{\tau}}\, e^{i\gamma k\cdot v_o} + \frac{1}{\tau}\,\frac{1}{s+\frac{1}{\tau}}\, e^{-\gamma^2 k^2/4\beta}\, L_W(\gamma^2 k,s) . \quad (2\text{-}3)$$

Equation (2-3) may be used to eliminate $L_W(\gamma k,s)$ from (2-2), yielding

$$L_W(k,s) = \frac{1}{s+\frac{1}{\tau}}\, e^{ik\cdot v_o} + \frac{1}{\tau}\,\frac{e^{i\gamma k\cdot v_o - k^2/4\beta}}{(s+\frac{1}{\tau})^2} + \frac{1}{\tau}\,\frac{e^{-\gamma^2 k^2/4\beta}}{s+\frac{1}{\tau}}\, L_W(\gamma^2 k,s). \quad (2\text{-}4)$$

Replacing k by $\gamma^2 k$ in (2-2), the resulting equation may be used to eliminate $L_W(\gamma^2 k,s)$ from (2-4). Continuing in this fashion yields the solution

$$L_W(k,s) = \sum_{n=0}^{\infty} \frac{(-1)^n}{\tau^n}\, \frac{e^{i\gamma^n k\cdot v_o - (k^2/4\beta)\Delta_n}}{(s+\frac{1}{\tau})^{n+1}} ,$$

where

$$\Delta_n = \frac{1-\gamma^{2n}}{1-\gamma^2} .$$

It is to be noted that the series is absolutely convergent.

$T_W(k,t)$ may readily be obtained from $L_W(k,s)$ by taking the inverse Laplace transform. Using a Bromwich contour it is apparent that

$$T_w(k,t) = \frac{1}{2\pi i}\int_{c-i\infty}^{c+i\infty} e^{st}\, L_w(k,s)ds$$

$$= \frac{1}{2\pi i}\int_{Br} e^{st}\, L_w(k,s)ds$$

$$= \text{Residue at } s = -\frac{1}{\tau} \text{ of } L_w(k,s)$$

$$= \sum_0^\infty (-1)^n \frac{e^{i\gamma^n k\cdot v_o - (k^2/4\beta)\Delta_n}}{\tau^n}$$

$$\times \left[\text{Residue at } s = -\frac{1}{\tau} \text{ of } \frac{e^{st}}{(s+\frac{1}{\tau})^{n+1}}\right]$$

$$= \sum_0^\infty (-1)^n \frac{e^{i\gamma^n k\cdot v_o - (k^2/4\beta)\Delta_n}}{\tau^n}\,\frac{(-t)^n}{n!}\, e^{-t/\tau}$$

$$= \left\{\sum_0^\infty \frac{1}{n!}\left(\frac{t}{\tau}\right)^n e^{i\gamma^n k\cdot v_o - (k^2/4\beta)\Delta_n}\right\} e^{-t/\tau} \quad . \qquad (2\text{-}5)$$

The inverse Fourier transform may now be performed and this yields

$$w(v,t) = \frac{1}{(2\pi)^3}\int e^{-ik\cdot v}\, T_w(k,t)\, dk$$

$$= \frac{e^{-t/\tau}}{(2\pi)^3}\sum_0^\infty \frac{1}{n!}\left(\frac{t}{\tau}\right)^n \int e^{i(\gamma^n v_o - v)\cdot k - \frac{k^2}{4\beta}\Delta_n}\, dk$$

$$= e^{-t/\tau}\left\{\delta(v-v_o) + \sum_1^\infty \frac{1}{n!}\left(\frac{t}{\tau}\right)^n \left(\frac{\beta}{\pi\Delta_n}\right)^{3/2} e^{-\frac{\beta}{\Delta_n}(v-\gamma^n v_o)^2}\right\}. \qquad (2\text{-}6)$$

This solution may be verified by substitution.

Equation (11) may also be solved in the following way.[4] Consider the sequence of equations,

$$\frac{\partial w_o(v,t)}{\partial t} = \frac{-w_o(v,t)}{\tau}$$

$$\frac{\partial w_1(v,t)}{\partial t} = \frac{-w_1(v,t)}{\tau} + \int w_o(v',t)\ \mathcal{Q}(v-\gamma v')dv' \qquad (2\text{-}7)$$

- - - - - - - -

$$\frac{\partial w_n(v,t)}{\partial t} = \frac{-w_n(v,t)}{\tau} + \int w_{n-1}(v',t)\ \mathcal{Q}(v-\gamma v')dv', \text{ etc.},$$

subject to the initial conditions

$$\begin{cases} w_o(v,0) = \delta(v-v_o) \\ w_i(v,0) = 0, \qquad i \neq 0\ . \end{cases} \qquad (2\text{-}8)$$

Plainly,

$$w(v,t) = \sum_{o}^{\infty} w_n(v,t) \qquad (2\text{-}9)$$

satisfies Eq. (11) and the condition $w(v,0) = \delta(v-v_o)$.

Then

$$w_o(v,t) = \delta(v-v_o)e^{-t/\tau}$$

and

$$w_n(v,t) = e^{-t/\tau} \int_o^t e^{s/\tau} \int w_{n-1}(v'',s)\ \mathcal{Q}(v-\gamma v'')dv''ds \qquad (2\text{-}10)$$

satisfy the equations and one need only evaluate the sequence of functions, $w_n(v,t)$. It is seen from this last equation that if $w_{n-1}(v,t)$ is a product of a function of v and a function of t, $w_n(v,t)$ is also such a product. Since w_o has such a form, all our w_n decompose in this way. Assume $w_n(v,t) = U_n(v)g_n(t)$. Then

$$g_n(t) = e^{-t/\tau} \int_0^t e^{s/\tau} g_{n-1}(s)ds$$

and

$$U_n(v) = \mathcal{a}_0 \int U_{n-1}(v')e^{-\beta(v-\gamma v')^2} dv' . \qquad (2\text{-}11)$$

It is now easy to see that

$$g_n(t) = \frac{t^n}{n!} e^{-t/\tau} . \qquad (2\text{-}12)$$

$U_n(v)$ has the form $\alpha_n e^{-\beta_n(v-\vec{\delta}_n)^2}$, and α_n, β_n, δ_n are connected by recursion relations derived from Eq. (2-11), stating

$$\alpha_{n+1} e^{-\beta_{n+1}(v-\delta_{n+1})^2} = \mathcal{a}_0 \int \alpha_n e^{-\beta_n(v'-\delta_n)^2} e^{-\beta(v-\gamma v')^2} dv'.$$

This gives

$$\alpha_{n+1} = \left[\frac{\pi}{\beta_n+\gamma^2\beta}\right]^{3/2} \mathcal{a}_0\alpha_n \qquad \text{with } \alpha_1 = \mathcal{a}_0$$

$$\beta_{n+1} = \frac{\beta\beta_n}{\beta_n+\gamma^2\beta} \qquad \beta_1 = B$$

$$\delta_{n+1} = \gamma\delta_n . \qquad \delta_1 = \gamma v_o \qquad (2\text{-}13)$$

$$\therefore \delta_n = \gamma^n v_o$$

$$\beta_n = \frac{\beta}{\Delta_n} \quad \text{where } \Delta_n = \frac{\gamma^{2n}-1}{\gamma^2-1} \qquad (2\text{-}14)$$

$$\alpha_n = \frac{1}{\tau^n} \left(\frac{\beta}{\pi\Delta_n}\right)^{3/2} .$$

When these are substituted into the series of Eq. (2-9), the solution is again obtained.

A POWER SERIES EXPANSION OF THE MASTER EQUATION[1]

N. G. van Kampen

ABSTRACT

In order to solve the master equation by a systematic approximation method, an expansion in powers of some parameter is needed. The appropriate parameter is the reciprocal size of the system, defined as the ratio of intensive and extensive variables. The lowest approximation yields the phenomenological law for the approach to equilibrium. The next approximation determines the mean square of the fluctuations about the phenomenological behavior. In equilibrium this approximation has the form of a linear Fokker–Planck equation. The higher approximations describe the effect of the non-linearity on the fluctuations, in particular on their spectral density. The method is applied to three examples: density fluctuations, Alkemade's diode, and Rayleigh's piston. The relation to the expansion recently given by Siegel is also discussed.

1. THE PROBLEM

Let a be a fluctuating physical quantity, $P(a,t)$ its probability distribution at time t. If a possesses Markov character, $P(a,t)$ obeys the equation

$$\frac{\partial P(a,t)}{\partial t} = \int \{W(a|a')P(a',t) - W(a'|a)P(a,t)\}\, da', \tag{1}$$

where $W(a|a')$ is the transition probability per unit time from a' to a. Equation (1) is the differential form of the Chapman–Kolmogorov equation and is often called the "master equation". The right-hand side can be formally expanded in a series,

$$\frac{\partial P(a,t)}{\partial t} = \sum_{n=1}^{\infty} \frac{1}{n!}\left(-\frac{\partial}{\partial a}\right)^n \alpha_n(a)P(a,t), \tag{2}$$

where α_n is the nth moment of the transition probability, or "derivate moment",

$$\alpha_n(a) = \int (a'-a)^n W(a'|a)\, da'.$$

The expansion (2) was given by Kramers (1940) and Moyal (1949). The Fokker–Planck equation (Fokker 1913, 1914; Planck 1917; Wang and Uhlenbeck 1945) is obtained from it by cutting off after $n = 2$,

$$\frac{\partial P(a,t)}{\partial t} = -\frac{\partial}{\partial a}\alpha_1(a)P(a,t) + \frac{1}{2}\frac{\partial^2}{\partial a^2}\alpha_2(a)P(a,t). \tag{3}$$

Clearly, this cutting off is not a systematic approximation procedure, because (2) is not a series expansion in powers of some small parameter.* Our purpose is to provide a systematic approximation method for (2).† For a special case

[1]Manuscript received November 28, 1960.
Contribution from the Institute for Theoretical Physics, Rijksuniversiteit te Utrecht, Nederland.

*For instance, one might just as well agree to rearrange the terms in (2) according to successive derivatives of P and omit all terms with higher derivatives than the second. Or one might, like Lax (1960), expand in "orders of non-linearity".

†The step from (1) to (2) is not entirely harmless. It can be shown for the example in Section 6 that certain terms are lost. These terms are not analytic in our expansion parameter and cannot, therefore, be treated by a power series expansion. Fortunately they are also extremely small.

(viz., the current fluctuations in a diode, see Section 6) such an expansion has been given previously (van Kampen 1959); we now treat the general case.

Note.—The name Fokker–Planck equation is sometimes confined to a more restricted kind of equation, to wit

$$\frac{\partial P(a,t)}{\partial t} = c_1 \frac{\partial}{\partial a} aP + c_2 \frac{\partial^2 P}{\partial a^2}, \tag{4}$$

with constants c_1 and c_2. We shall call this equation—first studied by Rayleigh (1891)—the *linear* Fokker–Planck equation.

2. POWER SERIES EXPANSION OF (2)

Rewrite the transition probability $W(a|a')$ as a function of the length of the jump, $\Delta a = a - a'$, and its starting point a',

$$W(a|a') = W(a';\Delta a). \tag{5}$$

So far the variable a has not been specified; it may be an extensive or an intensive quantity. We now stipulate that a is extensive. The corresponding intensive quantity X is related to it by

$$a = \Omega X,$$

where Ω is a measure for the size of the system. The essential point is that *the size of the jump, Δa, is properly expressed in terms of the extensive quantity, whereas the dependence on a' in (5) is properly expressed in terms of the intensive quantity X.* More precisely, if we write

$$W(a';\Delta a) = \Phi(X';\Delta a) = \Phi\left(\frac{a'}{\Omega};\Delta a\right),$$

the function Φ no longer involves Ω implicitly, in contrast with $W(a';\Delta a)$.* That this is so can be seen from the examples in Sections 4, 6, 7.

Since the dependence on Ω has now been made explicit, it is possible to expand in reciprocal powers of Ω. The moments α_n, when expressed as functions of X,

$$\int (\Delta a)^n \, \Phi(X;\Delta a) \, d\Delta a = \alpha_n(X),$$

no longer contain Ω.

Next one has to estimate how the successive derivatives of $P(a,t)$ depend on Ω. For this it is necessary to anticipate the kind of solution one is interested in. If one studies equilibrium alone, the interesting values of a are of order $\Omega^{\frac{1}{2}}$. It is then convenient to introduce a "*normalized*" *variable* $x = \Omega^{-\frac{1}{2}}a = \Omega^{\frac{1}{2}}X$. However, equation (2) can also be used to describe states that differ macroscopically from equilibrium. In such states a will have a mean value of order Ω, and a spread around this mean of order $\Omega^{\frac{1}{2}}$. Accordingly we put

$$a = \Omega\phi(t) + \Omega^{\frac{1}{2}}x.$$

ϕ is the function of time to be determined presently, x is the new variable.

*This requirement may be weakened: it is sufficient that Φ can be expanded in a power series of Ω^{-1}. Compare the example of Section 7.

After transforming from the variable a to the variable x, (2) becomes

$$\frac{\partial P(x,t)}{\partial t} - \Omega^{\frac{1}{2}}\phi'(t)\frac{\partial P(x,t)}{\partial x} = \sum_{n=1}^{\infty}\frac{\Omega^{-\frac{1}{2}n}}{n!}\left(-\frac{\partial}{\partial x}\right)^n \alpha_n\{\phi(t)+\Omega^{-\frac{1}{2}}x\}P(x,t).$$

The leading term on the right is

$$-\Omega^{-\frac{1}{2}}\alpha_1\{\phi(t)\}\frac{\partial P(x,t)}{\partial x}.$$

This can be made to cancel the second term on the left by subjecting $\phi(t)$ to the condition

(6) $$\phi'(t) = \frac{1}{\Omega}\alpha_1\{\phi(t)\}.$$

This is just the macroscopic, phenomenological law. The remaining equation is

(7) $$\frac{\partial P(x,t)}{\partial t} = -\Omega^{-\frac{1}{2}}\frac{\partial}{\partial x}[\alpha_1\{\phi(t)+\Omega^{-\frac{1}{2}}x\} - \alpha_1\{\phi(t)\}]P + \sum_{n=2}^{\infty}\frac{\Omega^{-\frac{1}{2}n}}{n!}\left(-\frac{\partial}{\partial x}\right)^n \alpha_n\{\phi(t)+\Omega^{-\frac{1}{2}}x\}P.$$

It is convenient to put $t/\Omega = \tau$, so that the new time scale increases with Ω, in agreement with the increase of the relaxation time. Then, expanding the right-hand side of (7) in powers of $\Omega^{-\frac{1}{2}}$, and rearranging terms

(8) $$\frac{\partial P(x,\tau)}{\partial \tau} = \sum_{m=2}^{\infty}\frac{\Omega^{-\frac{1}{2}(m-2)}}{m!}\sum_{n=1}^{m}\binom{m}{n}\alpha_n^{(m-n)}\{\phi(\tau)\}\left(-\frac{\partial}{\partial x}\right)^n x^{m-n}P.$$

In this equation all factors Ω are shown explicitly, so that the relevance of the several terms is indicated by their degree in $\Omega^{-\frac{1}{2}}$. The positive powers of Ω have been removed by means of (6).

For many purposes it is useful to find the successive *moments* rather than the distribution function itself. The general formula (8) gives

(9) $$\frac{d}{d\tau}\langle x^k\rangle = \sum_{m=2}^{\infty}\frac{\Omega^{-\frac{1}{2}(m-2)}}{m!}\sum_{n=1}^{m,k}\binom{m}{n}\alpha_n^{(m-n)}\{\phi(\tau)\}\frac{k!}{(k-n)!}\langle x^{m+k-2n}\rangle,$$

the upper limit of n being either m or k, whichever is smaller.

It is noteworthy that for many special cases exact results can be extracted from (9). For instance, if $\alpha_1(X)$ is a linear function,

$$\frac{d}{d\tau}\langle x\rangle = \alpha_1'.\langle x\rangle.$$

This can be solved to give $\langle x\rangle_\tau$ = const. $\phi(\tau)$. It will usually be convenient to choose $\phi(0)$ equal to $\langle a\rangle_0$, so that $\langle x\rangle_0 = 0$; our equation then shows that $\langle a\rangle_\tau = \phi(\tau)$ for all later τ.* Similarly, if $\alpha_2(X)$ happens to be linear, one finds a rigorous equation for $\langle x^2\rangle$. If all $\alpha_n(X)$ are linear, it is possible to find successive equations for all $\langle x^k\rangle$; such a case is the example in Section 4.

*More directly, from either (1) or (2) one has rigorously $(d/dt)\langle a\rangle = \langle\alpha_1(a)\rangle$; if α_1 is linear, this is equal to $\alpha_1(\langle a\rangle)$, which proves that $\langle a\rangle$ obeys the macroscopic equation.

3. THE ZEROTH ORDER APPROXIMATION

Neglecting all terms that vanish for $\Omega \to \infty$,

$$\frac{\partial P}{\partial \tau} = -\alpha_1'\{\phi(\tau)\}\frac{\partial}{\partial x}xP + \frac{1}{2}\alpha_2\{\phi(\tau)\}\frac{\partial^2 P}{\partial x^2}. \tag{10}$$

This would be a linear Fokker–Planck equation of the type (4) if the coefficients were not time dependent. Nevertheless the initial-value problem can be solved.

Suppose the initial distribution at $\tau = 0$ is

$$P(x,0) = \delta(x - x_0). \tag{11}$$

We define

$$s(\tau) = \log\frac{\alpha_1\{\phi(0)\}}{\alpha_1\{\phi(\tau)\}}, \tag{12}$$

so that $s(0) = 0$, and $ds/d\tau = -\alpha_1'\{\phi(\tau)\}$. Then (10) reduces to

$$\frac{\partial P}{\partial s} = \frac{\partial}{\partial x}xP - \frac{\alpha_2\{\phi(\tau)\}}{2\alpha_1'\{\phi(\tau)\}}\frac{\partial^2 P}{\partial x^2}.$$

Putting $x = ye^{-s}$ and $P(x,s) = e^s Q(y,s)$,

$$\frac{\partial Q}{\partial s} = -\frac{\alpha_2\{\phi(\tau)\}}{2\alpha_1'\{\phi(\tau)\}}e^{2s}\frac{\partial^2 Q}{\partial y^2}.$$

The solution of this equation is given by Chandrasekhar (1943)

$$Q(s) = [4\pi\gamma(s)]^{-\frac{1}{2}}\exp\left[-\frac{(y-x_0)^2}{4\gamma(s)}\right],$$

where

$$\gamma(s) = -\int_0^s \frac{\alpha_2\{\phi(\tau)\}}{2\alpha_1'\{\phi(\tau)\}}e^{2s}ds = \frac{1}{2}\int_0^\tau \alpha_2\{\phi(\tau)\}e^{2s(\tau)}d\tau.$$

Thus we finally find

$$P(x,\tau) = [2\pi\sigma^2]^{-\frac{1}{2}}\exp\left[-\frac{(x-x_0e^{-s})^2}{2\sigma^2}\right],$$

where

$$\sigma^2(\tau) = e^{-2s(\tau)}\int_0^\tau \alpha_2\{\phi(\tau)\}e^{2s(\tau)}d\tau. \tag{13}$$

In order to find the probability distribution of a with given initial distribution

$$P(a,0) = \delta(a-a_0) = \delta(a-\Omega X_0),$$

one first has to solve

$$d\phi/d\tau = \alpha_1(\phi), \qquad \phi(0) = X_0.$$

This can always be done by a quadrature,

$$\tau = \int_{x_0}^{\phi} \frac{d\phi}{\alpha_1(\phi)}\,.$$

One then has to determine $s(\tau)$ and $\sigma^2(\tau)$ from (12) and (13). The final result is

$$P(a,\tau) = [2\pi\Omega\sigma^2]^{-\frac{1}{2}} \exp\left[-\frac{\{a-\Omega\phi(\tau)\}^2}{2\Omega\sigma^2}\right]. \tag{14}$$

This is a Gaussian peak, whose width is of the order of the equilibrium fluctuations, and whose center moves according to the phenomenological law.* Higher approximations will destroy the Gaussian shape.

4. FIRST EXAMPLE: DENSITY FLUCTUATIONS

A box of volume Ω communicates through a small hole with an infinitely large volume, filled with a dilute gas with density ρ (Fig. 1). The number N of particles in Ω obeys the master equation (in suitable time units)

$$\frac{\partial P(N)}{\partial t} = \frac{N+1}{\Omega} P(N+1) - \frac{N}{\Omega} P(N) + \rho\{P(N-1)-P(N)\}. \tag{15}$$

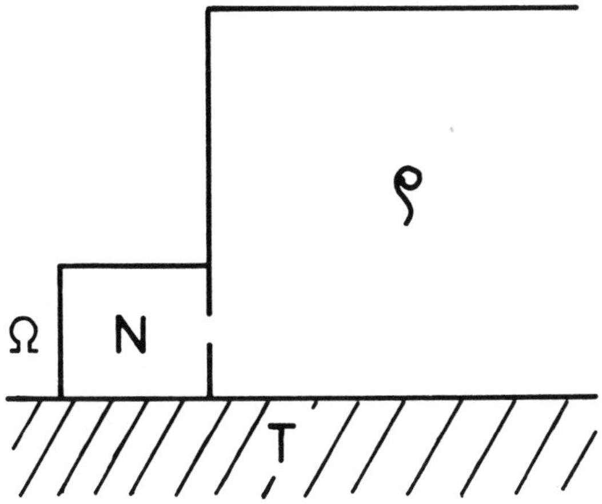

FIG. 1.

N is the extensive variable, the density $X = N/\Omega$ the intensive one. The transition probability is

$$W(N;\Delta N) = \frac{N}{\Omega}\,\delta_{\Delta N,-1} + \rho\,\delta_{\Delta N,+1} = \Phi(X;\Delta N).$$

The moments $\alpha_n(X)$ are given by

$$\alpha_n(X) = (-1)^n X + \rho.$$

*This result has been employed previously in evaluating the entropy of non-equilibrium states (van Kampen 1959).

The macroscopic equation (6) is

$$\frac{dX}{dt} = \frac{\rho - X}{\Omega},$$

with the solution

$$X = \phi(\tau) = X_0\, e^{-\tau} + \rho(1-e^{-\tau}),$$

where X_0 is the density at $\tau = 0$. The equilibrium value is $X^{eq} \equiv \phi(\infty) = \rho$, as was to be expected. One finds furthermore

$$s(\tau) = \tau,$$

$$\sigma^2(\tau) = (1-e^{-\tau})\,(\rho + X_0 e^{-\tau}).$$

Hence (14) takes the form

$$(16)\quad P(N,\tau) = [2\pi(1-e^{-\tau})(\Omega\rho + N_0 e^{-\tau})]^{-\frac{1}{2}} \exp\left[-\frac{\{N - N_0 e^{-\tau} - \Omega\rho(1-e^{-\tau})\}^2}{2(1-e^{-\tau})(\Omega\rho + N_0 e^{-\tau})}\right].$$

This example is sufficiently simple to check the result (16) in several ways. First one finds rigorously from either (1) or (2)

$$(d/dt)\langle N\rangle = \langle\alpha_1(N)\rangle = -\langle N\rangle/\Omega + \rho.$$

$$(d/dt)\langle N^2\rangle = 2\langle N\alpha_1(N)\rangle + \langle\alpha_2(N)\rangle$$

$$= -2\,\frac{\langle N^2\rangle}{\rho} + \left(2\rho + \frac{1}{\Omega}\right)\langle N\rangle + \rho.$$

These equations can be solved to give

$$\langle N\rangle_\tau = N_0 e^{-\tau} + \Omega\rho\,(1-e^{-\tau}).$$

$$\langle N^2\rangle_\tau - \langle N\rangle_\tau^2 = (N_0 e^{-\tau} + \Omega\rho)(1-e^{-\tau}).$$

This agrees exactly with the mean and variance of the Gaussian (16). The higher terms in $P(N,\tau)$, therefore, while destroying its Gaussian shape, do not affect the first two moments.

As a second check one may compare the equilibrium value of (16)

$$(17)\qquad P(N,\infty) = [2\pi\Omega\rho]^{-\frac{1}{2}} \exp\left[-\frac{(N-\Omega\rho)^2}{2\Omega\rho}\right],$$

with the correct expression

$$P^{eq}(N) = \frac{(\Omega\rho)^N}{N!}\, e^{-\Omega\rho}.$$

For this purpose we expand $\log P^{eq}(N)$ near its maximum, which is at $N_m = \Omega\rho - \frac{1}{2} + O(\Omega^{-1})$.

$$\log P^{eq}(N) = \log P^{eq}(N_m) - \frac{1}{2}\,\frac{(N-N_m)^2}{\Omega\rho + O(\Omega^{-1})} + \frac{1}{3!}\,\frac{(N-N_m)^3}{(\Omega\rho)^2} + \dots.$$

This agrees with (17) when terms of order $\Omega^{-\frac{1}{2}}$ are neglected.

As a final check it is possible to solve (15) exactly with the aid of the characteristic function, defined by

$$F(\xi,t) = \langle e^{\xi N}\rangle_t = \sum_{N=0}^{\infty} e^{\xi N} P(N,t).$$

The solution is

$$\log F(\xi,t) = \Omega\rho(1-e^{-\tau})(e^{\xi}-1)+N_0 \log(1-e^{-\tau}+e^{\xi-\tau}).$$

Let this be expanded to second order in ξ

$$\log F(\xi,t) = [\Omega\rho(1-e^{-\tau})+N_0 e^{-\tau}]\xi+\tfrac{1}{2}[\Omega\rho(1-e^{-\tau})+N_0 e^{-\tau}(1-e^{-\tau})]\xi^2.$$

This approximate expression is just the characteristic function of the Gaussian distribution (16).

5. EQUILIBRIUM FLUCTUATIONS

In order to describe the fluctuations in equilibrium one has to substitute for $\phi(\tau)$ in the general equation (8) its equilibrium value $\phi(\infty)$, so that the quantities $\alpha_n^{(m-n)}$ become constants.* The zeroth order equation (10) then reduces to

$$\frac{\partial P(x,\tau)}{\partial \tau} = -\alpha_1' \frac{\partial}{\partial x} xP + \frac{1}{2}\alpha_2 \frac{\partial^2 P}{\partial x^2}. \tag{18}$$

This is the linear Fokker–Planck equation (4). The solution with initial distribution (11) is well known,

$$P(x,\tau) = \left[\frac{\pi\alpha_2}{-\alpha_1'}(1-e^{\alpha_1'\tau})\right]^{-\frac{1}{2}} \exp\left[-\frac{(x-x_0 e^{\alpha_1'\tau})^2}{(-\alpha_2/\alpha_1')(1-e^{2\alpha_1'\tau})}\right]. \tag{19}$$

(Of course, for all realistic cases $\alpha_1' < 0$, $\alpha_2 > 0$.) The difference between this solution and (14) consists in the fact that the present one is only valid for initial distributions which are confined to a range of order 1 in the x-scale, i.e. of order $\Omega^{\frac{1}{2}}$ in the a-scale, around the equilibrium value.

The next approximation adds to the right-hand side of (18)

$$\frac{1}{3!}\Omega^{-\frac{1}{2}}\left\{-3\alpha_1'' \frac{\partial}{\partial x} x^2 P + 3\alpha_2' \frac{\partial^2}{\partial x^2} xP - \alpha_3 \frac{\partial^3 P}{\partial x^3}\right\}. \tag{20}$$

(Note that the resulting equation can also be obtained directly from (2) by breaking off $\alpha_1(a)$ after the second order of a, $\alpha_2(a)$ after the first order, $\alpha_3(a)$ after the zeroth order, and omitting all higher terms.) The first order equation can be solved to the order $\Omega^{-\frac{1}{2}}$ by treating the additional term (20) as a perturbation. That is, one substitutes the zeroth order solution (19) in (20), and solves the resulting inhomogeneous equation for P.

Alternatively one can find the successive moments, using (9), which for the equilibrium region simplifies to

$$\frac{d}{d\tau}\langle x^k\rangle = \sum_{m=2}^{\infty} \Omega^{-\frac{1}{2}(m-2)} \sum_{n=1}^{m,k} \binom{k}{n} \frac{\alpha_n^{(m-n)}}{(m-n)!}\langle x^{m+k-2n}\rangle. \tag{21}$$

*We write α_n for the nth moment of Δa, taken at the equilibrium. $\alpha_n^{(k)}$ denotes the kth derivative of $\alpha_n(X)$ *with respect to* X, also at the equilibrium value of X.

Explicitly,

$$(22a)\qquad (d/d\tau)\,\langle x\rangle = \alpha_1'\langle x\rangle + \tfrac{1}{2}\Omega^{-\frac{1}{2}}\alpha_1''\langle x^2\rangle + \tfrac{1}{6}\Omega^{-1}\alpha_1'''\langle x^3\rangle + O(\Omega^{-3/2}),$$

$$(22b)\qquad (d/d\tau)\,\langle x^2\rangle = 2\alpha_1'\langle x^2\rangle + \alpha_2 + \Omega^{-\frac{1}{2}}\alpha_1''\langle x^3\rangle + \Omega^{-\frac{1}{2}}\alpha_2'\langle x\rangle + O(\Omega^{-1}),$$

$$(22c)\qquad (d/d\tau)\,\langle x^3\rangle = 3\alpha_1'\langle x^3\rangle + 3\alpha_2\langle x\rangle + O(\Omega^{-\frac{1}{2}}).$$

Now let the initial distribution be given by (11), so that

$$\langle x\rangle_0 = x_0, \qquad \langle x^2\rangle_0 = x_0^2, \qquad \langle x^3\rangle_0 = x_0^3.$$

It is then possible to solve the set of equations (22) in successive orders and to find from it $\langle x\rangle$ as a function of τ to order Ω^{-1}. For convenience we choose such units that $\alpha_1 = -1$, $\alpha_2 = 2$. Then one has to zeroth order

$$\begin{aligned}
\langle x\rangle_\tau &= x_0 e^{-\tau},\\
\langle x^2\rangle_\tau &= x_0^2 e^{-2\tau} + 1 - e^{-2\tau},\\
\langle x^3\rangle_\tau &= x_0^3 e^{-3\tau} - 3x_0 e^{-3\tau} + 3x_0 e^{-\tau}.
\end{aligned}$$

Subsequently one finds to order $\Omega^{-\frac{1}{2}}$

$$\langle x^2\rangle_\tau = x_0^2 e^{-2\tau} + 1 - e^{-2\tau} + \Omega^{-\frac{1}{2}}(1-e^{-\tau})\{(a_2' + 3\alpha_1'')x_0 e^{-\tau} + \alpha_1''(x_0^3 - 3x_0)e^{-2\tau}\}.$$

Finally to order Ω^{-1},

$$\begin{aligned}
(23)\qquad \langle x\rangle_\tau = {} & x_0 e^{-\tau} + \tfrac{1}{2}\Omega^{-\frac{1}{2}}\alpha_1''\{1 - 2e^{-\tau} + e^{-2\tau} + x_0^2(e^{-\tau} - e^{-2\tau})\}\\
& + \tfrac{1}{2}\Omega^{-1}\{\alpha_1''^2(x_0^3 - 3x_0)(\tfrac{1}{2}e^{-\tau} - e^{-2\tau} + \tfrac{1}{2}e^{-3\tau})\\
& \qquad - (3\alpha_1''^2 + \alpha_1''\alpha_2')x_0(e^{-\tau} - e^{-2\tau})\}\\
& + \tfrac{1}{2}\Omega^{-1}(3\alpha_1''^2 + \alpha_1''\alpha_2' + \alpha_1''')x_0\tau e^{-\tau}\\
& + \tfrac{1}{12}\Omega^{-1}\alpha_1'''(x_0^3 - 3x_0)(e^{-\tau} - e^{-3\tau}).
\end{aligned}$$

As an example of the moment method we compute the *spectral density* of the fluctuations to order Ω^{-1}. The autocorrelation function is obtained from (23) by multiplying with x_0 and averaging over the equilibrium distribution $P^{\mathrm{eq}}(x_0)$. For this one needs the equilibrium moments, which can be found to any desired order from (21) by putting the left side equal to zero

$$\begin{aligned}
\langle x\rangle^{\mathrm{eq}} &= \tfrac{1}{2}\Omega^{-\frac{1}{2}}\alpha_1'' + O(\Omega^{-3/2}),\\
\langle x^2\rangle^{\mathrm{eq}} &= 1 + \tfrac{1}{2}\Omega^{-1}\{\tfrac{5}{2}\alpha_1''^2 + \tfrac{3}{2}\alpha_1''\alpha_2' + \tfrac{1}{3}\alpha_1''\alpha_3 + \alpha_1''' + \tfrac{1}{2}\alpha_2''\} + O(\Omega^{-2}),\\
\langle x^3\rangle^{\mathrm{eq}} &= \Omega^{-\frac{1}{2}}\{\tfrac{5}{2}\alpha_1'' + \alpha_2' + \tfrac{1}{3}\alpha_3\} + O(\Omega^{-3/2}),\\
\langle x^4\rangle^{\mathrm{eq}} &= 3 + O(\Omega^{-1}).
\end{aligned}$$

With the aid of these values one finds for the autocorrelation function

$$\begin{aligned}
\langle x(0)x(\tau)\rangle \equiv \langle x_0\langle x\rangle_\tau\rangle^{\mathrm{eq}} = {} & e^{-\tau}\\
& + \tfrac{1}{2}\Omega^{-1}[\tfrac{1}{2}\alpha_1''^2 + \{\alpha_1''^2 + \tfrac{3}{2}\alpha_1''\alpha_2' + \tfrac{2}{3}\alpha_1''\alpha_3 + \alpha_1''' + \tfrac{1}{2}\alpha_2''\}\,e^{-\tau}\\
& \qquad + \{\alpha_1''^2 - \tfrac{1}{3}\alpha_1''\alpha_3 e\}^{-2\tau}\\
& \qquad + \{3\alpha_1''^2 + \alpha_1''\alpha_2' + \alpha_1'''\}\tau e^{-\tau}].
\end{aligned}$$

The last term may be combined with the term $e^{-\tau}$ to give $e^{-\lambda\tau}$ with

$$\lambda = 1-\tfrac{1}{2}\Omega^{-1}(3\alpha_1''^2+\alpha_1''\alpha_2'+\alpha_1''').$$

The constant term $\frac{1}{4}\Omega^{-1}\alpha_1''^2$ exhibits the fact that $\langle x\rangle^{eq}$ does not vanish, so that

$$\lim_{\tau\to\infty}\langle x(0)x(\tau)\rangle = (\langle x\rangle^{eq})^2 = (\tfrac{1}{2}\Omega^{-\frac{1}{2}}\alpha_1'')^2.$$

Hence (24) may also be written

$$\langle\{x(0)-\langle x\rangle^{eq}\}\{x(\tau)-\langle x\rangle^{eq}\}\rangle =$$

$$[1+\tfrac{1}{2}\Omega^{-1}\{\alpha_1''^2+\tfrac{3}{2}\alpha_1''\alpha_2'+\tfrac{2}{3}\alpha_1''\alpha_3+\alpha_1'''+\tfrac{1}{2}\alpha_2'\}]\,e^{-\lambda\tau}+\{\alpha_1''^2-\tfrac{1}{3}\alpha_1''\alpha_3\}e^{-2\tau}.$$

From this follows the fluctuation spectrum

$$(25)\quad S_x(\omega) = \frac{2}{\pi}\left[\frac{1+\frac{1}{2}\Omega^{-1}\{-2\alpha_1''^2+\frac{1}{2}\alpha_1''\alpha_2'+\frac{2}{3}\alpha_1''\alpha_3+\frac{1}{2}\alpha_2''\}}{\lambda^2+\omega^2} + \frac{\Omega^{-1}\{\alpha_1''^2-\frac{1}{3}\alpha_1''\alpha_3\}}{4+\omega^2}\right].$$

Bernard and Callen (1960) arrived at the conclusion that the fluctuation spectrum is entirely independent of the non-linear terms. According to (25), this is only correct for $\Omega\to\infty$. (In fact it can be seen from our examples that both coefficients { } in (25) do not happen to vanish.)

6. SECOND EXAMPLE: ALKEMADE'S DIODE

(Alkemade 1958; van Kampen 1960)

Consider the circuit of Fig. 2. The whole system is kept at constant temperature T. The two electrodes have different work functions, and electrode 1 is supposed to operate under saturation conditions. The number of electrons N on the left condenser plate obeys the master equation (in suitable time units)

$$(26)\qquad \frac{\partial P(N)}{\partial t} = P(N+1)-P(N)+\zeta\{e^{-\epsilon(N-1)}P(N-1)-e^{-\epsilon N}P(N)\}.$$

Here $\epsilon = e^2/kTC$, and ζ is a constant depending on both work functions. The expansion parameter Ω is the capacity C of the condenser, or rather

$$\Omega = kTC/e^2 = \epsilon^{-1}.$$

For the extensive parameter we take $a = N$, so that $X = \epsilon N$. The Kramers–Moyal equation in terms of a is

$$\frac{\partial P(N,t)}{\partial t} = \sum_{n=1}^{\infty}\frac{1}{n!}\left(-\frac{\partial}{\partial N}\right)^n\{(-1)^n+\zeta e^{-X}\}P(N,t).$$

The macroscopic equation is

$$(27)\qquad \frac{dX}{d\tau} = -1+\zeta e^{-X}.$$

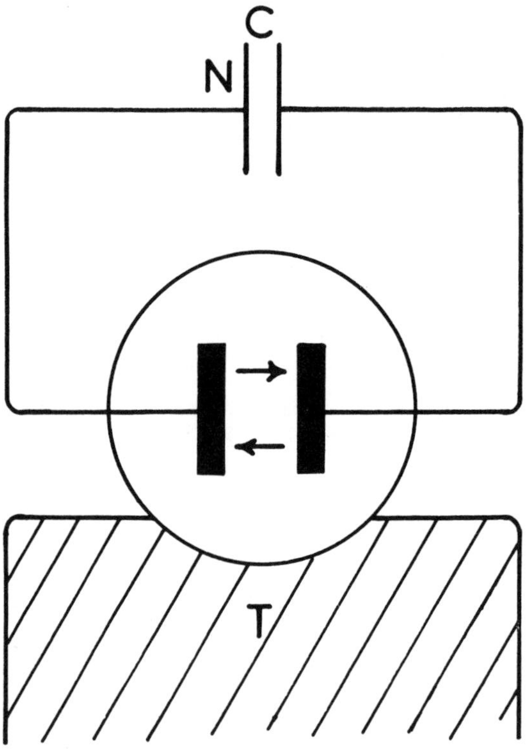

FIG. 2.

Equilibrium obtains for $X = \log \zeta$, which determines the contact potential X_c. Writing (27) in the form

$$\frac{dN}{dt} = e^{-(X-X_c)} - 1,$$

one recognizes the usual I–V characteristic of the diode.

The solution of (27) is

$$X(\tau) - X_c = \log(1+e^{\tau_0-\tau}), \tag{28}$$

where τ_0 is an integration constant given by

$$e^{\tau_0} = e^{X(0)-X_c} - 1.$$

One subsequently finds

$$s(\tau) = \log \frac{e^{\tau}+e^{\tau_0}}{1+e^{\tau_0}}$$

and

$$\sigma^2(\tau) = 1 + \frac{e^{\tau+\tau_0} + \tau e^{2\tau_0} - e^{2\tau_0} - 3e^{\tau_0} - 1}{(e^{\tau}+e^{\tau_0})^2}. \tag{29}$$

This is in agreement with the result obtained previously (van Kampen 1960). The distribution $P(N,t)$ is the Gaussian (14) with $\phi(\tau)$ taken from (28) and $\sigma^2(\tau)$ from (29). The equilibrium distribution is, to this order,

$$P^{eq}(N) = \frac{1}{\sqrt{2\pi\Omega}} \exp\left[-\frac{(N-\Omega X_c)^2}{2\Omega} \right]. \tag{30}$$

The spectral density of the equilibrium fluctuations is obtained from (25) by substituting

$$\alpha_1'' = \alpha_1''(X_c) = 1,$$

$$\alpha_2 = -1, \qquad \alpha_2'' = 1,$$

$$\alpha_3 = 0.$$

The result is again identical with the one obtained previously.

According to (30) the average number of electrons in equilibrium is $\langle N \rangle^{eq} = \Omega X_c$. According to (23), however, there is a higher order correction $\Omega^{\frac{1}{2}} \langle x \rangle^{eq} = \frac{1}{2}\alpha_1'' = \frac{1}{2}$. [Incidentally, this corrected value happens to be the exact one, because the exact equilibrium distribution is

$$P^{eq}(N) = [2\pi\Omega]^{-\frac{1}{2}} \exp\left[-\frac{(N-\Omega X_c - \frac{1}{2})^2}{2\Omega} \right], \tag{31}$$

as can easily be found from (26).] It means that the equilibrium distribution as found from the zeroth order equation, viz. (30), differs from the exact one through higher order terms. In this case these higher order terms do not destroy the Gaussian shape, but merely cause a shift. Accordingly the value for the contact potential, $X_c = \log \zeta$, is not the exact one. Using the explicit value for ζ, one finds for the exact contact potential

$$X_c + \tfrac{1}{2}\,\Omega = (W_1 - W_2)/kT,$$

where W_1 and W_2 are the work functions of both electrodes. The fact that the equilibrium value of the potential in lowest order, X_c, does not exactly correspond with the expected value, $W_1 - W_2$, has given rise to some discussion (MacDonald 1957; Alkemade 1958; Marek 1959).

7. THIRD EXAMPLE: RAYLEIGH'S PISTON

(Rayleigh 1891; Zernike 1929)

In one dimension a heavy particle with mass M is subjected to collisions with light gas molecules with mass m and velocity distribution $f(v)$ (Fig. 3). A collision of a molecule with velocity v changes the velocity V of the heavy particle into

$$V' = V + \frac{2m}{M+m}(v - V) = V + \Delta V.$$

One thus obtains a master equation for the velocity distribution $P(V,t)$, with transition probability

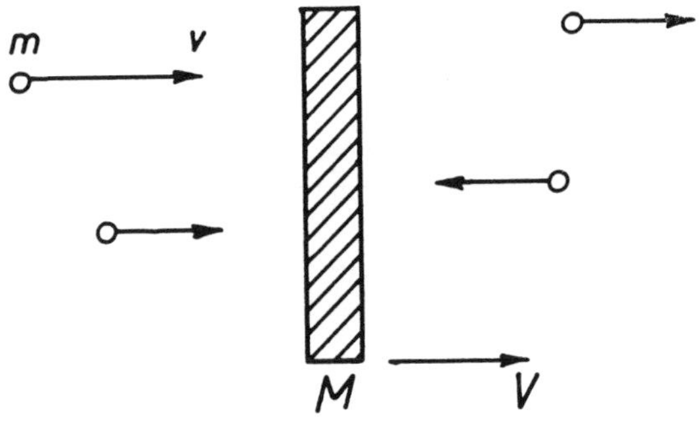

FIG. 3.

$$W(V'|V) = \nu\left(\frac{M+m}{2m}\right)^2 |V'-V| f\left(\frac{M+m}{2m}V' - \frac{M-m}{2m}V\right), \tag{32}$$

where ν is the number of molecules per unit length.

Although the concept of intensive and extensive variables is somewhat alien to this system, a similar distinction can be made. Indeed, the individual jumps in V are due to collisions with the gas molecules, and are therefore appropriately measured in terms of the momentum MV. On the other hand, the probability distribution of these jumps depends on the velocity V itself. Thus one is led to the following identification

$$a = \frac{M}{m}V, \qquad X = V, \qquad \Omega = \frac{M}{m}. \tag{33}$$

The transition probability (32) then takes the form

$$W(a;\Delta a) = \tfrac{1}{4}\nu(1+\Omega^{-1})^2|\Delta a| f\{X + \tfrac{1}{2}(1+\Omega^{-1})\Delta a\}.$$

This expression is not entirely independent of Ω. However, it can be expanded in powers of Ω^{-1}, so that the previous treatment can still be applied with minor modifications. This leads to the usual expansion in m/M.

A more elegant application of the previous treatment, however, is possible if one puts

$$a = \frac{M+m}{m}V, \qquad X = V, \qquad \Omega = \frac{M+m}{m}. \tag{34}$$

The transition probability now takes the form

$$W(a;\Delta a) = \tfrac{1}{4}\nu\,|\Delta a|\, f(X + \tfrac{1}{2}\Delta a) = \Phi\,(X;\Delta a),$$

entirely independent of Ω. Hence the results of the previous sections can now be literally applied.

The macroscopic equation, which describes the approach of $\langle V\rangle$ to its equilibrium value $\langle V\rangle^{\text{eq}} = 0$, is rather unmanageable. We therefore only study the equilibrium fluctuations. The zeroth approximation is the linear Fokker–Planck equation (18) with

$$\alpha_1' = \frac{1}{4}\nu \int_{-\infty}^{+\infty} (\Delta a)|\Delta a| f'(\tfrac{1}{2}\Delta a)\, d\Delta a,$$

$$\alpha_2 = \frac{1}{4}\nu \int_{-\infty}^{+\infty} (\Delta a)^2 |\Delta a| f(\tfrac{1}{2}\Delta a)\, d\Delta a.$$

If f is the Maxwell distribution,

$$\alpha_1' = -\frac{m}{2kT} \qquad \alpha_2 = 4\nu\left(\frac{2kT}{m}\right)^{\frac{1}{2}}.$$

After transforming back to the usual variables $V = \Omega^{-\frac{1}{2}}x$, $t = \Omega\tau$ the Fokker-Planck equation becomes

$$\frac{\partial P(V,t)}{\partial t} = \nu \frac{4m}{M+m}\left(\frac{2kT}{\pi m}\right)^{\frac{1}{2}} \frac{\partial}{\partial V}\left\{VP + \frac{kT}{M+m}\frac{\partial^2 P}{\partial V^2}\right\}. \tag{35}$$

This is the well-known Rayleigh equation, but for the appearance of $M+m$ in place of M. The reason is that we have expanded in powers of $m/(m+M)$ rather than in m/M. The lowest order of the expansion in m/M can, of course, easily be obtained from (35) by simply replacing $M+m$ with M.

The next term in the expansion is (20), but all coefficients vanish because of symmetry.* The term of order Ω^{-1} can also be found; if one puts $m/kT = 1$, and $m/M = \mu$, the result is, when written as an expansion in μ,

$$\frac{\partial P(V,t)}{\partial t} = \frac{\nu}{\sqrt{2\pi}}\left[\frac{8\mu}{1+\mu}\left\{\frac{\partial}{\partial V}V + \mu\frac{\partial^2}{\partial V^2}\right\}P\right. \tag{36}$$
$$\left. + 8\mu\left\{-\mu^2\frac{\partial^2}{\partial V^2} + \frac{1}{6}\frac{\partial}{\partial V}V^3 + \frac{3}{2}\mu\frac{\partial^2}{\partial V^2}V^2 + \frac{8}{3}\mu^2\frac{\partial^3}{\partial V^3}V + \frac{4}{3}\mu^2\frac{\partial^4}{\partial V^4}\right\}P\right].$$

The terms on the second line are of order μ relative to those on the first line. They agree with the result of Siegel (1960).

An advantage of expanding in μ rather than Ω^{-1} is that the equilibrium distribution, when expressed in the normalized co-ordinate $\mu^{\frac{1}{2}}V$ is independent of μ. Hence it must satisfy each term of the expansion separately. This can be verified from (36). The equilibrium distribution, when expressed in $\Omega^{\frac{1}{2}}V$, contains higher terms in Ω^{-1} and hence does not satisfy the separate terms of the expansion in Ω^{-1} (see next section). That is the reason why it is not an exact solution of (35).

8. RELATIONS FOLLOWING FROM THE KNOWN EQUILIBRIUM DISTRIBUTION

So far the equilibrium distribution has been found as a special solution of the master equation. It often happens, however, that it is known a priori. Then this known solution can be used, by substituting it in the general equation, to find relations between the α's. For instance, it follows from (18) that to lowest order the equilibrium distribution is given by

*Recently MacDonald (private communication) has suggested an asymmetric modification of the Rayleigh piston model, for which these coefficients do not vanish.

$$P^{\text{eq}}(x) = [-\pi\alpha_2/\alpha_1']^{-\frac{1}{2}} \exp\left[-\frac{x^2}{-\alpha_2/\alpha_1'}\right]. \tag{37}$$

The variance $(-\alpha_2/2\alpha_1')$ of this Gaussian is usually known from equilibrium statistical mechanics. Hence one obtains a relation between α_2 and α_1', which corresponds to the usual Einstein relation for Brownian movement.

Additional relations can only be obtained if higher order terms of the equilibrium distribution are also known. In some cases one knows that (37) is exact to all orders. Thus, taking again $\alpha_1' = -1$, $\alpha_2 = 2$,

$$\begin{aligned} \langle x^k \rangle &= 0, && \text{(odd } k) \\ \langle x^k \rangle &= (k-1)!!. && \text{(even } k) \end{aligned} \tag{38}$$

On substituting in (21) the separate powers of Ω yield

$$\sum_{n=1}^{m,k} \frac{(m+k-2n-1)!!}{n!(k-n)!(m-n)!} \alpha_n^{(m-n)} = 0,$$

where $m = 2, 3, \ldots$; $k = 1, 2, \ldots$; $m+k$ even; n running from 1 up to m or k, whichever is smaller. It is easily seen that for $m = 2$ these equations are identically satisfied. For $m = 3$ they lead to

$$\alpha_1'' = 0, \qquad 3\alpha_2'+\alpha_3 = 0. \tag{39}$$

For $m = 4$,

$$2\alpha_1'''+\alpha_2'' = 0, \qquad 4\alpha_2''+4\alpha_3'+\alpha_4 = 0. \tag{40}$$

More generally one finds that all even derivatives of α_1 vanish (at equilibrium) whereas for odd m,

$$2\alpha_1^{(m)}+\alpha_2^{(m-1)} = 0.$$

The Rayleigh piston furnishes an example of an equilibrium distribution that is known to all orders of Ω, but is not independent of Ω. Indeed

$$P^{\text{eq}} \propto \exp\left[-\frac{MV^2}{2kT}\right] \propto \exp\left[-\frac{m}{2kT}(1-\Omega^{-1})x^2\right].$$

Putting $m/kT = 1$ one has instead of (38),

$$\langle x^k \rangle = (k-1)!!(1-\Omega^{-1})^{-\frac{1}{2}k}.$$

Substituting in (21) and separating the various powers of Ω, one finds in the same way as before that (39) is still valid, but (40) must be replaced with

$$2\alpha_1'''+\alpha_2''+4\alpha_1' = 0,$$

$$4\alpha_2''+4\alpha_3'+\alpha_4+4\alpha_2 = 0.$$

These relations can be verified by explicit computations.

MacDonald (1957) has emphasized that (37) need not be correct to all orders. In fact, in our example of Section 4 it is certainly only correct in zeroth order. Yet it may be asserted that the equilibrium distribution must be Gaussian under the following conditions.

(i) The energy of the total system is quadratic in the fluctuating quantity a; and

(ii) the transition probability $W(a|a')$ contains some physical parameter (other than Ω) as a factor, which permits to scale down its magnitude without affecting its functional dependence on a and a'.

The first condition ensures that, according to equilibrium statistical mechanics, the probability distribution is

$$P^{\mathrm{eq}}(x) \propto e^{-E/kT} \propto e^{-cx^2}. \tag{41}$$

For instance, this condition rules out the case of density fluctuations. The second condition is necessary because (41) is only exact for infinitely small interaction. The scaling parameter permits to go to the limit of infinitely small interaction, so that (41) becomes exact. It may then be concluded that (41) must also be exact for all other values of the scaling parameter, because P^{eq} does not depend on the magnitude of $W(a|a')$, but only on its functional dependence of a and a'.

Both conditions are fulfilled in the case of Rayleigh's piston, if the variable a is defined according to (33) rather than (34), the scaling parameter being ν. Hence the Gaussian distribution is valid for all values of M/m—which of course is well known. In the case of the diode, the electrical energy of the circuit is proportional to the square of $X-(W_1-W_2)/kT$, so that equilibrium statistical mechanics indeed leads to (31). The exchange of electrical energy with thermal energy of the heat bath can be scaled down by decreasing the area of the electrodes. (This area should occur as a factor in the right-hand side of (26), although for convenience it was not written explicitly but absorbed in the time unit.) This explains why (31) has to be exact.

9. COMPARISON WITH SIEGEL'S EXPANSION

Recently Siegel (1960) has given an ingenious transformation of the Kramers–Moyal expansion (2), namely

$$\frac{\partial P}{\partial t} = e^{-\frac{1}{4}x^2} \sum_{m=1}^{\infty} \sum_{l=1}^{m} A_{lm}\left(\frac{1}{2}x-\frac{\partial}{\partial x}\right)^{m-l+1}\left(\frac{1}{2}x+\frac{\partial}{\partial x}\right)^{l} e^{\frac{1}{4}x^2}P. \tag{42}$$

The expansion only applies to the neighborhood of equilibrium (our Section 5). The units are chosen such that $-\alpha_2/2\alpha_1' = 1$, and it is assumed that the equilibrium distribution is rigorously $\propto e^{-\frac{1}{2}x^2}$. The coefficients A_{lm} are related to our moments $\alpha_n(X)$ by

$$A_{lm} = \frac{1}{\sqrt{2\pi}}\frac{1}{l!}\sum_{k=l}^{m}\frac{\Omega^{-\frac{1}{2}(m-k+1)}}{(k-l)!(m-k+1)!}\int_{-\infty}^{+\infty} H_k(x)\,\alpha_{m-k+1}(\Omega^{-\frac{1}{2}}x)e^{-\frac{1}{2}x^2}dx, \tag{43}$$

where $H_k(x)$ denotes a Hermite polynomial,

$$H_k(x) = e^{\frac{1}{2}x^2}\left(-\frac{d}{dx}\right)^k e^{-\frac{1}{2}x^2}. \tag{44}$$

Siegel suggests that in (43) the successive values of m should constitute successive orders of approximation, and he shows for the case of the Rayleigh piston that, indeed, A_{lm} is of order $\Omega^{-\frac{1}{2}m}$ (Ω being given by (33)). As we have obtained a power series expansion for *all* cases, it is now possible to check Siegel's suggestion generally. From (43) follows with the aid of (44),

$$A_{lm} = \frac{1}{\sqrt{2\pi}}\frac{1}{l!}\sum_{k=l}^{m}\frac{\Omega^{-\frac{1}{2}(m+1)}}{(k-l)!(m-k+1)!}\sum_{\nu=0}^{\infty}\frac{\Omega^{-\frac{1}{2}}}{\nu!}\alpha_{m-k+1}^{(k+\nu)}(0)\int_{-\infty}^{\infty}x^{\nu}e^{-\frac{1}{2}x^2}\,dx.$$

This shows that A_{lm} is of order $\Omega^{-\frac{1}{2}(m+1)}$, so that the successive terms of Siegel's expansion are indeed of increasing order of smallness.*†

Siegel's aim is rather to give a systematic expansion, which achieves two other purposes:

(i) in each approximation (i.e., if (42) is broken off at an arbitrary value m_0 of m) the correct equilibrium distribution $e^{-\frac{1}{2}x^2}$ should be an exact solution;

(ii) each approximation should be "semidefinite", meaning that all solutions should tend to $e^{-\frac{1}{2}x^2}$ for $t \to \infty$.

The first purpose seems to me a laudable one, albeit mainly for reasons of elegance. Admittedly, it is an awkward feature of the systematic expansion that it leads to the denominator $M+m$ rather than M in (35), although it is not incorrect. For practical purposes, however, such as the computation of the fluctuation spectrum, a systematic power series expansion is more convenient.

Siegel achieves the second purpose by a mathematical construction, which may be roughly described as follows. After cutting off the expansion of the master equation at m_0, suitably chosen higher order terms are added such as to make the resulting equation "semidefinite". This construction could also be applied to our expansion (8). However, I do not feel that one ought to impose the requirement (ii) of being "semidefinite" on the successive approximations. It is true that, for each m_0 for which the cutoff series is not "semidefinite", there exist solutions that do not tend to the equilibrium distribution $e^{-\frac{1}{2}x^2}$. Such a behavior, however, only shows that the approximation used is not good enough for these solutions.‡ By adding higher order terms to make the equation "semidefinite" one does not improve the approximation. One only replaces the incorrect solutions by other solutions, which do tend to $e^{-\frac{1}{2}x^2}$, but are otherwise equally incorrect.

REFERENCES

ALKEMADE, C. T. J. 1958. Physica, **24**, 1029.
BERNARD, W. and CALLEN, H. B. 1960. Phys. Rev. **118**, 1466.
CHANDRASEKHAR, S. 1943. Revs. Modern Phys. **15**, 1.
FOKKER, A. D. 1913. Thesis, Leiden.
——— 1914. Ann. Physik, **43**, 810.

*The reason we find $A_{lm} \sim \Omega^{-\frac{1}{2}(m+1)}$, as compared to Siegel's $A_{lm} \sim \Omega^{-\frac{1}{2}m}$, is that he writes an additional factor $\Omega^{\frac{1}{2}}$ in the transition probability, which simply amounts to using a different time scale.

†For example, as mentioned in Section 7, it can be verified that our next approximation (36) coincides with the next approximation given by Siegel, apart from terms of order $(m/M)^2$.

‡Generally speaking, that occurs when the initial distribution is a very rapidly varying function, such that the terms involving higher derivatives are not small. It follows that a solution with initial condition (11) must not be regarded as a physically possible one, but only as a means to represent solutions that are sufficiently smooth for the approximation to be valid.

van Kampen, N. G. 1959. Physica, **25**, 1294.
——— 1960. Physica, **26**, 585.
Kramers, H. A. 1940. Physica, **7**, 284 (Collected Scientific Papers (Amsterdam 1956), p. 754).
Lax, M. 1960. Revs. Modern Phys. **32**, 25.
MacDonald, D. K. C. 1957. Phys. Rev. **108**, 541.
Marek, A. 1959. Physica, **25**, 1358.
Moyal, J. E. 1949. J. Roy. Statist. Soc. (B) **11**, 150.
Planck, M. 1917. Sitzungsber. Preuss. Akad. Wissens. p. 324 (Physikalische Abhandlungen und Vorträge II (Braunschweig 1958) p. 435).
Lord Rayleigh. 1891. Phil. Mag. **32**, 424 (Scientific Papers (Cambridge 1902) **3**, 473).
Siegel, A. 1960. J. Math. Phys. **1**, 378.
——— Proceedings of the 2nd International Symposium on Rarified Gas Dynamics, Berkeley 1960. To be published.
Wang, M. C. and Uhlenbeck, G. E. 1945. Revs. Modern Phys. **17**, 323.
Zernike, F. 1929. Handbuch der Physik **3**, 419.

Approximation of the Linear Boltzmann Equation by the Fokker-Planck Equation

R. F. Pawula

Department of the Aerospace and Mechanical Engineering Sciences, University of California, San Diego, La Jolla, California

and

*Institute for Radiation Physics and Aerodynamics**

In general, transformation of the linear Boltzmann integral operator to a differential operator leads to a differential operator of infinite order. For purposes of mathematical tractability this operator is usually truncated at a finite order and thus questions arise as to the validity of the resulting approximation. In this paper we show that the linear Boltzmann equation can be properly approximated *only* by the first two terms of the Kramers-Moyal expansion; i.e., the Fokker-Planck equation, with the retention of a *finite number* of higher-order terms leading to a logical inconsistency.

I. INTRODUCTION

Frequently, in the study of relaxation phenomena described by the linearized Boltzmann (master) equation, the Boltzmann integral operator is approximated by a differential operator.[1–10] The differential operator is obtained by terminating the Kramers-Moyal expansion at a finite number of terms, with the lowest-order approximation (the first two terms) being the Fokker-Planck operator. Intuitively, one is tempted to expect that the degree of approximation is directly related to the number of terms retained in the expansion. However, even if the expansion can be made in terms of a small parameter (such as the mass ratio of a light to a heavy particle) difficulties arise when terms of order greater than two are retained. For example, additional boundary conditions must somehow be prescribed for the solution of the resulting partial differential equation. Furthermore, if the solution represents a distribution function, then terms must be retained in such a way as to render a non-negative answer. Although these well-known difficulties can to some extent be overcome,[8] the general procedure of passing from an integral operator to a differential operator has eluded mathematical justification.[7]

As a result of investigating generalizations of the Fokker-Planck-Kolmogorov equations to non-Markov processes,[11] we have obtained a partial solution to the above problem. As is shown in Sec. III, if one assumes that terms above a given order are zero in the Kramers-Moyal expansion, this assumption implies that *all* terms above second order are zero.

It is to be emphasized that the passage from the linear Boltzmann equation to the Kramers-Moyal expansion is not free from mathematical criticism. One must *assume* the existence of certain partial derivatives, the convergence properties of certain series, and the interchange of certain limits. For example, if a distribution function (probability density function) contains a Dirac δ function and is otherwise analytic, we cannot expect the Kramers-Moyal expansion to yield the δ function *even if an infinite number of terms are retained.*[12] In the following, *we assume* that the linear Boltzmann equation is equivalent to the Kramers-Moyal expansion as given by Moyal.[13]

II. THE LINEAR BOLTZMANN EQUATION AND THE KRAMERS-MOYAL EXPANSION

For the sake of clarity, we confine our attention to a one-dimensional random process $x(t)$ which can take on a continuous range of values as a function of the continuous time parameter t. $x(t)$ might represent, for example, the position, speed, or energy of a gas particle. Let $P(x,t)$ denote the probability density function of the random variable $x(t)$ at time t and let $P(x,t|x_0,t_0)$ denote the transition probability density function, i.e., the conditional probability density function of $x(t)$ at time t given that $x(t)=x_0$ at time $t=t_0$. $P(x,t)$ then satisfies the linearized Boltzmann equation

$$\frac{\partial P(x,t)}{\partial t}=\lim_{\Delta\to 0}\frac{1}{\Delta}\int_{-\infty}^{\infty}[P(x',t)P(x,t+\Delta|x',t)-P(x,t)P(x',t+\Delta|x,t)]dx'. \quad (1)$$

* This research was supported in part by the Advanced Research Projects Agency (Project DEFENDER) and was monitored by the U. S. Army Research Office (Durham) under Contract No. DA-31-124-ARO-D-257.

1 H. A. Kramers, Physica **7**, 284 (1940).
2 J. E. Moyal, J. Roy. Stat. Soc. (London) **B11**, 150 (1949).
3 J. Keilson and J. E. Storer, Quart. Appl. Math. **10**, 243 (1952).
4 A. Siegel, J. Math. Phys. **1**, 378 (1960).
5 M. Lax, Rev. Mod. Phys. **32**, 25 (1960).
6 N. G. van Kampen, Can. J. Phys. **39**, 551 (1961).
7 K. Anderson and K. E. Shuler, J. Chem. Phys. **40**, 633 (1963).
8 H. Akama and A. Siegel, Physica **31**, 1493 (1965).
9 N. G. van Kampen, in *Fluctuation Phenomena in Solids*, edited by R. E. Burgess (Academic Press Inc., New York, 1965), p. 139.
10 C. F. Eaton and L. H. Holway, Jr., Phys. Rev., **143**, 48 (1966).
11 R. F. Pawula, IEEE Trans. Inform. Theory **13**, 33 (1967).

12 This statement is made without proof and is based upon the fact that a number of regularity assumptions must be imposed in transforming the Boltzmann operator to a differential operator. However, Siegel and Kohlberg [A. Siegel and I. Kohlberg, Bull. Am. Phys. Soc. **8**, 30 (1963)] have shown in a special case that the eigenvalues of the differential operator converge to the eigenvalues of the integral operator.
13 See Ref. 2, p. 197, Eqs. (8.1.15) and (8.1.16).

We now assume that the right-hand side of this equation can be expanded in the Kramers-Moyal expansion; viz.,

$$\frac{\partial P(x,t)}{\partial t}=\sum_{n=1}^{\infty}\frac{(-1)^n}{n!}\frac{\partial^n}{\partial x^n}[A_n(x,t)P(x,t)], \quad (2)$$

where the derivate moments A_n are given by

$$A_n(x,t)=\lim_{\Delta\to 0}\frac{1}{\Delta}\int_{-\infty}^{\infty}(x'-x)^n P(x',t+\Delta|x,t)dx'. \quad (3)$$

It is common to *assume* that the limit and integration operations in (1) and (3) can be interchanged and to define a transition probability density per unit time as

$$B(x,x')=\lim_{\Delta\to 0}\frac{P(x,t+\Delta|x',t)}{\Delta}. \quad (4)$$

The limit in this definition of $B(x,x')$ is to be interpreted in the physical, rather than in the mathematical, sense. By this we mean, for example, that if $x(t)$ were some property of a gas particle, that we might require that Δ always be much larger than a characteristic interaction time between gas particles (see, for example, the discussion by Uhlenbeck[14]). However, even for a well-behaved process such as a continuous Gaussian process, $B(x,x')$ is poorly behaved mathematically, consisting of Dirac δ functions and their derivatives. We thus choose to retain the form (1) for the linearized Boltzmann equation and (3) for the derivate moments. In Sec. IV, we discuss the implications of the limit and integration interchanges.

In general, (2) is an infinite-order partial differential equation which, for purposes of mathematical tractability, we desire to truncate at a finite number of terms.

III. THE TRUNCATION LEMMA

In this section we consider conditions under which (2) reduces to a partial differential equation of finite order. These conditions follow directly from the following:

Lemma: If A_n, as defined by (3), exists for all n, and if $A_n=0$ for some even n, then $A_n=0$ for all $n\geq 3$.

Usually the derivate moments A_n will be nonzero for all values of n. However, this lemma tells us that if we *assume* $A_n=0$ for some even n, that we are in actuality assuming A_n to be zero for all $n\geq 3$. Thus we conclude that it is logically inconsistent to retain more than two terms in the Kramers-Moyal expansion unless *all* of the terms are retained.

The proof of a generalized form of the above lemma is given in Ref. 11 and is reproduced here as a matter of completeness. From (3), we have

$$A_n=\lim_{\Delta\to 0}\frac{1}{\Delta}\int_{-\infty}^{\infty}(x'-x)^{(n-1)/2}(x'-x)^{(n+1)/2} \times P(x',t+\Delta|x,t)dx'. \quad (5)$$

Assuming n odd and $n\geq 3$, and applying the Schwarz inequality to (5), we obtain

$$A_n^2\leq A_{n-1}A_{n+1} \qquad n \text{ odd } n\geq 3. \quad (6)$$

In a similar way, it follows that

$$A_n^2\leq A_{n-2}A_{n+2} \qquad n \text{ even, } n\geq 4. \quad (7)$$

Setting $n=r-1$, $r+1$ in (6) and $n=r-2$, $r+2$ in (7), where r is an even integer, we obtain the four equations

$$A_{r-2}^2\leq A_{r-4}A_r, \quad r\geq 6 \quad (8)$$

$$A_{r-1}^2\leq A_{r-2}A_r, \quad r\geq 4 \quad (9)$$

$$A_{r+1}^2\leq A_rA_{r+2}, \quad r\geq 2 \quad (10)$$

$$A_{r+2}^2\leq A_rA_{r+4}, \quad r\geq 2. \quad (11)$$

If $A_n<\infty$ for all n and if $A_r=0$ for some even $r\geq 6$, then (8)–(11) show that A_{r-2}, A_{r-1}, A_{r+1}, A_{r+2} must be zero. By repeated application of this argument it follows that $A_n=0$ for all $n\geq r$. Going in the other direction and taking cognizance of the limits on r in (8)–(11), it follows that $A_n=0$ for all $n\geq 3$.

Note that the above lemma *does not* guarantee that the Fokker-Planck equation will be a good approximation to the linear Boltzmann equation. We should in general expect to obtain different solutions from each equation. The lemma merely leads to the conclusion that the probability density function of a random process cannot be correctly described by a finite number, greater than two, of terms of the Kramers-Moyal expansion.

If the derivate moment inequalities (8)–(11) are ignored, the equation resulting from a finite number, say n, of terms of the Kramers-Moyal expansion can, in principle, be solved to yield a function $Q_n(x,t)$. This function will in certain cases approach $P(x,t)$ as $n\to\infty$. However, the approximations $Q_n(x,t)$ may possess undesirable properties. Using a rearrangement of terms of the Kramers-Moyal expansion called Siegel's *CD* expansion, Kohlberg and Siegel[15] have found, for example, that approximate solutions for $P(x,t)$ are not always non-negative.

Although we have restricted ourselves to the simplest case of a one-dimensional random variable $x(t)$ and a marginal probability density function $P(x,t)$, the above lemma is true under much broader conditions. The

[14] G. E. Uhlenbeck, in *Probability and Related Topics in Physical Sciences*, edited by M. Kac (Interscience Publishers, Ltd., London, 1959), Appendix I, p. 183.

[15] I. Kohlberg and A. Siegel, Boston University report, 1965 (unpublished).

general one-dimensional case and the multidimensional case are discussed in detail in Ref. 11.

IV. USE OF A TRANSITION PROBABILITY DENSITY PER UNIT TIME

Let us assume that the limit defining the transition probability per unit time exists in some appropriate sense so that the linearized Boltzmann equation can be written as

$$\frac{\partial P(x,t)}{\partial t}=\int_{-\infty}^{\infty} P(x',t)B(x,x')dx' - P(x,t)\int_{-\infty}^{\infty} B(x',x)dx', \quad (12)$$

and the Kramers-Moyal expansion as

$$\frac{\partial P(x,t)}{\partial t}=\sum_{n=1}^{\infty}\frac{(-1)^n}{n!}\frac{\partial^n}{\partial x^n}[a_n(x,t)P(x,t)], \quad (13)$$

where the derivate moments a_n are given by

$$a_n(x,t)=\int_{-\infty}^{\infty}(x'-x)^n B(x',x)dx'. \quad (14)$$

Since $B(x',x)$ is non-negative it follows that if $a_n(x,t)$ vanishes for some even n, then $B(x',x)$ must be zero for almost all x'. Thus for well-behaved $B(x',x)$, for example, as in the Keilson-Storer[3] model, the Kramers-Moyal expansion apparently becomes meaningless if an even derivate moment vanishes.[16] Thus for well-behaved $B(x',x)$ we conclude that no even derivate moment can vanish and that approximate solutions obtained from a finite number of terms of the Kramers-Moyal expansion will not necessarily represent probability density functions of random processes.

[16] This conclusion has been pointed out to the author by Professor J. Keilson of the University of Rochester.

If, on the other hand, the transition probability density $B(x',x)$ is allowed to contain certain singularities, then we are led to the results of Sec. III. For example, if

$$B(x',x)=\tfrac{1}{4}\delta''(x-x'), \quad (15)$$

then the Kramers-Moyal expansion becomes

$$\frac{\partial P(x,t)}{\partial t}=\frac{1}{4}\frac{\partial^2 P(x,t)}{\partial x^2}, \quad (16)$$

which for the initial condition $P(x,0)=\delta(x)$ has the solution

$$P(x,t)=(\pi t)^{-1/2}\exp[-x^2/t] \quad t\geq 0. \quad (17)$$

V. DISCUSSION

The derivate moment inequalities presented in Sec. III have led to the conclusion that the linear Boltzmann integral operator cannot properly be approximated by a finite number, greater than two, of terms of the Kramers-Moyal expansion. Although it is possible to construct approximate solutions by ignoring these inequalities, the validity of these approximations has not yet, to the author's knowledge, been established.

In our above treatment we have avoided a number of fundamental questions, such as the continuity of the random processes under consideration, the ability of a continuous random process to approximate a discontinuous random process, the validity of interchanging limiting operations, etc. These questions, as well as the all important problem of establishing error bounds on approximate solutions to the linear Boltzmann equation, remain areas for further investigation.

ACKNOWLEDGMENTS

The author is indebted to Dr. Kurt E. Shuler of the National Bureau of Standards for helpful discussions and for pointing out to the author a number of pertinent references. Thanks also are extended to Professor A. Siegel of Boston University for critical and informative comments on the original version of this manuscript.

Expansion of the Master Equation for One-Dimensional Random Walks with Boundary*

N. G. van Kampen† and Irwin Oppenheim
Department of Chemistry, Massachusetts Institute of Technology, Cambridge, Massachusetts
(Received 3 September 1971)

In order to understand the behavior of coarse-grained equations in the presence of a boundary, the following model is investigated. A homogeneous one-dimensional random walk is bounded on one side by some boundary conditions of rather arbitrary form. The corresponding master equation is approximated by the Fokker–Planck equation plus partial differential equations for the higher orders. The boundary condition for the Fokker–Planck approximation is well known; but those for the higher order terms are here derived. To the second order they amount to a virtual displacement of the boundary. The case of a two-step random walk, however, gives rise to an unexpected complication, inasmuch as nonpropagating solutions of the master equation cannot be ignored in the boundary condition, although they do not contribute to the differential equations themselves.

1. INTRODUCTION

In recent years much effort in nonequilibrium statistical mechanics has been devoted to the derivation of equations describing the time dependence of reduced distribution functions and macroscopic variables. These equations have been derived from the exact equations of motion and are particularly simple when some pertinent physical parameter is quite small. Examples are the Boltzmann equation for a low density gas, the Langevin and Fokker–Planck equations for heavy particles in a bath of light particles, and the hydrodynamic equations in systems with small spatial gradients. All of these equations can be extended to situations where the parameter (e.g., the density) is still small, but large enough so that the lowest order equations are no longer valid. These extensions have resulted in the Choh–Uhlenbeck equation[1] for dilute gases and the generalized hydrodynamic equations.[2]

All of the derivations mentioned above were carried out in the thermodynamic limit and in the absence of boundaries. It would be of great interest to derive the transport equations for a system with boundaries, but little progress has been made in this direction.[3(a)] Phenomenological considerations do supply a clue to the expected results of such analyses, at least to the lowest order. Thus, for example, the hydrodynamic equations describing the bulk properties for nonequilibrium fluids have the same form whether a boundary is present or not. The presence of the boundary merely imposes boundary conditions on these equations. We emphasize that these boundary conditions are often not derived from molecular theory but imposed on intuitive phenomenological grounds. While the intuitive approach may work for the lowest order equations, there is no assurance that it will, and even less assurance that it will be valid for the extended equations.

In this paper we study a number of discrete random walk models in the presence of boundaries. These models are of interest because they are fairly easy to analyze and because they have been widely used to describe stochastic processes in physics and chemistry. We expand the exact differential-difference equation (master equation) and find to lowest order the Fokker–Planck equation, and similar partial differential equations for the successive higher orders. The boundary conditions for these differential equations are also derived from the exact master equation. The form of the differential equations is not affected by the presence of boundaries. However, the boundary conditions which their solutions must satisfy are not intuitive and can be determined only from a detailed study of the solutions to the master equation itself.

2. THE RANDOM WALK MODEL

By "random walk" we mean a stationary, continuous time Markov process whose range of possible values consists of integral numbers n. The random walk is *unbounded* when n ranges from $-\infty$ to $+\infty$, and is *bounded* (on the left) when n takes on only the values

1, 2, 3, $\cdots$. For the unbounded case we assume *homogeneity*, i.e., the probability per unit time a_r, for making a jump of r units is independent of n. If the largest possible single jump is s units, we call it an s-step random walk. Then the master equation for the probability $p_n(t)$ to be at site n at time t is

$$\dot{p}_n(t) = \sum_{r=-s}^{s} a_r p_{n+r}(t). \tag{1}$$

If $a_r = a_{-r}$, the random walk is *symmetric*. We shall here restrict ourselves to the symmetric one-step and two-step cases. We assume $a_1 \neq 0$ to ensure that all sites can be reached. [See Ref. 3(b).]

For the bounded case we take the same Eq. (1) for all positive n excepting a finite number of them near the boundary. For the exceptional p_n special equations are required, whose precise form is descriptive of the physical properties of the boundary. Clearly an infinite variety of boundaries is possible. We shall call a boundary *reflecting* when the boundary equations are such that total probability is conserved, and *absorbing* if total probability decreases. Among all varieties of absorbing boundaries one may single out a special case, to be called purely absorbing, which is the kind of absorbing boundary usually considered. [See Ref. 3(c).]

It is not hard to choose the boundary equations such that the total probability *increases*, at a rate proportional to the probability already present at the neighboring sites; this kind of "stimulated emission," however, will be excluded. Moreover, for nonhomogeneous master equations, i.e., a_r depending on n, it often happens that the process is automatically bounded due to the vanishing of a_{-r} ($r = 1, 2, \ldots, s$) at $n = 0$; such "natural boundaries" are much easier to treat and are not the subject of this article.

The bounded s-step random walk problem can be solved exactly, in the sense that all $p_n(t)$ can be found for $t > 0$ when their initial values are prescribed. In particular, the reflecting and the purely absorbing case have been amply treated in the literature,[5] the latter often in connection with first passage problems. However, for reasons explained in the Introduction, we are interested in developing an approximation scheme based on the smallness of the individual jumps. The first step of this approximation is the Fokker–Planck or diffusion equation with the well-known boundary conditions, viz., the probability density vanishes on an absorbing boundary and has zero slope on a reflecting boundary. Higher order corrections to the Fokker–Planck equation have previously been obtained for the unbounded case.[6] This article is concerned with the higher orders in the presence of a boundary. It will appear that a nontrivial complication arises for multiple step random walks. Rather than trying to provide an exhaustive treatment of all possibilities, we shall confine ourselves to three special paradigms.

It is convenient to define a *pure boundary*, meaning that the recursion relation (1) is valid for *all* $n \geq s + 1$ (supposing that a_{-s} does not vanish), so that only the equations for $p_1, p_2, \ldots, p_s$ are modified. In the case of a pure boundary the following equivalent but more convenient way of formulating the boundary conditions is possible. Introduce s auxiliary variables $p_0, p_{-1}, p_{-2}, \ldots, p_{-s+1}$ and stipulate that (1) is valid for *all* $n \geq 1$, but that the auxiliary variables are connected with the actual variables $p_1, p_2, \cdots$ by a set of s suitably chosen linear relations. How to choose these linear relations will become clear in the applications. *Pure absorption* will be defined as the special case that these linear relations have the simple form $p_0 = p_{-1} = \cdots = p_{-s+1} = 0$.

3. ONE-STEP RANDOM WALK WITH PURE BOUNDARY—EXACT SOLUTION

With suitable choice of time unit, Eq. (1) reduces, in the symmetric one-step case, to

$$\dot{p}_n = p_{n+1} + p_{n-1} - 2p_n. \tag{2}$$

In the case of a pure boundary this is valid for $n = 2, 3, \cdots$. Inasmuch as only one-step jumps are permitted, the special equation for p_1 must have the form

$$\dot{p}_1 = p_2 - (1 + c)p_1, \tag{3}$$

with a single parameter c. For the total probability W one finds

$$\frac{dW}{dt} = \frac{d}{dt}\sum_{n=1}^{\infty} p_n = -cp_1.$$

Hence the boundary is reflecting for $c = 0$ and absorbing for $c > 0$. The case $c < 0$ describes stimulated emission and will therefore be excluded.

Equation (2) for $n \geq 2$ together with (3) for $n = 1$ may be expressed in an equivalent way by declaring (2) valid for $n = 1$ as well and putting

$$p_0 = (1 - c)p_1. \tag{4}$$

(Note that p_0 is merely an auxiliary quantity and is *not* equal to the probability that the particle has been absorbed.)

According to the definition in Sec. 2 the boundary is called purely absorbing when $c = 1$.

To solve this bounded random walk problem exactly first note that (2) is obeyed by

$$e^{-\lambda t} z^n \tag{5}$$

provided that z and λ are connected by the characteristic equation

$$\lambda = 2 - z - 1/z. \tag{6}$$

For fixed time constant λ there are two roots z_1, z_2 and hence two solutions of the form (5); thus the general solution of (2) with time factor $e^{-\lambda t}$ is

$$p_n^{(\lambda)}(t) = e^{-\lambda t}(C_1 z_1^n + C_2 z_2^n). \tag{7}$$

We require that $p_n^{(\lambda)}$ is bounded for $n \to \infty$; it will be shown that that is sufficient for obtaining a complete set of normal modes. Hence one must have $|z_1| \leq 1$ and $|z_2| \leq 1$ (see, however, Appendix A). As $z_1 z_2 = 1$ according to (6), we may write

$$z_1 = e^{i\vartheta}, \quad z_2 = e^{-i\vartheta}, \quad 0 \leq \vartheta \leq \pi. \tag{8}$$

Note that in order to obey (2) for $n = 2, 3, \cdots$, it is necessary that (7) holds for $n = 1, 2, \cdots$. Thus *owing to the fact that the boundary is pure, all $p_n(t)$ have to be of this form.* ϑ is related to λ by (6), or

$$\lambda = 2 - 2\cos\vartheta = 4\sin^2\tfrac{1}{2}\vartheta. \tag{9}$$

This covers the values $0 \leqslant \lambda \leqslant 4$, which determines the eigenvalue spectrum of our bounded problem.

The constants C_1, C_2 are now chosen such that (3) is satisfied:

$$(1 + c - \lambda)(C_1 z_1 + C_2 z_2) = C_1 z_1^2 + C_2 z_2^2. \tag{10a}$$

Equivalently one may use (4) to obtain

$$C_1 + C_2 = (1 - c)(C_1 z_1 + C_2 z_2). \tag{10b}$$

It follows that

$$-C_1/C_2 = [1 - (1 - c)e^{-i\vartheta}]/[1 - (1 - c)e^{i\vartheta}]. \tag{10c}$$

The normal mode solutions of the bounded random walk are now fully determined by (7), (9), and (10).

It is instructive to write the result in a more familiar form. Write $S(\vartheta) = e^{2i\eta(\vartheta)}$ for the right-hand side of (10c); note that $|S(\vartheta)| = 1$ and η is real. With suitable normalization the normal mode solution (7) may then be written

$$p_n^{(\lambda)}(t) = e^{-\lambda t}(2/\pi)^{1/2}\sin[\vartheta n + \eta(\vartheta)]. \tag{11}$$

Thus $\eta(\vartheta)$ is the phase shift due to the boundary and $S(\vartheta)$ is the "S matrix". Note, however, that this S-matrix is always unitary, regardless of whether total probability is conserved or not! The orthogonality and completeness of these normal modes is shown in Appendix A. Consequently the final solution may be written in terms of an evolution operator

$$p_n(t) = \sum_{m=1}^{\infty} U_{nm}(t)p_m(0),$$
$$U_{nm}(t) = \frac{2}{\pi}\int_0^{\pi}\sin[\vartheta n + \eta(\vartheta)] \times \sin[\vartheta m + \eta(\vartheta)]e^{-4t\sin^2(\vartheta/2)}d\vartheta. \tag{12}$$

4. ONE-STEP RANDOM WALK WITH PURE BOUNDARY—APPROXIMATE TREATMENT

Following the program outlined in the Introduction we now investigate the approximations that are applicable when the individual jumps may be treated as small compared to the distances that one is interested in. Accordingly we set

$$\epsilon n = x, \quad p_n(t) = \epsilon P(x, t), \tag{13}$$

and expand in ϵ. In other words, we are interested in an approximation method for solutions that vary slowly compared to the size of the jumps. Of course the result can be found directly by expanding the exact result (12); but our aim is to find an independent method for solving the problem in this approximation.

The master equation (2) gives for $P(x,t)$,

$$\dot{P}(x, t) = P(x + \epsilon, t) + P(x - \epsilon, t) - 2P(x, t)$$
$$= \epsilon^2\frac{\partial^2 P}{\partial x^2} + \frac{\epsilon^4}{12}\frac{\partial^4 P}{\partial x^4} + \cdots.$$

Using the new time variable $\tau = \epsilon^2 t$ and expanding P,

$$P = P^{(0)} + \epsilon P^{(1)} + \epsilon^2 P^{(2)} + \cdots,$$

one finds successively

$$\frac{\partial P^{(0)}}{\partial\tau} = \frac{\partial^2 P^{(0)}}{\partial x^2}, \tag{14a}$$

$$\frac{\partial P^{(1)}}{\partial\tau} = \frac{\partial^2 P^{(1)}}{\partial x^2}, \tag{14b}$$

$$\frac{\partial P^{(2)}}{\partial\tau} = \frac{\partial^2 P^{(2)}}{\partial x^2} + \frac{1}{12}\frac{\partial^4 P^{(0)}}{\partial x^4}, \quad \cdots \tag{14c}$$

These equations apply to all $x > 0$.

The boundary condition (4) translates into

$$P(0, \tau) = (1 - c)P(\epsilon, \tau),$$

which, on expanding in ϵ, amounts to

$$P^{(0)} = (1 - c)P^{(0)}, \tag{15a}$$

$$P^{(1)} = (1 - c)P^{(1)} + (1 - c)P^{(0)\prime}, \tag{15b}$$

$$P^{(2)} = (1 - c)P^{(2)} + (1 - c)P^{(1)\prime} + \tfrac{1}{2}(1 - c)P^{(0)\prime\prime}. \tag{15c}$$

(Here and in the future we abbreviate the notation by writing $P^{(0)}$ for $P^{(0)}(0, \tau)$, etc., and $P^{(0)\prime}$ for $\partial P^{(0)}(x, \tau)/\partial x$ evaluated at $x = 0$.) The conclusion from these equations depends on whether or not c vanishes.

First for $c \neq 0$, one concludes

$$P^{(0)} = 0, \quad P^{(1)} = [(1 - c)/c]P^{(0)\prime},$$
$$P^{(2)} = [(1 - c)/c]P^{(1)\prime}, \cdots. \tag{16a}$$

In the last equation we have used the fact that $P^{(0)\prime\prime} = 0$ as a consequence of $P^{(0)} = 0$ and Eq. (14a). Thus we find that *to lowest order, $P(x, \tau)$ obeys the familiar boundary condition for an absorbing boundary whenever c does not vanish.* The reason is that any absorption, however small, is sufficient on our slow time scale to absorb everything that reaches the boundary. The next two orders in (16a) may be written in the form of the boundary condition

$$P(0, \tau) = \epsilon[(1 - c)/c]P'(0, \tau) + \mathcal{O}(\epsilon^3).$$

That is, to first and second order the value of $P(x, \tau)$ on the boundary differs from zero by an amount proportional to its slope. The proportionality constant $l = \epsilon(1 - c)/c$ has been called the slip coefficient.[3,7] In the next order, however, the boundary condition can no longer be expressed in terms of a single slip coefficient; rather one finds

$$P(0, \tau) = \epsilon[(1 - c)/c]P'(0, \tau) + \epsilon^3[(1 - c)(3 - 2c)/6c^2] \times P'''(0, \tau) + \cdots.$$

For the reflecting boundary $c = 0$, Eq. (15a) is moot, while (15b) and (15c) yield

$$P^{(0)\prime} = 0, \quad P^{(1)\prime} = -\tfrac{1}{2}P^{(0)\prime\prime}. \tag{16b}$$

The same pattern continues in higher orders, e.g.,

$$P^{(2)\prime} = -\tfrac{1}{2}P^{(1)\prime\prime} - \tfrac{1}{6}P^{(0)\prime\prime\prime} = -\tfrac{1}{2}P^{(1)\prime\prime}. \qquad (16c)$$

Equations (14), together with the boundary conditions (16) constitute the desired approximate treatment in successive powers of ϵ. In order to compare this with the exact solution in Sec. 3 we write (11) in terms of the continuous variable x, setting $\vartheta = \epsilon k$,

$$\epsilon P_\lambda(x, \tau) = e^{-\lambda t}(2/\pi)^{1/2} \sin(kx + \eta). \qquad (17)$$

If one now expands λ as given in (9),

$$\lambda = \epsilon^2 k^2 - \tfrac{1}{12}\epsilon^2 k^4 + \cdots,$$

one finds that, indeed, (17) obeys (14) in successive orders.

To verify the boundary conditions, expand η as given by (10):

$$\eta = \operatorname{Im} \log\{1 - (1 - c)e^{-i\epsilon k}\}$$
$$= \epsilon[(1 - c)/c]k + \mathcal{O}(\epsilon^3).$$

Substituting this in (17), one finds again that the successive orders of (17) obey the boundary conditions (16a). Note that, to second order, η is proportional to k, which according to (17) simply amounts to a displacement of the point where $P(x, \tau)$ vanishes from $x = 0$ to $x = -l$. This is equivalent with the existence of a slip coefficient and does not go through in higher orders.

In the reflecting case one finds by taking $c = 0$ in (10),

$$\eta = \tfrac{1}{2}\pi - \tfrac{1}{2}\epsilon k + \mathcal{O}(\epsilon^3).$$

On substituting this in (17) one easily finds agreement with (16b). Here again one may represent the effect of the boundary in first and second order as a shift of the boundary. The probability density $P(x, \tau)$ has no longer zero slope at $x = 0$, but at $x = \frac{1}{2}\epsilon$. The concept of a slip coefficient does not apply to the reflecting boundary.

In spite of this agreement it is not true that the approximate treatment leads to a *convergent* power series expansion of the evolution operator (12). The solutions of (14) involve all wave numbers k, whereas the integral in (12) contains values of k only up to π/ϵ. The solutions that we have added incorrectly are functions that vary appreciably within a distance ϵ and are therefore meaningless; they decay in time like e^{-t} rather than $e^{-\tau}$. Of course this kind of error is inevitable in any approximation in terms of continuous functions, and presumably in any approximation based on coarse graining. It demonstrates that the expansion in powers of ϵ can only be an asymptotic expansion.

5. EXAMPLE OF ONE-STEP RANDOM WALK WITH IMPURE BOUNDARY

Suppose a particle jumps randomly between neighboring points of a one-dimensional lattice, but the transition probabilities between the endpoint $n = 1$ and its neighbor $n = 2$ differ from the others:

$$\dot{p}_n = p_{n+1} + p_{n-1} - 2p_n, \qquad n = 3, 4, \ldots, \qquad (18a)$$

$$\dot{p}_2 = p_3 + ap_1 - (1 + b)p_2, \qquad (18b)$$

$$\dot{p}_1 = bp_2 - ap_1. \qquad (18c)$$

Note that the total probability W is conserved.

To find the normal mode solutions we first solve (18a) by setting $p_n^{(\lambda)}$ equal to (7) *for all* $n \geqslant 2$. As to $p_1^{(\lambda)}$, we know its time dependence,

$$p_1^{(\lambda)}(t) = e^{-\lambda t}q;$$

but its amplitude q is not determined by (18a). The two boundary equations (18b) and (18c), however, state

$$(1 + b - \lambda)(C_1 z_1^2 + C_2 z_2^2) = C_1 z_1^3 + C_2 z_2^3 + aq, \qquad (19a)$$

$$(a - \lambda)q = b(C_1 z_1^2 + C_2 z_2^2). \qquad (19b)$$

These equations determine C_1, C_2, and q uniquely up to an arbitrary normalization constant.

In order to make the transition to the continuous limit we set

$$\epsilon n = x, \qquad \epsilon^2 t = \tau, \qquad p_n(t) = \epsilon P(x, \tau) \qquad \text{for } n \geqslant 2.$$

Equation (18a) leads again to (14). Equations (18b) and (18c) serve to determine $p_1(\tau)$ and also to find the boundary condition for $P(x, \tau)$. We shall now use the dot for differentiation with respect to τ, so that the time derivatives pick up a factor ϵ^2. Thus (18b) and (18c) become

$$\epsilon^2 \dot{P}(2\epsilon) = P(3\epsilon) + (a/\epsilon)p_1 - (1 + b)P(2\epsilon),$$
$$\epsilon \dot{p}_1 = bP(2\epsilon) - (a/\epsilon)p_1.$$

Order ϵ^{-1} merely states $p_1^{(0)} = 0$. Order ϵ^0 yields

$$ap_1^{(1)} = bP^{(0)}, \qquad (20)$$

but does not yet lead to a boundary condition for $P^{(0)}$. Order ϵ^1 yields

$$0 = -bP^{(1)} + (1 - 2b)P^{(0)\prime} + ap_1^{(2)},$$
$$0 = bP^{(1)} + 2bP^{(0)\prime} - ap_1^{(2)}. \qquad (21)$$

Hence $P^{(0)\prime} = 0$ and $ap_1^{(2)} = bP^{(1)}$. Thus *to zeroth order P obeys the diffusion equation with the boundary condition for a reflecting wall.* According to (20) the probability for occupying site 1 is proportional to the value of the continuous probability density at the wall. The proportionality factor contains a factor ϵ owing to the fact that site 1 is only one state among a dense aggregate of other states. To the next order one finds

$$P^{(1)\prime} = (b/a - \tfrac{3}{2})P^{(0)\prime\prime},$$
$$ap_1^{(3)} = bP^{(2)} + b(2b/a - b - 1/a)P^{(0)\prime\prime}.$$

6. TWO-STEP RANDOM WALK WITH PURE BOUNDARY

The symmetric two-step random walk is described by the master equation

$$\dot{p}_n = \alpha p_{n+2} + p_{n+1} + p_{n-1} + \alpha p_{n-2} - (2 + 2\alpha)p_n. \qquad (22)$$

In the presence of a pure boundary this recursion relation applies to $n = 3, 4, \cdots$ and has to be supplemented by two special equations for p_1 and p_2. We take

$$\dot{p}_2 = \alpha p_4 + p_3 + p_1 - bp_2, \tag{23a}$$

$$\dot{p}_1 = \alpha p_3 + p_2 - ap_1, \tag{23b}$$

with two positive constants a and b.

For the total probability one finds from (22) and (23),

$$\dot{W} = (\alpha + 2 - b)p_2 + (\alpha + 1 - a)p_1. \tag{24}$$

Hence the boundary is reflecting when

$$a = 1 + \alpha, \quad b = 2 + \alpha. \tag{25}$$

Smaller values of a or b will not be considered as they correspond to stimulated emission. Larger values correspond to an absorbing boundary. An equivalent formulation of the boundary conditions is obtained by declaring (22) valid for all $n \geqslant 1$, and setting

$$p_0 = [(2 + 2\alpha - b)/\alpha]p_2 \equiv Bp_2, \tag{26a}$$

$$p_{-1} = [(2 + 2\alpha - a)/\alpha]p_1 - [(2 + 2\alpha - b)/\alpha^2]p_2 \equiv Ap_1 - (B/\alpha)p_2. \tag{26b}$$

The constants A and B are simply abbreviations. Note that one must have

$$B \leqslant 1, \quad A \leqslant 1 + 1/\alpha.$$

The equality sign corresponds to the reflecting case. The purely absorbing boundary is characterized by $A = B = 0$.

Solutions of (23*a*) are obtained by taking $p_n(t)$ equal to (5), where λ and z are now related to each other by

$$\lambda = 2 + 2\alpha - z - (1/z) - \alpha z^2 - (\alpha/z^2). \tag{27}$$

To each λ correspond four roots z_1, z_2, z_3, z_4, so that the general solution is

$$e^{-\lambda t}(C_1 z_1^n + C_2 z_2^n + C_3 z_3^n + C_4 z_4^n).$$

When one of the roots has absolute value greater than unity this solution grows exponentially for $n \to \infty$, unless the corresponding coefficient is zero. On the other hand, there are two boundary conditions (23), or equivalently (26), which can be satisfied only if at least three constants C are available.

The full discussion of the roots of (27) is elementary but laborious (Appendix C). When studying the continuous limit, however, one is interested only in the low values of λ. In that case it is easy to see that, in addition to the roots $z_1 = e^{i\vartheta}$, $z_2 = e^{-i\vartheta}$ given by

$$\lambda = 2 + 2\alpha - 2\cos\vartheta - 2\alpha\cos 2\vartheta, \tag{28}$$

there is another pair of roots z_3, z_4 with $|z_3| < 1$, and $|z_4| = 1/|z_3| > 1$ (see Eq. 39 in Appendix D). Hence

$$p_n^{(\lambda)}(t) = e^{-\lambda t}(C_1 e^{i\vartheta n} + C_2 e^{-i\vartheta n} + C_3 z_3^n). \tag{29}$$

The term with z_3^n decreases exponentially with increasing n, and is therefore nonpropagating. The phase shift η is still determined by $e^{2i\eta} = -C_1/C_2$.

Going to the continuous limit as before one obtains from (23a)

$$\frac{\partial P^{(0)}}{\partial \tau} = (1 + 4\alpha)\frac{\partial^2 P^{(0)}}{\partial x^2}, \tag{30a}$$

$$\frac{\partial P^{(1)}}{\partial \tau} = (1 + 4\alpha)\frac{\partial^2 P^{(1)}}{\partial x^2}, \tag{30b}$$

$$\frac{\partial P^{(2)}}{\partial \tau} = (1 + 4\alpha)\frac{\partial^2 P^{(2)}}{\partial x^2} + \frac{1 + 16\alpha}{12}\frac{\partial^4 P^{(0)}}{\partial x^4}. \tag{30c}$$

The reason why the exponentially decreasing term does not show up here is that $z_3^n = z_3^{x/\epsilon}$ collapses into the origin. In this connection it is essential to note that $|z_3|$ remains less than unity for $\epsilon \to 0$. In fact, one has from (27) with $\vartheta = \epsilon k$,

$$\lambda = \epsilon^2 k^2 + \cdots,$$

so that

$$z_3(\lambda) = z_3^{(0)} + \epsilon^2 z_3^{(2)} + \cdots.$$

Here $z_0^{(0)} = z_3(0)$ is the third root of (27) for $\lambda = 0$, which is readily seen to lie between -1 and 0.

On the other hand, it is clear that one does not have the freedom to impose two boundary conditions at $x = 0$ on the solutions of (30). The nonpropagating mode associated with z_3 is necessary in order to satisfy these boundary conditions. Thus even in the limit one needs more detailed information concerning the recursion relation (22) than just the form of its continuous approximation (30).[8]

To find the boundary condition for $P(x, \tau)$ at $x = 0$, it is convenient to study the individual modes separately for each time factor $e^{-\lambda t}$. Any factor λ that may arise in the boundary condition will ultimately be replaced with $(-\partial/\partial t)$, so that the boundary condition applies to all normal modes and therefore to $P(x, \tau)$ itself. Accordingly we substitute in (26),

$$p_n^{(\lambda)}(t) = e^{-\lambda t}(\epsilon P_\lambda(n\epsilon) + C_3 z_3^n).$$

To lowest order this leads to two equations for $C_3^{(0)}$ alone:

$$C_3^{(0)} = BC_3^{(0)} z_3^{(0)2},$$

$$C_3^{(0)} z_3^{(0)-1} = AC_3^{(0)} z_3^{(0)} - (B/\alpha)C_3^{(0)} z_3^{(0)2}.$$

It follows that $C_3^{(0)} = 0$, because $Bz_3^{(0)2} < 1$.

The next order yields two equations for $P_\lambda^{(0)}$ and $C_3^{(1)}$:

$$(1 - B)P_\lambda^{(0)} + \{1 - Bz_3^{(0)2}\}C_3^{(1)} = 0, \tag{31a}$$

$$(1 - A + B/\alpha)P_\lambda^{(0)} + \{z_3^{(0)-1} - Az_3^{(0)} + (B/\alpha)z_3^{(0)2}\}C_3^{(1)} = 0. \tag{31b}$$

Unless the determinant vanishes, the only solution is $P_\lambda^{(0)} = C_3^{(1)} = 0$. Thus to this order the exponentially decreasing solution does not come in, and $P^{(0)}(x, \tau)$ obeys the diffusion equation with the usual boundary condition for an absorbing boundary. In the reflecting case (25), however, one has $1 - B = 0$ and $1 - A + B/\alpha = 0$, so that the determinant of (31) does vanish. Hence there is a nontrivial solution

$$P_\lambda^{(0)} \text{ arbitrary}, \quad C_3^{(1)} = 0.$$

In this case, therefore, (31) does not lead to a boundary condition. It is shown in Appendix D that the reflecting boundary is the only case in which the determinant is zero.

To second order (26) gives

$$(1-B)P_\lambda^{(1)} + \{1 - Bz_3^{(0)2}\}C_3^{(2)} = 2BP_\lambda^{(0)\prime}, \qquad (32a)$$

$$(1-A+B/\alpha)P_\lambda^{(1)} + [z_3^{(0)-1} - Az_3^{(0)} + (B/\alpha)z_3^{(0)2}]C_3^{(2)} = (1+A-2B/\alpha)P_\lambda^{(0)\prime}. \qquad (32b)$$

By solving these equations one obtains the boundary value of $P^{(1)}(x,\tau)$ in terms of the slope of $P^{(0)}(x,\tau)$ at $x=0$. For instance, in the case of pure absorption $A=B=0$, one finds successively

$$P^{(1)} = -[z_3^{(0)}/(1-z_3^{(0)})]P^{(0)\prime},$$
$$C_3^{(2)} = [z_3^{(0)}/(1-z_3^{(0)})]P_\lambda^{(0)\prime},$$
$$P^{(2)} = [z_3^{(0)}/(1-z_3^{(0)})]P^{(1)\prime},$$
$$C_3^{(3)} = [z_3^{(0)}/(1-z_3^{(0)})]P_\lambda^{(1)\prime}.$$

This may again be interpreted in terms of a slip coefficient

$$l = -\epsilon[z_3^{(0)}/(1-z_3^{(0)})],$$

or alternatively as a displacement of the boundary point to

$$x = -l = \epsilon[z_3^{(0)}/(1-z_3^{(0)})].$$

In the reflecting case both coefficients of $P_\lambda^{(1)}$ in (32) are zero and the only solution is $C_3^{(2)} = P_\lambda^{(0)\prime} = 0$ (see Appendix D). After somewhat lengthy computations one finds in the next order

$$P^{(1)\prime} = -\tfrac{1}{2}P^{(0)\prime\prime}, \quad C_3^{(3)} = P_\lambda^{(0)\prime\prime}[1-z_3^{(0)2}]^{-1}. \qquad (33)$$

7. THE TOTAL PROBABILITY

In this section we investigate the behavior of the total probability in the successive orders of ϵ. Provided that the boundary is pure, one may use the Euler–Maclaurin formula

$$W = \sum_{n=1}^{\infty} p_n(t) = \epsilon\sum_{n=1}^{\infty} P(n\epsilon,\tau)$$
$$= \int_0^\infty P(x,\tau)dx - \tfrac{1}{2}\epsilon P(0,\tau)$$
$$-\tfrac{1}{12}\epsilon^2 P'(0,\tau) + \tfrac{1}{720}\epsilon^4 P'''(0,\tau) + \cdots.$$

Hence the successive orders of W are

$$W^{(0)} = \int_0^\infty P^{(0)}(x,\tau)dx, \qquad (34a)$$

$$W^{(1)} = \int_0^\infty P^{(1)}(x,\tau)dx - \tfrac{1}{2}P^{(0)}, \qquad (34b)$$

$$W^{(2)} = \int_0^\infty P^{(2)}(x,\tau)dx - \tfrac{1}{2}P^{(1)} - \tfrac{1}{12}P^{(0)\prime}. \qquad (34c)$$

Consider the *one-step random walk* treated in Secs. 3 and 4. Using (14) one obtains, for the derivatives with respect to τ,

$$\dot W^{(0)} = -P^{(0)\prime},$$
$$\dot W^{(1)} = -P^{(1)\prime} - \tfrac{1}{2}P^{(0)\prime\prime},$$
$$\dot W^{(2)} = -P^{(2)\prime} - \tfrac{1}{2}P^{(1)\prime\prime} - \tfrac{1}{6}P^{(0)\prime\prime\prime}.$$

For the reflecting case each term duly vanishes when the boundary conditions (16b) and (16c) are inserted. For the absorbing case the equations reduce to

$$\dot W^{(0)} = -P^{(0)\prime}, \qquad \dot W^{(1)} = -P^{(1)\prime},$$
$$\dot W^{(2)} = -P^{(2)\prime} - [(3-2c)/6c]P^{(0)\prime\prime\prime}.$$

These equations may be combined into

$$\dot W = -P'(0,\tau) - \epsilon^2[(3-2c)/6c]P'''(0,\tau) + \mathcal{O}(\epsilon^3).$$

Consider the *two-step random walk* with pure boundary treated in Sec. 6. One now has, for each normal mode, using $\lambda = \epsilon^2 k^2$,

$$W_\lambda(\tau) = e^{-k^2\tau}\left(\epsilon\sum_{n=1}^{\infty} P_\lambda(n\epsilon) + C_3\frac{z_3}{1-z_3}\right).$$

From this one finds, taking into account that $C_3^{(0)} = C_3^{(1)} = 0$,

$$\dot W^{(0)} = -(1+4\alpha)P^{(0)\prime},$$
$$\dot W^{(1)} = -(1+4\alpha)P^{(1)} - \tfrac{1}{2}(1+4\alpha)P^{(0)\prime\prime},$$
$$\dot W_\lambda^{(2)} = e^{-k^2\tau}[-(1+4\alpha)P_\lambda^{(2)\prime} - \tfrac{1}{2}(1+4\alpha)P_\lambda^{(1)\prime\prime}$$
$$-\tfrac{1}{6}(1+10\alpha)P_\lambda^{(0)\prime\prime\prime} - k^2C_3^{(2)}z_3^{(0)}/(1-z_3^{(0)})].$$

For the *reflecting*, boundary it is clear that $\dot W^{(0)} = 0$ and $\dot W^{(1)} = 0$, see (33). It is now easier to find the next higher order boundary condition for the reflecting case from the conservation of probability, since we found already $C_3^{(2)} = 0$,

$$P^{(2)\prime} = -\tfrac{1}{2}P^{(1)\prime\prime}.$$

Again this can be interpreted by saying that to second order in ϵ the slope of $P(x,\tau)$ vanishes at $x = \tfrac{1}{2}\epsilon$ rather than at $x = 0$.

For the two-step random walk with *purely absorbing boundary*, one finds for the total probability

$$\dot W^{(0)} = -(1+4\alpha)P^{(0)\prime}, \qquad \dot W^{(1)} = -(1+4\alpha)P^{(1)\prime},$$

and to second order

$$\dot W_\lambda^{(2)} = e^{-k^2\tau}\left[-(1+4\alpha)P_\lambda^{(2)\prime} + \frac{1+4\alpha}{2}\,\frac{z_3^{(0)}}{1-z_3^{(0)}}\,P_\lambda^{(0)\prime\prime}\right.$$
$$\left. -\frac{1+10\alpha}{6}P_\lambda^{(0)\prime\prime\prime} - k^2\left(\frac{z_3^{(0)}}{1-z_3^{(0)}}\right)^2 P_\lambda^{(0)\prime}\right].$$

Hence,

$$\dot W^{(2)} = -(1+4\alpha)P^{(2)}$$
$$+\left(\frac{1+4\alpha}{2}\,\frac{z_3^{(0)}(1+z_3^{(0)})}{(1-z_3^{(0)})^2} - \frac{1+10\alpha}{6}\right)P^{(0)\prime\prime\prime}.$$

Finally consider the *impure boundary* case of Sec. 5. One now has

$$W = p_1 + \epsilon\sum_{n=2}^{\infty} P(n\epsilon)$$
$$= p_1 - \epsilon P(\epsilon) + \int_0^\infty P(x,\tau)dx$$
$$-\tfrac{1}{2}\epsilon P(0,\tau) - \tfrac{1}{12}\epsilon^2 P'(0,\tau).$$

From this one finds that, indeed, the total probability is conserved in successive orders of ϵ:

$$\dot W^{(0)} = \int_0^\infty \dot P^{(0)}(x,\tau)dx = -P^{(0)\prime} = 0,$$
$$\dot W^{(1)} = \dot p_1^{(1)} - P^{(1)\prime} - \tfrac{3}{2}\dot P^{(0)}$$
$$= (b/a)P^{(0)\prime\prime} - (b/a - \tfrac{3}{2})P^{(0)\prime\prime} - \tfrac{3}{2}P^{(0)\prime\prime} = 0.$$

8. CONCLUSIONS

(i) Expansion of the master equation of a random walk leads to the familiar diffusion equation for the zeroth and first order, and to diffusion equations with inhomogeneous term for the higher orders. These differential equations are valid at all interior points of the accessible interval.

(ii) In the case of a pure boundary, the continuous probability density $P(x, \tau)$ obeying these equations constitutes a full description of all probabilities; in the case of an impure boundary one or more separate quantities $p_1(\tau), p_2(\tau), \cdots$ are needed for the probabilities at the sites near the boundary.

(iii) For one-step processes the boundary conditions on $P(x, \tau)$ are obtained by substituting in the discrete master equation simply $p_n(\tau) = \epsilon P(n\epsilon, \tau)$, possibly with separate values for $p_1(\tau), p_2(\tau), \cdots$. For s-step processes, however, one has to add linear combinations of $s-1$ additional nonpropagating modes of the homogeneous master equation.

(iv) To zeroth order the result is: The slope of $P(x, \tau)$ vanishes at any boundary that conserves probability, but $P(x, \tau)$ itself vanishes as soon as there is some absorption. To second order the same boundary conditions apply but at a slightly displaced point. In higher orders the boundary condition cannot be stated in such simple terms.

ACKNOWLEDGMENTS

The authors would like to thank Professor K. E. Shuler and Dr. R. Trimble for helpful discussions.

APPENDIX A

In addition to the solutions (7) where z_1, z_2 are given by (8), there may be solutions with $|z_1| < 1, |z_2| > 1$, and $C_2 = 0$. According to (10) this requires that $z_1 = (1-c)^{-1}$, which is consistent with $|z_1| < 1$ only when $c > 0$. The corresponding time constant λ is

$$\lambda_* = c + 1 + 1/(c-1).$$

This is an isolated point of the spectrum because $\lambda_* > 4$. The corresponding normal mode solution is

$$p_n^{(*)}(t) = e^{-\lambda_* t}(1-c)^{-n}.$$

Note that one has $z_1 = e^{i\vartheta_*}$, where ϑ_* is a pole of the S matrix in the upper half of the complex ϑ plane. More explicitly, $\vartheta_* = \pi + i\kappa$ with $e^\kappa = c - 1$. This additional isolated solution has to be included when $c > 2$, in order that the solutions are complete. However, this fact is not relevant when studying the continuous limit, because that limit involves only low values of λ.

To investigate the completeness of the set of solutions (11) with $0 \leqslant \vartheta \leqslant \pi$, we verify the completeness relation, or to put it differently, we compute $U_{nm}(0)$:

$$\frac{2}{\pi}\int_0^\pi \sin[\vartheta n + \eta(\vartheta)]\sin[\vartheta m + \eta(\vartheta)]d\vartheta$$
$$= \frac{1}{2\pi}\int_0^{2\pi}(e^{i\vartheta(n-m)} - e^{i\vartheta(n+m)+2i\eta(\vartheta)})d\vartheta$$
$$= \delta_{nm} - \frac{1}{2\pi}\int_0^{2\pi} S(\vartheta)e^{i(n+m)\vartheta}\,d\vartheta.$$

The integration path may be extended by adding the line from $i\infty$ to 0, and the line from 2π to $2\pi + i\infty$, because their contributions cancel owing to the periodicity. Furthermore the integrand vanishes at $i\infty$ since $S(\vartheta) \sim e^{|\vartheta|}$, whereas $n + m \geqslant 2$. For $0 \leqslant c \leqslant 2$, there are no poles in the upper half plane, so that the integral vanishes and the desired completeness relation is obtained. For $c > 2$ there is a pole $\vartheta_* = \pi + i\kappa$, whose residue yields for the integral

$$c(c-2)e^{-\kappa(n+m)}.$$

This shows that for $c > 2$ one has to include the isolated solution

$$p_n^{(*)}(t) = e^{-\lambda_* t}[c(c-2)]^{1/2}(c-1)^{-n}.$$

APPENDIX B

From (19) follows on eliminating q and expressing λ through (9),

$$-\frac{C_1}{C_2} = -e^{-i\vartheta}\frac{1 + (a+b-2)e^{-i\vartheta} + (1-b)e^{-2i\vartheta}}{1 + (a+b-2)e^{i\vartheta} + (1-b)e^{2i\vartheta}}. \quad \text{(B1)}$$

Again denoting this quantity by $e^{2i\eta(\vartheta)}$, one sees that the normal mode solutions are given by (11) for $n \geqslant 2$, and in addition one has

$$p_1^{(\lambda)} = e^{-\lambda t}(2/\pi)^{1/2}\frac{b\sin[2\vartheta + \eta(\vartheta)]}{a-\lambda}.$$

The denominator vanishes for $\lambda = a$, but so does the numerator.

The denominator of (B1) has one zero in the upper half plane when $a + 2b > 4$ and otherwise none. The zero is given by

$$e^{i\vartheta} = \{2 - a - b + [(a+b)^2 - 4a]^{1/2}\}/2(1-\mathrm{b}).$$

When going to the limit one has $\vartheta = \epsilon k$, so that in terms of k the pole moves to $+i\infty$ when ϵ tends to zero. This is the reason why the corresponding normal mode does not appear in the expansion in powers of ϵ; but it also demonstrates that the expansion can only be an asymptotic one.

APPENDIX C

To solve (27) for z put $z + z^{-1} = 2w$:

$$4\alpha w^2 + 2w + (\lambda - 2 - 4\alpha) = 0. \quad \text{(C1)}$$

Either solution of this quadratic equation for w gives rise to two roots z to be found from

$$z^2 - 2wz + 1 = 0.$$

Thus the four roots of (27) consist of two pairs z_1, z_2 and z_3, z_4, such that $z_1 z_2 = 1$ and $z_3 z_4 = 1$. In order that there are at least three roots with $|z_j| \leqslant 1$, it is necessary that at least one pair has the form $e^{i\vartheta}, e^{-i\vartheta}$ with real ϑ. Hence all admissible values of λ are of the form (28), and lie therefore between 0 and

$$2 + 4\alpha + 1/4\alpha = [(4\alpha)^{1/2} + 1/\sqrt{4\alpha}]^2. \quad \text{(C2)}$$

This upper bound is also the condition that the two solutions of (C1)

$$w_\pm = \{-1 \pm [(4\alpha + 1)^2 - 4\alpha\lambda]^{1/2}\}/4\alpha$$

are real. In addition it is necessary that at least one of these has the form $\cos\vartheta$, i.e., that it lies between -1 and $+1$.

To investigate this condition we have to distinguish between the two cases $\alpha \leqslant \frac{1}{4}$ and $\alpha > \frac{1}{4}$. In the case $\alpha \leqslant \frac{1}{4}$, one readily finds that $w_- < -1$, while $|w_+| \leqslant 1$ holds only for $0 \leqslant \lambda \leqslant 4$. This determines the spectrum of the bounded two-step random walk with $\alpha \leqslant \frac{1}{4}$; to each λ in (0, 4) belongs a single normal mode solution.

In the case $\alpha > \frac{1}{4}$, the same nondegenerate spectrum exists; but in addition one finds that for λ between 4 and the upper bound (C2) both w_+ and w_- lie in the interval $(-1, +1)$. Hence for these values of λ there are *four* admissible roots z_j, and therefore *two* linearly independent solutions of the bounded two-step random walk with $\alpha > \frac{1}{4}$. For the low values of λ relevant for the continuous limit this complication cannot occur.

APPENDIX D

The determinant of (31) is

$$\begin{vmatrix} 1-B & 1-Bz^2 \\ 1-A+B/\alpha & z^{-1}-Az+(B/\alpha)z^2 \end{vmatrix},$$

where z stands for $z_3^{(0)}$. After subtracting the left column from the column on the right, a factor $1-z$ splits off. Subsequently multiplying the top row with $1/\alpha$ and adding it to the bottom row, and adding the left column to the right one, we get

$$(1-z)\begin{vmatrix} 1-B & 1+Bz \\ 1+(1/\alpha)-A & (1/z)+1+1/\alpha \end{vmatrix}.$$

The two elements of the left column are nonnegative and will be denoted by $R = 1-B$, $S = 1+1/\alpha - A$. In terms of R and S the value of the determinant is found to be

$$(1-z)\{RSz-(1+z)S+[(1/z)+1+1/\alpha]R\}.$$

The first and second terms in { } are negative (or zero when $S=0$). Hence, if we show that the coefficient of the third term is negative, it proves that the determinant cannot vanish unless $R=S=0$, i.e., unless the boundary is reflecting. It suffices to show

$$1+[1+(1/\alpha)]z < 0, \tag{D1}$$

where the explicit value of z is found from solving (27) for $\lambda = 0$,

$$z = z_3^{(0)} = -(1+1/2\alpha)+[(1/\alpha)+1/4\alpha^2]^{1/2}. \tag{D2}$$

On substituting this z in (D1), one easily finds that the inequality is true. This completes the proof.

APPENDIX E

Equations (32) reduce in the reflecting case to

$$(1-z^2)C_3^{(2)} = 2P_\lambda^{(0)\prime},$$

$$\{z^{-1}-[1+(1/\alpha)]z+(1/\alpha)]z^2\}C_3^{(2)} = [2-(1/\alpha)]P_\lambda^{(0)\prime}.$$

The determinant is

$$(1-z)\begin{vmatrix} 1+z & -2 \\ (1/z)+1-z/\alpha & -2+1/\alpha \end{vmatrix} = (2/z)(1-z)^2\{1+(1+1/2\alpha)z\}.$$

The first two factors are not zero and for the factor { } one has, using (D1),

$$1+z+(z/2\alpha) > 1+z+(z/\alpha) > 0.$$

This justifies the statement that the only solution of (32) in the reflecting case is $P^{(0)\prime} = C_3^{(2)} = 0$.

---

* Supported in part by the National Science Foundation.

† Permanent address: Institute for Theoretical Physics, Rijksuniversiteit, Utrecht, Netherlands.

1 S. T. Choh and G. E. Uhlenbeck, *The Kinetic Theory of Dense Gases* (University of Michigan 1958); E. G. D. Cohen, "The Boltzmann Equation and its Generalization to Higher Densities," in *Fundamental Problems in Statistical Mechanics* edited by E. G. D. Cohen (North-Holland, Amsterdam, 1962).

2 For literature see P. A. Selwyn and I. Oppenheim, Physica **54**, 161 (1971).

3 (a) J. C. Maxwell, Phil. Trans. [A] **170**, 231 (1879); H. A. Kramers, Nuovo Cimento Suppl. **6**, 297 (1949). (b) For the general equation (1) the precise condition is: Those r for which a_r does not vanish must not have a common divisor other than unity.[4] (c) Those who feel that total probability should always be equal to unity may either read "number of particles" instead of "probability," or add a limbo state containing all absorbed probability.

4 F. L. Spitzer and C. J. Stone, Illinois J. Math. **4**, 253 (1960).

5 (a) See e.g. W. Feller, *An Introduction to Probability Theory and Its Applications* (Wiley, New York, 1957), 2nd ed., Vol. 1, Chaps. 14 and 16; F. L. Spitzer, *Principles of Random Walk* (Van Nostrand, Princeton, N.J. 1964), Chap. 4; D. R. Cox and H. D. Miller, *The Theory of Stochastic Processes* (Wiley, New York, 1965), Chap. 2; K. Lakatos-Lindenberg and K. E. Shuler J. Math. Phys. **12**, 633 (1971). (b) This expansion is more elementary than the one in Ref. 6, and is only possible for a homogeneous case, which has no equilibrium distribution with finite variance.

6 N. G. van Kampen, Can. J. Phys. **39**, 551 (1961); also in *Fluctuation Phenomena in Solids*, edited by R. E. Burgess (Academic, New York, 1965).

7 H. Helmholtz and G. v. Piotrowski, Wien. Sitzungsber. **40**, 607 (1860)

8 This statement needs to be qualified inasmuch as for an s-step random walk the coefficients in the continuous approximation (30) together uniquely fix the a_r and thereby the recursion relation from which one started. In principle, therefore, they contain all the information needed for computing z_3, but in a rather recondite form.

4) First Passage Time Problems

First passage time problems are those that involve random processes that terminate when some condition on the random function is satisfied for the first time. For example, in the dissociation model of Montroll and Shuler[4.4.1] a diatomic molecule dissociates when the bond energy becomes sufficiently large. The earliest studies of first passage time problems appear to be those of Smoluchowski[4.4.2] and Schrodinger[4.4.3] who studied first passage problems for colloidal particles in order to analyze experiments of Svedberg[4.4.4] and Westgren [4.4.5] that have proved to be fundamental to our understanding of Brownian motion. The mathematical literature contains much earlier examples of first passage time problems particularly in the analysis of the "gambler's ruin" problem in which two gamblers with initially fixed amounts of capital play against one another until the capital of the loser is exhausted.[4.4.6]

The first paper, by Darling and Siegert gives an extensive analysis of first passage time problems for Markov processes in one dimension in continuous time. They show that when the random process can be described by a Fokker-Planck equation, the distribution of the first passage time satisfies the adjoint equation. In one dimension the analysis can be carried quite far because, for example, the spatial moments satisfy ordinary differential equations for which explicit solutions are available. Darling and Siegert's paper goes deeper into a subject first explored by Pontryagin, Andronow, and Witt (4.4.7). The second paper in this section is a short review of first passage time problems as applied to physical problems. Further useful material on the subject includes the approximations first given by Wald (4.4.8) for first passage time probabilities and moments and the use of renewal theoretic ideas (4.4.9) for non-Markovian problems in one dimension (4.4.10).

References

4.4.1. E. W. Montroll, K. E. Shuler, Adv. Chem. Phys. 1, 361 (1958).
4.4.2. M. von Smoluchowski, S. B. Akad. Wiss. Wien. 2a, 124, 339 (1915).
4.4.3. E. Schrodinger, Phys. Zeit. 16, 289 (1915).
4.4.4. T. Svedberg, Zeit. Phys. Chem. 77, 147 (1911).
4.4.5. A. Westgren, Arkiv. Math. Astron. Fys. 11, 8, 14 (1916); 13, 14 (1918).
4.4.6. I. Todhunter, History of the Mathematical Theory of Probability (Chelsea reprint, New York) 1949.
4.4.7. L. Pontryagin, A. Andronow, A. Witt. Zh. Eksp. i. Teor. Fiz. 3, 172 (1933).
4.4.8. A. Wald, Sequential Analysis (John Wiley & Sons, New York) 1947.
4.4.9. D. R. Cox, Renewal Theory (John Wiley & Sons, New York) 1962.
4.4.10. G. H. Weiss, Sep. Sci. 5, 51 (1970).

THE FIRST PASSAGE PROBLEM FOR A CONTINUOUS MARKOV PROCESS[1]

By D. A. Darling and A. J. F. Siegert

Columbia University, University of Michigan, and Northwestern University

Summary. We give in this paper the solution to the first passage problem for a strongly continuous temporally homogeneous Markov process $X(t)$. If $T = T_{ab}(x)$ is a random variable giving the time of first passage of $X(t)$ from the region $a > X(t) > b$ when $a > X(0) = x > b$, we develop simple methods of getting the distribution of T (at least in terms of a Laplace transform). From the distribution of T the distribution of the maximum of $X(t)$ and the range of $X(t)$ are deduced. These results yield, in an asymptotic form, solutions to certain statistical problems in sequential analysis, nonparametric theory of "goodness of fit," optional stopping, etc. which we treat as an illustration of the theory.

1. Introduction. There are certain generalizations of the classical gambler's ruin problem which appear in various guises in numerous applications—besides statistical problems there are physical applications in the theory of noise, in genetics, etc. The exact solution of the associated random walk (or Markov chain) problem is often analytically difficult, if not impossible to obtain, and one is usually content with asymptotic solutions. The nature of the asymptotic solution is generally such that it is the solution to a Markov chain problem in which the length of the steps, and the interval between them, approach zero and which may in the limit be regarded as some sort of continuous stochastic process.

This circumstance suggests we might solve directly the associated problem with regard to the stochastic process and so obtain the asymptotic solution to the Markov chain problem without the intervention of a limiting process. Aside from the difficulty of justifying the interchange of these limiting operations, it turns out that this procedure is often quite feasible and leads to simple solutions. Using this idea Doob [7] obtained in a direct way the Kolmogorov-Smirnov limit theorems and the principle was further exploited by Anderson and Darling [1]. The general principle is, of course, quite old, and in connection with random walk problems goes back at least to Rayleigh.

A general feature of this method is that the analytical difficulties, if any, are revealed as more or less classical boundary value problems, eigenvalue problems, etc.—this intrinsic nature of the problem often being masked by the discrete approach. On the other hand, it suffers from the serious defect of giving no

Received 2/3/52, revised 3/3/53.

[1] The major part of this work was done while the authors were consultants with The RAND Corporation.

information as to the difference between the actual solution and the asymptotic one—information which is essential in the numerical applications.

In the present paper we treat the first passage (or ruin, or absorption probability) problem for a general class of Markov processes (cf. Section 2) and obtain the solution in the form of a Laplace transform (Section 3). This Laplace transform is generally given as a simple function of the solutions to an ordinary differential equation (Section 4). The methods used are similar to those used in the discrete theory by Wald [17] (fundamental identity) and Feller [9] (renewal and generating function techniques), but the analysis is considerably simplified, at least in a formal way, and not restricted to additive processes. It turns out that there is an intimate relationship between the one- and two-sided absorption probabilities, and the probability of eventual absorption in one of the boundaries.

We illustrate the theory in Section 5 by solving a problem of Wald [17] in the sequential test of the mean of a normal population against a single alternative, the derivation of a nonparametric test used by Anderson and Darling [1] and the solution to the optional stopping problem (Robbins [15]). These problems are treated by solving the associated absorption problem with the Wiener-Einstein process and the Uhlenbeck process.

In Section 6 we study the first passage moments which can be obtained by an expansion of the Laplace transforms or again through differential equations which can be explicitly solved in quadratures. There are some quite interesting relations between the moments.

In Section 7 we develop the distribution of the range which has been used by Feller [10] in a statistical study.

2. Definitions, notations, assumptions, etc. Given a stochastic process $X(t)$ with $X(0) = x$, $a > x > b$, we define the *first passage time* $T_{ab}(x)$ as the random variable

$$T = T_{ab}(x) = \sup \{t \mid a > X(\tau) > b, 0 \leqq \tau \leqq t\}.$$

We make the following assumptions about the stochastic process $X(t)$.

A) $X(t)$ has a transition probability

$$P(x \mid y, t) = \Pr\{X(t + s) < y \mid X(s) = x\}, \qquad s > 0,$$

satisfying the Chapman-Kolmogorov equation

$$P(x \mid y, t_1 + t_2) = \int_{-\infty}^{\infty} P(z \mid y, t_2)\, d_z P(x \mid z, t_1), \quad t_1 > 0, t_2 > 0;$$

that is, $X(t)$ is temporally homogeneous and stochastically definite (e.g. Markovian).

B) $X(t)$ is continuous with probability one (or is *strongly continuous*).

If $X(t)$ satisfies A) sufficient conditions on P are known that it satisfy B), cf. Doeblin [5], Fortet [11], Ito [12]. These conditions generally imply further that

P satisfies the diffusion equation of Section 4. Note that A) and B) imply the *existence* of the random variable T, and we denote by $F_{ab}(x \mid t)$ the distribution function of T, $F_{ab}(x \mid t) = Pr\{T_{ab}(x) < t\}$.

In the work to follow we shall presume P and F have derivatives p, f; these being the densities

$$p(x \mid y, t) = \frac{\partial}{\partial y} P(x \mid y, t)$$

$$f_{ab}(x \mid t) = \frac{\partial}{\partial t} F_{ab}(x \mid t),$$

the modification of the results if these conditions are not met being more or less immediate. The existence of a density for T has been established by Fortet [11] under some circumstances. In this fundamental paper of Fortet on absorption probabilities there is just one absorbing barrier, but the modification of his results for two barriers is easy.

If $a = +\infty$ or $b = -\infty$ so that we have a one-sided absorption time we write $T_c(x)$ as the corresponding random variable. That is

$$T_c(x) = \begin{cases} T_{\infty, c}(x) \text{ if } x > c \\ T_{c, -\infty}(x) \text{ if } x < c \end{cases}$$

with a corresponding distribution function $F_c(x \mid t)$ and density $f_c(x \mid t)$.

It may happen of course that absorption is not a certain event and that T is not a proper random variable, that is $Pr\{T_c(x) < \infty\} = F_c(x \mid \infty) < 1$ (or similarly for $T_{ab}(x)$) and in this case we may still meaningfully treat the conditional density of T, under the condition $T < \infty$.

We need, in addition, the conditional distribution of $T_{ab}(x)$ under the condition that the absorption takes place into the barrier $\underline{a}$, which we denote by $F_{ab}^{+}(x \mid t)$:

$$F_{ab}^{+}(x \mid t) = \Pr\{T_{ab}(x) < t, T_{ab}(x) = T_a(x)\}$$

and $F_{ab}^{-}(x \mid t)$ will denote a similar expression for the lower barrier $\underline{b}$. Hence

$$F_{ab}(x \mid t) = F_{ab}^{+}(x \mid t) + F_{ab}^{-}(x \mid t)$$

and the corresponding densities are $f_{ab}^{+}(x \mid t)$ and $f_{ab}^{-}(x \mid t)$.

We denote by a circumflex over the corresponding function its Laplace transform on t; for example

$$\hat{p}(x \mid y, \lambda) = \int_0^\infty e^{-\lambda t} p(x \mid y, t)\, dt,$$

$$\hat{f}_{ab}^{+}(x \mid \lambda) = \int_0^\infty e^{-\lambda t} f_{ab}^{+}(x \mid t)\, dt,$$

etc. The continuity of the process $X(t)$ ensures the existence of these transforms.

3. The distribution of T. In this section we obtain the distribution of T in terms of the transition density p of the process. Theorem 3.1 for the one-sided barrier is due to Siegert [16] essentially.

THEOREM 3.1. *If $X(t)$ satisfies conditions* A) *and* B), *then $\hat{p}(x \mid y, \lambda)$ is a product*

$$\hat{p}(x \mid y, \lambda) = \begin{cases} u(x)u_1(y), & y > x \\ v(x)v_1(y). & y < x \end{cases}$$

and

$$\hat{f}_c(x \mid \lambda) = \begin{cases} \dfrac{u(x)}{u(c)}, & x < c \\ \dfrac{v(x)}{v(c)}, & x > c. \end{cases} \tag{3.1}$$

We note that absorption may be uncertain and $\hat{f}_c(x \mid 0) = Pr\{T_c(x) < \infty\}$ may be less than 1. A necessary and sufficient condition that absorption be certain is that $\hat{f}_c(x \mid 0) = 1$.

To prove the theorem we use a renewal principle which is very old. We have by A) and B) for $y > c > x$

$$p(x \mid y, t) = \int_0^t f_c(x \mid \tau) p(c \mid y, t - \tau)\, d\tau$$

by a direct enumeration of the paths going from x to y. On taking Laplace transforms we obtain

$$\hat{p}(x \mid y, \lambda) = \hat{f}_c(x \mid \lambda)\hat{p}(c \mid y, \lambda), \qquad y > c > x$$

and thus $\hat{p}(x \mid y, \lambda)$ is a function of x times a function of y, say $u(x)u_1(y)$ and hence for $y > c > x$ we get $\hat{f}_c(x \mid \lambda) = u(x)/u(c)$. Similarly, for $y < c < x$ we obtain $\hat{f}_c(x \mid \lambda) = v(x)/v(c)$ and hence for any c, x we obtain the conclusions to the theorem. Finally it follows by cancelling any factor which depends only on λ that $u(x)$ and $v(x)$ are uniquely determined.

THEOREM 3.2. *Let $X(t)$ satisfy* A) *and* B) *and let the functions $u(x)$ and $v(x)$ be as in Theorem* 3.1. *Then*

$$\hat{f}^{+}_{ab}(x \mid \lambda) = \frac{v(b)u(x) - u(b)v(x)}{u(a)v(b) - u(b)v(a)} \tag{3.2}$$

$$\hat{f}^{-}_{ab}(x \mid \lambda) = \frac{u(a)v(x) - v(a)u(x)}{u(a)v(b) - u(b)v(a)} \tag{3.3}$$

$$\hat{f}_{ab}(x \mid \lambda) = \frac{v(x)(u(a) - u(b)) - u(x)(v(a) - v(b))}{u(a)v(b) - u(b)v(a)}. \tag{3.4}$$

To prove the theorem we consider the two expressions

$$f_a(x \mid t) = f_{ab}^{+}(x \mid t) + \int_0^t f_{ab}^{-}(x \mid \tau) f_a(b \mid t - \tau)\, d\tau$$

$$f_b(x \mid t) = f_{ab}^{-}(x \mid t) + \int_0^t f_{ab}^{+}(x \mid \tau) f_b(a \mid t - \tau)\, d\tau$$

which are established by a direct enumeration. Considering f^+ and f^- as unknown this pair of simultaneous integral equations is solved immediately by taking Laplace transforms

$$\hat{f}_a(x \mid \lambda) = \hat{f}_{ab}^{+}(x \mid \lambda) + \hat{f}_{ab}^{-}(x \mid \lambda)\hat{f}_a(b \mid \lambda) \tag{3.5}$$

$$\hat{f}_b(x \mid \lambda) = \hat{f}_{ab}^{-}(x \mid \lambda) + \hat{f}_{ab}^{+}(x \mid \lambda)\hat{f}_b(a \mid \lambda) \tag{3.6}$$

which are 2 linear equations in 2 unknowns. On using the expressions in Theorem 3.1 for $\hat{f}_a$ and $\hat{f}_b$ we get (3.2) and (3.3) for $\hat{f}_{ab}^{+}$ and $\hat{f}_{ab}^{-}$ and the last expression (3.4) is obtained by noting $\hat{f}_{ab} = \hat{f}_{ab}^{+} + \hat{f}_{ab}^{-}$.

A random variable closely related to T is the maximum of $X(t)$, and we define

$$M(x, t) = \sup_{0 \leqq \tau \leqq t} | X(\tau) |, \qquad X(0) = x. \tag{3.7}$$

Denoting the distribution of M by $G(x \mid \xi, t)$ we have clearly

$$\begin{aligned} G(x \mid \xi, t) &= Pr\{M(x, t) < \xi\} = \Pr\{T_{\xi,-\xi}(x) > t\} \\ &= 1 - F_{\xi,-\xi}(x \mid t), \qquad \xi > | x |, \end{aligned} \tag{3.8}$$

so that the distribution of M is given directly through that of T. On taking Laplace transforms of (3.8) we obtain the following corollary

COROLLARY 3.3. $\hat{G}(x \mid \xi, \lambda) = 1/\lambda(1 - \hat{f}_{\xi,-\xi}(x \mid \lambda))$ *for* $\hat{f}_{\xi,-\xi}(x \mid \lambda)$ *as in Theorem* 3.2.

For a symmetrical process there is a specially simple formula for the Laplace transform of $T_{a,-a}(x)$. A process $X(t)$ is symmetrical if $p(x \mid y, t) = p(-x \mid -y, t)$ for all x, y, t. In this case $u(x) = v(-x)$ and Theorem 3.2 yields the following corollary.

COROLLARY 3.4. *For a symmetrical process*

$$\hat{f}_{a,-a}(x \mid \lambda) = \frac{u(x) + u(-x)}{u(a) + u(-a)}, \qquad | x | < a. \tag{3.9}$$

4. A differential equation. The function $p(x \mid y, t)$ will in most cases of interest satisfy the so-called diffusion equation

$$\frac{\partial p}{\partial t} = A(x) \frac{\partial p}{\partial x} + \tfrac{1}{2}B^2(x) \frac{\partial^2 p}{\partial x^2} \tag{4.1}$$

with initial and boundary conditions $p(\infty \mid y, t) = p(-\infty \mid y, t) = 0$, $p(x \mid y, 0) = \delta(x - y)$ (the Dirac function). Sufficient conditions on p, involving the infinitesimal transition moments, are known in order that p satisfy an equation of the type (4.1), and which generally further ensure the process is continuous with probability one (cf. Doeblin [5]). When A and B^2 are given a priori, conditions on them are known which ensure that (4.1) has a unique solution which is the transition density of a process continuous with probability 1. (Cf. Fortet [11]). But general necessary and sufficient conditions are not known, and it does not appear to be known under what conditions a process continuous with probability one satisfies a diffusion equation. However, for specific processes these points are generally easy to resolve.

The following theorem shows that for processes satisfying (4.1) $\underline{u}$ and $\underline{v}$ can be determined from a differential equation.

THEOREM 4.1. *If $p(x \mid y, t)$ uniquely satisfies (4.1) with the stated boundary conditions and $X(t)$ is continuous with probability one, the functions $u(x)$ and $v(x)$ can be chosen as any two linearly independent solutions of the differential equation*

$$\tfrac{1}{2}B^2(x)\frac{d^2w}{dx^2} + A(x)\frac{dw}{dx} - \lambda w = 0. \tag{4.2}$$

To prove the theorem we note that if p satisfies (4.1) its Laplace transform satisfies

$$\lambda\hat{p} = A\frac{d\hat{p}}{dx} + \tfrac{1}{2}B^2\frac{d^2\hat{p}}{dx^2} \tag{4.3}$$

and indeed $-\hat{p}$ is the Green's solution to this equation over the infinite interval $(-\infty < x < \infty)$. As a consequence, if $u(\infty) = v(-\infty) = 0$ and $u(x)$, $v(x)$ satisfy (4.3) we obtain to a constant factor

$$\hat{p}(x \mid y, \lambda) = \begin{cases} v(x)u(y) & y \geqq x \\ v(y)u(x) & y \leqq x \end{cases}$$

so that we obtain the previous expression (3.1) for $\hat{f}_c(x \mid \lambda)$ and consequently we obtain (3.2), (3.3) and (3.4). But since (3.2), (3.3) and (3.4) are invariant under any nonsingular linear transformation of $\underline{u}$ and $\underline{v}$ we obtain Theorem 4.1.

As for (3.9) we can choose for $u(x)$ any solution to (4.2) provided $u(x)$ and $u(-x)$ are linearly independent.

The customary way to obtain the first passage probability $f_{ab}(x \mid t)$ is to solve (4.1) with the boundary conditions $p(a \mid y, t) = p(b \mid y, t) = 0$, $p(x \mid y, 0) = \delta(x - y)$ and then we should have $F_{ab}(x \mid t) = 1 - \int_b^a p(x \mid y, t)\, dy$ (cf. Fortet [11] for a proof and Lévy [14] for a general discussion). By using the Laplace transform method this will give (3.4) for $\hat{f}_{ab}(x \mid \lambda)$, but it does not appear to give $\hat{f}^+$ and $\hat{f}^-$.

Since $\hat{f}^+_{ab}(x \mid 0)$ is the probability that absorption in the barrier $\underline{a}$ occurs before absorption in $\underline{b}$, we should expect that, putting $\lambda = 0$ in (4.2), the solution to

$$\tfrac{1}{2}B^2\frac{d^2\phi}{dx^2} + A\frac{d\phi}{dx} = 0$$

with $\phi(a) = 1, \phi(b) = 0$ should give this probability. Khintchine [13] has proved this result directly from the limiting case of a Markov chain without the use of a stochastic process. Barnard [3] has considered this result in connection with a sequential analysis problem.

5. A few examples.

a) *The Wiener-Einstein process.* Here $X(t)$ is the free Brownian motion; $X(t)$ is Gaussian with mean 0 and covariance $E(X(s)X(t)) = \min(s, t)$ and its transition density p satisfies the differential equation $\frac{\partial p}{\partial t} = \frac{1}{2}\frac{\partial^2 p}{\partial x^2}$ (i.e. $A = 0$, $B^2 = 1$). Two linearly independent solutions to $\frac{1}{2}W'' - \lambda W = 0$ are $u(x) = e^{-\sqrt{2\lambda}x}$ and $v(x) = e^{\sqrt{2\lambda}x} = u(-x)$ and hence we obtain from (3.9)

$$\hat{f}_{a,-a}(x \mid \lambda) = \frac{\cosh \sqrt{2\lambda}\, x}{\cosh \sqrt{2\lambda}\, a}, \qquad |x| < a. \tag{5.1}$$

The inversion of this Laplace transform is easy, and we obtain

$$f_{a,-a}(x \mid t) = \frac{\pi}{a^2}\sum_{j=0}^{\infty} (-1)^j (j + \tfrac{1}{2}) \cos\left\{(j + \tfrac{1}{2})\frac{\pi x}{a}\right\} e^{-(j+\frac{1}{2})^2\pi^2 t/2a^2}$$

and by integration on t

$$F_{a,-a}(x \mid t) = Pr\{T_{a,-a}(x) < t\}$$
$$= 1 - \frac{2}{\pi}\sum_{j=0}^{\infty} \frac{(-1)^j}{j + \frac{1}{2}} \cos\left\{(j + \tfrac{1}{2})\frac{\pi x}{a}\right\} e^{-(j+\frac{1}{2})^2\pi^2 t/2a^2}.$$

This completely solves the case of Brownian motion for general barriers, since

$$F_{a,b}(x \mid t) = F_{(a-b)/2,-(a-b)/2}\left(x - \frac{a+b}{2} \,\middle|\, t\right). \tag{5.2}$$

This result is well known (Bachelier [2], Lévy [14]) and alternatively can be obtained by the method of images with the heat equation $\frac{\partial p}{\partial t} = \frac{1}{2}\frac{\partial^2 p}{\partial x^2}$.

b) *The Uhlenbeck process.* Here $X(t)$ is stationary, Markovian, and Gaussian, with mean 0 and covariance $E(X(s)X(t)) = e^{-|s-t|}$ and the transition density satisfies (4.1) with $B^2 = 2$, $A = -x$. Solutions to

$$\frac{d^2w}{dx^2} - x\frac{dw}{dx} - \lambda w = 0$$

are $u(x) = e^{x^2/4}D_{-\lambda}(x)$ and $v(x) = e^{x^2/4}D_{-\lambda}(-x)$ where $D_\nu(z)$ is the Weber function, (cf. Whittaker and Watson [18]). Hence (3.9) gives

$$\hat{f}_{a,-a}(x \mid \lambda) = \exp\left\{\frac{x^2}{4} - \frac{a^2}{4}\right\}\frac{D_{-\lambda}(x) + D_{-\lambda}(-x)}{D_{-\lambda}(a) + D_{-\lambda}(-a)}, \tag{5.3}$$

but it appears very difficult to invert this transform. For the particular case $x = 0$ this result (5.3) was obtained from a limiting case of an Ehrenfest urn scheme describing molecular equilibrium by Bellman and Harris [4].

c) *A problem of Wald in sequential analysis.* Let $X_1, X_2, \cdots$, be independent random variables, normally distributed with an unknown mean θ and a known variance K^2. That is, the density of X_i is

$$\phi(x, \theta) = \frac{1}{\sqrt{2\pi}\, K} e^{-(x-\theta)^2/2K^2}.$$

According to the sequential likelihood ratio test of Wald, in order to test the hypothesis H_2 that $\theta = \theta_2$ against the hypothesis $H_1 : \theta = \theta_1$ we consider random variables

$$Z_i = \log \frac{\phi(X_i, \theta_1)}{\phi(X_i, \theta_2)} = \frac{\theta_1 - \theta_2}{K^2}\left(X_i - \frac{\theta_1 + \theta_2}{2}\right)$$

and let $S_j = Z_1 + Z_2 + \cdots + Z_j$. Then for $a > 0 > b$ we study the random variable N defined as the smallest integer for which either $S_N > a$ or $S_N < b$ and determine for this N the probabilities of these outcomes.

Now

$$E(Z_i) = \frac{\theta_1 - \theta_2}{K^2}\left(\theta - \frac{\theta_1 + \theta_2}{2}\right) = \mu, \tag{5.4}$$

$$\text{Var}\,(Z_i) = \left(\frac{\theta_1 - \theta_2}{K}\right)^2 = \sigma^2; \tag{5.5}$$

so that this suggests we study a Gaussian process $S(t)$ with independent increments and with $E(S(t)) = \mu t$ and Var $(S(t)) = \sigma^2 t$ (a linear transformation of the Wiener process). Then the joint distribution of $S_1, S_2, \cdots, S_j$ is the same as the joint distribution of $S(1), S(2), \cdots, S(j)$, and in place of finding the distribution of N we approximate to it by finding the distribution of the absorption time $T_{a,b}(0)$ in connection with the process $S(t)$. It should be remarked that the nature of this approximation is quite different from Wald's approximation of "neglecting the excess" since the process $S(t)$ may leave and re-enter one of the barriers between two consecutive integer time instants.

The differential equation satisfied by the transition density p of the process $S(t)$ is

$$\frac{\partial p}{\partial t} = \mu \frac{\partial p}{\partial x} + \frac{\sigma^2}{2} \frac{\partial^2 p}{\partial x^2};$$

that is, $A = \mu$, $B^2 = \sigma^2$, and (4.2) becomes

$$\frac{\sigma^2}{2} \frac{d^2 w}{dx^2} + \mu \frac{dw}{dx} - \lambda w = 0. \tag{5.6}$$

It is simple to solve this equation with constant coefficients and since the two roots of $\frac{\sigma^2}{2}\xi^2 + \mu\xi - \gamma = 0$ are

$$\xi_1 = \frac{-\mu + \sqrt{\mu^2 + 2\sigma^2\lambda}}{\sigma^2}, \qquad \xi_2 = \frac{-\mu - \sqrt{\mu^2 + 2\sigma^2\lambda}}{\sigma^2},$$

two linearly independent solutions to (5.6) are $u(x) = e^{\xi_1 x}$ and $v(x) = e^{\xi_2 x}$ and hence by Theorem 3.2 we immediately obtain $\hat{f}^+$, $\hat{f}^-$, and $\hat{f}$ and the problem is formally solved. The expressions are to be considered for $x = 0$, and (3.2) gives for $x = 0$, with this $u(x)$, $v(x)$,

$$\hat{f}^+(0 \mid \lambda) = \frac{e^{\xi_1 b} - e^{\xi_2 b}}{e^{\xi_2 a + \xi_1 b} - e^{\xi_1 a + \xi_2 b}}$$

and at $\lambda = 0$ this gives the probability of being absorbed into the barrier a before b, and we abbreviate $L^+(\theta) = \hat{f}_{ab}^+(0 \mid 0)$ for this probability. For $\lambda = 0$ we have $\xi_1 = 2\mu/\sigma^2$, $\xi_2 = 0$ so that

$$L^+(\theta) = \frac{e^{-(2\mu b/\sigma^2)} - 1}{e^{-(2\mu/\sigma^2)b} - e^{-(2\mu/\sigma^2)a}}. \tag{5.7}$$

According to the test of Wald we should choose the barriers a and b so that $L^+(\theta_1) = 1 - \beta$, $L^+(\theta_2) = \alpha$ where α, β, are given positive numbers with $\alpha + \beta < 1$. For $\theta = \theta_1$ we have $2\mu = \sigma^2$ and for $\theta = \theta_2$ we have $2\mu = -\sigma^2$ from (5.4) and (5.5). Hence from (5.7) we get as two equations for a and b

$$1 - \beta = \frac{e^{-b} - 1}{e^{-b} - e^{-a}}, \qquad \alpha = \frac{e^{b} - 1}{e^{b} - e^{a}},$$

which are easily solved to give

$$a = \log\frac{1-\beta}{\alpha}, \qquad b = \log\frac{\beta}{1-\alpha}.$$

These are the formulas of Wald.

From (5.2) and (5.3) we see that $2\mu/\sigma^2 = (2\theta - (\theta_1 + \theta_2))/(\theta_1 - \theta_2)$ which denote by $h(\theta)$. Then setting $A = (1 - \beta)/\alpha$, $B = \beta/(1 - \alpha)$ we obtain from (5.7)

$$L^+(\theta) = \frac{B^{-h(\theta)} - 1}{B^{-h(\theta)} - A^{-h(\theta)}},$$

the probability of absorption in the barrier a, which is the power of the test (i.e., the probability of rejecting $H_2 : \theta = \theta_2$ when θ is the true mean) and $1 - L^+(\theta) = L^-(\theta)$ is the expression given by Wald for the operating characteristic of test.

For the distribution of T (approximate number of observations necessary to terminate the test) we use the expression (3.4) with $x = 0$ to give

$$E(e^{-\lambda T_{ab}(0)}) = \hat{f}_{ab}(0 \mid \lambda) = \frac{(e^{\xi_1 b} - e^{\xi_2 b}) - (e^{\xi_1 a} - e^{\xi_2 a})}{e^{\xi_2 a + \xi_1 b} - e^{\xi_2 b + \xi_1 a}}$$

which can be inverted to give a rather complicated expression:

$$F_{ab}(0 \mid t) = \Pr\{T_{ab}(0) < t\}$$

$$= 1 - \frac{\sigma^2\pi^2}{(a-b)^2} \sum_{n=1}^{\infty} \frac{n(-1)^n}{\dfrac{\mu^2}{2\sigma^2} + \dfrac{\sigma^2 n^2 \pi^2}{2(a-b)^2}} \cdot \left\{ e^{\mu b/\sigma^2} \sin \frac{n\pi a}{a-b} - e^{\mu a/\sigma^2} \sin \frac{n\pi b}{a-b} \right\}$$

$$\cdot \exp\left(-t\left\{\frac{\mu^2}{2\sigma^2} + \frac{\sigma^2 n^2 \pi^2}{2(a-b)^2}\right\}\right).$$

But the moments are easy to obtain by expanding about $\lambda = 0$, since we have the moment generating function of T (note that T is a proper random variable, that is, absorption is a certain event since $\hat{f}_{ab}(0 \mid 0) = 1$). An alternative way is to use the result of the next section which gives the moments as the solutions to differential equations. If we let $m(x) = E(T_{ab}(x))$ then from (6.6) it follows that m satisfies the differential equation $\frac{1}{2}\sigma^2 m''(x) + \mu m'(x) = -1$ with $m(a) = m(b) = 0$.

Assuming first that $\mu \neq 0$ we obtain by solving this equation

$$m(0) = E(T) = \frac{1}{\mu}(aL^+(\theta) + bL^-(\theta))$$

while for $\mu = 0$

$$E(T) = -\frac{ab}{\sigma^2} = \log\left(\frac{1-\alpha}{\alpha}\right) \log\left(\frac{1-\beta}{\beta}\right) \frac{K^2}{(\theta_1 - \theta_2)^2}.$$

Here $L^-(\theta) = 1 - L^+(\theta)$ is the probability of absorption in the barrier b, as before, and a, b, μ, σ^2 have their former significance.

It is rather remarkable that despite the differing nature of the approximations of Wald and the approximations by presuming a continuous process as here, they should give the same formulas.

d) *A nonparametric test in "goodness of fit."* In a test related to the Kolmogorov-Smirnov tests the following important absorption probability problem arose. If $X(t)$ is the Uhlenbeck process (cf. example b) above) calculate the probability

$$b(\xi \mid t) = \Pr\{\mid X(\tau) \mid < \xi, 0 \leqq \tau \leqq t\}$$

where $X(0)$ has its stationary distribution. Thus we have the problem of finding the distribution of the random variable $M(x, t)$ defined by (3.7) whose distribution function is $G(x \mid \xi, t)$ as in (3.8).

For $|x| < \xi$ we have $G(x \mid \xi, t) = 1 - F_{\xi,-\xi}(x \mid t)$ and since for $|x| \geqq \xi$ we have $G = 1$ we define $F = 0$ for $|x| \geqq \xi$. The stationary distribution of $X(t)$ is $N(0, 1)$, that is, has a density $\phi(x) = (2\pi)^{-\frac{1}{2}}e^{-x^2/2}$ and hence

$$b(\xi \mid t) = \int_{-\infty}^{\infty} \phi(x)G(x \mid \xi, t)\, dx = \int_{-\infty}^{\infty} \phi(x)F_{\xi,-\xi}(x \mid t)\, dx.$$

On taking Laplace transforms we get

$$\hat{b}(\xi \mid \lambda) = \frac{1}{\lambda} - \int_{-\xi}^{\xi} \phi(x)\hat{F}_{\xi,-\xi}(x \mid \lambda)\, dx = \frac{1}{\lambda}\left\{1 - \int_{-\xi}^{\xi} \phi(x)\hat{f}_{\xi,-\xi}(x \mid \lambda)\, dx\right\}$$

and substituting from (5.3) we have

$$\hat{b}(\xi \mid \lambda) = \frac{1}{\lambda}\left\{1 - \sqrt{\frac{2}{\pi}} \frac{e^{-\xi^2/4}}{D_{-\lambda}(\xi) + D_{-\lambda}(-\xi)} \cdot \int_0^{\xi} e^{-x^2/4}(D_{-\lambda}(x) + D_{-\lambda}(-x))\, dx\right\}.$$

This result was given, without proof, in [1].

e) *The optional stopping problem.* In [15] Robbins outlined the optional stopping problem. Let, as in example c) above, the problem be that of testing the mean of a normal universe with known variance, say σ^2, but instead of testing the hypotheses H_1 and H_2 of example c) we have the hypothesis $H_1 : \theta = 0$ to test against $H_2 : \theta \neq 0$ (Robbins considers $H_2 : \theta > 0$). As sketched by Robbins the basic problem is to calculate the probability, when $\theta = 0$, $S_n = X_1 + X_2 +, \cdots, + X_n$,

$$g(n_1, n_2, \alpha) = \Pr\{|S_n| < \alpha\sigma\sqrt{n}, n_1 \leqq n \leqq n_2\}$$

for given α, n_1 and n_2. For the case of S_n instead of $|S_n|$ Robbins gave an inequality, and here we give an approximate and an asymptotic result.

The random variables $\{S_n/\sigma\sqrt{n}\}$, $n = n_1, n_1 + 1, \cdots, n_2$ have mean 0, variance 1, are normally distributed and form a Markov chain with covariance

$$E\left\{\frac{S_j}{\sigma\sqrt{j}} \cdot \frac{S_n}{\sigma\sqrt{n}}\right\} = \frac{\min(j, n)}{\sqrt{jn}} = e^{-|\frac{1}{2}\log j - \frac{1}{2}\log n|}.$$

Hence their joint distribution is the same as the joint distribution of $X(\frac{1}{2}\log n_1)$, $X(\frac{1}{2}\log(n_1 + 1))$, $\cdots$, $X(\frac{1}{2}\log n_2)$ where $X(t)$ is the Uhlenbeck process; (cf. examples b) and d) above). Hence we have, using approximations like those in example c),

$$g(n_1, n_2, \alpha) \cong \Pr\{|X(t)| < \alpha, \tfrac{1}{2}\log n_1 \leqq t \leqq \tfrac{1}{2}\log n_2\} = b\left(\alpha \mid \tfrac{1}{2}\log\frac{n_2}{n_1}\right),$$

where $b(\xi \mid t)$ is the function of example d) and of which we have the Laplace transform.

It is also possible to give an exact asymptotic result which is applicable even if the variables are not normally distributed, but merely have mean 0, variance σ^2, and obey the central limit theorem (e.g., if they are identically distributed). Let $n_1 \to \infty$, $n_2 \to \infty$, $n_1/n_2 \to t$, $0 < t < 1$, and consider a sequence $\{t_n\}$, $n = n_1, n_1 + 1, \cdots, n_2$ defined for fixed n_2 by $t_n = n/n_2$; this sequence depends upon n_2, $\{t_n\}_{n_2}$, and for $n_2 \to \infty$ becomes everywhere dense in the interval $t \leqq \tau \leqq 1$. That is, given any $\tau(t \leqq \tau \leqq 1)$ we can choose an element τ_k from $\{t_n\}_k$ such that $\lim_{k\to\infty}\tau_k = \tau$.

Then since

$$g(n_1, n_2, \alpha) = \Pr\left\{\frac{|S_{n_2 t_n}|}{\sigma\sqrt{n_2}} < \alpha\sqrt{t_n}, n_1 \leqq n \leqq n_2\right\}$$

it will follow from a theorem of Donsker [6] that the limiting distribution g can be expressed as the distribution of the corresponding Wiener functional. Hence for $n_1 \to \infty$, $n_2 \to \infty$, $n_1/n_2 \to t$, $0 < t < 1$,

$$g(n_1, n_2, \alpha) \to \Pr\{|W(\tau)| < \alpha\sqrt{\tau}, t \leqq \tau \leqq 1\},$$

where $W(t)$ is the Wiener-Einstein process (cf. example a) above).

Now if $X(t)$ is the Uhlenbeck process (cf. example b)) we can write $W(t) = \sqrt{t}X(\frac{1}{2}\log t)$ (Doob [8]) and thus

$$\begin{aligned}\lim g &= \Pr\{|X(\tfrac{1}{2}\log\tau)| < \alpha, t \leqq \tau \leqq 1\} \\ &= \Pr\left\{|X(\tau)| \leqq \alpha, 0 \leqq \tau \leqq \tfrac{1}{2}\log\frac{1}{t}\right\} = b\left(\alpha \,\middle|\, \tfrac{1}{2}\log\frac{1}{t}\right),\end{aligned}$$

and since $1/t = \lim n_2/n_1$ we obtain $g \sim b(\alpha \mid \frac{1}{2}\log n_2/n_1)$, the approximate expression deduced above. It seems somewhat striking that these two expressions should agree, being deduced from essentially distinct principles.

6. On the moments of T. In the preceding work the distributions were generally expressed as Laplace transforms which are often difficult to invert but which give immediate information about the moments of T.

In the present section we suppose that $\Pr\{T < \infty\} = 1$, that is, that T is a proper random variable, as otherwise the moments will not exist. If the corresponding Laplace transform is 1 for $\lambda = 0$ the variable is proper. Let us put

$$t_{ab}^{(n)}(x) = E(T_{ab}^n(x)), \qquad t_c^{(n)}(x) = E(T_c^n(x))$$

which we suppose to *exist* for $n \leqq n_0$. We have by a series expansion

$$\hat{f}_{ab}(x \mid \lambda) = \sum_{n=0}^{n_0} \frac{t_{ab}^{(n)}(x)}{n!}(-\lambda)^n + o(\lambda^{n_0}), \qquad \lambda \to 0 \tag{6.1}$$

$$\hat{f}_c(x \mid \lambda) = \sum_{n=0}^{n_0} \frac{t_c^{(n)}(x)}{n!}(-\lambda)^n + o(\lambda^{n_0}), \qquad \lambda \to 0$$

from which the moments are determined.

From equations (3.5) and (3.6) it is possible to express $\hat{f}_{ab} = \hat{f}_{ab}^+ + \hat{f}_{ab}^-$ in terms of the transforms $\hat{f}_c$ and from this fact we can express the moments $t_{ab}^{(n)}$ in terms of the one-sided first passage moments $t_c^{(n)}$. We get in fact from (3.5) and (3.6)

$$\hat{f}_{ab}(x \mid \lambda) = \frac{\hat{f}_a(x \mid \lambda)(\hat{f}_b(a \mid \lambda) - 1) + \hat{f}_b(x \mid \lambda)(\hat{f}_a(b \mid \lambda) - 1)}{\hat{f}_a(b \mid \lambda)\hat{f}_b(a \mid \lambda) - 1}, \tag{6.2}$$

and it follows that $t_{ab}^{(n)}(x)$ will be given by an algebraic combination of $t_a^{(k)}(x)$ and $t_b^{(j)}(x)$ for $k \leqq n$, $j \leqq n$, provided these moments exist. But it should be remarked that $t_{ab}^{(n)}(x)$ will exist in general for finite a, b, even though $t_c^{(k)}(x)$ may not, as the simple Wiener-Einstein process, for which $t_c^{(k)}(x) = \infty$ for $k \geqq 1$, shows.

In particular for $n = 1$, where we put $t^{(1)} = t$, we get for the mean first passage time by a simple expansion of (6.2),

$$t_{ab}(x) = \frac{t_a(x)t_b(a) + t_b(x)t_a(b) - t_a(b)t_b(a)}{t_a(b) + t_b(a)}. \tag{6.3}$$

This formula leads to interesting consequences. Let a and b be such that $t_a(b) = t_b(a)$. Then since $t_b(a) = t_x(a) + t_b(x)$ (6.3) becomes simply

$$t_{ab}(x) = \frac{t_a(x) - t_x(a)}{2}. \tag{6.4}$$

The right-hand side of (6.4) is independent of b, and since $t_{ab}(x) \geqq 0$ we have the result that *when* $a > x > b$ *and* $t_a(b) = t_b(a)$ *then* $t_a(x) \geqq t_x(a)$. Thus it is possible in a stationary process that the mean length of time it takes to go from a less probable state to a more probable state for the first time is longer than that it takes to reverse the journey. It is simple to construct processes for which this result obtains, for example, one in which the stationary density is symmetric and bimodal.

It is possible also to express the probability of absorption in the barrier a before b by means of the one-sided first passage moments. Since $\hat{f}_{ab}^{+}(x|0)$ is this probability we obtain from (3.5) and (3.6)

$$\hat{f}_{ab}^{+}(x \mid \lambda) = \frac{\hat{f}_a(b \mid \lambda)\hat{f}_b(x \mid \lambda) - \hat{f}_a(x \mid \lambda)}{\hat{f}_b(a \mid \lambda)\hat{f}_a(b \mid \lambda) - 1};$$

hence letting $\lambda \to 0$ we obtain the conclusion that *if the first passage moments exist the probability of absorption in a before b is given by* $P = (t_a(b) + t_b(x) - t_a(x))/(t_a(b) + t_b(a))$.

Since the expressions $\hat{f}$, $\hat{f}^{+}$, and $\hat{f}^{-}$ satisfy the differential equation (4.2) if the corresponding transition density p satisfies (4.1) it is possible to find the moments $t^{(n)}$ directly through a differential equation, and this often affords a method that is computationally more feasible than a direct evaluation of $\hat{f}$. We have in fact the following theorem.

THEOREM 6.1. *Let* $X(t)$ *satisfy the hypotheses of Theorem* 4.1. *Then if* $T = T_{ab}(x)$

is a proper random variable whose moments of order $n \leqq n_0$ *exist,* $t^{(n)} = t_{ab}^{(n)}(x)$ *satisfies the system*

$$\tfrac{1}{2}B^2 \frac{d^2t^{(n)}}{dx^2} + A\frac{dt^{(n)}}{dx} = -nt^{(n-1)}, \qquad n \leqq n_0$$

$$t^{(0)} \equiv 1 \tag{6.5}$$

$$t_{ab}^{(n)}(a) = t_{ab}^{(n)}(b) = 0, \qquad n > 0.$$

To prove the theorem we merely substitute the expansion (6.1) in the differential equation (4.2) and equate the coefficient of λ^n to zero.

The system (6.5) is particularly easy to solve since the substitution $Z^{(n)} = dt^{(n)}/dx$ renders each equation linear of the first order, and the solution can be written immediately in quadratures. Starting with $n = 1$ each $t^{(n)}$ can be obtained in turn in quadratures from the previous $t^{(k)}(k < n)$. In particular for $n = 1$ we have

$$\tfrac{1}{2}B^2 \frac{d^2t}{dx^2} + A\frac{dt}{dx} = -1, \qquad t = t_{ab}^{(1)}(x)$$

$$t(a) = t(b) = 0, \tag{6.6}$$

a result we have used already in Section 5, Example c).

7. The range of $X(t)$. In this section we develop a formula for the distribution of the random variable

$$R(x, t) = \sup_{0 \leqq \tau \leqq t} X(\tau) - \inf_{0 \leqq \tau \leqq t} X(\tau)$$

which is called the *range* of $X(t)$, or the *oscillation* of $X(t)$, and we denote its distribution by $\Phi(x \mid r, t) = \Pr\{R(x, t) < r\}$. Note that this probability exists if $X(t)$ satisfies conditions A) and B) of Section 2.

A treatment of the random variable R for the Wiener-Einstein case has been given by Feller [10] in a statistical application, and the present section solves a problem he posed on finding the distribution of R for other processes.

Again we presume the existence of a density for R, say $\phi(x \mid r, t) = \partial\Phi(x \mid r, t)/\partial r$ only to expedite the analysis. It is not difficult to show that the existence of a density for T implies that for R.

THEOREM 7.1. *Let* $X(t)$ *satisfy conditions A) and B) and let* $\phi(x \mid r, t)$ *be the density of* $R(x, t)$. *Then for* $\hat{f}_{ab}(x \mid \lambda)$ *as in* (3.4) *we have*

$$\hat{\phi}(x \mid r, \lambda) = -\frac{1}{\lambda}\frac{\partial^2}{\partial r^2}\int_{x-(r/2)}^{x+(r/2)} \hat{f}_{v+(r/2),v-(r/2)}(x \mid \lambda)\, dv. \tag{7.1}$$

We note that $\Phi(x \mid r, \lambda)$, being merely $\int_0^r \hat{\phi}(x \mid u, \lambda)\, du$, is given immediately since $\hat{\phi}$ is expressed as a derivative.

The starting point of the proof is the formula

$$\phi(x \mid r, t) = \int_{x-r}^{x} \left[-\frac{\partial^2}{\partial a\, \partial b}(1 - F_{ab}(x \mid t))\right]_{a=b+r} db$$

which is established readily by an enumeration of cases. The existence of the derivative (under the integral sign) follows from the existence of the density of $X(t)$ at a and b, for when $\delta > 0$

$$F_{ab}(x \mid t) - F_{a+\delta,b}(x \mid t) = \Pr\{a < X(t) < a + \delta,\ X(\tau) > b,\ 0 \leqq \tau \leqq t\}.$$

On taking the Laplace transform of the preceding expression (which can be done under the integration and differentiation operations) we obtain

$$\hat{\phi}(x \mid r, \lambda) = \frac{1}{\lambda} \int_{x-r}^{x} \left[\frac{\partial^2}{\partial a\, \partial b} \hat{f}_{ab}(x \mid \lambda) \right]_{a=b+r} db$$

and the conclusions to the theorem follow by noting the identity

$$\frac{\partial^2}{\partial a\, \partial b} \hat{f}_{ab}(x \mid \lambda) \bigg|_{a=b+r} \equiv \frac{\partial^2}{\partial b\, \partial r} \hat{f}_{b+r}\,(x \mid \lambda) - \frac{\partial^2}{\partial r^2} \hat{f}_{b+r,b}(x \mid \lambda).$$

As an application we consider the Wiener-Einstein process for which we have shown ((5.1) and (5.2))

$$\hat{f}_{ab}(x \mid \lambda) = \frac{\cosh \sqrt{2\lambda} \left(x - \dfrac{a+b}{2} \right)}{\cosh \sqrt{2\lambda} \left(\dfrac{a-b}{2} \right)},$$

and here (7.1) gives on performing the integration,

(7.2) $$\hat{\phi}(x \mid r, \lambda) = -\sqrt{\frac{2}{\lambda^3}} \frac{\partial^2}{\partial r^2} \tanh \sqrt{\frac{\lambda}{2}}\, r$$

independent of x since the process is spatially homogeneous. This latter transform is easy to invert, and we have

$$\phi(x \mid r, t) = \frac{2}{\pi^2} \frac{\partial^2}{\partial r^2} \left\{ r \sum_{j=0}^{\infty} \frac{1}{(j + \frac{1}{2})^2} \exp\left(\frac{-2\pi^2 t (j + \frac{1}{2})^2}{r^2} \right) \right\}$$
$$= \frac{8}{\sqrt{2\pi t}} \sum_{j=1}^{\infty} (-1)^{j-1} j^2 e^{-(j^2 r^2/2t)},$$

these two expressions being related by Theta function identities, and the second being given by Feller [10]. For the moments we get from (7.2) immediately $E(R^n) = c_n t^{n/2}$ where

$$c_n = -\frac{2^{n/2}}{\Gamma\left(\dfrac{n}{2} + 1\right)} \int_0^\infty \rho^n \frac{d^2}{d\rho^2} \tanh \rho\, d\rho$$

so that, for example, $E(R) = \sqrt{8t/\pi}$, $E(R^2) = 4t \log 2$, etc.

REFERENCES

[1] T. W. Anderson and D. A. Darling, "Asymptotic theory of certain 'goodness of fit criteria' based on stochastic processes," *Ann. Math. Stat.*, Vol. 23 (1952), pp. 193–212.

[2] L. BACHELIER, "Théorie mathématique du jeu," *Ann. Scuela Norm. Super. Pisa*, Vol. 18 (1901), pp. 143–210.
[3] G. A. BARNARD, "Sequential tests in industrial statistics," *J. Roy. Stat. Soc.*, Vol. 8 (1946), pp. 27–41.
[4] R. BELLMAN AND T. HARRIS, "Recurrence times for the Ehrenfest model," *Pacific J. Math.*, Vol. 1 (1951), pp. 179–193.
[5] W. DOEBLIN, "Sur l'équation de Kolmogoroff," *C. R. Acad. Sci. Paris*, Vol. 207 (1938), pp. 705–707.
[6] M. D. DONSKER, "An invariance principle for certain probability limit theorems," *Mem. Amer. Math. Soc.*, No. 6 (1951), 12 pp.
[7] J. L. DOOB, "Heuristic approach to the Kolmogoroff-Smirnov theorems," *Ann. Math. Stat.*, Vol. 20 (1949), pp. 393–403.
[8] J. L. DOOB, "The brownian motion and stochastic equations," *Ann. Math.*, Vol. 2 (1942), pp. 351–369.
[9] W. FELLER, *An Introduction to Probability Theory and Its Application*, John Wiley and Sons, 1950.
[10] W. FELLER, "The asymptotic distribution of the range of sums of independent random variables," *Ann. Math. Stat.*, Vol. 22 (1951), pp. 427–432.
[11] R. FORTET, Les fonctions aléatoire du type de Markoff associées à certaines équations linéares aux dérivées partielles du type parabolique," *J. Math. Pures Appl.*, Vol. 22 (1943), pp. 177–243.
[12] K. ITO, "On stochastic differential equations," *Mem. Amer. Math. Soc.* No. 4 (1951).
[13] A. KHINCHINE, "Asymptotische Gesetze der Wahrscheinlichkeitsrechnung," *Ergebnisse der Math.* Vol. 2 (1933).
[14] P. LÉVY, *Processus Stochastique et Mouvement Brownien*, Gauthier-Villars, Paris, 1948.
[15] H. E. ROBBINS, "Some aspects of the sequential design of experiments," *Bull. Amer. Math. Soc.*, Vol. 58 (1952), pp. 527–535.
[16] A. J. F. SIEGERT, "On the first passage time probability problem," *Physical Review*, Vol. 81 (1951), pp. 617–623.
[17] A. WALD, *Sequential Analysis*, John Wiley and Sons, 1947.
[18] E. T. WHITTAKER AND G. N. WATSON, *Modern Analysis*, Cambridge University Press, 1927.

FIRST PASSAGE TIME PROBLEMS IN CHEMICAL PHYSICS*

GEORGE H. WEISS, *National Cancer Institute, National Institutes of Health, Bethesda, Maryland, U.S.A.*

CONTENTS

I. INTRODUCTION

There are many processes in chemical physics that require the calculation of rates or rate constants. In several of these calculations one assumes that the underlying process can be described in terms of stochastic models, and more specifically in terms of the properties of random walks. Chandrasekhar[1] has given an excellent introduction to some of these topics. There is one class of problem, mentioned only briefly in his article, that has recently found increasing application in many fields of chemical physics, namely, first passage time problems. As an example, several authors[2–4] have proposed models for the dissociation of diatomic molecules in which dissociation occurs when the molecules acquire a certain critical energy E_C through collisions. If the changes in energy can be described in probabilistic terms, then the time to reach E_C is known as the first passage time. Since the theory of first passage times plays an integral role in the formulation of many models in chemical physics, and since no general account of the theory is to be found in chemical literature, various techniques useful for solving such problems are collected below.

In order to define a first passage time problem we consider a space Ω that can be decomposed into two non-overlapping sub-

* This paper was presented at the La Jolla Summer School on Chemical Physics, August 1965, University of California, San Diego, California, U.S.A.

spaces V and $\bar{V}$. It will be assumed that initially the random variable of interest lies in the subspace V: the first passage time is defined to be the time elapsing before passage to $\bar{V}$ for the first time. For example, in the dissociation model mentioned above Ω would consist of all energies, the space V would be all energies satisfying $E < E_C$, and $\bar{V}$ would consist of all energies satisfying $E \geq E_C$. In more elaborate dissociation models several quantum numbers might be involved, in which cases Ω would be multi-dimensional.

There are many specific models involving first passage times to be found in the literature. The first of these appears to have been developed in connection with Brownian motion studies by Schrödinger[5] and by Smoluchowski.[6] A first passage time model is relevant in the discussion of the Ehrenfest urn model,[7] in the theory of escape of stars from star clusters,[8] in the theory of reaction rates,[2–4] and in certain problems in the theory of poly-electrolytes.[9] Although there is a considerable mathematical literature on first passage time problems,[10] there have been few publications in the physical or chemical literature on the subject.

Although our principal concern in this article will be first passage problems for Markov processes it is well to point out that there are occasional applications which are non-Markovian in nature. As an example the recurrence of colloid counts in a fixed volume as in the experiments of Svedberg[11] and Westgren[12] analysed by Smoluchowski,[13] is a fundamentally non-Markovian process. Although Smoluchowski's early analysis did not make use of any assumption of Markovian evolution Chandrasekhar's later account[1] erred in deriving the fundamental relations. This error was pointed out by Bartlett, who gave a more complete account of the general problem.[14] Some related material is also to be found in a paper by Siegert.[15] Most applications, however, fall under the heading of Markov processes, which will be the subject of our future developments.

The plan of the present article is as follows: Section II contains a general account of the calculation of first passage time moments for Markov processes in continuous time, together with the specialization to the case of a discrete set of states. Section III contains the specialization to systems where the transition probabilities satisfy a Fokker–Planck equation.

II. GENERAL FORMULAE FOR MOMENTS OF THE FIRST PASSAGE TIME

We shall label the random variable of interest $X(t)$, which can be a vector or a scalar. It will be assumed that the development of $X(t)$ in time can be described as a stationary Markov process. To describe the statistical properties of $X(t)$ we choose a probability density $p(x,t|y)$, defined so that $p(x,t|y)dx$ is the probability that $x < X(t) \leq x + dx$ given that $X(0) = y$, i.e., y is the initial position. When $X(t)$ takes on integer values only $p(n,t|m)$ will be the probability that $X(t) = n$, given that $X(0) = m$. In both cases the variables x or n can refer to multidimensional vectors unless otherwise noted. The function $p(x,t|y)$ in general satisfies a linear operator equation of the form:

$$\partial p/\partial t = -L_x p, \tag{1}$$

where L_x operates on x only, and where $p(x,0|y) = \delta(x - y)$ for X continuous, and $p(x,0|y) = \delta_{xy}$ for X discrete. To encompass both cases we introduce the symbol $\Delta(x - y)$ which will denote a delta function or a Kronecker delta depending on whether x and y are continuous or discrete. When X is continuous the most commonly used form for L_x is the Fokker–Planck operator which in one dimension is

$$-L_x = \frac{1}{2}\cdot\frac{\partial}{\partial x}\left(a_2\frac{\partial}{\partial x}\right) - \frac{\partial}{\partial x}(a_1), \tag{2}$$

where a_1 and a_2 are the first and second moments of the infinitesimal transition rates.[1] When X is discrete, as is the case for many quantum problems, L_x is a difference operator. There appears to have been no study of the case of mixed discrete and continuum problems, although they do arise in the context of neutron thermalization.

The probability that $X(t)$ is still in V at time t given that it started at y in V will be denoted by $p_V(y,t)$ and is given by

$$p_V(y,t) = \int_V p(x,t|y)dx. \tag{3}$$

Let $\eta(y,t)$ be the probability density for the first passage time: that is, if T is the first time that $X(t)$ reaches $\bar{V}$, given that

$X(0) = y$, then $\eta(y,t)dt$ is the probability that $t < T \leq t + dt$ given that $X(0) = y$. An expression for $\eta(y,t)$ is obtained by noting that if $X(t)$ is in V at time t, then it either makes a first passage in $(t,t + dt)$ or it remains in V at $t + dt$. These two possibilities lead to

$$p_V(y,t) = \eta(y,t)dt + p_V(y,t + dt) \tag{4}$$

or

$$\eta(y,t) = -\partial p_V(y,t)/dt. \tag{5}$$

Moments of the first passage time are defined by

$$\langle t^n(y)\rangle = \int_0^\infty t^n \eta(y,t)dt = n \int_0^\infty t^{n-1} p_V(y,t)dt, \tag{6}$$
$$n = 1, 2, \ldots$$

where the last form is obtained by an integration by parts.

We can obtain other, formal, expressions for the moments in terms of the operator L_x by starting from the formal solution to eq. (1):

$$p(x,t|y) = e^{-L_x t} \Delta(x - y). \tag{7}$$

Substituting this expression into eqs. (3) and (6) we find

$$\begin{aligned}\langle t^n(y)\rangle &= n \int_0^\infty t^{n-1}dt \int_V e^{-L_x t} \Delta(x - y)dx \\ &= n! \int_V L_x^{-n} \Delta(x - y)dx,\end{aligned} \tag{8}$$

where we have freely interchanged orders of integration.

In order to use eq. (8) for computation we must introduce a more explicit representation of the quantities involved. When $X(t)$ ranges over the integers, the operator L_x can be represented as a matrix $\mathbf{L} = (L_{mn})$, Eq. (3) becomes

$$p(n,t|m) = (e^{-\mathbf{L}t})_{nr}\delta_{rm} = (e^{-\mathbf{L}t})_{nm}, \tag{9}$$

and the r'th moment of the first passage time conditional on $X(0) = m$ is

$$\langle t^r(m)\rangle = r! \sum_{n\varepsilon V} (\mathbf{L}^{-r})_{nj}\delta_{jm} = r! \sum_{n\varepsilon V} (\mathbf{L}^{-r})_{nm}. \tag{10}$$

Hence the r'th moment is simply related to the r'th power of the inverse of the rate matrix. When L_x is the Fokker–Planck operator, the operator L_x^{-1} can be identified with a Green's function and operators L_x^{-n} are iterates of the Green's function. To establish this fact we expand the Green's function associated with the operator L_x and the boundary conditions of the problem, in terms of the eigenfunctions of L_x. It will then be seen that the resulting expansion is that which arises in the evaluation of $\langle t(y)\rangle$.

The eigenfunctions of $L_x u_n(x)$, are defined by

$$L_x u_n(x) = \lambda_n u_n(x), \tag{11}$$

where it will be assumed that the λ_n are real, distinct, and positive. These conditions are fulfilled in most problems of physical interest, although the theory can be extended to deal with more complicated situations. Since L_x is not necessarily self-adjoint the $u_n(x)$ do not directly form an orthonormal set, but are orthogonal with respect to the eigenfunctions of the adjoint operator $\tilde{L}$ which satisfy[16]

$$\tilde{L} v_n(x) = \lambda_n v_n(x). \tag{12}$$

The relation of orthogonality can be expressed as

$$\int_\Omega u_m(x) v_n(x) dx = \delta_{nm}. \tag{13}$$

One can easily verify by means of this property that the Dirac delta function $\delta(x - y)$ has the representation

$$\delta(x - y) = \sum_{n=0}^{\infty} u_n(x) v_n(y). \tag{14}$$

Since $L_x^{-1} u_n(x) = u_n(x)/\lambda_n$ by application of eq. (8), we have

$$\langle t(y)\rangle = \int L_x^{-1} \sum_n u_n(x) v_n(y) dx = \int \sum_n \frac{u_n(x) v_n(y) dx}{\lambda_n}; \tag{15}$$

and, by the same argument

$$\langle t^r(y)\rangle = r! \int \sum_n \frac{u_n(x) v_n(y) dx}{\lambda_n^r}. \tag{16}$$

One can now verify that the expression for $\langle t(y)\rangle$ is an integral over the Green's function, $G(x,y)$, associated with L_x, and that

$\langle t^r(y)\rangle$ can be written in terms of iterates of the Green's function. The Green's function, $G(x,y)$, associated with the operator L_x is the solution to

$$L_x G(x,y) = \delta(x - y), \tag{17}$$

i.e., it is the continuous analogue of the matrix inverse. If one expands $G(x,y)$ in a series of eigenfunctions associated with L_x:

$$G(x,y) = \sum_n g_n u_n(x), \tag{18}$$

and uses the representation of eq. (14) in eq. (17), it is found that $g_n = v_n(y)/\lambda_n$ or

$$G(x,y) = \sum_n \frac{u_n(x)v_n(y)}{\lambda_n}. \tag{19}$$

A comparison of eqs. (15) and (19) shows that

$$\langle t(y)\rangle = \int G(x,y)dx. \tag{20}$$

Iterates of $G(x,y)$ are defined by

$$G_{n+1}(x,y) = \int G_n(x,z)G(z,y)dz \tag{21}$$
$$G_1(x,y) = G(x,y).$$

The expansion in terms of eigenfunctions of L or $G_n(x,y)$ reads

$$G_n(x,y) = \sum_m \frac{u_m(x)v_m(y)}{\lambda_m^n}, \tag{22}$$

as may readily be confirmed from eqs. (19) and (21). Hence $\langle t^r(y)\rangle$ can also be written

$$\langle t^r(y)\rangle = r!\int G_r(x,y)dx. \tag{23}$$

The expansions of eqs. (10) and (22) are analogues because of the formal relation between the matrix inverse and Green's functions.

So far our results have taken the form of a reduction of a problem stated in probabilistic terms to a purely computational problem. In practice, the applications treated in the literature contain further restrictions on the form of the operator L_x, so that analytic results are possible. In the case of a master equation defined over a discrete set of states it is most often assumed that **L** represents a nearest-neighbor system, that is $L_{ij} = 0$ for $|i - j| > 1$. A typical example of this is the Montroll–Shuler model of the dissociation of a diatomic molecule, in which the elements of L are

$$L_{j,j+1} = -\kappa(j+1)e^{-\Theta}, \quad L_{j,j-1} = -\kappa j,$$

and

$$L_{ii} = L_{j,j-1} + L_{j,j+1}, \tag{24}$$

where κ is a rate parameter, and $\Theta = h\nu/(kT)$ where ν is the characteristic oscillator frequency.[2] It is possible to derive formulae for the $\langle t^r(m)\rangle$ in closed form for these nearest-neighbor systems. Let us consider, as an example, a derivation of the formula for $\langle t(m)\rangle$ when states 0, 1, 2, . . ., N are non-reactant but state N + 1 is a reactant state. We can first observe that we need only calculate $\langle t(0)\rangle$ since if $T_{0,j}$ is the random variable representing the time for the system to reach state j for the first time starting from state 0, it follows from the assumption of a nearest-neighbor system that

$$T_{0,N+1} = T_{0,m} + T_{m,N+1}. \tag{25}$$

That is to say, since every state must necessarily be traversed in passing from state 0 to state N + 1, the total first passage time is made up of the time to reach state m for the first time plus the time to reach state N + 1 for the first time starting from state m. If we let $\langle t_k\rangle$ denote the mean first passage time for getting from state 0 to state k, eq. (25) implies that

$$\langle t(m)\rangle = \langle t_{N+1}\rangle - \langle t_m\rangle, \tag{26}$$

so that we need only calculate a formula for $\langle t(0)\rangle$ with an arbitrary upper reaction level.

In order to calculate $\langle t(0)\rangle$ most expeditiously we will start, not from the general formulation of eq. (10), but rather from eq. (1) which in the present case can be written:

$$\begin{aligned}
\dot{p}_0 &= L_{10}p_1 - L_{01}p_0 \\
\dot{p}_1 &= L_{01}p_0 - (L_{10} + L_{12})p_1 + L_{21}p_2 \\
\dot{p}_2 &= L_{12}p_1 - (L_{21} + L_{23})p_2 + L_{32}p_3 \\
&\ \ \vdots \\
\dot{p}_N &= L_{N-1,N}p_{N-1} - (L_{N,N-1} + L_{N,N+1})p_N,
\end{aligned} \tag{27}$$

where it is to be understood that $p_j(0) = \delta_{j0}$.

Since $p_V(t) = p_0(t) + p_1(t) + \ldots + p_N(t)$, eq. (6) indicates that

$$\langle t(0)\rangle = \int_0^\infty [p_0(t) + p_1(t) + \ldots + p_N(t)]dt; \tag{28}$$

or, introducing the Laplace transforms $p_j^*(s) = \int_0^\infty e^{-st}p_j(t)dt$, we can also write

$$\langle t(0)\rangle = \sum_{j=0}^{N} p_j^*(0). \tag{29}$$

It is in this form that the calculation becomes most convenient. The Laplace transform of eq. (27) with $s = 0$ is

$$\begin{aligned} L_{01}p_0^*(0) &= L_{10}p_1^*(0) + 1 \\ (L_{10} + L_{12})p_1^*(0) &= L_{01}p_0^*(0) + L_{21}p_2^*(0) \\ &\vdots \\ (L_{N,N-1} + L_{N,N+1})p_N^*(0) &= L_{N-1,N}p_{N-1}^*(0). \end{aligned} \tag{30}$$

These may be solved recursively to yield

$$p_j^*(0) = \theta_j p_0^*(0) - \eta_j, \tag{31}$$

where

$$\theta_0 = 1,\ \theta_j = \frac{L_{01}L_{12}L_{23}\ldots L_{j-1,j}}{L_{10}L_{21}L_{32}\ldots L_{j,j-1}} \tag{32}$$

$$\eta_0 = 0,\ \eta_j = \frac{1}{L_{j,j-1}}\left[1 + \frac{L_{j-1,j}}{L_{j-1,j-2}} + \frac{L_{j-1,j}L_{j-2,j-1}}{L_{j-1,j-2}L_{j-2,j-3}} + \ldots + \frac{L_{j-1,j}L_{j-2,j-1}\ldots L_{12}}{L_{j-1,j-2}L_{j-2,j-3}\ldots L_{10}}\right].$$

Eq. (31) is derived from the first N lines of eq. (30): if we now substitute eq. (31) into the last line of eq. (30) we obtain an expression for $p_0^*(0)$:

$$p_0^*(0) = \eta_{N+1}/\theta_{N+1}. \tag{33}$$

Thus the mean first passage time is given by

$$\langle t(0)\rangle = \frac{\eta_{N+1}}{\theta_{N+1}}\sum_{j=0}^{N}\theta_j - \sum_{j=0}^{N}\eta_j. \tag{34}$$

In a similar fashion it can be shown that if states 0 and N + 1 form absorbing barriers (that is, the reaction ends when either state is reached) the mean first passage time conditional on the initial state's being r is

$$\langle t(r) \rangle = \frac{\eta_{N+1}}{\theta_{N+1}} \sum_{j=0}^{N} \theta_j - \sum_{j=r+1}^{N} \eta_j. \tag{35}$$

For particular models which are appropriately described by the nearest-neighbor approximation the sums indicated in eqs. (32) and (34) may be rather simple to evaluate. For example, in the Montroll–Shuler treatment of dissociation characterized by the transition rates of eq. (24) it is easily verified that the parameters θ_j and η_j are, respectively:

$$\theta_j = e^{-j\theta},$$
$$\eta_j = \frac{1}{k}\left(\frac{1}{j} + \frac{e^{-\theta}}{j-1} + \frac{e^{-2\theta}}{j-2} + \cdots + \frac{e^{-(j-1)\theta}}{1}\right); \tag{36}$$

so that the mean first passage time to reach state N + 1 in the case of a single reactive state is

$$\langle t(0) \rangle = \frac{1}{k} e^{(N+1)\theta} \left(\frac{1}{N+1} + \frac{e^{-\theta}}{N} + \frac{e^{-2\theta}}{N-1} + \cdots + \frac{e^{-N\theta}}{1}\right) \left(\frac{1 - e^{-(N+1)\theta}}{1 - e^{-\theta}}\right)$$
$$- \frac{1}{k(1 - e^{-\theta})}\left[1 - e^{-N\theta} + \frac{1}{2}(1 - e^{-(N-1)\theta}) + \frac{1}{3}(1 - e^{-(N-2)\theta}) + \cdots + \frac{1}{N}(1 - e^{-\theta})\right] \tag{37}$$
$$= \frac{1}{k(1 - e^{-\theta})} \sum_{j=1}^{N+1} \frac{e^{j\theta} - 1}{j}.$$

One can derive expressions for higher moments of the first passage time by following the same line of proof as above. A general treatment of the theory of equations of the form of eq. (27), i.e., with nearest-neighbor transitions only, has been developed by Ledermann and Reuter,[17] and by Karlin and MacGregor.[18] Their results include a general solution to eq. (27) in terms of recursively defined orthogonal polynomials. In particular, Karlin and MacGregor are able to discuss the statistics of the first passage time problem in terms of the orthogonal polynomials appropriate to the particular set of equations. With their results it is possible to derive expressions for moments of the first passage time rather easily, but I have chosen to omit a discussion of the orthogonal polynomial technique from this article because the moments can be obtained directly. It would be of considerable interest to have results as simple as those of eqs. (34) and (35) for systems not restricted to the nearest-neighbor type, as there are indications[19] that the nearest-neighbor theory does not give results of the right order of magnitude for dissociation times.[1]

III. FIRST PASSAGE TIME AND THE FOKKER–PLANCK EQUATION

We have obtained a general formula for moments of the first passage time in terms of the Green's function of the Fokker–Planck operator. This formulation is useful for problems involving multidimensional geometries. The one-dimensional case can be solved in detail for moments of the first passage time. This is not surprising since the one-dimensional problem is the continuous analogue of the nearest-neighbor models just discussed. We shall present results for the one-dimensional case. These are due originally to Pontryagin, Andronow, and Witt,[20] although special cases were treated earlier by Schrödinger. Recent contributions to the mathematical theory have been made by Darling and Siegert.[21] Jackson and his collaborators have made extensive use of this theory in certain polymer problems.[9,22,23]

In what follows we use the mathematical terminology "absorbing" or "reflecting" barrier to describe the properties of a designated point or surface. A surface is said to be absorbing if it forms a boundary between V and $\bar{V}$, that is, if the process terminates when $X(t)$ reaches a point of the surface. A surface is called reflecting if, when $X(t)$ reaches a point of the surface, it is automatically transferred to a point in the interior of V. In the present article we consider only the case of infinitesimal reflection, i.e., when $X(t)$ is transferred to an infinitesimal neighborhood of the point of impingement.

We begin by considering the one-dimensional case in which X(t) is constrained to lie between x = 0 and x = A. It will be assumed that x = A is always an absorbing point and x = 0 is either reflecting or absorbing.

The probability density for the position of X(t) satisfies eq. (1) with L_x given in eq. (2). Theoretically one can find statistical properties of the first passage time by solving the Fokker–Planck equation for p(x,t) with appropriate boundary conditions. However, it proves considerably more convenient to derive an equation for a function $\phi(x,t)$ defined to be the probability that the first passage time is less than t, given that X(0) = x. If $W(y,dt|x)dy$ is the probability that $y \leq X(dt) \leq y + dy$, given that X(0) = x, then we may write the equation

$$\phi(x,t + dt) = \int_0^A W(y,dt|x)\phi(y,t)dt, \tag{38}$$

which expresses the fact that a transition $x \rightarrow (y,y + dy)$ took place in time dt, and the new position can be regarded as a starting point for the process. The next step is to expand $\phi(y,t)$ in a Taylor series around the point x and substitute into the last equation. This leads to

$$\begin{aligned}\phi(x,t + dt) = \phi(x,t) \int_0^A W(y,dt|x)dy \\ + \frac{\partial\phi(x,t)}{\partial x}\int_0^A (y - x)W(y,dt|x)dy \\ + \frac{1}{2}\cdot\frac{\partial^2\phi(x,t)}{\partial x^2}\int_0^A (y - x)^2W(y,dt|x)dy + \ldots\end{aligned} \tag{39}$$

The Fokker–Planck equation is derived on the assumption that

$$\lim_{dt\rightarrow 0} \frac{1}{dt}\int_0^A (y - x)^n W(y,dt|x)dy = 0 \tag{40}$$

for $n \geq 3$. On this assumption, eq. (39) implies that $\phi(x,t)$ is the solution to

$$\frac{\partial\phi}{\partial t} = \frac{a_2(x)}{2}\frac{\partial^2\phi}{\partial x^2} + a_1(x)\frac{\partial\phi}{\partial x}, \tag{41}$$

where the $a_j(x)$ are infinitesimal transition moments

$$a_j(x) = \lim_{dt\to 0} \frac{1}{dt} \int_0^A (y - x)^j W(y,dt|x)dy. \tag{42}$$

The boundary conditions for eq. (41) are $\phi(A,t) = 1$ and $\phi(0,t) = 1$ for $x = 0$ an absorbing point, or $\frac{\partial\phi(0,t)}{\partial x} = 0$ for $x = 0$ a reflecting point. A comprehensive discussion of requirements on $a_2(x)$ to ensure a finite first passage time has been given by Feller.[24] For all problems of physical interest there is no difficulty with this point since $a_2(x)$ is strictly positive and bounded away from zero. In dimensions greater than one, it is possible to derive the equation

$$\frac{\partial\phi}{\partial t} = \sum_i a_i \frac{\partial\phi}{\partial x_i} + \frac{1}{2}\sum_i\sum_j b_{ij} \frac{\partial^2\phi}{\partial x_i \partial x_j} \tag{43}$$

for the distribution of first passage time. In this equation the coefficients are defined, analogously to eq. (40), by

$$\begin{aligned} a_i(\mathbf{x}) &= \lim_{dt\to 0} \frac{1}{dt} \int_V (y_i - x_i) W(\mathbf{y},dt|\mathbf{x})d^n\mathbf{y} \\ b_{ij}(\mathbf{x}) &= \lim_{dt\to 0} \frac{1}{dt} \int_V (y_i - x_i)(y_j - x_j) W(\mathbf{y},dt|\mathbf{x})d^n\mathbf{y}; \end{aligned} \tag{44}$$

and it is assumed that higher infinitesimal transition moments are zero. Equation (43) is to be solved under the initial condition $\phi(\mathbf{x},0) = 0$ for $\mathbf{x}$ in V, and $\phi(\mathbf{x},t) = 0$ for $\mathbf{x}$ belonging to an absorbing portion of the boundary between V and $\bar{V}$, and $\partial\phi/\partial n = 0$ for $\mathbf{x}$ belonging to a reflecting part of the boundary, where $\partial/\partial n$ denotes a normal derivative.

Equation (41) together with the boundary conditions on $\phi(x,t)$ enables us to calculate moments fairly readily. The same argument as has led to eq. (6) implies that the j'th moment of the first passage time starting from a point x is expressible as

$$\mu_j(x) = \int_0^\infty t^j \frac{\partial\phi(x,t)}{\partial t} dt. \tag{45}$$

If we differentiate eq. (41) with respect to t, multiply by t^j, and integrate over all t, we find that the $\mu_j(x)$ are the solution to the set of equations

$$\frac{1}{2}b_2\frac{d^2\mu_j}{dx^2} + b_1\frac{d\mu_j}{dx} = -j\mu_{j-1} \qquad j = 2, 3, \ldots$$
$$\frac{1}{2}b_2\frac{d^2\mu_1}{dx^2} + b_1\frac{d\mu_1}{dx} = -1 \tag{46}$$

with boundary conditions $\mu_j(A) = 0$ for $x = a$ an absorbing point and $d\mu_j(a)/dx = 0$ for $x = A$ a reflecting point.

The equation for $\mu_1(x)$ can be solved in closed form. The general solution is

$$\mu_1(x) = -2\int_0^x e^{-U(y)}\,dy\int_0^y \frac{e^{U(z)}}{b_2(z)}\,dz + C_1\int_0^x e^{-U(y)}\,dy + C_2, \tag{47}$$

where U(x) is defined by

$$U(x) = 2\int_0^x [b_1(y)/b_2(y)]dy. \tag{48}$$

When both boundaries are absorbing,

$$C_1 = 2\int_0^A e^{-U(y)}\,dy\int_0^y [e^{U(z)}/b_2(z)]dz \Big/ \int_0^A e^{-U(y)}\,dy \tag{49}$$

and $\quad C_2 = 0;$

and when $x = A$ is absorbing and $x = 0$ reflecting we have

$$C_1 = 0$$
$$C_2 = 2\int_0^A e^{-U(x)}\,dx\int_0^x [e^{U(y)}/b_2(y)]dy. \tag{50}$$

Formulae for higher moments can be derived recursively from eq. (46), with $\mu_1(x)$ given by eq. (47).

The simplest illustration of the use of these formulae is in terms of simple Brownian motion in one dimension for which $a_1(x) = 0$

and $a_2(x) = 2D$ with D the diffusion constant. The equation for the mean first passage time reduces to $\mu_1'' = -1/D$. If $x = 0$ and $x = A$ are both absorbing, then the expression for mean first passage time is

$$\mu_1(x) = \frac{1}{2D} x(A - x). \tag{51}$$

It is instructive to calculate the statistical properties of one dimensional Brownian motion in a constant force field. Let us consider the case of a semi-infinite interval $(0, \infty)$ with absorption at $x = 0$. The coefficients in eq. (41) are $a_1(x) = -v$ and $a_2(x) = 2D$, where v is assumed to be a constant. The expression for $\mu_1(x)$ is found to be

$$\mu_1(x) = x/v, \tag{52}$$

independent of diffusion effects. The diffusion constant D does, however, appear in the expression for $\mu_2(x)$. An equation for the distribution of absorption times can be written, following eq. (41), as

$$\frac{\partial \phi}{\partial t} = D \frac{\partial^2 \phi}{\partial x^2} - v \frac{\partial \phi}{\partial x}, \tag{53}$$

subject to $\phi(x,0) = 0$ for $x > 0$ and $\phi(0,t) = 1$. The solution to this equation is

$$\phi(x,t) = \frac{x}{\sqrt{\pi D}} \int_0^t \frac{dt}{t^{3/2}} \exp\left[-\frac{1}{4Dt}(2x - vt)^2\right], \tag{54}$$

as given by Schrödinger.[5]

First passage time problems in spaces of dimension greater than one require a simple geometry if useful information is to be obtained in closed form. Thus, solutions for the distribution function of first passage times and the moments are readily obtained for force-free Brownian motion in cylinders or spheres, and between parallel plates when these form the absorbing boundaries.[25,26] More complicated geometries may require purely numerical techniques in the solution of associated first passage time problems.

An interesting example of the use of first passage time techniques in chemical physics is provided by Lifson and Jackson's attempt[9] to explain long association times of sodium ions and polymer macroions observed in transference experiments for polyelectrolytes. Their model is that of the Brownian motion of a counterion in the electrostatic field of a polyelectrolyte molecule. The simplest way to pose the problem in detail is to suppose that we have two bounding planes of $\pm X$. Within this region is a series of equidistant parallel charged planes at $x = 2jL$, $j = -n, -n+1, \ldots n-1, n$ where $X = 2nL$. Lifson and Jackson then go on to calculate the expected time for an ion initially at $x = 0$ to reach a boundary plane. The force on an ion of charge e will be written $\mathbf{F} = -e\nabla\Psi$. It is convenient to define a reduced potential $\phi = e\Psi/(kT)$. The equation for the mean first passage time for a particle performing Brownian motion in a force field derivable from a potential is

$$D\nabla^2\mu_1 + (1/\gamma)\mathbf{F}\cdot\nabla\mu_1 = -1, \tag{55}$$

where $\mathbf{F}$ is the force and γ is the hydrodynamic friction constant. Substituting the expression for $\mathbf{F}$ in terms of Ψ into this equation, and making use of the Einstein relation $D = kT/\gamma$, we find that eq. (55) can be rewritten

$$\nabla\cdot(e^{\phi}\nabla\mu_1) = -e^{\phi}/D \tag{56}$$

Since the system is homogeneous on a macroscopic scale, we might expect that the diffusion process is characterized by a diffusion constant D^* which depends on D and the local electrostatic field of the polyelectrolyte molecule. The object of Lifson and Jackson's analysis was to calculate D^* for a model system. One solves for D^* by calculating $\mu_1(0)$ and equating it to the expression

$$\mu_1(0) = X^2/(2D^*), \tag{57}$$

which is the expression relevant for the case of field-free Brownian motion. The average time $\mu_1(0)$ is given by

$$\mu_1(0) = \frac{1}{D}\int_0^X du\, e^{-\phi(u)}\int_0^u e^{\phi(v)}\,dv. \tag{58}$$

To obtain an expression for D^* we introduce the notation

$$\langle e^{\phi}\rangle = \frac{1}{2L}\int_0^{2L} e^{\phi(y)}\,dy. \tag{59}$$

Since $\phi(x)$ is periodic with period 2L, far from the boundaries, eq. (58) can be rewritten

$$\mu_1(0) = \frac{1}{D}\sum_{m=0}^{n-1}\int_{2mL}^{2(m+1)L} du\, e^{-\phi(u)}\int_0^u dv\, e^{\phi(v)}$$
$$= \frac{1}{D}\sum_{m=0}^{n-1}\int_{2mL}^{2(m+1)L} du\, e^{-\phi(u)}\left[2mL\langle e^{\phi}\rangle + \int_{2mL}^{u} dv\, e^{\phi(v)}\right]$$
$$= \frac{1}{2D}(2nL)^2\langle e^{-\phi}\rangle\langle e^{\phi}\rangle = \frac{X^2}{2D}\langle e^{\phi}\rangle\langle e^{-\phi}\rangle \quad (60)$$

A comparison between eqs. (57) and (60) yields the result

$$D^* = D/(\langle e^{\phi}\rangle\langle e^{-\phi}\rangle) \quad (61)$$

for the particular one-dimensional model under consideration. The Cauchy–Schwartz inequality can be used to show that $\langle e^{\phi}\rangle\langle e^{-\phi}\rangle \geq 1$, so that

$$D^* \leq D. \quad (62)$$

Jackson and Coriell[22] have shown that this inequality holds in any number of dimensions. Lifson and Jackson concluded from their study that the retardation of drift of counterions in polyelectrolyte solutions might be accounted for by their entrapment in the electrostatic fields of the macroions.

Another problem related to the first passage time problem is that of calculating the probability of absorption by a particular one of a set of absorbing barriers. For example, in one dimension we can consider the probability of termination at $x = 0$, given that $x = 0$ and $x = A$ both offer the possibility of absorption. Such a mathematical problem finds application in elucidation of the theory of competitive rate processes. If the transfer of energy can be described in terms of a stochastic process, e.g., if the energy distribution obeys a Fokker–Planck equation, then the probability of one of a set of outcomes can be calculated by the technique to be described.

Let us first consider the one-dimensional case in which absorption can occur at either $x = 0$ or $x = A$. Let $\phi_0(x)$ be the probability of absorption at $x = 0$ given a starting point x. We define a function $\phi_0(x,t)$ to be the probability of absorption at

$x = 0$ before time t, given the starting point x. The desired probability $\phi_0(x)$ is then

$$\phi_0(x) = \lim_{t \to 0} \phi_0(x,t). \tag{63}$$

The function $\phi_0(x,t)$ is a solution to eq. (41), the derivation being the same as given above. However, the boundary conditions are now to be changed to

$$\phi_0(0,t) = 1, \phi_0(A,t) = 0. \tag{64}$$

Setting $\partial\phi_0/\partial t = 0$ in eq. (41), we see that $\phi_0(x)$ is the solution to

$$\frac{1}{2} a_2(x) \frac{d^2\phi_0}{dx^2} + a_1(x) \frac{d\phi_0}{dx} = 0 \tag{65}$$

with boundary conditions $\phi_0(0) = 1$, $\phi_0(A) = 0$. The solution to eq. (65) under these circumstances is

$$\phi_0(x) = \int_x^A e^{-U(y)}\, dy \Big/ \int_0^A e^{-U(y)}\, dy, \tag{66}$$

where $U(x)$ is defined in eq. (48). For Brownian motion in a constant force field $\phi_0(x)$ is found to be

$$\phi_0(x) = \frac{1 - e^{-(v/D)(A-x)}}{1 - e^{-(vA/D)}}. \tag{67}$$

In the limit $v \to 0$ this reduces to $\phi_0(x) = 1 - (x/A)$, as is to be expected. The theory for more complicated geometries is similar. If the boundary between V and $\bar{V}$ is broken up into n segments $S_1, S_2, \ldots, S_n$ the probability of absorption by a particular segment S_j, $\phi_j(x)$, satisfies eq. (43) with $\partial\phi_j(x)/\partial t = 0$ and the boundary conditions $\phi_j(x) = 0$ for $\mathbf{x}$ in

$$S_1, S_2, \ldots, S_{j-1}, S_{j+1}, \ldots, S_n, \text{ and } \phi_j(\mathbf{x}) = 1 \text{ for } \mathbf{x} \text{ in } S_j.$$

There are many further applications of first passage time problems in chemical physics, particularly in the study of polymers.[27] Some of these involve first passages defined on Markov chains, a subject that has not been touched in the present article. However, the theory is similar to that developed here. The interested reader should refer to Kemeny and Snell's book[28] for a discussion of first passage time problems in discrete time.

Although it might appear at first glance that such problems would be completely unphysical, they have applications in polymer physics, where the analogue of time units are bond lengths of monomers.

References

1. Chandrasekhar, S., *Rev. Mod. Phys.*, **15**, 1 (1943).
2. Montroll, E. W., and Shuler, K. E., *Adv. Chem. Phys.* **1**, 361 (1958).
3. Ree, F. H., Ree, T. S., Ree, T., and Eyring, H., *Adv. Chem. Phys.* **4**, 1 (1962).
4. Widom, B., *J. Chem. Phys.* **30**, 238 (1959).
5. Schrödinger, E., *Phys. Z.* **16**, 289 (1915).
6. Smoluchowski, M. von, *Ann. Phys.* **21**, 756 (1906).
7. Kac, M., *Am. Math. Monthly* **54**, 369 (1947).
8. Chandrasekhar, S., *Astron. J.* **97**, 263 (1943).
9. Lifson, S., and Jackson, J. L., *J. Chem. Phys.* **36**, 2410 (1962).
10. Kemperman, J. H. B., *The Passage Problem for a Stationary Markov Chain*, Univ. of Chicago Press, 1961.
11. Svedberg, T., *Z. Physik. Chem.* **77**, 147 (1911).
12. Westgren, A., *Arkiv. Math. Astron. Fys.* **11**, 8, 14 (1916); **13**, 14 (1918).
13. Smoluchowski, M. von, *S. B. Akad. Wiss. Wien* **2a**, 124, 339 (1915).
14. Bartlett, M. S., *Proc. Cambridge Phil. Soc.* **49**, 263 (1953).
15. Siegert, A. J. F., *Phys. Rev.* **81**, 617 (1951).
16. Morse, P. M., and Feshbach, H., *Methods of Theoretical Physics*, McGraw-Hill, New York, 1953.
17. Ledermann, W., and Reuter, G. E. H., *Phil. Trans. Roy. Soc. A* **246**, 321 (1954).
18. Karlin, S., and MacGregor, J. L., *Trans. Am. Math. Soc.* **85**, 489 (1957).
19. Shuler, K. E., and Weiss, G. H., *J. Chem. Phys.* **38**, 505 (1963).
20. Pontryagin, L., Andronow, A., and Witt, A., *Zh. Eksperim. i Teor. Fiz.* **3**, 172 (1933).
21. Darling, D. A., and Siegert, A. J. F., *Ann. Math. Stat.* **24**, 624 (1953).
22. Jackson, J. L., and Coriell, S. R., *J. Chem. Phys.* **38**, 959 (1963).
23. Coriell, S. R., and Jackson, J. L., *J. Chem. Phys.* **39**, 2418 (1963).
24. Feller, W., *Ann. Math.* **65**, 527 (1957).
25. Stadie, F., *Ann. Phys.* **86**, 751 (1928).
26. Klein, G., *Proc. Roy. Soc. A* **211**, 431 (1952).
27. Mazur, J., *J. Chem. Phys.* **41**, 2256 (1964).
28. Kemeny, J. G., and Snell, J. L., *Finite Markov Chains*, D. van Nostrand, New York, 1960.

5) Gas Phase Relaxation Processes

The papers in this section discuss the application and validity of the master equation description of gas phase relaxation processes. In the first paper, the master equation for the vibrational degrees of freedom in a low density gas is derived from the Liouville equation. The conditions under which the master equation is valid are discussed in detail; the vibrational degrees of freedom must relax much more slowly than the other degrees of freedom of the system. When this is true, the other degrees of freedom can be considered to act as a heat bath in which there is a well defined temperature which either remains constant in time or varies slowly in time. In the second paper, the relaxation of a system of harmonic oscillators in a constant temperature heat bath is considered. The vibrational energy of the oscillators changes as a result of collisions with the heat bath molecules and by radiation. The third paper extends this treatment to the situation in which the temperature of the heat bath is time dependent owing to the exchange of energy with the harmonic oscillator sub-system.

The fourth paper discusses the relaxation of the translational energy of a group of hard-sphere test particles in a low density gas. The concentration of the test particles is taken to be much less than the concentration of the bath particles. Under these conditions, a master equation can be obtained for the energy distribution of the test particles. In the special cases of the hard-sphere Rayleigh and Lorentz gases, explicit solutions are obtained by use of the Fokker-Planck approximation to the master equation.

Vibrational Relaxation in Gases—Some Constraints on the Use of Master Equations*

PAUL K. DAVIS† AND IRWIN OPPENHEIM

Department of Chemistry, Massachusetts Institute of Technology, Cambridge, Massachusetts 02139

(Received 19 August 1971)

The master equation for vibrational relaxation of diatomic dilute gases is derived. The derivation begins with the quantum Liouville equation and uses projection operator techniques; it can be extended to dense fluids under certain conditions. A careful discussion of the master equation's range of validity reveals restrictions on the experimental initial conditions and the relevant scattering cross sections.

I. INTRODUCTION

Experimental data on molecular relaxation processes is almost invariably interpreted via rate equations with a very simple and characteristic structure:

$$d\mathbf{P}/dt = \mathbf{AP}(t), \tag{1.1}$$

or, more generally,

$$d\mathbf{P}/dt = \mathbf{A}[\mathbf{P}(t)]\mathbf{P}(t), \tag{1.2}$$

where $\mathbf{P}(t)$ is a stochastic vector and $\mathbf{A}$ is a transition matrix which may be a matrix functional of $\mathbf{P}(t)$, but is otherwise time independent. Such expressions are often referred to as "master equations," and it will be the purpose of this work to evaluate the basis some of them have in microscopic theory. We shall restrict ourselves to diatomic molecules and, in this paper, to the case of dilute gases. Although we treat only the vibrational master equation explicitly in this paper, we are able to draw conclusions of a more general nature.

As for specific conclusions, we find that there are at least four important cases for which the vibrational master equation is not valid:

(1) relaxation from an initial state in which a significant fraction of molecules have been optically excited to higher vibrational states (laser studies),

(2) relaxation to a final state in which a significantly larger fraction of molecules will be vibrationally excited (shock tube studies),

(3) relaxation of molecules which tend to have two or more vibrational transitions within a period of perhaps five or ten mean free times, and

(4) "V–V relaxation" (the development of a Maxwellian distribution within the vibrational manifold prior to the establishment of complete equilibrium between vibration and translation) in systems for which the relevant time scale is of the same order as that for rotational relaxation. The most obvious instances of this would involve hydrogenlike molecules.

We also find[1] that the master equation for rotational relaxation is valid only under very restrictive conditions. In consequence, we conclude that published "experimental" relaxation times have a dubious significance when they have been obtained through the use of a rotational master equation.

The procedures used in this work are based on the projection operator technique of Zwanzig,[2] a derivation of the classical Boltzmann equation by Mazur and Biel,[3] and the Wigner function approach to quantum statistical mechanics.[4] We begin with the N-molecule Schrödinger equation, but ignore quantum statistics and processes which affect the electronic and nuclear states. Our projection operators serve to isolate the two-body terms. The essential problem is then whether or not the "remainder terms" containing all the corrections for many-body effects and correlations can be ignored.

Some stylistic features of the present work which are thought to be desirable include a reliance on projection operators which allows us to extract the term of interest without introducing such artifices as power series in density, well-defined but physically structured assumptions, and a formalism which allows us to distinguish steps which are inherently quantum mechanical from those which are classically understandable.

Were we interested only in gases, we might have been inclined to begin with the Waldmann–Snider equation rather than the complete Schrödinger equation. However, in a subsequent paper we use the formalism developed in the first part of this paper to discuss vibrational relaxation in liquids.

II. EXACT N-BODY EQUATIONS

We consider a system consisting of N diatomic molecules in a volume V. The representative ensemble for the system of interest consists of the states $\{|\psi_\alpha^N(t)\rangle\}$, and the density operator is given by

$$\rho^N(t) = \sum_\alpha g_\alpha |\psi_\alpha^N(t)\rangle\langle\psi_\alpha^N(t)|,$$

where g_α is the statistical weight of $|\psi_\alpha^N(t)\rangle$.[5] It follows from the Schrödinger equation that

$$d\rho^N/dt = -(i/\hbar)[H^N, \rho^N(t)] = -iL^N\rho^N(t),$$

where the Liouville operator L^N is defined by this equation in terms of the Hamiltonian H^N.

For reduction of the many-body problem, it is convenient to introduce operators which are still abstract with respect to internal states, but explicit with respect to center-of-mass variables and expressed in terms of the phase space quantities $\mathbf{R}^N$ and $\mathbf{P}^N$. These we denote with a caret or circumflex $\hat{}$. Using the notation $\rho(\mathbf{R}^N, \bar{\mathbf{R}}^N) = \langle \mathbf{R}^N \mid \rho^N \mid \bar{\mathbf{R}}^N \rangle$, we have as a natural generalization of the usual Wigner function[4]

$$\hat{F}^N(\mathbf{R}^N, \mathbf{P}^N, t) = [1/(\pi\hbar)^{3N}] \int d\mathbf{Y}^{3N} \times \exp[(2i/\hbar)\mathbf{P}^N \cdot \mathbf{Y}^N] \rho^N(\mathbf{R}^N - \mathbf{Y}^N, \mathbf{R}^N + \mathbf{Y}^N)$$
$$\equiv \mathfrak{F}_{\mathbf{P}^N\mathbf{Y}^N} \rho^N(\mathbf{R}^N - \mathbf{Y}^N, \mathbf{R}^N + \mathbf{Y}^N), \qquad (2.1)$$

where the operator $\mathfrak{F}_{\mathbf{P}^N\mathbf{Y}^N}$ is defined by Eq. (2.1). The expectation value of an arbitrary observable χ at time t is given by

$$\langle \chi \rangle_t = \int d\mathbf{R}^N d\mathbf{P}^N \, \mathrm{Tr}_{Z^N} \hat{F}^N(\mathbf{R}^N, \mathbf{P}^N, t) \hat{\chi}(\mathbf{R}^N, \mathbf{P}^N)$$
$$= \int d\mathbf{R}^N d\mathbf{P}^N \sum_{Z^N, \bar{Z}^N} \langle Z^N \mid \hat{F}^N(\mathbf{R}^N, \mathbf{P}^N) \mid \bar{Z}^N \rangle \times \langle \bar{Z}^N \mid \chi(\mathbf{R}^N, \mathbf{P}^N) \mid Z^N \rangle,$$

where $\hat{\chi}^N(\mathbf{R}^N, \mathbf{P}^N)$ is defined by $\mathfrak{F}_{\mathbf{P}^N\mathbf{Y}^N}\chi(\mathbf{R}^N - \mathbf{Y}^N, \mathbf{R}^N + \mathbf{Y}^N)$, and $\mid Z^N \rangle$ denotes the eigenket for internal states.

It now follows from the equation given above for $\rho^N(t)$ that

$$\partial \hat{F}^N / \partial t = -i\hat{L}^N \hat{F}^N, \qquad (2.2)$$

where $\hat{L}^N$ is the representative of L^N appropriate for operation on careted functions or operators. Its explicit form is given below.

The Hamiltonian is decomposed into parts representing kinetic energy of translation, internal energy of noninteracting molecules, and intermolecular potential energy, i.e.,

$$H^N = H_0^N(\text{c.m.}) + H_0^N(\text{int}) + U^N,$$

where U^N is a sum of two molecule potentials. We shall assume that $H_0(\text{int})$ possesses eigenstates which can be well approximated by the states $\{\mid z \rangle\}$, where $\mid z \rangle = \mid v, J, M \rangle$ with v, J, and M denoting quantum numbers for vibration, rotational angular momentum, and the latter's z component, respectively. Thus, we neglect rotation–vibration coupling.

The operators $H_0(\text{c.m.})$, $H_0(\text{int})$, and U and the corresponding Liouville operators $L_0(\text{c.m.})$, $L_0(\text{int})$, and θ, are abstract. The counterparts of the latter group in the mixed representation are

$$\hat{L}_0^N(\text{c.m.}) = -iP^N/M \cdot (\partial/\partial \mathbf{R}^N), \qquad (2.3)$$

$$\hat{L}_0^N(\text{int}) = (1/\hbar)[H_0(\text{int}), -], \qquad (2.4)$$

and

$$\hat{\theta}^N \hat{\chi}^N(\mathbf{R}^N, \mathbf{P}^N) = \mathfrak{F}_{\mathbf{P}^N\mathbf{Y}^N} \langle \mathbf{R}^N - \mathbf{Y}^N \mid \theta^N \chi^N \mid \mathbf{R}^N + \mathbf{Y}^N \rangle, \qquad (2.5)$$

where

$$-i\theta^N \chi^N = -(i/\hbar)[U^N, \chi^N].$$

The reduced Wigner operator $\hat{F}^s$ is defined by

$$\hat{F}^s(\mathbf{R}^s, \mathbf{P}^s, t) = V^s \int d\mathbf{R}^{N-s} d\mathbf{P}^{N-s} \sum_{Z^{N-s}} \langle Z^{N-s} \mid \hat{F}^N \mid Z^{N-s} \rangle. \qquad (2.6)$$

It follows from (2.2), in the thermodynamic limit, that

$$(\partial/\partial t)\hat{F}^s(\mathbf{R}^s, \mathbf{P}^s, t) = -i\hat{L}^s \hat{F}^s - i(N/V) \sum_{j=1}^{s} \mathfrak{L}_{j,s+1} \hat{F}^{s+1}, \qquad (2.7)$$

where

$$\mathfrak{L}_{12}\hat{\chi}^2 \equiv \sum_{z_2} \int d\mathbf{R}_2 d\mathbf{P}_2 \langle z_2 \mid \hat{\theta}_{12}\hat{\chi}^2 \mid Z_2 \rangle. \qquad (2.8)$$

Equation (2.7) is the analog for the classical BBGKY equations for structureless particles.[3]

III. REARRANGEMENT OF THE MANY-BODY PROBLEM

Equation (2.7) is still formidably complex because of the $\hat{F}^{s+1}$ term. Fortunately, $\hat{F}^s$ often contains far more information than is relevant to a particular problem. When this is so, the part of $\hat{F}^s$ which is relevant may be extracted by projection operator methods. In some (but not all) cases, the projected part obeys an equation far simpler than (2.7).

We define some useful reduced probabilities by the relations

$$\phi^s(p^s) \equiv (1/V^s) \int d\mathbf{R}^s \sum_{Z^s} \langle Z^s \mid {}^0\hat{F}^s \mid Z^s \rangle, \qquad (3.1)$$

$$\mathcal{G}_{J^s} \equiv (1/V^s) \int d\mathbf{R}^s d\mathbf{P}^s \sum_{V^s, M^s} \langle Z^s \mid {}^0\hat{F}^s \mid Z^s \rangle, \qquad (3.2)$$

and

$$\hat{P}_{V^s}(t) \equiv (1/V^s) \int d\mathbf{R}^s d\mathbf{P}^s \sum_{J^s, M^s} \langle Z^s \mid \hat{F}^s(t) \mid Z^s \rangle. \qquad (3.3)$$

Here ${}^0\hat{F}^s$ denotes the equilibrium reduced Wigner operator of order s; $\phi^s(p^s)$ and $\mathcal{G}_{J^s}$ are, respectively, the probability density for the momenta $\mathbf{p}^s = \{\mathbf{P}_1 \cdots \mathbf{P}_s\}$ and the probability for the rotational states $J^s = \{J_1, \cdots, J_s\}$; and $P_{V^s}(t)$ is the time-dependent probability for vibrational states $V^s = \{V_1, \cdots, V_s\}$. The symbols $\mathcal{G}^s$ and P^s will denote diagonal operators with nonzero matrix elements $\mathcal{G}_{J^s}$ and P_{V^s}. We shall suppress the superscript for $s=1$; $P_V(t)$ is then the object for which we want a master equation.

The correlation operation operator g^s is defined by

$${}^0\hat{F}^s = \phi(p^s) \, \mathcal{G}^s \, {}^0\hat{P}^s \hat{g}^s, \qquad (3.4)$$

where

$${}^0\hat{P}^s = \hat{P}^s(t = \infty).$$

When the s points represented by $\mathbf{R}^s$ are widely separated (this means distances large compared to σ, the effective range of the potential, and λ, the mean de Broglie wavelength), $\hat{g}^s$ will factor into a product of singlet operators which we shall assume are diagonal. It then follows that if R_{ij} is large for all (i, j) from

1 to s, then

$$\langle Z^s \mid {}^0\hat{F}^s \mid \bar{Z}^s\rangle = \prod_{j=1}^{s} \phi(p)\mathcal{J}_{J_j}{}^0P_{V_j}g_{Z_j}\Delta(Z^s, \bar{Z}^s), \quad (3.5)$$

where Δ denotes the Kronecker delta function. The assumption that $\hat{g}$ is diagonal is consistent with our earlier assumption that $H_0(\text{int})$ has eigenstates $\mid vJM\rangle$ in which v, J, and M are the quantum numbers for vibration, rotation, and orientation of the rotational angular momentum, respectively. In this approximation it is still possible for the moment of inertia to depend upon vibrational state.

Our next task is to define the projection operators $\mathcal{S}^s$, D^s, and $\mathcal{P}^s$:

$$\mathcal{S}^s \equiv (1/V^s)\int dR^s, \quad (3.6)$$

$$\langle Z^s \mid D^s\chi^s \mid \bar{Z}^s\rangle = \langle Z^s \mid \chi^s \mid Z^s\rangle\Delta(Z^s, \bar{Z}^s), \quad (3.7)$$

$$\mathcal{P}^s \equiv \hat{\phi}^s(\mathbf{p}^s)\mathcal{J}^s\hat{g}^s\int d\mathbf{p}^s\, \mathrm{Tr}^s{}_{\text{rot}}\mathcal{S}^sD^s. \quad (3.8)$$

As usual, χ represents an arbitrary observable. With these definitions we find that

$$P_V(t) = [\phi(\mathbf{p})\mathcal{J}_Jg_Z]^{-1}\langle Z \mid \mathcal{P}\hat{F}(t) \mid Z\rangle. \quad (3.9)$$

Thus, an equation for $\mathcal{P}\hat{F}(t)$ contains all the information necessary for the description of vibrational relaxation. The problem is to discover whether or not this equation is closed with respect to $\mathcal{P}\hat{F}(t)$ under reasonable physical circumstances.

We now proceed in the manner usual to projection operator applications[2,3] by deriving two equations from (2.7); first we apply $\mathcal{P}$, and then $(1-\mathcal{P})$. The result is (here and in most of what follows we drop the carets)

$$\frac{\partial}{\partial t}f^s = -i\mathcal{P}^sL^s(f^s+h^s) - i\frac{N}{V}\mathcal{P}^s\sum_{j=1}^{s}\mathfrak{L}_{j,s+1}(f^{s+1}+h^{s+1}), \quad (3.10)$$

$$\frac{\partial}{\partial t}h^s = i(1-\mathcal{P}^s)L^s(f^s+h^s)$$
$$-i\frac{N}{V}(1-\mathcal{P}^s)\sum_{j=1}^{s}\mathfrak{L}_{j,s+1}(f^{s+1}+h^{s+1}), \quad (3.11)$$

where we have introduced f^s and h^s:

$$f^s \equiv \mathcal{P}^sF^s, \qquad h^s \equiv (1-\mathcal{P}^s)F^s, \qquad F^s = f^s + h^s. \quad (3.12)$$

Equations (3.10) and (3.11) can be considerably simplified by making use of the following properties, some of which are valid for any system and some of which are valid only in the thermodynamic limit ($N\to\infty$, $V\to\infty$, N/V finite) (we assume the forces have a finite effective range σ):

(1) $\mathcal{S}^sL_0{}^s(\text{c.m.}) = 0,$ (3.13)

(2) $\mathcal{S}^s\theta^s = 0,$ (3.14)

(3) $\mathcal{S}^sL^s = \mathcal{S}^sL_0{}^s(\text{int}) = L_0{}^s(\text{int})\mathcal{S}^s$ (3.15)

[Equation (3.15) follows from (3.13), (3.14), and the fact $H_0(\text{int})$ depends only upon variables of the internal state.],

(4) $L_0{}^s(\text{c.m.})\chi(P^s) = 0,$ (3.16)

(5) $\mathcal{P}^s(1-\mathcal{P}^s) = 0, \quad (\mathcal{P}^s)(\mathcal{P}^s) = \mathcal{P}^s,$ (3.17)

(6) $\mathcal{P}L_0(\text{int}) = 0$ (3.18)

[In (3.18), the commutation of a diagonal matrix and an arbitrary matrix has no diagonal elements.]

One advantage to the present phase space formalism is that it makes explicit the significance of a finite interaction length σ.

We can now write (3.10) and (3.11) more simply as

$$\frac{\partial}{\partial t}f^s = -i\frac{N}{V}\mathcal{P}^s\sum_{j=1}^{s}\mathfrak{L}_{j,s+1}f^{s+1} - i\frac{N}{V}\mathcal{P}^s\sum_{j=1}^{s}\mathfrak{L}_{j,s+1}h^{s+1}, \quad (3.19)$$

$$\frac{\partial}{\partial t}h^s = -iL^sh^s - iL^sf^s - i\frac{N}{V}(1-\mathcal{P}^s)\sum_{j=1}^{s}\mathfrak{L}_{j,s+1}F^{s+1}. \quad (3.20)$$

The two equations are coupled; but if we solve the second for $h^2(t)$ in terms of $f^2(t)$ and $h^2(0)$, the result can be used in the first equation for $f(t)$. We then have

$$\frac{\partial f}{\partial t} = -i\frac{N}{V}\mathcal{P}\mathfrak{L}_{12}f^2 - \frac{N}{V}\int_0^t ds\mathcal{P}\mathfrak{L}_{12}G(s)L^2f^2(t-s)$$
$$-\left(\frac{N}{V}\right)^2\int_0^t ds\mathcal{P}\mathfrak{L}_{12}G(s)(1-\mathcal{P}^2)(\mathfrak{L}_{13}+\mathfrak{L}_{23})F^3(t-s)$$
$$-i\frac{N}{V}\mathcal{P}\mathfrak{L}_{12}G(t)h^2(0). \quad (3.21)$$

The propagator $G(t)$ is defined as the solution of

$$(dG/dt) = -iL^2G, \qquad G(0) = 1.$$

It can be given the explicit form $G(t) = \exp(-itL^2)$. It is of some interest that $G(t)$ appears in our basic equation; in most applications of the projection operator technique a more complicated operator $\exp[-it(1-P^N)L^N]$ appears instead. Working from a hierarchy of reduced equations has allowed us to avoid this object altogether.[3]

As it stands, (3.21) is no simpler than the equation from which it was derived. Now, however, we can make use of the physical approximations at our disposal. The assumption of low density may well help to eliminate the term of second order in N/V, and a time-smoothing argument may allow us to effect further changes. For instance, we would not expect the "initial correlations term" [the term with $h^2(0)$] to be significant on a sufficiently coarse time scale. In fact, however, the proof of this is very subtle and dependent on the choice of projection operator.

For simplicity of notation we write (3.21) as

$$\partial f/\partial t = (\partial f/\partial t)^M + (\partial f/\partial t)^R + (\partial f/\partial t)^I \quad (3.22)$$

and refer to the master, remainder, and initial value

terms, respectively. By definition,

$$\left(\frac{\partial f}{\partial t}\right)^M = -i\frac{N}{V}\mathcal{P}\mathfrak{L}_{12}f^2(t) - \frac{N}{V}\int_0^t ds\mathcal{P}\mathfrak{L}_{12}G(s)L^2f^2(t-s),$$

$$\left(\frac{\partial f}{\partial t}\right)^R = -\left(\frac{N}{V}\right)^2\int_0^t ds\mathcal{P}\mathfrak{L}_{12}G(s)(1-\mathcal{P}^2)(\mathfrak{L}_{13}+\mathfrak{L}_{23})F^3(t-s),$$

$$\left(\frac{\partial f}{\partial t}\right)^I = -i\frac{N}{V}\mathcal{P}\mathfrak{L}_{12}G(t)h^2(0). \tag{3.23}$$

Again we observe that (3.22) is essentially exact; our only assumptions have been to assume pairwise additive and short-ranged forces, the thermodynamic limit, the separability of rotation and vibration in the Hamiltonian $H_0(\text{int})$, and the diagonality of the singlet reduced correlation operator g. In a subsequent paper we shall use (3.22) as the starting point for a discussion of vibrational relaxation in liquids.

The various terms of (3.22) can be given some physical interpretation, but caution is necessary. The master term represents the effect of binary collisions of molecules which are not correlated prior to their collision. The role of $\mathcal{S}^s$ was to isolate a term of this sort. As we shall see, the master term can be expressed in terms of the inelastic differential scattering cross section.

Since not all collisions involve molecules which were previously uncorrelated, the remainder term must contain information about triple collisions and even more complex events. In addition, the remainder term contains information of a rather different character; although the master term computes the results of binary events, it does so having made assumptions about the singlet operator F; the remainder term corrects for errors which result from this approximation. We shall see that the master term suffices only when the degrees of freedom other than the one of interest are uncorrelated and at equilibrium, the density is low, and an appropriate time scale is adopted.

The initial value term has much the same nature as the remainder term, but describes how properties of the system at time zero effect developments at time t. It should be emphasized that if the remainder term is not ignorable, the initial value term may also be important *if* a physically reasonable initial condition is set. It is not realistic to pretend that an experimentalist can prepare his system so that $h^2(0)$ is zero.

IV. THE DILUTE AND ISOTROPIC GAS

A. The Principle of Molecular Chaos

The formalism developed so far is general and impossible to use without some gross simplifications. For the remainder of this paper we shall treat the very important case of a dilute and isotropic gas. The singlet Wigner operator will be independent of $\mathbf{R}$ in such a system, and if t_c and t_f are the mean times of and between collisions, respectively, then diluteness implies $t_c \ll t_f$.

In order to use this constraint we assume the validity of "molecular chaos" in the following restricted form: If $R_{12}<\sigma$ and $s>\alpha t_c$, where $t_c \ll \alpha t_c \ll t_f$, then

$$G(s)F^2(\mathbf{R}_{12}, \mathbf{P}^2, t) = G(s)F(\mathbf{P}_1, t)F(\mathbf{P}_2, t) + \mathcal{O}(t_c/t_f). \tag{4.1}$$

This definition emphasizes that molecular chaos is a property of $F^2(\mathbf{R}_{12}, \mathbf{P}^2, t)$ in the "precollision region" of two-molecule phase space.[6] The distinction between pre and post collision phase spaces is important because molecular chaos is demonstrably invalid in the latter region except at equilibrium: Let the system be infinitely dilute, and choose $(\mathbf{R}_{12}, \mathbf{P}^2)$ such that $\mathbf{R}_{12} \gg \sigma$, $\mathbf{R}_{12}(-t_0) < \sigma$, and $\mathbf{R}_{12}(-T) \gg \sigma$, where $t_0 \ll T \ll t_f$ (t_f is infinite in this example). A two-molecule Liouville equation applies and $F^2(\mathbf{R}_{12}, \mathbf{P}^2, t) = F^2[\mathbf{R}_{12}(-T), \mathbf{P}_1', \mathbf{P}_2']$, where primes denote precollision momenta. If molecular chaos were assumed in the two-sided form, and applied to this equality, the result would be $F(\mathbf{P}_1)F(\mathbf{P}_2) = F(\mathbf{P}_1')F(\mathbf{P}_2')$—a result which is manifestly at variance with reality except at equilibrium.

Physically, the point of (4.1) is that if two molecules are about to collide, and if the density is quite low, then the probability is overwhelming that the events which caused them to move toward the collision region were unrelated; hence, the molecules should be moving independently prior to the collision. A discussion of molecular chaos at finite densities has been given by Andersen and Oppenheim,[7] who show by means of a binary collision expansion that the property (4.1) is consistent with the laws of motion for *at least* a time t where $0<t\ll (t_f/t_c)t_f$. For our purposes it is reasonable to merely accept (4.1) as an assumption, and to note that the correction term in (4.1) need not be analytic in density, but must only be small when t_c/t_f is small.

B. Time Scales

Vibrational relaxation in a pure gas of diatomic molecules is characterized by at least two relaxation times: t_{vv} and t_{tv}. The former estimates the time required for the establishment of a nearly Boltzmann-like distribution of energies within the vibrational manifold. Roughly speaking, we may say that a vibrational temperature $T_{\text{vib}}(t)$ will exist after a time t_{vv}, and will thereafter characterize the distribution of vibrational energies on the time scale of t_{vv}. This reflects the existence of nearly resonant collisions (V–V collisions) in which a vibrational quantum is transferred from one oscillator to another. Less frequent "T–V collisions" serve to convert a vibrational quantum to translational and/or rotational energy, or vice versa. These lead, after a time of the order of t_{tv}, to complete equilibrium—at which time $T_{\text{vib}}(t)=T$. The most important feature of all this is that a clear separation of time scales usually exists for the common diatomic molecules in a dilute gas:

$$t_c \ll t_f \ll t_{vv} \ll t_{tv}. \tag{4.2}$$

It is now useful to introduce some scaling parameters which will help to identify the time scales on which certain terms are important:

$$\xi^2 \equiv t_f/t_{vv}, \qquad \lambda^2 = \xi^2/t_f = 1/t_{vv}. \tag{4.3}$$

Here ξ^2 is a measure of the probability that a collision will lead to a V–V transition. The more important parameter is λ^2, which is directly proportional to $1/t_f$, and so goes to zero in the limit of low density for any finite value of ξ^2.

C. The Master Term

The master term was defined in (3.23) as

$$\left(\frac{\partial f}{\partial t}\right)^M = -\frac{N}{V}\int_0^t ds \mathcal{P}\mathcal{L}_{12} G(s) L^2 f^2(t-s) - i(N/V)\mathcal{P}\mathcal{L}_{12} f^2(t). \tag{4.4}$$

Because each operation $\mathcal{P}\mathcal{L}_{12}$ involves an integral over $\mathbf{R}_{12}$ of a function which effectively vanishes for $R_{12} > \sigma$, it is possible to write

$$(N/V)\mathcal{P}\mathcal{L}_{12} \propto (N\sigma^3/V) = t_c/t_f. \tag{4.5}$$

The operator θ_{12} can also be given a formal size, but we emphasize that to do so requires a careful study of context. It is valid to say, in terms of lowest order in density, and in our formalism in which θ_{12} is preceded always by a projection operator, that

$$\theta_{12} = \mathcal{O}(\xi/t_c). \tag{4.6}$$

Essentially, this reflects the fact that only a fraction ξ^2 of collisions involve vibrational transitions. Combining the above, we find

$$(N/V)\mathcal{P}\mathcal{L}_{12} = \mathcal{O}(\xi/t_f). \tag{4.7}$$

Next, we may define scaled operators $\mathcal{L}_{12}^{sc}$, $\tilde{\theta}_{12}$, and $\tilde{L}^2$:

$$\mathcal{L}_{12}^{sc} = (1/\xi\sigma^3)\mathcal{L}_{12}, \qquad \tilde{\theta}_{12} = (1/\xi)\theta_{12}, \qquad \tilde{L}^2 = (1/\xi)L^2. \tag{4.8}$$

The operator L^2 is scaled because in the context of (4.4), and in the limit of low density, only its interaction term contributes. It is now proper to write (4.4) as

$$\left(\frac{\partial f}{\partial t}\right)^M = -\lambda^2 t_c \int_0^t ds \mathcal{P}\mathcal{L}_{12}^{sc} G(s) \tilde{L}^2 f^2(t-s;\lambda) - \frac{i\lambda^2 t_c}{\xi}\mathcal{P}\mathcal{L}_{12}^{sc} f^2(t,\lambda), \tag{4.9}$$

where the order of the terms is explicit in the factors of λ.

Introducing reduced time scales

$$\tau \equiv \lambda^2 t, \qquad g(\tau) \equiv f(t), \tag{4.10}$$

it follows that

$$\left(\frac{\partial g}{\partial \tau}\right)^M = t_c \int_0^{\tau/\lambda^2} ds \mathcal{P}\mathcal{L}_{12}^{sc} G(s) \tilde{L}^2 g^2(\tau-\lambda^2 s;\lambda) - \frac{it_c}{\xi}\mathcal{P}\mathcal{L}_{12}^{sc} g^2(\tau,\lambda). \tag{4.11}$$

If we consider the "$\lambda^2 t$ limit" ($\lambda \to 0$, $t \to \infty$, τ finite), this becomes

$$\left(\frac{\partial g}{\partial \tau}\right)^M = -t_c\left[\int_0^\infty ds \mathcal{P}\mathcal{L}_{12}^{sc} G(s)\tilde{L}^2\right] g^2(\tau,0) - \frac{it_c}{\xi}\mathcal{P}\mathcal{L}_{12}^{sc} g^2(\tau,0); \tag{4.12}$$

or, equivalently, (4.9) becomes

$$\left(\frac{\partial f}{\partial t}\right)^M = -\frac{N}{V}\left[\int_0^\infty ds \mathcal{P}\mathcal{L}_{12} G(s) L^2\right] f^2(t,0) - i\frac{N}{V}\mathcal{P}\mathcal{L}_{12} f^2(t,0). \tag{4.13}$$

Noting that $G(s)L^2 = i(d/ds)G(s)$, and using this to integrate in (4.13), we obtain a cancellation and the result

$$(\partial f/\partial t)^M = -\lim_{T\to\infty} i(N/V)\mathcal{P}\mathcal{L}_{12} G(T) f(\mathbf{p}_1,t) f(\mathbf{p}_2,t). \tag{4.14}$$

It can be demonstrated that the first nonzero term arising from the expansion of $G(T)$ is $\mathcal{O}(\xi^2/t_c)$. Thus,

$$(\partial f/\partial t)^M = \mathcal{O}(t_c\xi^2/t_f t_c) = \mathcal{O}(\lambda^2) = \mathcal{O}(1/t_{vv}). \tag{4.15}$$

This equation is local in time and closed in the singlet Wigner operator if $[(\partial/\partial t)f]^M = (\partial/\partial t)f$. It predicts that $f(t)$ [and $P_v(t)$] is essentially constant over periods small with respect to $1/t_{vv}$. As we demonstrate in Sec. VI, (4.14) is a master equation for $P_v(t)$ if $\partial f/\partial t = (\partial f/\partial t)^M$. We also write the master equation in "standard form" there, identifying the transition matrix with an appropriate collision average.

The limiting procedure we have used is formally identical to that introduced by Van Hove,[8] and later exploited by Zwanzig,[2] and by Mazur and Biel[3] among others. Quite generally, if one takes the $\lambda^2 t$ limit of a differential equation of the form

$$\frac{\partial \rho}{\partial t} = -\lambda^2 \int_0^t ds K(s,\lambda)\rho(t-s),$$

one gets an equation which, at best, describes events on the coarse time scale of $1/\lambda^2$. All information about processes occurring on shorter time scales is eliminated. Such a procedure may be regarded as a modern version of the time-smoothing operations introduced by Kirkwood in his first papers on transport theory.[9] In our work, $(1/\lambda^2)$ corresponds to the shortest of the times characterizing vibrational relaxation. Thus, we have suppressed information about momentum relaxation (which occurs on the time scale of t_f), but we have retained all information concerning vibrational processes.

D. The Initial Value Term

We have seen that the projector formalism has served to isolate the so-called master term, and we have estimated the size of $(\partial f/\partial t)^M$ above. Now we must see whether or not the master term deserves its title.

The effect of initial "correlations" is contained in $(\partial f/\partial t)^I$, where

$$(\partial f/\partial t)^I = -(N/V)\mathcal{P}\mathcal{L}_{12}G(t)(1-\mathcal{P}^2)F^2(0). \quad (4.16)$$

The operator $\mathcal{L}_{12}$, defined by (2.8), includes an integration over R_{12} with integrand $I(t)$,

$$I(t) = \theta_{12}G(t)(1-\mathcal{P}^2)F^2(\mathbf{R}_{12}, \mathbf{P}^2, t=0). \quad (4.17)$$

The interaction operator θ_{12} is a function of R_{12}, and is zero for $R_{12}>\sigma$. The $F^2(\mathbf{R}_{12}, \mathbf{P}^2, 0)$ is also a function of $(\mathbf{R}_{12}, \mathbf{P}^2)$, but in context is to be evaluated at $[G(s)\mathbf{R}_{12}, G(s)\mathbf{P}^2]$ which, if θ_{12} is to be nonzero, must represent a point in the precollision region of phase space whenever s is large $(s>\alpha t_c)$. The principle of molecular chaos now applies and [see (4.1) and the related discussion]

$$I(t) = \theta_{12}G(t)[(1-\mathcal{P}^2)F(1,0)F(2,0)+\mathcal{O}(t_c/t_f)]. \quad (4.18)$$

Apparently, our task is to show that $(1-\mathcal{P}^2)F(1, 0) F(2, 0)$ is "small" or zero. If it is of order t_c/t_f or smaller, we then have [see (4.7) and (4.16)]

$$(\partial f/\partial t)^I = \mathcal{O}[(\xi/t_f)(t_c/t_f)] = \mathcal{O}(\lambda^4 t_c/\xi^3) \quad (4.19)$$

which can be ignored relative to $(\partial f/\partial t)^M$ in the $\lambda^2 t$ limit. Before discussing $(1-\mathcal{P}^2)F(1, 0)F(2, 0)$ in detail, we consider the nature of $(\partial f/\partial t)^R$.

E. The Remainder Term

The remaining contribution to $(\partial f/\partial t)^M$ is defined by

$$\left(\frac{\partial f}{\partial t}\right)^R = -i\left(\frac{N}{V}\right)^2 \int_0^t ds\mathcal{P}\mathcal{L}_{12}G(s)(1-\mathcal{P}^2) \times(\mathcal{L}_{13}+\mathcal{L}_{23})F^3(t-s). \quad (4.20)$$

Since this is apparently of second order in density we might naively dismiss it out of hand. A more careful analysis indicates, however, that

$$(\partial f/\partial t)^R = \mathcal{O}(\lambda^2 t_0/t_f), \quad (4.21)$$

where t_0 is the value of t (not necessarily finite) such that $(\partial f/\partial t)^R$ is essentially constant for $t>t_0$. Thus, the remainder term can be ignored compared with the master term if $t_0 \ll t_f$.

Introducing $\zeta(s)$,

$$\zeta(s) \equiv -i(N/V)\theta_{12}G(s)(1-\mathcal{P}^2)(\mathcal{L}_{13}+\mathcal{L}_{23})F^3(t-s), \quad (4.22)$$

and comparing it with (2.7), we see that

$$-i(N/V)(\mathcal{L}_{13}+\mathcal{L}_{23})F^3(t') = (\partial F^2/\partial t')_{\text{coll}},$$

where $(\partial F^2/\partial t')_{\text{coll}}$ is the change of F^2 due to collisions of molecules 1 and 2 with the reservoir molecules 3, $\cdots$, N, and $t'=t-s$. For the purpose of determining t_0, we use the BBGKY equation (2.7) to write $\zeta(s)$ as

$$\zeta(s) = \theta_{12}G(s)(1-\mathcal{P}^2)(\partial F^2/\partial t' + iL^2F^2). \quad (4.23)$$

The second term vanishes for large s $(s>\alpha t_c)$, and $(1-\mathcal{P}^2)$ and $G(s)$ commute with $\partial/\partial t'$, permitting application of the molecular chaos principle. We then find that for $s>\alpha t_c$, the properties of $\zeta(s)$ depend essentially on $(1-\mathcal{P}^2)F(1, t')F(2, t')$. If this quantity were no larger than $\mathcal{O}(t_c/t_f)$, then we could identify t_0 with αt_c and, in that case,

$$(\partial f/\partial t)^R = \mathcal{O}[\lambda^4(\alpha t_c/\xi^3)]. \quad (4.24)$$

The problem can be cast into the language of reduced time scales as follows:

$$g(\tau) = f(t), \qquad g^3(\tau) = F^2(t'), \qquad \lambda^2 t = \tau,$$

in which case (4.23) implies an equation of the form

$$\left(\frac{\partial g}{\partial \tau}\right)^R = -\lambda^2 \int_0^t ds K(s, \lambda) g^2(\tau - \lambda^2 s, \lambda).$$

Most of the serious difficulties which can arise in deriving a master equation via projectors are implicit here: does the kernel K decay quickly enough (it is sufficient that it decays at least as fast as an exponential independent of λ)? Since it depends on λ, it might decay more and more slowly as $\lambda \to 0$; this would correspond to the case $t_0 = \mathcal{O}(t)$. Some discussion of similar problems is contained in papers by Terwiel and Mazur,[10] and by Resibois, Brocas, and Decan.[11] The physical situations treated in these papers are, however, entirely different from that considered here.

Pulling together the results (4.19) and (4.24), we see that our basic task is to prove $(1-\mathcal{P}^2)F(1, t)F(2, t) \leq \mathcal{O}(t_c/t_f)$. If this condition is met, $f(t)$ will be correctly described by $(\partial f/\partial t)^M$, at least in the $\lambda^2 t$ limit.

V. CONDITIONS FOR THE VALIDITY OF MASTER EQUATIONS

A. Formal Conditions

As demonstrated in Sec. IV, our problem is to determine when, if ever,

$$(1-\mathcal{P}^2)F(1, t)F(2, t) = \mathcal{O}(\gamma), \quad (5.1)$$

where $\gamma \leq t_c/t_f$, and where (5.1) need only be valid on a coarse time scale with unit t_{vv}. If we refer back to the definitions of $F(1, t)$ and $\mathcal{P}^2$ given in Sec. III, we immediately see that (5.1) holds when

$$F(1, t) = \phi(\mathbf{p})\mathcal{G}gP(1, t), \quad (5.2)$$

i.e., $\langle Z \mid F(1, t) \mid Z' \rangle = \phi(\mathbf{p})\mathcal{G}_j g_z P_v(t)\Delta(Z, Z')$.

We now see that there are three *sufficient* conditions for a master equation to be valid:

(1) The nonvibrational degrees of freedom must be at equilibrium (that is, the *singlet* probabilities must be those at equilibrium),

(2) there must be no time-dependent correlations among the degrees of freedom, and

(3) the singlet reduced density matrix for vibration must be diagonal:

$$\langle v \mid \rho_{\text{vib}}(t) \mid v' \rangle \equiv \langle v \mid (1/V) \int d\mathbf{R} d\mathbf{P}\ \text{Tr}_{\text{rot}} F(1, t) \mid v' \rangle = P_v(t) \Delta(v, v'). \quad (5.3)$$

B. Physical Significance of the Conditions

The conditions listed above are logically distinct, as the following illustrates: Suppose a laser is used to excite 0.1% of the molecules to a state (v', J'), where $| v' \rangle$ is normally unpopulated. The *singlet* probability for rotational energy may still be very well approximated by $\mathcal{J}$, its equilibrium value. Clearly, however, the *joint* probability for rotation and vibration will be distinctly nonequilibrium in nature. It will not factor into $\mathcal{J} P_v(t)$ because rotation and vibration are correlated.

After a mean free time t_f, a new correlation appears. It is well known that $T-V$ transitions arise only from very hard collisions. Thus, at time $t \approx t_f$, any molecule in the state $v'-1$ is likely to be moving very rapidly (if this state is also unpopulated at equilibrium); this constitutes a correlation between a molecule's speed and its vibrational state. While this correlation builds, the other correlation tends to dissipate because rotational transitions occur readily and without much regard for vibrational state. If changes in the rotational state J occur in the fraction t_f/t_{rot} of collisions, then the correlation will be absent for $t \gg t_{\text{rot}}$. For nearly all diatomic molecules,

$$t_f \lesssim t_{\text{rot}} \ll t_{vv} \ll t_{tv}.$$

Thus, an initial rotation–vibration correlation would ordinarily play no role on the time scale of vibrational relaxation.

The case of H_2 is, however, quite exceptional. For this case t_{rot} is rather large $(t_{\text{rot}} \gg t_f)$[12]; indeed, it seems to be of the same general order as t_{vv}[12b]; thus, the master equation may not be valid for $V-V$ relaxation of H_2. This conclusion prevails for an arbitrary initial condition because there will in general exist some coupling between $V-V$ transitions and rotational transitions. The former involves some change in translational and/or rotational energy (the energy "defect") because of anharmonicity and the associated unequal level spacings. Some rotational spacings are comparable to this energy defect in the case of H_2. Thus, the possibility exists that some specific joint rotational and vibrational transitions would occur more readily than either or both of the separate transitions. On the other hand, other rotational transitions might represent a competing and somewhat easier process than a $V-V$ transition. The net result would be to create a rotational nonequilibrium if a vibrational nonequilibrium is present initially. The two relaxations would then proceed on the same time scale if $t_{\text{rot}} \approx t_{vv}$. Finally, we might mention that experimental evidence already exists[12c] which suggests that rotation plays a role in the vibrational relaxation of H_2 and D_2; certainly, the results do not fit into the usual empirical or theoretical schemes.

To analyze the significance of the correlation linking speed and vibrational state, we must first determine the time necessary for the thermalization of those highly energetic molecules which play such an important role in $T-V$ relaxation. Energies of 40 kT, or even more, are sometimes involved. It is always said that the momentum relaxation time is t_f, but in principle this could be very misleading because it is really an experimental conclusion about momentum transfer in ordinary molecules, those with nearly thermal energies.

The potential difficulty is this: Since the velocity distribution for molecules which have just had a $T-V$ transition is grossly skewed toward high energies, it is entirely plausible that such molecules would tend to have a second transition within a very few collisions—before their speeds had been brought down to thermal levels. If this were so, the relevant momentum distribution would be distinctly nonequilibrium in nature. Again, the distinction must be made between singlet and joint probabilities.

We can formulate a useful criterion as follows: Suppose the mean relative speed in collisions involving $T-V$ transitions is c^*, and suppose the transition probability per collision is a function $W(c)$. Clearly, the problem of correlations will be absent only if

$$W(c^*) \ll t_f / t_{\text{mom}}, \quad (5.4)$$

where $t_{\text{mom}}(c^*)/t_f$ is the mean number of collisions necessary to thermalize a molecule with speed c^*.

Some theoretical estimates of $t_f/t_{\text{mom}}(c^*)$ are available.[13a,13b] For collisions between molecules of equal mass, we may expect that even the very energetic molecules would thermalize within ten mean free times. If the molecules of interest (mass M) were dilutely dispersed in a reservoir of particles with mass m, the relaxation time would be somewhat longer for M/m much different from unity.[13b] We might add that the finite duration of translation relaxation has also proved important in chemical kinetics.[13c]

Rapp and Kassal[14] have commented on the size of $W(c^*)$ in connection with the validity of perturbation theory (p. 86 of their paper). They conclude that $W(c^*)$ may be large $[W(c^*) > 0.1]$ in many cases of interest, particularly when the reported value of t_{tv} is small $(t_{tv} < 10^3 t_f)$.

It would appear that the problem of correlations is

not one to be ignored. Again we emphasize that this is not merely a problem of an initial "incubation period"; it is a problem which, if present at all, will persist throughout the $T-V$ relaxation.

According to (5.3), another condition for the validity of a master equation is that $\rho_{\mathrm{vib}}(t)$ be diagonal. To better understand why this might be so, let us write $\rho_{\mathrm{vib}}(t)$ as

$$\rho_{\mathrm{vib}}(t) = \sum_{v,v'} \langle C_{v'}^*(t) C_v(t) \rangle^{\mathrm{ens}} \times \exp(i\omega_{vv'}t) \, |\langle v \mid \chi \mid v' \rangle|. \quad (5.5)$$

The $\langle \cdots \rangle^{\mathrm{ens}}$ denotes an ensemble average. This form is motivated by circumstances at zero density when $C_v(t)$ is a mere constant. This suggests that, more generally, $C_v(t)$ varies only on time scales at least as long as t_f. If this is so, $\langle v \mid \rho_{\mathrm{vib}} \mid v' \rangle$ oscillates about zero with period $2\pi/\omega_{vv'}$—roughly the duration of an individual vibration. Although $\langle v \mid \rho_{\mathrm{vib}}(t) \mid v' \rangle$ *may* not be zero, it is effectively so on the time scale t_{vv}. If we return briefly to Sec. IV.D, we find that $\zeta(s)$ may not vanish for $s > \alpha t_c$, but that t_0 is still $\mathcal{O}(\alpha t_c)$ because $\zeta(s)$ oscillates rapidly for larger values of s, and so makes little net contribution to $(\partial f/\partial t)^R$. An alternative view observes that the experimental initial state is usually ill defined on time scales as fine as $2\pi/\omega_{vv'}$; thus, the density operator for the representative ensemble should have no time dependence on such a scale. This and (5.5) can be reconciled if the "random phase" property holds:

$$\langle C_{v'}^*(t) C_v(t) \rangle^{\mathrm{ens}} = 0, \qquad v \neq v'.$$

The last condition we consider is that mentioned first in Sec. IV. The nonvibrational degrees of freedom must remain at equilibrium during the relaxation. Upon reflection, we see that if total energy is to be conserved, the condition cannot possibly be met—except perhaps in some good approximation.

To better understand the implications of this, let us consider an experiment in which a laser is used to excite molecules to some specific vibrational state at time zero. We shall assume that the level spacing is 3000°K, the initial temperature 300°K, and that 10% of the molecules are initially excited. Even if the vibrational transition is as small as possible ($\Delta v = 1$), a total of 300°K of energy must be dissipated among the five classical degrees of freedom in order to achieve equilibrium. This corresponds to a 20% change in the translation temperature on the same time scale as the relaxation process of interest.[15] Simple calculations indicate[16] that a 20% change in temperature leads to a similar change in the time t_{tv}. Thus, by forcing the master equation to fit experimental data under such circumstances one would lose accuracy, as well as potentially important information. We conclude that a master equation will not be valid if the translational temperature changes significantly during the experiment.

Fortunately, it is usually possible to replace the master equation with a more accurate expression in which the transition matrix has an explicit time dependence.[1] If the vibrational energy decays in approximately exponential fashion, a sufficient condition for which is that $P_v(t)$ maintain an approximately Maxwellian form throughout the relaxation,[17] then conservation of total energy leads to the expression

$$T(t) = T(\infty) + [T(0) - T(\infty)] \exp(-t/t_{tv}). \quad (5.6)$$

The value of t_{tv} can be estimated experimentally, after which the $\phi(p)\mathcal{G}_j$ form which appears in the master equation can be replaced by its analogue with the time-dependent temperature. The new equation would then allow one to make corrections to the original description.

In summary, we find that the constraints on the use of a master equation are both subtle and nontrivial. Indeed, we have indicated several situations for which use of a master equation is not justified:

(1) Experiments[15] in which the translational temperature changes significantly (this could include laser or shock-tube studies),

(2) relaxation of molecules which tend to have two or more $T-V$ transitions within a period of perhaps five or ten mean free times [see (5.4) for a more precise criterion],

(3) and $V-V$ relaxation of hydrogenlike molecules.

We have conducted a similar study for rotational relaxation by using a slightly different projection operator.[1] As can be readily anticipated from our discussion of vibration, the rotational master equation will usually not be valid because of strong coupling between translational and rotational relaxations. The case of H_2 dilute in a reservoir of atoms would be exceptional because, in this case, $t_{\mathrm{rot}} \gg t_f$. In pure H_2 there would possibly be some interference between rotational and $V-V$ relaxations. A master equation can also be justified for very specialized applications such as an experiment described theoretically by Gordon.[18] In Gordon's analysis, the master equation is postulated and used only for a period of about one mean free time; also, the initial state includes no correlations among the degrees of freedom. This would not be so, however, if polarized light were used to create the initial excitation; we hope to examine this case in more detail in a later publication. Mathematically, the problem would be more complicated because the appropriate kinetic equation would involve a reduced density operator with both diagonal and off-diagonal elements.

Finally, we would like to note some recent work by Keizer[19] in which a master equation is derived for "some" of the internal degrees of freedom in a single molecule immersed in a low density reservoir of structureless particles. Superficially, there is some resemblance between his work and our own: We both start with the Liouville equation, introduce a projection operator, and consider results in the limit of low density and long times. Keizer's model does not allow for V–V processes or nonequilibrium in the translational temperature. In our work such matters have been emphasized. Even apart from this, the general tone of the two papers is quite different: We have emphasized the possibility of correlations he does not mention; we have treated a single well-defined and physically realistic problem is great detail while he has adopted a more abstract approach in which the physical significance of various points is less obvious; finally, we have treated the matter of time scales (not just absolute value of the time elapsed since the initial state was formed) carefully and explicitly, and have demonstrated that the master equation is valid only on a time scale which is at once coarse with respect to t_{mom} and t_{rot}, and fine with respect to the shortest time characteristic of vibrational relaxation, t_{vr}.

VI. MASTER EQUATIONS: MICROSCOPIC AND MACROSCOPIC

The form of the master equation is given in (4.14). If we restore carets to avoid confusion below, this reads

$$\partial\hat{f}/\partial t=-i\lim_{T\to\infty}(N/V)\mathcal{P}\mathcal{L}_{12}\hat{G}(T)\hat{f}(1,t)\hat{f}(2,t). \quad (6.1)$$

By using the definitions introduced in Sec. I, it can be shown[1] that

$$\hat{\theta}_{12}\hat{G}(\infty)\hat{f}(1,t)\hat{f}(2,t)$$
$$=\mathfrak{F}_{\mathbf{P}^2\mathbf{Y}^2}\langle\mathbf{R}^2-\mathbf{Y}^2\mid\theta_{12}G(\infty)f(1,t)f(2,t)\mid\mathbf{R}^2+\mathbf{Y}^2\rangle. \quad (6.2)$$

Recalling that $P_v(t)$ and $f(1,t)$ are closely related [see (3.9)], we find

$$dP_v/dt=[-(2\pi)^3N/V]i$$
$$\times\sum_{a_2}\sum_{J_1,M_1,P_1}\langle a^2\mid\theta_{12}G(\infty)f(1,t)f(2,t)\mid a^2\rangle, \quad (6.3)$$

where $|a\rangle=|\mathbf{P}\rangle|VJM\rangle$ and $\sum_{\mathbf{P}}=(V/(2\pi)^3)\int d\mathbf{P}$.

The relationship of this to the scattering T matrix has been derived by Andersen and Oppenheim.[7] Using their result and the diagonality of $f(1,t)$, we obtain

$$dP_v/dt=(2\pi N/V)\sum_{a^2,J_1,M_1,P_1}\langle a^2\mid[Tf(1,t)f(2,t)T^{\dagger}$$
$$-TT^{\dagger}f(1,t)f(2,t)]\mid a^2\rangle. \quad (6.4)$$

The T matrix is to be considered diagonal in energy.

Let us now define transition rates by

$$D(V^2,V'^2)=(2\pi N/V)\sum_{J^2,M^2,P^2,b^2}T(a^2,b^2)T^{\dagger}(b^2,a^2)$$
$$\times\mathfrak{J}_{b^2}\phi(P_b{}^2)g_{b_1}g_{b_2}, \quad (6.5)$$

where $b^2=\{V'^2,J'^2,M'^2,\mathbf{P}'^2\}$. Now (6.4) can be written compactly as

$$dP_v/dt=\sum_{V_1',V_2,V_2'}[D(V^2,V'^2)P_{v_1'}(t)P_{v_2'}(t)$$
$$-D(V'^2,V^2)P_{v_1}(t)P_{v_2}(t)]. \quad (6.6)$$

This is the form of the phenomenological master equation; it is bilinear, and it can be easily demonstrated that **D** obeys detailed balance.

If the molecules were infinitely dilute in a reservoir of structureless particles, the corresponding master equation would be linear. And, if $P_v(t)$ is not too far from equilibrium, (6.5) can be linearized by writing $P_v(t)={}^0P_v+\Delta P_v(t)$, inserting this in (6.5), and keeping only terms of first order in $\Delta P_v(t)$.

* A portion of this work was supported by the National Science Foundation.

† Present address: Institute for Defense Analyses, Science and Technology Division, 400 Army–Navy Drive, Arlington, Va. 22202.

[1] P. K. Davis, Ph.D. thesis, MIT, 1970.

[2] R. Zwanzig, in *Lectures in Theoretical Physics*, edited by W. E. Brittin *et al.* (Interscience, New York, 1961), Vol. 3.

[3] P. Mazur and J. Biel, Physica **32**, 1633 (1966).

[4] H. Mori, I. Oppenheim, and J. Ross, in J. deBoer and G. E. Uhlenbeck, *Studies in Statistical Mechanics* (North-Holland, Amsterdam, 1962), Vol. 1.

[5] U. Fano, Rev. Mod. Phys. **29**, 74 (1957).

[6] N. Bogoliubov, J. Phys. (USSR) **10**, 265 (1946).

[7] H. C. Andersen and I. Oppenheim, Ann. Phys. (N.Y.) **48**, 1 (1968).

[8] L. Van Hove, Physica **21**, 517 (1955); **23**, 441 (1957).

[9] J. G Kirkwood, J. Chem. Phys. **14**, 180 (1946). See, however, the comments of Zwanzig in the reprint volume *Selected Topics in Statistical Mechanics*, edited by R. Zwanzig (Gordon and Breach, New York, 1967).

[10] R. Terwiel and P. Mazur, Physica **32**, 1813 (1966).

[11] P. Resibois, J. Brocas, and G. Decan, J. Math. Phys. **10**, 964 (1969).

[12] (a) R. M. Jonkman, G. J. Prangsma, I. Ertas, H. Knaap, and J. Beenaker, Physica **38**, 441 (1968). (b) Typical V–V relaxation times for diatomics are $\mathcal{O}(10^2\,\alpha t_f)$; we have no data on H_2 in particular. (c) J. Kiefer and R. Lutz, J. Chem. Phys. **44**, 658 (1966).

[13] (a) C. Rebick (private communication). (b) R. N. Porter, J. Chem. Phys. **45**, 2284 (1966). (c) R. Kapral, S. Hudson, and J. Ross, J. Chem. Phys. **53**, 4387 (1970).

[14] D. Rapp and T. Kassal, Chem. Rev. **69**, 61 (1969).

[15] Dr. J. Steinfeld has informed us that temperature changes have been observed in relaxation experiments in his laboratory; moreover, the assumption of 10% excitation is not beyond the possibilities of present techniques. Since this work was completed, Dr. G. Flynn has also reported such effects and, indeed, has found it necessary to use a modified master equation (seminar, University of Chicago, 1971).

[16] Crude but reasonable calculations can be performed using formulas given in Ref. 14, and in (a) K. Herzfeld and T. Litovitz, *Absorption and Dispersion of Ultrasonic Waves* (Academic, New York, 1959). (b) T. Cottrell and J. C. McCoubrey, *Molecular Energy Transfer in Gases* (Butterworths, London, 1961).

[17] H. C. Andersen, I. Oppenheim, K. Shuler, and G. Weiss, J. Math. Phys. **5**, 522 (1964).

[18] R. Gordon, J. Chem. Phys. **46**, 4399 (1967).

[19] J. Keizer, J. Chem. Phys. **53**, 4195 (1970).

Studies in Nonequilibrium Rate Processes.* I. The Relaxation of a System of Harmonic Oscillators

ELLIOTT W. MONTROLL, *Institute for Fluid Dynamics and Applied Mathematics, University of Maryland, College Park, Maryland*

AND

KURT E. SHULER, *National Bureau of Standards, Washington, D. C.*

(Received April 13, 1956)

As a part of an investigation of nonequilibrium phenomena in chemical kinetics a theoretical study has been made of the collisional and radiative relaxation of a system of harmonic oscillators contained in a constant temperature heat bath and prepared initially in a vibrational nonequilibrium distribution. An exact solution has been obtained for the general relaxation equation applicable to this system and expressions have been derived for the relaxation of initial Boltzmann distributions, Poisson distributions, and δ-function distributions as well as for the relaxation of the moments of the distributions. Using the latter result, explicit expressions are given for the relaxation of the internal energy of the system of oscillators and for the time dependence of the dispersion of the distributions.

1. INTRODUCTION

IT has been recognized for many years that by its very nature a chemical reaction must produce a perturbation in the initial Maxwell-Boltzmann distribution of the reactant species.[1] The extent of the departure from equilibrium will depend upon the relative magnitudes of the rates of the elementary chemical reactions (i.e., the rate of transformation of reactants

* This research was partially supported by the U. S. Air Force through the Office of Scientific Research of the Air Research and Development Command and by the U. S. Atomic Energy Commission.

[1] See, e.g., R. H. Fowler and E. A. Guggenheim, *Statistical Thermodynamics* (Cambridge University Press, New York, 1949), Chap. XII, or Eyring, Walter, and Kimball, *Quantum Chemistry* (John Wiley and Sons, Inc., New York, 1944), Chap. XVI.

to products) and the rates of energy exchange between the various atomic and molecular species in the reaction system. If the rate of the chemical transformation is small compared with the rate of energy exchange, the perturbation of the initial equilibrium distribution will be small and the reaction system can be discussed in terms of equilibrium statistical mechanics. If, however, the rate of the chemical transformation exceeds the rate of intra- and intermolecular energy exchange, there may develop a considerable perturbation of the equilibrium Maxwell-Boltzmann distribution of energy during the course of the chemical reaction. Under these conditions the equilibrium hypothesis underlying the present collision and absolute rate theories of chemical kinetics may no longer be tenable. Recent experimental work on various rapid high-temperature chemical reactions has shown quite clearly that there is indeed in many cases a considerable perturbation of the initial equilibrium distribution of energy during the course of the chemical reaction.[2] It therefore becomes important to study the distribution of energy in a reaction system during the course of a chemical reaction so that a foundation can be laid for the development of a non-equilibrium theory of chemical kinetics.

The specific problem which we wish to consider in the above context concerns the relaxation of the distribution of a system of harmonic oscillators prepared initially in nonequilibrium vibrational distributions. The oscillators are excited to these distributions either by external perturbations, such as irradiation with short duration, high intensity light or by the passage of a shock wave, or internally by some specific chemical reaction.[3] After the external perturbation has been removed (i.e., after the light has been turned off or after the passage of the shock wave) or after the cessation of the reaction, the system of oscillators will relax to its final equilibrium distribution by inelastic collisions and by radiative transitions. We wish to study in detail the dynamic behavior of the distribution and of the moments of the distribution of the oscillators among their energy levels for various initial nonequilibrium distributions.

Our study of the relaxation of a system of harmonic oscillators is based on the following model:

(a) The oscillators are contained in a large excess of (chemically) inert gas which acts as a *constant temperature* heat bath throughout the relaxation process. This implies that the concentration of the excited oscillators is sufficiently small and the energy absorbed by them during their excitation is sufficiently small that the heat bath remains at its initial equilibrium temperature T throughout the relaxation process.

(b) The total concentration of excited oscillators is sufficiently small so that the relaxation process is first order with respect to the concentration of oscillators. The energy exchange which controls the relaxation thus takes place primarily between the oscillators and the heat bath.

(c) The excited oscillators can transfer their vibrational energy both by collision and by radiation. In the collisional transfer of energy, the vibrational energy of the excited oscillators can be exchanged with both the translational and the vibrational degrees of freedom of the heat bath molecules.

(d) The collisional transition probabilities for transitions between the vibrational levels i and j of the harmonic oscillators are to be calculated according to the prescription of Landau and Teller.[4] According to this prescription, the perturbations which induce the transitions are linear in the normal coordinate (i.e., the internuclear separation in the case of a harmonic oscillator) and sufficiently small for a first order perturbation calculation. With these assumptions, the matrix elements for collisional transitions are identical, except for a constant factor, with those for the radiative transitions of a harmonic oscillator. The same "selection rules" will thus hold for collisional transitions as for radiative ones in that the collision induced transitions of the oscillators will take place only between adjacent vibrational levels. The collisional transition probabilities per collision, $P_{i,j}$ are thus given by

$$P_{i,j}=P_{j,i}, \quad P_{i,j}=0 \quad \text{for} \quad j\neq\begin{cases} i+1 \\ i-1 \end{cases}$$
$$P_{i,j+1}=(i+1)P_{10}, \tag{1.1}$$

where P_{10} is the collisional transition probability per collision for transitions between vibrational levels $i=1$ and $i=0$.

We now wish to derive the differential equations which govern the relaxation of the ensemble of harmonic oscillators in our model. It has been pointed out by Herzfeld[5] that an exact energy balance in a relaxation process of the type discussed here can be obtained when either (a) the excited system or the heat bath have a nearly continuous array of levels or (b) the excited system and the heat bath have equidistant energy levels and exchange only vibrational energy. Under either of these conditions, a transition $-\Delta E$ between two states in the excited system can be matched by a transition of a corresponding energy ΔE between two states of the heat bath. This latter case can readily be realized if one chooses for the relaxation system an

[2] For a more detailed discussion of this point see K. E. Shuler, J. Phys. Chem. **57**, 396 (1953); *5th Symposium (International) on Combustion* (Reinhold Publishing Corporation, New York, 1955), pp. 56–74.

[3] An example of the latter process is the formation of OH in the vibrational state $v=9$ in the reaction $H+O_3\rightarrow OH+O_2$ studied by A. B. Meinel, J. Astrophys. **111**, 207, 433, 555 (1950) and by McKinley, Garvin, and Boudart, J. Chem. Phys. **23**, 784 (1955).

[4] L. Landau and E. Teller, Physik. Z. Sowjetunion **10**, 34 (1936).

[5] K. F. Herzfeld in *Temperature, Its Measurement and Control in Science and Industry* (Reinhold Publishing Corporation, New York, 1955), p. 233.

ensemble of harmonic oscillators of which a small fraction are excited to an initial vibrational nonequilibrium distribution while the large excess of unexcited oscillators serves as the heat bath. If we let

$x_n(t)$ = fraction of excited oscillators in level n
y_i = concentration of heat bath oscillators in level i

$P_{n,\,n+1;\,i,\,i-1}$ = probability per collision for the energy transfer $n\rightarrow n+1$ as $i\rightarrow i-1$

the relaxation equation can be written as

$$\frac{dx_n(t)}{dt}=-Z[x_n(\sum_{i=0}^{\infty} y_iP_{n,\,n+1;\,i,\,i-1}+\sum_{i=0}^{\infty} y_iP_{n,\,n-1;\,i,\,i+1})$$
$$-x_{n+1}\sum_{i=0}^{\infty} y_iP_{n+1,\,n;\,i,\,i+1}$$
$$-x_{n-1}\sum_{i=0}^{\infty} y_iP_{n-1,\,n;\,i,\,i-1}] \quad (1.2)$$

where Z is the collision number, i.e., the number of collisions per second suffered by the oscillator in the level system $n=0, 1, \cdots$ when the gas density is one molecule per unit volume. Using Eq. (1.1), the probabilities $P_{n,\,n+1;\,i,\,i-1}$ for concurrent collisional transitions can be written as

$$P_{n,\,n+1;\,i,\,i-1}=(n+1)iP_{10}$$
$$P_{n,\,n-1;\,i,\,i+1}=n(i+1)P_{10} \quad (1.3)$$

so that Eq. (1.2) becomes

$$\frac{dx_n(t)}{dt}=-ZP_{10}[(n+1)x_n\sum_{i=0}^{\infty} iy_i+nx_n\sum_{i=0}^{\infty}(i+1)y_i$$
$$-(n+1)x_{n+1}\sum_{i=0}^{\infty}(i+1)y_i-nx_{n-1}\sum_{i=0}^{\infty} iy_i] \quad (1.4)$$
$$n=0, 1, 2, \cdots.$$

Since we assume that the heat bath remains in its initial Boltzmann distribution at temperature T throughout the relaxation process we can write, for all times t,

$$y_i=N(1-e^{-\theta})e^{-i\theta} \quad (1.5)$$

where N is the total concentration of oscillators in the heat bath and where $\theta=h\nu/kT$ and ν is the fundamental frequency of the oscillators. Substitution of (1.5) into (1.4) finally leads to

$$dx_n(t)/dt=k_{10}(1-e^{-\theta})^{-1}\{ne^{-\theta}x_{n-1}$$
$$-[n+(n+1)e^{-\theta}]x_n+(n+1)x_{n+1}\}$$
$$n=0, 1, 2, \cdots \quad (1.6)$$

where $k_{10}=ZP_{10}N$ is the collisional transition probability per second for transitions between levels 1 and 0 of the oscillators. The set of differential difference equations (1.6) governs the relaxation of a system of excited harmonic oscillators contained in a harmonic oscillator heat bath (with $\nu_n=\nu_i$) when there is only vibrational energy exchange between the excited oscillators and the heat bath.

It is not possible to follow the method used above to obtain Eq. (1.6) when the relaxation proceeds by the interchange of the vibrational energy of the excited oscillators with the translational energy of the heat bath. In this case it is not possible to establish internal equilibrium by considering only the energy transfer between the excited oscillators and the heat bath as was done previously since the oscillators will give up their excitation energy only in quanta of $h\nu$ while the heat bath has a nearly continuous array of translational energy states. Furthermore, it is not possible to write down simple explicit expressions for the joint transition probabilities $P_{n;\,i}$ as was done in (1.3), where i now refers to the translational energy levels of the heat bath, within the framework of the Landau-Teller approximation used in our model. It has been shown, however, by Rubin and Shuler[6] that the set of differential difference equations governing the relaxation process now under discussion can be obtained by the method used by Fowler in discussing the equilibrium relationship between collisions of the first and second kind.[7] Using properties (a) to (d) of our model and applying the principle of detailed balancing at equilibrium, Rubin and Shuler showed that the relaxation equation for the case when the relaxation proceeds by the interchange of the vibrational energy of the excited oscillators with the translational energy of the heat bath has the form of (1.6) except for the absence of the factor $(1-e^{-\theta})^{-1}$ in front of the braces (see Appendix II).

A third relaxation mechanism involves the interchange of radiation between the excited oscillators and the heat bath. The relaxation equations for this case have been derived by Rubin and Shuler[8] by considering the interaction of the oscillators with a radiation heat bath in equilibrium with the heat bath at the temperature T. Using the Einstein coefficients A and B for spontaneous and induced emission and for absorption and Planck's radiation law for the density of the radiation, it could readily be shown that the relaxation equation for radiative transitions is again of the form of (1.6) but with k_{10} replaced by A_{10}, the Einstein coefficient for spontaneous emission between vibrational levels 1 and 0 of the oscillators.[9]

[6] R. J. Rubin and K. E. Shuler, J. Chem. Phys. **25**, 59 (1956).
[7] R. H. Fowler, Phil. Mag. **47**, 257 (1924).
[8] R. J. Rubin and K. E. Shuler, J. Chem. Phys. **26**, 137 (1957).
[9] It should be noted that the case of radiative relaxation could also be discussed in terms of the transfer of photons with energy $h\nu$ between the excited oscillators and the heat bath oscillators by the method used above for the transfer of vibrational energy and without recourse to the radiation field. The exact correspondence between these two relaxation process explains the exact correspondence between the equations describing the two processes when the appropriate transition probabilities, i.e., k_{10} or A_{10}, are used in Eq. (1.1). It should also be noted that an internal energy balance can be maintained for this relaxation process.

The general relaxation equation applicable to the relaxation of a system of harmonic oscillators in a constant temperature heat bath can finally be written as

$$\frac{dx_n(t)}{dt}=\kappa\{ne^{-\theta}x_{n-1} - [n+(n+1)e^{-\theta}]x_n+(n+1)x_{n+1}\} \qquad n=0,\,1,\,2,\,\cdots \quad (1.7)$$

where

$$k=\begin{cases} k_{10}(1-e^{-\theta})^{-1} & \text{for collisional vibration-vibration exchange} \\ k_{10}' & \text{for collisional vibration-translation energy exchange} \\ A_{10}(1-e^{-\theta})^{-1} & \text{for radiative energy exchange.} \end{cases} \quad (1.7a)$$

It is the object of this paper to obtain an exact solution of (1.7) subject to the condition $\sum x_n(t)=1$ (closed system) for various initial distributions $x_n(0)$.

Rubin and Shuler[6] obtained a solution of (1.7) for the special case $\theta\ll 1$, which, in essence, corresponds to replacing the discrete set of energy levels by a quasi-continuum, by approximating the set of differential difference equations (1.7) by the related partial differential equation which then admitted of a solution in terms of a Fourier development in Laguerre polynomials. The choice of $\theta\ll 1$, made by Rubin and Shuler for mathematical convenience, is realized physically only for very few molecules and then only at rather high temperatures. Thus, for instance, one finds $\theta=h\nu/kT=60/T$ for Cs_2 so that the inequality $\theta\ll 1$ can be fulfilled for $T>10^3$ °K. For most diatomic species, however, $\theta>1$ at ordinary temperatures (300–1000°K). Some examples are $NO(\theta=2.73\cdot 10^3/T)$, $CO(\theta=2.13\cdot 10^3/T)$, and $OH(\theta=5.37\cdot 10^3/T)$, where the frequencies ν correspond to the electronic ground states. For a heat bath at 300°K, one thus finds $\theta\sim 10$ to 20. We will show in the appendix that the general solution of (1.7), valid for all θ, reduces to the solution of Rubin and Shuler when $\theta\rightarrow 0$. The qualitative characteristics of relaxation from various initial distributions as determined by the small θ theory are in general agreement with the exact results derived below.

We shall show in Sec. 2 that the exact solution of (1.7) can be written in terms of the generating function

$$G(z,t)=\sum_{n=0}^{\infty} z^n x_n(t) \quad (1.8)$$

and the dimensionless time $\tau=\kappa t(1-e^{-\theta})$ as

$$G(z,t)=\frac{(e^{\theta}-1)}{(z-1)e^{-\tau}-(z-e^{\theta})} \times G_0\left[\frac{(z-1)e^{-\tau}e^{\theta}-(z-e^{\theta})}{(z-1)e^{-\tau}-(z-e^{\theta})}\right] \quad (1.9)$$

where $G_0(y)\equiv G(y,0)$ is determined by the initial condition (distribution) $x_n(0)$ and where $x_n(t)$, the fraction of the molecules in level n is the coefficient of z^n in (1.9). We consider in this study the relaxation of three initial nonequilibrium distributions which could readily be obtained in a physical system and the relaxation of their moments:

(1) An initial Boltzmann distribution with temperature $T_0\neq T$ for which $x_n(0)$ is given by

$$x_n(0)=[1-\exp(-\theta_0)]\exp(-n\theta_0) \quad (1.10)$$

where $\theta_0=h\nu/kT_0$. Substitution of (1.10) into (1.9) yields the Boltzmann distribution (for details see Sec. 4)

$$x_n(t)=[1-\exp(-\Theta)]\exp(-n\Theta) \quad (1.11)$$

with

$$\Theta=\log\left[\frac{e^{-\tau}(1-e^{\theta-\theta_0})-e^{\theta}(1-e^{-\theta_0})}{e^{-\tau}(1-e^{\theta-\theta_0})-(1-e^{-\theta_0})}\right]. \quad (1.12)$$

At early times

$$\Theta\sim\theta+\tau\frac{(1-e^{\theta-\theta_0})}{(1-e^{\theta})}(e^{\theta_0}-1)+O(\tau^2) \quad (1.12a)$$

and as $\tau\rightarrow\infty$

$$\Theta\sim\theta+e^{-\tau}\left[\frac{(1-e^{-\theta})}{(1-e^{-\theta_0})}(1-e^{\theta-\theta_0})\right]+O(e^{-2\tau}).$$

The initial Boltzmann distribution (1.10) thus relaxes to a final equilibrium Boltzmann distribution via the continuous sequence of Boltzmann distributions (1.11). Since the transient distribution of the relaxing oscillators is always canonical in this case, it is possible to characterize it by a "temperature" $\mathcal{T}(t)=h\nu/k\Theta(t)$. To give an indication of the relaxation of this "temperature" we have plotted Θ^{-1} as a function of time for various initial and final temperatures T_0 and T in Fig. 1.[10]

An interesting feature of the curves in Fig. 1 is that the relaxation time associated with the temperature rise from $T_1\rightarrow T_2(\theta_2<\theta_1)$ is less than that for the corresponding temperature drop $T_2\rightarrow T_1$ (see curves A and B). Qualitatively this is not surprising because more levels are available for occupation at the higher temperature equilibrium than at the lower. The system becomes "disordered" faster than it can be ordered.

(2) An initial Poisson distribution with

$$x_n(0)=e^{-a}a^n/n! \quad (1.13)$$

where a is the mean value $\bar{n}$ of the level number n. This represents a "peaked" initial distribution, $x_n(0)$, in which most of the excited oscillators are found initially in levels near $n=a$. The level population $x_n(t)$ resulting

[10] The persistence of the form of the Boltzmann distribution is a consequence of the Landau-Teller transition probabilities. It has been shown by Rubin and Shuler, J. Chem. Phys. **24**, 68 (1956), that other choices of transition probabilities will lead to a different relaxation behavior.

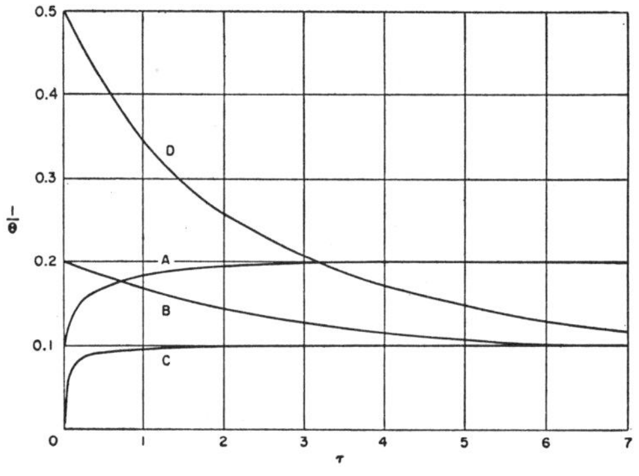

FIG. 1. The "temperature" $Tk/h\nu=\Theta^{-1}$ as a function of time τ for the relaxation of initial Boltzmann distributions [Eq. (1.12)]

Curve A: $\theta_0=10$, $\theta=5$, $T_0=\frac{1}{2}T$
Curve B: $\theta_0=5$, $\theta=10$, $T_0=2T$
Curve C: $\theta_0=20$, $\theta=10$, $T_0=\frac{1}{2}T$
Curve D: $\theta_0=2$, $\theta=10$, $T_0=5T$.

from an initial Poisson distribution is found to depend on the nth Laguerre polynomial [Eq. (5.6)]. We have plotted x_n as a function of τ for $\theta=3$ and $a=15$ in Fig. 2. Notice that the distribution narrows with time and shifts toward the equilibrium distribution as $t\rightarrow\infty$. We have also plotted $\log x_n$ vs t for the Poisson distribution (see Fig. 3) in order to gain some further information about the approach to equilibrium.

(3) An initial δ function distribution with all excited oscillators in state m:

$$x_n(0)=1 \quad \text{when} \quad n=m$$
$$x_n(0)=0 \quad \text{when} \quad n\neq m. \tag{1.14}$$

The level population $x_n(t)$ is given in terms of hypergeometric functions [see Eq. (6.5)]. The initially sharp distribution broadens and shifts to lower energy states

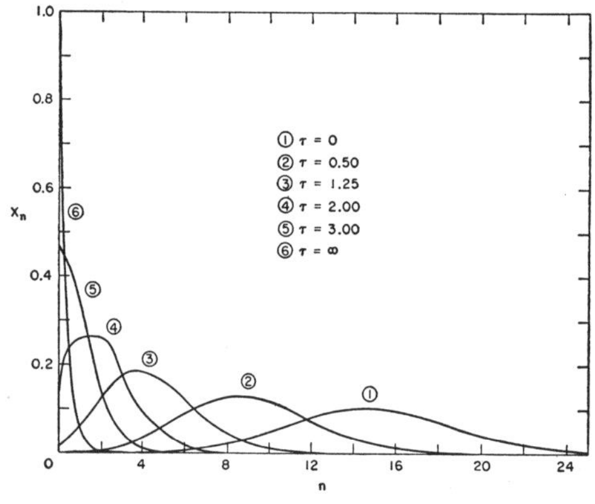

FIG. 2. The relaxation of an initial Poisson distribution $x_n(0)=e^{-a}a^n/n!$ with $a=\langle n\rangle=15$ to a final Boltzmann distribution with $\theta=3$. The ordinate x_n gives the fraction of oscillators in energy level n.

if $m>\bar{n}$ (in a manner similar to that plotted in reference 6 for the case $\theta\ll1$).

(4) We have also obtained (see Sec. 3) a solution for the relaxation of the moments of the distribution. The transient behavior of the factorial moments of $x_n(t)$ [Eq. (3.1a)] defined by

$$f_m(t)=\sum_{n=0}^{\infty} n(n-1)\cdots(n-m+1)x_n(t)$$
$$m=1,2,\cdots \tag{1.15}$$

is described by (3.7):

$$\frac{df_m}{\kappa dt}+m(1-e^{-\theta})f_m=m^2e^{-\theta}f_{m-1}. \tag{1.16}$$

The internal energy $E(t)$ of our system of excited oscillators is related to the first moment f_1 by

$$E(t)=h\nu\sum_{n=0}^{\infty} nx_n(t)=h\nu f_1. \tag{1.17}$$

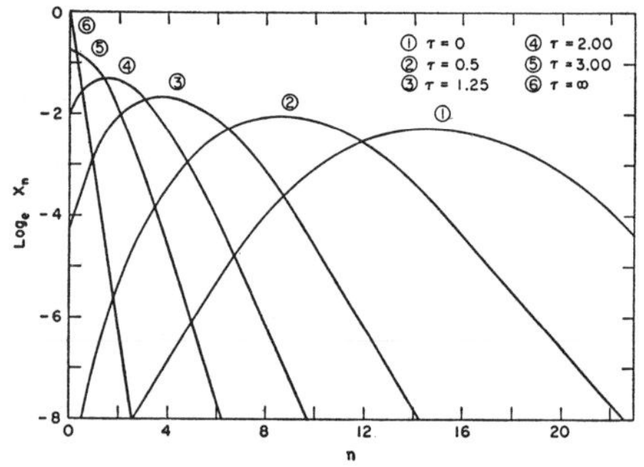

FIG. 3. A plot of $\log x_n$ vs n for the relaxation of the initial Poisson distribution shown in Fig. 2. The straight line portions of these curves for high n give a good indication of the adjustment of the initial Poisson distribution (at $\tau=0$) to the equilibrium Boltzmann distribution at $\tau=\infty$.

The combination of Eqs. (1.16) and (1.17) readily leads to the remarkably simple expression of Bethe and Teller[11]

$$\frac{E(t)-E(\infty)}{E(0)-E(\infty)}=e^{-\tau} \tag{1.18}$$

for the relaxation of the internal energy where $E(\infty)$ is the internal energy which corresponds to the final Boltzmann distribution. It should be noted that according to (1.18), the magnitude of the internal energy at any time t depends only on $E(0)$ and not on the form of the initial distribution $x_n(0)$. The relaxation of the mean energy of a system of harmonic oscillators is therefore determined solely by the amount of energy added to the system and not by its distribution. Hence

[11] H. A. Bethe and E. Teller, "Deviations from thermal equilibrium in shock waves," Ballistic Research Laboratory, Report X-117, 1941. See also Rubin and Shuler, reference 6.

any description of nonequilibrium distributions (at least for the case of harmonic oscillators) which is based solely on the magnitude of the internal energy $E(t)$ can never give any information (other than $\bar{n}$) about the distribution $x_n(t)$ associated with this energy. From Eq. (1.18) and the definition of τ in (1.8) one obtains

$$t_{\text{relax}}=[\kappa(1-e^{-\theta})]^{-1} \tag{1.19}$$

for the relaxation time of the internal energy.

It is also possible to obtain the time dependence of the dispersion $\sigma^2(t)$, from Eq. (1.16) where

$$\sigma^2(t)=\langle(n-f_1)^2\rangle_{\text{Av}}=f_2+f_1(1-f_1). \tag{1.20}$$

A knowledge of the dependence of the dispersion of the distribution with time is of particular interest in connection with "peaked" initial distributions such as the δ function, Poisson, or Gaussian distributions since it gives some information about the "spreading" or "contraction" of the distribution as it tends toward the equilibrium Boltzmann distribution.

The "easy" problems of nonequilibrium statistical mechanics are those associated with physical systems which can be divided into two parts, (a) the large heat bath with many degrees of freedom which remains at equilibrium (and has the fluctuations expected in a system at equilibrium) and (b) the small nonequilibrium part with relatively few degrees of freedom which relaxes through interactions with the heat bath without disturbing the heat bath equilibrium.

The Einstein theory of Brownian motion is the classical example of this type of situation. The large Brownian particle with an initial δ-function distribution interacts with the surrounding fluid which remains at equilibrium. The theory developed in the present paper follows in the same spirit.[12] Mathematically the relevant equations associated with these processes are linear and can be discussed in considerable detail.

Those processes which do not permit the postulation of an equilibrium heat bath usually lead to nonlinear equations (for example the Boltzmann equation for the transport theory of gases) and have not been discussed in any really satisfactory manner. If the x_n's and y_i's in our Eq. (1.2) were both in a nonequilibrium state (for example by setting $x_i=y_i$) our differential equations would become nonlinear and little could be done with them.

2. GENERAL SOLUTION OF FUNDAMENTAL EQUATION

We solve (1.7) through the introduction of the generating function

$$G(z,t)=\sum_{n=0}^{\infty} z^n x_n(t) \tag{2.1a}$$

which is defined so that the coefficient of z^n is the fraction of molecules in the nth state at time t. We note that

$$\frac{\partial G}{\partial z}=z^{-1}\sum_{n=0}^{\infty} n z^n x_n(t). \tag{2.1b}$$

If we multiply (1.7) by z^n and sum from $n=0$ to $n=\infty$ we find that G satisfies the first-order partial differential equation

$$\frac{1}{\kappa}\frac{\partial G}{\partial t}=(z-1)e^{-\theta}\left\{\frac{\partial G}{\partial z}[(z-1)+(1-e^{\theta})]+G\right\}. \tag{2.2}$$

This equation can be transformed into a somewhat simpler one by letting

$$y=z-1,\quad \lambda=\kappa t e^{-\theta},\quad c=1-e^{\theta}, \tag{2.3a}$$

and

$$H(y,\lambda)=(y+c)G. \tag{2.3b}$$

Then H satisfies

$$g\left(\frac{\partial H}{\partial \lambda},\frac{\partial H}{\partial y}\right)=\frac{\partial H}{\partial \lambda}-y(y+c)\frac{\partial H}{\partial y}=0 \tag{2.4}$$

or, if we let $p_1=\partial H/\partial\lambda$ and $p_2=\partial H/\partial y$

$$g(p_1,p_2)=p_1-y(y+c)p_2=0. \tag{2.5}$$

Following the method of characteristics we consider[13]

$$\frac{dy}{\partial g/\partial p_2}=\frac{d\lambda}{\partial g/\partial p_1} \tag{2.6a}$$

or

$$d\lambda=-dy/y(y+c) \tag{2.6b}$$

whose solution

$$ye^{\lambda c}[y+c]^{-1}=\text{constant} \tag{2.6c}$$

implies that the general solution of (2.4) is

$$H(y,\lambda)=f(y[y+c]^{-1}e^{\lambda c}) \tag{2.7}$$

where f is an arbitrary function which is to be determined from the initial distribution $\{x_n(0)\}$.

The general solution of (2.2) is then

$$G(z,t)=(z-e^{\theta})^{-1}\times f\{[z-1][z-e^{\theta}]^{-1}\exp[-t\kappa(1-e^{-\theta})]\}. \tag{2.8}$$

It is to be noted that the definition of (2.1a) implies $G(1,t)=\sum x_n(t)=1$. Hence

$$f(0)=(1-e^{\theta}). \tag{2.9}$$

The initial distribution $x_n(0)$ characterizes $G(z,0)$. The function f is related to $G_0(z)\equiv G(z,0)$ by setting

[12] The general theory of the "easy" problems has been discussed recently by many authors: M. Wang and G. E. Uhlenbeck, Revs. Modern Phys. 17, 323 (1945); H. B. Callen and T. A. Welton, Phys. Rev. 83, 34 (1951); P. Bergmann and J. Lebowitz, Phys. Rev. 99, 578 (1955); and others. A brief review and bibliography of this subject has been prepared by E. Montroll and M. S. Green, Ann. Rev. Phys. Chem. 5, 449 (1954).

[13] F. S. Woods, *Advanced Calculus* (Ginn and Company, New York, 1934), p. 292.

$t=0$ in (2.8) so that

$$f([z-1][z-e^\theta]^{-1})=(z-e^\theta)G_0(z) \quad (2.10)$$

or

$$f(\eta)=\left(\frac{e^\theta-1}{\eta-1}\right)G_0\left(\frac{\eta e^\theta-1}{\eta-1}\right). \quad (2.11)$$

Clearly, if we let

$$\tau=\kappa t(1-e^{-\theta}) \quad (2.12)$$

we can express $G(z,t)$ as

$$G(z,t)=\frac{1-e^\theta}{(z-e^\theta)-(z-1)e^{-\tau}} \times G_0\left(\frac{(z-1)e^{-\tau}e^\theta-(z-e^\theta)}{(z-1)e^{-\tau}-(z-e^\theta)}\right). \quad (2.13)$$

In principle our problem is solved because G_0 is determined by the initial conditions and our $x_n(t)$'s are coefficients of z^n in (2.13). As $t\to\infty$ (and hence as $\tau\to\infty$) we have, since $G(1,t)=1=G_0(1)$ for all t,

$$G(z,\infty)=(1-e^{-\theta})/(1-ze^{-\theta}). \quad (2.14a)$$

Hence by (2.1a) the equilibrium distribution is

$$x_n(\infty)=(1-e^{-\theta})e^{-n\theta} \quad (2.14b)$$

which is the Boltzmann distribution of a set of oscillators with $\theta=h\nu/kT$.

The more familiar type of expansion of the solution of (1.7) as a linear combination of orthogonal polynomials is presented for completeness in Appendix I.

3. MOMENTS OF THE LEVEL OCCUPATION DISTRIBUTION

We define the factorial moments of $x_n(t)$ by

$$f_m(t)=\sum_{n=0}^{\infty} n(n-1)\cdots(n-m+1)x_n(t) \qquad m=1, 2, \cdots \quad (3.1a)$$

$$f_0(t)=\sum_{n=0}^{\infty} x_n(t)=1. \quad (3.1b)$$

Clearly the internal energy of our system is

$$E(t)=h\nu\sum nx_n(t)=h\nu f_1(t). \quad (3.2)$$

In general

$$f_m(t)=\frac{\partial^m}{\partial z^m}G(z,t)\Big]_{z=1}. \quad (3.3)$$

A differential equation for f_m is readily derivable from (2.2), which we write as

$$\frac{1}{\kappa}\frac{\partial G}{\partial t}=(z-1)e^{-\theta}F \quad (3.4)$$

with

$$F=\frac{\partial G}{\partial z}[(z-1)+(1-e^\theta)]+G. \quad (3.5)$$

Then by differentiating (3.4) m times with respect to z and setting $z=1$ we find

$$\frac{1}{\kappa}\frac{\partial f_m}{\partial t}=me^{-\theta}\frac{\partial^{m-1}F}{\partial z^{m-1}}\Big]_{z=1}.$$

But (3.5) yields

$$\frac{\partial^{m-1}F}{\partial z^{m-1}}\Big]_{z=1}=m\frac{\partial^{m-1}G}{\partial z^{m-1}}\Big]_{z=1}+(1-e^\theta)\frac{\partial^m G}{\partial z^m}\Big]_{z=1}. \quad (3.6)$$

Hence

$$\frac{1}{\kappa}\frac{df_m}{dt}+m(1-e^{-\theta})f_m=m^2e^{-\theta}f_{m-1}. \quad (3.7)$$

By using the steady state value of f_1

$$f_1(\infty)=\sum_{n=0}^{\infty} ne^{-n\theta}(1-e^{-\theta}) =e^{-\theta}/(1-e^{-\theta}) \quad (3.8)$$

(3.7) is equivalent to

$$\frac{df_m}{d\tau}+mf_m=m^2f_{m-1}f_1(\infty) \quad (3.9)$$

and

$$f_m(\tau)=f_m(0)e^{-m\tau}+m^2f_1(\infty)e^{-\tau m}\int_0^\tau f_{m-1}(x)e^{mx}dx. \quad (3.10)$$

Since by (3.1b) $f_0(t)=1$ we find

$$f_1(t)=f_1(0)e^{-\tau}+f_1(\infty)(1-e^{-\tau}). \quad (3.11)$$

The variation of the internal energy with time is obtained by combining (3.2) with (3.11)

$$\frac{E(t)-E(\infty)}{E(0)-E(\infty)}=e^{-\tau} \quad (3.12)$$

as has been previously obtained by Bethe and Teller.[11]

Substitution of (3.11) into (3.10) when $m=2$ yields

$$f_2(t)=2f_1^2(\infty)+4f_1(\infty)[f_1(0)-f_1(\infty)]e^{-\tau} +[f_2(0)-4f_1(0)f_1(\infty)+2f_1^2(\infty)]e^{-2\tau}. \quad (3.13)$$

The dependence of the dispersion of our distribution on time is given by

$$\sigma^2(t)=\langle(n-f_1)^2\rangle_{Av}=f_2+f_1(1-f_1) =\sigma^2(\infty)+[\sigma^2(0)-\sigma^2(\infty)]e^{-2\tau} +[f_1(0)-f_1(\infty)][1+2f_1(\infty)]e^{-\tau}(1-e^{-\tau}) \quad (3.14)$$

where

$$\sigma^2(\infty)=f_1(\infty)[1+f_1(\infty)]. \quad (3.14a)$$

4. THE RELAXATION OF ONE BOLTZMANN DISTRIBUTION TO ANOTHER

It will now be shown that if our oscillators are initially distributed in their energy levels according to some Boltzmann distribution, then this distribution will persist during the relaxation but its effective temperature will vary monotonically until the temperature of the heat bath is achieved.

Let

$$x_n(0)=(1-e^{-\Theta_0})e^{-n\Theta_0} \tag{4.1}$$

where $\theta_0=h\nu/kT_0$, T_0 being the temperature corresponding to the initial distribution. Then

$$G_0(z)=\sum z^n x_n(0)=(1-e^{-\theta_0})/(1-ze^{-\theta_0}). \tag{4.2}$$

Hence, (2.13) yields

$$G(z,t)=\frac{(1-e^{\theta})(1-e^{-\theta_0})}{[(e^{-\tau}-e^{\theta})+e^{(\theta-\theta_0)}(1-e^{-\tau})]+z[(1-e^{-\tau})-e^{-\theta_0}(1-e^{\theta-\tau})]} \tag{4.3}$$

so that

$$x_n(t)=[1-\exp(-\Theta)]\exp(-n\Theta) \tag{4.4a}$$

where

$$\Theta=\log\left\{\frac{e^{-\tau}(1-e^{\theta-\theta_0})-e^{\theta}(1-e^{-\theta_0})}{e^{-\tau}(1-e^{\theta-\theta_0})-(1-e^{-\theta_0})}\right\} \tag{4.4b}$$

and τ is given by (2.12), $\tau=\kappa t[1-\exp(-\theta)]$.

Our distribution (4.4) is that of Boltzmann at all times and the "effective temperature" $\mathcal{T}$ varies with time as $\mathcal{T}=h\nu/k\Theta(\tau)$. We have plotted $1/\Theta$ as a function of time for various initial and final temperatures in Fig. 1.

5. INITIAL POISSON DISTRIBUTION

We now examine the relaxation of the Poisson distribution

$$x_n(0)=e^{-a}a^n/n! \tag{5.1}$$

where a is the mean value of n. We have

$$G_0(z)=\sum_{n=0}^{\infty} z^n x_n(0)=e^{-a(1-z)} \tag{5.2}$$

so that by (3.3)

$$f_1(0)=\bar{n}=\partial G_0/\partial z=a \tag{5.3a}$$

and

$$\sigma^2=\langle(n-\bar{n})^2\rangle_{\text{Av}}=a. \tag{5.3b}$$

Application of (2.13) yields

$$G(z,t)=\frac{1-e^{\theta}}{(z-e^{\theta})-(z-1)e^{-\tau}} \times\exp\left\{-\frac{ae^{-\tau}(1-z)(1-e^{\theta})}{(z-e^{\theta})-(z-1)e^{-\tau}}\right\}. \tag{5.4}$$

Since the generating function of Laguerre polynomials defined by[14]

$$n!L_n(y)=e^{y}(d/dy)^n(e^{-y}y^n)$$

$$=n!\sum_{\nu=0}^{n}\binom{n}{n-\nu}\frac{(-y)^{\nu}}{\nu!} \tag{5.5a}$$

is

$$(1-\alpha)^{-1}\exp\{y\alpha/(\alpha-1)\}=\sum_{n=0}^{\infty}\alpha^n L_n(y) \tag{5.5b}$$

we obtain

$$x_n(t)=\left[\frac{1-e^{\theta}}{e^{-\tau}-e^{\theta}}\right]\left[\exp\left\{\frac{ae^{-\tau}(e^{\theta}-1)}{e^{-\tau}-e^{\theta}}\right\}\right]\left[\frac{1-e^{-\tau}}{e^{\theta}-e^{-\tau}}\right]^n \times L_n\left\{\frac{ae^{-\tau}(e^{\theta}-1)^2}{(1-e^{-\tau})(e^{-\tau}-e^{\theta})}\right\}. \tag{5.6}$$

Since $L_n(0)=1$ it is easy to show that as $\tau\to\infty$, $x_n(t)$ tends to the final equilibrium Boltzmann distribution (2.14b). We have plotted $x_n(t)$ in Fig. 2 for $\theta=3$ and $a=15$.

6. ALL MOLECULES INITIALLY IN mTH STATE

Let

$$x_n(0)=\begin{cases}1 & \text{if } n=m\\ 0 & \text{otherwise.}\end{cases} \tag{6.1}$$

This sharp distribution broadens and its peak is displaced toward $n=0$ with increasing time. Then

$$G_0(z)=z^m \tag{6.2}$$

and

$$G(z,t)=\frac{(1-e^{\theta})[z(1-e^{-\tau+\theta})-e^{\theta}(1-e^{-\tau})]^m}{[z(1-e^{-\tau})+(e^{-\tau}-e^{\theta})]^{m+1}} =\frac{(1-e^{\theta})e^{m\theta}(e^{-\tau}-1)^m[1-\alpha z]^m}{(e^{-\tau}-e^{\theta})^{m+1}[1-\beta z]^{m+1}} \tag{6.3}$$

where

$$\alpha=\left(\frac{e^{-\tau}-e^{-\theta}}{e^{-\tau}-1}\right)\quad\text{and}\quad\beta=\left(\frac{e^{-\tau}-1}{e^{-\tau}-e^{\theta}}\right). \tag{6.4}$$

Now, it is well known[15] that if $|y|<1$ and $|y(1-s)|<1$

$$(1-y)^{a-1}(1-y+sy)^{-a}=\sum_{n=0}^{\infty}y^nF(-n,a,1;s) \tag{6.5}$$

F being the hypergeometric function. Hence if we let $a=-m$

$$x_n(t)=\frac{(1-e^{\theta})e^{m\theta}}{(e^{-\tau}-e^{\theta})}\left(\frac{e^{-\tau}-1}{e^{-\tau}-e^{\theta}}\right)^{m+n}F(-n,-m,1;u^2) \tag{6.5a}$$

$$u=\frac{\sinh\frac{1}{2}\theta}{\sinh\frac{1}{2}\tau}.$$

[14] G. Szego, *Orthogonal Polynomials* (American Mathematical Society, 1939), p. 96.

[15] Erdelyi, Magnus, Oberhettinger, and Tricomi, *Higher Transcendental Functions* (McGraw-Hill Book Company, Inc., New York, 1953), Vol. 1, p. 82.

A standard transformation formula yields

$$x_n(t)=\frac{(1-e^\theta)e^{m\theta}}{(e^{-\tau}-e^\theta)}\left(\frac{e^{-\tau}-1}{e^{-\tau}-e^\theta}\right)^{m+n}$$
$$\times[1-u^2]^{1+m+n}F(1+n,\,1+m,\,1;\,u^2). \quad (6.5b)$$

Note that as $\tau\to\infty$ (t large), $u\to 0$ and $F\to 1$ and the Boltzmann distribution develops as is required. When $\tau>\theta$, (6.5b) converges rapidly and is suitable for making calculations.

At early times we expect the distribution to be close to a Gaussian,

$$x_n(t)\simeq\frac{1}{(2\pi\sigma^2)^{\frac{1}{2}}}\exp(n-\bar{n})^2/2\sigma^2. \quad (6.6)$$

Initially $\bar{n}(0)=m$ and $\sigma^2(0)=0$. Hence from (3.11) and (3.14)

$$\bar{n}(\tau)=f_1(t)=me^{-\tau}+f_1(\infty)(1-e^{-\tau}) \quad (6.7a)$$

$$\sigma^2(\tau)=\sigma^2(\infty)[1-e^{-2\tau}]$$
$$+[m-f_1(\infty)][1+2f_1(\infty)]e^{-\tau}(1-e^{-\tau}) \quad (6.7b)$$

where

$$\sigma^2(\infty)=f_1(\infty)[1+f_1(\infty)]$$

and

$$f_1(\infty)=e^{-\theta}(1-e^{-\theta})^{-1}.$$

It is to be noted that the relaxation from any initial distribution can be expressed as a linear combination of the $x_n(t)$'s which result from initially sharp distributions.

APPENDIX I. FOURIER SERIES SOLUTION OF EQ. (1.7)

For completeness we shall discuss the solution of (1.7) as a linear combination of a certain set of eigenfunctions. This form of solution could be used as the basis of a perturbation theory or for the analysis of complicated initial distributions for which $G_0(z)=G(z,0)$ [see Eq. (2.1)] cannot be summed easily.

We seek solutions $x_n(t)$ which are a superposition of terms of the following type

$$a_\mu l_n(\mu)\exp\{-\mu t\kappa(1-e^{-\theta})\}=a_\mu l_n(\mu)e^{-\mu\tau}, \quad (I.1)$$

where μ is a positive number and a_μ a constant which is to be determined from the initial level population distribution $\{x_n(0)\}$. Direct substitution of (I.1) into (1.7) yields a difference equation in the numbers $\{l_n(\mu)\}$

$$(e^{-\theta}-1)\mu l_n=ne^{-\theta}l_{n-1}$$
$$-\{n+(n+1)e^{-\theta}\}l_n+(n+1)l_{n+1}$$
$$n=0,\,1,\,2,\,\cdots. \quad (I.2)$$

This set of equations can be solved through the introduction of the generating function

$$F(w,\mu)=\sum_{n=0}^{\infty}l_n(\mu)w^n. \quad (I.3a)$$

Multiplication of (I.2) by w^n and summation with respect to n yields, after application of (I.3a) and

$$w\frac{\partial F}{\partial w}=\sum_{n=0}^{\infty}nl_n(\mu)w^n, \quad (I.3b)$$

$$F'(w)/F(w)=\mu(w-1)^{-1}-(\mu+1)(w-e^\theta)^{-1} \quad (I.4)$$

whose solution is

$$F(w,\mu)=(1-w)^\mu(1-we^{-\theta})^{-\mu-1} \quad (I.5)$$

if l_0 is chosen to be 1. Our required functions $l_n(\mu)$ are the coefficients of w^n in $F(w,\mu)$. These functions have been studied by Gottlieb and can be shown to be[16]

$$l_n(\mu)=e^{\theta\mu}\Delta^n\left\{\binom{\mu}{n}e^{-\theta\mu}\right\} \quad (I.6a)$$

or

$$l_n(\mu)=e^{-n\theta}\sum_{\nu=0}^{\infty}(1-e^\theta)^\nu\binom{n}{\nu}\binom{\mu}{\nu} \quad (I.6b)$$

where the standard binomial coefficient notation is used with

$$\binom{n}{\nu}=\begin{cases}n!/\nu!(n-\nu)! & \text{if } n\geqslant\nu\\ 0 & \text{if } n<\nu \text{ or } \nu<0\\ 1 & \text{if } \nu=0.\end{cases} \quad (I.7)$$

The first few of these functions are

$$l_n(0)=e^{-n\theta} \quad (I.8a)$$

$$l_n(1)=e^{-n\theta}\{1+(1-e^\theta)n\}. \quad (I.8b)$$

These functions can be written in various ways in terms of hypergeometric functions; for example[17]

$$l_n(\mu)=F(-n,\,\mu+1,\,1;\,1-e^{-\theta}). \quad (I.9)$$

If μ is chosen to be an integer, it is clear by symmetry from (I.6b) that

$$e^{n\theta}l_n(\mu)=e^{\mu\theta}l_\mu(n). \quad (I.10)$$

A pair of useful orthogonality relations exist when μ and n are non-negative integers

$$\sum_{\nu=0}^{\infty}e^{-\theta\nu}l_n(\nu)l_m(\nu)=\begin{cases}0 & n\neq m\\ e^{-n\theta}(1-e^{-\theta})^{-1} & n=m\end{cases} \quad (I.11a)$$

$$\sum_{n=0}^{\infty}e^{n\theta}l_n(\nu)l_n(\mu)=\begin{cases}0 & \mu\neq\nu\\ e^{\nu\theta}(1-e^{-\theta})^{-1} & \mu=\nu.\end{cases} \quad (I.11b)$$

Gottlieb has also shown that for a fixed complex number x and large n

$$l_n(x)=(-1)^n(1-e^{-\theta})^{-x-1}\binom{x}{n}+O(n^{-Re\,x-2}). \quad (I.12)$$

---

[16] M. J. Gottlieb, Am. J. Math. 60, 455 (1938).

[17] Erdelyi, Magnus, Oberhettinger, and Tricomi, *Higher Transcendental Functions* (McGraw-Hill Book Company, Inc., New York, 1953), Vol. 2, p. 225.

Hence as $n \to \infty$, $l_n(x) \to 0$. This shows that our $l_n(x)$'s satisfy the physical requirement that the population of states characterized by the quantum number n decreases as $n \to \infty$ for θ fixed.

We can now construct a general solution of (1.7) by superposition of terms (I.1),

$$x_n(t) = \sum_{\mu=0}^{\infty} a_\mu l_n(\mu) e^{-\mu\tau} \quad \text{(I.13a)}$$

$$= \sum_{\mu=0}^{\infty} a_\mu e^{\theta(\mu-n)} l_\mu(n) e^{-\mu\tau} \quad \text{(I.13b)}$$

where the constants a_μ are determined in the usual manner through the orthogonality relations. We find

$$a_\mu = (1-e^{-\theta}) \sum_{n=0}^{\infty} x_n(0) l_\mu(n) \quad \text{(I.14a)}$$

or alternatively

$$a_\mu = (1-e^{-\theta}) e^{-\mu\theta} \sum_{n=0}^{\infty} e^{n\theta} l_n(\mu) x_n(0). \quad \text{(I.14b)}$$

When $x_n(0)$ is chosen to be a Boltzmann distribution, one can by repeated use of the generating function $F(w,\mu)$ of (I.5) readily obtain Eq. (4.4) for $x_n(t)$.

Rubin and Shuler[6] previously derived the following formula for $x(n,t) = x_n(t)$ which is to be valid when θ is very small; ($L_n(x)$ is the nth Laguerre polynomial)

$$x(n,t) = e^{-n\theta} \sum_{\mu=0}^{\infty} a_\mu L_\mu(n\theta) \exp(-\mu\theta t k_{10}) \quad \text{(I.15a)}$$

where

$$a_\mu = \theta \int_0^\infty x(y,0) L_\mu(\theta y) dy. \quad \text{(I.15b)}$$

This result can be obtained from (I.14a) and (I.13b) by letting $\theta \to 0$ in certain terms and retaining it in others. Gottlieb pointed out that

$$\lim_{\theta \to 0} l_n(x/\theta, \theta) = L_n(x). \quad \text{(I.16)}$$

If we write $l_\mu(n) = l_\mu(n\theta/\theta, \theta)$ and suppose $n\theta$ is fixed while $\theta \to 0$, and if we let $1-\exp(-\theta) \simeq \theta$, (I.14a) becomes

$$a_\mu \simeq \theta \sum_{y=0}^{\infty} x(y,0) L_\mu(y\theta) \simeq \theta \int_0^\infty x(y,0) L_\mu(\theta y) dy.$$

The conversion of the summation to an integration is valid if $x(y,0)$ and $L_\mu(\theta y)$ are slowly varying functions of y. Equation (I.13b) reduces to (I.15a) if one uses (I.16), sets $\exp\theta\mu = 1$ and writes $\tau = tk_{10}(1-e^{-\theta}) \simeq k_{10}\theta t$.

APPENDIX II. THE RELAXATION EQUATION FOR VIBRATIONAL-TRANSLATIONAL ENERGY EXCHANGE

We now wish to derive the analog of the relaxation equation (1.6) when the energy exchange is between the vibration of the harmonic oscillators and the translational degrees of freedom of the heat bath molecules. The formulation presented here is in principle analogous to that given by Rubin and Shuler[6] but is more detailed in that it takes explicit account of the matrix elements for the transitions between the translational "energy levels."

Let

$x_n(t)$ = fraction of excited oscillators in vibrational level n

y_i = concentration of heat bath molecules (or atoms) with momentum† p_i

$P_{n,\,n+1;\,i,\,i-1}$ = joint probability per collision for the energy transfer $nh\nu \to (n+1)h\nu$ as $p_i \to p_{i-1}$.

The relaxation equation can now be written as [see (1.2)]

$$\frac{dx_n(t)}{dt} = -Z\Big[x_n\Big(\sum_{i=0}^{\infty} y_i P_{n,\,n+1;\,i,\,i-1} + \sum_{i=0}^{\infty} y_i P_{n,\,n-1;\,i,\,i+1}\Big) - x_{n+1} \sum_{i=0}^{\infty} y_i P_{n+1,\,n;\,i,\,i+1} - x_{n-1} \sum_{i=0}^{\infty} y_i P_{n-1,\,n;\,i-1}\Big] \quad \text{(II.1)}$$

where Z is again the collision number. The joint transition probabilities can now be written as

$$P_{n,\,n+1;\,i,\,i-1} = P_{10}(n+1) Q_{i(-)}$$
$$P_{n,\,n-1;\,i,\,i+1} = P_{10} n Q_{i(+)} \quad \text{(II.2)}$$

where $(n+1)P_{10}$ and nP_{10} are the vibrational transition probabilities (see 1.1) and where the Q's are the translational transition probabilities to be determined. Substitution of (II.2) in (II.1) leads to

$$\frac{dx_n(t)}{dt} = ZP_{10}(n+1) \times \Big[x_{n+1} \sum_i y_i Q_{i(+)} - x_n \sum_i y_i Q_{i(-)}\Big] + ZP_{10} n \Big[x_{n-1} \sum_i y_i Q_{i(-)} - x_n \sum_i y_i Q_{i(+)}\Big]. \quad \text{(II.3)}$$

At equilibrium, (i.e., as $t \to \infty$), we have

$$\frac{dx_n(\infty)}{dt} = 0$$
$$\frac{x_{n+1}(\infty)}{x_n(\infty)} = \frac{x_n(\infty)}{x_{n-1}(\infty)} = e^{-\theta} \quad \text{(II.4)}$$

† The indices i, $i \pm 1$ do not represent successive translational energy levels but are used to indicate translational levels separated by $h\nu$.

and Eq. (II.3) becomes

$$0=ZP_{10}(n+1)x_n(\infty)[e^{-\theta}\sum_i y_iQ_{i(+)}-\sum_i y_iQ_{i(-)}]$$
$$-ZP_{10}nx_{n-1}(\infty)[e^{-\theta}\sum_i y_iQ_{i(+)}-\sum_i y_iQ_{i(-)}]. \quad \text{(II.5)}$$

Since the heat bath is assumed to remain at its Maxwell-Boltzmann equilibrium distribution at all times t, $y_i\neq f(t)$ and $y_i(\infty)=y_i$.

The first two terms in Eq. (II.5) refer to the rate of the (vibrational) transition $n\rightleftarrows n+1$ and the last two terms to the rate of the transition $n\rightleftarrows n-1$. By the principle of detailed balancing at equilibrium, the net rate of each of these transitions must be independently zero to satisfy (II.5). We thus find that

$$\sum_i y_iQ_{i(-)}=e^{-\theta}\sum_i y_iQ_{i(+)}. \quad \text{(II.5a)}$$

Substitution of (II.5a) into (II.3) leads to

$$\frac{dx_n(t)}{dt}=ZP_{10}\sum_i y_iQ_{i(+)}\{ne^{-\theta}x_{n-1}$$
$$-[n+(n+1)e^{-\theta}]x_n+(n+1)x_{n+1}\} \quad \text{(II.6)}$$

which is of the same form as (1.6) except for the replacement of $k_{10}(1-e^{-\theta})^{-1}$ by $ZP_{10}\sum_i y_iQ_{i(+)}$. To evaluate the term in front of the braces in (II.6) we write, with Landau and Teller[4]

$$y_i=y_w=2N(m/2kT)^2w^3\exp\left(-\frac{mw^2}{2kT}\right) \quad \text{(II.7)}$$

for the number of heat bath molecules with momentum mw undergoing collisions (N is the total number of heat bath molecules) and

$$Q_{i(+)}=Q_{w(+)}=Q_0e^{-2\pi\nu a/w} \quad \text{(II.8)}$$

where Q_0 is a constant, ν is the frequency of the oscillator, and where a is a length characteristic of the interaction forces in the collisions between the oscillators and the heat bath molecules. Replacing the summation over i indicated in (II.6) by integration over the velocities w one obtains

$$\frac{dx_n(t)}{dt}=k_{10}Q_0\left(\frac{6}{\pi}\frac{kT}{\epsilon}\right)^{-\frac{1}{2}}\exp\left[-\frac{3}{2}\left(\frac{\epsilon}{kT}\right)^{\frac{1}{3}}\right]$$
$$\times\{ne^{-\theta}x_{n-1}-[n+(n+1)e^{-\theta}]x_n$$
$$+(n+1)x_{n+1}\} \quad \text{(II.9)}$$

where $k_{10}=ZNP_{10}$ has been defined in connection with (1.6) and where $\epsilon=m(2\pi\nu a)^2$ with m equal to the effective mass of the collision system. One thus obtains finally

$$\frac{dx_n(t)}{dt}=k_{10}'\{ne^{-\theta}x_{n-1}$$
$$-[n+(n+1)e^{-\theta}]x_n+(n+1)x_{n+1}\} \quad \text{(II.9a)}$$

where

$$k_{10}'=k_{10}Q_0\left(\frac{6}{\pi}\frac{kT}{\epsilon}\right)^{-\frac{1}{2}}\exp\left[-\frac{3}{2}\left(\frac{\epsilon}{kT}\right)^{\frac{1}{3}}\right] \quad \text{(II.9b)}$$

is the transition probability used in (1.7a) for the vibration-translation energy exchange.

A more accurate evaluation of the joint vibrational-translational transition probabilities in (II.2) could be obtained from the quantum mechanical treatments of Jackson and Mott[18] and Herzfeld and his co-workers.[19]

[18] T. M. Jackson and N. F. Mott, Proc. Roy. Soc. (London) **A137**, 703 (1932).

[19] Slawsky, Schwartz, and Herzfeld, J. Chem. Phys. **20**, 1591 (1952); R. N. Schwartz and K. F. Herzfeld, *ibid.*, **22**, 767 (1954); see also K. F. Herzfeld in *Thermodynamics and Physics of Matter* (Princeton University Press, Princeton, 1955), Sec. H.

Relaxation of a Gas of Harmonic Oscillators*

C. C. RANKIN† AND J. C. LIGHT‡

Department of Chemistry and Institute for the Study of Metals, University of Chicago, Chicago, Illinois

(Received 28 September 1966)

The temporal evolution of the vibrational distribution function for a gas of harmonic oscillators undergoing binary collisions, in which they can exchange vibrational quanta among themselves as well as transfer energy between the vibrational and translational degrees of freedom, is determined exactly. The solution, given in terms of a generating function, involves only a double integral and is valid for transition probabilities due to an interaction potential linear in the oscillator coordinate. The relaxation toward "local equilibrium" of the vibrational distribution is found to be at least twice as fast as the relaxation of the average vibrational energy to the final equilibrium value. The solution is valid for an arbitrary initial vibrational distribution and for arbitrary "dilution" of the oscillators by inert collision partners.

INTRODUCTION

THE problem of the relaxation of the vibrational degrees of freedom of a system containing diatomic molecules has been treated extensively in the past.[1–6] The problem is simplified while retaining the essential physics of the problem by assuming that the molecules behave like harmonic oscillators and that the transition probabilities obey the selection rule $\Delta\nu=\pm1$ for the vibrational transitions of each oscillator. Since this selection rule is derivable in first-order perturbation theory from an interaction potential linear in the oscillator coordinate, the results of this investigation should approximate quite closely the behavior of real systems except in the very high temperature region. A knowledge of the kinetics of relaxation processes of this type is important not only for relaxation behind shock waves,[7] but also for such studies as the kinetics of dissociation,[8] and the relaxation of vibrational energy of excited reaction products.[9,10]

Heretofore, solutions of the equations governing the relaxation process have been restricted to (1) the case in which the initial distribution is canonical and characterized by a "vibrational temperature,"[1,3] and (2) the case in which the translational temperature remains constant throughout the relaxation either because the oscillators are diluted by a large excess of inert collision partners,[1,2] or because the vibrational degree of freedom is treated separately as an isolated system.[4] In this paper, we have extended the solution to include any arbitrary initial distribution and any degree of oscillator dilution. Such solutions should prove significant, for example, in the relaxation of the vibrational energy of excited reaction products, in which the translational temperature change during the relaxation may not be safely neglected.

RELAXATION EQUATIONS

The Boltzmann "master equation" governing the relaxation of the vibrational degree of freedom takes the following form in the event that the transition probabilities obey the selection rule $\Delta\nu=\pm1$ and that the rotational–translational degrees of freedom can be characterized by a temperature, T, throughout the relaxation process[1,4]:

$$\begin{aligned}\frac{1}{Z(T)}\frac{dx_n}{dt} = N^{(2)}\sum_{m=0}^{\infty}\{ &-x_nx_m[P'_{n,n+1;m,m-1}(T)+P'_{n,n-1;m,m+1}(T)]+x_{n+1}x_mP'_{n+1,n;m,m+1}(T)+x_{n-1}x_mP'_{n-1,n;m,m-1}(T)\} \\ &+x_{n+1}[N^{(1)}P^{(1)}_{n+1,n}(T)+N^{(2)}P^{(2)}_{n+1,n}(T)]+x_{n-1}[N^{(1)}P^{(1)}_{n-1,n}(T)+N^{(2)}P^{(2)}_{n-1,n}(T)] \\ &-x_n[N^{(1)}P^{(1)}_{n,n-1}(T)+N^{(2)}P^{(2)}_{n,n-1}(T)+N^{(1)}P^{(1)}_{n,n+1}(T)+N^{(2)}P^{(2)}_{n,n+1}(T)]. \qquad (1)\end{aligned}$$

* This research was supported by a grant from the National Science Foundation.

† NSF Predoctoral Fellow.

‡ Alfred P. Sloan Foundation Fellow.

1 E. W. Montroll and K. E. Shuler, J. Chem. Phys. **26**, 454 (1957).
2 K. E. Shuler, J. Phys. Chem. **61**, 849 (1957).
3 R. Herman and R. J. Rubin, Phys. Fluids **2**, 547 (1959).
4 K. E. Shuler, J. Chem. Phys. **32**, 1692 (1960).
5 K. A. Osipov, Russ. J. Phys. Chem. **35**, 748 (1961).
6 T. Carrington, J. Chem. Phys. **43**, 473 (1965).
7 C. C. Chow and E. F. Green, J. Chem. Phys. **43**, 324 (1965).
8 E. W. Montroll and K. E. Shuler, Advan. Chem. Phys. **1**, 361 (1958); K. E. Shuler, J. Chem. Phys. **21**, 1375 (1959).
9 P. Cashion, Proc. Roy. Soc. (London) **A258**, 529 (1960).
10 J. C. Polanyi, J. Quant. Spectry. Radiative Transfer **3**, 471 (1963).

x_n refers to the population of oscillators in the quantum state n; $Z(T)$ is the collision frequency,[3] T is the temperature characterizing the translational–rotational distribution, and $N^{(1)}$ and $N^{(2)}$ refer to the mole fractions of a dilutant species of inert gas and the oscillators, respectively. $P'(T)_{n,l;p,q}$ refers to the probability per collision that a molecule in the vibrational state n colliding with a molecule in the state p will be left in the state l and its partner in state $q(n\rightarrow l;\ p\rightarrow q)$. This transition probability represents the appropriate average over the translational Boltzmann velocity distribution, and hence its dependence on T. $P_{n,m}^{(1)}(T)$ and $P_{n,m}^{(2)}(T)$ denote a similarly averaged probability for the transfer of a quantum of vibrational energy into the translational–rotational degrees of freedom of the dilutant species and another oscillator, respectively; the subscripts refer to the $(n\rightarrow m)$ vibrational transition of the oscillator in question.

The transition probabilities involving the exchange of quanta between vibration and translation are

$$P_{n,m}=[n\delta_{n-1,m}+(n+1)\delta_{n+1,m}]P_{10}, \tag{2}$$

when the potential is linear in the oscillator coordinate.[11] P_{10} is the transition probability per collision for the transition $(1\rightarrow 0)$.

For joint transitions between oscillators, we have[4]

$$P'_{n,n+k;m,m-j}=(n+1)mP'_{10}\delta_{1,k}\delta_{1,j},$$
$$P'_{n,n-k;m,m+j}=n(m+1)P'_{10}\delta_{1,k}\delta_{1,j}, \tag{3}$$

where P'_{10} again refers to the transition $(1\rightarrow 0;\ 0\rightarrow 1)$.

Insertion of Eqs. (2) and (3) into Eq. (1), together with the relations,

$$P_{n-1,n}=P_{n,n-1}e^{-\theta}, \qquad \theta=h\nu/kT \text{ (detailed balance)},$$

$$\sum_{n=0}^{\infty}x_n(t)=1 \qquad \text{(normalization)},$$

and

$$\sum_{n=0}^{\infty}nx_n(t)=\langle n\rangle=\epsilon(t) \text{ (first-moment equation defining average number of quanta per oscillator)}, \tag{4}$$

yields the following simplified master equation[8]:

$$[1/Z(t)](dx/dt)n=P_{10}\{ne^{-\theta}x_{n-1}-[n+(n+1)e^{-\theta}]x_n+(n+1)x_{n+1}\}$$
$$+N^{(2)}P'_{10}(n\epsilon(T)x_{n-1}-\{n[1+\epsilon(T)]+\epsilon(T)(n+1)\}x_n+(n+1)[1+\epsilon(T)]x_{n+1}), \tag{5}$$

in which P_{10} is defined by Eq. (6),

$$P_{10}=N^{(1)}P_{10}^{(1)}+N^{(2)}P_{10}^{(2)}, \tag{6}$$

and where $P_{10}^{(1)}$ and $P_{10}^{(2)}$ now refer to the $(1\rightarrow 0)$ transfer of a quantum into translation in a collision with a dilutant particle and another oscillator, respectively. $\theta=h\nu/kT$, ν is the oscillator frequency; all other symbols are the same as in previous equations.

RELAXATION OF THE OSCILLATOR ENERGY

If it were not for the fact that θ, Z, P_{10}, P'_{10}, and ϵ varied with time, the relaxation would be governed by a set of first-order linear differential equations with constant coefficients which have been solved exactly[1,4] through the use of a generating function and the Gottlieb polynomials. Note, however, that the coefficients of the x_n's in Eq. (5) depends only implicitly on time through the translational temperture, T. $\epsilon(T)$ is governed by the conservation equation

$$h\nu\epsilon(T)+C_vT=E=h\nu\epsilon(T_0)+C_vT_0, \tag{7}$$

where T_0 is the initial translational temperature and E is the (fixed) total energy of the system. C_v is a combined translational–rotational heat capacity for the *entire* system, including the dilutant species. If the x_n's are known, T, and hence the coefficients, can be found as functions of time through the defining relation

$$\sum_{n=0}^{\infty}nx_n(t)=\epsilon(t) \tag{4}$$

for ϵ. It is evident that $h\nu\epsilon(t)$ is just the total energy above the ground state in the vibrational degree of freedom.

If the concentration of oscillators is large enough such that the variation in T during the relaxation cannot be neglected, the coefficients will change with time in a very complicated manner. Not only is prior knowledge of the x_n's required [Eqs. (4) and (7)] to find the parameters ϵ and T, but also P_{10} is generally a very complicated function of the temperature.[1,12–14]

[11] L. Landau and E. Teller, Phys. Z. Sowjetunion **10**, 34 (1936). (This is the first-order perturbation result, valid for $P_{10}<0.2$.)

[12] R. N. Schwartz, Z. I. Slawsky, and K. H. Herzfeld, J. Chem. Phys. **20**, 159 (1952).

[13] D. Rapp and T. E. Sharp, J. Chem. Phys. **38**, 2641 (1963).

[14] D. Rapp, J. Chem. Phys. **40**, 2813 (1964).

Fortunately, as shown as early as 1941 by Bethe and Teller,[15] there is a simpler differential equation for ϵ obtained by multiplying Eq. (5) by n and summing over n from 0 to ∞. With (4) in mind, the equation

$$d\epsilon/dt = Z(\epsilon)P_{10}(\epsilon)\{(1+\epsilon)\exp[-\theta(\epsilon)]-\epsilon\} \quad (8a)$$

obtains. The ϵ dependence arises from Eq. (7). $\epsilon(t)$ can then, in principle, be solved by the quatrature,

$$\int_{\epsilon_0}^{\epsilon(t)} \frac{d\epsilon}{Z(\epsilon)P_{10}(\epsilon)[(1+\epsilon)e^{-\theta}-\epsilon]} = t, \qquad \epsilon_0=\epsilon(0). \quad (8b)$$

For special cases, Eq. (8) has simple solutions. If the translational temperature remains constant for any reason, then Eq. (8) takes on the well-known[15] solution

$$\epsilon(t) = \epsilon(\infty) + [\epsilon_0 - \epsilon(\infty)]e^{-\tau},$$
$$\tau = (1-e^{-\theta})P_{10}Zt,$$
$$\epsilon(\infty) = e^{-\theta}/(1-e^{-\theta}). \quad (9)$$

Equation (9) will be valid for times so large that $|\epsilon(t)-\epsilon(\infty)|$ is small compared to the change in the integrand [Eq. (8b)]. Since $\exp(-\theta)$ is more slowly varying with respect to ϵ than the coefficient of $1/[(1+\epsilon)e^{-\theta}-\epsilon]$, $1/ZP_{10}$, then, for a rough estimate, the condition that $\epsilon(t)$ have the exponential dependence on time in Eq. (9) would be

$$1 \gg \frac{(d/dt)(1/ZP_{10})}{[\chi(\epsilon_t)-\chi(\epsilon_0)]}, \quad (10)$$

where

$$\chi \equiv [(1+\epsilon)e^{-\theta}-\epsilon]^{-1}; \qquad \epsilon_t \equiv \epsilon(t).$$

In the limit of large times, χ becomes singular, so that the relaxation of the energy must assume an exponential asymptotic form. The value of ϵ at the singularity is, or course, the equilibrium value of the vibrational energy, and is, in fact, identical to the value for the energy of a canonical distribution of oscillators at the equilibrium temperature.

GENERATING FUNCTION

With $\epsilon(t)$ known, Eq. (5) can be solved exactly through the use of the same generating function procedure used several times before[1,16] to solve the heat-bath problem (constant T). One seeks to express Eq. (5) in terms of a generating function alone, which is defined as follows:

$$\pi(y,t) \equiv \sum_{n=0}^{\infty} x_n(t)y^n, \quad (11)$$

from which come Eqs. (12) and (13):

$$\frac{\partial\pi}{\partial y} = \sum_{n=0}^{\infty} n x_n(t) y^{n-1}, \quad (12)$$

$$\frac{\partial\pi}{\partial t} = \sum_{n=0}^{\infty} y_n \frac{dx_n}{dt}. \quad (13)$$

Equation (5) becomes, on multiplying by y^n and summing over n and using Eqs. (12) and (13), the following partial differential equation containing only the generating function, its derivatives, function of time, and the factor $y-1$:

$$(y-1)^{-1}(\partial\pi/\partial t)+[g(t)-(y-1)f(t)](\partial\pi/dy)=f(t)\pi, \quad (14)$$

$$f(t)\equiv e^{-\theta}ZP_{10}+ZN^{(2)}P'_{10}\epsilon, \quad (15)$$

$$g(t)\equiv(1-e^{-\theta})ZP_{10}+ZN^{(2)}P'_{10}. \quad (16)$$

By virtue of the relation

$$\sum_{n=0}^{\infty} x_n(t) = 1 \quad (17)$$

expressing the conservation of the number of oscillators,

$$\pi(1,t)=1 \quad (18)$$

results. Equation (4) is translated through Eq. (12) into this equivalent of the first-moment equation:

$$(\partial\pi/\partial y)\,|_{y=1} = \epsilon(t). \quad (19)$$

Finally, $\pi(y, 0)$ must describe the initial oscillator distribution:

$$\pi(y,0)\equiv\pi_0(y). \quad (20)$$

Equation (14) reduces to Shuler's heat-bath equation[2] if $N^{(2)}=0$, or to his equation for the isolated ensemble of oscillators[4] if $P_{10}=0$, in both cases eliminating the time dependence of the coefficients of π, $\partial\pi/\partial y$, and $\partial\pi/\partial t$ because T remains constant.

The solution is obtained exactly as done by Shuler for the previously solved special cases, and which is spelled out specifically in his most recent paper.[16] The mechanics of the actual solution of Eq. (11), found through the method of characteristics,[17] can be found in the Appendix. The following solution can be verified by direct substitution into Eq. (11):

$$\pi(y,t) = \left\{\exp\left[\int_t g(t')dt'\right] \Big/ (y-1)\right\} \Phi\left(\left\{\exp\left[\int_t g(t')dt'\right]\Big/ (y-1)\right\} - \int_t f(t') \exp\left[\int_{t'} g(t'')dt''\right]dt'\right). \quad (21a)$$

Φ is some function of its argument suitably chosen to fit the initial oscillator distribution when limits are chosen for the indefinite integrals in Eq. (21a).

[15] H. A. Bethe and E. Teller, "Deviations from Thermal Equilibrium," Ballistic Research Laboratory Rept. X-117 (1941).
[16] K. E. Shuler and G. H. Weiss, J. Chem. Phys. **45**, 1105 (1966).
[17] R. Courant, *Methods of Mathematical Physics* (Interscience Publishers, Inc., New York, 1962), Vol. 2.

It becomes surprisingly simple to fit Φ to the initial conditions if we choose the limits of integration such that, if

$$A(t)\equiv \exp\left[\int_t g(t')dt'\right], \qquad A(0)=1,$$

i.e.,

$$A(t)\equiv \exp\left[\int_0^t g(t')dt'\right]. \tag{22a}$$

Similarly, we define $B(t)$ such that $B(0)=0$:

$$B(t)\equiv\int_0^t A(t')f(t')dt'. \tag{23}$$

Equation (21a) becomes

$$\pi(y,t)=[A(t)/(y-1)]\Phi\{[A(t)/(y-1)]-B(t)\}. \tag{21b}$$

The initial oscillator distribution is now sufficient to fix Φ.

An important relation between $A(t)$ and $B(t)$ can be obtained by comparison of the integrals for ϵ, B, and A:

$$B(t)=\epsilon(t)A(t)-\epsilon_0. \tag{24}$$

Equation (24) can be easily derived if one assumes that B can be expressed as a product of A with some arbitrary function $k(t)$ plus a constant:

$$B(t)=k(t)A(t)+\text{const.} \tag{25}$$

Differentiating (25), and using the relations $dB/dt=A(t)f(t)$ [see Eqs. (23) and (15)], $dA/dt=A(t)g(t)$ [see Eqs. (22a) and (16)], Eq. (25) becomes, after differentiation,

$$A(t)[f(t)-k(t)g(t)-(dk/dt)]=0. \tag{26}$$

Since A is always greater than zero (by inspection of its defining relations), Eq. (26) is equivalent to the differential equation

$$dk/dt=f-gk; \tag{27a}$$

insertion of the defining relations for f [Eq. (15)] and g [Eq. (16)] yields

$$(1/Z)(dk/dt)=[(1+k)e^{-\theta}-k]P_{10}+P'_{10}N^{(2)}(\epsilon-k). \tag{27b}$$

In view of the equation for ϵ [Eqs. (8)], Eq. (27) implies that ϵ and k are one and the same. Since $B(0)=0$, the constant becomes $-\epsilon_0$ and (24) results.

Therefore, if G and

$$\int_0^t Z[(1-e^{-\theta})P_{10}+N^{(2)}P'_{10}]dt'$$

are known, $\pi(y,t)$ can be found. Hence the problem of finding the populations reduces to the problem of calculating the energy $\epsilon(t)$, of solving an additional integral, and of fitting $\pi(y,0)$ to the initial distribution. Once $\pi(y,t)$ is known $x_n(t)$ is found from its Taylor's expansion in y.

With the inclusion of (24), (21b) assumes its final form:

$$\pi(y,t)=[A(t)/(y-1)]\times\Phi\{[A(t)/(y-1)]+\epsilon_0-\epsilon(t)A(t)\}, \tag{21c}$$

together with the definition

$$A(t)\equiv\exp\left\{\int_0^t Z[(1-e^{-\theta})P_{10}+N^{(2)}P'_{10}]dt'\right\} \tag{22b}$$

and the relations

$$\pi(1,t)=1, \tag{18}$$

$$(\partial\pi/\partial y)|_{y=1}=\epsilon(t) \tag{19}$$

$$\pi(y,0)=\pi_0(y). \tag{20}$$

Equation (21c) reduces to Montroll and Shuler's[1] form when the temperature is constant and (8) has the simple solution (9): A becomes $\exp(\tau)$, B becomes $\epsilon(\infty)(e^{\tau}-1)$, $\tau=(1-e^{-\theta})ZP_{10}t$.

INITIAL BOLTZMANN DISTRIBUTION

An excellent illustration of the use of $\pi(y,t)$ is its application to an initial Boltzmann distribution of oscillator populations. It is evident that the expression

$$\pi_0(y)=\frac{1-\exp(-\theta_v)}{1-y\exp(-\theta_v)}$$

$$=[1-\exp(-\theta_v)]\sum_{n=0}^{\infty}y^n\exp(-n\theta_v) \tag{28}$$

yields as the coefficient of y^n the population of the nth state of a canonical distribution of oscillators characterized by a "vibrational temperature" θ_v. Since $\epsilon_0=\exp(-\theta_v)/[1-\exp(-\theta_v)]$ for any canonical distribution of harmonic oscillators, Eq. (28) becomes

$$\pi_0(y) = [1+\epsilon_0(1-y)]^{-1} \qquad (29)$$

in terms of the initial oscillator energy. Applying (21c) to this case,

$$\pi(y, 0) = [1/(y-1)] \times\Phi[1/(y-1)] = \pi_0(y) = [1-\epsilon_0(y-1)]^{-1}. \qquad (30)$$

If we let $\eta(y, t)$ be the argument of Φ, then

$$\Phi = (y-1)/[1-\epsilon_0(y-1)].$$

At $t=0$, $\eta = 1/(y-1)$ $[A(0)=1,\ B(0)=0]$, so that, finally,

$$\phi(\eta) = 1/(\eta-\epsilon_0). \qquad (31)$$

At later times,

$$\eta = [A(t)/(y-1)]+\epsilon_0-\epsilon(t)A(t); \qquad (32)$$

insertion of this relation into (31), and using (21c),

$$\pi_{\mathrm{Boltz}}(y, t) = [1-\epsilon(t)(y-1)]^{-1} \qquad (33)$$

obtains, which is identical to Montroll and Shuler's result.[1] It follows from (33) that the initial Boltzmann distribution relaxes to equilibrium via a continuous sequence of intermediate Boltzmann distributions characterized by the (time-dependent) energy $\epsilon(t)$ and the vibrational temperature

$$\theta_v(t) = -\ln\{\epsilon(t)/[1+\epsilon(t)]\}.$$

Evidently, $A(t)$ characterizes a deviation from the canonical distribution since it does not even appear in (33).

ARBITRARY INITIAL DISTRIBUTION

If the master equation can be solved for an initial distribution in which every oscillator is in the state n $[x_m(0)=\delta_{mn}]$, then it can be solved for any arbitrary distribution in the following manner: Since $\epsilon(t)$ is determined by Eq. (8), it is evident that the vibrational energy is not dependent on the detailed nature of the oscillator distribution. By virtue of Eq. (7), the translational temperature and the coefficients of the x_n's in Eq. (5) which depend upon it are likewise insensitive to the distribution and may instead be considered pure functions of time. Equation (5) can now be interpreted as a set of linear differential equations with time-dependent coefficients. It follows that any solution can be constructed from a suitable superposition of a complete set of linearly independent solutions. In this case, if a solution exists for the distribution $x_m(0)=\delta_{mn}$ for *every* n, the arbitrary distribution is constructed by multiplying each solution, say $\pi_n(y, t)$, by the value of the actual initial population $[x_n(0)]$ and summing over the complete range of nonzero populations. In addition, if one is given a distribution at any known time t, one can find out what the distribution is at any time for $t=0$ to $t=\infty$ by taking the solutions $\pi_n(y, t)$ and finding what weight each must have to form the given distribution; this is equivalent to a matrix inversion. With the weights known, the distribution at any time may be found by substituting into the various π_n's the desired time.

It suffices to find $\pi_n(y, t)$ and the resulting populations. Obviously $\pi_n(y, 0)=y^n$. From (21c), $\Phi(y, 0) = y^n(y-1) = (1+\eta)/\eta^{n+1}$ by reasoning similar to that used to find π_{Boltz}. Putting in the value of $\eta(y, t)$,

$$\pi_n(y, t) = \frac{A}{y-1}\,\frac{\{1+[A/(y-1)]+\epsilon_0-\epsilon A\}^n}{\{[A/(y-1)]+\epsilon_0-\epsilon A\}^{n+1}},$$

which, on rearrangement, yields the relation

$$\pi_n(y, t) = \frac{A\{[(1+\epsilon_0-A\epsilon)y+[A(1+\epsilon)-(\epsilon_0+1)]\}^n}{[A(1+\epsilon)-\epsilon_0-(A\epsilon-\epsilon_0)y]^{n+1}}; \qquad (34a)$$

the subscript n labels the state initially filled. Note that, although $\pi_n(1, t)=1$, $\epsilon\neq\partial\pi_n/\partial y\,|_{y=1}=\epsilon(t)+[(n-\epsilon_0)/A(t)]$ because π_n characterizes only a *part* of the total (arbitrary) distribution, whereas ϵ represents the energy of the whole system and must enter in Eq. (5) as a "system parameter." [In the case that $\pi_n(y, t)$ represents the whole distribution, $n=\epsilon_0$ and $(\partial\pi_n/\partial y)\,|_{y=1}=\epsilon$.]

For general distribution, these relations hold:

$$\pi_{\mathrm{arb}}(y, t) = \sum_{n=0}^{\infty} x_n(0)\pi_n(y, t), \qquad (35)$$

$$\pi_{\mathrm{arb}}(1, t) = \sum_{n=0}^{\infty} x_n(0) = 1, \qquad (18)$$

$$\left.\frac{\partial\pi_{\mathrm{arb}}}{\partial y}\right|_{y=1} = \sum_{n=0}^{\infty} x_n(0)\left(\frac{n-\epsilon_0}{A}+\epsilon\right) = A^{-1}[\sum_{n=0}^{\infty} n x_n(0)-\epsilon_0]+\epsilon = \epsilon(t). \qquad (19)$$

We now investigate the role of A in the approach to "local" equilibrium, the canonical distribution at a non-equilibrium vibrational temperature. Remembering that [Eq. (22b)]

$$A(t)=\exp\left\{\int_0^t Z[(1-e^{-\theta})P_{10}+N^{(2)}P'_{10}]dt'\right\}, \tag{22b}$$

and noting that because Z, P_{10}, $N^{(2)}$, P'_{10}, and θ are inherently positive and reach finite limiting values, A must grow without bound as $t\to\infty$. If we let $\pi_{\mathrm{Boltz}}=[1-\epsilon(y-1)]^{-1}$, Eq. (34a) can be re-expressed as follows:

$$\pi(y,t)=\pi_{\mathrm{Boltz}}\frac{\{1+[(\epsilon_0+1)/A]\pi_{\mathrm{Boltz}}(y-1)\}^n}{[1+(\epsilon_0/A)\pi_{\mathrm{Boltz}}(y-1)]^{n+1}} \tag{34b}$$

which reduces to π_{Boltz} for very large times.

Relation (34b) confirms the notion that $A(t)$ is in some way a measure of the time dependence of the deviation from local equilibrium, since A^{-1} goes to zero exponentially with time as $t\to\infty$. We may examine the approach to local equilibrium for the general distribution by expanding Eq. (34b) about the time-dependent Boltzmann distribution in a power series in A^{-1} and performing the summation indicated in Eq. (35). The resulting expansion [see Eqs. (48) and (49) and the Appendix for the details] yields, to order A^{-2},

$$\pi_{\mathrm{arb}}(y,t)=\pi_{\mathrm{Boltz}}-\{[\epsilon_0(1+2\epsilon_0)-\langle n^2\rangle_0]/2A^2\}\pi^3_{\mathrm{Boltz}}(y-1)^2, \tag{36}$$

where

$$\langle n^2\rangle_0=\sum_{n=0}^{\infty}n^2x_n(0),$$

the "second moment" of the initial arbitrary distribution. Note that the deviation from the Boltzmann distribution for long times depends on two factors, one (A^{-2}) a pure function of time with an exponential asymptotic character, and the other a function of the initial distribution.

The coefficient $[\epsilon_0(1+2\epsilon_0)-\langle n^2\rangle_0]$ of the exponential term is a measure of the deviation of the initial distribution from the local equilibrium. However, the factor A^{-2} itself does not depend on the initial deviation from local equilibrium. Since the coefficient of the term A^{-1} is zero, it is clear that the relaxation proceeds at the faster rate of

$$A^{-2}=\exp\left[-2\int_0^t g(t')dt'\right].$$

The quantity

$$2\int_0^t g(t')dt'\left\{=2\int_0^t Z[(1-\theta^{-\theta})P_{10}+N^{(2)}P'_{10}]dt'\right\}$$

serves as a kind of relaxation constant analogous to Shuler's τ^1 [see Eqs. (9)]. Since it is well known that P'_{10} is as large (for T not too high) or larger than P_{10},[4,12,18] the relaxation to local equilibrium will be very much faster (for any appreciable $N^{(2)}$) than the final passage to equilibrium determined by Eq. (8). To illustrate this point, we may estimate the total relaxation by using Eq. (9) for $\epsilon(t)$. τ $[=ZP_{10}(1-e^{-\theta})t]$, then, serves as the relaxation constant for the approach to the true equilibrium. To be conservative, we could choose $N^{(2)}$ small enough such that $N^{(2)}P'_{10}\lessapprox\tau/t$ and take it as a constant [same approximation used in Eq. (9)]. 4τ would become the constant characterizing the approach to *local* equilibrium, or four times the final rate. In natural systems which consist of a small fraction of oscillators diluted by an inert collision species, two factors contribute to the observed fact that relaxation to local equilibrium is much faster: (1) the large value of P'_{10}/P_{10} tends to offset the smallness of $N^{(2)}$, and (2) relaxation even with $N^{(2)}=0$ is *twice* as fast ($\sim e^{-2\tau}$) because of the absence of the A^{-1} term.

The effect of these two factors is well illustrated if we take the gas O_2 under the experimental conditions of Lipscomb, Norrish, and Thrush.[19] They observed the relaxation of low concentrations of the excited oxygen at the temperatures of 288° and 1800°K. Shuler[2] calculated the relaxation, assuming that the relaxation was in a pure heat bath (N_2 in this case) and that only the N_2–O_2 collisions were important. His value of P_{10} is 2×10^{-8} for 288°K and 7×10^{-4} at 1800°K. Our calculations for P'_{10}, from Eqs. (A9) and (A11) in Ref. 12, yield 1.3×10^{-3} for 288°K and 8×10^{-3} for 1800°K. At 288°K, the value [ratio $N^{(2)}/N^{(1)}$] of $N^{(2)}$ must be less than 10^{-6} in order to neglect the effect of P'_{10} on the relaxation to local equilibrium.[1] Certainly, in this case, for any but very small values of $N^{(2)}$, the change in $\epsilon(t)$ can be ignored when speaking of relaxation to local equilibrium, i.e., $\epsilon(0)$ can be used to calculate A in this case. However, the situation at high temperatures is not so clear-cut. P_{10} is a very sensitive function of temperature, so that at 1800°K, $N^{(2)}$ need only be less

[18] K. F. Herzfeld, *Temperature, Its Measurement and Control in Science and Industry* (Reinhold Publ. Corp., New York, 1955), Chap. 15.

[19] F. S. Lipscomb, R. G. W. Norrish, and B. A. Thrush, Proc. Royal Soc. (London) **A233**, 455 (1956).

than 10^{-2} for P'_{10} to be ignored. However, because the relaxation rate to local equilibrium is twice that of final equilibrium in any case, it is still correct to say that the relaxation proceeds by two fairly well separated steps: relaxation to a canonical distribution characterized by a (time-dependent) nonequilibrium vibrational temperature, and the cooling or warming of this distribution as a whole.

To get the actual populations for the π_nth distribution, one must express the generating function as a power series in y. This expansion (see the Appendix for the actual derivation) takes the following form:

$$\pi_n(y,t)=\sum_{\lambda=0}^{\infty}\left\{A\left[\frac{A(1+\epsilon)-1-\epsilon_0}{A(1+\epsilon)-\epsilon_0}\right]^n [A(1+\epsilon)-\epsilon_0]^{-\lambda-1}\sum_{k=0}^{\lambda}\binom{n}{k}\binom{\lambda}{k}(A\epsilon-\epsilon_0)^{\lambda-k}\left[\frac{A}{A(1+\epsilon)-1-\epsilon_0}\right]^k\right\}y^\lambda. \quad (37)$$

The population $x_{\lambda n}(t)$ is the coefficient of y^λ,

$$x_{\lambda n}(t)=\frac{A}{[A(1+\epsilon)-\epsilon_0]^{\lambda+1}}\left[\frac{(1+\epsilon)A-1-\epsilon_0}{A(1+\epsilon)-\epsilon_0}\right]^n\sum_{k=0}^{\lambda}\binom{n}{k}\binom{\lambda}{k}(A\epsilon-\epsilon_0)^{\lambda-k}\left[\frac{A}{A(1+\epsilon)-1-\epsilon_0}\right]^k, \quad (38)$$

where λ refers to the level and n to the initial populated state. Note that the form of Eq. (38) is identical to Montroll and Shuler's[1] heat-bath equation in that it consists of a finite sum of $\lambda+1$ terms with functions to the same powers of λ and k. Equation (38) reduces to the heat-bath equation when the corresponding values of A and ϵ are inserted. As required, $x_{\lambda n}(t)$ reduces to $\delta_{\lambda n}$ at $t=0$ ($A=1$, $\epsilon=\epsilon_0$). As $A\to\infty$, only a single term ($k=0$) in the sum survives, so that, as $t\to\infty$,

$$x_{\lambda n}(t)\to[\epsilon/(1+\epsilon)]^\lambda[1/(1+\epsilon)], \quad (39)$$

$A\to\infty$ as $t\to\infty$. Thus $x_{\lambda n}$ assumes the Boltzmann form as soon as A becomes sufficiently large.

To see how the populations approach local equilibrium, again we must expand $x_{\lambda n}(t)$ in powers of A^{-k}. However, it is better to use Eq. (36) because the calculations are simple and result applicable to an arbitrary initial distribution. If $\pi_{\text{Boltz}}=(1-e^{-\theta})/(1-ye^{-\theta})$, then

$$\pi^3_{\text{Boltz}}=[1-\exp(-\theta_v)]^3[1-y\exp(-\theta_v)]^{-3}=[1-\exp(-\theta_v)]^3\sum_{k=0}^{\infty}\binom{2+k}{2}\exp(-k\theta_v)y^k.$$

Multiplying by $(y-1)^2$ and collecting powers of y,

$$(y-1)^2\pi^3_{\text{Boltz}}=[1-\exp(-\theta_v)]^3\sum_{k=0}^{\infty}\left[\binom{2+k}{2}\exp(-k\theta_v)-2\binom{1+k}{2}\exp[-(k-1)\theta_v]+\binom{k}{2}\exp[-(k-2)\theta_v]\right]y^k \quad (40)$$

results. Applying Eq. (36) and the definition of the generating function, the correction to the Boltzmann population for the stane λ second order in A^{-1} is

$$x_\lambda^{\text{arb}}-x_\lambda^{\text{Boltz}}\equiv\Delta_\lambda^{(2)}=-\{[1-\exp(-\theta_v)]^3/2A^2\}[\epsilon_0(1+2\epsilon_0)-\langle n^2\rangle_0]$$
$$\times\left\{\binom{2+\lambda}{2}-2\binom{1+\lambda}{2}\exp(\theta_v)+\binom{\lambda}{2}\exp(2\theta_v)\right\}\exp(-\lambda\theta_v) \quad (41a)$$

for large time and initial second moment $\langle n^2\rangle_0$. Equation (41a) applies to the general initial distribution. Equation (41a) can be written in terms of the oscillator energy to yield

$$\Delta_\lambda^{(2)}=-\frac{\epsilon_0(1+2\epsilon_0)-\langle n^2\rangle_0}{2A^2(1+\epsilon)^3}\left(\frac{\epsilon}{1+\epsilon}\right)^\lambda\left[\binom{2+\lambda}{2}-2\binom{1+\lambda}{2}\frac{\epsilon+1}{\epsilon}+\binom{\lambda}{2}\left(\frac{\epsilon+1}{\epsilon}\right)^2\right]. \quad (41b)$$

In Figs. 1(a) and 1(b) there is a plot of $\Delta_\lambda^{(2)}$ against λ along with deviations to higher order in A^{-1}, to be discussed later. Notice that the only information about the distribution at zero time needed is the second moment; presumably all other detail is "ironed out" by the time Eq. (41) sufficiently represents the total deviation. It should be clear from the foregoing discussion that A^{-2} is indeed a useful measure of the approach to local equilibrium.

The general distribution is found [see Eq. (11) and especially Eq. (35)] through the relation

$$x_\lambda(t)=\sum_{n=0}^{\infty}x_{\lambda n}(t)x_n(0), \quad (42)$$

where the $x_{\lambda n}$ are from Eq. (38), the $x_n(0)$ form the initial distribution, and $x_\lambda(t)$ is the population of the λth state at time t. Obviously, $x_{\lambda n}(0)=\delta_{\lambda n}$. If we denote the matrix of the $x_{\lambda n}$'s by $\mathfrak{U}(t,0)$ and the array

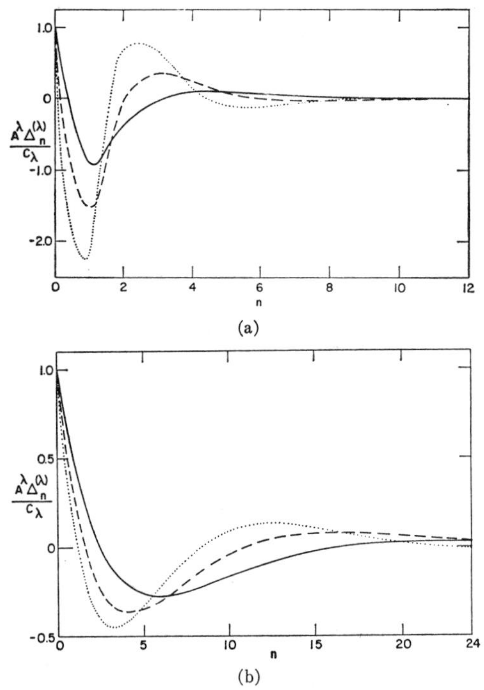

FIG. 1. Deviation of populations from the time-dependent Boltzmann distribution. Plots of $A^\lambda \Delta_n{}^{(\lambda)}/C_\lambda$ vs n. ——, $\lambda=2$; - - - -, $\lambda=3$; $\cdots$, $\lambda=4$. $\theta=1$ in (a); $\theta=0.2$ in (b).

of x_λ's and x_n's by the "vectors" $\mathbf{X}(t)$ and $\mathbf{X}(0)$, Eq. (42) becomes

$$\mathbf{X}(t)=\mathcal{U}(t,0)\mathbf{X}(0), \qquad [\mathcal{U}(t,0)]_{\lambda n}=x_{\lambda n}. \tag{43}$$

$\mathbf{X}(0)$ clearly represents the populations at $t=0$, whereas $\mathcal{U}(t,0)$ is the "time-evolution" operator which converts the initial distribution to the "final" one $\mathbf{X}(t)$. The inverse operation takes the form of Eq. (44):

$$\mathbf{X}(0)=\mathcal{U}^{-1}(t,0)\mathbf{X}(t), \tag{44}$$

which is the operation of finding the initial conditions from a given condition at a later time, as described previously.

There are several more general properties of π_n which are worthy of discussion in light of their connection with the properties of Eq. (5) discussed in Refs. 1 and 16. It can be shown by an argument similar to that used in Sec. 3 of Ref. 1, or by just taking Eq. (5), multiplying by n^k and summing over n and manipulating the sums, that differential equations for the kth moment,

$$\equiv\sum_{n=0}^{\infty} n^k x_n=\langle n^k\rangle,$$

contain no higher moments than the kth. If one can get the moments up to $k-1$, the kth can be solved for by an additional integration. Such an equation (5) is called a closed moment equation; this subject is taken up in greater detail in Ref. 16. The connection between π and $\langle n^k\rangle$ is obtained by noting that

$$\left.\frac{d^k\pi}{d\ln y^k}\right|_{y=1}=\langle n^k\rangle=\left.\frac{yd}{dy}\left[\frac{yd}{dy}\left(\cdots\frac{yd\pi}{dy}\right)\right]\right|_{y=1}$$

[see Eq. (11)]. In terms of the derivatives with y, this connection becomes

$$\langle n^k\rangle=\sum_{l=0}^{k-1}\binom{k-1}{l}y^l\left.\frac{\partial^{l+1}\pi}{\partial y^{l+1}}\right|_{y=1}. \tag{45}$$

If one can expand $\pi_{\text{arb}}(y,t)$ in powers of $(y-1)$—it appears natural to do so in light of the equations thus far derived for π—then the coefficients will be the sum of moments shown below. See the Appendix for the derivation of Eq. (46):

$$\pi_{\text{arb}}(y,t)=\sum_{k=0}^{\infty}\langle n(n-1)^{k-1}\rangle\frac{(y-1)^k}{k!}. \tag{46}$$

Equation (46) gives the direct link between the moments and the populations, showing the connection between the two different methods of solving Eq. (5). It should be clearly pointed out, however, that whereas the solution to the master equation by the step-by-step method of integration requires an infinite number of integrations, solution by means of a generating function requires only two. The first one is precisely the one needed to find the first moment, a prerequisite to finding any of the others; the second one is no more difficult, and can, with knowledge of $\epsilon(=\langle n\rangle)$, be done by machine. The trick which has enabled this drastic simplification lies in the fact that we have converted a set of ordinary differential equations into a *partial* differential equation containing all the arbitrariness of the original set. Such an equation is usually no easier to solve than Eq. (5), but luckily this partial differential equation has an exact, analytic, simple solution. The property of Eq. (5) which led to the simplification is precisely the "closed moment"—and the fact that the sums in Eq. (5) extend to infinity and that the coefficients of the $x_n(t)$ contain n up to the first power only. More discussion can be found in Ref. 16. The moments enter into another important relation governing the relaxation, the expansion of π_n into a power series in A^{-1}. The details used in reaching the expansion Eq. (47) can be found in the Appendix:

$$\pi_n(y,t)=\sum_{\lambda=0}^{\infty}\pi_{\text{Boltz}}{}^{\lambda+1}\frac{(y-1)^\lambda}{A^\lambda}\times\left[\sum_{l=0}^{\lambda}\binom{n}{l}\binom{\lambda}{l}(-1)^{\lambda+l}\epsilon_0{}^{\lambda-l}\right]. \tag{47}$$

Notice that each of the coefficients of $A^{-\lambda}$ can have no higher power in n than n^λ. Hence, on performing the summation in Eq. (35), no higher moment than the

λth can appear as a coefficient of $A^{-\lambda}$. The equation for the arbitrary distribution is

$$\pi_{\mathrm{arb}}(y,t)=\sum_{\lambda=0}^{\infty}\pi_{\mathrm{Boltz}}{}^{\lambda+1}\frac{(y-1)^{\lambda}}{A^{\lambda}}$$
$$\times\left[\sum_{l=0}^{\lambda}\left\langle\binom{n}{l}\right\rangle_0\binom{\lambda}{l}(-1)^{\lambda+l}\epsilon_0{}^{\lambda-l}\right],\quad(48)$$

where

$$\left\langle\binom{n}{l}\right\rangle_0\equiv\sum_{n=0}^{\infty}\binom{n}{l}\chi_n(0).$$

Thus it is clear that each order of the relaxation is characterized completely through moments of order λ or lower no matter what the character of the initial distribution. The higher the order, the more "detail" is required through the larger and larger number of moments needed. When calculating for the deviation from the Boltzmann form characterized by the λth order of relaxation, the y dependence enters through the power of π_{Boltz} and the $(y-1)^{\lambda}$ term; the first contributes a power series in y, whereas the second "shifts" their coefficients and groups them into polynominals of $\exp(-\theta_v)$ to order λ. Since the second summation in Eq. (49) depends only on the order of the relaxation and the initial distribution, $\pi_{\mathrm{Boltz}}{}^{\lambda+1}$ can be expanded, and the following general relation is found for the λth order of relaxation of the population in the nth state:

$$\Delta_n{}^{(\lambda)}=\left[\exp(-n\theta_v)\sum_{l=0}^{\lambda}\binom{\lambda-l+n}{\lambda}\binom{\lambda}{l}(-1)^l\exp(l\theta_v)\right]A^{-\lambda}C_{\lambda}[\mathbf{X}(0)],$$

$$C_{\lambda}[\mathbf{X}(0)]=[1-\exp(-\theta_v)]^{\lambda+1}(-1)^{\lambda}\sum_{l=0}^{\lambda}\left\langle\binom{n}{l}\right\rangle_0\binom{\lambda}{l}(-\epsilon_0)^{\lambda-l},\quad(49)$$

where θ_v is the vibrational temperature and $\mathbf{X}(0)$ is the array of the populations at zero time. Relation (50) points out explicitly the fact that each order of the relaxation is of a fixed form, dependent on the initial distribution only through the weight factor $C_{\lambda}[\mathbf{X}(0)]$ (θ_v is a system parameter independent of the detail of the distribution). Each of the deviations can be considered a polynomial of n to order λ times the factor $\exp(-n\theta_v)$, so that the behavior of a plot of $\Delta_n{}^{(\lambda)}(t)$ against n should look like a graph of that polynomial damped by an exponential factor. In Figs. 1(a) and 1(b) we have plotted $A^2\Delta_n{}^{(2)}/C_2$, $A^3\Delta_n{}^{(3)}/C_3$, and $A^4\Delta_n{}^{(4)}/C_4$ ($\Delta_n{}^{(1)}=0$) against the level number, n, for $\theta_v=1$ and 0.2, respectively.

It is interesting to see explicitly spelled out what is intuitively reasonable—that each order of the relaxation to local equilibrium depends only on moments of the same order or lower, and that the form of the contribution of that order to the entire distribution is independent of the initial distribution except in its magnitude.

In summary we have shown that the infinite set of nonlinear coupled equations [Eq. (1)] describing the vibrational relaxation of a gas of harmonic oscillators can be solved exactly for the case of the Landau–Teller transition probabilities. The solution, for arbitrary initial distribution and arbitrary dilution of the oscillators, involves only single and double integrals (and, to be sure, some heavy algebraic expressions). Furthermore, the solution can be used to generate the distribution at time t' from the distribution at t, and thus may be used to find the initial distribution from the observed distribution at a later time. The relaxation of an arbitrary distribution toward a time-dependent Boltzmann distribution proceeds with an asymptotic relaxation time at least twice as fast as the relaxation of the time-dependent Boltzmann distribution toward final equilibrium; the relaxation ia much faster than this when there is a significant concentration of oscillators present.

APPENDIX

The following manipulations are required to solve Eq. (14):

$$(y-1)^{-1}[\partial\pi(y,t)/\partial t]$$
$$+[g(t)-(y-1)f(t)][\partial\pi(y,t)/\partial y]=f(t)\pi(y,t),$$

where g and f are functions of time only. Following the method of characteristics as outlined in Ref. 17, we set up the relations

$$(y-1)dt=dy[g-(y-1)f]^{-1}=d\pi/f\pi.\quad(50)$$

Using the first two relations,

$$dy/dt=g(y-1)-f(y-1)^2.\quad(51)$$

We let $v=(y-1)^{-1}$ and differentiate:

$$-v^{-2}dv/dt=dy/dt.$$

Substituting this into Eq. (51) and multiplying the resulting expression by v^2 yields

$$(dv/dt)+vg=f,$$

whose solution is well known to be

$$\frac{1}{y-1}\equiv v=C\exp\left[-\int_t g(t')dt'\right]+\exp\left[-\int_t g(t')dt'\right]\int_t f(t')\exp\left[\int_{t'} g(t'')dt''\right]dt'. \tag{52a}$$

Solving for the constant of integration, C, we get

$$C=\frac{1}{y-1}\exp\left[\int_t g(t')dt'\right]-\int_t f(t')\exp\left[\int_{t'} g(t'')dt''\right]dt'. \tag{52b}$$

The general solution is gotten by taking the first and the last relations in Eq. (50) and substituting Eq. (52a) for $(y-1)$ and getting this differential equation:

$$\frac{d\pi}{\pi}=fdt\Big/\left[C\exp\left(-\int gdt\right)+\exp\left(-\int gdt\right)\int f\exp\left(\int gdt\right)\right]$$

$$=d\left\{\int_t f(t')\exp\left[\int_{t'} g(t'')dt''\right]dt'\right\}\Big/\left\{C+\int_t f(t')\exp\left[\int_{t'} g(t'')dt''\right]\right\}, \tag{53}$$

which has the solution

$$\pi=D\left\{C+\int_t f(t')\exp\left[\int_{t'} g(t'')dt''\right]dt'\right\}. \tag{54}$$

D is the second integration constant. Continuing the method, we insert Eq. (52b) for C, converting Eq. (54) into

$$\pi=D\exp\left[\int_t g(t')dt'\right]\Big/(y-1). \tag{55}$$

D now plays the role of an arbitrary function of the integration constant C, $D(C)$. Performing the substitutions turns Eq. (56) into Eq. (21a). Equation (21a) can be verified by direct substitution into Eq. (14).

Binomial Relations

The symbols $\binom{n}{k}$ used throughout this paper are the binomial coefficients defined as follows:

$$\binom{n}{k}=n!/k!(n-k)!, \tag{56}$$

from which follow the relations

$$\binom{n}{k}=\binom{n}{n-k},$$

$$\binom{n}{0}=1;\quad \binom{n}{1}=n. \tag{57}$$

In addition, the following relation holds:

$$\binom{n}{k}=0,\qquad k\rangle n, \tag{58}$$

which is used to automatically set the limits of many of the summations in this paper.

Derivation of Eq. (38)

Equation (38) is easily derived of one notes that the expression

$$\frac{(1+xu-u)^n}{(1-u)^{n+1}}\equiv\frac{1+[xu/(1-u)]}{1-u}$$

can be expanded as follows:

$$\frac{\{1+[xu/(1-u)]\}^n}{1-u}=\sum_{k=0}^{n}\binom{n}{k}x^ku^k(1-u)^{-(k+1)}$$

$$=\sum_{l=0}^{\infty}\sum_{k=0}^{n}\binom{n}{k}\binom{k+l}{k}x^ku^{k+l}.$$

This becomes, on letting $k+l=\lambda$,

$$\frac{1+[xu/(1-u)]}{1-u}=\sum_{\lambda=0}^{\infty}u^{\lambda}\sum_{k=0}^{\lambda}\binom{n}{k}\binom{\lambda}{k}x^{k}=\sum_{n=0}^{\infty}u^{\lambda}F(-n,-\lambda,1;x).^{20} \tag{59}$$

$\pi_n(y,t)$ [Eq. (34a)] can be rearranged to read

$$\pi_n(y,t)=\frac{A[A(1+\epsilon)-\epsilon_0-1]^n}{[A(\epsilon+1)-\epsilon_0]^{n+1}}\frac{(\{(1+\epsilon_0-A\epsilon)y/[A(1+\epsilon)-\epsilon_0-1]\}+1)^n}{(1-\{(A\epsilon-\epsilon_0)y/[A(1+\epsilon)-\epsilon_0]\})^{n+1}}. \tag{34c}$$

If we let

$$u=(A\epsilon-\epsilon_0)/[A(1+\epsilon)-\epsilon_0]y$$

and

$$x=A/\{[A(1+\epsilon)-\epsilon_0-1](A\epsilon-\epsilon_0)\},$$

then

$$\pi_n(y,t)=A\frac{[A(1+\epsilon)-\epsilon_0-1]^n}{[A(\epsilon+1)-\epsilon_0]^{n+1}}\frac{(1+xu-u)^n}{(1-u)^{n+1}},$$

from which Eq. (38) immediately follows by application of Eq. (59).

Derivation of Eqs. (46) and (47)

In order to derive Eqs. (46) and (47), we must first note the following sequence of equations:

$$(-x)^n\equiv[1-(1-x)]^n=\sum_{k=0}^{n}(-1)^k\binom{n}{k}(1+x)^k=\sum_{k=0}^{n}\sum_{l=0}^{k}(-1)^k\binom{n}{k}\binom{k}{l}x^l$$

$$=\sum_{l=0}^{n}x^l\sum_{k=l}^{n}(-1)^k\binom{n}{k}\binom{k}{l},$$

which yields Eq. (60) by comparison with $(-x)^n$

$$\sum_{k=l}^{n}\binom{n}{k}\binom{k}{l}(-1)^k=(-1)^n\delta_{nl}. \tag{60}$$

To derive Eq. (46), we start by assuming a Taylor's expansion of $\pi(y,t)$ is possible about the point $y=1$:

$$\pi(y,t)=\sum_{k=0}^{\infty}\pi_k\frac{(y-1)^k}{k!},\qquad \pi_k\equiv\left.\frac{\partial^k\pi}{\partial y^k}\right|_{y=1}.$$

If such a relation exists, we only have to prove the relation

$$\pi_k=\langle n(n-1)^{k-1}\rangle. \tag{61}$$

We begin by expanding $\langle n(n-1)^{k-1}\rangle$,

$$\langle n(n-1)^{k-1}\rangle=\sum_{l=0}^{k-1}\binom{k-1}{l}\langle n^{k-l}\rangle(-1)^l,$$

which becomes, using Eq. (45),

$$\langle n(n-1)^{k-1}\rangle=\sum_{l=0}^{k-1}\binom{k-1}{l}(-1)^l\sum_{t=0}^{k-l-1}\left(y^t\frac{\partial^{t+1}\pi}{\partial y^{t+1}}\right)_{y=1}\binom{k-l-1}{t}$$

and rearranged to read

$$\sum_{t=0}^{k-1}\left(y^t\frac{\partial^{t+1}\pi}{\partial y^{t+1}}\right)_{y=1}\sum_{l=0}^{k-1-t}\binom{k-l-1}{t}\binom{k-1}{l}(-1)^l. \tag{62}$$

By substituting the relations $k-1-l=m$ and Eq. (57), Eq. (62) becomes

$$\sum_{t=0}^{k-1}\left(y^t\frac{\partial^{t+1}\pi}{\partial y^{t+1}}\right)_{y=1}\sum_{m=k-1}^{t}\binom{m}{t}\binom{k-1}{m}(-1)^{k-1}(-1)^m,$$

[20] A. Erdelyi, W. Magnus, F. Oberhettinger, and F. G. Tricomi, *Higher Transcendental Functions* (McGraw-Hill Book Co., New York, 1953), Vol. 2, p. 255.

which by virtue of Eq. (60) becomes

$$\sum_{t=0}^{k-1}\left(y^t\frac{\partial^{t+1}\pi}{\partial y^{t+1}}\right)_{y=1}(-1)^{k-1}\delta_{t,k-1}(-1)^{k-1}=\pi_k$$

proving Eq. (61).

Equation (47) is derived by comparing Eq. (34b) with Eq. (59). If in Eq. (34b) we let

$$u=-(\epsilon_0/A)\pi_{\mathrm{Boltz}}(y-1)$$

and

$$x=-1/\epsilon_0,$$

then Eq. (48) follows through the use of Eq. (59). Equation (36) is obtained by taking only the first two non-zero terms of Eq. (47).

ACKNOWLEDGMENTS

The authors gratefully acknowledge the general support of the Institute for the Study of Metals by the Advanced Research Projects Agency. We would also like to thank Dr. K. E. Shuler for his preprint which was so helpful throughout this work.

On the Relaxation of the Hard-Sphere Rayleigh and Lorentz Gas

KNUD ANDERSEN*
H. C. Ørsted Institute, University of Copenhagen, Copenhagen Ø, Denmark
AND
KURT E. SHULER
National Bureau of Standards, Washington, D. C. 20234
(Received 22 April 1963)

As part of a study of the relaxation of nonequilibrium systems, the (translational) relaxation of a hard-sphere Rayleigh and Lorentz gas is investigated. From a detailed analysis of the collision dynamics an exact expression is derived for the kernel $A(x \mid x')$ of the collision integral which gives the probability per unit time for a change of the reduced kinetic energy from x' to x during a binary collision between a subsystem and a heat bath particle. A master equation, i.e., a linearized Boltzmann equation, incorporating this kernel is then formulated to represent the time variation of the distribution function of the subsystem particles. Making use of the special property of this kernel that it is a strongly peaked function around $x-x'=0$ for both the Rayleigh and Lorentz gas, a technique is developed for transforming this integral master equation into differential Fokker–Planck equations consistent in the order of the expansion parameter λ, the ratio of the mass of the heat bath particles to the subsystem particles. The Fokker–Planck equation for the Rayleigh gas is solved analytically and explicit solutions are presented for the relaxation of initial Maxwell and initial δ-function distributions of the energy. It is shown that an initial Maxwell distribution of the energy (or speed) of the subsystem particles relaxes to the final equilibrium Maxwell distribution via a continuous sequence of Maxwell distributions. The mean energy of the subsystem particles is shown to relax exponentially to its equilibrium value independent of the form of the initial distribution. For the hard-sphere Lorentz gas the Fokker–Planck equation is not susceptible of an analytical solution. Machine solutions are presented for various initial distributions which show that the Maxwell distribution is not preserved in the relaxation of the hard sphere Lorentz gas. Finally, a brief discussion is given of the relation between the hard-sphere model ($r^{-\infty}$) considered here and the more general model of the Rayleigh and Lorentz gas with a r^{-s} repulsive central force law.

I. INTRODUCTION

THIS investigation was stimulated by three closely connected problems which have been of interest to one of us (KES) for a number of years. These were the general problem of the relaxation of nonequilibrium systems, the problem of "inverse Brownian motion," i.e., relaxation in an ensemble where a subsystem particle undergoes a *large* change in some relevant parameter (position, momentum, etc.) as a consequence of each binary interaction with a heat-bath particle, and finally the problem of the passage from an integral (master) equation to its differential representation in the form of a Fokker–Planck equation.

In an attempt to obtain some clarification of these problems we have studied the relaxation of a three-dimensional hard-sphere Rayleigh and Lorentz gas. The Rayleigh gas[1] as studied here is a spatially isotropic two-species gas with a very dilute subsystem of heavy mass points dispersed in a heat bath of light particles. The Lorentz gas[2] considered here is the "inverse" of the Rayleigh gas in that the dilute subsystem of light particles is now dispersed in a heat bath of heavy particles. By limiting oneself to classical hard sphere interactions, it is possible to carry through the binary collision dynamics exactly in deriving the linearized Boltzmann equation which describes the relaxation of these systems. This in turn permits a very detailed analysis of the passage from the integral Boltzmann

* The major part of this work was carried out at the National Bureau of Standards (1961–1962) during the tenure of a NATO Post-Doctoral Fellowship.

[1] Lord Rayleigh, Phil. Mag. **32**, 424 (1891).
[2] H. A. Lorentz, Proc. Amsterdam Acad. **7**, 438, 585, 684 (1905).

equation (the master equation) to the differential Fokker–Planck equation.

The Lorentz gas serves as our model of the "inverse Brownian motion" problem. It is evident that the light subsystem particles in their interaction with the heavy heat-bath particles will suffer large displacements and undergo large changes in momentum in each binary encounter. From previous analyses of the validity of the Fokker–Planck equation[3–5] it is evident that the relaxation of a Lorentz gas in position and/or momentum space cannot be represented by a Fokker–Planck equation. If one, however, considers *speed* space (the scalar speed must be clearly distinguished here from the vectorial velocity) or *energy* space the picture changes. The change in speed and kinetic energy suffered by the light subsystem particles in binary encounters with a heat bath particle is very small compared to the average thermal speed or energy. There is thus good reason to believe that a Fokker–Planck equation in speed or energy space will form a valid description of the relaxation of a Lorentz gas. The development in this paper of the kinetic equations in speed and energy variables was undertaken to provide a parallel and unified treatment for both the Lorentz and the Rayleigh gas.[6a]

It is well known that the Lorentz gas can serve as a convenient model for a weakly ionized plasma where the electrons (light particles of the very dilute subsystem) interact, in uncorrelated binary collisions, predominantly with the neutral species (the heat bath). The hard-sphere interaction considered here is of course unrealistic as a representation of the electron–neutral species interactions in a plasma. The treatment developed here can, however, be extended to more realistic interactions and we plan to consider this in a subsequent publication.[6b]

In Sec. II, we present a brief discussion of the hard-sphere Rayleigh and Lorentz gas model used in the body of the paper. Starting from the Boltzmann equation we derive in Sec. III the transition probability kernel of the master equation which governs the time evolution of the subsystem distribution function. In Sec. IV we discuss the passage from the master equation to the Fokker–Planck equation and then derive the Fokker–Planck equations for the relaxation of the Rayleigh gas and the Lorentz gas. In Sec. V we develop the general analytical solution of the Fokker–Planck equation for the Rayleigh gas and then work out in some detail the specific examples of the relaxation of an initial δ-function distribution and an initial Maxwell distribution of reduced energies. In Sec. VI, we discuss the properties of the Fokker–Planck equation for the Lorentz gas and present machine calculations for the relaxation of an initial Maxwell distribution.

Finally we should point out that we have studied here the "how" and not the "why" of the irreversible relaxation of an initial nonequilibrium distribution to a final equilibrium state. By utilizing the Boltzmann equation we have built irreversibility into our model from the outset.

II. MODEL

Our model consists of a three-dimensional spatially isotropic system of identical, hard, smooth spheres—for further reference called the subsystem—in collisional interaction with another system of identical, hard smooth spheres of mass and radius different from those of the subsystem; the latter system we call the heat bath. For all variables referring to the particles of the *subsystem* we use the *subscript 1*; similarly the *subscript 2* refers to the *heat bath*. The ratio of the masses $m_2/m_1 \equiv \lambda^2$ of the two types of particles is taken to be either very small or very large ($m_2/m_1 \ll 1$ or $m_2/m_1 \gg 1$). The total system is taken sufficiently dilute as measured by the total number of particles N_1+N_2 per unit of volume that we are in the range of applicability of the simple Boltzmann equation based on uncorrelated binary collisions. The concentration of the subsystem is taken to be very much smaller than that of the heat bath ($N_1 \ll N_2$) so that the time behavior of the subsystem can be properly accounted for by considering only collisions between 1 and 2 particles, with 1–1 collisions occurring very infrequently. In addition, when $N_1 \ll N_2$ the presence of the subsystem can be assumed to perturb the distribution of the heat bath only to a very minor extent. Thus, when the heat bath is initially prepared with a Maxwellian distribution of velocities corresponding to a temperature T_2, this distribution will be assumed to remain invariant with time. From the model described here it follows that in the limit as the time $t \to \infty$ the subsystem, irrespective of its initial distribution, will relax to the equilibrium Maxwell distribution at the temperature T_2.

III. DERIVATION OF THE TRANSITION PROBABILITY KERNEL $A(x \mid x')$

We use a notation in which primes (′) indicate values of the collision variables prior to the collision and unprimed letters refer to values after the collision. We let $\mathbf{p}$ stand for momentum and m for mass. The relative velocity of a 2 particle with respect to a 1

[3] S. Chandrasekhar, Rev. Mod. Phys. **15**, 2 (1943).
[4] J. Keilson and J. E. Storer, Tech. Report No. 127, Cruft Laboratory, Harvard University, Cambridge, Massachusetts.
[5] N. G. van Kampen, Ned. Tijds. Natuurk. **26**, 225 (1960).
[6] (a) It is of course well known that the relaxation of a Rayleigh gas corresponds to a Brownian motion random walk and can be represented by a Fokker–Planck equation in velocity (or momentum) space; (b) Other applications of the hard-sphere model Lorentz gas can be found in a number of investigations on the thermalization of neutrons by heavy gaseous moderators. See, e.g., E. P. Wigner and J. E. Wilkins, AECD–2275 (1944); J. E. Wilkins, AEC Rept. CP–2481 (1944); N. Hurwitz, M. S. Nelkin, and G. J. Habetler, Nucl. Sci. Eng. **1**, 280 (1956); N. Corngold, Ann. Phys. (N.Y.) **6**, 368 (1959).

particle is called $\mathbf{g}=(\mathbf{p}_2/m_2)-(\mathbf{p}_1/m_1)$; during a collision process this vector is turned through the angle χ. The differential collision cross section is called $I(g',\chi)$ and we use the abbreviation $d\Omega$ for the corresponding differential $\sin\chi d\chi d\epsilon$. Finally we introduce the notation $N_1f_1(\mathbf{p}_1',t)d\mathbf{p}_1'$ for the number of 1 particles per unit volume with momenta between $\mathbf{p}_1'$ and $\mathbf{p}_1'+d\mathbf{p}_1'$ at time t; for 2 particles this number is called

$$N_2f_2(\mathbf{p}_2',t)d\mathbf{p}_2'.$$

As our starting point we take the Boltzmann equation

$$\frac{\partial f_1(\mathbf{p}_1',t)}{\partial t}=N_2\iint[f_2(\mathbf{p}_2,t)f_1(\mathbf{p}_1,t)-f_2(\mathbf{p}_2',t)f_1(\mathbf{p}_1',t)]g'I(g',\chi)d\Omega d\mathbf{p}_2' \tag{3.1}$$

from which we can deduce that the total number of collisions per unit volume in the time interval dt undergone by 1 particles of initial momentum around $\mathbf{p}_1'$ is given by

$$N_2N_1f_1(\mathbf{p}_1',t)d\mathbf{p}_1'dt\iint f_2(\mathbf{p}_2',t)g'I(g',\chi)d\Omega d\mathbf{p}_2'. \tag{3.2}$$

We want to express this collision number as an integral using a transition probability kernel:

$$N_2N_1f_1(\mathbf{p}_1',t)d\mathbf{p}_1'dt\iiint W(\mathbf{p}_1,\mathbf{p}_2\mid\mathbf{p}_1',\mathbf{p}_2')f_2(\mathbf{p}_2',t)d\mathbf{p}_1d\mathbf{p}_2d\mathbf{p}_2', \tag{3.3}$$

where

$$W(\mathbf{p}_1,\mathbf{p}_2\mid\mathbf{p}_1',\mathbf{p}_2')N_1f_1(\mathbf{p}_1',t)N_2f_2(\mathbf{p}_2',t)d\mathbf{p}_1d\mathbf{p}_2d\mathbf{p}_1'd\mathbf{p}_2'dt \tag{3.4}$$

gives the number of binary collisions per unit volume in time dt by which 1 particles change their momenta from around $\mathbf{p}_1'$ to around $\mathbf{p}_1$ while 2 particles change their momenta from around $\mathbf{p}_2'$ to around $\mathbf{p}_2$. Our problem is now to set up a formula for $W(\mathbf{p}_1,\mathbf{p}_2\mid\mathbf{p}_1',\mathbf{p}_2')$ which will make the two expressions (3.2) and (3.3) identical. As demonstrated by Waldmann[7a] the answer can be found from an analysis of the dynamics of a binary collision process; we obtain the formula

$$W(\mathbf{p}_1,\mathbf{p}_2\mid\mathbf{p}_1',\mathbf{p}_2')$$
$$=\mu^{-3}I(g',\chi)\delta_3(\mathbf{p}_1+\mathbf{p}_2-\mathbf{p}_1'-\mathbf{p}_2')\delta_1[\tfrac{1}{2}(g^2-g'^2)] \tag{3.5}$$

in which μ is the reduced mass $m_1m_2/(m_1+m_2)$ and

$$\chi=\arc\cos(\mathbf{g}\cdot\mathbf{g}'/|\mathbf{g}||\mathbf{g}'|) \tag{3.6}$$

and $\delta_3(\mathbf{a})$ is the three-dimensional Dirac delta function equal to $\delta_1(a_x)\delta_1(a_y)\delta_1(a_z)$.

Knowing the transition probability kernel

$$W(\mathbf{p}_1,\mathbf{p}_2\mid\mathbf{p}_1',\mathbf{p}_2')$$

we can evaluate the number of binary collisions per unit volume occurring in the time interval t, $t+dt$ which take subsystem particles with initial momenta around $\mathbf{p}_1'$ to final momenta around $\mathbf{p}_1$:

$$N_1Z(\mathbf{p}_1\mid\mathbf{p}_1')f_1(\mathbf{p}_1',t)d\mathbf{p}_1'd\mathbf{p}_1dt=N_1f_1(\mathbf{p}_1',t)d\mathbf{p}_1'dt\cdot N_2\iint W(\mathbf{p}_1,\mathbf{p}_2\mid\mathbf{p}_1',\mathbf{p}_2')f_2(\mathbf{p}_2',t)d\mathbf{p}_2'd\mathbf{p}_2. \tag{3.7}$$

Using Eq. (3.5) we obtain for $Z(\mathbf{p}_1\mid\mathbf{p}_1')$

$$Z(\mathbf{p}_1\mid\mathbf{p}_1')=N_2\mu^{-3}\iint I(g',\chi)\delta_3(\mathbf{p}_1+\mathbf{p}_2-\mathbf{p}_1'-\mathbf{p}_2')\delta_1[\tfrac{1}{2}(g^2-g'^2)]f_2(\mathbf{p}_2',t)d\mathbf{p}_2'd\mathbf{p}_2. \tag{3.8}$$

For the special case of hard, smooth spheres the differential cross section $I(g',\chi)$ is known to be equal to the $\frac{1}{4}(r_1+r_2)^2$, where r_i is the radius of an i sphere. If we now use the Maxwell distribution law for the heat bath particles

$$f_2(\mathbf{p}_2',t)=(2\pi m_2kT_2)^{-3/2}\exp[-p_2'^2/2m_2kT_2] \tag{3.9}$$

[7] (a) L. Waldmann, *Handbuch der Physik*, edited by S. Flügge (Springer-Verlag, Berlin, 1958), Vol. 12, p. 348 ff; (b) It is very important, in order to follow the development presented here, to make a clear distinction between vector and scalar quantities. For some physical quantities there are simple distinguishing terms, e.g., velocity and speed. This unfortunately is not the case for momenta, and we therefore have to use the awkward phrase "magnitude of the momentum" in several places to denote the scalar value; (c) The kernel $B(x\mid x')$ given in Eq. (3.25) is identical, except for notation and definition of variables, with the one derived by Wigner and Wilkins[6b] in their study of the thermalization of neutrons.

we obtain from (3.8)

$$Z(\mathbf{p}_1 \mid \mathbf{p}_1') = \frac{N_2(r_1+r_2)^2}{4\mu^3(2\pi m_2 kT_2)^{3/2}} \int \exp\left(-\frac{p_2'^2}{2m_2kT_2}\right)\delta_1\left[\frac{\alpha}{m_2\mu}(\mathbf{p}_2'\cdot\mathbf{n}-q)\right]d\mathbf{p}_2', \tag{3.10}$$

where

$$\alpha = |\mathbf{p}_1'-\mathbf{p}_1|, \tag{3.11}$$

$$\mathbf{n} = (\mathbf{p}_1'-\mathbf{p}_1)/\alpha; \qquad |\mathbf{n}| = 1, \tag{3.12}$$

and

$$q = m_2(\mathbf{p}_1'\cdot\mathbf{n}/m_1 - \alpha/2\mu). \tag{3.13}$$

To perform the integration over $\mathbf{p}_2'$ in Eq. (3.10) we shall make use of a Cartesian coordinate system for which the direction of the z-axis is given by the fixed vector $\mathbf{n}$, i.e.,

$$\mathbf{p}_2' = \mathbf{p}_{2,x}' + \mathbf{p}_{2,y}' + \mathbf{p}_2'\cdot\mathbf{n}\mathbf{n} \tag{3.14}$$

which yields

$$Z(\mathbf{p}_1 \mid \mathbf{p}_1') = \frac{N_2(r_1+r_2)^2}{4\mu^3(2\pi m_2kT_2)^{3/2}} \int_{-\infty}^{\infty}\int_{-\infty}^{\infty}\int_{-\infty}^{\infty} \exp\left[-\frac{p_{2,x}'^2+p_{2,y}'^2+(\mathbf{p}_2'\cdot\mathbf{n})^2}{2m_2kT_2}\right]\delta_1\left[\frac{\alpha}{m_2\mu}(\mathbf{p}_2'\cdot\mathbf{n}-q)\right]dp_{2,x}'dp_{2,y}'d(\mathbf{p}_2'\cdot\mathbf{n}). \tag{3.15}$$

Since $\mathbf{n}$ and q are constants with respect to the integration variables one finally obtains

$$Z(\mathbf{p} \mid \mathbf{p}_1') = \frac{N_2(r_1+r_2)^2}{4\alpha\mu^2}\left(\frac{m_2}{2\pi kT_2}\right)^{\frac{1}{2}} \exp\left[-\frac{q^2}{2m_2kT_2}\right]. \tag{3.16}$$

We now wish to integrate the expression (3.7) with $Z(\mathbf{p}_1 \mid \mathbf{p}_1')$ given by Eq. (3.16) over all possible orientations of $\mathbf{p}_1$ with respect to $\mathbf{p}_1'$. This will yield an expression for the number of collisions $N_1 Z(p_1 \mid \mathbf{p}_1')f_1(\mathbf{p}_1', t)\,dp_1 d\mathbf{p}_1' dt$ per unit volume in the time interval dt which take 1 particles with initial momenta around $\mathbf{p}_1'$ to momenta whose *magnitudes*[7b] lie between p_1 and p_1+dp_1. We now take $\mathbf{p}_1'$ as the polar axis of a spherically polar coordinate system in which $\mathbf{p}_1$ is defined by p_1 and the angles θ and ϕ, i.e., $d\mathbf{p}_1 = p_1^2 \sin\theta dp_1 d\theta d\phi$. We then have

$$Z(p_1 \mid \mathbf{p}_1') = \int_0^{\pi}\int_0^{2\pi} Z(\mathbf{p}_1 \mid \mathbf{p}_1')\,p_1^2 \sin\theta d\theta d\phi$$

$$= \frac{N_2(r_1+r_2)^2}{4\mu^2}\left(\frac{2\pi m_2}{kT_2}\right)^{\frac{1}{2}} p_1^2 \int_0^{\pi} \frac{1}{\alpha(\theta)} \exp\left[-\frac{q^2(\theta)}{2m_2kT_2}\right]\sin\theta d\theta, \tag{3.17}$$

where

$$\alpha(\theta) = (p_1'^2 + p_1^2 - 2p_1p_1'\cos\theta)^{\frac{1}{2}} \tag{3.18}$$

and

$$q(\theta) = m_2\left[\frac{p_1'}{m_1}\left(\frac{p_1'-p_1\cos\theta}{\alpha(\theta)}\right) - \frac{\alpha(\theta)}{2\mu}\right].$$

Introducing the new dimensionless variable z, defined by $z = \alpha(\theta)/(8m_2kT_2a)^{\frac{1}{2}}$ with $a = |p_1'^2 - p_1^2|/8m_1kT_2 \geq 0$ permits us to rewrite Eq. (3.17) as

$$Z(p_1 \mid \mathbf{p}_1') = \frac{N_2(r_1+r_2)^2 \pi^{\frac{1}{2}} m_2}{\mu^2}\frac{p_1}{p_1'} a^{\frac{1}{2}} \int_{l_2}^{l_1} \exp\left[-a\left(z \mp \frac{1}{z}\right)^2\right]dz, \tag{3.19}$$

where

$$l_1 = \frac{1}{\lambda}\left(\frac{p_1'+p_1}{|p_1'-p_1|}\right)^{\frac{1}{2}}, \qquad l_2 = \frac{1}{\lambda}\left(\frac{|p_1'-p_1|}{p_1'+p_1}\right)^{\frac{1}{2}} \qquad \text{and} \qquad \lambda = \left(\frac{m_2}{m_1}\right)^{\frac{1}{2}}.$$

Here and in all subsequent equations containing the combination $\pm$ or $\mp$ the upper sign should be applied to the case $p_1 < p_1'$ and the lower sign to the case $p_1 > p_1'$. Evaluation of the definite integral in Eq. (3.19) in terms of the error function $\mathrm{erf}(y)$ defined by

$$\mathrm{erf}(y) \equiv \int_0^y \exp(-t^2)\,dt$$

yields

$$Z(p_1\mid \mathbf{p}_1')=\frac{N_2(r_1+r_2)^2\pi^{\frac{1}{2}}m_2\,p_1}{2\mu^2\quad p_1'}[G(p_1,p_1')\mp H(p_1,p_1')],\tag{3.20}$$

with

$$G(p_1,p_1')=\exp\left(\frac{p_1'^2-p_1^2}{2m_1kT_2}\right)\operatorname{erf}\left[Q\frac{p_1'}{(2m_1kT_2)^{\frac{1}{2}}}+R\frac{p_1}{(2m_1kT_2)^{\frac{1}{2}}}\right]+\operatorname{erf}\left[R\frac{p_1'}{(2m_1kT_2)^{\frac{1}{2}}}+Q\frac{p_1}{(2m_1kT_2)^{\frac{1}{2}}}\right]$$

and

$$H(p_1,p_1')=\exp\left(\frac{p_1'^2-p_1^2}{2m_1kT_2}\right)\operatorname{erf}\left[Q\frac{p_1'}{(2m_1kT_2)^{\frac{1}{2}}}-R\frac{p_1}{(2m_1kT_2)^{\frac{1}{2}}}\right]+\operatorname{erf}\left[R\frac{p_1'}{(2m_1kT_2)^{\frac{1}{2}}}-Q\frac{p_1}{(2m_1kT_2)^{\frac{1}{2}}}\right].\tag{3.21}$$

The quantities Q and R are given by

$$Q=\frac{1}{2}\left(\frac{1}{\lambda}+\lambda\right)=\frac{1}{2}\left[\left(\frac{m_1}{m_2}\right)^{\frac{1}{2}}+\left(\frac{m_2}{m_1}\right)^{\frac{1}{2}}\right]$$

and

$$R=\frac{1}{2}\left(\frac{1}{\lambda}-\lambda\right)=\frac{1}{2}\left[\left(\frac{m_1}{m_2}\right)^{\frac{1}{2}}-\left(\frac{m_2}{m_1}\right)^{\frac{1}{2}}\right].\tag{3.22}$$

Finally, we ask for the number of collisions

$$N_1Z(p_1\mid p_1')f_1(p_1',t)\,dp_1dp_1'dt$$

per unit volume undergone by 1 particles in the time interval dt in which the initial momenta with magnitudes between p_1' and $p_1'+dp_1'$ are changed to momenta with magnitudes between p_1 and p_1+dp_1. This number is evaluated by integrating the expression

$$N_1Z(p_1\mid \mathbf{p}_1')f_1(\mathbf{p}_1')\,dp_1d\mathbf{p}_1'dt$$

with $Z(p_1\mid \mathbf{p}_1')$ given by Eq. (3.20) and with

$$f_1(\mathbf{p}_1',t)\,d\mathbf{p}_1'=p_1'^2f_1(p_1',t)\,\sin\theta_1dp_1'd\theta_1d\phi_1\tag{3.23}$$

over $0\leq\phi_1\leq 2\pi$ and $0\leq\theta_1\leq\pi$. Since $Z(p_1\mid \mathbf{p}_1')$ does not depend upon ϕ_1 or θ_1 we immediately obtain

$$Z(p_1\mid p_1')=[2\pi^{3/2}N_2(r_1+r_2)^2m_2/\mu^2]p_1p_1'\times[G(p_1,p_1')\mp H(p_1,p_1')].\tag{3.24}$$

We now introduce the reduced, dimensionless kinetic energy x for *the subsystem particles* as the state variable, i.e. we set $x'=p_1'^2/2m_1kT_2$ and $x=p_1^2/2m_1kT_2$ and define $N_1P(x',t)\,dx'$ as the number of 1 particles per unit volume with reduced energy between x' and $x'+dx'$. Similarly we use the notation

$$N_1B(x\mid x')P(x',t)\,dxdx'dt$$

for the number of collisions per unit volume in time dt which take 1 particles of initial reduced energy around x' to a final reduced energy around x. Substitution of the new dimensionless variables into Eq. (3.24) yields the following expression for $B(x\mid x')$:

$$B(x\mid x')=(C/x'^{\frac{1}{2}})[\operatorname{erf}(Qx^{\frac{1}{2}}+Rx'^{\frac{1}{2}})+e^{x'-x}\operatorname{erf}(Rx^{\frac{1}{2}}+Qx'^{\frac{1}{2}})\pm\{\operatorname{erf}(Qx^{\frac{1}{2}}-Rx'^{\frac{1}{2}})+e^{x'-x}\operatorname{erf}(Rx^{\frac{1}{2}}-Qx'^{\frac{1}{2}})\}]\tag{3.25}$$

with

$$C=Q^2N_2\pi(r_1+r_2)^2(2kT_2/\pi m_1)^{\frac{1}{2}}.\tag{3.25a}$$

The upper sign should be applied when $x<x'$, and the lower sign when $x>x'$.[7c]

When the equilibrium form of $P(x',t)$, i.e.,

$$P(x',\infty)=(2/\pi^{\frac{1}{2}})x'^{\frac{1}{2}}e^{-x'}\tag{3.26}$$

is inserted into the collision number as given above one obtains an expression for the number of collisions per unit volume in time *dt at equilibrium* which produce a change of the reduced energy of the 1 particles from around x' to around x. It is easy to show that the identical result is obtained for changes from around x to around x'. This equivalence proves that the collision number $B(x\mid x')$ satisfies the condition of detailed balancing at equilibrium.

We assume, as indicated previously, that our total system (subsystem+heat bath) is sufficiently dilute that any change in $N_1P(x,t)\,dx$ in a sufficiently short time interval dt is due to binary collisions only. We are then able to express $N_1dP(x,t)\,dx$, the change during the time interval dt in the number of 1 particles per unit volume having an energy between x and $x+dx$, as the difference between the number of binary collisions in dt taking 1 particles to an energy around x from all other energy "states" and the number of binary collisions in dt which remove 1 particles from the energy range around x to all other "states," i.e.,

$$N_1[P(x,t+dt)-P(x,t)]dx=N_1dxdt\left\{\int_0^\infty B(x\mid x')P(x',t)\,dx'-P(x,t)\int_0^\infty B(x'\mid x)\,dx'\right\}.\tag{3.27}$$

In the limit as $dt\to 0$, Eq. (3.27) leads to the master equation (linearized Boltzmann equation)

$$\frac{\partial P(x,t)}{\partial t}=\int_0^\infty A(x\mid x')P(x',t)\,dx'\tag{3.28}$$

which governs the time evolution of the energy distribution function $P(x,t)$. The transition probability kernel $A(x\mid x')$ is related to the $B(x\mid x')$ by

$$A(x\mid x')=B(x\mid x')-\delta(x-x')Z_{12}(x)\tag{3.29}$$

with $B(x \mid x')$ as given explicitly by Eq. (3.25), $\delta(x-x')$ being the Dirac delta function and

$$Z_{12}(x) = \int_0^\infty B(x' \mid x)\,dx'. \tag{3.30}$$

The collision number $Z_{12}(x)$ represents the total number of collisions in the time interval dt between 1 particles with reduced energy x and 2 particles. The explicit evaluation of $Z_{12}(x)$, which is presented in Appendix I, yields

$$Z_{12}(x) = \frac{C}{\lambda Q^2}\left[\left(2\lambda x^{\frac{1}{2}} + \frac{1}{\lambda x^{\frac{1}{2}}}\right)\operatorname{erf}(\lambda x^{\frac{1}{2}}) + \exp(-\lambda^2 x)\right]$$

$$= N_2 \pi (r_1+r_2)^2 \left(\frac{2kT_2}{\pi m_2}\right)^{\frac{1}{2}} \left\{\left[2\left(\frac{m_2 p_1^2}{2m_1^2 kT_2}\right)^{\frac{1}{2}} + \left(\frac{2m_1^2 kT_2}{m_2 p_1^2}\right)^{\frac{1}{2}}\right] \operatorname{erf}\left(\frac{m_2 p_1^2}{2m_1^2 kT_2}\right)^{\frac{1}{2}} + \exp\left(-\frac{m_2 p_1^2}{2m_1^2 kT_2}\right)\right\} \tag{3.31}$$

It can readily be verified that $Z_{12}(x)$ of Eq. (3.31) is identical with the expressions 5.4_5 and 5.4_6 given by Chapman and Cowling[8] for the probability per unit time of a collision between a particle of the "first type" moving with velocity p_1/m_1 and a particle of the "second type" in hard sphere collisions. This identity provides a convenient check on the validity of our calculations.

IV. DERIVATION OF THE FOKKER-PLANCK EQUATION

The master-equation in the form of the integral equation Eq. (3.28) does not appear amenable to an exact analytical solution. In this section we show how the master equation can be transformed through expansion in the mass ratios of the 1 and 2 particles into a partial differential equation of second order, i.e., a Fokker–Planck equation, which may be more readily subject to analytical solution and which, it is hoped, will provide a correct description of the time evolution of the distribution function $P(x, t)$. The procedure employed in the following derivation follows rather closely the method used by Keilson and Storer[9] and van Kampen.[10]

It should be stated here at the outset that in the attempt to pass from an integral equation (the master equation) to a second-order differential equation (the Fokker–Planck equation) one operates in a mathematical wilderness full of quicksand. What is involved here, as a broader and more general problem, is the representation of an integral operator by a differential operator. As far as the authors are aware, no general constructive theory exists which would permit one to state the conditions under which such a transformation is valid, its range of validity and the errors introduced. The specific case of the passage from a master equation to a Fokker–Planck equation has been studied by several authors[9–15] who have attempted to provide a systematic and valid prescription. These prescriptions are, however, in our view not free from objections as to lack of mathematical rigor and the resulting Fokker–Planck equations should be checked in each case, *a posteriori*, to verify whether they form a valid representation of the original integral equation.[16] Certainly, in the particular example to be considered here, we make no claims as to mathematical rigor; we only claim that the Fokker–Planck equation "derived" here appears to be a reasonable approximation to the master equation (3.28) under various restrictions to be discussed below.

We multiply Eq. (3.28) on both sides by an arbitrary, well-behaved function $g(x)$ defined in $0 \leq x < \infty$ and then integrate with respect to x over this interval

$$\int_0^\infty g(x)\frac{\partial P(x,t)}{\partial t}dx = \int_0^\infty \int_0^\infty g(x) A(x \mid x') P(x', t)\,dx'dx. \tag{4.1}$$

On the right-hand side we now replace $g(x)$ with its Taylor expansion around x'

$$g(x) = \sum_{n=0}^{\infty} \frac{(x-x')^n}{n!}\frac{d^n g(x')}{dx'^n} \tag{4.2}$$

to obtain

$$\int_0^\infty g(x)\frac{\partial P(x,t)}{\partial t}dx = \sum_{n=0}^{\infty} \int_0^\infty \frac{\partial^n g(x)}{\partial x^n} P(x,t) \frac{b_n(x)}{n!}dx \tag{4.3}$$

with

$$b_n(x) \equiv \int_0^\infty (x'-x)^n A(x' \mid x)\,dx'. \tag{4.4}$$

[8] S. Chapman and T. G. Cowling, *The Mathematical Theory of Non-Uniform Gases* (Cambridge University Press, New York, 1960).
[9] J. Keilson and J. E. Storer, Quart. Appl. Math. **10**, 248 (1952).
[10] N. G. van Kampen, Can. J. Phys. **39**, 551 (1961)
[11] H. A. Kramers, Physica **7**, 284 (1940).
[12] M. C. Wang and G. E. Uhlenbeck, Rev. Mod. Phys. **17**, 323 (1945).
[13] J. E. Moyal, J. Roy. Stat. Soc. (London) **B11**, 150 (1949).
[14] A. Siegel, J. Math. Phys. **1**, 378 (1960).
[15] M. Lax, Rev. Mod. Phys. **32**, 25 (1960).
[16] Only where the third- and higher-order moments (4.4) of the transition probability kernel $A(x \mid x')$ are *identically* equal to zero will the Fokker–Planck equation be *identically* equivalent to the master equation.

The $b_n(x)$ appearing in Eq. (4.3) are the nth order transition moments. They play a fundamental role in the derivation of the Fokker–Planck equation and we briefly digress at this point to discuss their properties.

It can readily be verified from Eq. (3.29) that $b_0(x)=0$. Exact expressions for $b_n(x)$ for $n=1$, 2 and 3 are presented in Appendix I; it is noted that the $b_n(x)$ are quite complicated functions of x. However, as is shown below, in the case of the Rayleigh gas $[\lambda=(m_2/m_1)^{\frac{1}{2}}\ll 1]$ and the Lorentz gas

$$[\lambda=(m_2/m_1)^{\frac{1}{2}}\gg 1]$$

there exist functions $\beta_n(x)$ which are either simple polynomials in x, or products of $x^{\frac{1}{2}}$ with polynomials in x, and which are very good approximations to $b_n(x)$ in the important part of the x-interval $0\leq x<\infty$. Furthermore we show that the moments $b_n(x)$ of order $n>2$ are smaller than $b_1(x)$ or $b_2(x)$ for all values of x by a factor of at least Q, with

$$Q=\tfrac{1}{2}[(m_1/m_2)^{\frac{1}{2}}+(m_2/m_1)^{\frac{1}{2}}]. \qquad (3.22)$$

Qualitatively, these properties of the moments can readily be understood from the "definition" of the transition probability kernel $A(x\mid x')dx$ as the probability per unit time that a 1 particle with a reduced energy x' will undergo, through a binary collision, a change in reduced energy x' to x. It then follows that the $b_n(x')$ are the nth-order moments around the origin of the distribution function of the "jumps" $(x'-x)$. In the case of the Rayleigh gas and the Lorentz gas, $A(x\mid x')$ considered as a function of the "jump" $x-x'$ is very peaked in a narrow region in the neighborhood of $x-x'=0$. This can readily be seen from Fig. 1 where $A(x\mid x')$ [from Eqs. (3.29) and (3.25)] has been plotted for the cases $\lambda^2=0.1$ (Rayleigh gas) and $\lambda^2=10$ (Lorentz gas) and compared with $A(x\mid x')$ for $\lambda^2=1.0$ corresponding to $m_1=m_2$. It can readily be verified that $Q\equiv\frac{1}{2}(\lambda+\lambda^{-1})$, is a measure of the sharpness of this peak of $A(x\mid x')$ around $x-x'=0$; i.e., as Q increases, the peak becomes sharper.[17] This behavior of the transition probability kernel corresponds to the fact that as the ratio λ^2 of the masses of the colliding partners becomes either very large or very small the amount of energy transferred in a collision becomes very small. Large changes in the energy of a subsystem particle can only be brought about by collisions with very energetic heat bath particles of which there are extremely few owing to the rapid decrease of the Maxwell distribution $f_2(\mathbf{p}_2')d\mathbf{p}_2'$ [Eq. (3.9)] as $\mathbf{p}_2'\to\infty$. Since the moments $b_n(x')$ are taken around the point $x-x'=0$ where $A(x\mid x')$ has a sharp peak, it is to be expected that their magnitude decreases with increasing order n.

[17] For very small x', $(0\leq x'<\frac{1}{2}Q)$, and very large x', $(2Q<x'<\infty)$, the plot of $A(x\mid x')$ versus $x-x'$ has a different shape. The strongly peaked curves are, however, characteristic for the whole physically important x' interval.

For the purpose of discussing the $b_n(x)$ quantitatively we divide up the x interval into three parts—I: $0\leq x<1/2Q$, II: $1/2Q\leq x\leq 2Q$, and III: $2Q\leq x<\infty$, and we show that in Interval II the simple functions $\beta_n(x)$ are good approximations to $Z_{12}(x)$ and $b_n(x)$ ($n=1$, 2, and 3) both for the Rayleigh gas and the Lorentz gas.

Rayleigh Gas

The Rayleigh gas corresponds to $\lambda=(m_2/m_1)^{\frac{1}{2}}\ll 1$. Therefore $2Q\simeq\lambda^{-1}$ and Interval II covers $\lambda\leq x\leq\lambda^{-1}$. This range of x spans the most important part of the energy interval $0\leq x<\infty$ since it excludes only particles with very small energy ($x<\lambda\ll 1$) and very high energy ($x>\lambda^{-1}\gg 1$) compared to the mean energy kT_2 of the heat-bath molecules.

In Interval II it follows from $\lambda x<1$ and $\lambda\ll 1$ that $\lambda^2x\ll 1$ and from $x/\lambda>1$ that $x/\lambda^2\gg 1$. Consequently we can introduce the Taylor expansion of the error function $\mathrm{erf}(\lambda x^{\frac{1}{2}})$ valid for $\lambda x^{\frac{1}{2}}\ll 1$ in the exact expressions (Appendix I) for $Z_{12}(x)$, $b_1(x)$, $b_2(x)$, and $b_3(x)$. We write

$$[\mathrm{erf}(\lambda x^{\frac{1}{2}})/x^{\frac{1}{2}}]=\lambda\exp(-\lambda^2x)[1+\tfrac{2}{3}\lambda^2x+\tfrac{4}{15}(\lambda^2x)^2+\tfrac{8}{105}(\lambda^2x)^3+\cdots] \qquad (4.5)$$

and then obtain

$$Z_{12}(x)=(2C/\lambda Q^2)\exp(-\lambda^2x)[1+O(\lambda^2x)], \qquad (4.6)$$

$$b_1(x)=(4C/3\lambda Q^4)\exp(-\lambda^2x)(\tfrac{3}{2}-x)[1+O(\lambda^2x)], \qquad (4.7)$$

$$b_2(x)=(2C/3\lambda^3Q^6)\exp(-\lambda^2x)(x+6\lambda^2)[1+O(\lambda^2x)]$$
$$\approx(2C/3\lambda^3Q^6)\exp(-\lambda^2x)[1+O(\lambda^2x)], \qquad (4.8)$$

$$b_3(x)=(C/\lambda^3Q^8)\times\exp(-\lambda^2x)(\tfrac{8}{5}x^2-4x-12\lambda^2)[1+O(\lambda^2x)]$$
$$\approx(4C/\lambda^3Q^8)\exp(-\lambda^2x)(\tfrac{2}{5}x^2-x)[1+O(\lambda^2x)]. \qquad (4.9)$$

Using $\exp(-\lambda^2x)=1+O(\lambda^2x)$, $\lambda^2Q^2\approx\frac{1}{4}$ and

$$C/\lambda Q^2=N_2\pi(r_1+r_2)^2(2kT_2/\pi m_2)^{\frac{1}{2}} \qquad (3.25a)$$

we finally obtain

$$Z_{12}\cong 2C/\lambda Q^2=2N_2\pi(r_1+r_2)^2(2kT_2/\pi m_2)^{\frac{1}{2}} \qquad (4.10)$$

$$b_n(x)=\beta_n(x)+O(\lambda^2x); \qquad n=1,2,3\cdots \qquad (4.11)$$

with

$$\beta_1(x)=\tfrac{8}{3}(m_2/m_1)(\tfrac{3}{2}-x)Z_{12} \qquad (4.12)$$

$$\beta_2(x)=\tfrac{16}{3}(m_2/m_1)xZ_{12} \qquad (4.13)$$

$$\beta_3(x)=128(m_2/m_1)(\tfrac{2}{5}x-1)Z_{12}\lambda^2x. \qquad (4.14)$$

From Eqs. (4.12–4.14) it follows that

$$\beta_3(x)/\beta_1(x)=O(\lambda^2x), \qquad \beta_3(x)/\beta_2(x)=O(\lambda^2x). \qquad (4.15)$$

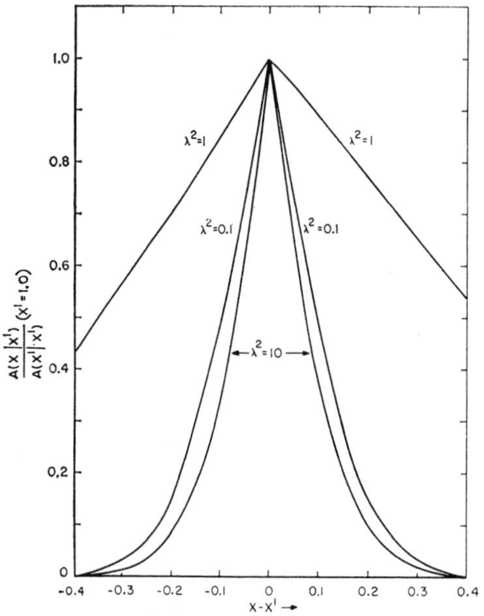

FIG. 1. The normalized transition probability kernel $[A(x\mid x')/A(x'\mid x')]$ as a function of the energy change per collision, $x-x'$. The curve for $\lambda^2=0.1$ corresponds to the Rayleigh gas, the curve for $\lambda^2=10$ corresponds to the Lorentz gas, the curve for $\lambda^2=1$ corresponds to $m_1=m_2$.

In general, from the definition of $b_n(x)$ and the form of $A(x\mid x')$ as displayed in Fig. 1, one can infer that the ratios $\beta_n(x)/\beta_1(x)$ and $\beta_n(x)/\beta_2(x)$ for $n>3$ are at least of $O(\lambda^2 x)$ or higher.[18] It should be noted that the collision number Z_{12} for the Rayleigh gas in the limit as $m_2/m_1\to 0$ as given by Eq. (4.10) is independent of the reduced energy x of the subsystem particles.

Lorentz Gas

In the case of the Lorentz gas we have $\lambda=(m_2/m_1)^{\frac12}\gg 1$. Therefore $2Q\simeq\lambda$ and Interval II extends over the energy range $\lambda^{-1}<x<\lambda$ which includes essentially all the subsystem particles. In this interval we find from $\lambda x>1$ that $\lambda^2 x\gg 1$ and from $x/\lambda<1$ that $x/\lambda^2\ll 1$. Substitution of the asymptotic expansion of the error function, valid for arguments $\lambda x^{\frac12}\gg 1$,

$$\operatorname{erf}(\lambda x^{\frac12})=\frac{\pi^{\frac12}}{2}-\frac{\exp(-\lambda^2 x)}{2\lambda x^{\frac12}}\left(1-\frac{1}{2}\frac{1}{\lambda^2 x}+\frac{3}{4}\frac{1}{(\lambda^2 x)^2}-\cdots\right) \tag{4.16}$$

into the exact expressions for $Z_{12}(x)$ and $b_n(x)$ (Appendix I), then leads to

$$Z_{12}(x)=\frac{C}{Q^2}\left\{(\pi x)^{\frac12}\left(1+\frac{1}{2\lambda^2 x}\right)+\frac{1}{\lambda}\exp(-\lambda^2 x)\,O\left[\frac{1}{(\lambda^2 x)^2}\right]\right\}, \tag{4.17}$$

$$b_1(x)=\frac{C}{Q^4}\left\{\frac{(\pi x)^{\frac12}}{2}(2-x)\left[1+O\left(\frac{1}{\lambda^2 x}\right)\right]+\frac{1}{\lambda}\exp(-\lambda^2 x)\,O\left[\frac{1}{(\lambda^2 x)^2}\right]\right\}, \tag{4.18}$$

$$b_2(x)=\frac{C\lambda^2}{Q^6}\left\{\frac{(\pi x)^{\frac12}}{4}x\left(1+\frac{4}{3}\frac{x}{\lambda^2}\right)\left[1+O\left(\frac{1}{\lambda^2 x}\right)\right]+\frac{1}{\lambda^3}\exp(-\lambda^2 x)\,O\left[\frac{1}{(\lambda^2 x)^2}\right]\right\}, \tag{4.19}$$

$$b_3(x)=\frac{C\lambda^2}{Q^8}\left\{\frac{(\pi x)^{\frac12}}{2}\left(\frac{1}{2}\frac{x^3}{\lambda^2}+x^2-3x\right)\left[1+O\left(\frac{1}{\lambda^2 x}\right)\right]+\frac{1}{\lambda^3}\exp(-\lambda^2 x)\,O\left[\frac{1}{(\lambda^2 x)^2}\right]\right\}. \tag{4.20}$$

From $\exp(-\lambda^2 x)\ll 1$ and $\lambda^2/Q^2\approx 4$ we then find

$$Z_{12}(x)\simeq\frac{C\pi^{\frac12}}{Q^2}(x)^{\frac12}=N_2\pi(r_1+r_2)^2\left(\frac{2kT_2 x}{m_1}\right)^{\frac12}$$

$$=N_2\pi(r_1+r_2)^2\frac{p_1}{m_1}, \tag{4.21}$$

$$b_n(x)=\beta_n(x)+O(1/\lambda^2 x);\qquad n=1,2,3 \tag{4.22}$$

with

$$\beta_1(x)=2(m_1/m_2)(2-x)Z_{12}(x), \tag{4.23}$$

$$\beta_2(x)=4(m_1/m_2)xZ_{12}(x), \tag{4.24}$$

$$\beta_3(x)=32(m_1/m_2)(x-3)Z_{12}(x)(x/\lambda^2). \tag{4.25}$$

From Eqs. (4.22)–(4.24) it follows that

$$\frac{\beta_3(x)}{\beta_1(x)}=O\left(\frac{x}{\lambda^2}\right);\qquad\frac{\beta_3(x)}{\beta_2(x)}=O\left(\frac{x}{\lambda^2}\right). \tag{4.26}$$

In general, from the definition of $b_n(x)$ and from the form of $A(x\mid x')$ as displayed in Fig. 1 one can infer that the ratios $\beta_n(x)/\beta_1(x)$ and $\beta_n(x)/\beta_2(x)$ for $n>3$ are at least of $O(x/\lambda^2)$ or higher.

It should be noted that $Z_{12}(x)$ for the hard-sphere Lorentz gas, in contrast to the situation in the Rayleigh gas, is an explicit function of the reduced energy x, being proportioned to $x^{\frac12}$, i.e., proportional to the speed v. Physically, this difference can readily be understood. In the case of the Rayleigh gas (heavy subsystem particles–light heat-bath particles), the speed of the heavy particles is small, on the average, compared to that of the light particles. The number of collisions per subsystem particle per unit time is thus

[18] For the one-dimensional Rayleigh gas, van Kampen[10] has shown, using a method similar to the one employed here, that the ratios $b_n(x)/b_1(x)$, $b_2(x)$ for $n\geq 3$ are at least of order $m_2/(m_1+m_2)\simeq O(\lambda^2)$.

to a good approximation (and exactly in the limit $m_2/m_1 \to 0$) independent of the heavy particle speed. In the case of the Lorentz gas (light subsystem particles–heavy heat-bath particles) the situation is reversed with the speed of the subsystem particles being, on the average, much greater than that of the heat-bath particles; in the limit $m_2/m_1 \to \infty$, the heat-bath particles can be considered to be stationary. The collision number $Z_{12}(x)$ for the hard-sphere Lorentz gas is therefore to a good approximation (and exactly in the above limit) proportional to the speed of the light particles. As is seen in Sec. VI, this dependence of $Z_{12}(x)$ on x gives rise to a Fokker–Planck equation for the Lorentz gas which, contrary to the one for Rayleigh gas, is not amenable to an analytical solution in terms of standard orthogonal functions.

As a check on the value of $Z_{12}(x)$ in Eq. (4.21) it may be noted that the mean free path l_1 for a light, hard sphere moving in a gas of heavy hard spheres is given, in terms of $Z_{12}(x)$, by the well-known expression[19]

$$l_1 = \frac{(p_1/m_1)}{Z_{12}(x)} = \frac{1}{N_2\pi(r_1+r_2)^2}. \tag{4.27}$$

We now return to the consideration of Eq. (4.3). This equation can be greatly simplified if the polynomials $\beta_n(x)$ are substituted for the moments $b_n(x)$ on the right-hand side of the equation. We have shown above that the error introduced by replacing $b_n(x)$ by $\beta_n(x)$ in Region II is at most of order $(1/Q)$ and thus well defined and small. We cannot make such a quantitative statement about the errors in Regions I and III. A qualitative analyses of terms on the right-hand side of Eq. (4.3), i.e.,

$$\int_0^\infty \frac{\partial g^n(x)}{\partial x^n} P(x,t) \frac{b_n(x)}{n!} dx,$$

leads to the conclusion that the contributions of Regions I and III to these integrals for all values of t are very small compared to the contribution of Region II provided that *the initial subsystem distribution $P(x, o)$ is not a δ-function distribution in either Region I or III*. Thus, even though we are not certain about the deviation of $\beta_n(x)$ from $b_n(x)$ in Regions I and III we can conclude, to a good approximation, that the error introduced in extending the range of integration of x from zero to infinity as $b_n(x)$ is replaced by $\beta_n(x)$ will be small.

Equation (4.3) can now be rewritten with the $\beta_n(x)$ as

$$\int_0^\infty g(x)\frac{\partial P(x,t)}{\partial t}dx = \sum_{n=1}^{\infty}\int_0^\infty \frac{d^n g(x)}{dx^n} P(x,t)\frac{\beta_n(x)}{n!}dx \tag{4.28}$$

with the term corresponding to $n=0$ equal to zero since $b_0=0$. We now investigate the dependence of the factors on the right-hand side of Eq. (4.28) on the "expansion parameter" Q. From the results contained in Eqs. (4.15) and (4.26) and the subsequent discussion, we know that $\beta_3(x) \simeq O(1/Q)\{\beta_1(x), \beta_2(x)\}$ and at least of the same or higher order in $(1/Q)$ for $n>3$. We have assumed [Eq. (4.2)] that $g(x)$ can be represented by its Taylor series expansion around x'. This places a limit on the possible increase of $d^n g(x)/dx^n$ with n. Since $g(x)$, being an arbitrary function, will be independent of Q, it is certainly possible to choose a value of $Q \gg 1$ such that

$$\left.\frac{\beta_n(x,Q)}{n!}\frac{d^n g(x)}{dx^n}\right|_{n\geq 3} \ll \left.\frac{\beta_n(x,Q)}{n!}\frac{d^n g(x)}{dx^n}\right|_{n=1,2}.$$

Under the above listed conditions and for $Q \gg 1$, all terms on the right-hand side of Eq. (4.28) corresponding to $n \geq 3$ can be neglected to within a good approximation, and one obtains

$$\int_0^\infty g(x)\frac{\partial P(x,t)}{\partial t}dx = \int_0^\infty \frac{dg(x)}{dx}\beta_1(x)P(x,t)dx + \frac{1}{2}\int_0^\infty \frac{d^2g(x)}{dx^2}\beta_2(x)P(x,t)dx. \tag{4.29}$$

In the limit as $Q\to\infty$, i.e., as $m_2/m_1 \to 0$, the above equation becomes exact. This is also the result obtained by Wang-Chang and Uhlenbeck.[20]

The two terms on the right-hand side of Eq. (4.29) are now integrated by parts one and two times, respectively, to yield

$$\int_0^\infty g(x)\frac{\partial P(x,t)}{\partial t}dx = \Big[g(x)\beta_1(x)P(x,t)\Big]_0^\infty + \frac{1}{2}\left[\frac{dg(x)}{dx}\beta_2(x)P(x,t)\right]_0^\infty - \frac{1}{2}\left[g(x)\frac{\partial}{\partial x}(\beta_2(x)P(x,t))\right]_0^\infty$$
$$-\int_0^\infty g(x)\frac{\partial}{\partial x}(\beta_1(x)P(x,t))dx + \frac{1}{2}\int_0^\infty g(x)\frac{\partial^2}{\partial x^2}(\beta_2(x)P(x,t))dx. \tag{4.30}$$

[19] See, e.g., the discussion in Ref. 8, p. 190.

[20] C. S. Wang Chang and G. E. Uhlenbeck, *The Kinetic Theory of a Gas in Alternating Outside Force Fields* (Engineering Research Institute, University of Michigan, 1956) Tech. Rept. 2457–3–T.

Since both $g(x)$ and $P(x, t)$ must be well behaved it is clear that the upper limits in each of the three square brackets vanish. The lower limits in the first two brackets will also give a zero contribution since $P(0, t)=0$ at all times t. [We have already excluded initial distribution functions of the type $\delta(x-0)$.] As far as the lower limit in the bracket

$$\tfrac{1}{2}[g(x)(\partial/\partial x)(\beta_2(x)P(x,t))]_0^{\infty}$$

is concerned, we know that $\beta_2(x)\propto x$ for the Rayleigh gas and $\beta_2\propto x^{3/2}$ for the Lorentz gas. This term will therefore give a zero contribution at the lower limit provided that either (i) $\partial P(x,t)/\partial x$ exists for $x=0$ or (ii) if $\partial P(x,t)/\partial x$ diverges for $x\rightarrow 0$, it will increase slower than x^{-1} as $x\rightarrow 0$. We shall assume that these conditions are met.[21] Equation (4.30) then reduces to

$$\int_0^{\infty} g(x)\frac{\partial P(x,t)}{\partial t}dx=-\int_0^{\infty} g(x)\frac{\partial}{\partial x}(\beta_1(x)P(x,t))dx$$
$$+\frac{1}{2}\int_0^{\infty} g(x)\frac{\partial^2}{\partial x^2}(\beta_2(x)P(x,t))dx. \quad (4.31)$$

Finally we make use of the fact that Eq. (4.31) must hold for any well behaved function $g(x)$; this condition implies that

$$\frac{\partial P(x,t)}{\partial t}=\frac{\partial}{\partial x}\left[-\beta_1(x)P(x,t)+\frac{1}{2}\frac{\partial}{\partial x}(\beta_2(x)P(x,t))\right]. \quad (4.32)$$

Equation (4.32) is the desired Fokker–Planck equation which is to serve as the differential approximation to the integral master Eq. (3.28).

From the general Fokker–Planck equation (4.32) we can now obtain the F–P equation for the Rayleigh gas and the Lorentz gas, respectively. For the Rayleigh gas we introduce the coefficient

$$k_R=\frac{8}{3}\frac{m_2}{m_1}Z_{12}=\frac{16}{3}\frac{m_2}{m_1}N_2\pi(r_1+r_2)^2\left(\frac{2kT_2}{\pi m_2}\right)^{\frac{1}{2}}. \quad (4.33)$$

Equation (4.32) then takes the form [see Eqs. (4.10) to (4.13)]:

$$\frac{\partial P(x,t)}{\partial t}=k_R\frac{\partial}{\partial x}\left[(x-\tfrac{3}{2})P(x,t)+\frac{\partial}{\partial x}(xP(x,t))\right]. \quad (4.34)$$

If in place of the reduced energy $x=\epsilon/kT_2$ one uses the translational energy $\epsilon=p_1^2/2m_1$ as the independent variable, one obtains the F–P equation

$$\frac{\partial P(\epsilon,t)}{\partial t}=k_R\frac{\partial}{\partial\epsilon}\left[(\epsilon-\tfrac{3}{2}kT_2)P(\epsilon,t)+kT_2\frac{\partial}{\partial\epsilon}(\epsilon P(\epsilon,t))\right]. \quad (4.35)$$

By arguments and techniques different from those employed here, a number of authors[20,22,23] have obtained Fokker–Planck equations for the time dependent distribution function $f_1(\mathbf{p}_1, t)$ of the momentum vector for Brownian motion problems and for the Rayleigh gas. Our Eq. (4.35) can easily be shown to be identical with their equations subject to our model in which $f_1(\mathbf{p}_1, t)$ depends only upon the magnitude p_1 of $\mathbf{p}_1$. The coefficient k_R in Eq. (4.33) is identical with the corresponding factor obtained by Green.[22]

For the Lorentz gas we introduce the coefficient

$$k_L=2(m_1/m_2)^{\frac{1}{2}}N_2\pi(r_1+r_2)^2(2kT_2/m_2)^{\frac{1}{2}}. \quad (4.36)$$

Equation (4.32) then takes the form [see Eqs. (4.21) to (4.24)]

$$\frac{\partial P(x,t)}{\partial t}=k_L\frac{\partial}{\partial x}\left[(x^{3/2}-2x^{1/2})P(x,t)+\frac{\partial}{\partial x}(x^{3/2}P(x,t))\right]. \quad (4.37)$$

If we again set $x=\epsilon/kT_2$ with $\epsilon=p_1^2/2m_1$ we obtain

$$\frac{\partial P(\epsilon,t)}{\partial t}=k_L(kT_2)^{-\frac{1}{2}}\frac{\partial}{\partial\epsilon}\left[(\epsilon^{3/2}-2kT_2\epsilon^{1/2})P(\epsilon,t)+kT_2\frac{\partial}{\partial\epsilon}(\epsilon^{3/2}P(\epsilon,t))\right]. \quad (4.38)$$

Fokker–Planck equations in "energy space" for the Lorentz gas can be found in papers by W. P. Allis,[24] and J. L. Delcroix and his co-workers.[25] The method of derivation employed in these papers is again different from ours. The equation obtained by Allis is identical with Eq. (4.38). Delcroix and his co-workers in their discussion of the relaxation of a Lorentz gas start from the equation [Ref. 8, pp. 346–350]

$$\frac{\partial f(v,t)}{\partial t}=\frac{1}{v^2}\frac{\partial}{\partial v}\left[\frac{kT_2}{m_2}\frac{v^3}{l_1(v)}\frac{\partial f(v,t)}{\partial v}+\frac{m_1}{m_2}\frac{v^4}{l_1(v)}f(v,t)\right] \quad (4.39)$$

with $v=p_1/m_1$ and where $l_1(v)$ is the mean free path of the light particle. If the hard-sphere value for l_1 [Eq. (4.27)] is substituted in the above expression, Eqs. (4.38) and (4.39) can readily be shown to be equivalent.

V. SOLUTION OF THE FOKKER–PLANCK EQUATION FOR THE RAYLEIGH GAS

In this section we present the solution of the Rayleigh gas F–P equation (4.34)

$$\frac{\partial P(x,t)}{\partial t}=k_R\frac{\partial}{\partial x}\left\{(x-3/2)P(x,t)+\frac{\partial}{\partial x}[xP(x,t)]\right\}.$$

First, we discuss the relaxation of the mean reduced

[21] For a Maxwell distribution, for example, $\partial P(x,t)/\partial x$ increases as $x^{-\frac{1}{2}}$ as $x\rightarrow 0$.

[22] M. S. Green, J. Chem. Phys. **19**, 1036 (1951).
[23] P. Mazur, Physica **25**, 149 (1959).
[24] W. P. Allis, *Handbuch der Physik*, edited by S. Flügge, (Springer-Verlag, Berlin, 1956), Vol. 21, p. 383 ff.
[25] M. Bayet, J. L. Delcroix, J. F. Denisse, J. Phys. Radium **15**, 795 (1954); **16**, 274 (1955); **17**, 923, 1005 (1956).

energy $\bar{x}(t)$ defined by

$$\bar{x}(t)=\int_0^\infty xP(x,t)dx. \tag{5.1}$$

On the basis of previously derived results[26] on the functional form of the moments $b_n(x)$ we can state immediately that the relaxation of the mean energy $\bar{x}$ of the subsystem particles obeys the exponential law

$$\frac{d\bar{x}(t)}{dt}=-k_R[\bar{x}(t)-3/2]=-k_R[\bar{x}(t)-\bar{x}(\infty)] \tag{5.2}$$

so that

$$[\bar{x}(t)-\bar{x}(\infty)]/[\bar{x}(0)-\bar{x}(\infty)]=\exp(-k_Rt) \tag{5.3}$$

independent of the form of the initial distribution $P(x,0)$. Note that $\bar{x}(\infty)$, the reduced energy at time $t=\infty$, equals 3/2 as can readily be verified from Eqs. (3.26) and (5.1). The above result follows directly from the form of the coefficient $\beta_1(x)$ and $\beta_2(x)$, i.e.,

$$\beta_1(x)=k_R(3/2-x), \tag{4.12}$$

$$\beta_2(x)=2k_Rx, \tag{4.13}$$

which satisfy the necessary and sufficient conditions given in Ref. 26 for the exponential relaxation of the first moment of a distribution function obeying a F–P equation. The relaxation time $t_r=k_R^{-1}$ is found to be [see Eq. (4.33)]

$$t_r=\tfrac{3}{8}(m_1/m_2)(1/Z_{12}), \tag{5.4}$$

and is thus proportional to the product of the mass ratio of the heavy subsystem particle to the light heat-bath particle and the average time, Z_{12}^{-1}, between collisions of Particles 1 and 2. The temperature dependence of t_r is of the form $t_r\propto(T_2)^{-\frac{1}{2}}$. For a heat-bath temperature $T_2=300°\mathrm{K}$ at atmospheric pressure and for a mass ratio $m_1/m_2=10^2$, the relaxation time t_r is of the order of 10^{-8} sec. In general, the number of collisions required to reduce the mean energy $\bar{x}(t)$ to e^{-1} of its initial value are of the order $\lambda^{-2}=m_1/m_2$. It must be noted that this result is valid only for $\lambda\ll 1$.

Equation (4.34) is a typical example of a F–P equation describing a Brownian motion random walk. Mazur in discussing the Brownian motion of a free particle in two-dimensional energy space obtained an analogous result [Ref. 23, Eq. (36)] as did Rubin and Shuler[27] in their treatment of the relaxation of an ensemble of harmonic oscillators in weak interaction with a heat bath.[28]

To solve Eq. (4.34), i.e., to obtain an explicit expression for the distribution function $P(x,t)$, we set $k_Rt=\tau$ and $P(x,t)=x^{\frac{1}{2}}e^{-x}\psi(x,\tau)$. We then find for $\psi(x,\tau)$:

$$\frac{\partial\psi(x,\tau)}{\partial\tau}=x\frac{\partial^2\psi(x,\tau)}{\partial x^2}+(\tfrac{3}{2}-x)\frac{\partial\psi(x,\tau)}{\partial x}. \tag{5.5}$$

The right-hand side of Eq. (5.5) defines the operator

$$\mathfrak{L}\equiv x(d^2/dx^2)+(\tfrac{3}{2}-x)(d/dx). \tag{5.6}$$

We make use of the eigenfunctions $\phi_\nu(x)$ of this operator given by

$$\mathfrak{L}\phi_\nu(x)=\gamma_\nu\phi_\nu(x)$$

$$x[d^2\phi_\nu(x)/dx^2]+(\tfrac{3}{2}-x)[d\phi_\nu(x)/dx]-\gamma_\nu\phi_\nu(x)=0 \tag{5.7}$$

to effect the solution. Equation (5.7) is satisfied by the confluent hypergeometric function

$$\phi_\nu(x)={}_1F_1(\gamma_\nu,\tfrac{3}{2};x). \tag{5.8}$$

Since we are going to expand $\psi(x,\tau)$ in the set of eigenfunctions $\phi_\nu(x)$ with $P(x,t)$ written as an infinite sum of functions of the type $\Omega_\nu(\tau)x^{\frac{1}{2}}e^{-x}\phi_\nu(x)$, we require that none of the functions $\phi_\nu(x)$ behave like e^x for large x in order for the integral of $\Omega_\nu(\tau)x^{\frac{1}{2}}e^{-x}\phi_\nu(x)$ from zero to infinity to remain finite. This implies that the infinite series defined by ${}_1F_1(\gamma_\nu,\frac{3}{2};x)$ must be broken off after a finite number of terms in order to obtain a physically meaningful solution. We therefore set $\gamma_\nu=-\nu$, with ν being integer. The confluent hypergeometric functions can be written in terms of generalized Laguerre polynomials as

$$L_\nu^{\frac{1}{2}}(x)=\frac{\Gamma(\nu+\frac{3}{2})}{\Gamma(\frac{3}{2})\cdot\Gamma(\nu+1)}\,{}_1F_1(-\nu,\tfrac{3}{2};x). \tag{5.9}$$

The Laguerre polynomials $L_\nu^{\frac{1}{2}}(x)$ are known to form a complete orthogonal set with the orthogonality relation

$$\int_0^\infty x^{\frac{1}{2}}e^{-x}L_\nu^{(\frac{1}{2})}(x)L_\mu^{(\frac{1}{2})}(x)dx=\frac{\Gamma(\nu+\frac{3}{2})}{\Gamma(\nu+1)}\delta_{\nu\mu} \tag{5.10}$$

and $\psi(x,\tau)$ can consequently be expanded in the set $L_\nu^{(\frac{1}{2})}(x)$ as

$$\psi(x,\tau)=\sum_{\nu=0}^\infty c_\nu\Omega_\nu(\tau)L_\nu^{(\frac{1}{2})}(x). \tag{5.11}$$

Substitution of this expansion into Eq. (5.5) and the use of (5.7) and (5.10) leads to

$$d\Omega(\tau)/d\tau=-\nu\Omega;\qquad \Omega(\tau)=e^{-\nu\tau} \tag{5.12}$$

for the time dependent factors $\Omega_\nu(\tau)$. The solution of the F–P Eq. (4.34) can thus be written in terms of the generalized Laguerre polynomials $L_\nu^{(\frac{1}{2})}(x)$ as

$$P(x,\tau)=x^{\frac{1}{2}}e^{-x}\sum_{\nu=0}^\infty c_\nu L_\nu^{(\frac{1}{2})}(x)e^{-\nu\tau} \tag{5.13}$$

[26] K. E. Shuler, G. H. Weiss, and K. Andersen, J. Math. Phys. 3, 550 (1962).

[27] R. J. Rubin and K. E. Shuler, J. Chem. Phys. 25, 59 (1956).

[28] In the work of Mazur and of Rubin and Shuler, the constant factor in $\beta_1(x)$ is 1 (or $kT\equiv 1$ in reduced coordinates) rather than $\frac{3}{2}$ as in Eq. (4.34). The factor $\frac{3}{2}$ reflects the three dimensional coordinate space used in this treatment.

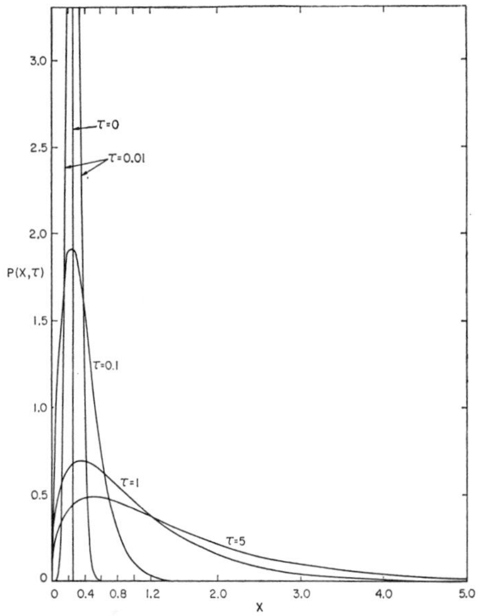

FIG. 2. The relaxation of an initial δ function distribution of reduced energy x with $P(x,0)=\delta(x_0-\frac{1}{4})$ for the Rayleigh gas. For $\tau=5$, the distribution function $P(x,\tau)$ coincides essentially with the equilibrium Maxwellian distribution $P(x,\infty)$.

with the coefficients c_ν determined by the initial distribution $P(x,0)$ through the use of the orthogonality relation (5.10) as

$$c_\nu=\frac{\Gamma(\nu+1)}{\Gamma(\nu+\frac{3}{2})}\int_0^\infty P(x,0)\,L_\nu^{(\frac{1}{2})}(x)\,dx. \qquad (5.14)$$

Relaxation of an Initial δ-Function Distribution

From Eqs. (5.13) and (5.14) we now derive an explicit closed expression for $P(x,\tau)$ for the relaxation of an initial δ-function distribution in the reduced energy x. In Eq. (5.14) we set $P(x,0)=\delta(x-x_0)$ where we specify that $\lambda<x_0<\lambda^{-1}$ (see the discussion in Sec. IV). For c_ν we obtain

$$c_\nu=[\Gamma(\nu+1)/\Gamma(\nu+\tfrac{3}{2})]L_\nu^{(\frac{1}{2})}(x_0) \qquad (5.15)$$

and for $P(x,t)$

$$P(x,\tau)=x^{\frac{1}{2}}e^{-x}\sum_{\nu=0}^{\infty}\frac{\Gamma(\nu+1)}{\Gamma(\nu+\frac{3}{2})}L_\nu^{(\frac{1}{2})}(x_0)\,L_\nu^{(\frac{1}{2})}(x)\,e^{-\nu\tau}. \qquad (5.16)$$

The summation indicated in (5.16) can be carried out explicitly[29] to yield

$$P(x,\tau)=\frac{1}{1-e^{-\tau}}\exp\left[-\frac{(x+x_0e^{-\tau})}{(1-e^{-\tau})}\right]\left(\frac{x}{x_0}\right)^{\frac{1}{4}}$$

$$\times e^{\tau/4}I_{1/2}\left[2\frac{(xx_0e^{-\tau})^{\frac{1}{2}}}{(1-e^{-\tau})}\right], \qquad (5.17)$$

[29] Bateman Manuscript Project, *Higher Transcendental Functions* (McGraw-Hill Book Company, Inc., New York, 1953), Vol. 2.

where $I_{\frac{1}{2}}$ is the modified Bessel function of order $\frac{1}{2}$. Since

$$I_{\frac{1}{2}}(z)=(2/\pi z)^{\frac{1}{2}}\sinh z=(2\pi z)^{-\frac{1}{2}}(e^z-e^{-z}), \qquad (5.18)$$

Eq. (5.17) reduces to

$$P(x,\tau)=\frac{e^{\tau/2}}{2[\pi x_0(1-e^{-\tau})]^{\frac{1}{2}}}\left\{\exp\left[-\frac{(x^{\frac{1}{2}}-(x_0e^{-\tau})^{\frac{1}{2}})^2}{1-e^{-\tau}}\right]-\exp\left[-\frac{(x^{\frac{1}{2}}+(x_0e^{-\tau})^{\frac{1}{2}})^2}{1-e^{-\tau}}\right]\right\}. \qquad (5.19)$$

It can readily be verified that $P(x,\tau)$ is normalized for all values of τ, that it approaches the initial δ-function distribution in the limit of $\tau\rightarrow0$ and the Maxwell distribution (3.26) as $\tau\rightarrow\infty$. The distribution function $P(x,\tau)$ of Eq. (5.19) has been plotted as a function of τ in Figs. 2 and 3 for $x_0=\frac{1}{4}$ and $x_0=4$, corresponding to $p_1^2(t=0)/2m_1=\frac{1}{4}kT_2$ and $p_1^2(t=0)/2m_1=4kT_2$.

For $\tau=5$, the value of $P(x,\tau)$ of Eq. (5.19) essentially coincides with the Maxwellian equilibrium value, $P(x,\infty)=2(x/\pi)^{\frac{1}{2}}\exp(-x)$, for both of the initial δ functions plotted in Figs. 2 and 3. From the definition of k_R in Eq. (4.33) it then follows that the number of collisions, tZ_{12}, between 1 particles (subsystem) and 2 particles (heat bath) required for equilibration of the initial δ-function distribution is given by the ratio $\frac{15}{8}(m_1/m_2)\simeq2(m_1/m_2)$. This is certainly of the proper functional form since one would expect an increase in the number of collisions required for the equilibration of the heavy-particle subsystem

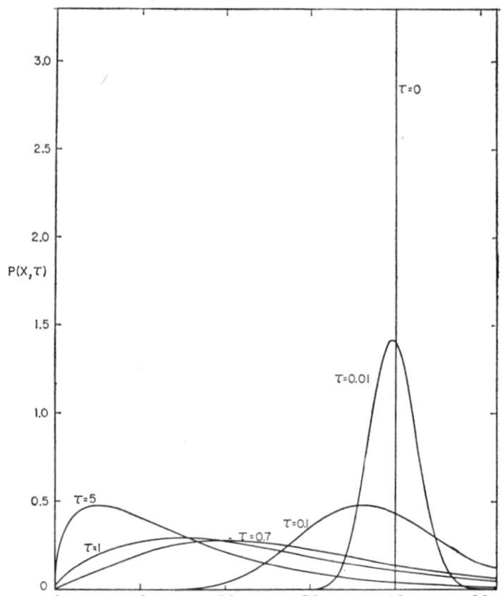

FIG. 3. The relaxation of an initial δ function distribution of reduced energy x with $P(x,0)=\delta(x_0-4)$ for the Rayleigh gas. For $\tau=5$, the distribution function $P(x,\tau)$ coincides essentially with the equilibrium Maxwellian distribution $P(x,\infty)$.

in collisional interaction with the light particle heat bath with an increase in the value of the mass ratio m_1/m_2. For a mass ratio $m_1/m_2=100$, about 200 collisions of 1 particles with 2 particles will thus be required for the equilibration of the initial δ-function distribution. This compares with about 4 collisions computed by Alder and Wainwright[30a] for the equilibration of an initial δ-function distribution of velocities in a hard-sphere gas (of identical atoms) with $m_1=m_2$. It would thus appear, at least for $m_1 \geq m_2$, that the number of collisions required for equilibration of an initial δ-function distribution of energy or velocity is proportional to the mass ratio m_1/m_2.[30b]

Relaxation of an Initial Maxwell Distribution

Next we study a Rayleigh gas undergoing relaxation from an initial Maxwell energy distribution with the temperature $T_1(0)$ of the subsystem of heavy particles different from the (time invariant) temperature T_2 of the heat bath. The normalized equilibrium distribution function $P(x, \infty)$ is given by

$$P(x, \infty)dx = 2(x/\pi)^{\frac{1}{2}}e^{-x}dx. \quad (5.20)$$

Introducing the ratio $\alpha_0 = (T_1(0)/T_2)$ we can write the normalized Maxwell distribution function of the subsystem at the temperature $T_1(0)$ as

$$P(x, 0)dx = 2(x/\pi)^{\frac{1}{2}}\alpha_0^{-3/2}\exp(-x/\alpha_0)dx. \quad (5.21)$$

From Eq. (5.14) we find for c_ν

$$c_\nu = \frac{\Gamma(\nu+1)}{\Gamma(\nu+\frac{3}{2})\Gamma(\frac{3}{2})}\alpha_0^{-3/2}\int_0^\infty x^{\frac{1}{2}}\exp(-x/\alpha_0)L_\nu^{(\frac{1}{2})}(x)dx. \quad (5.22)$$

The integral is evaluated by writing $L_\nu^{(\frac{1}{2})}(x)$ as[29]

$$L_\nu^{(\frac{1}{2})}(x) = \frac{x^{-\frac{1}{2}}}{\Gamma(\nu+1)}e^x\frac{\partial^\nu}{\partial x^\nu}(x^{\nu+\frac{1}{2}}e^{-x}) \quad (5.23)$$

and then integrating by parts. This yields

$$c_\nu = [1/\Gamma(\tfrac{3}{2})](-1)^\nu(\alpha_0-1)^\nu. \quad (5.24)$$

For $P(x, t)$ we then obtain from Eq. (5.13)

$$P(x, t) = 2(x/\pi)^{\frac{1}{2}}e^{-x}\sum_{\nu=0}^{\infty}[(1-\alpha_0)e^{-\tau}]^\nu L_\nu^{(\frac{1}{2})}(x). \quad (5.25)$$

The summation in Eq. (5.25) can again be carried out in closed form[29] to yield[31]

$$P(x, t) = 2(x/\pi)^{\frac{1}{2}}e^{-x}\frac{1}{[1+(\alpha_0-1)e^{-\tau}]^{3/2}} \times \exp\left[\frac{x(\alpha_0-1)e^{-\tau}}{1+(\alpha_0-1)e^{-\tau}}\right]. \quad (5.26)$$

If we now define the variable $\alpha(\tau)$ by

$$\alpha(\tau) \equiv 1+(\alpha_0-1)e^{-\tau}, \quad (5.27)$$

we can rewrite Eq. (5.26) as

$$P(x, \tau) = 2(x/\pi)^{\frac{1}{2}}\alpha^{-3/2}\exp(-x/\alpha). \quad (5.28)$$

From a comparison of Eq. (5.28) with Eq. (5.21) it is evident that Eq. (5.28) represents a Maxwell distribution of reduced energies x with

$$\alpha(\tau) = T_1(\tau)/T_2. \quad (5.29)$$

Combining Eq. (5.29) with Eq. (5.27) and recalling the definition $\alpha_0 = T_1(0)/T_2$, we obtain

$$[T_1(\tau)-T_2]/[T_1(0)-T_2] = e^{-\tau} \quad (5.30)$$

for the relaxation of the temperature $T_1(\tau)$ of the subsystem of heavy particles.

We have thus been led to the interesting result [Eq. (5.28)] that a subsystem of heavy hard spheres with an initial Maxwell distribution of kinetic energy [corresponding to a temperature $T_1(0)$] in contact with a heat bath of light particles with a Maxwellian distribution of energy [at a temperature $T_2 \neq T_1(o)$] will relax to its equilibrium Maxwellian distribution [with a temperature $T_1(\infty) = T_2$] via a sequence of Maxwellian distributions. Since all the intermediate distributions are Maxwell distributions, they can be characterized exactly by a (translational) temperature $T_1(\tau)$, whose time dependence is given by Eq. (5.30). It may be pointed out that this result is quite analogous to the result of Montroll and Shuler[32] on the vibrational relaxation of a subsystem of harmonic oscillators in weak interaction with a heat bath. In Figs. 4 and 5 we have plotted $P(x, \tau)$ vs x for various values of τ for initial Maxwell distributions with $\alpha_0 \equiv T_1(0)/T_2 = \frac{1}{6}$ and $\alpha_0 \equiv T_1(0)/T_2 = \frac{8}{3}$. These initial distributions correspond, respectively, to distributions with initial mean reduced energies $\bar{x}(0) = \frac{1}{4}$ and $\bar{x}(0) = 4$.

It will be noted that Eqs. (5.3) and (5.30) are of identical form. This is as should be and provides a check for the development presented in this section. For a Maxwellian distribution, $\bar{x}(\tau) \equiv \bar{\epsilon}(\tau)/kT_2 =$

---

[30] (a) B. J. Alder and T. Wainwright in *Transport Processes in Statistical Mechanics*, edited by I. Prigogine (Interscience Publishers Inc., New York, 1958); (b) The solution (5.19) for the F–P Eq. (4.34) for an initial δ-function distribution has also been obtained recently by F. H. Ree and R. E. Kidder, Phys. Fluids **6**, 857 (1963) in connection with their discussion of the thermalization of a fast ion in a plasma. Their condition (47) on the validity of the analytical results is equivalent to our condition $\lambda < x_0 < \lambda^{-1}$. On the basis of their model, the thermalization of an ion in a plasma is an example of a Rayleigh gas relaxation process.

[31] It should be pointed out that the series in Eq. (5.25) converges only for $(1-\alpha_0)e^{-\tau} > -1$. However, since the solution obtained [Eq. (5.26)] satisfies both the partial differential equation [Eq. 4.34)] and the initial condition [Eq. (5.21)], the solution (5.26) is valid for all values of α.

[32] E. W. Montroll and K. E. Shuler, J. Chem. Phys. **26**, 454 (1957).

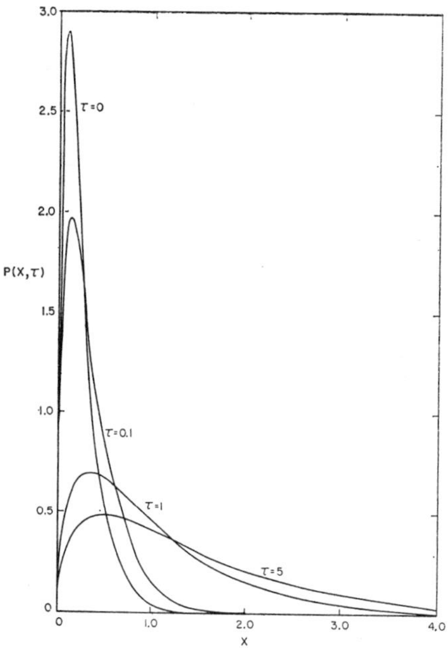

FIG. 4. The relaxation of an initial Maxwell distribution of reduced energy x with $T_1(0)=(1/6)T_2$ for the Rayleigh gas. For $\tau=5$, the distribution function $P(x,\tau)$ coincides essentially with the equilibrium Maxwell distribution $P(x,\infty)$ at the heat bath temperature T_2.

$3/2[T_1(\tau)/T_2]$ with $\bar{\epsilon}(\tau)=3/2kT_1(\tau)$. Substitution of this value of $\bar{x}$ into Eq. (5.3) then yields Eq. (5.30).

Finally we should like to point out that Eq. (5.30) is the integrated form of "Newton's law of cooling"

$$dT_1(t)/dt=-k_R[T_1(t)-T_2]. \tag{5.31}$$

It is usually stated that this law is valid only if the temperature difference $[T_1(0)-T_2]$ is "not too large." As we have shown previously,[26] such simple exponential laws for the moments of the distribution function will always hold *near equilibrium*, i.e., for $(T_1(0)-T_2)\simeq T_2\pm\Delta T$, for rate processes which can be represented by a linear master equation. For the hard-sphere Rayleigh gas studied here, this "law of cooling" is however valid, as shown above, for *all* temperature differences $T_1(0)-T_2$.

VI. FOKKER-PLANCK EQUATION FOR THE LORENTZ GAS

The partial differential equation (4.37) describing the relaxation of the Lorentz gas in terms of the reduced kinetic energy x is

$$\frac{\partial P(x,t)}{\partial t}=k_L\frac{\partial}{\partial x}\left\{(x^{3/2}-2x^{1/2})P(x,t)+\frac{\partial}{\partial x}[x^{3/2}P(x,t)]\right\}. \tag{4.37}$$

The appearance of $x^{\frac{1}{2}}$ in this equation suggests that we choose the quantity $y=x^{\frac{1}{2}}$ as a new independent variable. Since $x\equiv\epsilon/kT_2=m_1v_1^2/2kT_2$, the variable $y=v_1(m_1/2kT_2)^{\frac{1}{2}}$ is proportional to the reduced speed of the subsystem particle. The distribution function for $y(0\leq y<\infty)$ will be called $S(y,t)$, and from $S(y,t)dy=P(x,t)dx$ we find $S(y,t)=2yP(y^2,t)$. For $S(y,t)$ we then obtain

$$\frac{\partial S(y,t)}{\partial t}=\frac{k_L}{4}\frac{\partial}{\partial y}\left\{(2y^2-3)S(y,t)+\frac{\partial}{\partial y}[yS(y,t)]\right\}. \tag{6.1}$$

First of all, it will be of some interest to examine whether the mean energy, or some other moment of the distribution function, has a simple exponential time behavior similar to that of the mean energy of the Rayleigh gas [Eq. (5.2)]. If we define the mean value $\langle y^n\rangle_{\text{Av}}$ by

$$\langle y^n(t)\rangle_{\text{Av}}\equiv\int_0^\infty y^nS(y,t)dy \tag{6.2}$$

we obtain from Eq. (6.1), using the procedure indicated in Ref. 26, the following set of differential difference equations:

$$d\langle y^n\rangle_{\text{Av}}/dt=k_L[-\tfrac{1}{2}n\langle y^{n+1}\rangle_{\text{Av}}+\tfrac{1}{2}n(\tfrac{1}{2}n+1)\langle y^{n-1}\rangle_{\text{Av}}],$$
$$n=0,1,2\cdots. \tag{6.3}$$

We have been unable so far to obtain a closed analytical solution of Eqs. (6.3). Clearly, they are not satisfied by the simple exponential expressions obtained for the Rayleigh gas. An interesting relationship between the (time-dependent) mean speed $\langle y^1(t)\rangle_{\text{Av}}$ and the mean energy $\langle y^2(t)\rangle_{\text{Av}}$ can, however, be established from the Eqs. (6.3). For $n=1$, one finds:

$$\frac{1}{k_L}\frac{d\langle y^1(t)\rangle_{\text{Av}}}{dt}=-\tfrac{1}{2}\langle y^2(t)\rangle_{\text{Av}}+\tfrac{3}{4}, \tag{6.4}$$

where we have made use of the normalization of $S(y,t)$ which yields $\langle y^0\rangle_{\text{Av}}=1$. As the mean kinetic energy $\langle\tfrac{1}{2}m_1v_1^2\rangle_{\text{Av}}(=kT_2\langle y^2\rangle_{\text{Av}})$ of the subsystem particles at equilibrium $(t=\infty)$ has the value $\tfrac{3}{2}kT_2$, we

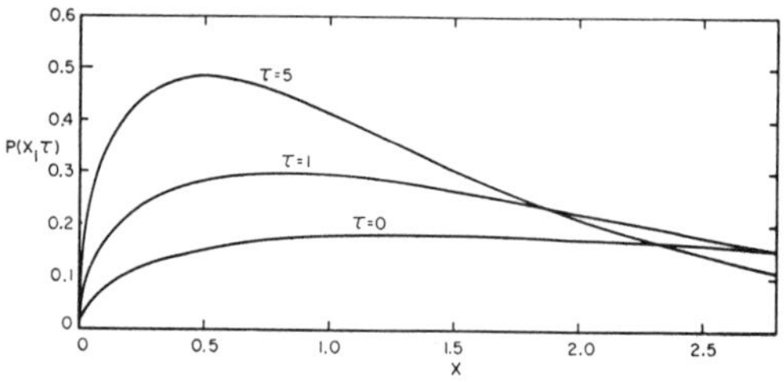

FIG. 5. The relaxation of an initial Maxwell distribution of reduced energy x with $T_1(0)=(8/3)T_2$ for the Rayleigh gas. For $\tau=5$, the distribution function $P(x,\tau)$ coincides essentially with the equilibrium Maxwell distribution $P(x,\infty)$ at the heat bath temperature T_2.

can put $\langle y^2(\infty)\rangle_{Av}=\frac{3}{2}$ and rewrite Eq. (6.4) as

$$d\langle y^1(t)\rangle_{Av}/dt=-\tfrac{1}{2}k_L[\langle y^2(t)\rangle_{Av}-\langle y^2(\infty)\rangle_{Av}]. \quad (6.5)$$

Equation (6.5) states that for the hard-sphere Lorentz gas, the rate of change of the mean speed, $d\langle y^1\rangle_{Av}/dt$, of the light subsystem particles is proportional to the difference between their mean energy $\langle y^2(t)\rangle_{Av}$ at time t and the final equilibrium mean energy, $\langle y^2(\infty)\rangle_{Av}$. This is to be compared with Eq. (5.2) for the relaxation of Rayleigh gas where it is the mean energy, rather than the mean speed, whose time rate of change is proportional to the energy difference $\bar{x}(t)-\bar{x}(\infty)$.

In an attempt to solve Eq. (6.1) by a method analogous to that used for the Rayleigh gas we set $S(y,\ t)=y^2\ \exp(-y^2)\chi(y,\ t)$ and write $\tau=k_Lt$. For $\chi(y,\ \tau)$ we then obtain

$$\frac{\partial\chi(y,\tau)}{\partial\tau}=\frac{1}{4}\left[y\frac{\partial^2\chi(y,\tau)}{\partial y^2}+(3-2y^2)\frac{\partial\chi(y,\tau)}{\partial y}\right]. \quad (6.6)$$

A comparison of Eq. (6.6) with the corresponding F–P equation for the Rayleigh gas [Eq. (5.5)] shows that the linear coefficient $(3/2-x)$ of the Rayleigh gas is replaced by the nonlinear coefficient $(3-2y^2)$. We now try to express the solution to Eq. (6.6) in terms of eigenfunctions $\eta_\nu(y)$ to the operator

$$\mathfrak{L}=\tfrac{1}{4}[y(d^2/dy^2)+(3-2y^2)(d/dy)] \quad (6.7)$$

so that the $\eta_\nu(y)$ are the solutions to

$$\mathfrak{L}\eta_\nu(y)=\mu_\nu\eta_\nu(y),$$

$$y[d^2\eta_\nu(y)/dy^2]+(3-2y^2)[d\eta_\nu(y)/dy]-4\mu_\nu\eta_\nu(y)=0. \quad (6.8)$$

Equation (6.8) is identical with the eigenvalue equation [Eq. (22)] obtained by Delcroix *et al.*[33] and by Wilkins[6b] in his study of the thermalization of neutrons in a heavy gaseous moderator. It can easily be transformed into a Sturm–Liouville form with a self-adjoint operator; we can therefore conclude that all the eigenvalues μ_ν are real. The eigenvalue μ_0 is zero corresponding to the equilibrium solution $\chi(y,\ \infty)=\eta_0=$constant.

It has not been possible to affect an analytical solution of Eq. (6.8). We have therefore carried out a numerical solution on the GIER digital computer at the University of Copenhagen.[34] The first 10 (nonzero) eigenvalues μ_ν were computed to eight significant figures by repeated trial integrations of Eq. (6.8) with the boundary condition $d\eta_\nu(y)/dy\rightarrow0$ as $y\rightarrow\infty$.[35] For the integration we used the Merson method which is an improved Runge–Kutta fourth-order procedure involving continuous adjustment of the interval length to keep the truncation error below a predetermined small value along the entire integration range. The eigenfunctions $\eta_\nu(y)$ were obtained by integrating Eq. (6.8) after substitution of the corresponding eigenvalues μ_ν. The distribution function $S(y,\ \tau)$ was then approximated by the finite series

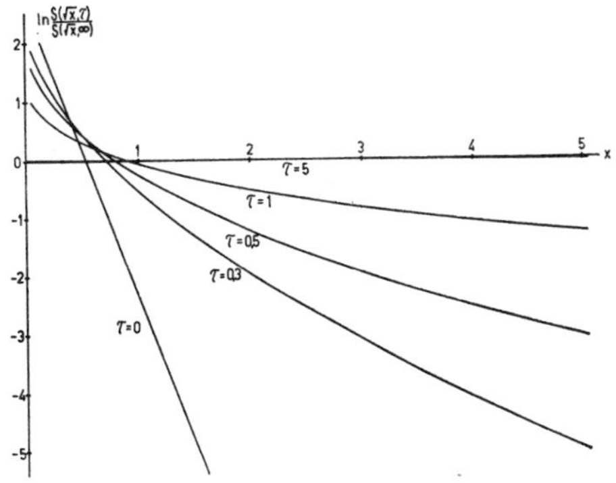

FIG. 6. The relaxation of an initial Maxwell distribution with $T_1(0)=(1/6)T_2$ for the Lorentz gas for various values of τ. The deviation from a straight-line relationship indicates the departure from a Maxwellian distribution during the relaxation.

$$S(y,\tau)=y^2\exp(-y^2)\sum_{\nu=0}^{10}c_\nu\eta_\nu(y)\ \exp(\mu_\nu\tau), \quad (6.9)$$

which is valid in the interval $0<y<5$ and for $\tau>0.3$. Equation (6.9) is the analog of the exact infinite series representation (5.13) for the distribution function $P(x,\ \tau)$ of the Rayleigh gas. The coefficients c_ν were evaluated numerically from

$$c_\nu=\frac{1}{\Gamma_\nu}\int_0^\infty S(y,0)\eta_\nu(y)dy, \quad (6.10)$$

where $S(y,\ 0)$ is the normalized initial distribution and where the Γ_ν were determined from the equation

$$\int_0^\infty y^2\eta_\nu(y)\eta_\mu(y)\ \exp(-y^2)dy=\Gamma_\nu\delta_{\nu\mu} \quad (6.11)$$

by numerical integration. The integrations indicated in (6.11) also served to verify the orthogonality of the computed eigenfunctions $\eta_\nu(y)$.

The evaluation of Eqs. (6.9) through (6.11) now permit us to study the relaxation of the hard-sphere Lorentz gas for various initial distributions $S(y,\ 0)$. We are particularly interested here in the relaxation of an initial Maxwell distribution with

$$S(y,0)=4(\pi)^{-\frac{1}{2}}y^2(\alpha_0)^{-\frac{3}{2}}\exp(-y^2/\alpha_0) \quad (6.12)$$

corresponding to an initial temperature $T_1(0)=\alpha_0T_2$.

[33] A. Arsac, L. Basquin, J. F. Denisse, J. L. Delcroix, and J. Salmon, J. Phys. Radium **19**, 624 (1958).

[34] A numerical solution of Eq. (6.8) has also been worked out by Delcroix *et al.*[33] We have extended their calculations both in range and accuracy primarily in order to investigate the extent of the departure from equilibrium in the relaxation of initial Maxwell distributions.

[35] The eigenvalues μ_ν for the hard sphere Lorentz gas $(s=\infty)$ as well as for some other values of $s(2\leq s\leq\infty)$ will be tabulated in a subsequent publication.

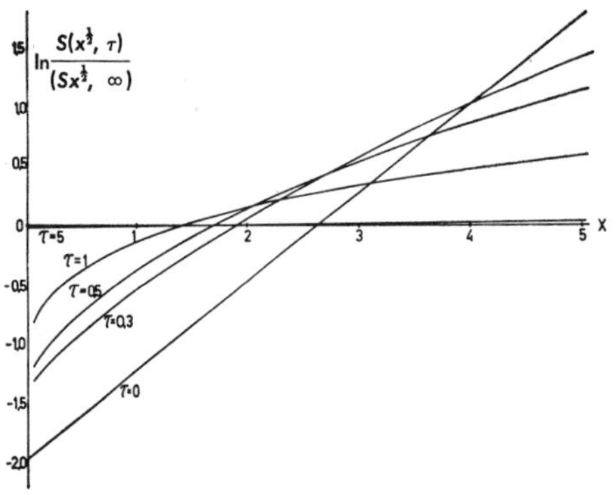

FIG. 7. The relaxation of an initial Maxwell distribution with $T_1(0) = (8/3) T_2$ for the Lorentz gas for various values of τ. The deviation from a straight-line relationship indicates the departure from a Maxwellian distribution during the relaxation.

In the case of the Rayleigh gas, which undergoes a "Maxwellian relaxation" in the sense that $P(x, \tau)$ as given by Eq. (5.28) is a Maxwellian distribution with a time-dependent temperature $T_1(\tau) = \alpha(\tau) T_2$ for an initial Maxwellian distribution $P(x, 0)$, a plot of $\ln[P(x, \tau)/P(x, \infty)]$ vs x will yield, for any given value of τ, a straight line with slope $[\alpha(\tau)-1]/\alpha(\tau) = [T_1(\tau)-T_2]/T_1(\tau)$. For the Lorentz gas, the analogous plot of $\ln[S(x^{\frac{1}{2}}, \tau)/S(x^{\frac{1}{2}}, \infty)]$ vs x for various values of τ shows considerable deviation from the above straight-line relationship. This is clearly exhibited in Figs. 6 and 7 which present, respectively, the relaxation of a hard-sphere Lorentz gas with an initial Maxwell distribution with $\alpha_0 \equiv T_1(0)/T_2 = \frac{1}{6}$ and $\alpha_0 \equiv T_1(0)/T_2 = \frac{8}{3}$. The deviation from the straightline relationship is a convenient measure of the deviation from equilibrium during the relaxation.

It must be recalled that all the results of this paper pertain to the *hard-sphere* Rayleigh and Lorentz gas. As has been pointed out, for the hard-sphere Rayleigh gas the collision frequency Z_{12} of the heavy subsystem particles, in the limit $m_2/m_1 \rightarrow 0$, is independent of the speed, $x^{\frac{1}{2}}$. This is true, within the limit $m_2/m_1 \rightarrow 0$, for any central field force law. For the Lorentz gas, however, in the limit as $m_2/m_1 \rightarrow \infty$, the collision frequency $Z_{12}(y)$ is a function of the speed $y(\equiv x^{\frac{1}{2}})$ of the light subsystem particles; for the repulsive force law r^{-s} it is of the form[8,25]

$$Z_{12}(y) = Ay^{(s-5)/(s-1)} \tag{6.13}$$

with A a constant independent of y. For hard-sphere interaction, $s=\infty$, and Eq. (6.13) is identical with Eq. (4.21). For $s=5$, i.e., for a "Maxwell" gas, $Z_{12}=A$ and the collision frequency is independent of the speed y. It is for this Maxwellian force law with $s=5$, and for this force law only, that the Fokker–Planck equation for the Lorentz gas assumes the same form as the F–P equation for the Rayleigh gas and is thus amenable to an analytical solution of the form presented in Sec. V. As a corollary, it is for this force law only that the mean reduced energy of the light subsystem particles ("the electrons") follows a simple exponential relaxation. The (identical) analytical solutions for the relaxation of the Lorentz gas given by Delcroix,[33] Salmon,[36] Kahalas and Kashian,[37] and Osipov[38] all pertain to the Maxwellian ($s=5$) Lorentz gas. The assumption made by some of these authors that the mean free path $l_1(y)$ is proportional to the speed y is equivalent [see, e.g., Eq. (4.27)] to setting $Z_{12}=A$.

The use of the generalized repulsive force law $r^{-s}(s \geq 2)$ in place of the hard-sphere interaction $(s=\infty)$ leads to some interesting extensions of the results presented in this paper:

(1) For the Rayleigh gas, i.e., in the limit $m_2/m_1 \rightarrow 0$, the differential Fokker–Planck collision operator is independent of the value of s and only the coefficient k_R [see e.g., Eq. (4.34)] depends upon s. This coefficient enters into the expression for the eigenvalues but not for the eigenfunctions in the solution (5.13) of the Fokker–Planck equation. Therefore, the *time scale* of the relaxation depends upon the value of s, but the functional form of the relaxation is independent of s.

(b) For the Lorentz gas, i.e., in the limit $m_2/m_1 \rightarrow \infty$, the differential Fokker–Planck collision operator as well as the coefficient k_L [Eq. (5.37)] are explicit functions of s. For the Lorentz gas then, both the time scale of the relaxation and the functional form of the relaxation depend explicitly on the value of s.

One of us (KES) in collaboration with G. Weiss is presently engaged in carrying out numerical computations for the solution of the Lorentz gas F–P equation as a function of the force law parameter s. The results, to be presented shortly, give a clear indication of the dependence of the distribution function $S(y, \tau)$ on s.

ACKNOWLEDGMENTS

We wish to acknowledge many helpful conversations with our colleagues M. S. Green, N. van Kampen, I. Oppenheim, J. Ross, and G. Weiss whose comments and suggestions have contributed greatly to the writing of this paper. Particular thanks are due to Miss Shirley G. Brown for her skill and devotion in the preparation of this manuscript for publication.

APPENDIX I

We derive here exact expressions for $Z_{12}(x)$ and for the moments $b_n(x)$ with $n=1$, 2, and 3 from the

[36] J. Salmon, J. Phys. Radium **17**, 931 (1956).
[37] S. L. Kahalas and H. C. Kashian, Phys. Fluids **2**, 100 (1959).
[38] O. I. Osipov, Bull. Moscow Univ. **1**, 13 (1961).

transition probability kernel $A(x' \mid x)$, where

$$Z_{12}(x)=\int_0^\infty B(x' \mid x)\,dx', \tag{A1}$$

$$b_n(x)=\int_0^\infty (x'-x)^n A(x' \mid x)\,dx' \qquad n=1,2,3\cdots, \tag{A2}$$

$$A(x' \mid x)=B(x' \mid x)-\delta(x'-x)Z_{12}(x), \tag{A3}$$

and

$$B(x' \mid x)=(C/x^{\frac{1}{2}})[\operatorname{erf}(Qx'^{\frac{1}{2}}+Rx^{\frac{1}{2}})+e^{x-x'}\operatorname{erf}(Rx'^{\frac{1}{2}}+Qx^{\frac{1}{2}})\pm\{\operatorname{erf}(Qx'^{\frac{1}{2}}-Rx^{\frac{1}{2}})+e^{x-x'}\operatorname{erf}(Rx'^{\frac{1}{2}}-Qx^{\frac{1}{2}})\}]. \tag{A4}$$

In Eq. (A4) the upper sign (+) applies to the interval $x'<x$, the lower sign (−) to the interval $x'>x$.

We have to evaluate the two indefinite integrals

$$J_n(x',x)=\int (x'-x)^n[\operatorname{erf}(Qx'^{\frac{1}{2}}+Rx^{\frac{1}{2}})+e^{x-x'}\operatorname{erf}(Rx'^{\frac{1}{2}}+Qx^{\frac{1}{2}})]dx' \tag{A5}$$

$$I_n(x',x)=\int (x'-x)^n[\operatorname{erf}(Qx'^{\frac{1}{2}}-Rx^{\frac{1}{2}})+e^{x-x'}\operatorname{erf}(Rx'^{\frac{1}{2}}-Qx^{\frac{1}{2}})]dx' \tag{A6}$$

for $n=0, 1, 2,$ and 3.

By partial integration and substitution of the new variable $y=Qx'^{\frac{1}{2}}-Rx^{\frac{1}{2}}$ in the remaining integrals we obtain for I_n:

$$I_n=[1/(n+1)](x'-x)^{n+1}\operatorname{erf}(Qx'^{\frac{1}{2}}-Rx^{\frac{1}{2}})-\exp(x-x')\,n!\sum_{\nu=0}^{n}\frac{1}{\nu!}(x'-x)^{\nu}\operatorname{erf}(Rx'^{\frac{1}{2}}-Qx^{\frac{1}{2}})$$
$$-\frac{1}{(n+1)Q^{2n+2}}\int[y(y+2Rx^{\frac{1}{2}})-x]^{n+1}\exp(-y^2)\,dy+\frac{R}{Q}n!\sum_{\nu=0}^{n}\frac{1}{\nu!Q^{2\nu}}\int[y(y+2Rx^{\frac{1}{2}})-x]^{\nu}\exp(-y^2)\,dy. \tag{A7}$$

Integrals of the type

$$\int[y(y+2Rx^{\frac{1}{2}})]^m\exp(-y^2)\,dy$$

can easily be expressed by standard integrals[39] and can be written in the form

$$P_{m/2}(R^2x)\operatorname{erf}(y)+S_m(y,\,y+2Rx^{\frac{1}{2}})\exp(-y^2)$$

with $P_{m/2}$ a polynomial of degree $(m/2)$ for m even or $(m-1)/2$ for m odd and S_m a linear combination of products of the two arguments raised to powers ranging from 0 to m. For I_n we then obtain

$$I_n=\frac{\operatorname{erf}(Qx'^{\frac{1}{2}}-Rx^{\frac{1}{2}})}{n+1}\Bigg[(x'-x)^{n+1}-\frac{1}{Q^{2n+2}}\sum_{\mu=0}^{n+1}\binom{n+1}{\mu}P_\mu(R^2x)(-x)^{n+1-\mu}$$
$$+(n+1)!\frac{R}{Q}\sum_{\nu=0}^{n}\frac{1}{\nu!Q^{2\nu}}\sum_{\mu=0}^{\nu}\binom{\nu}{\mu}P_\mu(R^2x)(-x)^{\nu-\mu}\Bigg]-n!e^{x-x'}\operatorname{erf}(Rx'^{\frac{1}{2}}-Qx^{\frac{1}{2}})\sum_{\nu=0}^{n}\frac{1}{\nu!}(x'-x)^{\nu}+\frac{1}{n+1}\exp[-(Qx'^{\frac{1}{2}}-Rx^{\frac{1}{2}})^2]$$
$$\times\Bigg[\frac{-1}{Q^{2n+2}}\sum_{\mu=0}^{n+1}\binom{n+1}{\mu}S_\mu(Qx'^{\frac{1}{2}}-Rx^{\frac{1}{2}},\,Qx'^{\frac{1}{2}}+Rx^{\frac{1}{2}})(-x)^{n+1-\mu}$$
$$+(n+1)!\frac{R}{Q}\sum_{\nu=0}^{n}\frac{1}{\nu!Q^{2\nu}}\sum_{\mu=0}^{\nu}\binom{\nu}{\mu}S_\mu(Qx'^{\frac{1}{2}}-Rx^{\frac{1}{2}},Qx'^{\frac{1}{2}}+Rx^{\frac{1}{2}})(-x)^{\nu-\mu}\Bigg]$$
$$=T_n(x',x)\operatorname{erf}(Qx'^{\frac{1}{2}}-Rx^{\frac{1}{2}})-U_n(x',x)\operatorname{erf}(Rx'^{\frac{1}{2}}-Qx^{\frac{1}{2}})$$
$$+[(Qx'^{\frac{1}{2}}+Rx^{\frac{1}{2}})V_n(x',x)+Rx^{\frac{1}{2}}W_n(x',x)]\exp[-(Qx'^{\frac{1}{2}}-Rx^{\frac{1}{2}})^2]. \tag{A8}$$

In a completely analogous manner we find for J_n:

$$J_n=T_n(x',x)\operatorname{erf}(Qx'^{\frac{1}{2}}+Rx^{\frac{1}{2}})-U_n(x',x)\operatorname{erf}(Rx'^{\frac{1}{2}}+Qx^{\frac{1}{2}})$$
$$+[(Qx'^{\frac{1}{2}}-Rx^{\frac{1}{2}})V_n(x',x)-Rx^{\frac{1}{2}}W_n(x',x)]\exp[-(Qx'^{\frac{1}{2}}+Rx^{\frac{1}{2}})^2], \tag{A9}$$

[39] W. Gröbner and N. Hofreiter, *Integraltafeln* (Springer-Verlag, Vienna, 1949).

where the polynomials T_n, U_n, V_n, and W_n are identical with those appearing in Eq. (A8) for I_n. We have worked out the explicit expressions for these polynomials, but do not reproduce them here owing to their excessive length.

In the integral (A2) defining $b_n(x)$ the δ-function term in $A(x' \mid x)$ vanishes owing to the factor $(x'-x)^n$. For $b_n(x)$ we then obtain

$$b_n(x) = (C/x^{\frac{1}{2}})[J_n(\infty, x) - I_n(\infty, x) - J_n(0, x) - I_n(0, x) + 2I_n(x, x)]. \tag{A10}$$

Since

$$\lim_{x'\to\infty} \operatorname{erf}(Qx'^{\frac{1}{2}} + Rx^{\frac{1}{2}}) = \lim_{x'\to\infty} \operatorname{erf}(Qx'^{\frac{1}{2}} - Rx^{\frac{1}{2}}) = 1$$

$$\lim_{x'\to\infty} \exp[-(Qx'^{\frac{1}{2}} + Rx^{\frac{1}{2}})^2] = \lim_{x'\to\infty} \exp[-(Qx'^{\frac{1}{2}} - Rx^{\frac{1}{2}})^2] = 0$$

and $\operatorname{erf}(z) = -\operatorname{erf}(-z)$, we note that Eq. (A10) reduces to

$$b_n(x) = (2C/x^{\frac{1}{2}}) I_n(x, x) \tag{A11}$$

$$Z_{12}(x) = (2C/x^{\frac{1}{2}}) I_0(x, x). \tag{A12}$$

We can now write the following exact expressions for the collision number $Z_{12}(x)$ and the moments $b_n(x)$:

$$Z_{12}(x) = (C/\lambda Q^2)\{[2\lambda x^{\frac{1}{2}} + (1/\lambda x^{\frac{1}{2}})] \operatorname{erf}(\lambda x^{\frac{1}{2}}) + \exp(-\lambda^2 x)\}, \tag{A13}$$

$$b_1(x) = \frac{C}{Q^4}\left\{\left[-x^2 + \left(2 - \frac{1}{\lambda^2}\right)x + \frac{1}{\lambda^2}\left(1 + \frac{1}{4\lambda^2}\right)\right]\frac{\operatorname{erf}(\lambda x^{\frac{1}{2}})}{x^{\frac{1}{2}}} + \left(-\tfrac{1}{2}x + 1 - \frac{1}{4\lambda^2}\right)\frac{\exp(-\lambda^2 x)}{\lambda}\right\}, \tag{A14}$$

$$\begin{aligned} b_2(x) = \frac{C}{Q^6}\Bigg\{&\left[\tfrac{2}{3}x^3 + \left(\tfrac{1}{2}\lambda^2 - 3 + \frac{3}{2}\frac{1}{\lambda^2}\right)x^2 + 3\left(\frac{3}{2} - \frac{1}{\lambda^2}\right)x + \frac{15}{8}\frac{1}{\lambda^2} + \frac{3}{4}\frac{1}{\lambda^4} + \frac{1}{8}\frac{1}{\lambda^6}\right]\frac{\operatorname{erf}(\lambda x^{\frac{1}{2}})}{x^{\frac{1}{2}}} \\ &+ \left[\tfrac{1}{3}x^2 + \left(\tfrac{1}{4}\lambda^2 - \tfrac{3}{2} + \frac{7}{12}\frac{1}{\lambda^2}\right)x + \frac{17}{8} - \frac{3}{4}\frac{1}{\lambda^2} - \frac{1}{8}\frac{1}{\lambda^4}\right]\frac{\exp(-\lambda^2 x)}{\lambda}\Bigg\}, \end{aligned} \tag{A15}$$

$$\begin{aligned} b_3(x) = -\frac{C}{Q^8}\Bigg\{&\left[\tfrac{1}{2}x^4 + \left(\lambda^2 - 4 + \frac{2}{\lambda^2}\right)x^3 + 3\left(-\lambda^2 + \frac{9}{2} - \frac{3}{\lambda^2} + \frac{1}{4}\frac{1}{\lambda^4}\right)x^2 - 15\left(1 - \frac{3}{4}\frac{1}{\lambda^2}\right)x - 3\left(\frac{7}{4}\frac{1}{\lambda^2} + \frac{7}{8}\frac{1}{\lambda^4} + \frac{1}{4}\frac{1}{\lambda^6} + \frac{1}{32}\frac{1}{\lambda^8}\right)\right]\frac{\operatorname{erf}(\lambda x^{\frac{1}{2}})}{x^{\frac{1}{2}}} \\ &+ \left[\tfrac{1}{4}x^3 + \left(\tfrac{1}{2}\lambda^2 - 2 + \frac{7}{8}\frac{1}{\lambda^2}\right)x^2 + \left(-\tfrac{3}{2}\lambda^2 + \frac{13}{2} - \frac{7}{2}\frac{1}{\lambda^2} + \frac{1}{16}\frac{1}{\lambda^4}\right)x + 3\left(-\frac{9}{4} + \frac{7}{8}\frac{1}{\lambda^2} + \frac{1}{4}\frac{1}{\lambda^4} + \frac{1}{32}\frac{1}{\lambda^6}\right)\right]\frac{\exp(-\lambda^2 x)}{\lambda}\Bigg\}, \end{aligned} \tag{A16}$$

6) Chemical Kinetics

The use of the master equation for the description of chemical rate processes has a long history which is detailed in McQuarrie's review paper[4.6.1]. These models make use of phenomenological assumptions that are not necessarily rigorous consequences of physical laws. The major unsolved problem in this area of research is the rigorous derivation of the master equation description of chemical kinetics from the Liouville equation. The first paper presents a stochastic model for the dissociation of a subsystem of diatomic molecules in a heat bath. The second paper is addressed to a stochastic model for reactions of species firmly attached to sites, a problem which is of importance in polymer chemistry. The master equation is used to determine the time dependence of the process. Papers three and four discuss the differences between the stochastic and deterministic model for chemical reactions. It is shown for a wide variety of reactions that in the thermodynamic limit the fluctuations in concentration predicted by the stochastic theory are unobservably small. Thus the results for the stochastic and deterministic theories are essentially identical.

References

4.6.1. D. A. McQuarrie, J. Appl. Prob. 4, 413 (1967).

THE APPLICATION OF THE THEORY OF STOCHASTIC PROCESSES TO CHEMICAL KINETICS *

ELLIOTT W. MONTROLL, *Institute for Fluid Dynamics and Applied Mathematics, University of Maryland*

and

KURT E. SHULER, *National Bureau of Standards*

CONTENTS

I. INTRODUCTION

The standard theories of chemical kinetics are equilibrium theories in which a Maxwell-Boltzmann energy (or momentum or internal coordinate) distribution of reactants is postulated to persist during a reaction. In the collision theory, mainly due to Hinshelwood,[7] the number of energetic, reaction producing collisions is calculated under the assumption that the molecular velocity distribution always remains Maxwellian. In the absolute

* The research on which this review is based was supported in part by the United States Air Force through the Office of Scientific Research of the Air Research and Development Command and by the United States Atomic Energy Commission.

rate theory, developed particularly by Eyring,[4] the equilibrium assumption is stated (in its most refined form) in terms of a time invariant Boltzmann distribution of reactant molecules in the reactant valley far removed from the potential barrier. It has long been recognized that the equilibrium postulate is only an approximation of reality, since any process with a nonvanishing rate disturbs an initial equilibrium. Hence it is of interest to examine various models of chemical reactions to estimate the extent of departure from equilibrium and to develop techniques for analyzing reactions in which the departure is significant. "Microscopic" chemical kinetics or the adjustment of the energy distributions induced by reactions is the main theme of this paper.

This section is a review of previous treatments of this subject. Subsequent sections are an application of the theory of stochastic processes to chemical rate phenomena; the harmonic oscillator model of a diatomic molecule is used to obtain explicit results by the general formalism.

A. The Curtiss-Prigogine-Takayanagi Model

Curtiss,[3] Prigogine[15,16] and collaborators, and Takayanagi[21] investigated the perturbation of an initial Maxwell velocity distribution by a chemical reaction. The perturbation of the energy distribution of internal degrees of freedom was neglected. A generalized Boltzmann equation was derived which incorporated effects such as (a) addition of energy to the system through the heat of reaction, and (b) the removal of highly energetic molecules from the system of colliding reactants by inelastic collisions (those which result in a chemical reaction). After postulating the probability of an inelastic collision to be small, the generalized Boltzmann equation was solved by the Chapman-Enskog method of using a perturbed distribution function f of the form

$$f = f_0(1+\Phi) \qquad \text{(I.1)}$$

f_0 being the equilibrium Maxwellian velocity distribution, and Φ the perturbation induced by the chemical reaction.

Prigogine considered two relations between the probability (cross section) of inelastic collisions σ and the relative velocity modulus g; namely

$$\sigma = 1 - \exp(-\alpha g^2) \tag{I.2a}$$

and

$$\sigma = \begin{cases} 0 \text{ if } g < g_0 \\ 1 \text{ if } g \geqq g_0 \end{cases} \tag{I.2b}$$

where α and g_0 are related to the activation energy E_{act} of the reaction by

$$E_{act} = m/4\alpha = \tfrac{1}{4}mg_0^2 \tag{I.3}$$

where m is the mass of a molecule of the reacting species. Φ was found to depend only on E_{act}/kT in both of these cases. Quantitative results were very sensitive to the choice of σ. For $E_{act}/kT = 5$, Eq. I.2a produced a perturbation of the Maxwellian distribution which reduced the reaction rate to about 99 per cent of the equilibrium rate whereas Eq. I.2b resulted in a decrease to about 80 per cent. Curtiss, by employing Eq. I.2b found a reduction of the rate by about 20 per cent for $E_{act}/kT = 4$, and 10 per cent for $E_{act}/kT = 5$, in good agreement with Prigogine.

An important result of these studies is that the perturbation of the equilibrium distribution and the corresponding deviation from the equilibrium rate is quite small when $E_{act}/kT \geqq 5$ and that the use of the equilibrium theory of chemical reactions is justified under this condition. The other investigations described below are in essential agreement with this result.

The extension of the kinetic theory approach to include large values of σ (and hence large deviations from equilibrium) requires higher order perturbations for the solution of the Boltzmann equation. It is probably unprofitable to proceed in this difficult and laborious direction until one understands the detailed analytical dependence of the transition probability σ on the mechanism of molecular energy exchange and redistribution on collision. Currently available information on intermolecular forces is insufficient to establish this dependence.

B. The Model of Zwolinski and Eyring

Zwolinski and Eyring[24] schematically describe the reactants of a chemical reaction by one set of quantum energy states and the reaction product by another. These levels are not especially identified with translational states or those of internal degrees of freedom but are left quite general. Molecules in the reactant states are postulated to pass by collision with other molecules into the product states. Reactions which are first order in a species A were studied. In this case linear rate equations of the form (prime denoting omission of terms with $n = m$)

$$\frac{dx_n}{dt} = \sum_m{}' \{W_{nm} x_m - x_n W_{mn}\} \quad n = 0, 1, 2, \ldots \qquad \text{(I.4)}$$

can be constructed to describe the rate of transition for species A between the various reactant and product levels. The W_{nm}'s are probabilities per unit time of transitions through collision from states m to states n. Note that the first subscript refers to the final state. The $x_n(t)$ is the fraction of molecules in state n at time t.

The rate constants W_{nm} in the Zwolinski-Eyring model are analogous to the inelastic collision cross section σ of the Prigogine-Curtiss model. They are calculable, in principle, from the quantum-mechanical theory of collisions, but, as pointed out above, our ignorance of intermolecular forces and interactions prevent us from deducing their analytical form or numerical values. In the absence of this information Zwolinski and Eyring assumed certain relations between the various rate constants and assigned plausible numerical values to enough of them so that all could be determined.

The solution of Eq. I.4 is given by

$$x_n(t) = \sum_j B_{nj} \exp t\lambda_j \qquad \text{(I.5)}$$

where the λ'_j's are the characteristic roots of the matrix of the coefficients B_{nj} obtained by setting $x_n = B_n \exp \lambda t$. The B'_{nj}'s are components of the characteristic vectors of the matrix. Zwolinski and Eyring evaluated Eq. 1.5 numerically for a 4-level model to obtain the time dependent concentration $x_n(t)$. The rate of reaction, i.e., the rate of passage between the energy levels, was then obtained by computing the appropriate products $W_{nm} x_m$ and summing these products over the reactant and product levels.

Zwolinski and Eyring expressed the deviation of the actual rate from the equilibrium rate by forming the ratio Γ of the above calculated rate to the rate determined on the assumption that the concentration of the reacting species was given at all times by the equilibrium Maxwell-Boltzmann distribution. They found $\Gamma < 1$, the extent of the departure from equilibrium depending upon the numerical choice for the transition probabilities W_{nm}. The maximum deviation from the equilibrium rate found for a 4-level model was about 20 per cent.

Zwolinski and Eyring, and one of us[10] have suggested that an extension of this 4-level model to one of N levels with a systematic treatment of transitions between these levels would be instructive in the formulation of microscopic, nonequilibrium chemical kinetics. Such an extension is presented in Sections III to V of this paper. We have generalized the Zwolinski-Eyring discrete energy level model to N levels and have characterized the transitions of the molecules between levels as a one-dimensional random walk. This general formulation is developed in terms of the properties of stochastic matrices whose elements involve the transition probabilities. Chemical reactions, i.e., the removal of reactant species from the reaction system, are introduced into this level system via an absorbing barrier at level $N+1$. The rate of the chemical reaction is then given in terms of the mean first passage time $\bar{l}$, the average time required for a species to pass level N and reach the absorbing barrier for the first time.

C. The Model of Kramers

Kramers[11] proposed a Brownian motion model for a chemical reaction. In this model reactant molecules become activated through collisions with other molecules of the surrounding medium which acts as a constant temperature heat bath. After many collisional exchanges of energy, some of the reactant molecules acquire sufficient energy (the energy of activation) to cross over a potential barrier. This crossing constitutes the chemical reaction, and the rate of crossing is equal to the rate of the reaction. The interaction of the reactant molecules with the heat bath is analogous to the Brownian motion of a particle in a viscous medium under the action of a force whose potential is the potential energy surface along the reaction coordinate. The theory of reaction rates is then treated as a Brownian motion problem. The interaction (coupling) of the molecules with the heat bath is expressed through a viscosity coefficient η, such that a large value of η corresponds to a strong interaction between the molecules and the heat bath.

The rate of reaction is given by the diffusion current over the potential barrier, and the energy distribution of the reacting species along the reaction coordinate is given through the density distribution in momentum space. The calculation rests, as remarked by Kramers, on the construction and solution of the equation of diffusion obeyed by a density distribution of particles in phase space. A very clear presentation of the Kramers diffusion equation has been given by Chandrasekhar.[1,2]

The coefficient of viscosity η plays the same rôle in the Kramers model as the cross section σ in the theory of Curtiss and Prigogine and the transition probabilities W_{nm} in the Zwolinski-Eyring treatment. Neither its value nor analytical form can be determined from our present knowledge of intermolecular forces. It is interesting to see how this factor enters into all the theories and models of "microscopic" nonequilibrium chemical kinetics. Its absence from equilibrium chemical kinetics is, of course, due to the fact that the properties of the equilibrium state are independent of the manner of its establishment.

Mathematical difficulties forced Kramers to restrict his discussion to the case in which the barrier height $Q = E_{act}$ is large compared to the mean thermal energy of the molecules kT and in which the diffusion over the barrier can be treated as a quasi-stationary process. Kramers showed that under these conditions the calculated reaction rate is very close to the equilibrium rate, as given by absolute rate theory, and that for $E/kT > 10$ the rate calculated from his model agrees with the equilibrium rate to within about 10 per cent over a rather wide range of η.

The results of Kramers and of Curtiss and Prigogine are in good agreement in showing that for $E_{act}/kT \geqq 10$ the departure of the actual rate from the equilibrium rate is quite small and can be neglected for all practical purposes. Our own results, which are developed in the following sections, are in excellent agreement with these findings. This does not answer, however, the question as to the deviation of the rate from the equilibrium rate when

$E_{act}/kT < 10$. Conceptual and mathematical difficulties prevent the application of the methods developed and described in Sections I.A to I.C to an analysis of this problem. The application of the theory of stochastic processes, however, permits the evaluation of reaction rates under these conditions and will be shown in Section VII to lead to results that may deviate considerably from the equilibrium rate when $E_{act}/kT \ll 10$.

II. DISCRETE ENERGY-LEVEL MODEL FOR UNIMOLECULAR REACTIONS

Our approach to the study of the departure from equilibrium in chemical reactions and of the "microscopic" theory of chemical kinetics is a discrete quantum-mechanical analog of the Kramers-Brownian-motion model. It is most specifically applicable to a study of the energy-level distribution function and of the rate of activation in unimolecular (dissociation)reactions. Our model is an extension of one which we used in a discussion of the relaxation of vibrational nonequilibrium distributions.[14, 18, 20]

We consider an ensemble of reactant molecules with quantized energy levels to be immersed in a large excess of (chemically) inert gas which acts as a *constant temperature* heat bath throughout the reaction. The requirement of a constant temperature T of the heat bath implies that the concentration of reactant molecules is very small compared to the concentration of the heat bath molecules. The reactant molecules are initially in a Maxwell-Boltzmann distribution appropriate to a temperature T_0 such that $T_0 < T$. By collision with the heat bath molecules the reactants are excited in a stepwise processs into their higher-energy levels until they reach "level" $(N+1)$ where they are removed irreversibly from the reaction system. The collisional transition probabilities per unit time W_{mn} which govern the rate of transfer of the reactant molecules between levels with energies E_n and E_m are functions of the quantum numbers n and m and can, in principle, be calculated in terms of the interaction of the reactant molecules with the heat bath.

The model described above corresponds to an unimolecular activation followed by dissociation

$$M+A \underset{k_{\text{deact}}}{\overset{k_{\text{act}}}{\rightleftarrows}} M+A^* \tag{II.1}$$

$$A^* \xrightarrow[k_D]{} \text{dissociation product}$$

where M represents the heat bath molecules, A the reactant molecules, and A^* the reactants with energy greater than E_N. Since we have assumed that the concentration $[M] \gg [A]$, the activating and deactivating collisions suffered by the molecules A will take place predominantly with the heat bath molecules M, so that the activation process is of first order in $[A]$ and the differential equations governing the rate of activation will be linear in the concentration of the reacting species A.

The relevant stochastic process corresponding to our model of a chemical reaction is that of a one-dimensional random walk with an absorbing barrier. In this random walk, the probability per unit time W_{mn} that a "walker" will take a step from level n to m is a function of the distance from the origin, $n = 0$. The time dependent distribution of the reactant molecules among the energy levels $n = 0, 1, \ldots, N$ is then given by the fraction $x_n(t)$ of walkers $(\sum x_n(0) = 1)$ n levels from the origin at time t. The rate of activation, v_{act} is inversely proportional to the mean first passage time, which is the average time required (appropriately weighted for the initial distribution of walkers) for a walker to reach the absorbing level $(N+1)$ (i.e., to pass the level N) for the first time. The mean first passage time l also gives directly the time lag for activation for our ensemble of reactants.

Generally our transition probabilities W_{mn} are as difficult to obtain as the corresponding parameters used by Curtiss, Prigogine, Zwolinski and Eyring, and Kramers. However, if the reactant molecules can be treated as simple harmonic oscillators, and if *only weak interactions* exist between the oscillators and heat bath molecules, an explicit calculation of the collisional transition probabilities W_{mn} can be carried out. This was first done by Landau and Teller[6, 12] by using the following argument. The classical interaction energy of one of our oscillator molecules with a heat bath molecule depends on the displacement r of the atomic separation distance between the atoms of our diatomic molecule from its equilibrium value r_0. If the interaction with the heat bath is weak (and the adiabatic condition which requires the vibration period of the diatomic molecule to be short compared with collision times is satisfied), the interaction energy can be expanded as a power series in r and terms of order r^2 or higher can be neglected. The coefficients appropriate to a given collision depend on collision angles and relative velocities but these must be averaged over all possible collisions with each weighted according to a Maxwellian velocity distribution of heat bath molecules.

If the transition probability per collision, P_{10}, for the transition $0 \rightarrow 1$ can be determined, Landau and Teller show that the linear perturbation of the vibrations of the diatomic molecules by the heat bath induce other transitions with probabilities:

$$P_{mn} = [(m+1)\delta_{n-1,m} + m\delta_{n+1,m}]P_{10} = P_{nm} \tag{II.2}$$

per collision. Here

$$\delta_{nm} = \begin{cases} 1 & \text{if } m = n \\ 0 & \text{otherwise} \end{cases} \tag{II.3}$$

It is to be noted that these transition probabilities are exactly those associated with optical transitions of a harmonic oscillator in a radiation bath. Transitions are only possible between adjacent levels.

The transition probabilities per collision P_{mn} are related to the transition probabilities per unit time W_{mn} (for $n \rightarrow m$) by[18]

$$W_{n+1,n} = Z^* N^* e^{-\theta} P_{n,n+1} \tag{II.4}$$

$$W_{n,n+1} = Z^* N^* P_{n+1,n} \tag{II.5}$$

so that $W_{n,n+1}/W_{n+1,n} = \exp \theta$ as required by the principle of detailed balancing. The quantity Z^* is the collision number, i.e., the number of collisions per unit time suffered by the oscillator when the gas density is one molecule per unit volume, and N^* is the total concentration of heat bath molecules, and $\theta = h\nu/kT$.

We postulate the potential energy curve of a "dissociating harmonic oscillator" reactant as that shown in Fig. 1. It is a truncated harmonic oscillator potential with a finite number of equally spaced energy levels such that the level N is the last discrete level of the oscillator and the energy $E_{N+1} = h\nu(N+1)$ is the dissociation energy and the activation energy for the reaction. Here ν is the natural frequency of the oscillator.

In Sections III to VI we will discuss some of the general stochastic properties of an ensemble of molecules undergoing stepwise

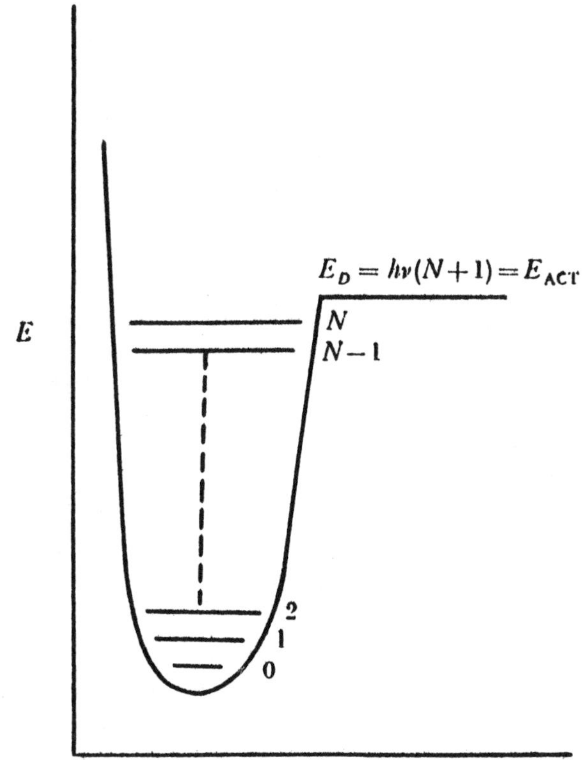

Fig. 1. Potential energy for a harmonic oscillator reactant with a dissociation "cut-off." The energy $E_D = h\nu(N+1)$ is the dissociation energy

transitions in a discrete energy-level system with and without absorbing barriers. The development presented in these sections is quite general and pertains to any ensemble of reactant molecules in the quantized energy levels. We will derive general expressions for the time-dependent distribution of the molecules among these energy levels, for mean first passage times, and for the perturbation introduced into an initial equilibrium distribution by chemical reactions. In Section VII we will then apply some of these results to the study of a system of harmonic oscillator reactants with analytical forms for the transition probabilities given in the foregoing.

III. TRANSITIONS WITHOUT REACTION

Let us assume (a) that the energy levels of our molecules are $E_0, E_1, E_2, \ldots, E_N$; (b) that the fraction of molecules in the mth state at time t is $x_m(t)$; (c) that the transition probabilities per unit time W_{nm} from state m to n can be computed in terms of the interaction of the molecules with a heat bath (which is postulated to remain at temperature T) by application of quantum-mechanical time dependent perturbation theory (the W_{nm}'s being proportional to squares of absolute values of the matrix elements of the interaction energy); and (d) that the temporal variations of the level concentrations are described through the transport equation

$$\frac{dx_n}{dt} = \sum_{\substack{m=0,1,2,\ldots \\ m \neq n}} \{W_{nm} x_m - x_n W_{mn}\} = \sum_{m=0,1,\ldots} A_{nm} x_m \qquad \text{(III.1)}$$

The positive terms represent the increase in number of occupants of the nth level by transitions $m \to n$ while the negative sum corresponds to the loss associated with transitions $n \to m$. Critical discussions of the derivation and validity of transport equations such as Eq. III.1 have been made by van Hove[9] and Luttinger and Kohn.[10] We hope sometime to make an analogous analysis of the validity of these equations as they are applied to problems in chemical kinetics.

Various properties of the matrices $W = (W_{nm})$ and $A = (A_{nm})$ and of the vector $x = \{x_0(t), x_1(t), \ldots\}$ are immediately apparent in a system which suffers no loss of molecules (i.e., *when no reaction occurs*). The equilibrium distribution of x_n is

$$x_n(\infty) = Z^{-1} \exp(-\beta E_n), \; Z \doteq \sum_n \exp(-\beta E_n) \qquad \text{(III.2)}$$

For detailed balance at equilibrium

$$W_{nm} \exp(-\beta E_m) = W_{mn} \exp(-\beta E_n) \qquad \text{(III.3)}$$

We note that

$$A_{nm} = W_{nm}(1-\delta_{nm})-\delta_{nm}\sum_l W_{ln}(1-\delta_{nl}) \qquad \text{(III.4)}$$

so that for fixed m

$$\begin{aligned}\sum_n A_{nm} &= \sum_n W_{nm}(1-\delta_{nm})-\sum_l (1-\delta_{nl})\sum_n \delta_{nm} W_{ln} \\ &= \sum_n W_{nm}(1-\delta_{nm})-\sum_l W_{lm}(1-\delta_{lm})\end{aligned} \qquad \text{(III.5)}$$

and

$$\sum_n A_{nm} = 0,\ m = 0, 1, \ldots \qquad \text{(III.6)}$$

This is equivalent to the statement that particles are conserved during transitions.

If the property, Eq. III.6, is common to two matrices A and B, then it is also characteristic of their product $C = AB$, for

$$C_{nm} = \sum_l A_{nl}\, B_{lm} \qquad \text{(III.7a)}$$

and

$$\sum_n C_{nm} = \sum_l B_{lm} \sum_n A_{nl} = 0 \qquad \text{(III.7b)}$$

The matrix A can be symmetrized through the introduction of the new variables

$$y_n = x_n \exp\left(\tfrac{1}{2}\beta E_n\right) \qquad \text{(III.8)}$$

for then Eq. III.1 becomes

$$\frac{dy_n}{dt} = \sum_{m=0,1,\ldots} A_{nm}\, e^{\frac{1}{2}\beta(E_n-E_m)}\, y_m = \sum_{m=0,1,\ldots} B_{nm}\, y_m \qquad \text{(III.9)}$$

where

$$B_{nm} = A_{nm} \exp \tfrac{1}{2}(E_n-E_m)\beta \qquad \text{(III.10)}$$

Since the off-diagonal elements of A_{nm} (see Eq. III.4) are W_{nm} the off-diagonal elements of B are, in view of detailed balance, Eq. III.3,

$$B_{nm} = W_{nm} \exp\{-\tfrac{1}{2}\beta(E_m-E_n)\} = B_{mn} \qquad \text{(III.11)}$$

Solutions of the symmetrized transport equation exist in the form

$$y_n(t) = \sum_j c_j \psi_j(n) \exp \lambda_j t \qquad \text{(III.12)}$$

where the $\psi_j(n)$'s and λ_j's (with $-\lambda_0 \leqq -\lambda_1 \leqq -\lambda_2 \leqq \ldots$) are, respectively, the normalized characteristic vectors and characteristic values of the matrix B, and the c_j's are related to the initial level distribution by

$$c_j = \sum_m y_m(0)\psi_j(m) = \sum_{m=0,1,\ldots} x_m(0)\psi_j(m) \exp \tfrac{1}{2}\beta E_m \qquad \text{(III.13)}$$

Also

$$x_n(t) = \exp(-\tfrac{1}{2}\beta E_n) \sum_j c_j \psi_j(n) \exp \lambda_j t \qquad \text{(III.14)}$$

The conservation conditions Eqs. III.6 and III.10 imply that

$$\sum_n B_{nm} \exp(-\tfrac{1}{2}\beta E_n) = 0; \qquad \text{(III.15)}$$

which means that

$$\psi_0(n) = \frac{\exp(-\tfrac{1}{2}\beta E_n)}{\{\sum_n \exp(-\beta E_n)\}^{1/2}} \qquad \text{(III.16)}$$

is a characteristic vector of B which has a characteristic value $\lambda_0 = 0$. Furthermore, since $\sum x_n(0) = 1$

$$c_0 = Z^{-1/2} \qquad \text{(III.17)}$$

Hence

$$x_n(t) = Z^{-1} \exp(-\beta E_n) + \sum_{j>0} c_j \psi_j(n) \exp(\lambda_j t) \qquad \text{(III.18)}$$

Since $x_n(t) \leqq 1$, all λ_j must be negative (we assume $\lambda_0 = 0$ is the only vanishing λ).

IV. REACTION AS AN ABSORBING BARRIER

The discussion of the last section can be generalized to include the possibility of a chemical reaction.[13] Consider the case in which the achievement of the $(N+1)$st level represents the completion of the reaction and in which the reaction occurs only by a molecule passing into the $(N+1)$st level. Any molecule which reaches this level is absorbed or "dies." The reaction rate is determined by the rate at which molecules in their "random walk" from level to level reach the $(N+1)$st level *for the first time*. In the language of the theory of stochastic processes the mean time for level $(N+1)$ to be reached is the *mean first passage time* for the Nth level (the time required to *pass N for the first time*).

We let $F(t)$ be the fraction of molecules which have not yet reached $(N+1)$ in the time interval $(0, t)$. Then

$$F(t) = \sum_{n=0}^{N} x_n(t) \tag{IV.1}$$

The fraction of molecules which dissociate in an infinitesimal time interval $(t, t+\delta t)$ is

$$-[F(t+\delta t)-F(t)] = -(dF/dt)\delta t \tag{IV.2}$$

If $P(t)$ is the distribution of first passage times for transitions past level N, the number of molecules which pass N in the interval $(t, t+\delta t)$ is $P(t)\delta t$. Since all passing molecules are immediately "absorbed" by the barrier at $(N+1)$

$$P(t) = -\frac{dF}{dt} = -\frac{d}{dt}\sum_{0}^{N} x_n(t) \tag{IV.3}$$

The mean first passage time is

$$\begin{aligned} \bar{t} &= \int_0^\infty tP(t)dt = -\int_0^\infty t\frac{d}{dt}\sum_0^N x_n(t)dt \\ &= \int_0^\infty \sum_0^N x_n(t)\,dt \end{aligned} \tag{IV.4}$$

We first consider in detail the case of molecules with simple selection rules which allow transitions between neighboring levels

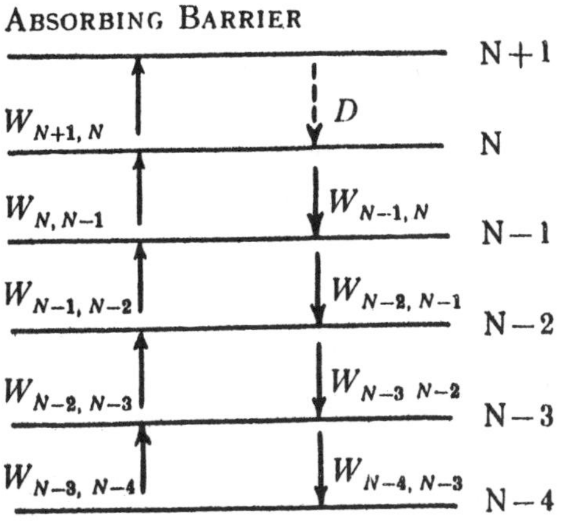

Fig. 2. Schematic diagram of nearest neighbor transitions.

only (see Fig. 2) and include the possibility that the "level" $(N+1)$ is only partially absorbing.

The transport equations, III.1, are now

$$dx_0/dt = -W_{10}x_0 + W_{01}x_1 \tag{IV.5a}$$

$$dx_j/dt = W_{j,j-1}x_{j-1} - (W_{j-1,j} + W_{j+1,j})x_j + W_{j,j+1}x_{j+1}, \quad j = 1, 2, \ldots, (N-1) \tag{IV.5b}$$

$$dx_N/dt = W_{N,N-1}x_{N-1} - (W_{N-1,N} + W_{N+1,N})x_N + D \tag{IV.5c}$$

where D represents the downflow rate from the state $(N+1)$. The characterization of D depends on the decomposition mechanism of molecules which leave the Nth state. We postulate the state $(N+1)$ to be a partially absorbing barrier with an absorption coefficient α such that of the normal rate of jump from level N to $N+1$, a fraction $\alpha W_{N+1,N}x_N$ is absorbed and "decomposes" while the remaining fraction $(1-\alpha)W_{N+1,N}x_N = \beta W_{N+1,N}x_N$ remains at the Nth level. To an observer at the Nth level this is equivalent to a flow rate $W_{N+1,N}x_N$ upward and $D = \beta W_{N+1,N}x_N$ downward, so that our required equation, IV.5c becomes

$$dx_N/dt = W_{N,N-1}x_{N-1} - (W_{N-1,N} + \alpha W_{N+1,N})x_N$$

where $\beta = 1-\alpha$ is the reflection coefficient of the state $(N+1)$. We shall usually be concerned with the perfect absorber $\alpha = 1$.

When $\alpha = 1$ the matrix A of Eq. IV.5 becomes

$$A = \begin{bmatrix} -W_{10} & W_{01} & 0 & 0 & 0\ldots & 0 & 0 \\ W_{10} & -(W_{01}+W_{21}) & W_{12} & 0 & 0\ldots & 0 & 0 \\ 0 & W_{21} & -(W_{12}+W_{32}) & W_{23} & 0\ldots & 0 & 0 \\ \cdot & \cdot & \cdot & \cdot & \cdot & & \\ \cdot & \cdot & \cdot & \cdot & \cdot & & \\ \cdot & \cdot & \cdot & \cdot & \cdot & & \\ 0 & 0 & 0 & 0 & 0\ldots & -(W_{N-2,N-1} + W_{N,N-1}) & W_{N-1,N} \\ 0 & 0 & 0 & 0 & 0\ldots & W_{N,N-1} & -(W_{N-1,N} + W_{N+1,N}) \end{bmatrix} \tag{IV.6}$$

The conservation condition, Eq. III.6, is violated in the last column since

$$\sum_n A_{nN} = -W_{N+1,N} \neq 0 \qquad \text{(IV.7)}$$

Hence the characteristic value $\lambda_0 = 0$ is perturbed to have a non-zero value. We can express A as

$$A = A_0 + \delta A \qquad \text{(IV.8)}$$

where A_0 satisfies the conservation condition, and δA is a matrix whose elements are all zero except for $(\delta A)_{N,N} = -W_{N+1,N}$.

The distribution of first passage times $P(t)$, Eq. IV.3, can be expressed in terms of $x_n(t)$ by summing Eqs. IV.5a, IV.5b, and IV.5c:

$$P(t) = -(d/dt)\sum x_n(t) = W_{N+1,N}\,x_N(t) \qquad \text{(IV.9)}$$

Then the mean first passage time is

$$\bar{t} = \int_0^\infty t\,P(t)dt = W_{N+1,N}\int_0^\infty t x_N(t)dt \qquad \text{(IV.10)}$$

These formulas are immediately generalizable to the case of transition to both nearest and next nearest neighbor levels (see Fig. 3). Here the conservation law is violated in the last two

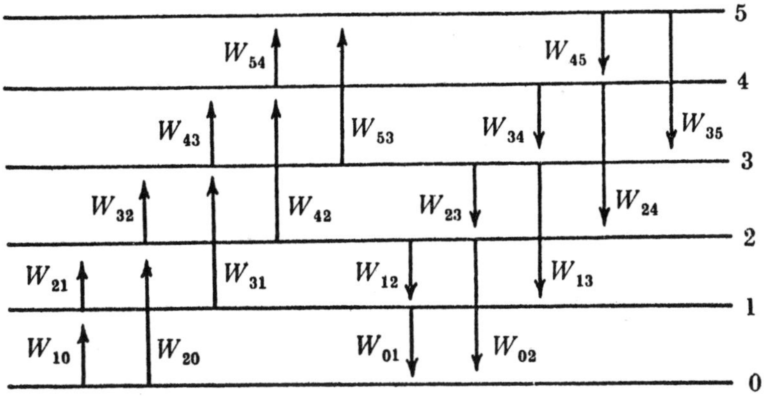

Fig. 3. Schematic diagram of nearest and next-nearest neighbor transitions.

columns in such a way that if A_0 represents the part of A which satisfies the conservation conditions, Eq. III.6, then

$$\delta A = - \begin{bmatrix} 0 & 0 & \dots & 0 & 0 & 0 \\ 0 & 0 & \dots & 0 & 0 & 0 \\ \cdot & \cdot & \dots & \cdot & \cdot & \cdot \\ 0 & 0 & \dots & 0 & 0 & 0 \\ 0 & 0 & \dots & 0 & W_{N+1,N-1} & 0 \\ 0 & 0 & \dots & 0 & 0 & W_{N+1,N} \end{bmatrix} \tag{IV.11}$$

Then

$$P(t) = \{W_{N+1,N}x_N(t) + W_{N+1,N-1}x_{N-1}(t)\} \tag{IV.12}$$

Under the more general conditions in which transitions can occur between levels n and $n\pm1$, $n\pm2, \dots, n\pm k$, the matrix δA has vanishing off-diagonal elements, and its diagonal elements are

$$[0, 0, \dots, 0, W_{N+1,N-k}, W_{N+1,N-k+1} \dots, W_{N+k,N}] \tag{IV.13}$$

Then

$$P(t) = \sum_{i=0}^{k} W_{N+1,N-i}\, x_{N-i}(t) \tag{IV.14}$$

V. GENERAL THEORY OF MEAN FIRST PASSAGE TIME

Let A be the transition probability matrix of the set of differential equations (analogous to Eq. IV.5) which describe the variation of level concentration with time in the presence of an absorbing barrier. Then the set of equations in matrix form is

$$dX(t)/dt = AX(t) \tag{V.1}$$

where $X(t)$ is a vector with components $x_0(t), x_1(t), \dots, x_N(t)$. The solution of this equation is

$$X(t) = e^{At}X(0) \tag{V.2}$$

$X(0)$ being the initial concentration vector. If all molecules are initially in the ground state

$$X(0) = \{1, 0, 0, \dots, 0\} \tag{V.3}$$

It is convenient to express the exponential matrix $\exp(At)$ as a linear combination of the characteristic matrices of A, the $f_j(A)$'s, which satisfy the relations[19]

$$Af_j(A) = \lambda_j f_j(A), j = 0, 1, 2, \dots, N \tag{V.4a}$$

$$f_k(A)f_j(A) = \delta_{kj}f_j(A) \tag{V.4b}$$

$$\sum_{j=0}^{N} f_j(A) = I \tag{V.4c}$$

where I is the identity matrix, and the λ_j's are the characteristic values of A. An explicit representation of $f_j(A)$ is

$$f_j(A) = \frac{(\lambda_0-A)(\lambda_1-A)\ldots(\lambda_{j-1}-A)(\lambda_{j+1}-A)\ldots(\lambda_N-A)}{(\lambda_0-\lambda_j)(\lambda_1-\lambda_j)\ldots(\lambda_{j-1}-\lambda_j)(\lambda_{j+1}-\lambda_j)\ldots(\lambda_N-\lambda_j)} \tag{V.5}$$

Since Eq. V.4a implies $G(A)f_j(A) = G(\lambda_j)f_j(A)$, when $G(\lambda)$ can be expressed as a power series in λ, we have

$$\begin{aligned} e^{At}I &= e^{At}\sum_{j=0}^{N} f_j(A) \\ &= \sum_{j=0}^{N} e^{At} f_j(A) = \sum_{j=0}^{N} e^{\lambda_j t} f_j(A) \end{aligned} \tag{V.5}$$

and

$$\begin{aligned} \int_0^\infty X(t)dt &= \sum_{j=0}^{N} \left\{\int_0^\infty e^{\lambda_j t}\, dt\right\} f_j(A)X(0) \\ &= -\sum_{0}^{N} \lambda_j^{-1} f_j(A)X_0(0) = -\sum_{0}^{N} A^{-1} f_j(A)X(0) \\ &= -A^{-1}X(0). \end{aligned} \tag{V.6}$$

Let the characteristic equation whose roots are λ_j be

$$\lambda^{N+1} + a_1\lambda^N + \ldots + a_N\lambda + a_{N+1} = 0 \tag{V.7a}$$

Then, since A satisfies its own characteristic equation:

$$A^N + a_1 A^{N-1} + \ldots + a_N I = -a_{N+1}A^{-1} \tag{V.7b}$$

Also, since the determinant of A is $(-1)^{N+1}a_{N+1}$

$$\int_0^\infty X(t)dt = \frac{(-1)^{N+1}}{\det A}\left\{a_N I + a_{N-1}A + \ldots + A^N\right\} X(0) \tag{V.8}$$

The a_j's are, of course, the invariants of the matrix A. For example $a_1 = -$ trace A and

$$\begin{aligned} a_N &= (-1)^N \sum_i \lambda_0\lambda_1\ldots\lambda_{i-1}\lambda_{i+1}\ldots\lambda_N \\ &= (-1)^N \det A \sum_{i=0}^{N} \lambda_0^{-1} = (-1)^N \det A \text{ trace } A^{-1} \end{aligned} \tag{V.9}$$

Let $(i|A^m|j)$ represent the ,elements in the ith row and jth column of the matrix A^m. Then, from Eq. IV.4 which states that the mean first passage time from an initial jth state is

$$\begin{aligned} \bar{t} &= \int_0^\infty [x_0(t) + \ldots + x_N(t)]dt = -\sum_{i=0}^{N} (i|A^{-1}|j) \\ &= \frac{(-1)^{N+1}}{\det A} \sum_{l=0}^{N} a_l \sum_{i=0}^{N} (i|A^{N-l}|j) \end{aligned} \tag{V.10}$$

This formula simplifies considerably when all molecules are initially in the ground state and when det A is a continuant, as it is when only transitions between nearest neighbor states occur (for example in the case of the simple harmonic oscillator model). We recall that A as defined in Eq. IV.6 has the property

$$\sum_{i=0}^{N} (i|A|j) = 0 \text{ for } j = 0, 1, \ldots, N-1 \tag{V.11}$$

but not for $j = N$. Now

$$\begin{aligned} \sum_{i=0} (i|A^2|j) &= \sum_{i,k}^{N} (i|A|k)(k|A|j) \\ &= \sum_{k=0}^{N} (k|A|j) \sum_{i} (i|A|k) \\ &= -(N|A|j)W_{N+1,N} \\ &= 0 \text{ unless } j = N \text{ or } N-1 \end{aligned} \tag{V.12}$$

By repeating this argument we find

$$\sum_{i=0}^{N} (i|A^3|j) = 0 \text{ unless } j = N, \ N-1, \text{ or } N-2, \text{ etc.} \tag{V.13a}$$

until

$$\sum_{i=0}^{N} (i|A^N|j) = 0 \text{ only for } j = 0. \tag{V.13b}$$

Since $\sum_i (i|I|j) = 1$ we finally obtain (after combining Eqs. V.10 and V.13)

$$\bar{t} = -\text{ trace } A^{-1} \tag{V.14}$$

If transitions are made from a given level to nearest and next nearest neighboring levels and if only the ground state is occupied initially, an argument similar to that used above implies that the terms $l = 0$ and $l = N$ contribute to $\bar{t}$.

An alternative formulation of the solution of Eq. V.1 is that given by Eq. III.4:

$$x_n(t) = e^{-\frac{1}{2}\beta E_n} \sum_{j=0}^{N} c_j \psi_j(n) e^{t\lambda_j} \tag{V.15}$$

where the ψ_j's and λ_j's are now the characteristic vectors and values of the symmetrized form of Eq. IV.6. The c_j's are given by Eq. III.13. The characteristic value $\lambda_j = 0$ is perturbed by the matrix δA which describes the absorbing barrier. We show in the next section that the equilibrium theory for our model of a chemical reaction can be obtained by applying perturbation theory to the determination of the corrected λ_0.

Equation IV.14 implies the mean first passage time to be

$$\begin{aligned} \bar{t} &= \int_0^\infty tP(t)dt \\ &= \sum_{j=0}^{N} \sum_{i=0}^{k} W_{N+1,N-i}\, e^{-\frac{1}{2}\beta E_{N-i}}\, c_j \psi_j(N-i) \lambda_j^{-2} \end{aligned} \tag{V.16}$$

VI. PERTURBATION THEORY AND EQUILIBRIUM THEORY OF CHEMICAL KINETICS

Let us suppose that $E_{N+1} \gg kT$. Then if level $(N+1)$ were not an absorbing barrier but merely the level of a typical highly excited state, the fraction of molecules with energy E_{N+1} would be very small at equilibrium. When level $(N+1)$ is an absorbing barrier we would expect rate processes associated with phenomenon such as propagation of shock waves through a diatomic gas to be governed by two time scales. First the initial Boltzmann distribution of the diatomic molecules would be transformed into the new Boltzmann distribution which corresponds to the final temperature of the gas after the passage of the shock wave. The characteristic time for this process is $\simeq -1/\lambda_1$. The molecules occasionally are excited to the $(N+1)$st level where a reaction occurs. The time associated with this process is determined by the variation of $-\lambda_0$ from 0 associated with the perturbation δA (Eq. IV.8) which arises from making level $(N+1)$ into an absorbing barrier.

The perturbed λ_0 is given by

$$\lambda_0 = \psi_0^{(0)} \cdot \delta A \cdot \psi_0^{(0)} \tag{VI.1}$$

where the squares of the elements of the unperturbed characteristic vector $\psi_0^{(0)}$, $[\psi_0^{(0)}(n)]^2$, form a Boltzmann distribution. Then in the case of nearest neighbor transitions only we find by combining Eqs. VI.1, IV.8, and III.16

$$\lambda_0 = -W_{N+1,N}Z^{-1}\exp(-\beta E_N) = -W_{N,N+1}Z^{-1}\exp(-\beta E_{N+1}) \quad \text{(VI.2)}$$

The time scale associated with the reaction processes is of the order of $-1/\lambda_0$, and the reaction rate is proportional to $-\lambda_0$.

The λ's other than λ_0 and the ψ's are changed but little (of $0(\exp -\beta E_{N+1})$) by the perturbation. Hence when $n \ll N$ the value of x_n at times $\gg -\lambda_1^{-1}$ is

$$x_n \simeq Z^{-1}\exp(-\beta E_n)\exp\{-tW_{N,N+1}Z^{-1}\exp(-\beta E_{N+1})\} \quad \text{(VI.3)}$$

so that

$$x_n/x_m \simeq \exp\beta(E_m - E_n) \text{ for } m, n \ll N \quad \text{(VI.4)}$$

and the Boltzmann ratio of level concentrations is preserved during the reaction. This is the usual hypothesis of the equilibrium theory of chemical kinetics.

The generalization of Eqs. IV.8 and IV.13 which is appropriate when transitions between more distant neighbors can occur is

$$\lambda_0 \simeq -Z^{-1}\{\sum_j W_{j,N+1}\}\exp(-\beta E_{N+1}) \quad \text{(VI.5)}$$

Unfortunately the mean first passage time is more difficult to deduce than λ_0 for, if we refer to Eq. V.16 (first in the case $k = 0$), we see that even when $|\lambda_0^{-1}| \gg |\lambda_1^{-1}|$

$$l \simeq W_{N+1,N}\, e^{-\frac{1}{2}\beta E_N}\, \psi_0(N)\lambda_0^{-2}Z^{-\frac{1}{2}} \quad \text{(VI.6)}$$

The unperturbed $\psi_0^{(0)}(N)$ is $Z^{-\frac{1}{2}}\exp(-\frac{1}{2}\beta E_N)$ which is already a small number. Hence it is quite possible for the perturbation in $\psi_0(N)$ to be of the same order of magnitude as the unperturbed $\psi_0^{(0)}(N)$ itself. All other $\psi_j^{(0)}(n)$'s would be necessary in order to ascertain whether or not this is the case. The rough order of magnitude of l can be obtained however from

$$\begin{aligned} l &\simeq W_{N,N+1}Z^{-1}\exp(-\beta E_{N+1})\lambda_0^{-2} \\ &= ZW_{N,N+1}^{-1}\exp(\beta E_{N+1}) \end{aligned} \quad \text{(VI.7)}$$

The error obtained in using this formula for the harmonic oscillator molecule will be discussed in the next section.

VII. THE HARMONIC OSCILLATOR

The general theory developed in the previous sections can be applied immediately to the harmonic oscillator model of a diatomic molecule. The quantum-mechanical transition probabilities given in Section II yield the transport equation

$$dx_n(t)/dt = \varkappa\{ne^{-\theta}x_{n-1} - [n+(n+1)e^{-\theta}]x_n + (n+1)x_{n+1}\} \qquad \text{(VII.1a)}$$

where $\varkappa$ depends only on the coupling between molecules and heat bath and where $\theta = h\nu/kT$. These equations have been solved by the authors in the absence of an absorbing barrier.[14] In the presence of an absorbing barrier at level $(N+1)$ the top boundary condition is

$$dx_N/dt = \varkappa\{Ne^{-\theta}x_{N-1} - [N+\alpha(N+1)e^{-\theta}]x_N\} \qquad \text{(VII.1b)}$$

Here α is the absorption coefficient (which in the next few paragraphs we set equal to 1).

The symmetrical matrix B is obtained by letting

$$y_n = x_n \exp(\tfrac{1}{2}\theta n) \qquad \text{(VII.2)}$$

Then

$$dy_n/dt = \varkappa\{ne^{-\frac{1}{2}\theta}y_{n-1} - [n+(n+1)e^{-\theta}]y_n + (n+1)y_{n+1}e^{-\frac{1}{2}\theta}\};$$
$$n = 0, 1, \ldots, N-1 \qquad \text{(VII.3a)}$$

$$dy_N/dt = \varkappa\{Ne^{-\frac{1}{2}\theta}y_{N-1} - [N+\alpha(N+1)e^{-\theta}]y_N\} \qquad \text{(VII.3b)}$$

and B has the form (exhibited for $N = 3$)

$$B = \varkappa\begin{bmatrix} -e^{-\theta} & e^{-\frac{1}{2}\theta} & 0 & 0 \\ e^{-\frac{1}{2}\theta} & -(1+2e^{-\theta}) & 2e^{-\frac{1}{2}\theta} & 0 \\ 0 & 2e^{-\frac{1}{2}\theta} & -(2+3e^{-\theta}) & 3e^{-\frac{1}{2}\theta} \\ 0 & 0 & 3e^{-\frac{1}{2}\theta} & -(3+4\alpha e^{-\theta}) \end{bmatrix} \qquad \text{(VII.4)}$$

Det B, having nonvanishing elements only along the diagonal and first off-diagonal, is a continuant. Trace B^{-1} and hence the mean first passage time $\bar{t}$ from the ground state, past the Nth level is found in Eq. A1.15 of Appendix I to be

$$
\begin{aligned}
\varkappa l &= -\text{ trace } B^{-1} \\
&= \sum_{j=1}^{N+1} e^{j\theta}\left\{\frac{1}{j}+\frac{1}{j+1}+\ldots+\frac{1}{N+1}\right\} \\
&= e^{(N+1)\theta}\sum_{k=0}^{N} e^{k\theta}\left\{\frac{1}{N+1}+\frac{1}{N}+\ldots+\frac{1}{N+1-k}\right\}.
\end{aligned}
\tag{VII.5}
$$

This corresponds to a totally absorbing barrier ($\alpha = 1$) at level $(N+1)$.

It is shown in Eq. A1.17 that $N \to \infty$

$$
\begin{aligned}
\varkappa l &= -\varkappa \text{ trace } B^{-1} \\
&= \frac{e^{(N+1)\theta}}{(N+1)(e^{-\theta}-1)^2}\left\{1+\frac{e^{-\theta}}{N(1-e^{-\theta})}+\frac{e^{-2\theta}}{2N(N-1)(1-e^{-\theta})^2}+0(N^{-3})\right\}
\end{aligned}
\tag{VII.6}
$$

The equilibrium theory first passage time is applicable in limit as $N \to \infty$. Hence corrections to it are to be expected when the second term in the product above is not negligible, i.e., when N is not much greater than $e^{-\theta}(1-e^{-\theta})^{-1}$. The mean first passage time and the rate of activation, $v_{\text{act}} \propto l^{-1}$, deviate from their equilibrium value by more than 10 per cent when

$$N(1-e^{-\theta}) < 10e^{-\theta} \tag{VII.7a}$$

In the high-temperature limit this corresponds approximately to

$$Nh\nu/kT = E_{\text{act}}/kT < 10 \tag{VII.7b}$$

which is in agreement with the results discussed in Section I.

The ratio of the equilibrium first passage time to that calculated on the basis of Eqs. VII.5 is plotted in Fig. 4 as a function of N for various values of θ. We shall see later that this ratio is exactly that of the reaction rate, based on Eq. VII.5, to the equilibrum rate.

The complete characteristic vector analysis of the harmonic oscillator model can be effected through the aid of Gottlieb polynomials.[5] Let us assume $x_n(t)$ to be a linear combination of products

$$l_n \exp\{\varkappa\mu t(e^{-\theta}-1)\} \tag{VII.8}$$

Substitution of this quantity into Eqs. VII.1a and VII.1b shows that l_n satisfies the difference equation

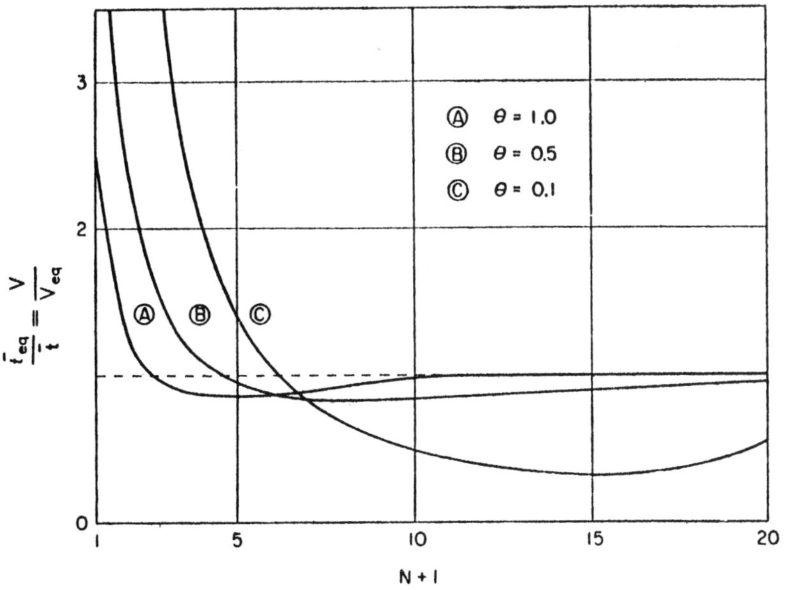

Fig. 4. The ratio of the equilibrium theory first passage time to exact first passage time as a function of number of levels for various values of $\theta = h\nu/kT$.

$$(e^{-\theta}-1)\mu l_n = ne^{-\theta} l_{n-1} - \{n+(n+1)e^{-\theta}\} l_n + (n+1) l_{n+1}, \quad n = 0, 1, \ldots, N \qquad \text{(VII.9a)}$$

where μ is a root of the end-condition equation

$$l_{N+1}(\mu) = (1-\alpha)e^{-\theta} l_N(\mu) \qquad \text{(VII.9b)}$$

The symmetrizing transformation

$$l_n = y_n e^{-\frac{1}{2}n\theta} \qquad \text{(VII.10)}$$

allows us to express the set, Eq. VII.9a in matrix form

$$By = \lambda y \qquad \text{(VII.11)}$$

where B is an $(N+1) \times (N+1)$ generalization of Eq. VII.4,

$$\lambda = \mu\varkappa(e^{-\theta}-1) \qquad \text{(VII.12)}$$

and

$$y = \{y_0, y_1, \ldots, y_N\}$$

In view of the symmetrical nature of the matrix B operating on the y vector we see that the characteristic vectors

$$y^{(k)} = \{y_0^{(k)}, y_1^{(k)}, \ldots, y_N^{(k)}\} \tag{VII.13}$$

which correspond to different characteristic values $\lambda_k = \mu_k(e^{-\theta}-1)$ must be orthogonal

$$y^{(k)} \cdot y^{(l)} = 0 \text{ if } k \neq l \tag{VII.14}$$

(the characteristic values being degenerate). Hence the solutions $l_n(\mu_j)$ which correspond to characteristic values $\mu_j(e^{-\theta}-1)$ satisfy the orthogonality condition

$$\sum_{n=0}^{N} e^{n\theta} l_n(\mu_j) l_n(\mu_k) = 0 \text{ if } j \neq k \tag{VII.15}$$

The Gottlieb polynomials $l_n(\mu)$ which are defined by

$$l_n(\mu) = e^{\theta\mu}\Delta^n\left\{\binom{\mu}{n} e^{-\theta\mu}\right\} \tag{VII.16}$$

or

$$l_n(\mu) = e^{-n\theta} \sum_{\nu=0}^{N} (1-e^{\theta})^{\nu} \binom{n}{\nu}\binom{\mu}{\nu} \tag{VII.17}$$

satisfy the recursion formula, Eq. VII.9a. Hence the end condition, Eq. VII. 1b, is satisfied by choosing μ to be a root of Eq. VII.9b.

The general solution of Eq. VII.1a is then

$$x_n(t) = \sum_{j=0}^{N} a_j l_n(\mu_j) \exp\{\mu_j(e^{-\theta}-1)t\varkappa\} \tag{VII.18}$$

where the coefficients a_j are given in terms of the initial distribution $x_n(0)$ through the aid of the orthogonality relation, Eq. VII.15

$$a_j = \sum_{n=0}^{N} x_n(0) l_n(\mu_j) e^{n\theta} \Bigg/ \sum_{n=0}^{N} l_n^2(\mu_j) e^{n\theta} \tag{VII.19}$$

Two special cases, the initial delta distribution and initial Boltzmann distribution are of interest. If

$$x_n(0) = \delta_{nm} \tag{VII.20}$$

$$\begin{aligned} a_j &= \sum_{n=0}^{N} \delta_{nm} l_n(\mu_j) e^{n\theta} \Big/ \sum_{n=0}^{N} l_n^2(\mu_j) e^{n\theta} \\ &= l_m(\mu_j) e^{m\theta} \Big/ \sum_{n=0}^{N} l_n^2(\mu_j) e^{n\theta} \end{aligned} \tag{VII.21}$$

Hence

$$x_n(t) = \sum_{j=0}^{N} \left\{ \frac{l_m(\mu_j)\, l_n(\mu_j)\, e^{m\theta}}{\sum_{s=0}^{N} l_s^2(\mu_j) e^{s\theta}} \right\} \exp\{\mu_j(e^{-\theta}-1)t\varkappa\} \tag{VII.22}$$

If $x_n(0)$ is a Boltzmann distribution with $\theta_0 \gg \theta$

$$x_n(0) = e^{-\theta_0}(1-e^{-\theta_0})$$

(we assume θ_0 to be sufficiently large so that $e^{-n\theta_0} \simeq 0$ when $n \geqq N$), then

$$\begin{aligned} a_j &\simeq \frac{[1-e^{-\theta_0}] \sum_{n=0}^{\infty} e^{-n(\theta_0-\theta)}\, l_n(\mu_j)}{\sum_{n=0}^{N} l_n^2(\mu_j) e^{n\theta}} \\ &= \left\{ \frac{1-e^{-(\theta_0-\theta)}}{1-e^{-\theta_0}} \right\}^{\mu_j} \Big/ \sum_{n=0}^{N} l_n^2(\mu_j) e^{n\theta} \end{aligned} \tag{VII.23}$$

and

$$x_n(t) = \sum_{j=0}^{N} \left\{ l_n(\mu_j) \left[\frac{1-e^{-(\theta_0-\theta)}}{1-e^{-\theta_0}} \right]^{\mu_j} \Big/ \sum_{m=0}^{N} l_m^2(\mu_j) e^{m\theta} \right\} e^{-\mu_j(1-e^{-\theta})t\varkappa} \tag{VII.24}$$

The distribution of mean first passage times $P(t)$ is obtained from Eq. IV.9:

$$\begin{aligned} P(t) &= -\frac{d}{dt} \sum x_n(t) = W_{N+1,N}\, x_N(t) \\ &= \alpha e^{-\theta}(N+1) x_N(t) \end{aligned}$$

so that

$$P(t) = \varkappa\alpha(N+1)e^{-\theta} \sum_{j=0}^{N} a_j l_N(\mu_j) \exp\,[-\mu_j(1-e^{-\theta})\varkappa t] \tag{VII.25}$$

In particular when all molecules are initially in the ground state

$$P(t) = \varkappa\alpha(N+1)e^{-\theta}\sum_{j=0}^{N} l_N(\mu_j)\exp(\lambda_j t)\{\sum_{n=0}^{N} l_n^2(\mu_j)e^{n\theta}\}^{-1} \tag{VII.26}$$

The mean first passage time in this case is then

$$\begin{aligned} \bar{t} &= \int_0^\infty tP(t)dt \\ &= \varkappa\alpha(N+1)e^{-\theta}\sum_{j=0}^{N}\lambda_j^{-2} l_N(\mu_j)\{\sum_{n=0}^{N} l_n^2(\mu_j)e^{n\theta}\}^{-1} \end{aligned} \tag{VII.27}$$

We can derive the equilibrium theory estimate of $\bar{t}$ by letting $N \to \infty$ and using some limit properties of $l_N(\mu)$ as $N \to \infty$. No chemical reaction takes place in this limit. It was shown by Gottlieb that as $N \to \infty$, the roots μ of $l_N(\mu) = 0$ approach $0, 1, 2, \ldots, N-1$. This means that the roots of Eq. VII.9b also approach nonnegative integers. Hence $\lambda_0 \to 0$ while $|\lambda_j|\varkappa > (1-e^{-\theta})$ for $j > 0$ so that the first term in Eq. VII.27 dominates all others and as $N \to \infty$

$$\bar{t} \sim \varkappa\,\alpha e^{-\theta}(N+1)\lambda_0^{-2} l_N(\mu_0)\Big/\sum_{n=0}^{N} l_n^2(\mu_0)e^{n\theta} \tag{VII.28}$$

We have shown in Appendix II, Eqs. A2.9 and A2.10, that

$$\mu_0 \sim \frac{(N+1)\alpha(1-e^{-\theta})e^{-(N+1)\theta}}{[1-(1-\alpha)e^{-\theta}]} \tag{VII.29a}$$

and

$$l_N(\mu_0) \sim e^{-N\theta}\left\{\frac{1-e^{\theta}}{1-(1-\alpha)e^{-\theta}}\right\} \tag{VII.29b}$$

while (see Eq. A2.8)

$$l_n(\mu_0) \sim e^{-n\theta} + 0(e^{-N\theta}) \text{ for } n \ll N \tag{VII.29c}$$

The denominator of Eq. VII.28 is, as $N \to \infty$

$$\sum_{n=0}^{N} l_n^2(\mu_0)e^{n\theta} \sim \sum_{n=0}^{N} l_n^2(0)e^{n\theta} = (1-e^{-\theta})^{-1} \tag{VII.29d}$$

The first passage time given by Eq. VII.28 is that of equilibrium chemical kinetics since it corresponds to the expression for $x_n(t)$ (Eq. VII.18) which in the limit as $N \to \infty$ would be the Boltzmann distribution. Substitution of Eqs. VII.29a to VII.29d and VII.12 into Eq. VII.28 yields

$$\varkappa l \sim \frac{[1-(1-\alpha)e^{-\theta}]e^{(N+1)\theta}}{(N+1)\alpha(1-e^{-\theta})^2} \qquad \text{(VII.30)}$$

which, when $\alpha = 1$ is exactly the same as the leading term in Eq. VII.6. The remainder of this article will concern this choice of α.

The relations $\beta E_{N+1} = (N+1)\theta = \beta E_{act}$, Eqs. II.2, II.5, the expression for the harmonic oscillator partition function $Z = (1-e^{-\theta})^{-1}$, and the definition of $\varkappa$:

$$\varkappa = Z^* P_{10} N^* \qquad \text{(VII.31)}$$

(where Z^* and N^* are the collision numbers and heat bath molecule concentration defined in Section II) allow us to rewrite Eq. VII.30

$$l = Z^2(W_{N,N+1})^{-1} \exp(\beta E_{N+1}) = Z^2(W_{N+1,N})^{-1} \exp(\beta E_N) \qquad \text{(VII.32)}$$

This differs by a factor Z from the estimate derived in Eq. VI.7 by the assumption that the characteristic value λ_0 of the transition probability matrix A can be determined by perturbation theory while the corresponding characteristic vector remains unchanged by the existence of an absorbing barrier. The significance of the error will be discussed presently.

The rate of activation v_{act} (in mole/cm^3/sec) is inversely proportional to the mean first passage time l:

$$v_{act} = \varkappa(N+1)X(1-e^{-\theta})^2 e^{-(N+1)\theta} = X/l \qquad \text{(VII.33)}$$

where X is the concentration of diatomic oscillator reactants in mole/cm^3. The rate constant for activation, k_{act} is, according to its usual definition, and from Eqs. VII.31 and VII.33

$$k_{act} = v_{act}/XN^* = Z^* P_{10}(N+1)(1-e^{-\theta})^2 \exp(-\beta E_{act}) \qquad \text{(VII.34)}$$

If we compare this equation with the Arrhenius equilibrium expression

$$k_{act} = A \exp(-\beta E_{act}) \qquad \text{(VII.35)}$$

we find the frequency factor A of our model to be

$$A = Z^* P_{10}(N+1)(1-e^{-\theta})^2 \qquad \text{(VII.36)}$$

Equations VII.33 and VII.36 correspond to the equilibrium values of l i.e., as $N \to \infty$. When $N\theta \ll 10$, v_{act} becomes (see Eq. VII.5)

$$v_{\text{act}} = \varkappa X \left[\sum_{j=1}^{N+1} e^{j\theta} \left(\frac{1}{j} + \frac{1}{j+1} + \ldots + \frac{1}{N+1} \right) \right]^{-1} \quad \text{(VII.37)}$$

Additional insight into stepwise activation processes is gained by expressing Eqs. VII.34 and VII.33 in the alternative form

$$\begin{aligned} v_{\text{act}} &= Z^{-2} X\, W_{N+1,N} \exp(-\beta E_N) \\ &= X_N^{(0)} (1-e^{-\theta}) W_{N+1,N} \end{aligned} \quad \text{(VII.38)}$$

where $X_N^{(0)} = XZ^{-1} \exp(-\beta E_N)$ is the equilibrium concentration of oscillators at the Nth level. This equation is to be interpreted as follows. If there were no reaction, the equilibrium concentration of oscillators in level N would be given by $X_N^{(0)}$. When there is a reaction, i.e., when the oscillators are removed irreversibly through passage upward from level N, the concentration X_N of oscillators in level N is less than the (closed system) Boltzmann equilibrium concentration $X_N^{(0)}$. In our model, this reduction in the concentration of oscillators in level N is given by the factor $(1-e^{-\theta})$ of Eq. VII.38. The product $(1-e^{-\theta})X_N^{(0)}$ thus represents the actual concentration of oscillators X_n in level N in a reacting system, and the product of this quantity with the transition probability per unit time $W_{N+1,N}$, gives the rate of activation to level $(N+1)$. Hirschfelder,[8] in discussing a simpler and more schematic model of stepwise activation in unimolecular reactions assumed the absorption coefficient α ($k_F/k_n = 1$ in Hirschfelder notation) to be 0.5 and obtained the equation

$$X_N = \left(\frac{1-e^{-\theta}}{2-e^{-\theta}} \right) X_N^{(0)} \quad \text{(VII.39)}$$

For $\theta = 1$ Hirschfelder thus finds $X_N/X_N^{(0)} = 0.387$ while we obtain $X_N/X_N^{(0)} = 0.632$ from Eq. VII.38. The difference in the numerical values is not important, being due to different models and different assumptions used by Hirschfelder and by us. The important point, as already expressed by Hirschfelder, is that in unimolecular reactions the concentration of molecules in the "activated state" is less than the value calculated for statistical equilibrium.

According to the "collision theory" of chemical kinetics, the rate constant for activation is given by the expression

$$k_{act} = P_c Z^* \exp(-E_{act}\beta) \qquad \text{(VII.40)}$$

where $Z^* \exp(-\beta E_{act})$ is the number of collisions per unit time per unit gas density in which the relative kinetic energy of the collision partners along the line of centers at contact exceeds the energy E_{act} and where P_c is the probability per collision that such a collision will actually leave one of the collision partners with an energy $E \geqq E_{act}$. A comparison of Eq. VII.40 with Eq. VII.35 yields

$$A = Z^* P_c \qquad \text{(VII.41)}$$

for the frequency factor of the rate of activation in the standard collision theory treatment.

As far as we are aware, no *a priori* calculations have even been carried out to evaluate P_c from the molecular properties of the collision partners. Its order of magnitude has only been found *a posteriori* for reaction systems. It is, therefore, impossible to make a direct numerical comparison between the frequency factors A found for the process of stepwise activation Eq. VII.36 and for the "all or nothing" kinetic theory activation Eq. VII.41. The following indirect comparison is, however, instructive. Measurements on the rate of activation of I_2 at about 300°K in various inert gases such as He, Ne, A, Kr, and N_2 have shown[22] that the frequency factor A is of the order of about 5×10^{16} cm³/mole/sec. Since the collision number Z^* is only about 10^{14} cm³/mole/sec at 300°K, even a value of $P_c = 1$, which corresponds to unit efficiency in direct collisional activation, could not raise the calculated A-value for the standard collision theory (Eq. VII.41) to the observed one. This is, of course, one of the old and vexing problems in chemical kinetics.[17]

We can also make an approximate calculation of the frequency factor A for the stepwise activation process (Eq. VII.36). The transition probability per collision is not known for I_2 in the various gases just listed. We can, however, estimate it to be about 10^{-4} on the basis of the results obtained by Herzfeld[6] for Cl_2 in Cl_2 and Br_2 in Br_2. At 300°K, $\theta \simeq 1$ and $N+1 \simeq 60$, based on

$\omega_e(I_2) \simeq 200$ cm^{-1} and $D_e(I_2) \simeq 12{,}000$ cm^{-1}. Using these values and $Z^* = 10^{14}$ cm^3/mole/sec, we obtain $A \simeq 2 \times 10^{11}$cm^3/mole/sec.

It appears from these calculations that the frequency factor calculated for a stepwise activation deviates even more from the experimental value than that calculated by the usual collision theory. Two points are to be noted however. If we had taken $P_{10} = 1$, as we did for P_c, we would have found $A \simeq 2 \times 10^{15}$ cm^3/mole/sec which is closer to the experimental value of A $\simeq 10^{16}$ cm^3/mole/sec. This comparison is, however, quite unrealistic since neither P_e nor P_{10} are unity; both quantities probably being several orders of magnitude smaller. The second, and significant point is that we have used, for mathematical convenience, a *harmonic* oscillator molecule in our model of a stepwise activation process. Owing to the selection rule $P_{n,m} = 0$ for $m \neq n \pm 1$ which limits transitions to adjacent levels only, we are considering the slowest possible process of stepwise activation. One would expect, intuitively, that the rate of stepwise activation would be more rapid for an *anharmonic* oscillator where transition can occur between any two energy levels and where in addition the level density increases with increasing energy.

In conclusion we summarize some of the principal points of this review:

(1) A formal treatment has been developed for the application of the theory of stochastic processes to reaction rate problems.

(2) On the basis of the model employed here, it has been shown that the rate of a chemical reaction will deviate from the equilibrium rate, as calculated from collision or absolute rate theory, when $E_{act}/kT < 10$. The calculated deviation of about 20 per cent for $E_{act}/kT = 5$ (see Fig. 4) is in good agreement with the results obtained previously by other authors. For $E_{act}/kT \geqq 10$ the error in the rate as calculated by equilibrium theory is < 10 per cent.

(3) The rate of activation in a unimolecular dissociation reaction has been calculated for a stepwise process of activation involving transition only between adjacent energy levels (harmonic oscillator model with weak interactions). The experimental rate of activation is higher by several orders of magnitude than the rate obtained from these calculations. A process of stepwise activation involving

only weak collisional interaction is thus inadequate to account for the observed high rate of activation.

In conclusion the authors wish to thank Mr. George Weiss for several useful discussions concerning Section VII and Appendix II.

Note added in proof: In view of the failure of the harmonic oscillator model to account for the observed rate of activation in unimolecular dissociation reactions (the dissociation lag problem) these calculations have been repeated for a Morse anharmonic oscillator with transition between nearest and next-nearest neighbor levels [S. K. Kim, *J. Chem. Phys.* (to be published)]. The numerical evaluation of the analytical results obtained by Kim has not yet been carried out. From the results obtained by us and our co-workers [Bazley, Montroll, Rubin, and Shuler, *J. Chem. Phys.* (in press)] on the relaxation of vibrational nonequilibrium distributions of a system of Morse anharmonic oscillators it seems clear, however, that the anharmonic oscillator model *with weak interactions* (i.e., adiabatic perturbation type matrix elements) does not constitute much of an improvement on the harmonic oscillator model in giving the observed rates of activation. The answer to this problem would seem to lie in a recalculation of the collisional matrix elements for translational-vibrational energy exchange which takes account of the strong interactions in highly energetic collisions which can lead to direct dissociation.

APPENDIX 1

Trace of the Inverse of the Matrix of a Continuant

Let us consider the matrix B of a continuant

$$B = \begin{vmatrix} a_1 & b_1 & 0 & 0 & \ldots & 0 & 0 \\ b_1 & a_2 & b_2 & 0 & \ldots & 0 & 0 \\ 0 & b_2 & a_3 & b_3 & \ldots & 0 & 0 \\ \cdot & \cdot & \cdot & \cdot & \cdot\cdot\cdot & \cdot & \cdot \\ & & & & & a_{n-1} & b_{n-1} \\ 0 & 0 & 0 & 0 & \ldots & b_{n-1} & a_n \end{vmatrix} \tag{A1.1}$$

The mth diagonal element of B^{-1} is (since the elements of B^{-1} are the cofactors of the corresponding elements of A):

$$(m|B^{-1}|m) = D_{m+1}A_{m-1}/\det A \qquad \text{(A1.2)}$$

where

$$A_m = \begin{vmatrix} a_1 & b_1 & 0 & 0 \ldots 0 \\ b_1 & a_2 & b_2 & 0 \ldots 0 \\ \cdot & \cdot & \cdot & \cdots \cdot \\ 0 & 0 & 0 & 0 \ldots a_m \end{vmatrix} \qquad \text{(A1.3)}$$

and

$$D_m = \begin{vmatrix} a_m & b_m & 0 & 0 & \ldots 0 \\ b_m & a_{m+1} & b_{m+1} & 0 & \ldots 0 \\ 0 & b_{m+1} & a_{m+2} & b_{m+2} & \ldots 0 \\ \cdot & \cdot & \cdot & \cdot & \cdots \cdot \\ 0 & \cdot & \cdot & \cdot & \ldots a_m \end{vmatrix} \qquad \text{(A1.4)}$$

By expanding A_m with respect to the mth row, we find

$$A_m = a_m A_{m-1} - b_{m-1}^2 A_{m-2} \qquad \text{(A1.5)}$$

and by expanding D_m with respect to the first row we have

$$D_m = a_m D_{m+1} - b_m^2 D_{m+2} \qquad \text{(A1.6)}$$

We define

$$A_0 = D_{n+1} = 1 \qquad \text{(A1.7)}$$

and note that

$$A_n = D_1 = \det B \qquad \text{(A1.8)}$$

Finally

$$\text{trace } B^{-1} = D_1^{-1} \sum_{m-1}^{n} D_{m+1} A_{m-1} \qquad \text{(A1.9)}$$

The rest of this appendix is devoted to the calculation of trace B^{-1} when (Eq. VII.4)

$$b_m = me^{-\frac{1}{2}\theta} \quad \text{and} \quad a_m = -(m-1+me^{-\theta}) \qquad \text{(A1.10)}$$

Then Eqs. A1.5 and A1.6 are, respectively,

$$A_m = -(m-1+me^{-\theta})A_{m-1} - (m-1)^2 e^{-\theta} A_{m-2} \qquad \text{(A1.11)}$$

and

$$D_m = -(m-1+me^{-\theta})D_{m+1} - m^2 e^{-\theta} D_{m+2} \tag{A1.12}$$

We can now show by induction that

$$A_m = (-1)^m m!\, e^{-m\theta} \tag{A1.13}$$

Clearly this is true of $m = 0$ and $m = 1$ since $A_0 = 1$ and $A_1 = -e^{-\theta}$. If it is true for A_{m-1} and A_{m-2}, it is true for A_m since direct substitution of Eq. A1.13 into the right-hand side of Eq. A1.11 yields

$$A_m = (-1)^m (m-1+me^{-\theta})(m-1)!\, e^{-(m-1)\theta} - (-1)^m (m-1)(m-1)!\, e^{-(m-1)\theta}$$
$$= (-1)^m m!\, e^{-m\theta}$$

as is required.

We can also show by induction that

$$D_{n-k} = (-1)^{k+1} n(n-1)\ldots(n-k-1)$$
$$\cdot\left\{\frac{1}{n} + \frac{1}{n-1}e^{-\theta} + \frac{1}{n-2}e^{-2\theta} + \ldots + \frac{1}{n-(k+1)}e^{-(k+1)\theta}\right\} \tag{A1.14}$$
$$k = 0, 1, \ldots, n$$

By definition $D_{n+1} = 1$. Equation A1.4 implies that $D_n = -(n-1+ne^{-\theta})$ which is the form given by Eq. A1.14. If Eq. A1.14 is valid for $D_{n-(k-1)}$ and $D_{n-(k-2)}$, then Eq. A1.12) implies

$$D_{n-k} = (-1)^{k+1}[(n-k-1)+(n-k)e^{-\theta}]n(n-1)\ldots(n-k)\left\{\frac{1}{n}+\ldots+\frac{1}{n-k}e^{-k\theta}\right\}$$
$$-(n-k)^2 e^{-\theta}(-1)^{k+1} n(n-1)\ldots(n-[k-1])\left\{\frac{1}{n}+\ldots+\frac{e^{-(k-1)\theta}}{n-(k-1)}\right\}$$
$$= (-1)^{k+1} n(n-1)\ldots(n-k-1)\left\{\frac{1}{n}+\ldots+\frac{1}{n-k}e^{-k\theta}\right\}$$
$$+(-1)^{k+1} n(n-1)\ldots(n-k)e^{-(k+1)\theta}$$

which is exactly Eq. A1.14 as is required to prove its validity by induction.

We now substitute Eqs. A1.14 and A1.13 into Eq. A1.9 to find

$$-\text{ trace } B^{-1} = \sum_{k=0}^{n-1} e^{(k+1)\theta}\left\{\frac{1}{n} + \frac{e^{-\theta}}{n-1} + \ldots + \frac{e^{-k\theta}}{n-k}\right\} \quad \text{(A1.15)}$$

$$= \sum_{j=1}^{n} e^{j\theta}\left\{\frac{1}{j} + \frac{1}{j+1} + \ldots + \frac{1}{n}\right\} = e^{n\theta}\sum_{k=0}^{n-1} e^{-k\theta}\left\{\frac{1}{n} + \frac{1}{n-1} + \ldots + \frac{1}{n-k}\right\}$$

The values of $-$ trace B^{-1} for the first few integers n are given in Table I.

TABLE I. Mean First Passage Time $\bar{t}$ as a Function of N.

N	$\bar{t}$
0	e^{θ}
1	$\frac{e^{2\theta}}{2}(1 + 3e^{-\theta})$
2	$\frac{e^{3\theta}}{6}(2 + 5e^{-\theta} + 11e^{-2\theta})$
3	$\frac{e^{4\theta}}{12}(3 + 7e^{-\theta} + 13e^{-2\theta} + 25e^{-3\theta})$
4	$\frac{e^{5\theta}}{60}(12 + 27e^{-\theta} + 47e^{-2\theta} + 77e^{-3\theta} + 137e^{-4\theta})$
5	$\frac{e^{6\theta}}{60}(10 + 22e^{-\theta} + 37e^{-2\theta} + 57e^{-3\theta} + 87e^{-4\theta} + 147e^{-5\theta})$

The asymptotic value of $-$ trace B^{-1} as $n \to \infty$ is easily obtained by noting that

$$\left\{\frac{1}{n}+\frac{1}{n-1}+\ldots+\frac{1}{n-k}\right\} = \frac{1}{n}(k+1)+\frac{1}{n}\left\{\frac{1}{n-1}+\frac{2}{n-2}+\ldots+\frac{k}{n-k}\right\}$$

$$=\frac{1}{n}(k+1)+\frac{1}{2}\frac{k(k+1)}{n(n-1)}+\frac{1}{n(n-1)}\left\{\frac{2.1}{n-2}+\frac{3.2}{n-3}+\ldots+\frac{k(k-1)}{n-k}\right\}$$

$$=\frac{1}{n}(k+1)+\frac{1}{2}\frac{k(k+1)}{n(n-1)}+\frac{(k+1)k(k-1)}{3n(n-1)(n-2)}$$

$$+\frac{1}{n(n-1)(n-2)}\left\{\frac{3.2.1}{n-3}+\frac{4.3.2}{n-4}+\ldots+\frac{k(k-1)(k-2)}{n-k}\right\}$$

Hence

$$-\text{ trace } B^{-1} = e^{n\theta}\left\{\frac{1}{n}\sum_{k=0}^{n-1}(k+1)e^{-k\theta} + \frac{1}{2n(n-1)}\sum_{k=1}^{n-1} e^{-k\theta}k(k+1)\right.$$

$$\left. + \frac{1}{3n(n-1)(n-2)}\sum_{k=2}^{n-1} e^{-k\theta}(k+)k(k-1) + \ldots\right\} \qquad \text{(A1.16)}$$

But

$$\sum_{k=0}^{n-1}(k+1)e^{-k\theta} = (1-e^{-\theta})^{-2} + 0(e^{-n\theta});$$

$$\sum_{k=2}^{n-1}(k+1)k(k-1)e^{-k\theta} = \frac{6e^{-2\theta}}{(1-e^{-\theta})^4} + 0(e^{-n\theta})$$

$$\sum_{k=1}^{n-1}(k+1)ke^{-k\theta} = 2e^{-\theta}(1-e^{-\theta})^{-3} + 0(e^{-n\theta})$$

Therefore

$$-\text{ trace } B^{-1} = \frac{e^{n\theta}}{n(e^{-\theta}-1)^2}$$

$$\left\{1 + \frac{e^{-\theta}}{(1-e^{-\theta})(n-1)} + \frac{2e^{-2\theta}}{(1-e^{-\theta})^2(n-1)(n-2)} + 0(n^{-3})\right\} \qquad \text{(A1.17)}$$

APPENDIX II

On the Smallest Zero of $l_{N+1}(\mu) = (1-\alpha)\, e^{-\theta}\, l_N(\mu)$ as $N \to \infty$

As $N \to \infty$ the smallest zero of this equation approaches 0. Hence we must find the behavior of $l_N(\mu)$ as $\mu \to 0$ and $N \to \infty$.

By definition

$$l_n(\mu) = e^{-n\theta} + e^{-n\theta}\sum_{\nu=1}^{n}(1-e^{\theta})^{\nu}\binom{n}{\nu}\binom{\mu}{\nu} \qquad \text{(A2.1)}$$

As $\mu \to 0$

$$\binom{\mu}{\nu} = \frac{\mu(\mu-1)\ldots(\mu-\nu+1)}{\nu!} \sim \frac{\mu}{\nu}(-1)^{\nu-1}$$

Hence

$$
\begin{aligned}
l_n(\mu) &\sim e^{-n\theta}-\mu e^{-n\theta}\sum_{\nu=1}^{n}\nu^{-1}(e^{\theta}-1)^{\nu}\binom{n}{\nu} \\
&= e^{-n\theta}-\mu e^{-n\theta}\sum_{\nu=1}^{n}\int_0^{e^{\theta}-1}\binom{n}{\nu}x^{\nu-1}\,dx \\
&= e^{-n\theta}-\mu e^{-n\theta}\int_0^{e^{\theta}-1}x^{-1}\left\{\sum_{\nu=0}^{n}x^{\nu}\binom{n}{\nu}-1\right\}dx \qquad \text{(A2.2a)} \\
&= e^{-n\theta}-\mu e^{-n\theta}\int_0^{e^{\theta}-1}x^{-1}\{(1+x)^{n}-1\}dx \\
&= e^{-n\theta}-\mu e^{-n\theta}I_n
\end{aligned}
$$

where

$$I_n=\int_1^{e^{\theta}}\frac{y^n-1}{y-1}\,dy \qquad \text{(A2.2b)}$$

Since

$$I_{n+1}-I_n=\int_1^{e^{\theta}}y^n dy=\frac{e^{(n+1)\theta}-1}{n+1}$$

and

$$I_0=0,\ I_1=e^{\theta}-1,$$

we have

$$
\begin{aligned}
e^{-(n+1)\theta}I_{n+1} &= \sum_{m=0}^{n}\left\{\frac{e^{(m+1)\theta}-1}{m+1}\right\}e^{-(n+1)\theta} \\
&= \sum_{m=0}^{n}\frac{e^{(m-n)\theta}}{m+1}+0(e^{-(n+1)\theta}\log n)
\end{aligned} \qquad \text{(A2.3)}
$$

Since we are interested in asymptotic values of $e^{-n\theta}I_n$ as $n\to\infty$, we neglect the term of $0(e^{-(n+1)\theta}\log n)$ and find that

$$e^{-(n+1)\theta}I_{n+1}\sim\sum_{m=0}^{n}\frac{e^{(m-n)\theta}}{(m+1)}. \qquad \text{(A2.4)}$$

An immediate application of Abels' partial summation formula yields

$$\sum_{m=0}^{n}\frac{e^{m\theta}}{m+1}=\frac{1}{n+1}\left\{\frac{e^{(n+1)\theta}-1}{e^{\theta}-1}\right\}+\sum_{m=0}^{n}\frac{1}{(m+1)(m+2)}\left\{\frac{e^{(m+1)\theta}-1}{e^{\theta}-1}\right\} \qquad \text{(A2.5)}$$

But

$$\sum_{m=0}^{n} [(m+1)(m+2)]^{-1} < \sum_{m=0}^{\infty} (m+1)^{-2} = \pi^2/6$$

and

$$J_n = \sum_{m=0}^{n} \frac{e^{(m+1)\theta}}{(m+1)(m+2)} < \sum_{m=0}^{n} \frac{e^{(m+1)\theta}}{(m+1)^2} < \int_1^{n+1} \frac{e^{x\theta}}{x^2} dx$$

$$= \frac{1}{\theta} \frac{(e^{(n+1)\theta} - 1)}{(n+1)^2} + \frac{2}{\theta} \int_1^{n+1} \frac{e^{x\theta}}{x^3} dx$$

also

$$\int_1^{n+1} \frac{e^{\theta x}}{x^3} dx < e^{(n+1)\theta} \int_1^{n+1} \frac{dx}{x^3} < \frac{e^{(n+1)\theta}}{2(n+1)^2}$$

Hence

$$0 < J_n < 2(n+1)^{-2} \exp{(n+1)\theta} \tag{A2.6}$$

so that as $n \to \infty$

$$e^{-(n+1)\theta} I_{n+1} \sim e^{\theta}/(n+1)(e^{\theta}-1) \tag{A2.7}$$

and as $\mu \to 0$

$$l_n(\mu) \sim e^{-n\theta} - \frac{\mu e^{\theta}}{n(e^{\theta}-1)} \tag{A2.8}$$

In this limit our characteristic equation

$$l_{N+1}(\mu) = (1-\alpha)e^{-\theta} l_N(\mu)$$

yields

$$\mu_0 \sim \frac{\alpha e^{-(N+1)\theta} (N+1)(1-e^{-\theta})}{[1-(1-\alpha)e^{-\theta}]} \tag{A2.9}$$

which approaches zero exponentially as $N \to \infty$. Substitution of Eq. A2.9 into Eq. (A2.8) yields

$$l_N(\mu) \sim e^{-N\theta} \left\{1 - \frac{\alpha e^{-\theta}}{[1-(1-\alpha)e^{-\theta}]}\right\} \tag{A2.10}$$

References

1. See also: Brinkman, H. C., *Physica* **22**, 29, 149 (1956).
2. Chandrasekhar, S., *Revs. Modern Phys.* **15**, 1 (1943).
3. Curtiss, C. F., "The Equilibrium Assumption in the Theory of Absolute Reaction Rates," University of Wisconsin, Report CM-476, June 1948.

4. Glasstone, S., Laidler, K. J., and Eyring, H., *The Theory of Rate Processes*, McGraw-Hill, New York, 1941.
5. Gottlieb, M. J., *Am. J. Math.* **60**, 455 (1938).
6. Herzfeld, K. F., in *Thermodynamics and Physics of Matter* (High Speed Aerodynamics and Jet Propulsion, Vol. I), Section H, Princeton Univ. Press, Princeton, New Jersey, 1955.
7. Hinshelwood, C. N., *Kinetics of Chemical Change*, Oxford Univ. Press, London, 1940.
8. Hirschfelder, J. O., *J. Chem. Phys.* **16**, 22 (1948).
9. Van Hove, L., in *Proceedings of the International Symposium on Transport Processes in Statistical Mechanics*, ed. I. Prigogine, Interscience, New York, 1958; *Physica* **23**, 441 (1957).
10. Kohn, W., and Luttinger, J., private communication.
11. Kramers, H. A., *Physica* **7**, 284 (1940).
12. Landau, L., and Teller, E., *Physik. Z. Sowjetunion* **10**, 34 (1936).
13. Montroll, E. W., in *Proceedings of the International Symposium on Transport Processes in Statistical Mechanics*, ed. I. Prigogine, Interscience, New York, 1958.
14. Montroll, E. W., and Shuler, K. E., *J. Chem.Phys.* **26**, 454 (1957).
15. Prigogine, I., and Xhrouet, E., *Physica* **15**, 913 (1949).
16. Prigogine, I., and Mahieu, M., *Physica* **16**, 51 (1950).
17. Rice, O. K., *J. Chem. Phys.* **9**, 258 (1941).
18. Rubin, R. J., and Shuler, K. E., *J. Chem. Phys.* **25**, 59, 68 (1956); *ibid.* **26**, 137 (1957).
19. Shuler, K. E., in *5th Symposium on Combustion, Pittsburgh*, 1954, Rheinhold, New York, 1955, pp. 56–74.
20. Shuler, K. E., *J. Phys. Chem.*, **61**, 849 (1957).
21. Takayanagi, K., *Progr. Theoret. Phys. (Japan)* **6**, 486 (1951).
22. For a summarizing review on these measurements see A. F. Trotman-Dickenson, *Gas Kinetics*, Butterworth, London, 1955, Chap. 2.
23. Wedderburn, J. H. M., *Am. Math. Soc. Colloq. Series* **17**, 25 (1934).
24. Zwolinski, B. J., and Eyring, H., *J. Am. Chem. Soc.* **69**, 2702 (1947).

Kinetics of Reactant Isolation. I. One-Dimensional Problems*

E. Richard Cohen and H. Reiss

North American Aviation Science Center, Canoga Park, California

(Received 18 October 1962)

This paper treats the time-dependent statistics of bond formation (both irreversible and reversible) between fixed sites. This model corresponds to a number of real physical processes. Among these are reaction of adjacent functional groups on a linear polymer, matrix isolation of free radicals, and chemisorption. Because the sites are fixed, certain of them become isolated from others, not yet reacted, and must survive to infinite time. A certain amount of survival occurs even when bonds can be undone.

Exact descriptions of the kinetics of these processes are given for one-dimensional systems. The mean probability of site survival as a function of time is calculated for linear arrays and for rings. The probability of site survival as a function of position on the chain (end effects) is treated in Sec. IV. A generating function is introduced in order to calculate higher moments of the survival probability distribution function and to treat the effect of a random diluent. An approximate method of solution is developed for the time dependence of the number of reacted particles in the reversible case.

I. INTRODUCTION

THERE are examples of systems in which species firmly attached to sites can react with one another. Reaction is usually confined to entities occupying adjacent sites. An early example[1] involves the polymer of methyl vinyl ketone. The methyl ketone groups distributed at equal intervals along the polymer chain can be made to condense (when the polymer is heated to 300°C) with one another so as to produce rings. The typical reaction (limited to nearest neighbors) is the following:

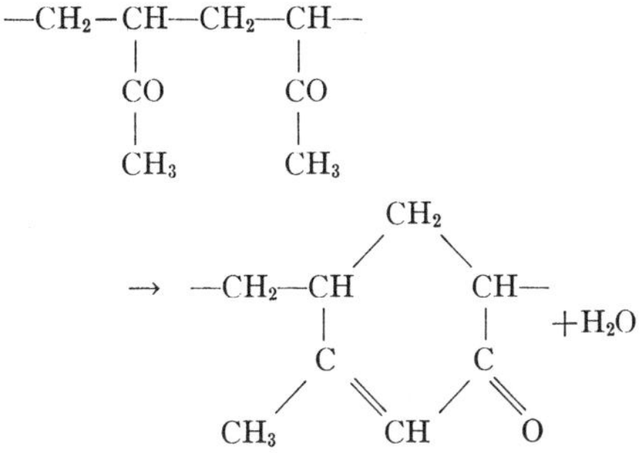

$$-CH_2-CH-CH_2-CH-$$
$$\quad\quad |\quad\quad\quad |$$
$$\quad\quad CO\quad\quad CO$$
$$\quad\quad |\quad\quad\quad |$$
$$\quad\quad CH_3\quad\quad CH_3$$

$$\rightarrow\ -CH_2-CH \quad CH-\ +H_2O \quad (1.1)$$

A longer segment of the chain may be represented schematically as follows:

with the vertical lines designating the methyl ketone groups. After reaction the situation may be represented by

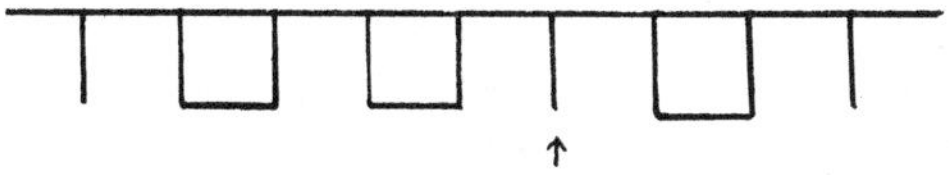

where the rings are now evident. It is also clear that certain methyl ketone groups such as the one indicated by the arrow become isolated and are prevented from reacting. At infinite time a definite residue of such groups is isolated. It is of interest to compute the number of unreacted groups remaining after any time *t*. This sort of calculation forms the basis of the present investigation.

The question of condensation in polyvinyl methyl ketone has been examined previously by Flory[1] whose results are entirely correct. However, he limits his calculation to the number of methyl ketone groups which survive ultimately and does not attack the problem of rates nor specify the time dependence of the surviving groups.

There are other systems besides the one described in which similar processes take place. One which has received attention consists of a crystal of free radicals[2] formed by directing a stream of such particles against a cold surface. Neighboring radicals in the solid are able to combine to form diatomic molecules, and eventually certain of them become isolated because all their nearest neighbors have combined with other radicals. This is a three-dimensional system in contrast to the one-dimensional system discussed above, but the process involved is of the same class. Sometimes the situation is modified because the radicals are codeposited with an inert diluent. In this case it is also desirable to know how many radicals survive after a time *t*.

In the free-radical system the real situation is complicated because the radicals[3] may be able to diffuse and therefore are not strictly fixed to their sites. Calculations have been made by Jackson and Montroll[4] and others[5,6] to estimate how many radicals remain if dif-

* Work supported in part by U. S. AEC under contract AT-(11-1)-GEN-8.

[1] P. J. Flory, J. Am. Chem. Soc. **61**, 1518 (1939).

[2] A. M. Bass and H. P. Broida, Phys. Rev. **101**, 1740 (1956).
[3] J. L. Jackson, J. Chem. Phys. **31**, 154, 722 (1959).
[4] J. L. Jackson and E. W. Montroll, J. Chem. Phys. **28**, 1101 (1958).
[5] P. L. Chessin, J. Chem. Phys. **31**, 159 (1959).
[6] S. Golden, J. Chem. Phys. **29**, 61 (1958).

TABLE I. Possible configurations for short chains. For a fixed chain length N, the possible ultimate bond configurations are listed. Mirror-image configurations are considered to be equivalent and are not individually listed. C_N is the relative frequency of occurrence of equivalent patterns; S_N is the number of survivors in each pattern; n_N is the average number of survivors, $\Sigma S_N C_N / \Sigma C_N$; f_N is the average survival probability per site, n_N/N; $\langle n^2 \rangle = \Sigma S_N^2 C_N / \Sigma C_N$; $\sigma_f = [\langle n^2 \rangle - n_N^2]^{\frac{1}{2}}/N$.

N		C_N	S_N	n_N	$\langle n^2 \rangle$	f_N	σ_f
1	○	1	1	1	1	1	0
2	●—●	1	0	0	0	0	0
3	●—● ○	1	1	1	1	1/3	0
4	●—● ●—●	2	0	(2/3)	(4/3)	(1/6)	$(1/6)(2)^{\frac{1}{2}}$
	○ ●—● ○	1	2				
5	●—● ●—● ○	3	1	1	1	(1/5)	0
	●—● ○ ●—●	1	1				
6	●—● ●—● ●—●	7	0				
	●—● ○ ●—● ○	5	2	(16/15)	(32/15)	(8/45)	$(2/45)(14)^{\frac{1}{2}}$
	○ ●—● ●—● ○	3	2				
7	●—● ●—● ●—● ○	37	1				
	●—● ●—● ○ ●—●	27	1	(11/9)	(17/9)	(11/63)	$(4/63)(2)^{\frac{1}{2}}$
	○ ●—● ○ ●—● ○	8	3				

fusion is prohibited. These authors predict the survival of a much larger fraction of radicals than is actually observed so that diffusion seems to be an important factor.

The manner in which previous workers have formulated the problem leaves something to be desired (even in the absence of diffusion). Their mathematical analysis is correct, but the physical assumptions upon which their developments are based involve the reversibility of the reactions. Even though the back reaction rate may be vanishingly small there is a quantitative difference in the steady-state configuration between the reversible and irreversible situation. We illustrate the difficulty by reference to a simple system consisting of four radicals arranged into a linear chain:

1 2 3 4

If we restrict bonding to nearest neighbors, two ultimate configurations are possible. These are:

1——2 3——4 (A)

1 2——3 4 (B)

Configuration A contains no survivors while B contains two. If after deposition of the free radicals, configurations A and B are equally probable, the *average* number of survivors is 1. In essence this represents the technique employed previously to compute the average number of survivors in what is of course the much larger (and three-dimensional) array of free radicals. The various ultimate bonding configurations are enumerated together with the number of survivors in each, and the average is computed with the assumption of equal *a priori* probability for each configuration.

It seems reasonable, however, that all configurations are not equally probable because they are formed by sequences of events. For example, there is only one path leading to B above while there are two leading to A. In the case of B a bond forms between particles 2 and 3 and the process is over. In the case of A a bond can form first between 1 and 2 and then between 3 and 4. Alternatively the bond can form first between 3 and 4 and then between 1 and 2. Either sequence leads to the same final configuration. Thus, if each sequence is equally probable, A is twice as probable as B since it can be formed in two ways while the method of formation of B is unique. Employing the weighted average, the mean number of survivors now becomes 2/3 rather than 1. The possible configurations for short chains (up to seven particles) are listed in Table I.

Thus it appears as though the first method overestimates the number of survivors. This may constitute part of the reason for the disparity between theory and experiment, although the diffusion of free radicals undoubtedly plays a role.

Another process which falls into the general class under discussion is the irreversible chemisorption of diatomic molecules upon a regular array of surface sites. Each diatomic molecule spans two sites, effectively producing a bond between them. Eventually certain sites, although unoccupied, are isolated because each neighboring site has one of the atoms in a diatomic

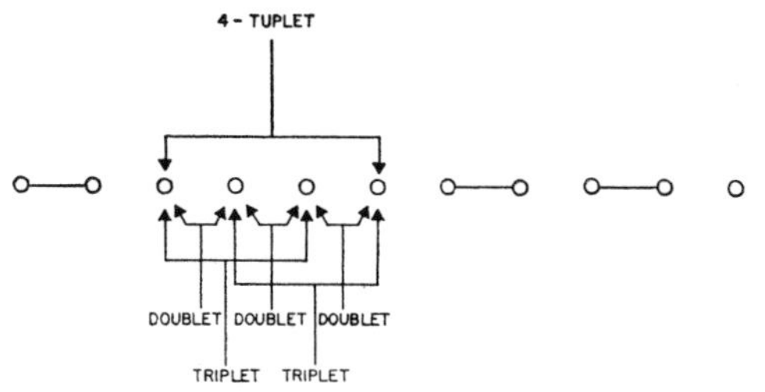

FIG. 1. An illustration of site multiplets. The definition is nonexclusive and a given multiplet contains all multiplets of lower order.

molecule fixed to it. Since a pair of unoccupied neighbor sites is required for the adsorption of a single diatomic molecule, the rate of adsorption of molecules is related to the rate of consumption of sites so that calculation of the number of sites surviving at time t makes it possible to know the degree of adsorption at t. This is just another problem in site isolation (this time two-dimensional) and can be solved in common with the two problems described above.

Much work has been done[7] on this aspect of adsorption but unfortunately the issue has been obscured by the misinterpretation found in the free-radical case. This is particularly true in the so-called "immobile" film[7] case. Because of the importance of the phenomenon of adsorption, we do not treat it in the present text but reserve it for the second paper in this series.

In the present paper we confine our attention to one-dimensional arrays, both finite and infinite, and determine survival as it depends on time. It is possible to supply exact solutions for many problems. This seems to be characteristic of the one-dimensional case, and we indicate the reason at an appropriate juncture.

Solutions to the problems of radical isolation and adsorption are of course, included in our results. In these instances, however, the one-dimensional array represents an idealization. In the case of radical isolation we also consider the effect of diluent.

II. LINEAR ARRAY

Consider a chain of N sites between which bonds are formed. A single unbonded site is referred to as a "singlet," a run of two sites as a "doublet," and a run of three a "triplet." In general, a run of n sites termed an "n-tuplet." It is clear that an "n-tuplet" may contain two distinct "$(n-1)$-tuplets," three distinct "$(n-2)$-tuplets," etc. This fact is illustrated in Fig. 1 with respect to a "4-tuplet."

Consider an ensemble of M identical chains of N sites. Denote the number of n-tuplets in the jth chain at time t by $C_n^{(j)}(t)$. Let $k(t)dt$ be the probability that a bond forms between two unreacted neighbors in the time interval $(t, t+dt)$, the same for all pairs of neighbors. In most cases k may be assumed independent of time, but since it is no more difficult to perform the analysis with k dependent on time we do so. Then the rate of change of $C_n^{(j)}(t)$ with time is

$$-dC_n^{(j)}/dt=k(t)\{(n-1)C_n^{(j)}+2C_{n+1}^{(j)}\}. \qquad (2.1)$$

The origin of this equation is simple. The minus sign appears because the reaction is irreversible and n-tuplets can only be destroyed, never created. The first term on the right corresponds to the destruction of n-tuplets by the formation of a bond within the n-tuplet itself. Since there are $n-1$ possible bonds within an n-tuplet, the rate of destruction is proportional to $k(t)(n-1)C_n^{(j)}$. The second term on the right corresponds to the destruction of n-tuplets by the formation of a bond between either of its terminal sites and a site not belonging to the n-tuplet. The site does, however, belong to either of the two $(n+1)$-tuplets which contain the n-tuplet. Thus we arrive at the term $k(t)2C_{n+1}^{(j)}$.

The mean number of n-tuplets averaged over the M identical chains is

$$\bar{C}_n(t)=M^{-1}\sum_{j=1}^{M}C_n^{(j)}(t). \qquad (2.2)$$

Summing (2.1) over j and dividing by M yields

$$-d\bar{C}_n/dt=k(t)\{(n-1)\bar{C}_n+2\bar{C}_{n+1}\}. \qquad (2.3)$$

Initially, there are $N-n+1$ n-tuplets so that the set of equations (2.3) must be solved subject to boundary condition

$$\bar{C}_n(0)=N-n+1. \qquad (2.4)$$

It is convenient to introduce the variable

$$z=\int_0^t k(t)dt, \qquad (2.5)$$
$$dz=k(t)dt.$$

With this transformation (2.3) becomes

$$-d\bar{C}_n/dz=(n-1)\bar{C}_n+2\bar{C}_{n+1}. \qquad (2.6)$$

The solution of (2.6) subject to (2.4) is

$$\bar{C}_n=\exp[-(n-1)z]\sum_{s=0}^{N-n}(N-n-s+1)\frac{[2e^{-z}-2]^s}{s!}. \qquad (2.7)$$

The fraction of n-tuplets which survive, or the probability of survival of an n-tuplet, is

$$P_n(z)=\frac{\bar{C}_n}{N-n+1}$$
$$=\exp[-(n-1)z]\sum_{s=0}^{N-n}\left(1-\frac{s}{N-n+1}\right)\frac{[2e^{-z}-2]^s}{s!}. \qquad (2.8)$$

[7] A. R. Miller, *The Adsorption of Gases on Solids* (Cambridge University Press, New York, 1949), pp. 30–37.

In the special case of the infinite chain ($N\rightarrow\infty$), we have for all *finite n*

$$P_n(z)=\exp[-(n-1)z]\exp\{-2[1-\exp(-z)]\}. \tag{2.9}$$

At infinite time ($z=\infty$), P_1 is therefore

$$P_1(\infty)=e^{-2}. \tag{2.10}$$

This agrees with Flory's[1] result for the fraction of singlets which ultimately survive.

An examination of formula (2.9) for the infinite chain establishes the physical reason underlying the ready acquisition of an exact solution in the one-dimensional case. Notice that

$$P_1=\exp\{-2[1-\exp(-z)]\} \tag{2.11}$$

and that

$$P_n=\exp[-(n-1)z]P_1(z). \tag{2.12}$$

Consider P_2. The probability that a doublet survives is equivalent to the probability P_1 that one site of the doublet survives multiplied by the conditional probability p that the adjacent site survives, given that the first has survived. Thus

$$P_2=P_1p. \tag{2.13}$$

Comparison with (2.12) reveals that

$$p=e^{-z}. \tag{2.14}$$

Similarly the probability that a triplet survives is the probability P_2 that one of the doublets in the triplet survives multiplied by the conditional probability p' that the site adjacent to the doublet survives when it is known that doublet has survived. Thus

$$P_3=P_2p'=(\exp(-z)\exp\{-2[1-\exp(-z)]\})\times\exp(-z), \tag{2.15}$$

from which it is apparent that

$$p'=e^{-z}=p. \tag{2.16}$$

In general,

$$P_n=P_{n-1}e^{-z}. \tag{2.17}$$

Thus the conditional probability of survival for a site adjacent to one which is known to have survived is independent of the number of sites which has survived on the other side. The reason for this is clear from Fig. 2. Suppose X is a site known to survive in an infinite chain. Then site 1 to the right of X has not interacted with X and X had not interacted with site -1 to its left. As a result the segment of the chain 1, 2, 3, 4, etc., to the right is effectively isolated from the segment -1, -2, -3, -4 to the left and no information can be transmitted from left to right or vice versa through X. Thus it is of no account to the survival of 1 that 1, 2, 3, or any number of particles has survived in the left segment. Thus $p=p'$, etc.

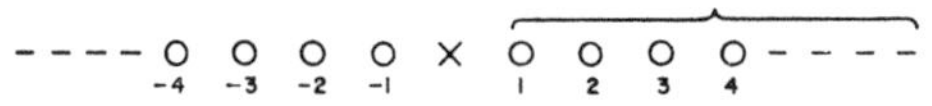

FIG. 2. Isolation of chain by site survival. If site X is known to have survived, the survival of sites 1, 2, 3, etc., is independent of the survival of any of the sites -1, -2, -3, etc.

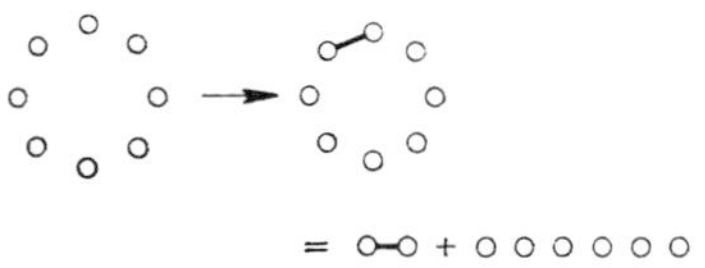

FIG. 3. The correspondence between chains and rings. A ring configuration of N sites becomes topologically equivalent to a linear chain of N-2 sites as soon as the first bond is formed.

In a two- or three-dimensional array this simple barrier to the flow of information past a surviving site does not exist as a circuitous path around the site over which information can pass can usually be found. Thus in the multidimensional case we may expect a more difficult situation.

In the one-dimensional case for the infinite chain the equations [obtained from (2.3) and (2.8)]

$$-dP_1/dz=2P_2,$$

$$-dP_2/dz=P_2+2P_3, \tag{2.18}$$

together with the relations

$$P_2=P_1p,$$

$$P_3=P_1p^2, \tag{2.19}$$

constitute a determinate set from which P_1 can be derived. Thus the infinite chain of equations can be broken after the second. This is a direct consequence of the foregoing considerations.

III. RINGS

The linear chain exhibits end effects. Thus the probability of survival of a site depends upon how far it is from either end of the chain. These effects have entered implicitly in Sec. II and we treat them explicitly later.

The end effects can be eliminated by bending the chain into a closed ring, and it is therefore of interest to study the survival of n-tuplets in an N-membered ring. A connection with the linear problem can be made immediately through use of the following device. After the first bond is formed in an N-membered ring the residue is a linear chain containing $N-2$ members. Thus for an eight-membered ring the formation of a bond (Fig. 3) leaves a linear chain of six members.

It is convenient to employ the transformed time z defined by (2.5). In what follows, for convenience we refer to z as "time." No confusion should result from this terminology.

TABLE II.

N	Chain	Ring
2	$P_1(z)=\exp(-z)$ $P_1(\infty)=0$	
3	$P_1(z)=(1/3)+(2/3)\exp(-2z)$ $P_1(\infty)=1/3$	$\Pi_1(z)=(1/3)+(2/3)\exp(-3z)$ $\Pi_1(\infty)=(1/3)$
4	$P_1(z)=(1/6)+(1/6)[3\exp(-z)+2\exp(-3z)]$ $P_1(\infty)=1/6$	$\Pi_1(z)=(2/3)\exp(-z)+(1/3)\exp(-4z)$ $\Pi_1(\infty)=0$
5	$P_1(z)=(1/5)+(2/5)[(2/3)\exp(-z)+\exp(-2z)$ $+(1/3)\exp(-4z)]$ $P_1(\infty)=1/5$	$\Pi_1(z)=(1/5)+(2/3)\exp(-2z)$ $+(2/15)\exp(-5z)$ $\Pi_1(\infty)=1/5$
6	$P_1(z)=(8/45)+(1/3)\exp(-z)+(2/9)\exp(-2z)$ $+(2/9)\exp(-3z)+(2/45)\exp(-5z)$ $P_1(\infty)=8/45$	$\Pi_1(z)=(1/9)+(2/5)\exp(-z)$ $+(4/9)\exp(-3z)+(2/45)\exp(-6z)$ $\Pi_1(\infty)=1/9$

An N-membered ring possesses N potential bonds (N doublets). Until the first bond is formed all of these doublets are equivalent. The conditional probability that a specific doublet is annihilated by bond formation in the interval z' to $z'+dz'$ is just dz' (or $k(t')dt'$). Since any of the N doublets may annihilate in the interval dz', the chance that some doublet annihilates is

$$Ndz'. \tag{3.1}$$

The probability that the N-particle ring survives to time z' without any doublet annihilating is therefore seen to be

$$\exp(-Nz'). \tag{3.2}$$

The chance that all doublets survive until z' and then some one of them is annihilated in dz' is just the product of (3.1) and (3.2)

$$N\exp(-Nz')dz'. \tag{3.3}$$

This is therefore the chance that an $N-2$ membered chain is formed in dz' at z'.

The survival of n-tuplets is now governed by $\bar{C}_n$ written for a linear chain of $N-2$ sites. Thus survival until time z is governed by ${}_{(N-2)}\bar{C}_n(z-z')$ where $(N-2)$ is written as a pre-subscript as a reminder that the linear chain created at $z-z'=0$ contains $(N-2)$ sites. To obtain the average number of n-tuplets surviving from the original ring (we shall denote this number by ${}_NR_n$), it is necessary to average ${}_{(N-2)}\bar{C}_n(z-z')$ over all possible z' prior to z since the first bond could have formed at any one of these times. The weight factor is obviously (3.3). We must also include the case in which no bonds form in the internal 0 to z. In this case the number of n-tuplets per ring which will have survived will be N and the probability e^{-Nz}. This contribution leads to the term Ne^{-Nz}. Thus,

$${}_NR_n(z)=N\exp(-Nz)+\int_0^z N\exp(-Nz')_{N-2}\bar{C}_n(z-z')dz'. \tag{3.4}$$

This can be simplified through the introduction of

$$y=z-z'. \tag{3.5}$$

The result is

$${}_NR_n=Ne^{-Nz}\left[1+\int_0^z e^{Ny}{}_{(N-2)}\bar{C}_n(y)dy\right]. \tag{3.6}$$

Substituting from (2.7) we have

$${}_NR_n(z)=Ne^{-Nz}\left[1+\sum_{s=0}^{N-n-2}\frac{N-n-s-1}{s!}\times\int_0^z\exp[(N-n+1)y][2e^{-y}-2]^s dy\right]. \tag{3.7}$$

The probability of survival in an infinite ring should be indistinguishable from that in an infinite linear chain since end effects are negligible at all sites save a finite few near either end. Let us test this for the special case of the singlet survival probability. This is

$$\lim_{N\to\infty}\frac{{}_NR_1}{N}=\lim_{N\to\infty}\left[e^{-Nz}\int_0^z e^{Ny}{}_{N-2}\bar{C}_1(y)dy+e^{-Nz}\right],$$

$$=\int_0^z P_1(y)\{\lim_{N\to\infty}N\exp[-N(z-y)]\}dy, \tag{3.8}$$

where $P_1(y)$ refers to the infinite chain. Now

$$\lim_{N\to\infty}N\exp[-N(z-y)]=\delta(y-z) \tag{3.9}$$

TABLE III. End effects in a semi-infinite chain.

$P_{(1)}^{(\infty)}=\exp[\exp(-z)-1]$	$P_{(1)}^{(\infty)}(\infty)=1/e$
$P_{(2)}^{(\infty)}=\exp(-z)\exp[\exp(-z)-1]$	$P_{(2)}^{(\infty)}(\infty)=0$
$P_{(3)}^{(\infty)}=[(1/2)+(1/2)\exp(-2z)]\exp[\exp(-z)-1]$	$P_{(3)}^{(\infty)}(\infty)=1/(2e)$
$P_{(4)}^{(\infty)}=[(1/3)+(1/2)\exp(-z)+(1/6)\exp(-3z)]\exp[\exp(-z)-1]$	$P_{(4)}^{(\infty)}(\infty)=1/(3e)$
$P_{(5)}^{(\infty)}=[(3/8)+(1/3)\exp(-z)+(1/4)\exp(-2z)+(1/24)\exp(-4z)]\exp[\exp(-z)-1]$	$P_{(5)}^{(\infty)}(\infty)=(3/8e)$

since the exponential is zero when $y<z$ and infinite when $y=z$ while

$$\lim_{N\to\infty}\int_0^z N\exp[-N(z-y)]dy=\lim_{N\to\infty}(1-e^{-Nz})=1. \quad (3.10)$$

Thus

$$\lim_{N\to\infty}\frac{{}_NR_1}{N}=\int_0^z P_1(y)\,\delta(y-z)\,dy=P_1(z). \quad (3.11)$$

Thus the survival probability for singlets in an infinite ring is identical with that for an infinite chain.

It is instructive to compute the singlet survival probabilities for a few short chains and rings. Using (2.8) and (3.7) we arrive at the formulas in Table II in which Π_1 stands for ${}_NR_1/N$.

All P's and Π's are initially unity as they should be. As N increases both $P(\infty)$ and $\Pi(\infty)$ oscillate, the oscillation being more pronounced in the ring case. This is understandable. For example, in the four-membered ring no particle can survive since whichever two particles bond first, the remaining two are juxtaposed and eventually annihilate one another. Thus in this case $\Pi_1(\infty)$ is zero and the oscillation carries clear to zero. By contrast some survival is possible in the four-membered chain. The rate of convergence to the limit $1/e^2$ as $N\to\infty$ is more rapid among the rings than the chains. Thus for $N=6$, $\Pi_1(\infty)=1/9=0.111$ as compared to $1/e^2=0.135$, whereas $P_1(\infty)=8/45=0.178$.

IV. END EFFECTS

The discussion of end effects requires consideration of the dependence of site survival on location. For this purpose we introduce a new notation in which P always stands for the singlet survival probability and the subscript (now in brackets) denotes the location of the site in respect to the end of the chain. Thus $P_{(n)}^{(N)}$ is the survival probability of the nth site from the end of an N-particle chain. Consider the third site from the left end of the six particle chain in Fig. 4.
Its survival probability would be denoted by $P_{(3)}^{(6)}$.

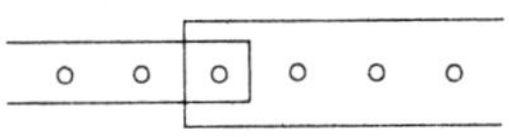

FIG. 4. The survival of a site in a linear chain is equivalent to the simultaneous survival of an end site in two other chains.

Now, if this site survives it isolates (according to Sec. II) the left segment of the chain from the right. In fact, it is effectively simultaneously an end site in an isolated three-membered chain to the left and an isolated four-membered chain to the right. Its survival requires the simultaneous survival of end particles in three- and four-membered chains. Thus

$$P_{(3)}^{(6)}=P_{(1)}^{(3)}P_{(1)}^{(4)}. \quad (4.1)$$

By the same argument, in general,

$$P_{(n)}^{(N)}=P_{(1)}^{(n)}P_{(1)}^{(N+1-n)}. \quad (4.2)$$

Further progress depends upon the evaluation of $P_{(1)}^{(n)}$. Let N be infinity. Then (4.2) yields

$$P_{(1)}^{(n)}=P_{(n)}^{(\infty)}/P_{(1)}^{(\infty)}. \quad (4.3)$$

Now,

$$-dP_{(1)}^{(\infty)}/dz=P_{(1)}^{(\infty)}p=P_{(1)}^{(\infty)}e^{-z}, \quad (4.4)$$

where $P_{(1)}^{(\infty)}p$ is the survival probability of the doublet consisting of the first two particles in an infinite chain (p is the same conditional probability defined in Sec. II). This equation merely expresses the fact that the first site can only be destroyed by destruction of the first doublet.

For the second site we have

$$-dP_{(2)}^{(\infty)}/dz=P_{(1)}^{(\infty)}p+P_{(2)}^{(\infty)}p=[P_{(1)}^{(\infty)}+P_{(2)}^{(\infty)}]e^{-z} \quad (4.5)$$

which expresses the fact that the second site can be destroyed in the first or second doublet. In general,

$$\frac{dP_{(n)}^{(\infty)}}{dz}=[P_{(n-1)}^{(\infty)}+P_{(n)}^{(\infty)}]e^{-z} \quad (4.6)$$

$$P_{(n)}^{(\infty)}(0)=1, \qquad n\geq 1, \qquad P_{(0)}^{(\infty)}=0.$$

The general solution of this set of equations is

$$P_{(n)}^{(\infty)}=\exp_n[\exp(-z)-1]\exp[\exp(-z)-1], \quad (4.7)$$

where $\exp_n[x]$ is the sum of the first n terms in the expansion of e^x. That (4.7) is the correct solution can be verified by direct substitution in (4.6). For $n\to\infty$ (deep within the chain)

$$P_{(\infty)}^{(\infty)}=\exp[2\exp(-z)-2], \quad (4.8)$$

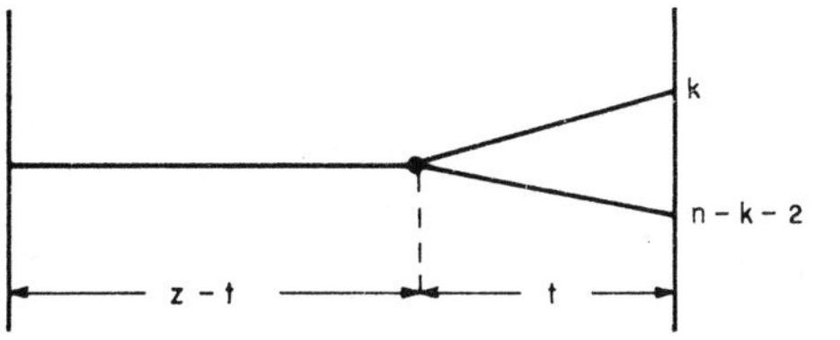

FIG. 5. Schematic representation of linear chain survival. Starting with an n-site linear chain, the formation of a bond at z-t produces a k-site chain and a (n-k-2)-site chain.

which is in agreement as it should be with the singlet survival probability specified by (2.9).

According to (4.7),

$$P_{(1)}^{(\infty)}=\exp[\exp(-z)-1]. \qquad (4.9)$$

Use of (4.7) and (4.9) in (4.3) then yields

$$P_{(1)}^{(n)}=\exp_n[\exp(-z)-1]. \qquad (4.10)$$

Therefore (4.2) requires

$$P_{(n)}^{(N)}=\exp_n[\exp(-z)-1]\exp_{N+1-n}[\exp(-z)-1]. \qquad (4.11)$$

Table III contains formulas for $P_{(n)}^{(\infty)}$ for several small values of n. Now 1/3 and 3/8 are already reasonably good approximations to $1/e$ so that the end effect only extends to the fourth or fifth site from the end.

The function $P_{(n)}^{(\infty)}$ is oscillatory in its dependence upon n. $P_{(2)}^{(\infty)}$ is zero and the reason for this is clear. The first particle can either survive or be annihilated. If it survives, it is because the second particle has been annihilated by the third. If it is annihilated, it is annihilated by the second which must then be annihilated also. Thus in all possible cases the second particle is annihilated and $P_{(2)}^{(n)}(\infty)$ is zero.

V. GENERATING FUNCTION FOR PARTICLE SURVIVAL

We now consider the probability $p_{r,n}(z)$ that exactly r particles survive to "time" z in a chain initially of length n. From this fairly complete description of the probabilities we may easily generate less detailed distributions such as the survival probability at $z=\infty$, the mean number of surviving sites, the variance in this number, etc.

We can write down directly an integral equation for the probability $p_{r,n}(z)$:

$$p_{r,n}(z)=\delta_{r,n}\exp[-(n-1)z]+\int_0^z \exp[-(n-1)(z-t)]\sum_{k=0}^{n-2}\sum_{s=0}^{k}p_{s,k}(t)p_{r-s,n-k-2}(t)dt. \qquad (5.1)$$

The meaning of this equation is clear. The first term which contains the Kronecker delta function ($\delta_{r,n}=0$ for $r\neq n$, $\delta_{r,n}=1$ for $r=n$) represents the probability that none of the $n-1$ bonds of the initial chain have been formed by time z. The second term represents the summation (integral) of all of the processes which involve at least one bond formation. The first factor of the integrand represents the survival of all n sites to time $z-t$ at which time the chain is decomposed into two chains of k and $n-k-2$ sites (two sites are destroyed by each bond) (see Fig. 5). The probability of each mode of decay from $k=0$ to $k=n-2$ is equally likely and each is equal to dt. In order to have r sites survive to time z we must then have s sites on one chain and $r-s$ sites on the other chain surviving at a time t after formation of those chains. The exact mode of division does not concern us, and hence we sum over all of the possible independent alternate modes which still lead to r particles surviving at time z.

In order to simplify the solution of Eq. (5.1) we define a generating function $p_n(y,z)$ which is by definition a polynomial of order n in y which is so defined that the coefficient of y^r is the probability $p_{r,n}(z)$

$$p_n(y,z)=\sum_{r=0}^{n}p_{r,n}(z)y^r. \qquad (5.2)$$

Now, from the structure of its derivation Eq. (5.1) is obviously valid only for $n\geq 2$ since otherwise the integral term cannot appear. In order to complete the definition we must look separately at the cases for $n=0$ and $n=1$. For $n=0$ it is trivially true that

$$p_{0,0}(z)=1, \qquad p_0(y,z)=1, \qquad (5.3)$$

i.e., a zero-site chain is certain to have zero survival at all times. For $n=1$ the single site survives indefinitely since sites can annihilate only in pairs. Thus a single initial site can never have zero survival. Therefore

$$p_{1,1}(z)=1, \qquad p_{0,1}(z)=0, \qquad p_1(y,z)=y. \qquad (5.4)$$

Equation (5.1) is multiplied by y^r and summed over all values of r. This gives

$$p_n(y,z)\exp[(n-1)z]=y^n+\int_0^z \exp[(n-1)t]\times\sum_{k=0}^{n-2}p_k(y,t)p_{n-k-2}(y,t)dt. \qquad (5.5)$$

This equation may easily be converted to a differential equation

$$\frac{\partial p_n(y,z)}{\partial z}+(n-1)p_n(y,z)=\sum_{k=0}^{n-2}p_k(y,z)p_{n-k-2}(y,z). \qquad (5.5')$$

Another generating function $G(x,y,z)$ can now be defined such that in its Taylor expansion the coefficient of x^n is $p_n(y, z)$

$$G(x, y, z) = \sum_{n=0}^{\infty} p_n(y, z)x^n = 1 + xy + \sum_{n=2}^{\infty} p_n(y, z)x^n \quad (5.6)$$

and

$$\sum_{n=2}^{\infty}(n-1)p_n(y, z)x^n = x^2\frac{\partial}{\partial x}\left[\frac{G(x, y, z)-1}{x}\right]. \quad (5.7)$$

Equation (5.5) is multiplied by x^n and summed over all values of n from $n=2$ to $n=\infty$, and after some manipulation we obtain

$$\frac{\partial G}{\partial z} + x\frac{\partial G}{\partial x} = x^2G^2 + G - 1. \quad (5.8)$$

This equation must be solved with the boundary condition which can be developed from the initial conditions on the original probabilities. These conditions are simply that there are certainly n sites at $z=0$ in an n-site chain and hence

$$p_n(y, 0) = y^n,$$

$$G(x, y, 0) = 1 + xy + x^2y^2 + \cdots = 1/(1-xy). \quad (5.9)$$

It is easy to verify that the solution of (5.8) with the boundary condition (5.9) is

$$G(x, y, z) = \{1 - x\tanh[x(1-e^{-z}) + \tanh^{-1}y]\}^{-1}$$

$$= \frac{1 + y\tanh[x(1-e^{-z})]}{1 - xy + (y-x)\tanh[x(1-e^{-z})]}. \quad (5.10)$$

The probability function $p_{r,n}(z)$ can then be extracted from the solution (5.10) by expanding $G(x, y, z)$ as a power series and extracting the coefficient of x^ny^r. This calculation is a bit cumbersome, and it is perhaps just as convenient to calculate $p_n(y, z)$ successively from (5.5). The first few functions are

$$p_0(y, z) = 1,$$

$$p_1(y, z) = y,$$

$$p_2(y, z) = y^2e^{-z} + 1 - e^{-z},$$

$$p_3(y, z) = y^3e^{-2z} + y(1-e^{-2z}),$$

$$p_4(y, z) = y^4e^{-3z} + y^2(1+3e^{-z}-4e^{-3z})/3 + (2-3e^{-z}+e^{-3z})/3,$$

$$p_5(y, z) = y^5e^{-4z} + y^3(2e^{-z}+3e^{-2z}-5e^{-4z})/3 + y(3-2e^{-z}-e^{-2z}+2e^{-4z})/3. \quad (5.11)$$

In the limit of $z \to \infty$ we obtain the ultimate survival probabilities studied by Flory.[1] For $z \gg 1$, $p_n(y, z)$ becomes independent of z and (5.5) reduces to

$$p_n(y, \infty) = \frac{1}{n-1}\sum_{k=0}^{n-2} p_k(y, \infty)p_{n-k-2}(y, \infty). \quad (5.12)$$

Although (5.5) or (5.12) may be more convenient for calculation than (5.10), we use this latter expression for the generation of other probability generating functions. From the definition the mean number of sites surviving to time z is given by

$$N_n(z) = \sum_{r=0}^{n} rp_{r,n}(z) = \left.\frac{\partial p_n(y, z)}{\partial y}\right|_{y=1}, \quad (5.13)$$

and hence the generating function for $N_n(z)$ can be obtained by partial differentiation of $G(x, y, z)$. If $G_1(x, z)$ is then defined to be the generating function for $N_n(z)$, it follows immediately that

$$G_1(x, z) = \sum_{n=0}^{\infty} N_n(z)x^n = \left.\frac{\partial G(x, y, z)}{\partial y}\right|_{y=1}$$

$$= \frac{x}{(1-x)^2}\exp\{2x[\exp(-z)-1]\}. \quad (5.14)$$

The mean survival probability for sites in an n-site chain is $N_n(z)$, so that by expansion of (5.14) we find

$$P_1^{(n)}(z) = N_n(z)/n = \exp_{n+1}\omega - (\omega/n)\exp_n\omega, \quad (5.15)$$

where $\omega = 2(e^{-z}-1)$. For long chains, $n \gg 1$, an excellent approximation is

$$P_1(z) = (1/n)[n+2-2\exp(-z)] \times \exp\{-2[1-\exp(-z)]\} \quad (5.16)$$

and

$$P_1(\infty) = [(n+2)/n]e^{-2}. \quad (5.16')$$

The second moment, or mean-square number of surviving sites, is given by

$$M_n(z) = \sum_{r=0}^{n} r^2p_{r,n}(z), \quad (5.17)$$

and hence we obtain

$$G_2(x, z) = \sum M_n(z)x^n = \left.\frac{\partial}{\partial y}\left\{y\frac{\partial G(x, y, z)}{\partial y}\right\}\right|_{y=1}$$

$$= \frac{x(1+x)}{(1-x)^3}\exp\{-4x[1-\exp(-z)]\}. \quad (5.18)$$

From this we find

$$M_n(z) = \sum_{s=0}^{n}\frac{(n-s)^2(2\omega)^s}{s!} = n^2\exp_{n+1}(2\omega) - 2(2n-1)\omega\exp_n(2\omega) + 4\omega^2\exp_{n-1}(2\omega) \quad (5.19)$$

and in the limit of large n

$$M_n(z) \approx [(n-2\omega)^2 + 2\omega]\exp\{-4[1-\exp(-z)]\}. \quad (5.20)$$

The variance in the number of surviving sites is

TABLE IV. Statistics of site survival.

n	$p_{0,n}$	$p_{1,n}$	$p_{2,n}$	$p_{3,n}$	$p_{4,n}$	N_n	M_n	p_n[a]	π_n[b]
1	0	1				1	1	1.00000	...
2	1	0				0	0	0	...
3	0	1				1	1	0.33333	0.33333
4	2/3	0	1/3			2/3	4/3	0.16667	0
5	0	1	0			1	1	0.20000	0.20000
6	7/15	0	8/15			16/15	32/15	0.17778	0.11111
7	0	8/9	0	1/9		11/9	17/9	0.17460	0.14286
8	34/105	0	71/105	0		142/105	284/105	0.16905	0.13333
9	0	34/45	0	11/45		67/45	133/45	0.16543	0.13580
10	638/2835	0	2092/2835	0	105/2835	4604/2835	10048/2835	0.16240	0.13524
11	0	977/1575	0	598/1575	0	2771/1575	6359/1575	0.15994	0.13535
12	4876/31185	0	23075/31185	0	3234/31185	59086/31185	144044/31185	0.15789	0.13533

[a] p_n, mean survival probability of a site in an n-site linear chain.
[b] π_n, mean survival probability of a site in an n-site ring; $\pi_n=[(n-2)/n]p_{n-2}$.

$M_n - N_n^2$ and in the limit of large n this becomes

$$\sigma_n^2 = 4[1-\exp(-z)][n+2-3\exp(-z)] \times \exp\{-4[1-\exp(-z)]\}. \quad (5.21)$$

For $z \gg 1$ we have

$$\sigma_n^2(\infty) = 4(n+2)e^{-4}, \quad (5.22)$$

and the relative standard deviation in the number of surviving sites σ_n/N_n becomes $2/(n+2)^{\frac{1}{2}}$. Table IV gives the exact probabilities for chains of up to 12 sites.

VI. EFFECT OF DILUENTS

If selected sites in an array are rendered inert so that they cannot participate in the reaction, it is actually possible, depending upon the conditions, to increase the number of sites which ultimately survive. Both Flory[1] and Jackson and Montroll[4] have recognized this possibility. In the free-radical case sites can be rendered inert by substituting a chemically inert atom (a diluent) for free radicals. The effect is achieved in essence by increasing the number of short chains which increases thereby the variability of site survival probabilities. The maximum effect which can be achieved in this way is an enhancement of the survival probability from $1/e^2$ to $1/2$ by introducing a diluent uniformly onto alternate sites, thus isolating completely the remaining sites and guaranteeing their survival. On the other hand, a one-third diluent concentration uniformly trapped on every third site would produce isolated pairs which could not survive. In this case the survival rate is reduced to zero. It is therefore important to determine the effect of introducing a diluent molecule randomly into an infinite array of sites.

With the background of the preceding section this problem is readily solved. If the diluent concentration is ϵ the probability that any given site is occupied by a diluent molecule is ϵ, and the probability that it is not occupied by diluent is $1-\epsilon$. If the diluent is randomly deposited the occupation state of any site is independent of the state of its neighbors. To obtain a run of exactly n empty sites requires a diluent at each end with n free sites between. The probability of this configuration is simply $\epsilon(1-\epsilon)^n\epsilon$, and the probability of survival of sites is therefore

$$P(\epsilon, z) = \sum_{n=0}^{\infty} \epsilon^2(1-\epsilon)^n n P_1^{(n)}(z), \quad (6.1)$$

where $P_1^{(n)}(z)$ is the singlet survival probability for a chain of length n; the factor n accounts for the fact that there are n possible sites in the run and $\epsilon^2(1-\epsilon)^n$ is the probability of the run. In terms of the functions $N_n(z)$ of (5.13), this can be written

$$P(\epsilon, z) = \epsilon^2 \sum_{n=0}^{\infty} (1-\epsilon)^n N_n(z), \quad (6.2)$$

and therefore

$$P(\epsilon, z) = \epsilon^2 G_1(1-\epsilon, z) = (1-\epsilon)\exp\{-2(1-\epsilon)[1-\exp(-z)]\} \quad (6.3)$$

from (5.14). The ultimate survival probability is therefore given by

$$P(\epsilon, \infty) = (1-\epsilon)e^{-2(1-\epsilon)}. \quad (6.4)$$

The maximum value of this is obtained for $\epsilon = 1/2$ in which case we have $P = 1/2e$ as calculated by Flory.[1] In Table V and Fig. 6 we compare the cases of random diluent with the case of uniform diluent. In the case of a uniform diluent one diluent molecule every $(n+1)$th site decomposes the system into subsystems consisting of chains of length $n=(1-\epsilon)/\epsilon$. The survival probability in the uniform system is therefore given by

$$P(\epsilon, z)=(1-\epsilon)P_1^{(1-\epsilon/\epsilon)}(z), \quad (6.5)$$

which is of course defined only if $(1-\epsilon)/\epsilon$ is an integer. It is clear from the data that there is practically no difference between the uniform and random diluent cases for diluent concentrations of less than 12%. This is a reflection of the fact that the end effects are negligible for chains of more than seven sites.

VII. REVERSIBLE REACTIONS

If we permit bonds to be undone so that the processes leading to reactant isolation become reversible, the problem becomes considerably more difficult. The fact that a given site is unoccupied no longer guarantees that information has not flowed past it. Because bonds can be undone an unoccupied site may have been occupied at an earlier time. As a result of these considerations it proves more expedient to introduce a new method of solution.

It is convenient to speak of the occupation of *bonds* rather than *sites*. In an infinite array of sites there are effectively as many bonds as sites. In the case of no reversibility we have defined the fraction of unoccupied sites by P_1 so that the fraction of occupied sites is $(1-P_1)$; and since each bond occupies two adjacent sites the fraction of bonds occupied by molecules is

$$S=(1-P_1)/2. \quad (7.1)$$

Thus the bonding process is described by the function P_1 derived earlier.

Order the bonds from left to right by an integer index j, and define an occupation number S_j which is unity if the bond is occupied and zero otherwise. If bonds $j-1$, j, and $j+1$ are all unoccupied, then it is possible for bond j to become occupied since neither of its sites has been consumed. The chance that it is occupied in time dt is $kdt=dz$. In the process S_j goes

TABLE V.

Mean chain length n	Diluent concentration ϵ	$P(\epsilon,\infty)$ random	$P(\epsilon,\infty)$ uniform
1	0.5000	0.183940	0.5000
2	0.3333	0.175731	0
3	0.2500	0.167348	0.2500
4	0.2000	0.161517	0.133333
5	0.1667	0.157297	0.166667
6	0.1428	0.154365	0.152381
7	0.1250	0.152052	0.152776
8	0.1111	0.150234	0.150265
9	0.1000	0.148769	0.148889
10	0.0909	0.147564	0.147635

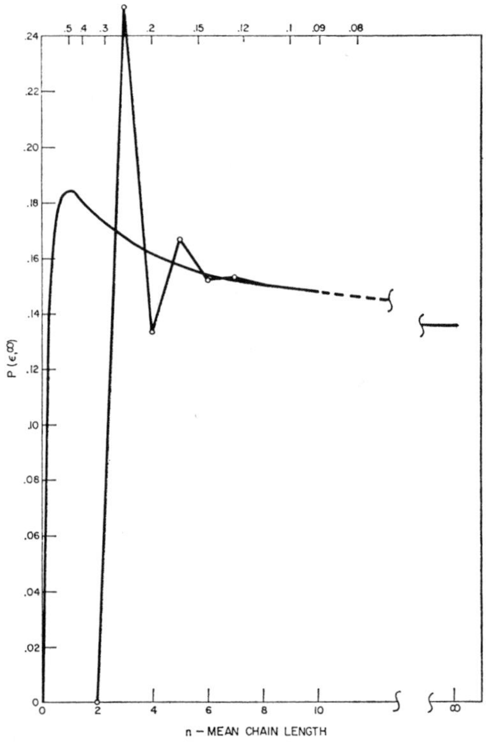

FIG. 6. Ultimate site survival probability as a function of diluent concentration. The smooth curve refers to a random distribution of occupied sites by the diluent. The segmented curve actually exists only at the circled points and corresponds to a regular array of diluent (n:1, free sites:diluent sites). The mean chain length between diluent sites is plotted along the bottom and α, the diluent concentration, is plotted along the top.

from zero to unity and the probability of this occurring can be denoted by

$$p_{0\to1}(j)dt=k(1-S_{j-1})(1-S_j)(1-S_{j+1})dt. \quad (7.2)$$

From this expression it can be seen that if either S_{j-1}, S_j, or S_{j+1} is unity (either of the three bonds is occupied, $p_{0\to1}(j)=0$ and that if S_{j-1}, S_j, and S_{j+1} are all zero (none of the three bonds is occupied), then $p_{0\to1}(j)=kdt$. Thus $p_{0\to1}(j)$ has the required properties.

If the chance of unbonding in time dt is βdt, then the chance of the occurrence in which S_j goes from unity to zero is

$$p_{1\to0}(j)dt=S_j\beta dt. \quad (7.3)$$

This has the requisite behavior because if S_j is zero (unoccupied bond), $p_{1\to0}(j)$ is zero—undoing of bond j cannot occur. On the other hand, if the bond is occupied ($S_j=1$) then $p_{1\to0}(j)=\beta dt$.

Now the change of $\langle S_j\rangle$, the average value of S_j

TABLE VI.

z	S Eq. (7.22) ($\alpha=0$)	s' Eq. (7.24)
0	0	0
0.5	0.270	0.272
1.0	0.346	0.369
1.5	0.370	0.394
2.0	0.378	0.411
2.5	0.381	0.420
∞	0.38196	0.43233

(averaged over an ensemble of arrays) in time dt must be

$$d\langle S_j\rangle = \langle p_{0\to1}(j)\rangle dt - \langle p_{1\to0}(j)\rangle dt. \quad (7.4)$$

Substitution from (7.2) and (7.3) then yields [expanding the factors in (7.2)]

$$d\langle S_j\rangle/dz = 1-\langle S_{j-1}\rangle - \langle S_j\rangle - \langle S_{j+1}\rangle + \langle S_{j-1}S_j\rangle + \langle S_jS_{j+1}\rangle + \langle S_{j-1}S_{j+1}\rangle - \langle S_{j-1}S_jS_{j+1}\rangle - \alpha\langle S_j\rangle, \quad (7.5)$$

where

$$\alpha = \beta/k. \quad (7.6)$$

Since two adjacent bonds cannot be occupied at the same time,

$$S_jS_{j+1} = 0 = S_jS_{j-1} \quad (7.7)$$

and (7.5) can be written

$$d\langle S_j\rangle/dz = 1-\langle S_{j-1}\rangle - \langle S_j\rangle - \langle S_{j+1}\rangle + \langle S_{j-1}S_{j+1}\rangle - \alpha\langle S_j\rangle \quad (7.8)$$

If we restrict ourselves entirely to infinite systems, then, by symmetry, $\langle S_j\rangle$ is independent of the specific site location,

$$\langle S_j\rangle = S = \langle S_{j-1}\rangle = \langle S_{j+1}\rangle \quad (7.9)$$

and the pair correlation $\langle S_{j-1}S_{j+1}\rangle$ is also independent of the index j

$$\langle S_{j-1}S_{j+1}\rangle = \langle S_jS_{j+2}\rangle = \phi_2. \quad (7.10)$$

With these relations, (7.8) becomes

$$dS/dz = 1-(3+\alpha)S + \phi_2. \quad (7.11)$$

In order to proceed further we must either approximate ϕ_2 in terms of S or develop an equation from which ϕ_2 may be calculated. We define, in general,

$$\phi_k = \langle S_jS_{j+k}\rangle = \langle S_jS_{j-k}\rangle \quad (7.12)$$

to be the pair correlation function of rank k. We have already made use of the fact that $\phi_1=0$.

Arguments similar to those that led to (7.4) allow us to write

$$d\phi_k/dt = (d/dt)\langle S_jS_{j+k}\rangle = \langle S_j[p_{0\to1}(j+k) - p_{1\to0}(j+k)] + [p_{0\to1}(j) - p_{1\to0}(j)]S_{j+k}\rangle. \quad (7.13)$$

We substitute (7.2) and (7.3) into (7.13) and simplify the resulting expressions, making use of (7.7), (7.9), and (7.10). This gives

$$d\phi_k/dt = 2[S - \phi_{k-1} - (1+\alpha)\phi_k - \phi_{k+1} + \langle S_{j-1}S_{j+1}S_{j+k}\rangle], \quad (7.14)$$

and we now have an expression which involves a triplet correlation function

$$\psi_{r,t} = \langle S_jS_{j+r}S_{j+t}\rangle. \quad (7.15)$$

From this definition we find several symmetry relationships

$$\psi_{r,t} = \psi_{t,r} = \psi_{-r,-t} = \psi_{-r,t-r} = \psi_{t-r,t} = \text{etc.} \quad (7.16)$$

These symmetry relations allow us to restrict consideration to subscripts for which $0\le r\le t$ and $r\le t/2$. In addition we also have the restriction that

$$\psi_{r,t} = 0 \quad \text{if} \quad r=1 \quad \text{or} \quad t-r=1. \quad (7.17)$$

From (7.14) we can then write

$$d\phi_0/dt = d\langle S_j^2\rangle/dt = 2[S-(1+\alpha)\phi_0], \quad (7.18a)$$

$$d\phi_1/dt = 0 = 2[S-\phi_0-\phi_2+\psi_{0,2}], \quad (7.18b)$$

$$d\phi_2/dt = 2[S-(1+\alpha)\phi_2-\phi_3], \quad (7.18c)$$

$$d\phi_3/dt = 2[S-\phi_2-(1+\alpha)\phi_3-\phi_4+\psi_{2,4}]. \quad (7.18d)$$

We are therefore led to an infinite hierarchy of equations which involve successively higher rank pair correlation functions as well as triplet correlation functions.

In order to solve such a system we must introduce an approximation which closes the hierarchy. As a start let us break the hierarchy with the first equation by the use of a superposition approximation for ϕ_2

$$\phi_2 \approx S^2. \quad (7.19)$$

This is equivalent to assuming that the occupancies of alternate bonds are independent of one another. Substitution of (7.19) with (7.11) yields

$$dS/[1-(3+\alpha)S+S^2] = dz \quad (7.20)$$

which, when integrated subject to the boundary condition

$$S=0, \qquad z=0 \quad (7.21)$$

yields

$$S = 2\tanh\tfrac{1}{2}\gamma z/[\gamma+(3+\alpha)\tanh\tfrac{1}{2}\gamma z], \quad (7.22)$$

where

$$\gamma = [(1+\alpha)(5+\alpha)]^{\frac{1}{2}}. \quad (7.23)$$

It is of interest to compare this with the value of S obtained by substitution of our earlier result [(2.11)] for the infinite chain into (7.1). In this irreversible case $\alpha=0$ so γ in (7.22) must be set equal to $\sqrt{5}$ before the latter relation can be compared with the result obtained from (2.11). Denoting the result obtained from (2.11) as s' we have

$$s'=\tfrac{1}{2}\{1-\exp[2\exp(-z)-2]\}. \qquad (7.24)$$

Table VI compares S and s' for several values of z. The table demonstrates that for early times S and s' are almost the same. This is to be expected because the occupancies of alternate bonds are less correlated for small z and (7.15) is more nearly valid. In the limit of zero time there is of course no correlation. The table demonstrates, however, that even at infinite time the error is not great. If reversibility is introduced ($\alpha\neq 0$) then this should reduce even further the degree of correlation (i.e., the degree of memory) so that (7.16) should be an even better approximation to the exact result than it is in Table VI for the case of $\alpha=0$.

We now turn to the second approximation in the hierarchy of equations. This time we make no approximation in (7.11) but rather we determine the pair function ϕ_2 from (7.18c). This equation involves ϕ_3 which we approximate by

$$\phi_3\approx S^2. \qquad (7.25)$$

The pair correlation function ϕ_3 involves bonds which are separated by two bond sites and hence the superposition approximation (7.25) should be more accurate than for ϕ_2 in which the bonds are separated by only a single bond site.

The set of simultaneous differential equations (7.11) and (7.18c) becomes, upon the introduction of (7.25), a closed set but one which is unfortunately nonlinear. This nonlinear system apparently has no solution in closed form. To obtain solution it is therefore necessary to use approximation techniques or numerical integration. It is, however, rather easy to obtain the equilibrium solution which is after all the quantity of primary physical interest. It is easy to show that for $z\gg 1$, both S and ϕ_2 approach constant values. Hence we can set

$$dS/dz=0=d\phi_2/dz, \qquad (7.26)$$

and solve the resulting algebraic equations. The solution is

$$S(\infty)=[2+5\alpha+5\alpha^2+2\alpha^3+\tfrac{1}{4}\alpha^4]^{\frac{1}{2}}-1-2\alpha-\tfrac{1}{2}\alpha^2 \qquad (7.27a)$$

$$\phi_2(\infty)=(3+\alpha)S(\infty)-1. \qquad (7.27b)$$

For $\alpha=0$, which corresponds to the irreversible case of Table VI, we obtain

$$S(\infty)=\sqrt{2}-1=0.4142, \qquad (7.28a)$$

$$\phi_2(\infty)=3\sqrt{2}-4=0.2426, \qquad (7.28b)$$

which is to be compared with the first-order approximation, $S=0.3820$, $\phi_2(\infty)=0.1459$, and the exact solution $S(\infty)=0.4323$ given in Table VI.

The second-order approximation of the hierarchy therefore leads to a significant improvement in accuracy. Furthermore the accuracy of the superposition approximation for ϕ_3 should improve for $\alpha>0$ since the introduction of reversibility into the system decreases the extent of correlation which exists between bonds. Use of additional equations in the hierarchy improves the accuracy still further.

Stochastic and Deterministic Formulation of Chemical Rate Equations

I. Oppenheim*

Department of Chemistry, Massachusetts Institute of Technology, Cambridge, Massachusetts

K. E. Shuler†

Department of Chemistry, University of California, San Diego, La Jolla, California

AND

G. H. Weiss

National Institutes of Health, Bethesda, Maryland

(Received 5 July 1968)

It is shown, on the basis of some examples, that the commonly used stochastic theory of gas-phase chemical rate equations reduces to the deterministic formulation in the thermodynamic limit, $N\rightarrow\infty$, $V\rightarrow\infty$, N/V fixed. Since the commonly used deterministic collision theory of chemical kinetics is derivable from the Boltzmann equation with reactive scattering terms, the stochastic formulation of chemical kinetics is thus shown to be equivalent to the results of the Boltzmann equation in the thermodynamic limit. However, since the stochastic theory has not been derived from the Liouville equation for finite systems, the validity of the calculations of the deviation of the stochastic mean from the deterministric results and the fluctuations about that mean for finite N has not been established.

I. INTRODUCTION

Much work has been carried out over the past fifteen years on what has been called the "stochastic" formulation of chemical rate equations. This work has been surveyed in an extensive review paper by McQuarrie.[1]

There are two essentially different conceptual approaches to a stochastic formulation of chemical kinetics which in the past have not been clearly differentiated. In one approach one phrases, in an *ad hoc* manner, rate problems in terms of probability theory (stochastic theory of birth and death processes) on the basis of intuititive feelings about the nature and mechanism of the rate process and without reference to deterministic dynamics. Examples of this formulation can be found in work on chemistry,[1] genetics,[2,3] and epidemology and ecology.[4] In the other approach, one considers the stochastic formulation as an *improvement* over a deterministic theory in the sense that the stochastic theory presumably takes account of the concentration fluctuations of reactants and products in small subvolumes of the over-all reaction volume and thus presumably yields some information on the dispersion of the reacting species about their deterministic mean. It is to the investigation of the validity of this second approach that this paper is addressed.

The usual transcription of the deterministic to the stochastic version of chemical kinetics involves essentially the following steps:

(1) Replacement of the deterministic particle number $n_A(t)\equiv$number of A particles at time t by the distribution $P_n(t)=\text{Prob}\{n_A(t)=n\}$. For nonconservative, bimolecular reaction systems such as the Weiss–Dietz model discussed below, the number densities $n_A(t)$, $n_B(t)$ are replaced by the joint probability $P_{nm}(t)=\text{Prob}\{n_A(t)=n, n_B(t)=m\}$.

(2) Representation of the dynamics of chemical change by a birth and death process with a transition probability $A(a, a\pm j)dt$ for the transition $a\rightarrow a\pm j$ during the time interval $(t, t+dt)$ of the form

$$A(a, a\pm j)dt=kadt+o(dt), \qquad (1.1)$$

where k is the same rate constant that appears in the deterministic theory and $o(dt)/dt\rightarrow 0$ as $dt\rightarrow 0$.

(3) Assumption of the validity of molecular chaos according to which collisions and the concomitant changes due to collisions can be treated as *uncorrelated binary* events. This assumption implies that a joint occupation number change $n\rightarrow n+1$, $m\rightarrow m-1$ is correctly represented by the *product* transition probability

$$A_2(n, n+1; m, m-1)dt=knmdt+o(dt). \qquad (1.2)$$

The combination of these operations then leads to "master equations" of the type used in Ref. 1 and in this paper.

To make any judgment as to whether this presently used stochastic theory, as detailed for instance in Ref. 1, is an improvement over the presently used deterministic theory of chemical rate processes,[5] it is first necessary to determine whether this determinsitic theory is a fundamentally correct description of a

* Part of this work was supported by the National Science Foundation.

† Part of this work was carried out while the author was at the National Bureau of Standards.

[1] D. A. McQuarrie, J. Appl. Probl. **4**, 413 (1967).

[2] P. A. P. Moran, *The Statistical Processes of Evolutionary Theory* (Clarendon Press, Oxford, England, 1962).

[3] S. Karlin and J. McGregor, *Stochastic Models in Medicine and Biology* (University of Wisconsin Press, Madison, Wisc., 1964), pp. 245–279.

[4] M. S. Bartlett, *Stochastic Population Models in Ecology and Epidemology* (John Wiley & Sons, Inc., New York, 1961).

[5] See, e.g., H. S. Johnston, *Gas Phase Reaction Theory* (Ronald Press Co., New York), 1966.

chemical rate process. To derive the correct deterministic rate theory for chemical reactions one would have to start with the Liouville equation (either classical or quantal) and then take proper account of the dynamics of the interactions, including correlations, to obtain the transport equation appropriate to the system under investigation (i.e., liquid, plasma, high-density gas, low-density gas, etc). There has been some work along these lines,[6] but the answer is certainly not yet at hand.

All one can say with confidence on the basis of a derivation from the Liouville equation is that in the thermodynamic limit: $N\rightarrow\infty$, $V\rightarrow\infty$, N/V is constant (where N is the total number of particles in the system with volume V), the Boltzmann equation is the correct dynamical description of a transport process in a dilute gas. As has been shown by Ross and Mazur,[7] the Boltzmann equation extended to include reactive scattering terms is equivalent to the usual deterministic uncorrelated binary collision theory of chemical kinetics. The Boltzmann equation has the important property that

$$\lim_{N\to\infty} f_k^{(N)}(v_1,\cdots,v_k;t)=\prod_{j=1}^{k}\lim_{N\to\infty} f_1^{(N)}(v_j;t), \quad (1.3)$$

i.e., the distribution function is factorized at all times t, $0\leq t\leq\infty$. For a stochastic formulation of chemical kinetics to be at all correct, it must reduce to the stochastic version of the Boltzmann equation in the limit $N\rightarrow\infty$, i.e., it must reduce to a Pauli equation[8] (lowest-order master equation) with the factorization property[9–11]

$$\lim_{N\to\infty} P_k^{(N)}(r_1,\cdots,r_k;t)=\prod_{j=1}^{k}\lim_{N\to\infty} P_1^{(N)}(r_j;t) \quad (1.4)$$

for all t, $0<t\leq\infty$, for the initial condition

$$P_k^{(N)}(\mathbf{r};0)=\prod_{j=1}^{k} P_1^{(N)}(r_j;0),$$

where $P_k^{(N)}(r_1,\cdots,r_k{:}\,t)$ is the joint probability that there are r_1 molecules of species $1,\cdots$, and r_k molecules of species k in the reaction system at time t.

Equivalently, since the Boltzmann equation implies deterministic kinetics, the stochastic formulation must reduce to the deterministic formulation in the thermodynamic limit in order for the stochastic version to be at least a correct limiting theory.

This is a *necessary* condition for the validity of a stochastic formulation of gas-phase chemical kinetics but it is by no means a sufficient one. In the absence of an exact dynamic theory of chemical rate processes for $N\neq\infty$, it is not possible to compare the stochastic theory for $N\neq\infty$ with the dynamic, deterministic theory for $N\neq\infty$ and it is therefore not possible to make any statements at all about the validity of a stochastic theory for finite N. Thus, the results obtained from the stochastic theory of gas-phase kinetics on concentration fluctuations about the deterministic mean for small N (or small V) may or may not be correct. We tend to believe that these results on fluctuations[1] are not correct since the generalized master equation for chemical reactions for N, $V\neq\infty$, which still needs to be derived from the Liouville equation, probably has a much more complicated structure than the simple birth and death process stochastic equations now being used.

We shall show in this paper that solutions of the presently used stochastic formulation of chemical kinetics indeed become identical with the solutions of the deterministic theory in the limit as $N\rightarrow\infty$. Also the coefficients of variation go to zero in this limit. In the absence of a "general" stochastic equation covering all possible types of chemical reactions, we shall demonstrate this limiting equivalence for two specific examples of chemical reactions. On the basis of the simple structure of the birth- and death-type stochastic equations for chemical rate processes and on the basis of our results given below there is every reason to believe that this equivalence in the limit $N\rightarrow\infty$ is a general property of such stochastic equations.

To demonstrate the equivalence of the stochastic formulation with the deterministic formulation in the thermodynamic limit, we shall show that (a) the stochastic mean $\mu_1(t)\equiv\sum nP(n,t)$ for the reactant (and product) molecules agrees with the deterministic results to order $(1/N)$ for all t, and (b) that $\mu_r\equiv\sum n^rP(n,t)=(\mu_1)^r$, $r\ll N$, to order $(1/N)$ for all t. These two results together imply that for all practical purposes the singlet probabilities $P_1^{(N)}(n,t)$ are delta functions to order $(1/N)$ for all t.

II. ANALYSIS OF A GENERAL SECOND-ORDER REACTION

We consider here the general second-order reaction

$$\mathrm{A+B}\underset{k_2}{\overset{k_1}{\rightleftarrows}}\mathrm{C+D}. \quad (\mathrm{I})$$

This reaction was first studied in detail in the above context by Darvey, Ninham, and Staff[12] who were able to compare, after lengthy calculations, the stochastic and deterministic results for the special cases $t=\infty$ (i.e., at equilibrium) and for all time t for $K\simeq1$ and

[6] T. Yamanoto, J. Chem. Phys. **33**, 281 (1960); H. Aroeste, Advan. Chem. Phys. **6**, 1 (1964); F. C. Andrews, J. Chem. Phys. **47**, 3170 (1967).

[7] J. Ross and P. Mazur, J. Chem. Phys. **35**, 19 (1961); see also J. C. Light, J. Ross and K. E. Shuler in *Kinetic Processes in Gases and Plasmas*, A. Hochstim, Ed. (Academic Press Inc., New York, to be published).

[8] I. Oppenheim and K. E. Shuler, Phys. Rev. **B138**, 1007 (1965).

[9] M. Kac, in *Third Berkeley Symposium on Mathematical Statistics and Probability* (University of California Press, Berkeley, Calif., 1956), Vol. 3, p. 171.

[10] J. A. McLennan and R. J. Swenson, J. Math. Phys. **4**, 1527 (1963).

[11] G. V. Chester and J. Sykes, J. Math. Phys. **7**, 2243 (1966).

[12] I. G. Darvey, B. W. Ninham, and P. J. Staff, J. Chem. Phys. **45**, 2145 (1966); I. G. Darvey and B. W. Ninham, *ibid.* **46**, 626 (1967).

$K \simeq 0$, where

$$K = k_2/k_1 \tag{2.1}$$

is the equilibrium constant. The present analysis will provide general results by a much simpler method.

Let the initial number of reactants and products in the system of fixed volume V be N_A, N_B, N_C, N_D, respectively. We shall specifically assume that N_A and N_B (or N_C and N_D) are large and will obtain results in the limit $N_A \to \infty$, where it should be remembered that N_A is of the order of 10^{18}–10^{23} for gas-phase reactions. We define parameters

$$\beta = N_B/N_A, \qquad \gamma = N_C/N_A, \qquad \delta = N_D/N_A, \tag{2.2}$$

where, β, γ, δ are numbers which are of order unity. Owing to the conservation of the number of molecules in Reaction I, the state of the system is completely determined by specifying the number of molecules of any one species, i.e.,

$$N_A - n_A(t) = N_B - n_B(t) = -(N_C - n_C(t)) = -(N_D - n_D(t)), \tag{2.3}$$

where the $n_i(t)$ are the number of molecules of species i at time t. In what follows, we denote by $n(t)$ the number of species A at time t, and by $P_n(t)$ the probability that there are n molecules of species A in the system at time t. The principal assumption of the presently used stochastic theory of chemical kinetics is that the probability of an A–B collision and reaction during the time interval $(t, t+dt)$ is $k_1 n_A(t) n_B(t) dt$, where k_1 is the rate constant which appears in Reaction I. An analogous assumption is made for the C–D reaction. The assumption just stated allows us to write the following set of equations for the $P_n(t)$:

$$\begin{aligned} \dot{P}_n(t) \equiv dP_n(t)/dt = {} & k_1(n+1)[n+1+(\beta-1)N_A] \\ & \times P_{n+1}(t) + k_2[(1+\gamma)N_A-(n-1)] \\ & \times[(1+\delta)N_A-(n-1)]P_{n-1}(t) - \{k_1 n[n+(\beta-1)N_A] \\ & + k_2[(1+\gamma)N_A-n][(1+\delta)N_A-n]\}P_n(t), \\ & n=0,1,\cdots, \qquad P_{-1}(t)=0 \end{aligned} \tag{2.4}$$

with the initial condition $P_n(0) = \delta_{n,N_A}$. Rather than dealing with this set of equations we shall go directly to the equations for the moments $\mu_r(t)$, defined by

$$\mu_r(t) = \sum_{n=0}^{\infty} n^r P_n(t). \tag{2.5}$$

If we define a dimensionless time $\tau = k_1 t$ and $\dot{\mu}_r(\tau)$ to be $d\mu_r/d\tau$, then the $\mu_r(\tau)$ are easily shown to satisfy the equations

$$\begin{aligned} \dot{\mu}_r(\tau) = {} & \sum_{j=0}^{r-1} (-1)^{r-j} \binom{r}{j} [\mu_{j+2} + (\beta-1) N_A \mu_{j+1}] \\ & + K \sum_{j=0}^{r-1} \binom{r}{j} [\mu_{j+2} - (2+\gamma+\delta) N_A \mu_{j+1} \\ & + (1+\gamma)(1+\delta) N_A^2 \mu_j], \qquad r = 0, 1, 2, \end{aligned} \tag{2.6}$$

where K is defined by Eq. (2.1). This set of equations can be derived in straightforward fashion by multiplying Eq. (2.4) by n^r and summing from 0 to ∞. The initial conditions are $\mu_r(0) = N_A{}^r$ for all r.

To proceed further in the calculation we rewrite the right-hand side of Eq. (2.6) as the sum of two terms, one of which is "small" in comparison to the other, and carry out an iterative solution of the resulting equation. In this way we will show that the solutions for $\mu_1(\tau)$ and $\mu_2(\tau)$ can be expressed as a dominant term plus second-order terms that are negligible in comparison. By way of motivating the division of Eq. (2.6) into two sets of terms, we note that for any set of positive random variables it can be shown that[13]

$$\mu_{r-1}/\mu_r \geq \mu_r/\mu_{r+1}. \tag{2.7}$$

By induction it follows that

$$1/\mu_1 \geq \mu_r/\mu_{r+1}. \tag{2.8}$$

Hence, if $\mu_1(\tau)$ is very large compared to unity, $\mu_{r+1}(\tau)$ is much larger than $\mu_r(\tau)$. We will assume that $\mu_1(\tau)$ remains of order N_A for all time. This assumption will be justified *a posteriori*. We now rewrite Eq. (2.6) as

$$\begin{aligned} \dot{\mu}_r(\tau) = {} & r\{(K-1)\mu_{r+1} - [K(2+\gamma+\delta)+(\beta-1)]N_A\mu_r \\ & + K(1+\gamma)(1+\delta)N_A{}^2\mu_{r-1}\} \\ & + \sum_{j=0}^{r-2} \binom{r}{j} \{(-1)^{r-j}[\mu_{j+2}+(\beta-1)N_A\mu_{j+1}] \\ & + K[\mu_{j+2} - (2+\gamma+\delta)N_A\mu_{j+1} + (1+\gamma)(1+\delta)N_A{}^2\mu_j]\}. \end{aligned} \tag{2.9}$$

If it is assumed that μ_r is of order $N_A{}^r$ for $r \ll N_A$ then each term in the first set of braces is $0(N_A{}^{r+1})$ while each term in the summation is of lower order. Let us denote the sum in Eq. (2.9) by $H_r(\boldsymbol{\mu})$ where $\boldsymbol{\mu}$ is the vector of moments. Then Eq. (2.9) can be expressed as

$$\begin{aligned} \dot{\mu}_r(\tau) = {} & r\{(K-1)\mu_{r+1} - [K(2+\gamma+\delta)+(\beta-1)]N_A\mu_r \\ & + K(1+\gamma)(1+\delta)N_A{}^2\mu_{r-1}\} + H_r(\boldsymbol{\mu}) \end{aligned} \tag{2.10}$$

in which $H_r(\boldsymbol{\mu})$ is of lower order than the term in braces.

We will now solve Eq. (2.10) by successive approximations. To summarize the detailed calculations to follow we rewrite Eq. (2.10) in matrix form as

$$\dot{\boldsymbol{\mu}} = \mathbf{L}_0\boldsymbol{\mu} + \lambda \mathbf{L}_1 \boldsymbol{\mu}, \tag{2.11}$$

where the L_0 corresponds to the braces in Eq. (2.10), the λL_1 to the H terms, and λ is an ordering parameter which we will later set equal to one. Let us further assume the expansion

$$\boldsymbol{\mu} = \boldsymbol{\mu}^{(0)} + \lambda \boldsymbol{\mu}^{(1)} + \cdots \tag{2.12}$$

to be valid, and substitute this into Eq. (2.11). When powers of λ are equated, it is found that the first two

[13] H. Cramer, *Mathematical Methods of Statistics* (Princeton University Press, Princeton, N.J., 1946).

terms of this expansion satisfy

$$\dot{\mathbf{u}}^{(0)} = \mathbf{L}_0\mathbf{u}^{(0)},$$

$$\dot{\mathbf{u}}^{(1)} = \mathbf{L}_0\mathbf{u}^{(1)} + \mathbf{L}_1\mathbf{u}^{(0)}. \quad (2.13)$$

The initial conditions are $\mu_r(0) = N_A{}^r$. We will assume that the components $\mu_r{}^{(0)}(0)$, $\mu_r{}^{(1)}(0), \cdots$, satisfy the initial conditions

$$\mu_r{}^{(0)}(0) = N_A{}^r,$$

$$\mu_r{}^{(j)}(0) = 0, \qquad \text{for } j \geq 1. \quad (2.14)$$

consistent with the initial conditions on $\mu_r(0)$. It will be shown that the zeroth approximation $\mathbf{u}^{(0)}$ is equivalent to the results of the deterministic theory and that $\mathbf{u}^{(1)}$ is of lower order than $\mathbf{u}^{(0)}$ by a factor $N_A{}^{-1}$ so that the iteration procedure is convergent for large N_A.

The moments in the zeroth-order approximation satisfy the equation

$$\dot{\mu}_r{}^{(0)}(\tau) = r\{(K-1)\mu_{r+1}{}^{(0)} - [K(2+\gamma+\delta)+(\beta-1)] \times N_A\mu_r{}^{(0)} + K(1+\gamma)(1+\delta)N_A{}^2\mu_{r-1}{}^{(0)}\}. \quad (2.15)$$

One can readily verify that a solution of this set of equations satisfying the initial conditions of Eq. (2.14) is

$$\mu_r{}^{(0)}(\tau) = [\mu_1{}^{(0)}(\tau)]^r, \quad (2.16)$$

where $\mu_1{}^{(0)}(\tau)$ is the solution to

$$\dot{\mu}_1{}^{(0)} = (K-1)(\mu_1{}^{(0)})^2 - [K(2+\gamma+\delta)+(\beta-1)]N_A\mu_1{}^{(0)} + K(1+\gamma)(1+\delta)N_A{}^2. \quad (2.17)$$

This equation is identical with the equation for the time rate of change of the number of A molecules, $n_A(\tau)$, obtained from the deterministic theory if one sets $\mu_1{}^{(0)}(\tau) = n_A(\tau)$.

The solution to Eq. (2.17) for $K \neq 1$ is

$$\mu_1{}^0(\tau) = N_A\left(\frac{\epsilon_-(1-\epsilon_+)e^{aN_A\tau} - \epsilon_+(1-\epsilon_-)}{(1-\epsilon_+)e^{aN_A\tau} - (1-\epsilon_-)}\right) \quad (2.18)$$

in which a, ϵ_+, and ϵ_- are numbers of order 1 defined by

$$a = \{[K(2+\gamma+\delta)+(\beta-1)]^2 - 4K(K-1)(1+\delta)(1+\gamma)\}^{1/2},$$

$$\epsilon_\pm = [1/2(K-1)][K(2+\gamma+\delta)+(\beta-1)\pm a]. \quad (2.19)$$

For $K=1$, the solution of Eq. (2.17) is

$$\mu_1{}^0(\tau) = N_A\{(\Psi/\chi) + [1-(\Psi/\chi)]\exp(-\chi N_A\tau)\} \quad (2.20)$$

with

$$\chi = 1+\beta+\gamma+\delta$$

$$\Psi = (1+\gamma)(1+\delta). \quad (2.21)$$

The moment $\mu_1{}^{(0)}(\tau)$ is thus clearly of order N_A and, by Eq. (2.16), $\mu_r{}^{(0)}(\tau)$ is of order $N_A{}^r$ for all values of τ.

The next approximation, as indicated by Eq. (2.13), is one in which the second matrix operator $\mathbf{L}_1$ is taken into account. Because of the simple expression for $\mu_r{}^{(0)}$ shown in Eq. (2.16), the matrix $\mathbf{L}_1\mathbf{u}^{(0)}$ also has a simple form. The components of $\mathbf{u}^{(1)}$ are found to satisfy

$$\dot{\mu}_r{}^{(1)}(\tau) = r\{(K-1)\mu_{r+1}{}^{(1)} - [K(2+\gamma+\delta)+(\beta-1)]N_A\mu_r{}^{(1)} + K(1+\gamma)(1+\delta)N_A{}^2\mu_{r-1}{}^{(1)}\}$$
$$+[(\mu_1{}^{(0)})^2 + (\beta-1)N_A\mu_1{}^{(0)}][(\mu_1{}^{(0)}-1)^r - (\mu_1{}^{(0)})^r + r(\mu_1{}^{(0)})^{r-1}] + K[(\mu_1{}^{(0)})^2 - (2+\gamma+\delta)N_A\mu_1{}^{(0)} + (1+\gamma)(1+\delta)N_A{}^2]$$
$$\times[(\mu_1{}^{(0)}+1)^r - (\mu_1{}^{(0)})^r - r(\mu_1{}^{(0)})^{r-1}] = r\{(K-1)\mu_{r+1}{}^{(1)} - [K(2+\gamma+\delta)+(\beta-1)]N_A\mu_r{}^{(1)}$$
$$+K(1+\delta)(1+\gamma)N_A{}^2\mu_{r-1}{}^{(1)}\} + H_r(\mathbf{u}^{(1)}), \qquad r=1,2,\cdots, \quad (2.22)$$

with $\mu_0{}^{(1)}(\tau) = 0$. The quantity $H_r[\mathbf{u}^{(0)}(\tau)]$ is a known function of τ defined by the second and third lines of the above equation.

The solution of Eq. (2.22), which is given in the appendix, leads to the following bound for $\mu_1{}^{(1)}(\tau)$:

$$\mu_1{}^{(1)}(\tau) < (M/a)[1-(\epsilon_-/\epsilon_+)], \quad (2.23)$$

where the constant M is defined by Eq. (A.13). This is a bounded function of order 1 so that $\mu_1{}^{(1)} = 0(1)$. We thus find that

$$\mu_1{}^{(1)}(\tau)/\mu_1{}^{(0)}(\tau) = 0(1/N_A). \quad (2.24)$$

Proceeding as indicated above, it can be shown generally that $\mu_1{}^{(r)}$ is of order $N_A{}^{-(r-1)}$. We thus verify, as assumed earlier, that $\mu_1(\tau) = \mu_1{}^{(0)} + \mu_1{}^{(1)} + \cdots$ is of order N_A for all times τ.

A similar calculation for $\mu_2{}^{(1)}$ leads to

$$\mu_2{}^{(1)}(\tau) < M'N_A, \quad (2.25)$$

where M' is a different constant. The fluctuations in the reaction system, as measured by the coefficient of variation C are thus given by

$$C = \frac{\sigma(\tau)}{\mu_1(\tau)} = \left[\frac{\mu_2(\tau)}{[\mu_1(\tau)]^2} - 1\right]^{1/2} \sim 0\left(\frac{1}{N_A{}^{1/2}}\right), \quad (2.26)$$

which clearly vanishes in the limit as $N_A \to \infty$.

We have thus established that, in the limit as $N_A \to \infty$, the mean $\mu_1(\tau)$ of the stochastic theory becomes identical with the deterministic result [Eqs. (2.17), (2.18), and (2.20)] and that the fluctuations go to zero [Eq. (2.26)].

III. ANALYSIS OF A COMBINED FIRST- AND SECOND-ORDER REACTION

We now consider the simultaneous first- and second-order reactions

$$A+B\xrightarrow{k_1}A+C$$
$$A\xrightarrow{k_2}D. \qquad (II)$$

The stochastic theory of this set of reactions has been studied in the context of the theory of epidemics by Weiss,[14] Dietz,[15] and Downton.[16] The set of reactions (II) differ from the simple bimolecular reaction (I) treated in Sec. II in that the state of the system can no longer be specified by the number $n_i(t)$ of any one of the species i at time t, but requires instead the simultaneous specification of the number of both A and B molecules, i.e., $n_A(t)$ and $n_B(t)$. This in turn leads naturally to the use in the stochastic formulation of the joint probabilities $P_{nm}(t)$ that $n_A(t)=n$ and $n_B(t)=m$. Exact results for these joint probabilities and their moments are available from the work referred to above, Refs. 15 to 16. This problem differs also from that discussed in Sec. II in that $n_A(t)$ approaches zero as $t\to\infty$. We shall thus restrict our consideration here to times for which $n_A(t)\gg 1$.

Let N_A and N_B be the initial numbers of A and B molecules, respectively, V the (fixed) volume in which the reaction takes place, and α and τ the dimensionless parameters

$$\alpha=k_1/k_2V, \qquad \tau=k_2t. \qquad (3.1)$$

We again assume that N_A, N_B are of the order 10^{18}–10^{23} and will obtain our results in the limit as $N_A, N_B\to\infty$. The deterministic theory predicts that the numbers of A and B molecules at time τ, $n_A(\tau)$, $n_B(\tau)$, are

$$n_A(\tau)=N_A\exp(-\tau)$$
$$n_B(\tau)=N_B\exp[-\alpha N_A(1-e^{-\tau})]. \qquad (3.2)$$

The stochastic equations which describe reactions II can be written in terms of the joint probability $P_{n,m}(\tau)$, as[15]

$$\dot{P}_{n,m}(\tau)=(n+1)P_{n+1,m}+\alpha n(m+1)P_{n,m+1}$$
$$-n(1+\alpha m)P_{n,m}, \qquad (3.3)$$

where $\dot{P}_{n,m}(\tau)=dP_{n,m}(\tau)/d\tau$. The quantities which are to be compared with $n_A(\tau)$ and $n_B(\tau)$ are the first moments $\langle n_A(t)\rangle$ and $\langle n_B(\tau)\rangle$, where, for example,

$$\langle n_B(\tau)\rangle=\sum_{n=0}^{\infty}\sum_{m=0}^{\infty}mP_{nm}(\tau). \qquad (3.4)$$

We will also be interested in the variances $\sigma_A^2(\tau)$ and $\sigma_B^2(\tau)$, where $\sigma_B^2(\tau)=\langle n_B^2(\tau)\rangle-\langle n_B(\tau)\rangle^2$. One can show immediately by multiplying Eq. (3.3) by n and summing over all n and m that

$$\langle n_A(\tau)\rangle=N_A\exp(-\tau) \qquad (3.5)$$

which agrees exactly with the deterministic result (3.2). The coefficient of variance, $C_A(\tau)$, is readily evaluated as

$$C_A(\tau)=\sigma_A/\langle n_A(\tau)\rangle=[(e^{\tau}-1)/N_A]^{1/2} \qquad (3.6)$$

and goes to zero as $N_A\to\infty$ for a fixed time τ. The exact agreement between the stochastic mean and the deterministic number and the vanishing of the coefficient of variance in the limit $N_A\to\infty$ are well known results for unimolecular reactions.[1]

The calculation of the analogous quantities for the B molecules from the stochastic equation (3.3) requires considerable ingenuity, but exact analytic results have been obtained by Dietz[15] and Downton.[16] The mean number of B particles $\langle n_B(\tau)\rangle$ is given by

$$\langle n_B(\tau)\rangle=N_B\left[\frac{(1+\alpha e^{-(1+\alpha)\tau})}{(1+\alpha)}\right]^{N_A}. \qquad (3.7)$$

At a fixed pressure and temperature, the reaction volume V of Eq. (3.1) is proportional to the total number of molecules N in the reaction system so that the parameter α is of order $(1/N)$ where N is of order N_A, N_B. In the limit $N\to\infty$, we can thus expand Eq. (3.7) around $\alpha=0$, after taking logarithms, to obtain

$$\ln(\langle n_B(\tau)\rangle/N_B)=N_A\{\ln[1+\alpha e^{-(1+\alpha)\tau}]-\ln(1+\alpha)\}$$
$$=-N_A[\alpha(1-e^{-\tau})+0(\alpha^2)]. \qquad (3.8)$$

Neglect of the term $0(\alpha^2)\equiv 0(1/N_A^2)$ then yields

$$\langle n_B(\tau)\rangle=N_B\exp[-\alpha N_A(1-e^{-\tau})] \qquad (3.9)$$

which is identical with the deterministic result (3.2). The variance of the number of B molecules has been shown to be[15]

$$\sigma_B^2(\tau)=N_B(N_B-1)[F(\tau)]^{N_A}+\langle n_B(\tau)\rangle-\langle n_B(\tau)\rangle^2, \qquad (3.10)$$

where

$$F(\tau)=(1+2\alpha e^{-(1+2\alpha)\tau})/(1+2\alpha). \qquad (3.11)$$

Expanding $F(\tau)$ around $\alpha=0$ and neglecting terms of $0(\alpha^2)$ we find

$$[F(\tau)]^{N_A}=\exp[N_A\ln F(\tau)]\sim\exp[-2\alpha N_A(1-e^{-\tau})]. \qquad (3.12)$$

In the approximation represented by Eqs. (3.9) and (3.12), we can write for the coefficient of variation of the number of B molecules

$$C_B(\tau)\sim\left(\frac{1-\exp[-N_A\alpha(1-e^{-\tau})]}{N_B\exp[-N_A\alpha(1-e^{-\tau})]}\right)^{1/2}=0(1/N_B)^{1/2} \qquad (3.13)$$

which vanishes for all τ in the limit $N_B=\infty$. Hence, again no deviations from the results of the deterministic theory are found in the limit as N_A, $N_B\to\infty$.

[14] G. H. Weiss, Biometrics 21, 481 (1965).
[15] K. Dietz, J. Appl. Probl. 3, 375 (1966).
[16] F. Downton, J. Appl. Probl. 4, 264 (1967).

APPENDIX

Evaluation of $\mu_1^{(1)}$

To solve the set of equations (2.22) we introduce the generating functions

$$U(s,t)=\sum_{r=1}^{\infty}\mu_r^{(1)}(\tau)\frac{s^r}{r},$$

$$H(s,\tau)=\sum_{r=1}^{\infty}H_r(\boldsymbol{\mu}^{(0)})\frac{s^r}{r}-(K-1)\mu_1^{(1)}(\tau)=H^*(s,\tau)-(K-1)\mu_1^{(1)}(\tau), \tag{A1}$$

where $H^*(s,\tau)$ can be obtained explicitly from Eq. (2.22)

$$H^*(s,\tau)=[(\mu_1^{(0)})^2+(\beta-1)N_A\mu_1^{(0)}]\left[\ln\left(\frac{1-\mu_1^{(0)}s}{1-(\mu_1^{(0)}-1)s}\right)+\frac{s}{1-\mu_1^{(0)}s}\right]$$
$$+K[(\mu_1^{(0)})^2-(2+\gamma+\delta)N_A\mu_1^{(0)}+(1+\gamma)(1+\delta)N_A^2]\left[\ln\left(\frac{1-\mu_1^{(0)}s}{1-(\mu_1^{(0)}+1)s}\right)-\frac{s}{1-\mu_1^{(0)}s}\right]. \tag{A2}$$

Multiplying Eq. (2.22) by s^r/r and summing from 1 to ∞, one finds that $U(s,\tau)$ is the solution to

$$\partial U/\partial\tau=\{(K-1)-[K(2+\gamma+\delta)+(\beta-1)]N_As+K(1+\gamma)(1+\delta)N_A^2s^2\}(\partial U/\partial s)+H(s,\tau) \tag{A3}$$

subject to the initial condition $U(s,0)=0$. If we regard $H(s,\tau)$ as a known function we can solve this first order partial differential equation by the method of characteristics.[17] The solution which satisfies the initial condition is

$$U(s,\tau)=\int_0^{\tau}H(s,\tau)\left\{\frac{b}{N_A}\left(\frac{\epsilon_+[s-(b\epsilon_-/N_A)]-\epsilon_-[s-(b\epsilon_+/N_A)]\exp(aN_A\tau')}{[s-(b\epsilon_-/N_A)]-[s-(b\epsilon_+/N_A)]\exp(aN_A\tau')}\right),\tau-\tau'\right\}d\tau', \tag{A4}$$

where b is the constant

$$b=(K-1)/K(1+\gamma)(1+\delta). \tag{A5}$$

The quantities of immediate interest are $\mu_1^{(1)}(\tau)$ and $\mu_2^{(1)}(\tau)$. These can be found in terms of $U(s,\tau)$ by

$$\mu_1^{(1)}(\tau)=\partial U/\partial s\,|_{s=0+},\qquad \mu_2^{(1)}(\tau)=\partial^2U/\partial s^2\,|_{s=0+}. \tag{A6}$$

Let us first consider $\mu_1^{(1)}(\tau)$. From Eqs. (A4) and (A1) we find that

$$\mu_1^{(1)}(\tau)=\left.\frac{\partial U(s,\tau)}{\partial s}\right|_{s=0}=\int_0^{\tau}\left.\frac{\partial H^*(\sigma,\tau-\tau')}{\partial\sigma}\right|_{\sigma=f(\tau')}\left(\frac{\partial\sigma}{\partial s}\right)_{s=0}d\tau', \tag{A7}$$

where

$$f(\tau')=\frac{b}{N_A}\frac{\epsilon_+\epsilon_-[\exp(aN_A\tau')-1]}{\epsilon_+\exp(aN_A\tau')-\epsilon_-}=\frac{g(\tau')}{N_A} \tag{A8}$$

and

$$(\partial\sigma/\partial s)_{s=0}=[(\epsilon_+-\epsilon_-)^2\exp(aN_A\tau')]/[\epsilon_+\exp(aN_A\tau')-\epsilon_-]^2. \tag{A9}$$

The function $g(\tau')$ is a bounded function of order 1. The function $f(\tau')$ tends to 0 as N_A goes to ∞. The expansion of $H^*(s,\tau)$ of Eq. (A2) in the neighborhood of $s=0$ is

$$H^*(s,\tau)\sim\{(K+1)(\mu_1^{(0)})^2-[K(2+\gamma+\delta)+(1-\beta)]N_A\mu_1^{(0)}+(1+\gamma)(1+\delta)N_A^2\}\{[s^2/2(1-\mu_1^{(0)}s)]+0(s^3)\}. \tag{A10}$$

Since $\mu_1^{(0)}(\tau)$ is proportional to N_A each term in the braces is proportional to N_A^2. Hence we can rewrite this equation as

$$H^*(s,\tau)\sim[N_A^2F(\tau)s^2/2(1-\mu_1^{(0)}(\tau)s)^2]+\cdots, \tag{A11}$$

where $F(\tau)$, defined by Eq. (A10), is of order 1. The derivative appearing in Eq. (A7) can now be written as

$$\left.\frac{\partial H^*(\sigma,\tau-\tau')}{\partial\sigma}\right|_{\sigma=f(\tau)}=\left.\frac{N_A^2F(\tau-\tau')\sigma}{[1-\mu_1^{(0)}(\tau-\tau')\sigma]^3}\right|_{\sigma=f(\tau')}=\frac{N_AF(\tau-\tau')g(\tau')}{\{1-[\mu_1^{(0)}(\tau-\tau')g(\tau')/N_A]\}^3}. \tag{A12}$$

---

[17] G. F. D. Duff, *Partial Differential Equations* (University of Toronto Press, Toronto, Canada, 1956).

By some tedious algebra and checking in turn the various cases $K<1$, $K=1$, and $K>1$, it can be shown that the denominator of Eq. (A12) never vanishes, i.e., is bounded by a number that is $0(1)$. Since the functions $F(\tau-\tau')$, $g(\tau')$, and $\mu_1^{(0)}(\tau-\tau')$ are bounded, we can set

$$\left|\frac{F(\tau)g(\tau)}{\{1-[\mu_1^{(0)}g(\tau)/N_A]\}^3}\right| \leq M, \tag{A13}$$

where M is the maximum value of the function. Using Eqs. (A7) and (A9) we therefore obtain the following, bound for $\mu_1^{(1)}(\tau)$:

$$\mu_1^{(1)}(\tau) \leq MN_A\left(1-\frac{\epsilon_-}{\epsilon_+}\right)^2 \int_0^\tau \frac{\exp(aN_A\tau')\,d\tau'}{[\exp(aN_A\tau)-(\epsilon_-/\epsilon_+)]^2} < \frac{M}{a}\left(1-\frac{\epsilon_-}{\epsilon_+}\right). \tag{A14}$$

The Relationship between Stochastic and Deterministic Models for Chemical Reactions*

THOMAS G. KURTZ
Department of Mathematics, University of Wisconsin, Madison, Wisconsin 53706
(Received 30 April 1971)

The Markov chain and ordinary differential equation models for chemical reaction systems are compared. It is shown that if the volume of the reaction system is taken into account in an appropriate way in the formulation of the Markov chain model, then the o.d.e. model is the infinite volume limit of the Markov chain model. A central limit theorem is also given for the deviation of the Markov chain model from the o.d.e. model.

I. INTRODUCTION

A number of authors have considered Markov chain models for chemical reactions (see McQuarrie[1] for a survey of work done in this area), and the question has been raised as to the relationship between these models and the classical deterministic ordinary differential equation models. Oppenheim, Shuler, and Weiss[2] have shown in certain special cases that the deterministic model is the infinite volume limit of the Markov chain models and conclude that the same must be the case for more complex systems of reactions. The present paper is a restatement of results obtained in Refs. 3–5 in terms of the chemical reaction models and proves that this conclusion is indeed correct.

II. FORMULATION OF THE MODEL

The only difference between our formulation of the Markov chain models and earlier formulations is that we take explicitly into account the volume of the reaction system. In particular, for a reaction involving two molecules we assume that the chance of a *particular* pair of molecules reacting during a short interval of time $[t, t+\Delta t]$ is inversely proportional to the volume V of the reaction system. Similarly, for a reaction involving l molecules we assume that the chance of l particular molecules reacting during a short interval of time is inversely proportional to V^{l-1}. The reason for this assumption can be seen by considering the probability of l balls placed at random in n boxes all ending up in the same box.

The chance of having some l molecules react in a short interval of time is then inversely proportional to V^{l-1} and proportional to the number of different ways of selecting the l molecules.

For the simple reaction $A+B\rightarrow C$ the chance of the reaction occurring in the time interval $[t, t+\Delta t]$ is approximately $\alpha(i_1i_2/V)\Delta t\equiv V\alpha(i_1/V)(i_2/V)\Delta t$, where i_1 is the number of molecules of A present, i_2 is the number of molecules of B, and α is some constant.

In general consider a system of M reactants, R_1, $R_2, \cdots, R_M$, undergoing N reversible reactions

$$\sum_{m=1}^{M} c_{nm}R_m \leftrightarrows \sum_{m=1}^{M} d_{nm}R_m, \qquad n=1, 2, \cdots, N. \quad (2.1)$$

The Markov chain model for this system may be formulated either in terms of the number of molecules of each of the reactants present at time t, $X^V(t)=[X_1^V(t)X_2^V(t)\cdots X_M^V(t)]$, or in terms of the number of times each of the reactions has occurred in the forward direction minus the number of times it has occurred in the reverse direction,

$$Y^V(t)=[Y_1^V(t), Y_2^V(t)\cdots Y_N^V(t)].$$

Letting C denote the matrix $[(c_{nm})]$ and D the matrix $[(d_{nm})]$, $X^V(t)$ and $Y^V(t)$ are related by

$$X^V(s)+[Y^V(t+s)-Y^V(s)](D-C)=X^V(t+s). \quad (2.2)$$

Let

$$c_n=\sum_{m=1}^{M} c_{nm}$$

and

$$d_n=\sum_{m=1}^{M} d_{nm}.$$

(Of course c_{nm} and d_{nm} are nonnegative integers.) The chance of the nth reaction occurring in the forward direction during the interval $[t, t+\Delta t]$ is approximately

$$\alpha_n(V^{c_n-1})^{-1}\left[\prod_{m=1}^{M}\binom{i_m}{c_{nm}}\right]\Delta t$$

$$=V\alpha_n\left[\prod_{m=1}^{M}(V^{c_{nm}})^{-1}\binom{i_m}{c_{nm}}\right]\Delta t\equiv Vf_n^V(\mathbf{i}) \quad (2.3)$$

and in the reverse direction

$$\beta_n(V^{d_n-1})^{-1}\left[\prod_{m=1}^{M}\binom{i_m}{d_{nm}}\right]\Delta t$$

$$=V\beta_n\left[\prod_{m=1}^{M}(V^{d_{nm}})^{-1}\binom{i_m}{d_{nm}}\right]\Delta t\equiv Vg_n^V(\mathbf{i}), \quad (2.4)$$

where $\mathbf{i}=X^V(t)$.

For $\mathbf{x}=(x_1x_2\cdots x_M)$ define

$$f_n(\mathbf{x})=\alpha_n\prod_{m=1}^{M}\frac{x_m^{c_{nm}}}{c_{nm}!}$$

and

$$g_n(\mathbf{x})=\beta_n\prod_{m=1}^{M}\frac{x_m^{d_{nm}}}{d_{nm}},$$

and observe that

$$f_n^V(\mathbf{i})=f_n(V^{-1}\mathbf{i})+O(V^{-1})$$

and

$$g_n^V(\mathbf{i})=g_n(V^{-1}\mathbf{i})+O(V^{-1}).$$

If c_{nm} and d_{nm} are either 0 or 1 for all n and m, then equality holds without $O(V^{-1})$.

Finally, define

$$F^V(\mathbf{i})=[F_1^V(\mathbf{i}), \cdots, F_M^V(\mathbf{i})],$$

where

$$F_M^V(\mathbf{i})=\sum_{n=1}^{N}(d_{nm}-c_{nm})[f_n^V(\mathbf{i})-g_n^V(\mathbf{i})]$$

and

$$F(\mathbf{x}) = [F_1(\mathbf{x}) \cdots F_M(\mathbf{x})],$$

where

$$F_m(\mathbf{x}) = \sum_{n=1}^{N} (d_{nm} - c_{nm})[f_n(\mathbf{x}) - g_n(\mathbf{x})].$$

It can be shown that

$$\frac{dE(V^{-1}X^V(t))}{dt} = E(F^V[X^V(t)])$$

$$= E(F[V^{-1}X^V(t)]) + O(V^{-1}). \quad (2.5)$$

($E(\ )$ denotes the expectation of a random variable.) The system of differential equations $\dot{X} = F(X)$ is just the classical, deterministic model for our reaction system. Let $X(t, \mathbf{x}_0)$ denote the solution of the initial value problem

$$\frac{\partial X(t, \mathbf{x}_0)}{\partial t} = F(X(t, \mathbf{x}_0)), \qquad X(0, \mathbf{x}_0) = \mathbf{x}_0.$$

If

$$\lim_{V\to\infty} V^{-1}X^V(0) = \mathbf{x}_0,$$

then the theorems of Refs. 3–5 allow us to conclude that

$$\lim_{V\to\infty} P\{\sup_{s\le t} | V^{-1}X^V(s) - X(s, x_0) | > \epsilon\} = 0 \quad (2.6)$$

for every t and $\epsilon > 0$.

We can obtain estimates on the probabilities in (2.6) in two different ways. The first is similar to the Chebychev Inequality of elementary probability and the second is similar to the Central Limit Theorem.

III. AN INEQUALITY

Let

$$\Gamma(\mathbf{x}) = \sum_{n=1}^{N} [\sum_{m=1}^{M} (d_{nm} - c_{nm})^2][f_n(\mathbf{x}) + g_n(\mathbf{x})],$$

and

$$K_\epsilon = \{\mathbf{x} : \inf_{s\le t} | \mathbf{x} - X(s, \mathbf{x}_0) | \le \epsilon\};$$

i.e., K_ϵ is the set of points within a distance ϵ of the trajectory $X(s, x_0)$, $s \le t$. Define

$$\Gamma = \sup_{x \in K_\epsilon} \Gamma(\mathbf{x}),$$

$$M = \sup_{\mathbf{x}_1, \mathbf{x}_2 \in K_\epsilon} [| F(\mathbf{x}_1) - F(\mathbf{x}_2) | / | \mathbf{x}_1 - \mathbf{x}_2 |],$$

and

$$\eta = \sup_{(1/V)\mathbf{i} \in K_\epsilon} | F(V^{-1}\mathbf{i}) - F^V(\mathbf{i}) |.$$

Lemma (1.2) and the inequalities in Sec. 2 of Ref. 5 imply

$$P\{\sup_{s\le t} | V^{-1}X^V(s) - X(s, \mathbf{x}_0) | \ge \epsilon\} \le t\Gamma/(V\delta^2) \quad (3.1)$$

provided $\delta \equiv \epsilon e^{-Mt} - | V^{-1}X^V(0) - \mathbf{x}_0 | - t\eta > 0$.

It is reasonable to assume that $Y^V(0) = 0$. Under this assumption define $\hat{F}^V(\mathbf{j}) = [\hat{F}_1^V(\mathbf{j}) \cdots \hat{F}_N^V(\mathbf{j})]$, where

$$\hat{F}_n^V(\mathbf{j}) = f_n^V[X^V(0) + \mathbf{j}(D-C)] - g_n^V[X^V(0) + \mathbf{j}(D-C)],$$

$$\hat{F}(\mathbf{y}) = [\hat{F}_1(\mathbf{y}) \cdots \hat{F}_N(\mathbf{y})]$$

where

$$\hat{F}_n(\mathbf{y}) = f_n[\mathbf{x}_0 + \mathbf{y}(D-C)] - g_n[\mathbf{x}_0 + \mathbf{y}(D-C)],$$

and

$$\hat{\Gamma}(\mathbf{y}) = \sum_{n=1}^{N} \{f_n[\mathbf{x}_0 + \mathbf{y}(D-C)] + g_n[\mathbf{x}_0 + \mathbf{y}(D-C)]\}.$$

Let $Y(t)$ denote the solution of

$$\partial Y(t)/\partial t = \hat{F}[Y(t)],$$

$$Y(0) = 0.$$

Then with $\hat{\Gamma}$, $\hat{M}$ and $\hat{\eta}$ defined in a manner similar to Γ, M, and η we have

$$P\{\sup_{s\le t} | V^{-1}Y^V(s) - Y(s) | \ge \epsilon\} \le t\hat{\Gamma}/(V\delta^2) \quad (3.2)$$

provided $\delta \equiv \epsilon \exp(-\hat{M}t) - t\hat{\eta} > 0$.

Note: If $| \mathbf{x}_0 - V^{-1}X^V(0) | = O(V^{-1})$ then $\hat{\eta} = O(V^{-1})$.

IV. A CENTRAL LIMIT THEOREM

Let

$$\gamma_{ij}(\mathbf{x}) = \sum_{n=1}^{N} (d_{ni} - c_{ni})(d_{nj} - c_{nj})[f_n(\mathbf{x}) + g_n(\mathbf{x})].$$

Theorem (3.5) of Ref. 5 implies the following: If

$$\lim_{V\to\infty} V^{1/2}[V^{-1}X^V(0) - \mathbf{x}_0] = 0$$

then

$$\lim_{V\to\infty} P\{V^{1/2}[V^{-1}X^V(t) - X(t, \mathbf{x}_0)] \in (a_1, b_1) \times (a_2, b_2) \times \cdots \times (a_M, b_M)\}$$

$$= P\{W(t) \in (a_1, b_1) \times (a_2, b_2) \times \cdots \times (a_M, b_M)\},$$

where $W(t)$ has a multivariate normal distribution with a characteristic function $\psi(t, \boldsymbol{\theta}) \equiv E(\exp\{i\boldsymbol{\theta} \cdot W(t)\})$ satisfying

$$(\partial/\partial t)\psi(t, \boldsymbol{\theta}) = -\tfrac{1}{2} \sum_{j,k} \theta_j \theta_k \gamma_{jk}[X(t, \mathbf{x}_0)]\psi(t, \boldsymbol{\theta})$$

$$+ \sum_{j,k} \theta_j \partial_k F_j[X(t, \mathbf{x}_0)](\partial/\partial\theta_k)\psi(t, \boldsymbol{\theta}). \quad (4.1)$$

[$\mathbf{z} \in (a_1, b_1) \times (a_2, b_2) \times \cdots \times (a_M, b_M)$ means $a_m < z_m < b_m$, $m = 1, 2, \cdots, M$.] Letting

$$h_{ij}(\mathbf{x}) = \partial_i F_j(\mathbf{x}) \equiv \partial F_j(\mathbf{x})/\partial x_i,$$

$G(\mathbf{x})$ be the matrix $\{[\gamma_{ij}(\mathbf{x})]\}$, $H(\mathbf{x})$ the matrix $\{[h_{ij}(\mathbf{x})]\}$ and $H^*(\mathbf{x})$ its adjoint, (4.1) implies $W(t)$

has mean zero and covariance matrix given by

$$\int_0^t \exp\left(\int_s^t H^*[X(u, \mathbf{x}_0)]du\right) G[X(s, \mathbf{x}_0)]$$

$$\times \exp\left(\int_s^t H[X(u, \mathbf{x}_0)]du\right) ds. \quad (4.2)$$

The corresponding quantities for Y^V are

$$\hat{\gamma}_{ij}(\mathbf{y}) = 0 \qquad i \neq j$$

$$= f_i[\mathbf{x}_0 + \mathbf{y}(D-C)] + g_i[\mathbf{x}_0 + \mathbf{y}(D-C)] \qquad \text{for } i=j$$

and

$$h_{ij}(\mathbf{y}) = (\partial/\partial y_i)\hat{F}_j(\mathbf{y}).$$

If $|V^{-1}X^V(0) - \mathbf{x}_0| = O(V^{-1})$ then

$$\lim_{V\to\infty} P\{V^{1/2}[V^{-1}Y^V(t) - Y(t)] \in (a_1, b_1) \times \cdots \times (a_N, b_N)\}$$

$$= P\{Z(t) \in (a_1, b_1) \times \cdots \times (a_N, b_N)\}$$

where $Z(t)$ is multivariate normal with mean zero and covariance matrix given by

$$\int_0^t \exp\left(\int_s^t \hat{H}^*[Y(u)]du\right) \hat{G}[Y(s)]$$

$$\times \exp\left(\int_s^t \hat{H}[Y(u)]du\right) ds. \quad (4.3)$$

V. EXAMPLE

Consider a single reaction $A+B \rightleftarrows C$. The Y^V model is clearly the appropriate model to consider. We than have

$$f(\mathbf{x}) = \alpha x_1 x_2$$

$$g(\mathbf{x}) = \beta x_3,$$

$$\hat{F}(y) = \alpha(x_1^0 - y)(x_2^0 - y) - \beta(x_3^0 + y),$$

$$\hat{\Gamma}(y) = \hat{G}(y) = \alpha(x_1^0 - y)(x_2^0 - y) + \beta(x_3^0 + y),$$

and

$$\hat{H}(y) = 2\alpha y - \beta - \alpha(x_1^0 + x_2^0).$$

Suppose $\alpha x_1^0 x_2^0 - \beta x_3^0 = 0$, that is $\mathbf{x}_0 = (x_1^0, x_2^0, x_3^0)$ is the equilibrium value. Then $Y(t) \equiv 0$ and the variance of the normal random variable $Z(t)$ is

$$\int_0^t \exp\{-(t-s)[\beta + \alpha(x_1^0 + x_2^0)]\}(\alpha x_1^0 x_2^0 + \beta x_3^0)$$

$$\times \exp\{-(t-s)[\beta + \alpha(x_1^0 + x_2^0)]\} ds$$

$$= \frac{\alpha x_1^0 x_2^0}{\beta + \alpha(x_1^0 + x_2^0)} (1 - \exp\{-2t[\beta + \alpha(x_1^0 + x_2^0)]\}).$$

* Research supported in part by the NIH at the University of Wisconsin, Madison, Wisconsin.

[1] D. A. McQuarrie, J. Appl. Prob. **4**, 413 (1967).

[2] I. Oppenheim, K. E. Schuler, and G. H. Weiss, J. Chem. Phys. **50**, 460 (1969).

[3] T. G. Kurtz, "Convergence of Operator Semigroups with Applications to Markov Processes," dissertation, Stanford University, 1967.

[4] [illegible]. [illegible]. MMMMMM, J. Appl. Prob. **7**, 49 (1970).

[5] MMMMMMM, "Limit Theorems for Sequences of Jump Markov Processes Approximating Ordinary Differential Equations," J. Appl. Prob. (to be published).

7) Spin Relaxation Processes

The master equation can be used to describe spin relaxation processes in a large variety of systems. The spin degrees of freedom interact weakly with other degrees of freedom of the systems under consideration. The other degrees of freedom can be considered to remain in equilibrium while the spins relax slowly to their equilbrium distribution by their interactions with the time-independent lattice or heat bath. Master equations for spin systems have been obtained by Redfield[4.7.1] and by Sher and Primakoff[4.7.2] starting with the exact equations of motion.

An extremely simple stochastic model for spin relaxation has been considered by Glauber in the first paper. He obtains a master equation for the spin distribution function for a many spin system in which there are only nearest neighbor interactions. Glauber's model differs from all the systems in the papers reprinted here in that the equilibrium distribution is that for a set of interacting particles and not for a set of independent particles. Glauber obtains explicit solutions for the time dependent joint spin distribution functions for his model.

In the second paper, Glauber's model and results are used to investigate the decay of correlations in spin systems. The results obtained are similar to those obtained in other systems: namely, the m particle distribution functions decay asymptotically to functionals of m-1 particle distribution functions at a rate proportional to the m'th power of the decay of the one particle distribution to its equilibrium value.

References

4.7.1 A. G. Redfield, IBM. J. Res. Devel. 1, 19 (1951).

4.7.2 A. Sher and J. Primakoff, Phys. Rev. 119, 178 (1960).

Time-Dependent Statistics of the Ising Model*

Roy J. Glauber
Lyman Laboratory of Physics, Harvard University, Cambridge, Massachusetts

The individual spins of the Ising model are assumed to interact with an external agency (e.g., a heat reservoir) which causes them to change their states randomly with time. Coupling between the spins is introduced through the assumption that the transition probabilities for any one spin depend on the values of the neighboring spins. This dependence is determined, in part, by the detailed balancing condition obeyed by the equilibrium state of the model. The Markoff process which describes the spin functions is analyzed in detail for the case of a closed N-member chain. The expectation values of the individual spins and of the products of pairs of spins, each of the pair evaluated at a different time, are found explicitly. The influence of a uniform, time-varying magnetic field upon the model is discussed, and the frequency-dependent magnetic susceptibility is found in the weak-field limit. Some fluctuation–dissipation theorems are derived which relate the susceptibility to the Fourier transform of the time-dependent correlation function of the magnetization at equilibrium.

INTRODUCTION

The statistical study of systems of strongly interacting particles is beset by many problems, largely mathematical in nature. These difficulties have motivated theorists to devote a great deal of effort to devising and studying the simplest sorts of model systems which show any resemblance to those occurring in nature. The property most desired in these models is mathematical transparency. The deeper insights offered by the possibility of exact treatment are intended to compensate for any unrealistic simplifications in the formulation. The first, and most successful of these models is one introduced by Ising[1] in an attempt to explain the ferromagnetic phase transition. While many generalizations of this model have been studied, we may note that the first true understanding of a phase transition in an interacting system was reached by Onsager[2] for the case of the two-dimensional Ising model.

If the mathematical problems of equilibrium statistical mechanics are great, they are at least relatively well-defined. The situation is quite otherwise in dealing with systems which undergo large-scale changes with time. The principles of nonequilibrium statistical mechanics remain in largest measure unformulated. While this lack persists, it may be useful to have in hand whatever precise statements can be made about the time-dependent behavior of statistical systems, however simple they may be. We have attempted, therefore, to devise a form of the Ising model whose behavior can be followed exactly, in statistical terms, as a function of time. While certain of the assumptions underlying the model are to a degree arbitrary, it is surely one of the simplest ones involving N coupled particles for which exact time-dependent solutions can be found.

The model we shall discuss is a stochastic one. The spins of N fixed particles are represented as stochastic functions of time $\sigma_j(t)$, $(j = 1, \cdots N)$, which are restricted to the values ± 1, and make transitions randomly between these two values. These transitions take place because of the interaction of the spins with an external agency which may be regarded as a heat reservoir. The transition probabilities of the individual spins, however, are assumed to depend on the momentary values of the neighboring spins as well as on the influence of the heat bath. It is for this reason that statistical correlations arise between the values of neighboring spins. The coupling of the spins through their transition probabilities makes it necessary, in mathematical terms, to deal with the entire N-spin system as a unit. The spin functions form a Markoff process of N discrete random variables with a continuous time variable as argument. Fortunately, if the coupling of the spins is not too complicated, the differential equations governing the probabilities may be simplified greatly, making it possible to solve for all of the quantities of immediate physical interest by elementary means.

In the sections that follow, we introduce first the individual spins interacting with the heat bath, then the means by which they are coupled to one another. The description of the behavior of the model

* A brief account of this work was given at the Washington, D. C. meeting of the American Physical Society, 1960 [R. J. Glauber, Bull. Am. Phys. Soc. 5, 296 (1960)].

[1] E. Ising, Z. Physik **31**, 253 (1925).
[2] L. Onsager, Phys. Rev. **65**, 117 (1944).

is then formulated as a matter of solving for the expectation values of the spin functions and of their products. We center the subsequent discussion largely upon explicit solutions for the single-spin and two-spin averages, since most of the interesting properties of the system may be constructed in terms of these. In addition we find the time-delayed spin correlation function, i.e. the average product of two spin variables, each evaluated at a different time. We then describe the model in the presence of a uniform, time-varying magnetic field. Two results of this generalization are a derivation of the complex frequency-dependent magnetic susceptibility for weak fields, and a discussion of fluctuation–dissipation relations which hold when the field-induced departures from equilibrium are small. Our efforts, in the present paper, are confined to treating a one-dimensional model which, as already indicated by the treatment of the Ising model at equilibrium,[2] appears to be a great deal simpler than dealing with the model in two or more dimensions.

SINGLE-SPIN SYSTEM

It may be helpful in introducing our model to begin by discussing the most simple of such systems: a single particle whose interaction with a heat reservoir of some sort causes its spin to flip between the values $\sigma = 1$ and $\sigma = -1$ randomly, but at a known rate. We assume that no magnetic field is present so that neither of the states $\sigma = \pm 1$ is preferred. Then, if the rate per unit time at which the particle makes transitions from either state to the opposite one is written as $\alpha/2$, the probability $p(\sigma, t)$ that the spin takes on the value σ at time t obeys the equation

$$(d/dt)p(\sigma, t) = -\tfrac{1}{2}\alpha p(\sigma, t) + \tfrac{1}{2}\alpha p(-\sigma, t). \quad (1)$$

This equation, or more properly, this pair of equations for $\sigma = \pm 1$, preserves the normalization condition

$$p(1, t) + p(-1, t) = 1. \quad (2)$$

The pair of equations is therefore immediately reducible to a single equation for a single unknown function. A convenient choice of the latter function is the difference of the two probabilities

$$\begin{aligned} q(t) &= p(1, t) - p(-1, t) \\ &= \sum_{\sigma=\pm 1} \sigma p(\sigma, t), \end{aligned} \quad (3)$$

which is simply the expectation value of the spin as a function of time, i.e. if we think of the time-dependent spin variable as a stochastic function $\sigma(t)$ taking on the values $\sigma = \pm 1$ we have

$$q(t) = \langle \sigma(t) \rangle. \quad (4)$$

The equation obeyed by the mean spin is seen from (1) to be

$$(d/dt)q(t) = -\alpha q(t), \quad (5)$$

so that the mean spin simply decays exponentially with a relaxation time $1/\alpha$ from whatever value it is known to have initially,

$$q(t) = q(0)e^{-\alpha t}. \quad (6)$$

We may regain the individual probabilities $p(\pm 1, t)$ from a knowledge of $q(t)$ by means of the identities (2) and (3) which together yield

$$p(\sigma, t) = \tfrac{1}{2}[1 + \sigma q(t)]. \quad (7)$$

MANY-SPIN SYSTEM

Particles such as the one we have just discussed, each of them responding to a random spin-flipping agency, will form the basic units of the model we wish to describe. We shall assume that these particles are arranged in a regularly spaced linear array which may be closed to form an N-particle ring. The dynamical resemblance between this model and the Ising model rests on the assumption that the individual spins of the ring are not wholly independent stochastic functions. We may, for example, introduce a tendency for a particular spin σ_j ($j = 1 \cdots N$) to correlate with its neighboring spins by assuming that its transition probabilities between the states $\sigma_j = \pm 1$ depend appropriately on the momentary spin values of the other particles. To treat any such model we must consider the entire ring as a unit and introduce a set of 2^N probability functions $p(\sigma_1, \cdots \sigma_N t)$, one for each complexion, i.e. each set $\sigma_1, \cdots \sigma_N$ for the ring.

If we let $w_j(\sigma_j)$ be the probability per unit time that the jth spin flips from the value σ_j to $-\sigma_j$, while the others remain momentarily fixed, then we may write the time derivative of the function $p(\sigma_1, \cdots \sigma_N t)$ as

$$\begin{aligned} \frac{d}{dt} p(\sigma_1, \cdots \sigma_N t) &= -\left[\sum_j w_j(\sigma_j)\right] p(\sigma_1, \cdots \sigma_N t) \\ &\quad + \sum_j w_j(-\sigma_j) p(\sigma_1, \cdots -\sigma_j, \cdots \sigma_N t), \end{aligned} \quad (8)$$

i.e., the complexion $\sigma_1, \cdots \sigma_N$ is destroyed by a flip of any of the spins σ_j, but it may also be created by spin flip from any complexion of the form $\sigma_1, \cdots -\sigma_j, \cdots \sigma_N$. We shall refer to Eq. (8) as the master equation since its solution would contain the most complete description of the system available.

CORRESPONDENCE WITH THE ISING MODEL

We have already mentioned that the transition probabilities $w_j(\sigma_j)$ may be chosen to depend on neighboring spin values as well as on σ_j. If we want, for example, to describe a tendency for each spin to align itself parallel to its nearest neighbors we may choose the probabilities $w_j(\sigma_j)$ to be of the form

$$w_j(\sigma_j) = \tfrac{1}{2}\alpha\{1 - \tfrac{1}{2}\gamma\sigma_j(\sigma_{j-1} + \sigma_{j+1})\}, \tag{9}$$

which may be seen to take on three possible values

$$w_j(\sigma_j) = \tfrac{1}{2}\alpha(1 - \gamma), \quad \tfrac{1}{2}\alpha, \quad \tfrac{1}{2}\alpha(1 + \gamma). \tag{10}$$

The value $\frac{1}{2}\alpha$ corresponds to the case in which the neighboring spins are antiparallel, $\sigma_{j-1} = -\sigma_{j+1}$. When the neighboring spins are parallel to each other the transition probability takes on the value $\frac{1}{2}\alpha(1 - \gamma)$ for σ_j parallel to the two of them or $\frac{1}{2}\alpha(1 + \gamma)$ for σ_j antiparallel. Clearly as long as γ is positive the parallel configurations will be longer-lived than the antiparallel ones and we shall be dealing with a model having ferromagnetic tendencies. Conversely negative γ will mean a tendency of neighboring spins to remain aligned oppositely, and will describe the antiferromagnetic case. We note, incidently, that $|\gamma|$ may not exceed unity.

The parameter α which occurs in the transition probabilities simply describes the time scale on which all transitions take place. It has, of course, no analog in the familiar discussions of the Ising model at equilibrium. The parameter γ, however, describes the tendency of spins toward alignment and thereby determines the equilibrium state of the present model much as the exchange interaction does in the Ising model. To indicate the quantitative correspondence between the models we write the Hamiltonian for the linear Ising model as

$$\mathcal{H} = -J \sum_l \sigma_l\sigma_{l+1}. \tag{11}$$

When the Ising model has reached equilibrium at temperature T, the probability that the jth spin will take on the value σ_j as opposed to $-\sigma_j$ (for a given set of values of the neighboring spins) is just proportional to the Maxwell–Boltzmann factor $\exp(-\mathcal{H}/kT)$. The ratio of the probabilities $p_j(-\sigma_j)$ and $p_j(\sigma_j)$ corresponding to the two states for the jth spin is therefore

$$\frac{p_j(-\sigma_j)}{p_j(\sigma_j)} = \frac{\exp[-(J/kT)\sigma_j(\sigma_{j-1} + \sigma_{j+1})]}{\exp[(J/kT)\sigma_j(\sigma_{j-1} + \sigma_{j+1})]}. \tag{12}$$

If the spins other than σ_j are considered as fixed, the stochastic model described by (8) and (9) will approach an equilibrium in which

$$\frac{p_j(-\sigma_j)}{p_j(\sigma_j)} = \frac{w_j(\sigma_j)}{w_j(-\sigma_j)} \tag{13}$$

$$= \frac{1 - \frac{1}{2}\gamma\sigma_j(\sigma_{j-1} + \sigma_{j+1})}{1 + \frac{1}{2}\gamma\sigma_j(\sigma_{j-1} + \sigma_{j+1})}. \tag{14}$$

The exponentials which occur in the ratio (12) may be written in the forms

$$\exp[\pm(J/kT)\sigma_j(\sigma_{j-1} + \sigma_{j+1})]$$

$$= \cosh\left[\frac{J}{kT}(\sigma_{j-1} + \sigma_{j+1})\right] \pm \sigma_j \sinh\left[\frac{J}{kT}(\sigma_{j-1} + \sigma_{j+1})\right] \tag{15}$$

$$= \cosh\left[\frac{J}{kT}(\sigma_{j-1} + \sigma_{j+1})\right] \times \left\{1 \pm \tfrac{1}{2}\sigma_j(\sigma_{j-1} + \sigma_{j+1}) \tanh\frac{2J}{kT}\right\}, \tag{16}$$

the latter of which is readily checked for the three values the function can take on. The correspondence between the ratios of the equilibrium probabilities (12) and (14) may evidently be made precise by identifying the constant γ as

$$\gamma = \tanh(2J/kT). \tag{17}$$

We should mention that the particular choice we have made for the way in which the transition probabilities (9) depend on neighboring spin values is motivated more by the desire for simplicity than for generality. There exist other, but less simple, coupling schemes which also yield the same equilibrium states as the Ising model with nearest-neighbor interactions. Some of these are discussed in the Appendix. There exists, furthermore, the possibility that each spin is coupled through the transition probabilities to some or all of its more distant neighbors. We shall mention this possibility further at a later point. For the present we shall continue to deal with the transition probabilities (9) and discuss the mathematical treatment of the master equation based on them.

REDUCTION OF THE PROBABILITY FUNCTION

The functions $p(\sigma_1, \cdots \sigma_N t)$ which satisfy the master equation (8) furnish, as we have noted earlier, the fullest possible description of the system. While we cannot deny that it would be desirable to know these functions in their entirety we must nevertheless point out that, for N large, they contain vastly more information than we usually require in practice. To answer the most familiar physical questions about the system, in fact, it suffices to know just the probabilities that individual spins or pairs of spins occupy specified states. Alternatively, we need know only the expectation values of spins or the average products of pairs of spins. Most of our attention in the present paper will be devoted to discussing just these functions. However before proceeding to the discussion, it may be helpful to indicate some general relations between the probability functions and the expectation values of products of spin variables.

We define the functions $q_j(t)$ to be the expectation values of the spins $\sigma_j(t)$ regarded as stochastic functions of time:

$$\begin{aligned} q_j(t) &= \langle\sigma_j(t)\rangle \\ &= \sum_{\{\sigma\}} \sigma_j p(\sigma_1, \cdots \sigma_N t). \end{aligned} \tag{18}$$

Here and in future work we designate by a sum over $\{\sigma\}$, a sum carried out over the 2^N values of the set $\sigma_1, \cdots \sigma_N$. The functions $r_{j,k}(t)$ are defined, likewise, as the expectation values of the products $\sigma_j(t)\sigma_k(t)$:

$$\begin{aligned} r_{j,k}(t) &= \langle\sigma_j(t)\sigma_k(t)\rangle \\ &= \sum_{\{\sigma\}} \sigma_j\sigma_k p(\sigma_1, \cdots \sigma_N t). \end{aligned} \tag{19}$$

We note in particular that the "diagonal" expectation values $r_{j,j}$ are identically unity:

$$r_{j,j}(t) = 1. \tag{20}$$

We next construct a general identity relating the probability to the expectation values as follows: Let σ_j and σ_j' be two possibly different values of the jth spin. Then the function $\frac{1}{2}(1 + \sigma_j\sigma_j')$ equals unity for $\sigma_j' = \sigma_j$ and zero for $\sigma_j' = -\sigma_j$. We may therefore construct an identity expressing $p(\sigma_1, \cdots \sigma_N t)$ as a sum over all spins by writing

$$\begin{aligned} p(\sigma_1, \cdots \sigma_N t) &= \frac{1}{2^N} \\ &\times \sum_{\{\sigma'\}} (1 + \sigma_1\sigma_1') \cdots (1 + \sigma_N\sigma_N')p(\sigma_1', \cdots \sigma_N', t). \end{aligned} \tag{21}$$

If we expand the product in the summand of this relation and carry out the indicated summations, we find

$$\begin{aligned} p(\sigma_1, \cdots \sigma_N, t) = \frac{1}{2^N} \{1 &+ \sum_j \sigma_j q_j(t) \\ &+ \sum_{j \neq k} \sigma_j\sigma_k r_{j,k}(t) + \cdots\}, \end{aligned} \tag{22}$$

which exhibits a general expansion of the probability functions in terms of the expectation values of the spins and their products taken two at a time, three at a time, etc., i.e. the functions 1 and σ form a complete orthogonal basis for the expansion of any function of σ, and (22) is just such an expansion with N independent variables. The relation (7) for a single spin is a trivial example of the expansion.

The reduced probability functions which furnish the probabilities that individual spins or pairs of spins occupy specified states, whatever may be the states of the remaining spins, are defined by

$$p_j(\sigma_j, t) = \sum_{\{\sigma \neq \sigma_j\}} p(\sigma_1, \cdots \sigma_N, t), \tag{23}$$

$$p_{jk}(\sigma_j, \sigma_k, t) = \sum_{\{\sigma \neq \sigma_j, \sigma_k\}} p(\sigma_1, \cdots \sigma_N, t), \tag{24}$$

where the notation is intended to indicate summation over all the spin variables save σ_j in (23) and σ_j and σ_k in (24). If these summations are carried out upon the form (22) for $p(\sigma_1, \cdots \sigma_N, t)$ we find

$$p_j(\sigma_j, t) = \tfrac{1}{2}\{1 + \sigma_j q_j(t)\}, \tag{25}$$

$$\begin{aligned} p_{jk}(\sigma_j, \sigma_k, t) = \tfrac{1}{4}\{1 &+ \sigma_j q_j(t) \\ &+ \sigma_k q_k(t) + \sigma_j\sigma_k r_{j,k}(t)\}. \end{aligned} \tag{26}$$

It should be clear that by solving for the expectation values of the spins and their products we are beginning a systematic expansion of the probability functions as well as finding the quantities of greatest physical interest.

As a preliminary step to finding the time-dependent equations satisfied by the expectation values, we may write the master equation (8) in the more compact form

$$\begin{aligned} &\frac{d}{dt} p(\sigma_1, \cdots \sigma_N, t) \\ &= -\sum_m \sigma_m \sum_{\sigma_m'} \sigma_m' w_m(\sigma_m') p(\sigma_1, \cdots \sigma_m', \cdots \sigma_N, t). \end{aligned} \tag{27}$$

If we multiply both sides of this relation by σ_k and sum over all values of the σ variables we obtain

$$\begin{aligned} (d/dt)q_k(t) &= -2 \sum \sigma_k w_k(\sigma_k) p(\sigma_1, \cdots \sigma_N, t) \\ &= -2\langle\sigma_k(t) w_k[\sigma_k(t)]\rangle. \end{aligned} \tag{28}$$

Similarly, if both sides of (27) are multiplied by the product $\sigma_j\sigma_k$ (where $j \neq k$) and summed over the σ variables we obtain

$$\frac{d}{dt} r_{j,k}(t)$$
$$= -2 \sum_{\{\sigma\}} \sigma_j \sigma_k \{w_j(\sigma_j) + w_k(\sigma_k)\} p(\sigma_1, \cdots \sigma_N, t)$$
$$= -2\langle \sigma_j(t)\sigma_k(t)\{w_j[\sigma_j(t)] + w_k[\sigma_k(t)]\}\rangle. \quad (29)$$

If we substitute the form (9) for the transition probabilities in (28) we obtain a recursive system of differential equations for the expectation values $q_k(t)$:

$$(d/d\alpha t)q_k(t) = -q_k(t) + \tfrac{1}{2}\gamma\{q_{k-1}(t) + q_{k+1}(t)\}. \quad (30)$$

An analogous system of equations for the expectation values of products of pairs of spins results from the substitution of (9) in (29). For $j \neq k$ we have

$$(d/d\alpha t)r_{j,k}(t) = -2r_{j,k}(t) + \tfrac{1}{2}\gamma\{r_{j,k-1}(t) + r_{j,k+1}(t) + r_{j-1,k}(t) + r_{j+1,k}(t)\}, \quad (31)$$

while for $j = k$, the functions obey the identity (20).

These equations, as we shall see, may be solved quite readily. It is worth noting, however, that the assumption of forms different from (9) for the transition probabilities leads, in many cases, to systems of equations in which the expectation values of products of differing numbers of spins are coupled in each equation. Such systems are considerably less tractable than the present one.

SOLUTION FOR THE AVERAGE SPINS: INFINITE RING

The coupled differential equations (30) are particularly easy to solve for the case of an infinite ring, $N \to \infty$. It is convenient, for this case, to alter slightly the scheme for numbering the spins by labeling a particular spin as the zeroth and designating those to one side of it with positive integers and those to the other side with negative ones. We then construct the generating function

$$F(\lambda, t) = \sum_{k=-\infty}^{\infty} \lambda^k q_k(t), \quad (32)$$

which, according to Eq. (30), satisfies the differential equation

$$(\partial/\partial\alpha t)F(\lambda, t) = -F(\lambda, t) + \tfrac{1}{2}\gamma(\lambda + \lambda^{-1})F(\lambda, t). \quad (33)$$

The solution for the generating function is evidently

$$F(\lambda, t) = F(\lambda, 0) \exp[-\alpha t + \tfrac{1}{2}\gamma(\lambda + \lambda^{-1})\alpha t], \quad (34)$$

which furnishes us an implicit solution for the $q_k(t)$ in terms of the initial values $q_k(0)$. To make the solution an explicit one we note that one of the factors in (34) is just the generating function for the Bessel functions of imaginary argument,[3]

$$\exp[\tfrac{1}{2}x(\lambda + \lambda^{-1})] = \sum_{n=-\infty}^{\infty} \lambda^n I_n(x), \quad (35)$$

where

$$I_n(x) = i^{-n} J_n(ix). \quad (36)$$

Hence the time-dependent generating function is given by

$$F(\lambda, t) = F(\lambda, 0)e^{-\alpha t} \sum_{k=-\infty}^{\infty} \lambda^k I_k(\gamma\alpha t). \quad (37)$$

We consider first the case in which all of the spin expectations q_k vanish initially except for one, which we may choose to be the one at the origin

$$q_k(0) = \delta_{k,0}. \quad (38)$$

Then the initial value of the generating function is just unity, and at later times it is

$$F(\lambda, t) = e^{-\alpha t} \sum_{k=-\infty}^{\infty} \lambda^k I_k(\gamma\alpha t), \quad (39)$$

from which we conclude, by comparing with (32), that the spin expectations are given by

$$q_k(t) = e^{-\alpha t} I_k(\gamma\alpha t). \quad (40)$$

An examination of the functions I_k shows that q_0 decreases steadily to zero as time increases, while the neighboring spin expectations rise from zero to positive values for a while as a form of transient polarization induced by the positive spin at the origin. The functions q_k for spins neighboring the origin rise for times $t \ll k/\gamma\alpha$ as

$$q_k(t) \approx (1/|k|!)(\tfrac{1}{2}\gamma\alpha t)^{|k|} e^{-\alpha t}. \quad (41)$$

They then reach a maximum[4] at a time given, for $k \gg 1$, by $\alpha t \approx k(1 - \gamma^2)^{-\frac{1}{2}}$, and, for much larger times, decrease as

$$q_k(t) \sim (2\pi\gamma\alpha t)^{-\frac{1}{2}} e^{-\alpha(1-\gamma)t}. \quad (42)$$

The most general solution for the spin expectation values, corresponding to an arbitrary set of initial values $q_k(0)$, may clearly be obtained from (40) by linear superposition,

$$q_k(t) = e^{-\alpha t} \sum_{m=-\infty}^{\infty} q_m(0) I_{k-m}(\gamma\alpha t), \quad (43)$$

where we note that the functions I_n for negative

[3] See, for example, G. N. Watson, *Bessel Functions* (Cambridge University Press, Cambridge, England, 1958), pp. 14 and 77.

[4] The locations of the maxima and various other properties of the functions $e^{-ax}I_n(x)$ for $a \geq 1$ are discussed by E. W. Montroll, J. Math and Phys. **25**, 37 (1946).

order are the same as those for positive order, $I_n = I_{-n}$.

AVERAGE SPINS: FINITE RING

A somewhat more general means of treating the set of equations (30) for arbitrary N may be based on a system of normal modes for the spin expectation values q_k. If we seek solutions to Eqs. (30) in the form

$$q_k(t) = A\zeta^k e^{-\nu t}, \tag{44}$$

where A is a constant, then we have

$$\nu = \alpha\{1 - \tfrac{1}{2}\gamma(\zeta^{-1} + \zeta)\}. \tag{45}$$

The closure of the N-spin ring requires that the solution (44) be periodic in k with period N, i.e., that $\zeta^N = 1$. Hence there are N roots for ζ of the form

$$\zeta_m = \exp(2\pi i m/N), \quad m = 0, 1, \cdots N - 1, \tag{46}$$

and for these the eigenvalues ν_m are

$$\nu_m = \alpha\{1 - \gamma \cos(2\pi m/N)\}. \tag{47}$$

The system of mode functions $q_k^{(m)} = \exp(2\pi imk/N)$ forms a complete orthogonal basis on the ring. Hence any solution to (30) may be written in the form

$$q_k(t) = \sum_{m=0}^{N-1} A_m e^{(2\pi imk/N) - \nu_m t}, \tag{48}$$

where the constants A_m may be solved for in terms of the $q_k(0)$ by using the orthogonality theorem. These constants are

$$A_m = \frac{1}{N}\sum_{l=1}^{N} q_l(0)e^{-2\pi iml/N}. \tag{49}$$

The solution for the spin expectation values in terms of their initial values is thus

$$q_k(t) = \frac{1}{N}\sum_{l,m} q_l(0)e^{(2\pi im/N)(k-l) - \nu_m t}$$
$$= e^{-\alpha t}\sum_{l=1}^{N}\sum_{j=-\infty}^{\infty} q_l(0)I_{k-l+jN}(\gamma\alpha t). \tag{50}$$

The latter form of the solution is obtained from the former by carrying out the summation over m explicitly. That the solutions may be expressed in this way is obvious from the fact that the problem for a finite ring may be solved by inserting periodic initial values in (43).

A particular consequence of the solution (50) is the fact that the total magnetization always decreases exponentially,

$$\sum_k q_k(t) = e^{-\alpha(1-\gamma)t}\sum_l q_l(0), \tag{51}$$

a result which corresponds to the known absence of permanent magnetization in the linear Ising model (with interactions restricted to a finite number of neighbors). The net effect of the spin interactions is to reduce the coefficient in the exponent from the α of Eq. (6) to $\alpha(1 - \gamma)$.

SOLUTION FOR ONE SPIN FIXED

It is interesting to investigate the behavior of the spin system when one of the spins is assumed somehow to be fixed or frozen. We shall, for simplicity, consider the infinite ring and let the zeroth spin, the one at the origin, take on the fixed value $\sigma_0 = 1$. Then the differential equations derived earlier for the $q_k(t)$ still hold for $k \neq 0$. In particular, for $k = 1$, we have

$$(d/d\alpha t)q_1(t) = -q_1(t) + \tfrac{1}{2}\gamma\{1 + q_2(t)\}, \tag{52}$$

while the equations for $k > 1$ assume precisely the form (30). This sequence of equations for $k \geq 1$ is an inhomogeneous one because of the constant term on the right-hand side of (52). It possesses a nonvanishing equilibrium solution, which satisfies the recursion relation

$$q_k = \tfrac{1}{2}\gamma\{q_{k-1} + q_{k+1}\}, \qquad k \neq 0, \tag{53}$$

where $q_0 = 1$. The solution to such a linear difference equation may be written as

$$q_k = \eta^{|k|}, \tag{54}$$

where η satisfies the quadratic equation

$$\eta^2 - 2\gamma^{-1}\eta + 1 = 0. \tag{55}$$

It is worth noting that the same quadratic equation for η holds for negative values of k as for positive values of k, i.e., the equation is unchanged by the substitution of η^{-1} for η. The roots of (55), which are always real, form a reciprocal pair. One member of the pair, $\gamma^{-1}\{1 + (1 - \gamma^2)^{\frac{1}{2}}\}$, always has absolute value greater than unity for $|\gamma| \leq 1$ and therefore is of no use in solving the problem for an infinite ring. The correct root for η has absolute value less than unity and is given by

$$\eta = \gamma^{-1}\{1 - (1 - \gamma^2)^{\frac{1}{2}}\}. \tag{56}$$

For this value, using the correspondence (17) with the static Ising model, we find

$$\eta = \tanh(J/kT). \tag{57}$$

The solution (54) exhibits clearly the tendency of any spin, in this case a fixed one, to surround itself with a "polarization cloud." (In the antiferromagnetic case, $\gamma < 0$, the signs of the induced spins will alternate.) The value of η given by (57) is just the familiar short-range order parameter of the Ising model.

To complete the solution of the time-dependent equations for the $q_k(t)$, with the zeroth spin fixed, we need only note that (54) constitutes a particular solution of the inhomogeneous system. We may add to it any solution to the homogeneous system of equations obtained by requiring q_0 to vanish at all times. Such a boundary condition may easily be satisfied by using the method of images, since the requirement $q_0 = 0$ separates the system into two halves which do not influence each other. (The infinite ring need not be imagined as closed.) If we seek a solution to the homogeneous system of equations in which the q_k assume a particular set of initial values, say v_k for $k > 0$, we may reach a solution for the positive-k half of the system by using the general solution (43) and imagining that the initial values of the q_k at the negative sites are given by $q_{-k}(0) = -v_k$ for $k > 0$, and that we have $q_0(0) = 0$. Interpreted in this way for $k > 0$, the solution (43) may be made to fit the correct initial conditions and yet, since it remains odd at k at all times, meet the boundary condition $q_0(t) = 0$ as well. An analogous imaging procedure solves the equations for negative k as well.

To find the general solution to the time-dependent equations with the zeroth spin fixed we must add together the particular solution (54) for the inhomogeneous system and the general solution, constructed by the method of images, for the homogeneous system, i.e., we add to the solution η^k the solution to the homogeneous system which corresponds for $k > 0$ to the set of initial values $q_k(0) - \eta^k$. The resulting solution for $k > 0$ is

$$q_k(t) = \eta^k + e^{-\alpha t} \sum_{l=1}^{\infty} (q_l(0) - \eta^l) \times \{I_{k-l}(\gamma\alpha t) - I_{k+l}(\gamma\alpha t)\}. \quad (58)$$

An analogous solution exists for negative k values. For times $t \gg (\gamma\alpha)^{-1}$, the solutions in all cases decay exponentially to the equilibrium form.

SOLUTION FOR THE SPIN CORRELATIONS

We next turn our attention to the average values of products of pairs of spin variables. The functions $r_{j,k}(t)$ which express these averages obey the two-index system of Eqs. (31) for $j \neq k$, and for $j = k$ obey the identity $r_{j,j} = 1$. We can secure a rapid insight into the behavior of these functions by simplifying the problem so that they depend, in effect, on only one index. It often happens, in fact, that our knowledge of the initial state of the system is characterized by translational invariance, i.e., our initial knowledge about all of the spins is the same. Then $r_{j,k}(0)$ can only depend on $j - k$, and no other dependence on j or k can be present at later times. In that case it becomes convenient to introduce the abbreviation

$$r_m = r_{k,k+m} \quad (59)$$

for the spin correlation functions. We shall consider this translationally invariant situation first and then return to the more general one presently.

In the uniform case the functions r_m are seen to obey the relations

$$(d/d\alpha t)r_m(t) = -2r_m(t) + \gamma\{r_{m-1}(t) + r_{m+1}(t)\} \quad (60)$$

for $m \neq 0$, and

$$r_0(t) = 1. \quad (61)$$

Aside from a trivial change of a factor of two in the coefficients, this is precisely the sequence of equations we solved in the preceding section, for the single-spin averages with the zeroth spin fixed. The factor of two in the coefficients affects only the time scale in which the functions change. In particular, the equilibrium solution on the infinite chain is again given by

$$r_m = \eta^{|m|}, \quad (62)$$

where η is the short-range order parameter mentioned earlier. The time-dependent solution for arbitrary initial correlations may be constructed immediately from (58). For $m > 0$ we have

$$r_m(t) = \eta^m + e^{-2\alpha t} \sum_{l=1}^{\infty} [r_l(0) - \eta^l] \times \{I_{m-l}(2\gamma\alpha t) - I_{m+l}(2\gamma\alpha t)\}. \quad (63)$$

As a particular example of the type of problem to which this result is applicable, we may suppose that the spin system is suddenly subjected to a change of temperature; i.e., after coming to equilibrium with a heat reservoir at temperature T_0, it is suddenly placed in contact with another heat bath at a different temperature T. In that case the initial values of the r_l are given by

$$r_l(0) = \eta_0^l = [\tanh(J/kT_0)]^l, \quad (64)$$

and the way these relax into the equilibrium values at temperature T is shown by (63).

We return now to the general problem of solving

the two-index system of differential equations (31) without the simplifying assumption of translational invariance. The system is an inhomogeneous one because of the condition $r_{k,k}(t) = 1$, which plays a role similar to that of the fixed spin in the preceding section. The translationally invariant equilibrium solution $r_{k,l} = \eta^{|k-l|}$, which we have just discussed, clearly satisfies the system of equations. It can be used as a particular solution to the inhomogeneous system. To this particular solution, we must add a general solution to the homogeneous system obtained by supplementing (31) with the conditions $r_{k,k}(t) = 0$. The solutions to these equations may be obtained and the boundary conditions met by generalizing the methods of the preceding sections to deal with a two-index array $r_{j,k}(t)$, i.e. a matrix, rather than a linear sequence $q_j(t)$.

If, for the moment, we ignore the boundary condition on $r_{k,k}(t)$ and assume that Eqs. (31) hold even for $j = k$, it becomes a simple matter to solve the equations by using a two-parameter generating function analogous to (39). We then find that if all of the initial values of $r_{j,k}(0)$ vanish except one, which is unity, i.e.,

$$r_{j,k}(0) = \delta_{jl}\delta_{km}, \tag{65}$$

the solution for $r_{j,k}(t)$ is

$$r_{j,k}(t) = e^{-2\alpha t} I_{j-l}(\gamma\alpha t) I_{k-m}(\gamma\alpha t). \tag{66}$$

Such solutions may be superposed to secure the appropriate initial values and to meet the condition $r_{k,k}(t) = 0$. To satisfy the latter condition, we must generalize to a two-index array the method of images used earlier.

The matrix $r_{j,k}(t)$ is, or course, symmetric. However, it is quite convenient to think of it as if it were antisymmetric. What we shall do is fix our attention, for the moment, on the values of $r_{j,k}(t)$ for $j > k$ and only attempt to deal correctly with these. We assume that these matrix elements take on their correct initial values but that the elements $r_{k,j}(0)$ are given by $-r_{j,k}(0)$ for $j > k$, and that $r_{j,j}(0) = 0$. The matrix $r_{j,k}$ which is thus assumed initially antisymmetric, maintains its antisymmetry at later times and, therefore, always meets the condition $r_{j,j}(t) = 0$. In fact, it satisfies the sequence of equations (31) including, in virtue of its antisymmetry, the equation of the same form for $j = k$. We need not be embarrassed, therefore, by our inclusion of the $j = k$ equations in the arguments leading to (66).

The basic set of solutions we seek, which meets the initial condition (65) and the boundary condition $r_{j,j}(t) = 0$, is just the solution (66) antisymmetrized in the two indices l and m, i.e., for $j \geq k$ and $l \geq m$

$$r_{j,k}(t) = e^{-2\alpha t}\{I_{j-l}(\gamma\alpha t)I_{k-m}(\gamma\alpha t) - I_{j-m}(\gamma\alpha t)I_{k-l}(\gamma\alpha t)\}. \tag{67}$$

The general solution to the homogeneous system is obtained by superposing the solutions (67). In order to solve the inhomogeneous system with which we began, we must add the particular solution η^{j-k} to the solutions we have just found. The form which satisfies the correct initial conditions for $j \geq k$ is

$$r_{j,k}(t) = \eta^{j-k} + e^{-2\alpha t}\sum_{l>m}[r_{l,m}(0) - \eta^{l-m}] \times \{I_{j-l}(\gamma\alpha t)I_{k-m}(\gamma\alpha t) - I_{j-m}(\gamma\alpha t)I_{k-l}(\gamma\alpha t)\}, \tag{68}$$

which is the general solution for the expectation values of the spin products. When translational invariance holds, this solution may be seen to reduce to (63) by applying the relation

$$I_k(2x) = \sum_{m=-\infty}^{\infty} I_{k+m}(x)I_m(x), \tag{69}$$

which is a special case of the addition theorem for Bessel functions.[5]

TIME-DELAYED SPIN CORRELATION FUNCTIONS

The functions $r_{j,k}(t)$, which we have discussed up to this point, describe whatever tendency the pairs of spins σ_j and σ_k may have to be correlated in direction, on the average, at a particular instant of time t. Not all of the spin correlations of interest, however, have this instantaneous character. In particular, variation of any one spin at a given instant induces polarizations among its neighbors which only become appreciable after finite intervals of time. To describe correlation effects extending over an interval of length t', we shall discuss the functions $\langle\sigma_j(t)\sigma_k(t+t')\rangle$, i.e. the expectation values of the products of the stochastic spin functions σ_j evaluated at time t, and σ_k evaluated at time $t + t'$.

To evaluate these more general correlation functions we represent the values assumed by the spins at time t as $\sigma_1, \cdots \sigma_N$ and at the later time $t + t'$ as $\sigma_1', \cdots \sigma_N'$. The probability associated with the spin values $\sigma_1, \cdots \sigma_N$ at time t is $p(\sigma_1, \cdots \sigma_N, t)$, i.e., the solution to the master equation which satisfies whatever initial conditions our physical knowledge imposes. In order to carry out the averaging correctly, we must also know the probability associated with the final configuration $\sigma_1', \cdots \sigma_N'$ at time $t + t'$. The question we ask in determining that

[5] Reference 3, p. 361.

probability is rather different from the one answered by $p(\sigma_1, \cdots \sigma_N, t)$, since we assume that the spins are known to have the values $\sigma_1, \cdots \sigma_N$ at time t. The values $\sigma_1, \cdots \sigma_N$ are thus to be regarded as initial spin values in determining the probability of finding $\sigma_1', \cdots \sigma_N'$ at a time t' later. We shall write this conditional probability for finding $\sigma_1', \cdots \sigma_N'$ as $p(\sigma_1, \cdots \sigma_N \mid \sigma_1', \cdots \sigma_N' t)$. The expectation value we seek for the product of two spins may then be constructed by summing over all possible values of the sets $\sigma_1, \cdots \sigma_N$ and $\sigma_1', \cdots \sigma_N'$ as follows:

$$\langle\sigma_j(t)\sigma_k(t+t')\rangle = \sum_{\{\sigma\}\{\sigma'\}} p(\sigma_1, \cdots \sigma_N t)\sigma_j p(\sigma_1, \cdots \sigma_N \mid \sigma_1', \cdots \sigma_N' t')\sigma_k'. \tag{70}$$

The part of this summation which is to be carried out over the variables $\sigma_1', \cdots \sigma_N'$ may be regarded simply as the expectation value of the kth spin when the spins are initially $\sigma_1, \cdots \sigma_N$. We may then write

$$\sum_{\{\sigma'\}} p(\sigma_1, \cdots \sigma_N \mid \sigma_1', \cdots \sigma_N', t')\sigma_k' = q_k(t'), \tag{71}$$

where it is understood that the initial values of the q_k are given by $q_k(0) = \sigma_k$. For the case of an infinite chain, the functions $q_k(t')$ are given in terms of these initial values by the general solution (43) as

$$q_k(t') = e^{-\alpha t'} \sum_l \sigma_l I_{k-l}(\gamma\alpha t'). \tag{72}$$

By substituting (71) and (72) into (70) we find

$$\langle\sigma_j(t)\sigma_k(t+t')\rangle = e^{-\alpha t'} \sum_{l=-\infty}^{\infty} I_{k-l}(\gamma\alpha t') \sum_{\{\sigma\}} p(\sigma_1, \cdots \sigma_N t)\sigma_j\sigma_l. \tag{73}$$

The summation over $\sigma_1, \cdots \sigma_N$, however, is just the instantaneous correlation $r_{j,k}(t)$ defined by (19). The time-delayed correlation function, therefore, reduces to

$$\langle\sigma_j(t)\sigma_k(t+t')\rangle = e^{-\alpha t'} \sum_{l=-\infty}^{\infty} r_{j,l}(t) I_{k-l}(\gamma\alpha t'), \tag{74}$$

where the functions $r_{j,l}(t)$ are given, in general, by the results of the preceding section.

For the particular case of a system in thermal equilibrium at temperature T, the correlation function depends only on the interval t', i.e.,

$$\langle\sigma_j(t)\sigma_k(t+t')\rangle_T = e^{-\alpha t'} \sum_{l=-\infty}^{\infty} \eta^{|j-k+l|} I_l(\gamma\alpha t'). \tag{75}$$

The term corresponding to $l = k - j$ is the only contribution which would be present if there were no correlations between spins in the initial state, as would be true, for example, for infinite temperature. The remaining terms of the series describe the stabilizing effects upon the kth spin of the polarizations which exist about it in the initial state. For either sign of γ, the addition of the effects of neighboring spins in (75) makes the correlation function decrease in magnitude more slowly with increasing t'.

In all of our work to date, we have assumed that we are in possession of some knowledge about the system at an initial time $t = 0$, and have sought, in a probabilistic sense, to answer questions about the behavior of the system at later times. Of course, the same questions may be asked in a reversed sense. What may we say, on the basis of knowledge at $t = 0$, about the behavior of the system at negative times? Since the dynamical properties of our model are presumably reversible, there is no need to construct or solve a new master equation. The probabilities are simply even functions of time. The time t is to be construed more generally as $|t|$ in all of the probability functions we have calculated thus far. In particular, the time-dependent spin correlation function (75) may be written for $t = 0$ and arbitrary t' as

$$\langle\sigma_j(0)\sigma_k(t')\rangle_T = e^{-\alpha|t'|} \sum_{l=-\infty}^{\infty} \eta^{|j-k+l|} I_l(\gamma\alpha\,|t'|). \tag{76}$$

SINGLE SPIN IN A MAGNETIC FIELD

It is not difficult to formulate the equations which describe the behavior of our model when it is placed in a uniform magnetic field. The influence of the magnetic field H, which we suppose is parallel to the axis of spin quantization, is to introduce a preference of the spins for either the $\sigma = 1$ or the $\sigma = -1$ state. For the most simple case, in which only a single spin is present, the transition probability from σ to $-\sigma$ may be written as

$$w(\sigma) = \tfrac{1}{2}\alpha(1 - \beta\sigma).$$

If we equate the ratios of the equilibrium probabilities calculated according to the stochastic model and according to statistical mechanics, we find

$$\frac{p(-\sigma)}{p(\sigma)} = \frac{w(\sigma)}{w(-\sigma)} = \frac{1-\beta\sigma}{1+\beta\sigma} = \frac{\exp[-(\mu H/kT)\sigma]}{\exp[(\mu H/kT)\sigma]} = \frac{1 - \sigma\tanh(\mu H/kT)}{1 + \sigma\tanh(\mu H/kT)}, \tag{77}$$

where μ is the magnetic moment associated with the spins or, more concisely, we find the correspondence

$$\beta = \tanh(\mu H/kT). \tag{78}$$

The equation satisfied by the expectation value of the spin is then

$$(d/d\alpha t)q(t) = \beta - q(t). \tag{79}$$

In the work that follows, it will be interesting to be able to discuss the behavior of the spins in time-dependent magnetic fields. Since the arguments of statistical mechanics used in treating the Ising model deal only with constant magnetic fields, we are free in defining the stochastic model to choose any time-dependence of the parameter β which yields (78) when H is constant. The simplest way of defining a time-dependent β is to retain the relation (78) when H depends on time. The solution for the average spins is then

$$q(t) = q(t_0)e^{-\alpha(t-t_0)} + \int_{t_0}^{t} e^{-\alpha(t-t')}\beta(t')\alpha\, dt', \tag{80}$$

where t_0 is a time at which q is known initially.

SPIN SYSTEM IN A MAGNETIC FIELD

To construct a stochastic analog of the Ising model in a magnetic field, we must first find an appropriate set of transition probabilities. To this end we note that the Hamiltonian of the Ising model is

$$\mathcal{H} = -\mu H \sum_m \sigma_m - J \sum_m \sigma_m\sigma_{m+1}, \tag{81}$$

so that, if the spins other than σ_j are considered as fixed, the ratio of equilibrium probabilities for the states $-\sigma_j$ and σ_j is

$$\begin{aligned}\frac{p_j(-\sigma_j)}{p_j(\sigma_j)} &= \frac{\exp\{-(1/kT)\sigma_j[J(\sigma_{j-1}+\sigma_{j+1})+\mu H]\}}{\exp\{(1/kT)\sigma_j[J(\sigma_{j-1}+\sigma_{j+1})+\mu H]\}}\\ &= \frac{w_j(\sigma_j)\exp[-(\mu H/kT)\sigma_j]}{w_j(-\sigma_j)\exp[(\mu H/kT)\sigma_j]},\end{aligned} \tag{82}$$

where the identities (12) and (13) were used in securing the latter relation. If we write the transition probabilities for the model in a magnetic field as $w_j'(\sigma_j)$, the detailed balancing condition at equilibrium requires

$$\begin{aligned}\frac{w_j'(\sigma_j)}{w_j'(-\sigma_j)} &= \frac{p_j(-\sigma_j)}{p_j(\sigma_j)}\\ &= \frac{w_j(\sigma_j)[1-\sigma_j\tanh(\mu H/kT)]}{w_j(-\sigma_j)[1+\sigma_j\tanh(\mu H/kT)]}.\end{aligned} \tag{83}$$

Hence our model will approach the same equilibrium state as the Ising model if we choose

$$\begin{aligned}w_j'(\sigma_j) &= w_j(\sigma_j)[1-\sigma_j\tanh(\mu H/kT)]\\ &= w_j(\sigma_j)(1-\beta\sigma_j)\\ &= \tfrac{1}{2}\alpha\{1-\beta\sigma_j+\tfrac{1}{2}\gamma(\beta-\sigma_j)(\sigma_{j-1}+\sigma_{j+1})\}.\end{aligned} \tag{84}$$

The difference-differential equations satisfied by the average spins and the average products are easily constructed by means of (28) and (29). For the average spins we find the sequence of equations

$$\begin{aligned}(d/d\alpha t)q_k(t) = &-q_k(t)+\beta\\ &+\tfrac{1}{2}\gamma[q_{k-1}(t)+q_{k+1}(t)]\\ &-\tfrac{1}{2}\beta\gamma[r_{k-1,k}(t)+r_{k,k+1}(t)],\end{aligned} \tag{85}$$

which differs from the sequence (30) considered earlier by the inclusion of the inhomogeneous term β and, more importantly, through the inclusion of the pair-correlation terms $r_{k-1,k}$ and $r_{k,k+1}$. The equations for the pair correlation are likewise found to contain terms proportional to other correlation functions, i.e., the single-spin expectations and the expectation of the product of three spins. Such equations appear, because of their mixed structure, to be essentially more difficult to solve than those treated earlier. It is not difficult, however, to solve them in the limit of weak magnetic fields, $\mu H \ll kT$, and by doing so we are able to discuss the time-dependent magnetic susceptibility of the system.

In the weak-field limit, the parameter β is proportional to the magnetic field, $\beta = \mu H/kT$. The first-order changes of the averages $q_k(t)$ may be found from Eqs. (85) by using as a zeroth approximation for the functions $r_{k-1,k}$ and $r_{k,k+1}$ the solution (68) derived for them in our earlier work. The equations for the $q_k(t)$ become in this way an inhomogeneous sequence, with the inhomogeneous terms proportioned to H. The solution of these equations is simplified considerably if we assume that the model is in thermal equilibrium to zeroth order in H, i.e., that the field induces only small departures from equilibrium. In that case we have

$$r_{k-1,k} = r_{k,k+1} = \eta, \tag{86}$$

which is independent of k, and Eqs. (85) reduce to the sequence

$$\frac{d}{d\alpha t}q_k = -q_k + \tfrac{1}{2}\gamma(q_{k-1}+q_{k+1}) + \beta(1-\gamma\eta). \tag{87}$$

We shall assume, as before, that the definition of β holds for time-dependent magnetic fields as well as stationary ones. The inhomogeneous term in (87) may also be written, by using Eq. (55), as

$$\beta(1 - \gamma\eta) = \frac{\mu H(t)}{kT} \frac{1 - \eta^2}{1 + \eta^2}. \tag{88}$$

The sequence of Eqs. (87) differs from the sequence (52), which we solved earlier, only by the inclusion of this inhomogeneous term. Since the term is independent of k, the particular solution required may be chosen independent of k as well. Finding the particular solution is then a matter of treating the simplest of first-order linear differential equations. The general solution to the sequence (87) for an infinite chain is

$$q_k(t) = e^{-\alpha(t-t_0)} \sum_l q_l(t_0) I_{k-l}[\gamma\alpha(t - t_0)]$$
$$+ \frac{\mu}{kT} \frac{1 - \eta^2}{1 + \eta^2} \int_{t_0}^{t} e^{-\alpha(1-\gamma)(t-t')} H(t')\alpha \, dt', \tag{89}$$

where again we have let t_0 be the initial time. Since the model is assumed to be in thermal equilibrium before the magnetic field is turned on at time t_0, the initial values of the q_l may be taken to vanish. The spin expectations therefore all have the value given by the integral term of (89).

We now introduce the stochastic magnetization function

$$M(t) = \mu \sum_k \sigma_k(t), \tag{90}$$

whose average value is given by the sum

$$\langle M(t) \rangle = \mu \sum_k q_k(t). \tag{91}$$

If we let the initial time recede into the past, $t_0 \to -\infty$, the average magnetization obtained by summing (89) becomes

$$\langle M(t) \rangle = \frac{\mu^2 N}{kT} \frac{1 - \eta^2}{1 + \eta^2} \int_{-\infty}^{t} e^{-\alpha(1-\gamma)(t-t')} H(t')\alpha \, dt'. \tag{92}$$

For the case of a magnetic field which varies harmonically, $H(t) = H_0 e^{-i\omega t}$, we may define a complex, frequency-dependent magnetic susceptibility $\chi(\omega)$ via the relation

$$\langle M(t) \rangle = \chi(\omega) H_0 e^{-i\omega t}. \tag{93}$$

The susceptibility is then given by

$$\chi(\omega) = \frac{\mu^2 N}{kT} \frac{1 - \eta^2}{1 + \eta^2} \frac{\alpha}{\alpha(1 - \gamma) - i\omega}$$
$$= \frac{\mu^2 N}{kT} \frac{1 + \eta}{1 - \eta} \frac{\alpha(1 - \gamma)}{\alpha(1 - \gamma) - i\omega}. \tag{94}$$

In particular, in the low-frequency limit $\omega \to 0$, we find the static susceptibility

$$\chi(0) = \frac{\mu^2 N}{kT} \frac{1 + \eta}{1 - \eta} = \frac{\mu^2 N}{kT} \exp \frac{2J}{kT}, \tag{95}$$

which is the familiar result furnished by the Ising model.

FLUCTUATION-DISSIPATION THEOREMS

It is interesting to note that our result (94) for the magnetic susceptibility is closely related to the result (76) for the time-dependent correlation function. If we sum the correlation functions (76) over the indices j and k by means of the generating function (35), and multiply by μ^2, we find the time-dependent correlation function for the magnetization,

$$\langle M(0)M(t') \rangle_T = \mu^2 N \frac{1 + \eta}{1 - \eta} e^{-\alpha(1-\gamma)|t'|}. \tag{96}$$

The Fourier transform of this function is

$$\int_{-\infty}^{\infty} \langle M(0)M(t') \rangle_T e^{i\omega t'} \, dt'$$
$$= \mu^2 N \frac{1 + \eta}{1 - \eta} \frac{2\alpha(1 - \gamma)}{\alpha^2(1 - \gamma)^2 + \omega^2}$$
$$= \frac{2kT}{\omega} \operatorname{Im} \chi(\omega), \tag{97}$$

i.e., the imaginary, or dissipative part of the magnetic susceptibility is proportional to the Fourier transform of the time-dependent magnetization correlation function. We thus have in hand a particularly simple example of a fluctuation–dissipation relation. Although the derivation we have given depends on the explicit evaluation of the functions involved, analogous relations are known to hold for a wide class of mechanical systems. These relations are derived from statistical mechanics by discussing the way in which perturbations of the Liouville equation affect the distribution function or density matrix and the expectation values derived from them. Since the model we are discussing, on the other hand, is a stochastic one, our equations do not follow the dynamics of the spin variables in detail. In place of the quantum-mechanical Liouville equation we have the master equation, which has altogether different properties. Our model, nevertheless, does permit the statement of a number of simple identities analogous to the fluctuation–dissipation theorems of statistical mechanics, but differing from them slightly in form. Since these relations may be of use in finding the effect of a weak field upon the average values of quite general functions of the spin variables, we shall derive them here.

We denote the change of any quantity A induced

by the presence of the weak magnetic field by the increment symbol ΔA. The change of the transition probabilities according to (84) is then

$$\Delta w_j(\sigma_j) = w_j'(\sigma_j) - w_j(\sigma_j) = -(\mu H/kT)\sigma_j w_j(\sigma_j). \tag{98}$$

The first-order changes of the quantities involved in the master equation (27) are related by

$$\frac{d}{dt}\Delta p(\sigma_1', \cdots \sigma_N', t) = -\sum_l \sigma_l' \sum_{\sigma_l''} \sigma_l''\{\Delta w_l(\sigma_l'')p(\sigma_1', \cdots \sigma_l'', \cdots \sigma_N', t) + w_l(\sigma_l'')\Delta p(\sigma_1', \cdots \sigma_l'', \cdots \sigma_N', t)\}. \tag{99}$$

Now if $p(\sigma_1', \cdots \sigma_N' \mid \sigma_1, \cdots \sigma_N t)$ is a conditioned probability function in the sense described earlier, i.e., it satisfies the unperturbed master equation and reduces to $\prod_j \delta_{\sigma_j\sigma_j'}$ for $t = 0$, then it constitutes a Green's function for the sequence of Eqs. (99). If the initial time is $-\infty$, the solution to (99) may be written as

$$\Delta p(\sigma_1, \cdots \sigma_N, t) = -\sum_{\{\sigma'\}} \sum_l \sigma_l' \sum_{\sigma_l''} \sigma_l'' \times \int_{-\infty}^t \Delta w_l(\sigma_l'', t')p(\sigma_1', \cdots \sigma_l'', \cdots \sigma_N', t') \times p(\sigma_1', \cdots \sigma_N' \mid \sigma_1, \cdots \sigma_N, t - t')\, dt'. \tag{100}$$

We next substitute the expression (98) for the increment of the transition probabilities into (100) and sum explicitly over the values of σ'', finding

$$\Delta p(\sigma_1, \cdots \sigma_N, t) = \frac{\mu}{kT} \sum_{\{\sigma'\}} \sum_l \sigma_l' \times \int_{-\infty}^t H(t')\{w_l(\sigma_l', t')p(\sigma_1', \cdots \sigma_l', \cdots \sigma_N', t') + w_l(-\sigma_l', t')p(\sigma_1', \cdots -\sigma_l', \cdots \sigma_N, t')\} \times p(\sigma_1', \cdots \sigma_N' \mid \sigma_1, \cdots \sigma_N, t - t')\, dt'. \tag{101}$$

The detailed balancing relation (83) assures us that the two products within the curly brackets of (101) are equal, i.e. that the probability increment may be simplified to the form

$$\Delta p(\sigma_1, \cdots \sigma_N, t) = \frac{2\mu}{kT} \sum_l \sum_{\{\sigma'\}} \int_{-\infty}^t H(t')p(\sigma_1', \cdots \sigma_N', t')\sigma_l' w_l(\sigma', t') \times p(\sigma_1', \cdots \sigma_N' \mid \sigma_1, \cdots \sigma_N, t - t')\, dt'. \tag{102}$$

To evaluate the change induced by the magnetic field in the expectation value of any function of the σ variables, $F(\sigma_1, \cdots \sigma_N)$, we have only to multiply Eq. (102) through by F and sum over spins σ_j. The integrand on the right-hand side may then be recognized as an equilibrium-state average of a product of three stochastic functions. Expressed in this way, the change of the average value of F becomes

$$\Delta\langle F[\sigma_1(t), \cdots \sigma_N(t)]\rangle = \frac{2\mu}{kT}\int_{-\infty}^t \sum_l \langle \sigma_l(t')w_l[\sigma_l(t)]F[\sigma_1(t), \cdots \sigma_N(t)]\rangle_T \times H(t')\, dt'. \tag{103}$$

In particular, when the transition probabilities are given by (9) we find more simply

$$\Delta\langle F[\sigma_1(t), \cdots \sigma_N(t)]\rangle = \frac{\mu}{kT}\alpha(1-\gamma)\int_{-\infty}^t \sum_l \langle \sigma_l(t')F[\sigma_1(t), \cdots \sigma_N(t)]\rangle_T \times H(t')\, dt'. \tag{104}$$

If the function F is taken to be the magnetization, we find that it obeys the relation

$$\Delta\langle M(t)\rangle = \langle M(t)\rangle = \frac{1}{kT}\alpha(1-\gamma)\int_{-\infty}^t \langle M(t')M(t)\rangle_T H(t')\, dt'. \tag{105}$$

Since the equilibrium state is stationary, the thermal average in the integrand can only depend on $t - t'$. Hence for the case of a harmonic field $H(t) = H_0 e^{-i\omega t}$, we find

$$\chi(\omega) = \frac{1}{kT}\alpha(1-\gamma)\int_0^\infty \langle M(0)M(t)\rangle_T e^{i\omega t}\, dt. \tag{106}$$

The foregoing relations are rather similar in structure to the complex forms of the fluctuation–dissipation theorems of statistical mechanics, and furnish us with similar information. They differ from those relations, however, in two respects illustrated by comparing (97) and (106). The former equation relates the imaginary part of the susceptibility to the transform of the correlation function, while the latter relates the real part to it with a different proportionality constant. Although both types of relation hold true for the model at hand, it is interesting to see how the difference between them arises. For this purpose let us consider the stochastic function

$$L(t) = -2\mu \sum_m \sigma_m(t)w_m(\sigma_m, t). \tag{107}$$

The expectation value of L is the time derivative of the average magnetization. To see this, we use

(28) to write

$$\langle L(t)\rangle = \mu \sum_m \frac{d}{dt} q_m(t) = \frac{d}{dt} \langle M(t)\rangle. \quad (108)$$

The function $L(t)$ itself, however, is not the time derivative of $M(t)$. If it were, the substitution of $\dot{M}(t)$ for it in (103) would lead to precisely the relations furnished by discussions based on the Liouville equation. The relations we find instead are evidently quite similar in content.

ALTERNATIVE METHOD AND GENERALIZATION

It may be of interest to mention briefly another way of studying Markoff processes, one rather different from the preceding discussion. The 2^N values of the probability function $p(\sigma_1, \cdots \sigma_N, t)$ may be regarded as the components of a vector $\mathbf{p}$. Then by suitably defining the elements of a matrix M, we may write the master equation (27) in the form

$$(d/dt)\mathbf{p} = M\mathbf{p}, \quad (109)$$

which suggests that p is a superposition of eigenvectors $\mathbf{p}^{(s)}$ which satisfy

$$M\mathbf{p}^{(s)} = -\nu_s \mathbf{p}^{(s)}. \quad (110)$$

One eigenvector, at least, is quite well-known to us. The probability distribution for the Ising model at equilibrium, the normalized Maxwell–Boltzmann distribution, corresponds to the eigenvalue $\nu = 0$. It is

$$p^{(0)}(\sigma_1, \cdots \sigma_N) = Z^{-1} \exp\,[(J/kT) \sum_l \sigma_l \sigma_{l+1}], \quad (111)$$

where Z, the normalizing factor, is the partition function.

Other eigenvectors may be sought by multiplying $\mathbf{p}^{(0)}$ by sums of products of spin variables with undetermined coefficients. For example, if we write

$$p^{(1)}(\sigma_1, \cdots \sigma_N, t) = \sum_i a_i(t) \sigma_i p^{(0)}(\sigma_1, \cdots \sigma_N), \quad (112)$$

we find that the condition that this form satisfy (109) is that the functions $a_i(t)$ satisfy the same sequence of equations (30) as we discussed earlier in connection with $q_i(t)$. The mode functions ζ_m^k, where ζ_m is given by (46) therefore furnish us with N different eigenvectors corresponding to roots ν_m given by (47).

The eigenvectors which are constructed by multiplying $\mathbf{p}^{(0)}$ by higher-order polynomials in $\sigma_1, \cdots \sigma_N$, are somewhat more complicated in form, and will be discussed in a later publication. The eigenvalues to which they correspond are fairly simple, however. The eigenvectors which are formed from the products of rth degree polynomials with $\mathbf{p}^{(0)}$ have eigenvalues

$$\nu = \nu_{m_1} + \nu_{m_2} + \cdots + \nu_{m_r}, \quad (113)$$

where the ν_{m_j} are given by (47), and the set of integers $m_1, \cdots m_r$ is selected from $0, 1 \cdots N-1$ with no repetitions. The number of such eigenvalues is given by the binomial coefficient $\binom{N}{r}$. The full set of 2^N eigenvalues is obtained by allowing r to range from 0 to N.

In particular, the largest eigenvalue is obtained for $r = N$ and is $\nu = N$. The eigenvector for this case is simply proportional to

$$p^{(N)}(\sigma_1, \cdots \sigma_N, t) = \prod_{j=1}^{N} \sigma_j e^{-N\alpha t}. \quad (114)$$

All of the foregoing discussion has been restricted to the case of nearest-neighbor coupling among spins in order to make contact with the familiar studies of the Ising model. The coupling may be extended to include the first n nearest neighbors by introducing the transition probability

$$w_j(\sigma_j) = \tfrac{1}{2}\alpha\left\{1 - \tfrac{1}{2}\sigma_j \sum_{l=1}^{n} \gamma_l(\sigma_{j-l} + \sigma_{j+l})\right\}, \quad (115)$$

where $\sum_l |\gamma_l| \leq 1$. The methods of the preceding sections deal equally with the equations which follow from this more general type of coupling. The only significant change is that the quadratic equation, (55), for the short-range order is replaced by an equation of $2n$th degree which has n roots $\eta_1, \cdots \eta_n$ with absolute value less than unity. The equilibrium solution for the average spins, when the zeroth spin is fixed, is then an expression of the form

$$q_k = \sum_{j=1}^{n} c_j \eta_j^k, \quad (116)$$

where the coefficients c_j must be determined from the condition $q_0 = 1$ and the equations for $q_1, \cdots q_{n-1}$. These spin averages then determine the equilibrium spin correlations $r_{j,k}$ in precisely the way described earlier.

ACKNOWLEDGMENT

The author would like to thank the Bell Telephone Laboratories, Murray Hill, New Jersey, for their hospitality during a period of several summer weeks in which this work was carried out.

APPENDIX

We have already noted that other forms of the transition probability than (9) are capable of bringing the stochastic model to the same equilibrium state as the Ising model. The condition that such a transition probability $w_j(\sigma_j)$ must satisfy is that the ratio $w_j(\sigma_j)/w_j(-\sigma_j)$ be equal to the equilibrium probability ratio (12). If we assume that $w_j(\sigma_j)$ depends symmetrically on the two neighboring spins σ_{j-1} and σ_{j+1} as well as on σ_j, then the condition just mentioned may be regarded as a functional equation for the transition probability. Its most general solution is given by the form

$$w_j(\sigma_j) = \tfrac{1}{2}\alpha\{1 + \delta\sigma_{j-1}\sigma_{j+1} - \tfrac{1}{2}\gamma(1 + \delta)\sigma_j(\sigma_{j-1} + \sigma_{j+1})\}, \quad (117)$$

in the absence of any magnetic field. In this form the parameter γ must still be identified with the constant (17), but the parameter δ has no analog in the discussions of the Ising model at equilibrium, and may evidently be chosen arbitrarily. It was assumed to vanish in our discussions of the time-dependent model since its presence materially complicates the equations for the spin expectation values.

Decay of Correlations. III. Relaxation of Spin Correlations and Distribution Functions in the One-Dimensional Ising Lattice

Dick Bedeaux,[1] Kurt E. Shuler,[1] and Irwin Oppenheim[2]

Received February 3, 1970

We have studied the relaxation of the n-spin correlation function $\langle\sigma^{(n)}\rangle$ and distribution function $P_n(\sigma^{(n)}; t)$ for the Glauber model of the one-dimensional Ising lattice. We find that new combinations of correlation functions (C-functions) and distribution functions (Q-functions) are more useful in discussing the relaxation of this system from initial nonequilibrium states than the usual cumulants and Ursell functions used in our papers I and II. The asymptotic behavior of the P, C, and Q functions are: $P_n(\sigma^{(n)}; t) - P_n^{(0)}(\sigma^{(n)}) \sim P_1(\sigma; t) - P_1^{(0)}(\sigma)$; $C_n(\sigma^{(n)}; t) - C_n^{(0)}(\sigma^{(n)}) \sim \langle\sigma\rangle^n$; $Q_n(\sigma^{(n)}; t) - Q_n^{(0)}(\sigma^{(n)}) \sim [P_1(\sigma; t) - P_1^{(0)}(\sigma)]^n$; where the superscript zero denotes the equilibrium function. These results imply that $P_n(\sigma^{(n)}; t)$, $n > 2$, decays to a functional of lower-order distribution functions as $[P_1(\sigma; t) - P_1^{(0)}(\sigma)]^n$ and that the n-spin correlation function $\langle\sigma^{(n)}\rangle$ with $n > 2$ decays to a functional of lower-order correlation functions as $\langle\sigma\rangle^n$. This result for the distribution function $P_n(\sigma^{(n)}; t)$, $n > 2$, is identical with the results obtained in papers I and II for initially correlated, non-interacting many-particle systems in contact with a heat bath and for an infinite chain of coupled harmonic oscillators. As a special example, we study the relaxation of the spin system when the heat-bath temperature is changed suddenly from an initial temperature T_0 to a final temperature T. We obtain the interesting result that the spin system is not canonically invariant, i.e., it can *not* be characterized by a time-dependent "spin temperature."

KEY WORDS: Ising lattice; spin correlations; spin distribution function; dynamics of correlations; master equation.

The work of two of the authors (D. B. and K. E. S.) was supported in part by the Advanced Research Projects Agency of the Department of Defense as monitored by the U.S. Office of Naval Research under Contract N00014-67-A-0109-0010. A portion of this work (I. O.) was supported by the National Science Foundation.

[1] Department of Chemistry, University of California, San Diego, La Jolla, California.

[2] Department of Chemistry, Massachusetts Institute of Technology, Cambridge, Massachusetts.

1. INTRODUCTION

In this paper, we continue our discussion of the decay of correlations in systems relaxing from initial nonequilibrium states to their final equilibrium states. In two previous papers[1,2] (hereafter referred to as I and II, respectively), we developed the theory for noninteracting, initially correlated many-particle systems and for an infinite chain of coupled harmonic oscillators. We found that the initial correlations as measured by the Ursell function U_n decayed to their zero equilibrium value faster than the distribution functions relaxed to their equilibrium values. In particular, the n-particle distribution functions relaxed to their equilibrium forms $P_n^{(0)}$ asymptotically as

$$P_n(t) - P_n^{(0)} \sim P_1(t) - P_1^{(0)}, \qquad n \geqslant 1 \tag{1}$$

and the Ursell functions relaxed asymptotically as

$$U_n(t) \sim [P_1(t) - P_1^{(0)}]^n \tag{2}$$

Equation (2) implies the important result that $P_n(t)$, $n \geqslant 1$, relaxes to a functional of lower-order distribution functions $[P_{n-1}(t), P_{n-2}(t),..., P_1(t)]$ as $[P_1(t) - P_1^{(0)}]^n$.

In this paper, we discuss the relaxation of the n-spin correlations and distribution function of the infinite one-dimensional Ising system with nearest-neighbor interactions using the stochastic dynamical model of Glauber.[3] For this system, the n-spin equilibrium distribution function factorizes into a product of two-spin distribution functions rather than into a product of singlet distribution functions. Furthermore, the dynamical variables of the Ising model, the spins σ_i, can assume only the values ± 1, so that $\sigma_i^2 = 1$ for all i. Thus, for example, $\langle \sigma_i^2 \rangle = 1$ for all i and all times t. We shall see that these properties make it desirable to construct new functions, analogous to the cumulant and Ursell functions used in I and II, in order to discuss the relaxation of the n-spin correlation and distribution functions.

An important result of this paper is that the n-spin distribution function $P_n(\sigma^n; t)$, $n > 2$, decays to a functional of lower-order distribution functions $[P_{n-1}, P_{n-2},..., P_1]$ as $[P_1(\sigma; t) - P_1^{(0)}(\sigma)]^n$ and that the n-spin correlation function $\langle \sigma^{(n)} \rangle$, $n > 2$, decays to a functional of lower-order correlation functions $[\langle \sigma^{(n-1)} \rangle, \langle \sigma^{(n-2)} \rangle,..., \langle \sigma \rangle]$ as $\langle \sigma \rangle^n$. This result is identical with our findings for the systems considered in I and II. Some previous work[4] on spin relaxation in the one-dimensional Ising model which employed the usual cumulant and Ursell functions has led to some incorrect results. Application of the usual cumulants to the two-dimensional Ising model[5] probably does not lead to valid results either.

We consider an infinite, one-dimensional lattice with a spin $\sigma_i = \pm 1$ on each site i. The state of the system is specified by the spin vector $\{\sigma\} = (..., \sigma_{i-1}, \sigma_i, \sigma_{i+1}, ...)$. The probability of finding the system in the state $\{\sigma\}$ at time t is $P(\{\sigma\}; t)$. The n-spin The n-spin reduced probability $P_n(\sigma^{(n)}; t)$ is given by

$$P_n(\sigma^{(n)}; t) \equiv P_n(\sigma_{i_1}, \sigma_{i_2},..., \sigma_{i_n}; t) = \sum_{\{\sigma\} \neq \sigma^{(n)}} P(\{\sigma\}; t) \tag{3}$$

where the summation is over all spin variables except σ_{i_1} through σ_{i_n}. The time-dependent spin correlation functions are defined as

$$\langle \sigma_{i_1}\sigma_{i_2} \cdots \sigma_{i_n}\rangle = \sum_{\{\sigma\}} \sigma_{i_1}\sigma_{i_2} \cdots \sigma_{i_n} P(\{\sigma\}; t) = \sum_{\sigma^{(n)}} \sigma_{i_1}\sigma_{i_2} \cdots \sigma_{i_n} P_n(\sigma^{(n)}; t) \tag{4}$$

where the time dependence of $\langle \sigma_{i_1}\sigma_{i_2} \cdots \sigma_{i_n}\rangle$ is implicit. The reduced probabilities can be expressed in terms of the correlation functions as [3]

$$P_n(\sigma_{i_1}, \ldots, \sigma_{i_n}; t)$$

$$= 2^{-n} \left\{ 1 + \sum_{j=1}^{n} \sigma_{i_j}\langle \sigma_{i_j}\rangle + \sum_{j<k}^{n} \sigma_{i_j}\sigma_{i_k}\langle \sigma_{i_j}\sigma_{i_k}\rangle + \cdots + \sigma_{i_1}\sigma_{i_2} \cdots \sigma_{i_n}\langle \sigma_{i_1}\sigma_{i_2} \cdots \sigma_{i_n}\rangle \right\} \tag{5}$$

Transitions of the spins between their possible values ± 1 are due to their interactions with an external heat reservoir. The transition rate for the flip of the ith spin from the value σ_i to the value $-\sigma_i$, while the other spins remain momentarily fixed, is assumed to be.[3]

$$w_i(\sigma_i) = \tfrac{1}{2}\alpha[1 - \tfrac{1}{2}\gamma\sigma_i(\sigma_{i-1} + \sigma_{i+1})] \tag{6}$$

with $\alpha > 0$ and $0 \leqslant \gamma \leqslant 1$. The significance of the parameters α and γ has been discussed by Glauber. It is clear from the form of Eq. (6) that there is a correlation at all times between nearest-neighbor spins in that $w_i(\sigma_i)$ depends upon the values σ_{i+1} and σ_{i-1} of the $(i + 1)$th and $(i - 1)$th spins.

The equilibrium properties of the Ising spin systems are described by the Hamiltonian

$$H(\{\sigma\}) = -J \sum_i \sigma_i\sigma_{i+1} \tag{7}$$

Using detailed balance, the relation

$$\gamma = \tanh(2J/kT) \tag{8}$$

where T is the fixed temperature of the heat bath, can readily be derived. The equilibrium form for the distribution function is

$$P^{(0)}(\{\sigma\}) = e^{-H(\{0\})/kT} \Big/ \sum_{\{\sigma\}} e^{-H(\{\sigma\})/kT} \tag{9}$$

where the superscript zero denotes the equilibrium value. From Eqs. (4), (7), and (9), it then follows that the equilibrium correlation functions are

$$\begin{aligned} \langle \sigma_{i_1}\sigma_{i_2} \cdots \sigma_{i_n}\rangle^{(0)} &= 0 && \text{if } n \text{ is odd} \\ &= \langle \sigma_{i_1}\sigma_{i_2}\rangle^{(0)} \langle \sigma_{i_3}\sigma_{i_4}\rangle^{(0)} \cdots \langle \sigma_{i_{n-1}}\sigma_{i_n}\rangle^{(0)} && \text{if } n \text{ is even} \end{aligned} \tag{10}$$

where

$$\langle \sigma_{i_1} \sigma_{i_2} \rangle^{(0)} = \eta^{i_2 - i_1} \tag{11}$$

and

$$\eta = \tanh(J/kT) \tag{12}$$

In Eq. (10) and in all subsequent equations, the spin indices are ordered such that $i_1 \leqslant i_2 \leqslant \cdots \leqslant i_n$. It follows from Eqs. (5), (10), and (11) that the reduced equilibrium distribution functions are

$$P_n^{(0)}(\sigma_{i_1}, \sigma_{i_2}, \ldots, \sigma_{i_n}) = 2^{n-2} P_2^{(0)}(\sigma_{i_1}, \sigma_{i_2}) \, P_2^{(0)}(\sigma_{i_2}, \sigma_{i_3}) \cdots P_2^{(0)}(\sigma_{i_{n-1}}, \sigma_{i_n}), \quad n \geqslant 2 \tag{13}$$

$$P_2^{(0)}(\sigma_{i_1}, \sigma_{i_2}) = \tfrac{1}{4}(1 + \sigma_{i_1} \sigma_{i_2} \eta^{i_2 - i_1}) \tag{14}$$

and

$$P_1^{(0)}(\sigma_{i_1}) = \tfrac{1}{2} \tag{15}$$

Using Eq. (4) and the master equation for $P(\{\sigma\}; t)$ derived by Glauber, the dynamic equations for the correlations functions for $n \geqslant 1$ can be written as

$$\begin{aligned} \frac{d}{dt} \langle \sigma_{i_1} \sigma_{i_2} \cdots \sigma_{i_n} \rangle = {} & -n\alpha \langle \sigma_{i_1} \sigma_{i_2} \cdots \sigma_{i_n} \rangle + \frac{\alpha\gamma}{2} \{ \langle \sigma_{i_1+1} \sigma_{i_2} \cdots \sigma_{i_n} \rangle + \langle \sigma_{i_1-1} \sigma_{i_2} \cdots \sigma_{i_n} \rangle \\ & + \langle \sigma_{i_1} \sigma_{i_2+1} \cdots \sigma_{i_n} \rangle + \langle \sigma_{i_1} \sigma_{i_2-1} \cdots \sigma_{i_n} \rangle \\ & + \cdots + \langle \sigma_{i_1} \sigma_{i_2} \cdots \sigma_{i_n+1} \rangle + \langle \sigma_{i_1} \sigma_{i_2} \cdots \sigma_{i_n-1} \rangle \} \end{aligned} \tag{16}$$

where all indices $i_1 \cdots i_n$ are different. If any of the indices are the same, Eq. (16) does not apply. For instance, if $i_1 = i_2$, then $\langle \sigma_{i_1} \sigma_{i_2} \cdots \sigma_{i_n} \rangle$ reduces to $\langle \sigma_{i_3} \sigma_{i_4} \cdots \sigma_{i_n} \rangle$ since $\sigma_i^2 = 1$ for all i. In this case, we find from Eq. (4)

$$\begin{aligned} \frac{d}{dt} \langle \sigma_{i_3} \sigma_{i_4} \cdots \sigma_{i_n} \rangle = {} & -(n-2)\alpha \langle \sigma_{i_3} \sigma_{i_4} \cdots \sigma_{i_n} \rangle + \frac{\alpha\gamma}{2} \{ \langle \sigma_{i_3+1} \sigma_{i_4} \cdots \sigma_{i_n} \rangle + \langle \sigma_{i_3-1} \sigma_{i_4} \cdots \sigma_{i_n} \rangle \\ & + \langle \sigma_{i_3} \sigma_{i_4} \cdots \sigma_{i_n+1} \rangle + \langle \sigma_{i_3} \sigma_{i_4} \cdots \sigma_{i_n-1} \rangle \} \end{aligned} \tag{17}$$

This leads to difficulties in the solution of Eq. (16) since, for example, $i_1 + 1$ may be equal to i_2, even though $i_1 \neq i_2$.

For $n = 1, 2$, the differential difference equations for the spin correlation functions are, for $i < j$,

$$\frac{d}{dt} \langle \sigma_i \rangle = -\alpha \langle \sigma_i \rangle + \frac{\alpha\gamma}{2} [\langle \sigma_{i+1} \rangle + \langle \sigma_{i-1} \rangle] \tag{18}$$

$$\frac{d}{dt} \langle \sigma_i \sigma_j \rangle = -2\alpha \langle \sigma_i \sigma_j \rangle + \frac{\alpha\gamma}{2} [\langle \sigma_{i+1} \sigma_j \rangle + \langle \sigma_{i-1} \sigma_j \rangle + \langle \sigma_i \sigma_{j+1} \rangle + \langle \sigma_i \sigma_{j-1} \rangle] \tag{19}$$

The solution of these equations has been given by Glauber[3]:

$$\langle\sigma_i\rangle = e^{-\alpha t} \sum_{m=-\infty}^{\infty} \langle\sigma_m\rangle_0 \, I_{i-m}(\gamma\alpha t) \tag{20}$$

$$\langle\sigma_i\sigma_j\rangle = \langle\sigma_i\sigma_j\rangle^{(0)} + e^{-2\alpha t} \sum_{\substack{m<n \\ -\infty}}^{\infty} [\langle\sigma_m\sigma_n\rangle_0 - \langle\sigma_m\sigma_n\rangle^{(0)}]$$

$$\times\, [I_{i-m}(\gamma\alpha t)\, I_{j-n}(\gamma\alpha t) - I_{i-n}(\gamma\alpha t)\, I_{j-m}(\gamma\alpha t)] \tag{21}$$

where the subscript zero denotes the initial value at $t = 0$ of the correlation function, the superscript zero again denotes the equilibrium value at $t = -\infty$, and where the $I_n(x)$ are the modified Bessel function $I_n(x) = i^{-n} J_n(ix)$.[6]

For $n = 0$, the function $e^{-\alpha t} I_n(\gamma\alpha t)$ tends to zero monotonically as t increases. For $n > 0$, the function increases for times $t \ll n/\gamma\alpha$ as

$$e^{-\alpha t} I_n(\gamma\alpha t) \approx (n!)^{-1} (\tfrac{1}{2}\gamma\alpha t)^n \, e^{-\alpha t} \tag{22}$$

For $n \gg 1$, it reaches a maximum for $t \approx (n/\alpha)(1 - \gamma^2)^{1/2}$. For long times, the asymptotic behavior for all values of n is given by

$$e^{-\alpha t} I_n(\gamma\alpha t) \sim (2\pi\gamma\alpha t)^{-1/2} \, e^{-\alpha(1-\gamma)t} \left\{ 1 + \frac{4n^2 - 1}{8\gamma\alpha t} + \frac{(4n^2 - 1)(4n^2 - 9)}{2!\,(8\gamma\alpha t)^2} + \cdots \right\} \tag{23}$$

Various properties of the function $e^{-\alpha t} I_n(\gamma\alpha t)$ are discussed in detail by Montroll.[7]

The asymptotic behavior of the spin correlation functions $\langle\sigma_i\rangle$ and $\langle\sigma_i\sigma_j\rangle$ are easily obtained from Eqs. (20)–(23) under the conditions that a finite set of initial correlation functions $\langle\sigma_i\rangle_0$ and $\langle\sigma_i\sigma_j\rangle_0$ has nonequilibrium values, i.e., $\langle\sigma_i\rangle_0 \neq 0$ for some i and $\langle\sigma_i\sigma_j\rangle_0 \neq \langle\sigma_i\sigma_j\rangle^{(0)}$ for some i, j. The case of $\langle\sigma_i\sigma_j\rangle_0 \neq \langle\sigma_i\sigma_j\rangle^{(0)}$ for *all* i, j is considered in Section 5, and in the appendix. The results are

$$\langle\sigma_i\rangle \sim k_1(i)[(2\pi\gamma\alpha t)^{-1/2} \, e^{-\alpha(1-\gamma)t}] \equiv k_1(i)[A(t)] \tag{24}$$

and

$$\langle\sigma_i\sigma_j\rangle - \langle\sigma_i\sigma_j\rangle^{(0)} \sim k_2(i, j)\, t^{-1}[A(t)]^2 \tag{25}$$

where $[A(t)]$ is defined by Eq. (24) and where k_1 and k_2 are independent of time and depend only on the initial conditions. We note that $\langle\sigma_i\sigma_j\rangle$ approaches its equilibrium value somewhat faster than $\langle\sigma_i\rangle^2$. The factor of t^{-1} in Eq. (25) arise due to the cancellation of the first term in the Bessel-function expansion when Eq. (23) is substituted into Eq. (21).

The explicit form fors $k_1(i)$ and $k_2(i, j)$follow immediately from Eqs. (20), (21), and (23) and are

$$k_1(i) = \sum_{m=-\alpha}^{\infty} \langle\sigma_m\rangle_0 \tag{26}$$

$$k_2(i, j) = \frac{(i - j)}{\gamma\alpha} \sum_{\substack{m<n \\ -\infty}}^{\infty} [\langle\sigma_m\sigma_n\rangle_0 - \langle\sigma_m\sigma_n\rangle^{(0)}](n - m) \tag{27}$$

It is clear that the asymptotic expansions used here and below are valid only if the $k_n(i_1, ..., i_n)$ are finite. If the k_n are zero because of special initial conditions, additional factors of t^{-1} will occur in the asymptotic form.

In the next sections, we develop methods which permit us to obtain exact and asymptotic results for the time dependence of the n-spin correlation functions.

2. THE C-FUNCTIONS AND THEIR DYNAMICS

As we have discussed in Section 1, the solution of Eq. (16) for the dynamics of the n-spin correlation function presents difficulties owing to the possible occurrence of spin correlation functions of order $n - 2$ on the right-hand side of the equation when two or more spin indices are the same. In other words, Eq. (16) is then not a closed set of equations for the nth order correlation functions. In order to overcome this difficulty, we introduce a new set of functions, the C_n-functions,

$$C_n(i_1, i_2, ..., i_n; t) \equiv C_n(\sigma^{(n)}; t),$$

defined for $n > 2$ with $i_1 \leqslant i_2 \leqslant \cdots \leqslant i_n$, which are combinations of the correlation functions. These functions have the following properties:

(a) The C_n-function satisfies the same differential equation (16) as the n-spin correlation function,

$$\begin{aligned}\frac{d}{dt} C_n(i_1, i_2, ..., i_n; t) = &-n\alpha C_n(i_1, i_2, ..., i_n; t)\\ &+ \frac{\alpha\gamma}{2}\{C_n(i_1 + 1, i_2, ..., i_n; t) + C_n(i_1 - 1, i_2, ..., i_n; t)\\ &+ \cdots + C_n(i_1, i_2, ..., i_n + 1; t) + C_n(i_1, i_2, ..., i_n - 1; t)\}\end{aligned} \tag{28}$$

(b) The C_n-function is zero if two adjacent indices are the same,

$$C_n(i_1, ..., i_n; t) = 0 \qquad \text{for} \quad i_j = i_{j+1}, \quad 1 \leqslant j \leqslant n-1 \tag{29}$$

The differential equations (28) for the C_n-functions clearly form a closed set owing to the property (29). The equilibrium solution for the C_n-function is

$$C_n^{(0)}(i_1, ..., i_n; t) = 0, \qquad n > 2 \tag{30}$$

which can readily be seen from Eq. (28). The general solution of Eq. (28) is

$$\begin{aligned}C_n(i_1, ..., i_n; t)\\ = e^{-n\alpha t} \sum_{m_1 < m_2 < \cdots < m_n} C_n(m_1, ..., m_n; 0) \sum_{\mathscr{P}} (-1)^{\mathscr{P}} I_{i_1 - m_1'}(\gamma\alpha t) \cdots I_{i_n - m_n'}(\gamma\alpha t)\end{aligned} \tag{31}$$

where the sum over $\mathscr{P}$ is over all permutations $(m_1', m_2', ..., m_n')$ of $(m_1, m_2, ..., m_n)$. It will be noted that if two adjacent indices i_j, i_{j+1} are equal, the sum over the permutation makes the right-hand side of Eq. (31) equal to zero, in agreement with condition (29). It is interesting to note that $C_n(i_1, i_2, ..., i_n; t)$ will be zero for all times t if $C_n(m_1, m_2, ..., m_n; 0)$ is zero for all m_j, $j = 1, 2, ..., n$.

Using the asymptotic properties of the Bessel function $I_n(x)$ as given in Eq. (23) and the solution (31) of the C_n-function, we find for the asymptotic behavior of the C_n-function

$$C_n \sim K_n(\sigma^{(n)})\, t^{(1-n)}[A(t)]^n \tag{32}$$

where the factor $t^{(1-n)}$ arises from cancellations in the sum over permutations and where K_n is independent of time and depends only on the initial conditions. The asymptotic form (32) is valid if a finite number of $C_n(\sigma^{(n)}; 0)$ are nonzero. It follows directly from Eq. (32) that $C_n(\sigma^{(n)}; t)$ approaches zero faster than $\langle\sigma_i\rangle^n$, as can be seen from a comparison with Eq. (24).

We shall now relate the C_n-functions to the spin correlation functions. We define $C_n(\sigma^{(n)}; t)$ by

$$C_n(i_1, i_2, ..., i_n; t) \equiv \sum_{\xi} (-1)^{\mathscr{P}} (k-1)!\,(-1)^{k-1}\, \mathscr{P}\langle i_1 i_2 \cdots i_{n_1}\rangle \cdots \langle i_{n-n_k+1} \cdots i_n\rangle \tag{33}$$

where $\mathscr{P}$ is the permutation operator. The summation over ξ denotes a summation over all even partitions of the n spins into subgroups in which the indices in the subgroups are ordered. A partition of n spins into k subgroups containing n_1 spins in subgroup 1, n_2 spins in subgroup 2,..., n_k spins in subgroup k is called even if n_j, where $j = 1, 2, ...k$. is even except for at most one value of j. The notation $\langle i_1 i_2 \cdots i_n\rangle$, etc. In Eq. (33) is shorthand for the n-spin correlation function $\langle\sigma_{i_1}\sigma_{i_2} \cdots \sigma_{i_n}\rangle$. Performing the indicated operations in Eq. (33) leads to the following relations between the C-functions and the spin correlation functions:

$$\begin{aligned}
C_1(i_1; t) &= \langle\sigma_{i_1}\rangle \\
C_2(i_1, i_2; t) &= \langle\sigma_{i_1}\sigma_{i_2}\rangle \\
C_3(i_1, i_2, i_3; t) &= \langle\sigma_{i_1}\sigma_{i_2}\sigma_{i_3}\rangle - \langle\sigma_{i_1}\rangle\langle\sigma_{i_2}\sigma_{i_3}\rangle \\
&\quad - \langle\sigma_{i_3}\rangle\langle\sigma_{i_1}\sigma_{i_2}\rangle + \langle\sigma_{i_2}\rangle\langle\sigma_{i_1}\sigma_{i_3}\rangle \\
C_4(i_1, i_2, i_3, i_4; t) &= \langle\sigma_{i_1}\sigma_{i_2}\sigma_{i_3}\sigma_{i_4}\rangle - \langle\sigma_{i_1}\sigma_{i_2}\rangle\langle\sigma_{i_3}\sigma_{i_4}\rangle \\
&\quad - \langle\sigma_{i_1}\sigma_{i_4}\rangle\langle\sigma_{i_2}\sigma_{i_3}\rangle + \langle\sigma_{i_1}\sigma_{i_3}\rangle\langle\sigma_{i_2}\sigma_{i_4}\rangle
\end{aligned} \tag{34}$$

where, as always, $i_1 \leqslant i_2 \leqslant i_3 \cdots \leqslant i_n$. Note that the definition of Eq. (33) enables us to define $C_1(i_1; t)$ and $C_2(i_1, i_2; t)$. The properties of these two functions have been discussed by Glauber[3] and in Section 1 of this paper. We shall show below why the C-functions defined here are more useful than the usual cumulants (see, e.g., Gnedenko[8]) in discussing the decay of the n-spin correlation functions for $n > 2$.

We shall now demonstrate that the definition of the C_n-function given in Eq. (33) satisfies the conditions of Eqs. (28) and (29) for $n > 2$. That $C_n(\sigma^{(n)}; t)$ satisfies the differential equation (28) follows from the fact that each term in the sum of Eq. (33) satisfies Eq. (28). That $C_n(\sigma^{(n)}; t)$ is zero for $i_l = i_{l+1}$ can be proved by induction. We invert Eq. (33) to obtain an expression for the n-spin correlation function in terms of the C-functions

$$\langle \sigma_{i_1}\sigma_{i_2} \cdots \sigma_{i_n} \rangle = \sum_{\xi} (-1)^{\mathscr{P}} \mathscr{P} C_{n_1}(i_1, i_2, ..., i_{n_1}; t)\, C_{n_2}(i_{n_1+1}, ..., i_{n_1+n_2}; t) \cdots C_{n_k}(i_{n-n_k+1}, ..., i_n; t) \tag{35}$$

In the sum on the r.h.s. of Eq. (35) are the following contributions:

(a) $C_n(i_1, ..., i_n; t)$.

(b) Terms in which i_l and i_{l+1} are in different subgroups. These terms cancel in pairs due to the fact that the interchange of i_l and i_{l+1} is an odd permutation

(c) Terms in which i_l and i_{l+1} are in the same subgroups j and $n > n_j > 2$. These terms are zero, using the induction hypothesis that $C_{n_j} = 0$, $n > n_j > 2$, if two adjacent spin indices are the same.

(d) Terms in which i_l and i_{l+1} are in the same two-spin subgroup. Since

$$C_2(i_l, i_{l+1}; t) = \langle \sigma_{i_l}\sigma_{i_{l+1}} \rangle = 1 \tag{36}$$

these terms add up to $\langle \sigma_{i_1} \cdots \sigma_{i_{l-1}}\sigma_{i_{l+2}} \cdots \sigma_{i_n} \rangle$. By inspection, $C_3(i_1, i_2, i_3; t)$ is zero if $i_1 = i_2$ or $i_2 = i_3$. This finishes the proof of property (29) that $C_n(i_1, ..., i_n; t) = 0$ for $i_l = i_{l+1}$ and $n > 2$. Thus, the C_n-function as defined by Eq. (33), for $n > 2$, satisfy Eq. (29).

A cumulantlike property of the C-function is that

$$C_n(i_1, ..., i_n; t) = 0 \tag{37}$$

if two adjacent spins, i_l and i_{l+1}, are uncorrelated to the rest of the spin variables $i_1, i_2, ..., i_{l-1}, i_{l+2}, ..., i_n$.

It should be emphasized here that the definition of the C-functions in Eq. (33) in terms of the spin correlation functions is a convenient one but not a unique one. Other functions could be developed which possess the desirable properties (28) and (29).

In the next section, we shall use the asymptotic properties of the C_n-function to discuss the time-dependent behavior of the spin correlation functions.

In a subsequent paper, we will demonstrate that there is a close and interesting relation between the C-functions and Pfaffians. That such a relation exists can readily be seen from the expression for C_4 in Eq. (34), in that

$$\langle \sigma_{i_1}\sigma_{i_2}\sigma_{i_3}\sigma_{i_4} \rangle - C_4(i_1, i_2, i_3, i_4; t) = \begin{vmatrix} \langle i_1 i_2 \rangle & \langle i_1 i_3 \rangle & \langle i_1 i_4 \rangle \\ & \langle i_2 i_3 \rangle & \langle i_2 i_4 \rangle \\ & & \langle i_3 i_4 \rangle \end{vmatrix} \tag{38}$$

where the expression on the right-hand side is the Pfaffian.

3. RELAXATION OF THE SPIN CORRELATION FUNCTIONS

The dynamical behavior of the correlation functions is easily obtained from Eq. (35), which expresses the spin correlation functions in terms of the C-functions, and from Eq. (31), which gives the explicit dynamical behavior of the C-function. Explicit expressions for the time dependence of the one-and two-spin correlation functions have already been given in Eqs. (20) and (21). For example, the time dependence of the three-spin correlation function can be obtained from Eq. (35) in the form

$$\langle \sigma_{i_1}\sigma_{i_2}\sigma_{i_3} \rangle = C_3(i_1, i_2, i_3; t) + C_1(i_1; t)\, C_2(i_2, i_3; t) + C_1(i_3; t)\, C_2(i_1, i_2; t) - C_1(i_2; t)\, C_2(i_1, i_3; t) \tag{39}$$

Use of Eqs. (31), (20), and (21) then leads to an explicit but complicated expression in terms of Bessel functions.

The *asymptotic* time dependence of the spin correlation functions can readily be obtained from Eqs. (35), (32), (24), and (25). We shall discuss the asymptotic time dependence for the three- and four-spin correlation functions in detail and then give some general properties for the n-spin correlation function. From Eq. (39) it follows immediately that

$$\langle \sigma_{i_1}\sigma_{i_2}\sigma_{i_3} \rangle \sim at^{-2}[A(t)]^3 + bt^{-1}[A(t)]^3 + c[A(t)] \tag{40}$$

where

$$a = K_3(i_1, i_2, i_3),$$

$$b = k_1(i_1)\, k_2(i_2, i_3) + k_1(i_3)\, k_2(i_1, i_2) - k_1(i_2)\, k_2(i_1, i_3)$$

$$c = k_1(i_1)\langle \sigma_{i_2}\sigma_{i_3} \rangle^{(0)} + k_1(i_3)\langle \sigma_{i_1}\sigma_{i_2} \rangle^{(0)} - k_1(i_2)\langle \sigma_{i_1}\sigma_{i_3} \rangle^{(0)}$$

and where $k_1(i)$ and $k_2(i, j)$ are given by Eqs. (26) and (27). In Eq. (40), we have used the leading asymptotic term for each term on the right-hand side of Eq. (39). It is clear from the form of Eq. (40) that the relaxation of the three-spin correlation function proceeds in two stages[3]: in the first stage, $\langle \sigma_{i_1}\sigma_{i_2}\sigma_{i_3} \rangle$ becomes a functional of the two- and one-particle correlation functions

$$\langle \sigma_{i_1}\sigma_{i_2}\sigma_{i_3} \rangle \rightarrow F_3[\langle \sigma_{i_j}\sigma_{i_k} \rangle^{(0)}, \langle \sigma_{i_j} \rangle] \tag{41}$$

as[3] $[A(t)]$; in the second stage, the functional F_3 of Eq. (41) decays to its equilibrium value

$$F_3[\langle \sigma_{i_j}\sigma_{i_k} \rangle^{(0)}, \langle \sigma_{i_j} \rangle] \rightarrow F_3[\langle \sigma_{i_j}\sigma_{i_k} \rangle^{(0)}, \langle \sigma_{i_j} \rangle^{(0)}] = 0 \tag{42}$$

as $[A(t)]$. Overall, $\langle \sigma_{i_1}\sigma_{i_2}\sigma_{i_3} \rangle$ decays to its zero equilibrium value as $[A(t)]$.

In a completely analogous manner, we can write the four-spin correlation function in terms of the C-functions as

$$\langle \sigma_{i_1}\sigma_{i_2}\sigma_{i_3}\sigma_{i_4} \rangle = C_4(i_1, i_2, i_3, i_4; t) + C_2(i_1, i_2; t)\, C_2(i_3, i_4; t) + C_2(i_1, i_4; t)\, C_2(i_2, i_3; t) - C_2(i_1, i_3; t)\, C_2(i_2, i_4; t) \tag{43}$$

[3] We shall frequently neglect the slowly varying time factors of the form t^{-n} in front of the $[A(t)] \equiv [(2\pi\gamma\alpha t)^{-1/2}\, e^{-\alpha(1-\gamma)t}]$ when discussing the asymptotic behavior of various functions.

The asymptotic form of $\langle\sigma_{i_1}\sigma_{i_2}\sigma_{i_3}\sigma_{i_4}\rangle$ is then found to be

$$\langle\sigma_{i_1}\sigma_{i_2}\sigma_{i_3}\sigma_{i_4}\rangle = dt^{-3}[A(t)]^4 + et^{-2}[A(t)]^4 + ft^{-1}[A(t)]^2 + \langle\sigma_{i_1}\sigma_{i_2}\rangle^{(0)}\langle\sigma_{i_3}\sigma_{i_4}\rangle^{(0)} \quad (44)$$

where $d = K_4(i_1, i_2, i_3, i_4)$,

$$e = k_2(i_1, i_2)\,k_2(i_3, i_4) + k_2(i_1, i_4)\,k_2(i_2, i_3) - k_2(i_1, i_3)\,k_2(i_2, i_4)$$

and

$$f = k_2(i_1, i_2)\langle\sigma_{i_3}\sigma_{i_4}\rangle^{(0)} + k_2(i_3, i_4)\langle\sigma_{i_1}\sigma_{i_2}\rangle^{(0)} + k_2(i_1, i_4)\langle\sigma_{i_2}\sigma_{i_3}\rangle^{(0)}$$
$$+ k_2(i_2, i_3)\langle\sigma_{i_1}\sigma_{i_4}\rangle^{(0)} - k_2(i_1, i_3)\langle\sigma_{i_2}\sigma_{i_4}\rangle^{(0)} - k_2(i_2, i_4)\langle\sigma_{i_1}\sigma_{i_3}\rangle^{(0)}$$

Again we have used only the leading asymptotic terms of each term on the r.h.s. of Eq. (43). The relaxation of the four-spin correlation function also proceeds in two stages: in the first stage, $\langle\sigma_{i_1}\sigma_{i_2}\sigma_{i_3}\sigma_{i_4}\rangle$ becomes a function of the two-particle correlation functions,

$$\langle\sigma_{i_1}\sigma_{i_2}\sigma_{i_3}\sigma_{i_4}\rangle \rightarrow F_4[\langle\sigma_{i_j}\sigma_{i_k}\rangle] \quad (45)$$

as $[A(t)]^4$; in the second stage, the functional F_4 of Eq. (45) decays to its equilibrium value

$$F_4[\langle\sigma_{i_j}\sigma_{i_k}\rangle] \rightarrow F_4[\langle\sigma_{i_j}\sigma_{i_k}\rangle^{(0)}] = \langle\sigma_{i_1}\sigma_{i_2}\rangle^{(0)}\langle\sigma_{i_3}\sigma_{i_4}\rangle^{(0)} \quad (46)$$

as $[A(t)]^2$ with $\langle\sigma_{i_j}\sigma_{i_k}\rangle^{(0)}$ given by Eq. (11). The overall relaxation of $\langle\sigma_{i_1}\sigma_{i_2}\sigma_{i_3}\sigma_{i_4}\rangle$ to its equilibrium value $\langle\sigma_{i_1}\sigma_{i_2}\rangle^{(0)}\langle\sigma_{i_3}\sigma_{i_4}\rangle^{(0)}$ goes as $[A(t)]^2$.

The asymptotic behavior of $\langle\sigma_{i_1}\cdots\sigma_{i_n}\rangle$ depends upon whether n is even or odd. For odd n, $n > 3$, we find that in the first stage of the relaxation

$$\langle\sigma_{i_1}\sigma_{i_2}\cdots\sigma_{i_n}\rangle \rightarrow F_n[\langle\sigma^{(n-2)}\rangle] \quad (47)$$

as $[A(t)]^n$. The overall relaxation to the zero equilibrium value goes as $[A(t)]$. If n is even, $n > 2$, we find that in the first stage of the relaxation

$$\langle\sigma_{i_1}\sigma_{i_2}\cdots\sigma_{i_n}\rangle \rightarrow F_n[\langle\sigma^{(n-2)}\rangle] \quad (48)$$

as $[A(t)]^n$. The overall relaxation to the equilibrium value

$$\langle\sigma_{i_1}\sigma_{i_2}\cdots\sigma_{i_n}\rangle^{(0)} = \prod_{j=1}^{n/2}\langle\sigma_{i_{2j-1}}\sigma_{i_{2j}}\rangle^{(0)}$$

goes as $[A(t)]^2$.

We shall now discuss the time dependence of the cumulants.[8] The first-order cumulant, defined by

$$\langle\sigma_i\rangle_c \equiv \langle\sigma_i\rangle \quad (49)$$

has the asymptotic time behavior

$$\langle\sigma_i\rangle_c \sim k_1(i)[A(t)] \quad (50)$$

The second-order cumulant, defined by

$$\langle\sigma_{i_1}\sigma_{i_2}\rangle_c \equiv \langle\sigma_{i_1}\sigma_{i_2}\rangle - \langle\sigma_{i_1}\rangle\langle\sigma_{i_2}\rangle \tag{51}$$

has the asymptotic time behavior

$$\langle\sigma_{i_1}\sigma_{i_2}\rangle_c \sim \langle\sigma_{i_1}\sigma_{i_2}\rangle^{(0)} + \{k_2(i_1, i_2)\, t^{-1} - k_1(i_1)\, k_1(i_2)\}[A(t)]^2 \tag{52}$$

The third-order cumulant is given by

$$\begin{aligned}\langle\sigma_{i_1}\sigma_{i_2}\sigma_{i_3}\rangle_c &\equiv \langle\sigma_{i_1}\sigma_{i_2}\sigma_{i_3}\rangle - \langle\sigma_{i_1}\sigma_{i_2}\rangle\langle\sigma_{i_3}\rangle - \langle\sigma_{i_2}\sigma_{i_3}\rangle\langle\sigma_{i_1}\rangle \\ &\quad - \langle\sigma_{i_1}\sigma_{i_3}\rangle\langle\sigma_{i_2}\rangle + 2\langle\sigma_{i_1}\rangle\langle\sigma_{i_2}\rangle\langle\sigma_{i_3}\rangle \\ &= C_3(i_1, i_2, i_3; t) - 2\langle\sigma_{i_2}\rangle\langle\sigma_{i_1}\sigma_{i_3}\rangle + 2\langle\sigma_{i_1}\rangle\langle\sigma_{i_2}\rangle\langle\sigma_{i_3}\rangle\end{aligned} \tag{53}$$

Using some of our previous results, we find for the asymptotic time behavior of $\langle\sigma_{i_1}\sigma_{i_2}\sigma_{i_3}\rangle_c$

$$\langle\sigma_{i_1}\sigma_{i_2}\sigma_{i_3}\rangle_c \sim -2\langle\sigma_{i_1}\sigma_{i_3}\rangle^{(0)}\, k_1(i_2)[A(t)] \tag{54}$$

For the asymptotic behavior of the nth-order cumulant, we find

$$\langle\sigma_{i_1}\sigma_{i_2}\cdots\sigma_{i_n}\rangle_c \sim \langle\sigma_{i_1}\sigma_{i_2}\cdots\sigma_{i_n}\rangle_c^{(0)} + k[A(t)] \tag{55}$$

We note that the asymptotic time dependence of the cumulants is quite different from that of the C-functions. In fact, the cumulants relax to their equilibrium values even slower than the correlation functions for all n, $n > 1$. Because of this property, their application to the Ising spin model can give rise to incorrect deductions about the relaxation of the n-spin correlation functions.

4. RELAXATION OF THE Q-FUNCTIONS AND THE PROBABILITY DISTRIBUTIONS

We now wish to study the time-dependent behavior of the n-spin distribution function $P_n(\sigma^{(n)}; t)$. In order to do so, it is useful to define a function $Q_n(\sigma^{(n)}; t)$ which, for the Ising spin model considered here, is a convenient function for studying the relaxation of $P_n(\sigma^{(n)}; t)$. It is used here in the same fashion that the Ursell function $U_n(x^{(n)}; t)$ was used in papers I and II.

We define $Q_n(\sigma^{(n)}; t)$, for $n \geqslant 1$, by

$$Q_n(i_1, i_2, \ldots, i_n; t) \equiv 2^{-n}\sigma_{i_1}\sigma_{i_2}\cdots\sigma_{i_n}C_n(i_1, i_2, \ldots, i_n; t) \tag{56}$$

where $i_1 \leqslant i_2 \leqslant \cdots \leqslant i_n$. The properties of this function are:

(a) $Q_n(\sigma^{(n)}; t)$ satisfies the same differential equation, Eq. (16), as the n-spin correlation function.

(b) $Q_n(\sigma^{(n)}; t)$ is zero for $n > 2$ when two adjacent spin indices are equal. This follows from Eq. (29).

(c)
$$Q_2^{(0)}(i_1, i_2) = \tfrac{1}{4}\sigma_{i_1}\sigma_{i_2}\eta^{i_1-i_2}$$
$$Q_n^{(0)}(\sigma^{(n)}) = 0 \qquad \text{for all} \quad n, \quad n \neq 2 \tag{57}$$

This follows from Eqs. (10), (11), and (30).

(d) For $n > 2$, $Q_n(\sigma^{(n)}; t) = 0$ if two adjacent spins i_l and i_{l+1} are uncorrelated with the rest of the spin variables $i_1, i_2, \ldots, i_{l-1}, i_{l+2}, \ldots, i_n$. This property, which follows from Eq. (37), is analogous to an important property of the Ursell function discussed in I and II.

(e)
$$\sum_{\sigma_{i_j}} Q_n(\sigma^{(n)}; t) = 0, \qquad 1 \leqslant j \leqslant n \tag{58}$$

This follows immediately from the definition in Eq. (56) and the fact that the spin variables σ_{i_j} have the two values ± 1. This is another important property which is also possessed by the Ursell functions.

The asymptotic properties of $Q_n(\sigma^{(n)}; t)$ can readily be obtained from definition (56) and Eqs. (24), (25), and (32). They are

$$Q_1(i) \sim \tfrac{1}{2}\sigma_i k_1(i)[A(t)] \tag{59}$$
$$Q_2(i_1, i_2) \sim \tfrac{1}{4}\sigma_{i_1}\sigma_{i_2}[\langle\sigma_{i_1}\sigma_{i_2}\rangle^{(0)} + k_2(i_1, i_2)\, t^{-1}[A(t)]^2] \tag{60}$$
$$Q_n(\sigma^{(n)}) \sim 2^{-n}\sigma_{i_1}\sigma_{i_2}\cdots\sigma_{i_n}K_n(\sigma^{(n)})\, t^{(1-n)}[A(t)]^n, \qquad n > 2 \tag{61}$$

The n-spin probability distribution $P_n(\sigma^{(n)}; t)$ can be expressed in term of the Q-function as

$$P_n(\sigma^{(n)}; t) = \sum_{n'=0}^{n} 2^{n'-n} \sum_{\xi} (-1)^{\mathscr{P}}\, \mathscr{P} Q_{n_1}(i_1, i_2, \ldots, i_{n_1}; t) \cdots Q_{n_k}(i_{n'-n_k+1}, \ldots, i_{n'}; t) \tag{62}$$

where the notation is the same as in Eq. (33). The convention $Q_0 = 1$ is used. Equation (62) can be obtained from the definition (56) for the Q-function, the definition (33) for the C-function, and Eq. (5), which relates the $P_n(\sigma^{(n)}; t)$ to the spin correlation functions. The first few expressions for $P_n(\sigma^{(n)}; t)$ are

$$P_1(\sigma_i; t) = Q_1(i; t) + \tfrac{1}{2}$$
$$P_2(\sigma_{i_1}, \sigma_{i_2}; t) = Q_2(i_1, i_2; t) + \tfrac{1}{2}Q_1(i_1; t) + \tfrac{1}{2}Q_1(i_2; t) + \tfrac{1}{4}$$
$$\begin{aligned} P_3(\sigma_{i_1}, \sigma_{i_2}, \sigma_{i_3}; t) = {} & Q_3(i_1, i_2, i_3; t) + Q_1(i_1; t)\, Q_2(i_2, i_3; t) \\ & + Q_1(i_3; t)\, Q_2(i_1, i_2; t) - Q_1(i_2; t)\, Q_2(i_1, i_3; t) \\ & + \tfrac{1}{2}[Q_2(i_1, i_2; t) + Q_2(i_1, i_3; t) + Q_2(i_2, i_3; t)] \\ & + \tfrac{1}{4}[Q_1(i_1; t) + Q_1(i_2; t) + Q_1(i_3; t)] + \tfrac{1}{8} \end{aligned} \tag{63}$$

We have not succeeded in finding the analytical inversion of Eq. (62) to obtain a general expression for Q_n in terms of the P_n. We will, however, display here the first few explicit forms of Q_n in terms of the P_n:

$$
\begin{aligned}
Q_1(i; t) &= P_1(\sigma_i\,; t) - \tfrac{1}{2} \\
Q_2(i_1\,, i_2\,; t) &= P_2(\sigma_{i_1}, \sigma_{i_2}\,; t) - \tfrac{1}{2}[P_1(\sigma_{i_1}\,; t) + P_1(\sigma_{i_2}\,; t)] + \tfrac{1}{4} \\
Q_3(i_1\,, i_2\,, i_3\,; t) &= P_3(\sigma_{i_1}, \sigma_{i_2}, \sigma_{i_3}\,; t) - P_1(\sigma_{i_1}\,; t)\, P_2(\sigma_{i_2}, \sigma_{i_3}\,; t) \\
&\quad - P_1(\sigma_{i_3}\,; t)\, P_2(\sigma_{i_1}, \sigma_{i_2}\,; t) + P_1(\sigma_{i_2}\,; t)\, P_2(\sigma_{i_1}, \sigma_{i_3}\,; t) \\
&\quad - P_2(\sigma_{i_1}, \sigma_{i_3}\,; t) + P_1(\sigma_{i_1}\,; t)\, P_1(\sigma_{i_3}\,; t)
\end{aligned}
\tag{64}
$$

We now discuss the asymptotic relaxation of the n-spin distribution functions $P_n(\sigma^{(n)}; t)$. This discussion can be based either on the relaxation of the n-spin correlation functions or the relaxation of the Q_n functions. It follows from Eqs. (59)–(63) that

$$P_1(\sigma_i\,; t) \sim P_1^{(0)}(\sigma_i) + \tfrac{1}{2}\sigma_i k_1(i)[A(t)] \tag{65}$$

$$
\begin{aligned}
P_2(\sigma_{i_1}, \sigma_{i_2}\,; t) \sim{}& P_2^{(0)}(\sigma_{i_1}, \sigma_{i_2}) + \tfrac{1}{4}\sigma_{i_1}\sigma_{i_2} k_2(i_1\,, i_2)\, t^{-1}[A(t)]^2 \\
&+ \tfrac{1}{4}\{\sigma_{i_1} k_1(i_1) + \sigma_{i_2} k_1(i_2)\}[A(t)]
\end{aligned}
\tag{66}
$$

$$
\begin{aligned}
P_3(\sigma_{i_1}, \sigma_{i_2}, \sigma_{i_3}\,; t) \sim{}& P_3^{(0)}(\sigma_{i_1}, \sigma_{i_2}, \sigma_{i_3}) + \tfrac{1}{8}\sigma_{i_1}\sigma_{i_2}\sigma_{i_3} K_3(i_1\,, i_2\,, i_3)\, t^{-2}[A(t)]^3 \\
&+ \tfrac{1}{8}\sigma_{i_1}\sigma_{i_2}\sigma_{i_3}\{k_2(i_2\,, i_3)\, k_1(i_1) + k_2(i_1\,, i_2)\, k_1(i_3) \\
&- k_2(i_1\,, i_3)\, k_1(i_2)\}\, t^{-1}[A(t)]^3 \\
&+ \tfrac{1}{8}\sigma_{i_1}\sigma_{i_2}\sigma_{i_3}\{\langle\sigma_{i_2}\sigma_{i_3}\rangle^{(0)}\, k_1(i_1) + \langle\sigma_{i_1}\sigma_{i_2}\rangle^{(0)}\, k_1(i_3) \\
&- \langle\sigma_{i_1}\sigma_{i_3}\rangle^{(0)}\, k_1(i_2)\}[A(t)] \\
&+ \tfrac{1}{8}\,\{\sigma_{i_1}\sigma_{i_2} k_2(i_1\,, i_2) + \sigma_{i_1}\sigma_{i_3} k_2(i_1\,, i_3) + \sigma_{i_2}\sigma_{i_3} k_2(i_2\,, i_3)\}\, t^{-1}[A(t)]^2 \\
&+ \tfrac{1}{8}\{\sigma_{i_1} k_1(i_1) + \sigma_{i_2} k_1(i_2) + \sigma_{i_3} k_1(i_3)\}[A(t)]
\end{aligned}
\tag{67}
$$

where we have used the leading asymptotic term of each term on the r.h.s. of Eq. (63). The relaxation of $P_1(\sigma_i; t)$ to its equilibrium value $P_1^{(0)}(\sigma_i)$ proceeds in one stage,

$$P_1(\sigma_i\,; t) \to P_1^{(0)}(\sigma_i) \tag{68}$$

as $[A(t)]$. The relaxation of $P_2(\sigma_{i_1}, \sigma_{i_2}; t)$ proceeds in two stages[4] in the first stage:

$$P_2(\sigma_{i_1}, \sigma_{i_2}\,; t) \to G_2[P_2^{(0)}, P_1] \tag{69}$$

as $[A(t)]^2$, where G_2 is a functional of the equilibrium two-spin distribution function and the time-dependent one-spin distribution function; in the second stage,

$$G_2[P_2^{(0)}, P_1] \to G_2[P_2^{(0)}, P_1^{(0)}] = P_2^{(0)}(\sigma_{i_1}, \sigma_{i_2}) \tag{70}$$

[4] See footnote 3.

as $[A(t)]$. The overall relaxation to the equilibrium distribution function thus proceeds as $[A(t)]$. The relaxation of $P_3(\sigma_{i_1}, \sigma_{i_2}, \sigma_{i_3}; t)$ proceeds in three stages. In the first stage,

$$P_3(\sigma_{i_1}, \sigma_{i_2}, \sigma_{i_3}; t) \to G_3[P_2, P_1] \tag{71}$$

as $[A(t)]^3$. In the second stage,

$$G_3[P_2, P_1] \to G_3[P_2^{(0)}, P_1] \tag{72}$$

as $[A(t)]^2$. In the third stage,

$$G_3[P_2^{(0)}, P_1] \to G_3[P_2^{(0)}, P_1^{(0)}] = 2P_2^{(0)}(\sigma_{i_1}, \sigma_{i_2})\, P_2^{(0)}(\sigma_{i_2}, \sigma_{i_3}) \tag{73}$$

as $[A(t)]$. The overall relaxation to the factorized equilibrium distribution function, Eq. (13), again proceeds as $[A(t)]$. The asymptotic properties of $P_n(\sigma^{(n)}; t)$, $n > 3$, are most easily obtained from the relation between the P_n and the spin correlation functions, Eq. (5). It follows from Eqs. (47) and (48) that in the first stage

$$P_n(\sigma^{(n)}; t) \to G_n[P_{n-1}], \qquad n > 3 \tag{74}$$

as $[A(t)]^n$. In the last stage,

$$G_n[P_2^{(0)}, P_1] \to G_n[P_2^{(0)}, P_1^{(0)}] = P_n^{(0)}(\sigma^{(n)}) \tag{75}$$

as $[A(t)]$, where $P_n^{(0)}(\sigma^{(n)})$ is given by Eq. (13). It is evident from the above analysis that the n-spin distribution function decays very rapidly to a functional of lower-order distribution functions, with the slowest stage of the relaxation being the relaxation of the one-spin distribution function $P_1(\sigma_{i_1}; t)$ to its equilibrium value $P_1^{(0)}(\sigma_i)$.

It can readily be verified from the definition of the Ursell function given in papers I and II that the Ursell function $U_n(\sigma^{(n)}; t)$ for $n > 2$ does not decay to its equilibrium value any faster than the n-spin distribution functions. Thus, for instance,

$$U_3(\sigma_{i_1}, \sigma_{i_2}, \sigma_{i_3}; t) \to U_3^{(0)}(\sigma_{i_1}, \sigma_{i_2}, \sigma_{i_3}) = 0 \tag{76}$$

as $[A(t)]$. It is this undesirable property of the Ursell function that led us to develop the Q-functions in this section.

5. EXAMPLES

5.1. Relaxation of Spin Functions from Lattice Temperature T_0 to T

It is of interest to study the relaxation of the spin functions when the lattice is subjected to a sudden change in temperature from T_0 to T. The spin system is assumed to be in equilibrium with the heat bath at temperature T_0 at time $t \leqslant 0$. At time $t = 0$, the temperature of the heat bath is suddenly changed to T.

For $t \leqslant 0$, the C-functions are equal to their equilibrium values at temperature T_0,

$$C_n(i_1, i_2, \ldots, i_n; 0) = 0 \qquad \text{for} \quad n \neq 2$$
$$= \eta_0^{i_2 - i_1} \qquad \text{for} \quad n = 2 \tag{77}$$

where $\eta_0 = \tanh(J/kT_0)$. The time dependence of the C_n-functions is given by Eq. (31). It follows immediately that

$$C_n(i_1, i_2, \ldots, i_n; t) = 0 \qquad \text{for} \quad n \neq 2 \tag{78}$$

for all times t. For $n = 2$, it follows from Eq. (21) that

$$C_2(i_1, i_2; t) \equiv \langle \sigma_{i_1} \sigma_{i_2} \rangle = \eta^{i_2 - i_1} + e^{-2\alpha t} \sum_{\substack{m_1 < m_2 \\ -\infty}}^{\infty} (\eta_0^{m_2 - m_1} - \eta^{m_2 - m_1})$$
$$\times \{I_{i_1 - m_1}(\gamma\alpha t)\, I_{i_2 - m_2}(\gamma\alpha t) - I_{i_1 - m_2}(\gamma\alpha t)\, I_{i_2 - m_1}(\gamma\alpha t)\} \tag{79}$$

where $\eta = \tanh(J/kT)$ and $\gamma = \tanh(2J/kT)$. Setting $i_1 = i$, $i_2 = j + i_1$, $m_1 = m$, and $m_2 = m + n$ yields

$$C_2(i, i + j; t) = \eta^j + e^{-2\alpha t} \sum_{n=0}^{\infty} (\eta_0{}^n - \eta^n)\{I_{j-n}(2\gamma\alpha t) - I_{j+n}(2\gamma\alpha t)\} \tag{80}$$

were we have used the relation

$$I_k(2x) = \sum_{m=-\infty}^{\infty} I_{k+m}(x)\, I_m(x) \tag{81}$$

Since the sum in Eq. (80) involves an infinite number of nonzero terms, we must perform the asymptotic analysis in a somewhat different manner from that employed in the preceding sections. Substitution of the identity[6]

$$I_k(z) = (1/2\pi) \int_{-\pi}^{\pi} e^{z\cos\theta} e^{-ik\theta}\, d\theta \tag{82}$$

into Eq. (80) yields

$$C_2(i, i + j; t) = \eta^j + (2/\pi)\, e^{-2\alpha t} \int_0^{\pi} e^{2\alpha\gamma t\cos\theta} \sin j\theta \sin\theta$$
$$\times [(\eta_0 + 1/\eta_0 - 2\cos\theta)^{-1} - (\eta + 1/\eta - 2\cos\theta)^{-1}\, d\theta] \tag{83}$$

For $t \gg j/2\alpha\gamma$, the main contributions of the integral will be in the neigborhood of $\theta = 0$. The asymptotic form of Eq. (83) then becomes

$$C_2(i, i + j; t) \equiv \langle \sigma_i \sigma_{i+j} \rangle$$
$$\sim \eta^j + j\left(\frac{\pi}{\alpha\gamma}\right)^{1/2} \left[\frac{\eta_0}{(1 - \eta_0)^2} - \frac{\eta}{(1 - \eta)^2}\right] t^{-1/2}[A(t)]^2 \tag{84}$$

An inspection of Eq. (84) shows that the two-spin correlation function relaxes to its equilibrium values $\langle\sigma_i\sigma_{i+j}\rangle^{(0)} = \eta^j$ by a factor $t^{-1/2}$ slower than shown in the result obtained in Eq. (25). This difference is due to the fact that in the example studied here, $C_2(i_1, i_2; 0)$ differs from the equilibrium value $C_2^{(0)}(i_1, i_2)$ for *all* values of i_1 and i_2.

From the above analysis and Eq. (35) we find that:

$$\begin{aligned} &\text{for } n \text{ odd:} \quad \langle\sigma_{i_1}\sigma_{i_2}\cdots\sigma_{i_n}\rangle = 0 \\ &\text{for } n \text{ even:} \quad \langle\sigma_{i_1}\sigma_{i_2}\cdots\sigma_{i_n}\rangle = \sum_{\mathscr{P}}(-1)^{\mathscr{P}}\mathscr{P}C_2(i_1, i_2; t)\cdots C_2(i_{n-1}, i_n; t) \end{aligned} \tag{85}$$

The sum in Eq. (85) is over all permutations with the restriction that no two terms in the sum are the same and that the indices in each C_2 are ordered. Thus, the odd-order spin correlation functions retain their zero equilibrium form at all times, while the even-order spin correlation functions relax to their equilibrium value

$$\langle\sigma_{i_1}\sigma_{i_2}\cdots\sigma_{i_n}\rangle^{(0)} = \langle\sigma_{i_1}\sigma_{i_2}\rangle^{(0)}\langle\sigma_{i_3}\sigma_{i_4}\rangle^{(0)}\cdots\langle\sigma_{i_{n-1}}\sigma_{i_n}\rangle^{(0)}$$

as $t^{-1/2}[A(t)]^2$. The explicit coefficients for this relaxation can be obtained by substituting the result of Eq. (84) into Eq. (85).

The initial time behavior of C_2 is

$$C_2(i, i+j; t) \equiv \langle\sigma_i\sigma_{i+j}\rangle = \eta_0{}^j + 2\alpha t[G(\eta, \gamma)] + O(t^2) \tag{86}$$

with

$$G(\eta, \gamma) = \tfrac{1}{2}[(\eta_0^{i-1} - \eta^{i-1}) + (\eta_0^{j+1} - \eta^{j+1})] - (\eta^i - \eta_0{}^j) \tag{87}$$

which can readily be found by developing the exponentials in Eq. (83) in a Taylor series around $t = 0$. The correlation between two spins thus grows (or decays) linearly with time for $t \ll j/2\alpha\gamma$.

It is interesting to note from the analysis given below that the Ising spin system considered here is not canonically invariant. A system is called "canonically invariant" if it relaxes from an initial canonical distribution to its final canonical distribution via a continuous (in time) sequence of canonical distributions.[9] It is only for canonically invariant systems that a temperature can be defined exactly for the relaxing system. The results found here for the Glauber Ising spin system and by Anderson *et al.*[9] for noncorrelated spins in contact with a heat bath indicate that the widely used practiec of characterizing relaxing spin systems by a "spin temperature" needs to be reexamined in more detail.

If the spin system is to be canonically invariant, it is clear from the initial and final equilibrium forms of C_2, i.e., $C_2(i, i+j; 0) = \eta_0{}^j$ and $C_2^{(0)}(i, i+j) = \eta^j$, that C_2 must be of the form

$$C_2(i, i+j; t) \equiv \langle\sigma_i\sigma_{i+j}\rangle = \eta^j(t) \tag{88}$$

with

$$\eta(t) = \tanh[J/kT(t)] \tag{89}$$

where $T(t)$ is the time-dependent spin temperature. Let us now check whether the form (88) is a solution of the differential equation (19) for the two-spin correlation function. This yields

$$j \frac{d}{dt} \eta(t) = -2\alpha\eta(t) + \alpha\gamma[1 + \eta^2(t)] \tag{90}$$

Since this differential equation has no solution that is independent of j, except for the equilibrium solution at $t = \infty$, and since, according to Eq. (89), $\eta(t)$ must be independent of j, we have shown that the Ising spin system is not canonically invariant and thus cannot be described in terms of a "spin temperature."

5.2. Relaxation of an Initial Spin Fluctuation from Equilibrium

It is of interest to see how a local fluctuation from equilibrium relaxes to the final equilibrium state. We consider an initial state where all the C_n have their equilibrium values except for $C_1(0; 0)$ which we set equal to Δ, i.e.,

$$\begin{aligned} C_n(i_1, \ldots, i_n; 0) &= C_n^{(0)} && \text{for} \quad n > 1 \\ C_1(i; 0) &= \langle \sigma_i \rangle_0 = C_1^{(0)}(i) = 0 && \text{for} \quad i \neq 0 \\ C_1(0; 0) &= \langle \sigma_0 \rangle_0 = \Delta \end{aligned} \tag{91}$$

The time dependence of the C-functions is given by Eqs. (20), (21), and (31). It follows that

$$\begin{aligned} C_n(i_1, \ldots, i_n; t) &= C_n^{(0)} && \text{for} \quad n > 1 \\ C_1(i; t) &\equiv \langle \sigma_i \rangle = \Delta e^{-\alpha t} I_i(\gamma\alpha t) && \text{for all} \quad i \end{aligned} \tag{92}$$

Hence, for $t \ll |i|/\alpha\gamma$, the initial behavior as given by Eq. (22) is

$$\langle \sigma_i \rangle \approx \frac{\Delta}{|i|!} \left(\frac{\gamma\alpha t}{2}\right)^{|i|} e^{-\alpha t} \tag{93}$$

where we note that $I_n = I_{-n}$. For $|i| \gg 1$, $\langle \sigma_i \rangle$ reaches a maximum for

$$t \approx (|i|/\alpha)(1 - \gamma^2)^{1/2}.$$

For long times, the asymptotic behavior is given by Eq. (23),

$$\langle \sigma_i \rangle \sim \Delta(2\pi\gamma\alpha t)^{1/2} e^{-\alpha(1-1)t} = \Delta[A(t)] \tag{94}$$

This is in agreement with the general result obtained in Eq. (24), with $k_1(i) = \Delta$. It will be noted that $\langle \sigma_i \rangle$ for large t is independent of the distance i of the spin from the local disturbance at lattice site zero if only the constant term is retained in the expansion of the Bessel function.

The nth-order correlation function can be calculated using Eq. (35). This yields

$$\begin{aligned}\langle \sigma_{i_1} \cdots \sigma_{i_n} \rangle &= 0 && \text{for } n \text{ even} \\ &= \sum_{\mathscr{P}} (-1)^{\mathscr{P}} \mathscr{P} \langle \sigma_{i_1}\sigma_{i_2} \rangle^{(0)} \langle \sigma_{i_3}\sigma_{i_4} \rangle^{0} \cdots \langle \sigma_{i_{n-2}}\sigma_{i_{n-}} \rangle^{(0)} \langle \sigma_{i_n} \rangle && \text{for } n \text{ odd}\end{aligned} \tag{95}$$

where the sum is over all permutations, with the restriction that no two terms in the sum are the same and that the indices in each $\langle \sigma_i \sigma_j \rangle$ are ordered. Hence, the nth-order correlation function (for n is odd) decays to the zero equilibrium value as $[A(t)]$, which is in agreement with the general result stated below Eq. (47).

APPENDIX. Bounds on the Relaxation of C_n

We present here a simple argument for obtaining the upper and lower bounds for the time dependence of the C_n-functions. We define $u_n(t)$ to be equal to the *maximum value* of $C_n - C_n^{(0)}$ at time t. Since $C_n - C_n^{(0)}$ is identically equal to zero when two spin indices are equal, $u_n \geqslant 0$. The time dependence of $u_n(t)$ can be obtained from Eq. (28). The time derivative fulfills

$$du_n(t)/dt \leqslant -n\alpha(1 - \gamma)\, u_n(t) \tag{A.1}$$

Equation (A.1) is easily solved to yield

$$u_n(t) \leqslant u_n(0)\, e^{-n\alpha(1-\gamma)t} \tag{A.2}$$

The function $u_n(0)\, e^{-n\alpha(1-\gamma)t}$ provides an upper limit to the value of C_n at time t.

We define $v(t)$ to be equal to the *minimum value* of $C_n - C_n^{(0)}$ at time t. The time dependence of $v_n(t)$ can be obtained from Eq. (28). The time derivative fulfills

$$dv_n(t)/dt \geqslant -n\alpha(1 - \gamma)\, v_n(t) \tag{A.3}$$

with the solution

$$v_n(t) \geqslant v_n(0)\, e^{-n\alpha(1-\gamma)t} \tag{A.4}$$

The function $v_n(0)\, e^{-n\alpha(-\gamma)t}$ provides a lower limit to the value of C_n at time t.

It is clear that C_n at all times t must lie between the values of the functions on the r.h.s. of Eqs. (A.2) and (A.4). Thus, asymptotically. the function C_n must go to zero at least as fast as $e^{-n\alpha(1-\gamma)t}$. This argument, of course, only provides bounds for the asymptotic time dependence of C_n and cannot be expected to reproduce the pre-exponential time factors obtained in the bodu of the paper.

ACKNOWLEDGMENT

We wish to express our thanks to Dr. George Weiss for several helpful discussions during the writing of this paper.

REFERENCES

1. I. Oppenheim, K. E. Shuler, and G. H. Weiss, *J. Chem. Phys.* **46**:4100 (1967).
2. I. Oppenheim, K. E. Shuler, and G.H. Weiss, *J. Chem. Phys.* **50**:3662 (1969).

3. R. Glauber, *J. Math. Phys.* **4**:294 (1963).
4. G. H. Weiss, in: *Advances in Chemical Physics*, K. E. Shuler, ed. (1969), Vol. 15, p. 199.
5. N. Matsudaira, *Can. J. Phys.* **45**:2091 (1967); *J. Phys. Soc. Japan* **23**:232 (1967); K. Kawasaki and T. Yamada, *Progr. Theoret. Phys. (Japan)* **39**:1 (1968).
6. M. Abramowitz and I. Stegun (eds.), *Handbook of Mathematical Functions*, National Bureau of Standards, AMS 55, U.S. Government Printing Office, Washington, D.C. (1964), p. 374.
7. E. W. Montroll, *J. Math. Phys.* **25**:34 (1946).
8. B. V. Gnedenko, *The Theory of Probability*, Chelsea Publ. Co., New York (1963), Ch. VII.
9. H. C. Anderson, I. Oppenheim, K. E. Shuler, and G. H. Weiss, *J. Math. Phys.* **5**:522 (1964).

8) Dynamics of Chain Molecules

Recent years have seen an interest in the application of the theory of stochastic processes, and in particular the application of the master equation to the study of time dependent conformational properties of polymers. The first paper contains a discussion of the kinetics of a sequence of first order reactions which has application in the description of the kinetics of the helix-coil transition in polypeptides. The second and third paper use stochastic models to study chain dynamics. These models include a freely jointed chain in any number of dimensions and a one-dimensional chain with nearest neighbor correlations, which is equivalent to the Glauber spin model (see paper 1 of section 7). The third paper removes the restriction to free joints as treated in the second paper.

Kinetics of a Sequence of First-Order Reactions

Barry Ninham,* Ralph Nossal, and Robert Zwanzig†

Physical Sciences Laboratory, Division of Computer Research and Technology, National Institutes of Health, Department of Health, Education, and Welfare, Bethesda, Maryland 20014

(Received 21 May 1969)

Solutions are obtained for the finite set of coupled rate equations $\partial C_i/\partial t=\alpha_{i,i-1}C_{i-1}+\alpha_{i,i}C_i+\alpha_{i,i+1}C_{i+1}$ ($i=0, \cdots, N$), where $\alpha_{i,j}$ are given in general as $\alpha_{i,i-1}=A$, $\alpha_{i,i+1}=B$, $\alpha_{i,i}=-(A+B)$, except that $\alpha_{0,0}=-\alpha_{1,0}=-a$, $\alpha_{NN}=-\alpha_{N-1,N}=-b$, $\alpha_{0,-1}=\alpha_{N,N+1}=0$. Asymptotic expressions are given for the approach to equilibrium as a function of the various rate parameters and the chain length N. For large N, we find that if $A<B$, the eigenvalue spectrum approaches a continuum, and the approach to equilibrium is described by a simple relaxation time $\lambda_1\simeq(A^{1/2}-B^{1/2})^2$. However, if $A(1-a/A)^2>B$, the system exhibits a peculiar eigenvalue spectrum, and the relaxation is characterized by two distinct and well-separated relaxation times, λ_1 and $\lambda_2=-a\{1-B[A(1-a/A)^2]^{-1}\}$.

I. INTRODUCTION

The system of equations,

$$\dot{C}_0=-aC_0+BC_1,$$
$$\dot{C}_1=aC_0-(A+B)C_1+BC_2,$$
$$\dot{C}_j=AC_{j-1}-(A+B)C_j+BC_{j+1},\qquad j=2,3,\cdots,N-1,$$
$$\dot{C}_{N-1}=AC_{N-2}-(A+B)C_{N-1}+bC_N,$$
$$\dot{C}_N=AC_{N-1}-bC_N,\tag{1}$$

arise in a number of applications of mathematics and mathematical physics. Special cases of these equations have already received considerable attention. For example, when $a=A$, $b=B$, and N is infinite, the system appears in the theory of queues.[1,2] If, further, $A=B$ and periodic boundary conditions are imposed, the equations describe a one-dimensional symmetric continuous-time random walk on a circle. In the latter form, the equations have been studied by Montroll[3] and Glauber[4] in connection with the relaxation of a one-dimensional Ising model and by Goel[5] in describing the denaturation of DNA.

When $A=B$ and the rate constants a and b in the equations for $\dot{C}_1$ and $\dot{C}_{N-1}$ are replaced by A, the eigenvalues of the system are identical with those of a one-dimensional system of coupled harmonic oscillators with either free or fixed ends. This problem has been solved by Louck,[6] and the associated dynamical properties of a crystal with free ends have been investigated by Clem and Godwin.[7] Some related continuant determinants which arise in both one- and two-dimensional problems have been evaluated by Rutherford.[8]

* Permanent address: Department of Applied Mathematics, University of New South Wales, Kensington, N. S. W., Australia.

† Permanent address: Institute for Fluid Dynamics, University of Maryland, College Park, Md.

[1] N. U. Prabhu, *Queues and Inventories* (John Wiley & Sons, Inc., N. Y., 1965).

[2] A. Wragg, Proc. Cambridge Phil. Soc. 59, 117 (1963).

[3] E. W. Montroll, in *Energetics in Metallurgical Phenomena* (Gordon & Breach, New York, 1967) Vol. 3.

[4] R. J. Glauber, J. Math. Phys. **4**, 294 (1963).

[5] N. S. Goel, Biopolymers **6**, 55 (1968).

[6] J. D. Louck, Am. J. Phys. **30**, 585 (1962).

[7] J. R. Clem and R. P. Godwin, Am. J. Phys. **34**, 460 (1966).

[8] D. E. Rutherford, Proc. Roy. Soc. Edinburgh **A63**, 232 (1954); **A62**, 229 (1947).

In this paper we consider a generalization to the case where N is finite and where a, b are different from A, B. In this form the equations pertain to one of several models which may be used to characterize the kinetics of binding between cationic polypeptides and DNA. The equations also find application in descriptions of the kinetics of the helix–coil transition in polypeptides, when it is assumed that helical regions nucleate at one end of the coil and thereafter grow towards the other end of the molecule. The $C_j(t)$ represent the concentrations of molecules containing j "helix" and $N-j$ "coil" units.[9–12]

For the helix–coil problem it is of interest to study the solutions of Eqs. (1) as a function of the nucleation parameters a and b, and as a function of the chain length N.[10] In the case of an infinite chain the system of equations can be solved by the method of generating functions. However, except for special values of the rate constants, the solution given by this method is not particularly convenient. Nor does this method permit an easy generalization to include effects which arise from defective helix or coil units, i.e., a situation where one or more of the "propagation" rate constants associated with the jth transition differ from A or B. Similar remarks apply to the finite chain; a generating function

$$P(w,t)=\sum_{j=0}^{N} w^j C_j(t), \qquad w=\exp(2\pi i/N+1) \tag{2}$$

can be obtained, but the resulting solution for $C_j(t)$ is a complicated integral equation involving generalized hyperbolic functions[13] which are difficult to handle.

We use instead the alternative method of Laplace transforms and find explicit solutions to Eqs. (1). In principle our method can be generalized to obtain information concerning the relaxation spectrum for an arbitrary set of rate constants in the limit of large N, or for a system with several defect units. Thus, e.g., when $b=B$ so that the equations describe a stochastic birth and death process, we shall show that for large N the eigenvalue spectrum has the following curious form:

(1) $\xi=(A-a)/(AB)^{1/2}>1$. There exist two isolated eigenvalues $\lambda_0=0$, $\lambda_1>0$, and a continuum starting with $\lambda_2>\lambda_1$ which lies in a band of finite width.

(2) $\xi<1$, λ_1 merges into the continuum.

In addition to other possible application to the helix–coil transition problem such as the effect of defects, the explicit solution to Eqs. (1) may also be of value in so far as it provides a model calculation in the theory of multistate relaxation processes; errors introduced by the replacement of Eqs. (1) by the continuum analog (the Fokker–Planck equation) can be bounded precisely. The model can also be used to test the utility of the concept of a "mean relaxation time" introduced by Schwarz[14] in his review of kinetic analysis by chemical relaxation methods.

To avoid unnecessarily cumbersome algebra, we consider in detail only the case $b=B$, and the special initial condition $C_j=\delta_{j0}$. Corresponding results for general values of b are given subsequently.

II. FORMULATION: $b=B$

It is convenient to introduce a new variable $\sigma_j(\tau)$ defined by the transformation

$$C_j(t)=(A/B)^{j/2}\exp[-(A+B)t]\sigma_j(\tau);\qquad \tau=2(AB)^{1/2}t. \tag{3}$$

Substitution of this expression into Eqs. (1) yields

$$\begin{aligned}
2\dot\sigma_0&=\alpha\sigma_0+\sigma_1,\\
2\dot\sigma_1&=\beta\sigma_0+\sigma_2,\\
2\dot\sigma_j&=\sigma_{j-1}+\sigma_{j+1}, \qquad j=2,3,\cdots,N-1,\\
2\dot\sigma_N&=\sigma_{N-1}+\gamma\sigma_N,
\end{aligned} \tag{4}$$

where the parameters α, β, and γ are given by

$$\begin{aligned}
\gamma&=(A/B)^{1/2}, \qquad \beta=a/A,\\
\alpha&=\gamma+1/\gamma-\beta\gamma\\
&=(A/B)^{1/2}+(B/A)^{1/2}-[a(AB)^{-1/2}].
\end{aligned} \tag{5}$$

We rewrite Eqs. (4) as

$$2\dot{\boldsymbol\sigma}=L\boldsymbol\sigma, \tag{6}$$

where L is an $(N+1)\times(N+1)$ matrix, and require the solution of Eq. (6) subject to the initial condition

$$\sigma_j(0)=\delta_{j0}. \tag{7}$$

Writing

$$\hat\sigma(\epsilon)=\int_0^\infty \boldsymbol\sigma(\tau)\exp(-\epsilon\tau)\,d\tau, \tag{8}$$

we have

$$2\epsilon\hat{\boldsymbol\sigma}-2\boldsymbol\sigma(0)=L\hat\sigma, \tag{9}$$

an equation whose solution is

$$\hat{\boldsymbol\sigma}=2(2\epsilon-L)^{-1}\boldsymbol\sigma(0). \tag{10}$$

With the initial condition given by Eq. (7) we have for the Laplace transform of $\sigma_j(\tau)$ the result

$$\hat\sigma_j(\epsilon)=2[(2\epsilon-L)^{-1}]_{j0} \tag{11}$$

and require the $(j0)$ element of the inverse of the

---

[9] M. Saunders and P. D. Ross, Biochem. Biophys. Res. Commun. **3**, 314 (1960).

[10] J. A. Ferretti, Polymer Preprints **10**, 29 (1969).

[11] G. Schwarz, J. Mol. Biol. **11**, 64 (1965).

[12] A. C. Pipkin and J. H. Gibbs, Biopolymers **4**, 3 (1966).

[13] A. Erdelyi, W. Magnus, F. Oberhettinger, and F. G. Tricomi, *Higher Transcendental Functions*, (McGraw-Hill Book Co., New York, 1955), Vol. 3, p. 212.

[14] G. Schwarz, Rev. Mod. Phys. **40**, 206 (1968).

matrix

$$(2\epsilon-L)=\begin{bmatrix} 2\epsilon-\alpha & -1 & 0 & & & \\ -\beta & 2\epsilon & -1 & & & \\ 0 & -1 & 2\epsilon & -1 & & \\ \cdots & \cdots & \cdots & \cdots & \cdots & \cdots \\ & & & -1 & 2\epsilon & -1 \\ & & & & -1 & 2\epsilon-\gamma \end{bmatrix}_{N+1} \quad (12)$$

We define the determinant

$$D_j=\det\begin{bmatrix} 2\epsilon & -1 & 0 & & & \\ -1 & 2\epsilon & -1 & 0 & & \\ 0 & -1 & 2\epsilon & -1 & & \\ \cdots & \cdots & \cdots & \cdots & \cdots & \cdots \\ & & & -1 & 2\epsilon & -1 \\ & & & & -1 & 2\epsilon-\gamma \end{bmatrix}_j ,$$

with (13)

$$D_0\equiv 1,\qquad D_1=2\epsilon-\gamma,\qquad D_2=2\epsilon D_1-1. \quad (14)$$

In general these determinants satisfy the recurrence relations

$$D_j=2\epsilon D_{j-1}-D_{j-2};\qquad j=3,4,\cdots,N, \quad (15)$$

$$D_{N+1}\equiv\Delta(\epsilon)=\det(2\epsilon-L)=(2\epsilon-\alpha)D_N-\beta D_{N-1}. \quad (16)$$

The relations in Eq. (15) are the recurrence relations for the Tchebycheff polynomials; Eq. (15) will therefore be satisfied by any sum of these polynomials. To satisfy Eq. (14) we must take

$$D_j=U_j(\epsilon)-\gamma U_{j-1}(\epsilon), \quad (17)$$

where

$$U_j(\cos\theta)=\sin(j+1)\theta/\sin\theta;\qquad U_{-j}(\cos\theta)\equiv 0 \quad (18)$$

are polynomials of the second kind. Substitution of Eq. (17) into Eq. (16) then gives

$$\Delta(\epsilon)=(2\epsilon-\gamma-1/\gamma)[U_N(\epsilon)-\xi U_{N-1}(\epsilon)], \quad (19)$$

where

$$\xi=\gamma(1-\beta)=(A-a)/(AB)^{1/2}. \quad (20)$$

Equations (11)–(20) imply that

$$\hat{\sigma}_j(\epsilon)=[2/\Delta(\epsilon)][\delta_{j0}+\beta(1-\delta_{j0})]D_{N-j}(\epsilon), \quad (21)$$

so that

$$\sigma_j(\tau)=(2\pi i)^{-1}\int_C d\epsilon\,\exp(\epsilon\tau)\hat{\sigma}_j(\epsilon), \quad (22)$$

where the contour lies to the right of all the singularities of $\hat{\sigma}_j(\epsilon)$.

Finally, using Eq. (3), we have

$$C_j(t)=\exp[-(A+B)t]\gamma^j(2\pi i)^{-1}\int_C d\epsilon\hat{\sigma}_j(\epsilon)\times\exp[2(AB)^{1/2}\epsilon t]. \quad (23)$$

In this form it is straightforward to verify the conservation condition

$$\sum_{j=0}^{N} C_j(t)=1. \quad (24)$$

III. EXPLICIT SOLUTION

Most interest centers on the mean number of helix units, given by

$$G_1(t)=\sum_{j=0}^{N} jC_j(t); \quad (25)$$

the analysis for $C_j(t)$ proceeds in the same manner. Substituting Eq. (23) into Eq. (25) and interchanging orders of summation and integration, we can use Eq. (17) to rearrange the sum to give

$$G_1(t)=\exp[-(A+B)t](2\beta/2\pi i)\times\int_C\frac{d\epsilon\,\exp[2(AB)^{1/2}t\epsilon]}{\Delta(\epsilon)}\sum_{j=1}^{N}\gamma^j U_{N-j}(\epsilon). \quad (26)$$

The poles of the integrand are the zeros of $\Delta(\epsilon)$ defined by Eq. (19). This function has one isolated zero at $\epsilon=\epsilon_0=\frac{1}{2}(\gamma+1/\gamma)\geq 1$, which yields the equilibrium solution. The remaining zeros are those of the polynomial

$$P_N(\epsilon)=U_N(\epsilon)-\xi U_{N-1}(\epsilon). \quad (27)$$

By a well known theorem in the theory of orthogonal polynomials (Szego[15]), this polynomial has N distinct real zeros. These zeros lie in the range $-1<\epsilon<1$ with the exception of the greatest (least) zero which lies in $[-1, 1]$ only if

$$\xi\leq(N+1)/N,\quad\text{if}\quad\xi>0;$$

$$\xi\geq-[(N+1)/N],\quad\text{if}\quad\xi<0. \quad (28)$$

In the limit of large N, with the exception of this greatest (or least) zero, the zeros of $P_N(\epsilon)$ merge into a continuum.

We evaluate first the equilibrium solution $G_1(\infty)$, due to the pole of the integrand of Eq. (26) at $\epsilon_0=\frac{1}{2}(\gamma+\gamma^{-1})$. The contribution of this pole is

$$G_1(\infty)=\{\beta/[U_N(\epsilon_0)-\xi U_{N-1}(\epsilon_0)]\}\sum_{j=1}^{N}\gamma^j U_{N-j}(\epsilon_0). \quad (29)$$

This expression can be simplified by writing $\gamma=e^{\phi}$, $\epsilon_0=\cosh\phi$, and using the definition Eq. (18) in the form

$$U_j(\cosh\phi)=[\sinh(j+1)\phi]/\sinh\phi. \quad (30)$$

[15] G. Szego, *Orthogonal Polynomials* (American Mathematical Society Colloquium Publications, 1934), Vol. 23, Theorem 334.

This yields

$$G_1(\infty)=\frac{a/B[1-(N+1)(A/B)^N+N(A/B)^{N+1}]}{(1-A/B)\{1-A/B+a/B[1-(A/B)^N]\}},\tag{31}$$

a result which could have been obtained directly from Eq. (1).

To exhibit the time dependence of G_1 explicitly, we first use the trivial identity

$$\sum_{j=1}^{N}\gamma^j U_{N-j}(\epsilon)=\frac{\gamma U_N(\epsilon)-U_{N-1}(\epsilon)-\gamma^{N+1}}{(2\epsilon-\gamma-1/\gamma)};\qquad \epsilon\neq\tfrac{1}{2}(\gamma+1/\gamma)\tag{32}$$

to write

$$\Delta G_1(t)=G_1(t)-G_1(\infty)\tag{33}$$

$$=\exp[-(A+B)t](\beta/\pi i)$$

$$\times\int_{C_1}\frac{d\epsilon\exp[2(AB)^{1/2}t\epsilon](\gamma U_N-U_{N-1}-\gamma^{N+1})}{(2\epsilon-\gamma-1/\gamma)^2(U_N-\xi U_{N-1})},\tag{34}$$

where the contour C_1 now excludes the poles at $\epsilon=\frac{1}{2}(\gamma+\gamma^{-1})$.

IV. ASYMPTOTIC BEHAVIOR

In principle, the complete time dependence of $G_1(t)$ can be determined by evaluation of the integral which appears in Eq. (34). However, for many applications it is sufficient to know only the approach to equilibrium, and it is to this question that all subsequent analysis is directed.

The poles required for the evaluation of the integral Eq. (34) are the roots of

$$U_N(\epsilon_k)=\xi U_{N-1}(\epsilon_k);\qquad k=1,2,\cdots,N.\tag{35}$$

Several cases arise:

$$(1)\ \xi=(A-a)/(AB)^{1/2}>0;\qquad \xi>(N+1)/N\simeq 1:\tag{36}$$

In this case by Eq. (28) the *greatest* zero lies outside the interval $-1\leq\epsilon\leq 1$, while the remaining $N-1$ zeros lie inside this region. The relaxation to equilibrium will be determined by this greatest zero.

(2) $\xi>0$, $\xi<1$: All zeros lie in the interval $[-1, 1]$.
(3) $\xi<0$, $\xi>-1$: Again all zeros lie in $[-1, 1]$.
(4) $\xi<0$, $\xi<-1$: The *least* zero lies outside $[-1, 1]$.

For most applications cases (2)–(4) will be indistinguishable. We consider first case (1). To determine the greatest zero, ϵ^*, we note that $\epsilon^*>1$, and use Eq. (30) to write Eq. (35) as

$$\sinh(N+1)\theta^*=\xi\sinh N\theta^*,\tag{37}$$

with $\cosh\theta^*=\epsilon^*$. For large N, since $\theta^*>0$, terms in $\exp(-N\theta^*)$ are negligible, and Eq. (37) reduces to

$$\exp[(N+1)\theta^*]\simeq\xi\exp(N\theta^*).\tag{38}$$

One thus finds ϵ^* to be given as

$$\exp(\theta^*)\simeq\xi,\qquad \epsilon^*=\cosh\theta^*\simeq\tfrac{1}{2}(\xi+1/\xi).\tag{39}$$

We note that $\epsilon^*<\epsilon_0$. Evaluating that residue of the integrand of Eq. (34) which is associated with ϵ^*, we find a contribution to the integral $G_1^*(t)$ given as

$$G_1^*(t)=-\frac{4\beta\exp[-(A+B)t]\exp[2(AB)^{1/2}t\epsilon^*]\sinh^2\theta^*[\gamma^{N+1}\sinh\theta^*-(\gamma\xi-1)\sinh N\theta^*]}{(\xi^2-1)(2\epsilon^*-\gamma-1/\gamma)^2\sinh N\theta^*},$$

$$\simeq-\frac{\beta\exp[-a(1-1/\xi^2)t](\xi^2-1)[\gamma^{N+1}(\xi^2-1)-(\gamma\xi-1)\xi^{N+1}]}{(\xi+1/\xi-\gamma-1/\gamma)^2\xi^{N+3}}.\tag{40}$$

For larger values of γ and moderately small values of β, the latter may be expressed simply as

$$G_1^*(t)\simeq-\exp(-at)/\beta(1-\beta)^{N-1}.\tag{41}$$

The $(N-1)$ remaining roots of Eq. (35) may be found by using Eq. (18) to write

$$\sin(N+1)\theta_k=\xi\sin N\theta_k,\tag{42}$$

with $\epsilon_k=\cos\theta_k$. We take (cf. Szego[15])

$$\theta_k=k\pi/N+\delta_k/N;\qquad 0<\delta_k<\pi,$$

$$k=0,1,\cdots,N-2.\tag{43}$$

Then

$$\sin(N+1)\theta_k=\cos k\pi[\sin\delta_k\cos(k\pi/N)+\cos\delta_k\sin(k\pi/N)]+O(1/N),$$

$$\sin N\theta_k=\cos k\pi\sin\delta_k.\tag{44}$$

Substituting these expressions into Eq. (42) we have

$$\tan\delta_k=\frac{\sin(k\pi/N)}{\xi-\cos(k\pi/N)}+O\left(\frac{1}{N}\right),$$

$$\epsilon_k=\cos\theta_k=\cos(k\pi/N)+O(1/N),$$

$$\sin\delta_k=\frac{\sin(k\pi/N)}{|[1+\xi^2-2\xi\cos(k\pi/N)]^{1/2}|}+O\left(\frac{1}{N}\right).\tag{45}$$

With these expressions, it is easy to show that

$$[(d/d\epsilon)(U_N-\xi U_{N-1})]_{\epsilon=\epsilon_k}=\frac{N(-1)^k}{\sin^2(k\pi/N)}\,|\,[1-2\xi\cos(k\pi/N)+\xi^2]^{1/2}\,|\,[1+O(1/N)] \tag{46}$$

nd

$$[\gamma U_N-U_{N-1}]_{\epsilon=\epsilon_k}=(-1)^k(\gamma\xi-1)\,|\,[1-2\xi\cos(k\pi/N)+\xi^2]^{-1/2}\,|\,[1+O(1/N)]. \tag{47}$$

Hence, from Eq. (34) we have

$$\Delta G_1(t)=G_1^*(t)+G_1^{(1)}(t)+G_1^{(2)}(t), \tag{48}$$

here $G_1^{(1)}$ and $G_1^{(2)}$ are given as

$$G_1^{(1)}(t)=\exp[-(A+B)t]\frac{2\beta}{N}\sum_{k=0}^{N-2}\frac{\exp[2(AB)^{1/2}t\cos(k\pi/N)]}{[2\cos(k\pi/N)-\gamma-1/\gamma]^2}\,\frac{(\gamma\xi-1)\sin^2(k\pi/N)}{[1+\xi^2-2\xi\cos(k\pi/N)]}, \tag{49}$$

$$G_1^{(2)}(t)=-\exp[-(A+B)t]\frac{2\beta}{N}\sum_{k=0}^{N-2}\frac{(-1)^k\gamma^{N+1}\exp[2(AB)^{1/2}t\cos(k\pi/N)]\sin^2(k\pi/N)}{[2\cos(k\pi/N)-\gamma-1/\gamma]^2\,|\,[1-2\xi\cos(k\pi/N)+\xi^2]^{1/2}\,|}. \tag{50}$$

n the limit of large N, the sums may be replaced by integrals to give

$$G_1^{(1)}(t)=\exp[-(A+B)t]\frac{2\beta(\gamma-1/\xi)}{\pi}\int_0^\pi\frac{d\theta\exp[2(AB)^{1/2}t\cos\theta]\sin^2\theta}{(2\cos\theta-\gamma-1/\gamma)^2(1/\xi+\xi-2\cos\theta)} \tag{51}$$

nd

$$G_1^{(2)}(t)=-\exp[-(A+B)t]\frac{2\beta\gamma^{N+1}}{\pi}\int_0^\pi\frac{d\theta\exp[2(AB)^{1/2}t\cos\theta]\sin^2\theta\cos N\theta}{(2\cos\theta-\gamma-1/\gamma)^2\,|\,[1-2\xi\cos\theta+\xi^2]^{1/2}\,|}. \tag{52}$$

or the remaining cases (2)–(4) above, the approach to quilibrium is given by essentially the same expression, xcept that $G_1^*(t)$ is to be taken as zero.

V. EVALUATION OF INTEGRALS

Before further evaluating Eqs. (51) and (52), we rst remark on the effect of finite chain length N. everal sources of N dependence can be distinguished: 1) In evaluating the residue of the integrand of Eq. 34) at $\epsilon=\epsilon_k$ we have put

$$\exp[2(AB)^{1/2}t\epsilon_k]$$
$$=\exp\{2(AB)^{1/2}t[\cos(k\pi/N)-\sin(\delta_k/N)\sin(k\pi/N)]\}$$
$$\simeq\exp[2(AB)^{1/2}t\cos(k\pi/N)].$$

he exponent in Eqs. (51) and (52) is thus in error by term $O(1/N)$. Similar approximations in the rest of he residues are also $O(1/N)$. Finally the error in ap-roximating the sums of Eqs. (49) and (50) by inte-rals is $O(1/N)$. All of these approximations give rise to orrections to Eqs. (51) and (52) which may be gnored for large N. (2) Finite chain length is impor-ant in the relaxation process only if $\gamma=(A/B)^{1/2}$ is reater than 1. This remark is intuitively clear, and ollows explicitly from Eqs. (31), (51), and (52) and he conditions necessary in order that the term $G_1^*(t)$ e nonzero.

We now evaluate the integrals, and show that ssentially only a single relaxation time is involved. Let s first consider $G_1^{(1)}(t)$, which may be written as

$$G_1^{(1)}(t)=\exp[-(A+B)t](\beta/4\pi)(\gamma-1/\xi)I_1 \tag{53}$$

here

$$I_1=\int_0^\pi\frac{d\theta\exp[2(AB)^{1/2}t\cos\theta]\sin^2\theta}{[\frac{1}{2}(\gamma+1/\gamma)-\cos\theta]^2[\frac{1}{2}(\xi+1/\xi)-\cos\theta]}. \tag{54}$$

We shall be interested in the behavior of I_1 for large alues of t. The major contribution comes from the region $\theta\simeq0$. The change of variable $y=\frac{1}{2}(1-\cos\theta)$ gives

$$I_1=\tfrac{1}{2}\{\exp[2(AB)^{1/2}t]\}\times\int_0^1\frac{dy[y(1-y)]^{1/2}\exp[-4(AB)^{1/2}ty]}{(p+y)^2(q+y)}, \tag{55}$$

where

$$p=\tfrac{1}{2}[\tfrac{1}{2}(\gamma+\gamma^{-1})-1]\geq0,\qquad q=\tfrac{1}{2}[\tfrac{1}{2}(\xi+\xi^{-1})-1]\geq0. \tag{56}$$

For long times such that $p\gg[4(AB)^{1/2}t]^{-1}$, $q\gg[4(AB)^{1/2}t]^{-1}$, Eq. (55) becomes

$$I_1\simeq\frac{\exp2(AB)^{1/2}t}{2p^2q}\int_0^{4(AB)^{1/2}t}\frac{dz}{[4(AB)^{1/2}t]^{3/2}}z^{1/2}\times\left(1-\frac{z}{4(AB)^{1/2}t}\right)^{1/2}e^{-z}$$
$$=\{\pi^{1/2}\exp[2(AB)^{1/2}t]/4p^2q\}[4(AB)^{1/2}t]^{-3/2}\times[1-O(1/t)]. \tag{57}$$

Thus, substituting Eq. (57) into Eq. (53), we see that the contribution $G_1^{(1)}(t)$ relaxes towards equilibrium as

$$G_1^{(1)}(t)\simeq\frac{\beta}{2(\pi)^{1/2}}\times\frac{(\gamma-1/\xi)\exp[-(A^{1/2}-B^{1/2})^2t]}{(\gamma+1/\gamma-2)^2(\xi+1/\xi-2)[(AB)^{1/2}t]^{3/2}}. \tag{58}$$

The integral $G_1^{(2)}(t)$ is somewhat more complicated. For large (but finite) N and large t the major contribution to the integral comes from the region $\theta\simeq0+$. Hence we may write

$$G_1^{(2)}(t)\simeq-\exp[-(A^{1/2}-B^{1/2})^2t](2\beta/\pi)\times[\gamma^{N+1}/(\gamma^{1/2}-\gamma^{-1/2})^2\,|\,1-\xi\,|]\times\int_0^\pi\theta^2\exp[-(AB)^{1/2}t\theta^2]\cos N\theta d\theta. \tag{59}$$

The upper limit may be replaced by ∞ to a very good approximation, and the remaining integral is elementary. The result is

$$G_1^{(2)}(t)=\frac{-\beta\gamma^{N+1}\exp[-(A^{1/2}-B^{1/2})^2t]}{2(\pi)^{1/2}(\gamma^{1/2}-\gamma^{-1/2})^2\,|\,1-\xi\,|}\exp\left(-\frac{N^2}{4(AB)^{1/2}t}\right)\left([(AB)^{1/2}t]^{-3/2}-\frac{N^2}{2[(AB)^{1/2}t]^{5/2}}\right), \tag{60}$$

for large t.

When the rate constants for dissociation are greater than those for association, i.e., if $A/B<1$, then $\gamma^{N+1}\ll 1$ so that $G_1^{(2)}(t)$ is unimportant. In this case, the only term governing the approach to equilibrium is that given by Eq. (58), viz.,

$$\Delta G_1(t)=\frac{-\beta(1/\xi-\gamma)}{2(\pi)^{1/2}(\gamma+1/\gamma-2)^2(\xi+1/\xi-2)}\,\frac{\exp[-(A^{1/2}-B^{1/2})^2t]}{[(AB)^{1/2}t]^{3/2}}\quad\text{if}\quad B\gtrsim A$$
$$\approx-[\beta\gamma^{1/2}/2(\pi)^{1/2}][\exp(-Bt)/(Bt)^{3/2}]\quad\text{if}\quad B\gg A. \tag{61}$$

On the other hand, if $A>B$ the approach to equilibrium is characterized by two relaxation times. In general, the terms G_1^* and $G_1^{(2)}$ are dominant and the relaxation times are those which appear above in Eqs. (40) and (60). In the special case that $\gamma\gg 1$, a particularly simple result is obtained, viz.,

$$\Delta G_1(t)\simeq G_1^*+G_1^{(2)}$$
$$\simeq-\left(\frac{\exp(-at)}{\beta(1-\beta)^{N-1}}+\frac{\beta\gamma^{N+1/2}}{2\pi^{1/2}(1-\beta)}\,\frac{\exp(-At)}{(At)^{3/2}}\right),$$
$$[a\neq 0,\quad \gamma(1-\beta)>1,\quad t>N^2/4(AB)^{1/2}]. \tag{62}$$

Note that, when $A=B$, the analysis leading to Eqs. (58)–(62) must be modified. In this case the zero of the integrand of Eq. (26) which occurs at $\epsilon=\epsilon_0$ now occurs at $\epsilon_0=\frac{1}{2}(\gamma+\gamma^{-1})=1$ and is no longer well separated from the band of zeros of $U_N(\epsilon)-\xi U_{N-1}(\epsilon)$. Thus, the decomposition of Eq. (33) is no longer useful and, further, the identity given in Eq. (32) cannot be used. Although a somewhat similar analysis can be applied to this special case, we have not considered this point in detail.

VI. SOLUTION FOR ARBITRARY b

The solution of the more general system Eq. (1) proceeds in precisely the same manner. The last two equations of the set Eq. (4) are replaced by

$$2\dot{\sigma}_{N-1}=\sigma_{N-2}+\beta'\sigma_N,$$
$$2\dot{\sigma}_N=\sigma_{N-1}+\alpha'\sigma_N, \tag{63}$$

where now

$$\alpha'=\gamma+\gamma^{-1}-\beta'/\gamma;\qquad \beta'=b/B. \tag{64}$$

In the last column of Eqs. (12), (13) the entries -1, $2\epsilon-\gamma$ are replaced by $-\beta'$, $2\epsilon-\alpha'$, respectively and the equation corresponding to Eq. (14) is

$$D_0=1,\qquad D_1=2\epsilon-\alpha',\qquad D_2=2\epsilon D_1-\beta'. \tag{65}$$

The additional nucleation parameter β' requires that D_j must be modified to include a further Tchebycheff polynomial. Consequently, corresponding to Eqs. (17) and (19) we find

$$D_j=(U_j-\gamma U_{j-1})-[(1-\beta')/\gamma](U_{j-1}-\gamma U_{j-2}) \tag{66}$$
$$\Delta(\epsilon)=(2\epsilon-\gamma-\gamma^{-1})\{U_N(\epsilon)-\xi U_{N-1}(\epsilon)-[(1-\beta')/\gamma][U_{N-1}(\epsilon)-\xi U_{N-2}(\epsilon)]\}. \tag{67}$$

In general, if in addition the chain contains n defects, so that n of the propagation constants A or B are modified, it is clear that $\Delta(\epsilon)$ will involve an additional n polynomials.

Corresponding to Eq. (25) we have

$$G_1(t)=\exp[-(A+B)t]\frac{2\beta}{2\pi i}\int_C\frac{d\epsilon\,\exp[2(AB)^{1/2}t\epsilon]}{\Delta(\epsilon)}\sum_{j=1}^{N}\gamma^j[U_{N-j}-(1-\beta')\gamma^{-1}U_{N-j-1}], \tag{68}$$

which yields the equilibrium solution for the mean as

$$G_1(\infty)=\frac{\beta\{N\gamma^N[1-(1-\beta')/\gamma^2]-(\gamma^N-\gamma^{-N})-[(1-\beta')/\gamma][(\gamma^{N-1}-\gamma^{-(N-1)})/(\gamma-1/\gamma)]\}}{\{[\gamma^{N+1}-\gamma^{-(N+1)}]-[\xi+(1-\beta')/\gamma](\gamma^N-\gamma^{-N})+[(1-\beta')/\gamma][\gamma^{N-1}-\gamma^{-(N-1)}]\}} \tag{69}$$

and

$$\Delta G_1(t)=G_1(t)-G_1(\infty)$$
$$=\exp[-(A+B)t]\frac{2\beta}{2\pi i}\int_{C_1}d\epsilon\,\exp 2(AB)^{1/2}t\epsilon$$
$$\times\frac{[1-(1-\beta')/\gamma^2]\{[\gamma U_N(\epsilon)-U_{N-1}(\epsilon)-\gamma^{N+1}]/(2\epsilon-\gamma-1/\gamma)^2\}+[(1-\beta')/\gamma]\{[U_{N-1}(\epsilon)]/(2\epsilon-\gamma-1/\gamma)\}}{U_N(\epsilon)-[\xi+(1-\beta')/\gamma]U_{N-1}(\epsilon)+(1-\beta')\xi/\gamma U_{N-2}}. \tag{70}$$

A thorough investigation of the eigenvalue spectrum is a matter of some complexity. However, in the limit $A/B\gg 1$ it is clear that the previous conclusions remain substantially unaltered.

STOCHASTIC MODELS FOR CHAIN DYNAMICS*

R. A. ORWOLL and W. H. STOCKMAYER

Department of Chemistry, Dartmouth College, Hanover, New Hampshire

CONTENTS

I. INTRODUCTION

During the past fifteen years a familiar bead-and-spring model often associated with the name of Rouse[1] has dominated discussions of the dynamic behavior of flexible long-chain molecules in solution or bulk. Independently developed in more or less equivalent degree by others,[2,3] the model has since been improved and embellished in several physically important respects, such as (*a*) the inclusion of hydrodynamic interactions between chain segments in either scalar[4] or tensor[5] approximation, (*b*) the inclusion of long-range "excluded volume" effects in some approximation[5–7] and, (*c*) the *ad hoc* introduction of an "internal viscosity."[8,9] The phenomenological success of the model, as applied to dynamic viscoelastic,[10,11,12] dielectric,[12,13,14] optical,[4,14,15] and nuclear magnetic relaxation[16] experiments, is noteworthy, except at high frequencies[17,18] or low temperatures, or under nonlinear (e.g., non-Newtonian viscoelastic) conditions. Further extensions[19] of the model may yet be offered.

The molecular basis of the Rouse model (and its more sophisticated relatives), however, has remained somewhat obscure. The more complete and realistic treatment of Kirkwood,[20] though itself rarely tractable save in a formal sense, could in principle offer a starting point for the derivation

* Supported by the National Science Foundation.

of the Rouse model, in rough analogy with the reduction of the Liouville equation to the Boltzmann equation in gas kinetic theory; but recorded attempts[21] along this line have not been wholly satisfactory. In particular, it is not clear whether a realistic molecular model would *demand* (as seems required by various experimental indications[17,18,22]) the introduction of some form of "internal viscosity" into the Rouse formalism, or whether and how the shape of the relaxation spectrum in the high-frequency region would deviate from the Rouse-Zimm[1,4,6] form.

As a contribution to the study of these problems, stochastic models are here developed for two cases: a freely-jointed chain in any number of dimensions, and a one-dimensional chain with nearest-neighbor correlations. Our work has been directly inspired by two different sources: the Monte Carlo studies by Verdier[23,24] of the dynamics of chains confined to simple cubic lattices, and the analytical treatment by Glauber[25] of the dynamics of linear Ising models. No attempt is made in this work to introduce the effects of excluded volume or hydrodynamic interactions.

II. FREELY JOINTED CHAIN

Let a linear chain comprise $N+1$ beads which are indexed serially along the chain from 0 to N. The centers of adjoining beads are separated by a constant bond distance b. The vector drawn from bead $i-1$ to bead i is designated $b\boldsymbol{\sigma}_i$, where $\boldsymbol{\sigma}_i$ is a unit vector. Thus, the configuration of a chain is specified by the set of N vectors $\{\boldsymbol{\sigma}_1, \boldsymbol{\sigma}_2, \ldots, \boldsymbol{\sigma}_N\}$.

The chain configuration is made to vary in time by allowing the beads to move one at a time. When an interior bead i ($i = 1, 2, \ldots N-1$) moves (or "flips"), the vectors $\boldsymbol{\sigma}_i$ and $\boldsymbol{\sigma}_{i+1}$ before the flip are transformed to the vectors $\boldsymbol{\sigma}_i'$ and $\boldsymbol{\sigma}_{i+1}'$ after the flip according to the equations

$$\boldsymbol{\sigma}_i' = \boldsymbol{\sigma}_{i+1} \quad \text{and} \quad \boldsymbol{\sigma}_{i+1}' = \boldsymbol{\sigma}_i, \tag{1}$$

as illustrated in Figure 1. This process is essentially the same as that used by Verdier[24] for chains confined to a simple cubic lattice. For convenience, the terminal beads 0 and N are specified to move so that

$$\boldsymbol{\sigma}_1' = -\boldsymbol{\sigma}_1 \quad \text{and} \quad \boldsymbol{\sigma}_N' = -\boldsymbol{\sigma}_N \tag{2}$$

Let w_i denote the probability per unit time that the ith bead executes a flip. We consider a class of stochastic models for which this probability is given by

$$w_i = \alpha(1 - a\boldsymbol{\sigma}_i \cdot \boldsymbol{\sigma}_{i+1}) \tag{3}$$

with $|a| \leqq 1$. By choice of a suitable value of a, the mobility of a bead can

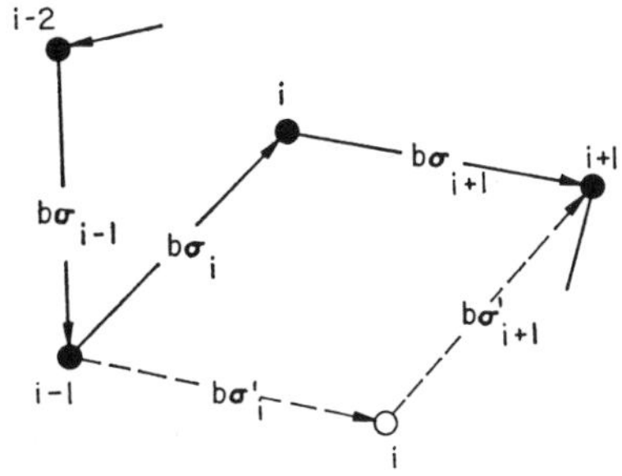

Fig. 1. The basic stochastic process. Position of bead i is indicated by a filled circle before it has flipped and an open circle after it has flipped. Solid lines represent the bond vectors $b\boldsymbol{\sigma}_{i-1}$, $b\boldsymbol{\sigma}_i$ and $b\boldsymbol{\sigma}_{i+1}'$ before the ith bead flips; dashed lines indicate bond vectors $b\boldsymbol{\sigma}_i'$ and $b\boldsymbol{\sigma}_{i+1}'$ after the flip.

be weighted in favor of acute or obtuse bond angles as desired. The flip probability for an end bead is assigned the constant value

$$w_0 = w_N = \alpha\gamma \tag{4}$$

The probability that at time t the chain has the configuration $\{\boldsymbol{\sigma}_1, \boldsymbol{\sigma}_2, \ldots, \boldsymbol{\sigma}_N\} \equiv \{\boldsymbol{\sigma}^N\}$ is designated $p(\boldsymbol{\sigma}_1, \boldsymbol{\sigma}_2, \ldots, \boldsymbol{\sigma}_N, t) \equiv p(\boldsymbol{\sigma}^N, t)$. Let this probability be normalized, so that the ensemble average value of any chosen bond vector $\boldsymbol{\sigma}_i$ at time t is given by

$$\langle \boldsymbol{\sigma}_i(t) \rangle = \int_{\text{all}\{\boldsymbol{\sigma}^N\}} \cdots \int \boldsymbol{\sigma}_i\, p(\boldsymbol{\sigma}^N, t)\, d\{\boldsymbol{\sigma}^N\} \equiv \mathbf{q}_i(t) \tag{5}$$

The time derivative of the probability function is expressed by the master equation[25]

$$\begin{aligned} dp(\boldsymbol{\sigma}^N, t)/dt = & -p(\boldsymbol{\sigma}^N, t)\left[w_0 + \sum_{i=1}^{N-1} w_i(\boldsymbol{\sigma}_i, \boldsymbol{\sigma}_{i+1}) + w_N\right] \\ & + w_0\, p(-\boldsymbol{\sigma}_1, \boldsymbol{\sigma}_2, \ldots, \boldsymbol{\sigma}_N, t) \\ & + \sum_{i=1}^{N-1} w_i(\boldsymbol{\sigma}_{i+1}, \boldsymbol{\sigma}_i) p(\boldsymbol{\sigma}_1, \boldsymbol{\sigma}_2, \ldots, \boldsymbol{\sigma}_{i+1}, \boldsymbol{\sigma}_i, \ldots, \boldsymbol{\sigma}_N, t) \\ & + w_N\, p(\boldsymbol{\sigma}_1, \boldsymbol{\sigma}_2, \ldots, \boldsymbol{\sigma}_{N-1}, -\boldsymbol{\sigma}_N, t) \end{aligned} \tag{6}$$

The first term on the right-hand side accounts for the destruction of the configuration $\{\boldsymbol{\sigma}^N\}$ which results from the flipping of any one of the $N+1$ beads; and the remaining terms describe the formation of $\{\boldsymbol{\sigma}^N\}$ from other chain configurations.

The time derivative of $\mathbf{q}_i$ is obtained after both sides of Eq. (6) have been multiplied by $\boldsymbol{\sigma}_i$ and integrated over all possible configurations. For $2 \leqq i \leqq N-1$,

$$d\mathbf{q}_i/dt = \int_{\text{all}\{\boldsymbol{\sigma}^N\}} \cdots \int \{[-\boldsymbol{\sigma}_i w_{i-1}(\boldsymbol{\sigma}_{i-1}, \boldsymbol{\sigma}_i) - \boldsymbol{\sigma}_i w_i(\boldsymbol{\sigma}_i, \boldsymbol{\sigma}_{i+1})]p(\boldsymbol{\sigma}^N, t)$$
$$+ \boldsymbol{\sigma}_i w_{i-1}(\boldsymbol{\sigma}_i, \boldsymbol{\sigma}_{i-1})p(\boldsymbol{\sigma}_1, \boldsymbol{\sigma}_2, \ldots, \boldsymbol{\sigma}_i, \boldsymbol{\sigma}_{i-1}, \ldots, \boldsymbol{\sigma}_N, t) \tag{7}$$
$$+ \boldsymbol{\sigma}_i w_i(\boldsymbol{\sigma}_{i+1}, \boldsymbol{\sigma}_i)p(\boldsymbol{\sigma}_1, \boldsymbol{\sigma}_2, \ldots, \boldsymbol{\sigma}_{i+1}, \boldsymbol{\sigma}_i, \ldots, \boldsymbol{\sigma}_N, t)\}\, d\{\boldsymbol{\sigma}^N\}$$

In writing Eq. (7), we have noted that the following integral vanishes when $j \neq i-1, i$:

$$\int_{\text{all}\{\boldsymbol{\sigma}^N\}} \cdots \int [-\boldsymbol{\sigma}_i w_j(\boldsymbol{\sigma}_j, \boldsymbol{\sigma}_{j+1})p(\boldsymbol{\sigma}^N, t)$$
$$+ \boldsymbol{\sigma}_i w_j(\boldsymbol{\sigma}_{j+1}, \boldsymbol{\sigma}_j)p(\boldsymbol{\sigma}_1, \boldsymbol{\sigma}_2, \ldots, \boldsymbol{\sigma}_{j+1}, \boldsymbol{\sigma}_j, \ldots, \boldsymbol{\sigma}_N, t)]\, d\{\boldsymbol{\sigma}^N\}$$

Substitution of w_j from Eq. (3) leads to

$$\alpha^{-1}\, d\mathbf{q}_i/dt = -2\mathbf{q}_i + \mathbf{q}_{i-1} + \mathbf{q}_{i+1} + a\langle(\boldsymbol{\sigma}_i - \boldsymbol{\sigma}_{i-1})(\boldsymbol{\sigma}_{i-1} \cdot \boldsymbol{\sigma}_i)\rangle$$
$$+ a\langle(\boldsymbol{\sigma}_i - \boldsymbol{\sigma}_{i+1})(\boldsymbol{\sigma}_i \cdot \boldsymbol{\sigma}_{i+1})\rangle; \qquad i = 2, \ldots, N-1 \tag{8}$$

Equations for the terminal bonds can be derived in a similar manner

$$\alpha^{-1}\, d\mathbf{q}_1/dt = -(1+2\gamma)\mathbf{q}_1 + \mathbf{q}_2 + a\langle(\boldsymbol{\sigma}_1 - \boldsymbol{\sigma}_2)(\boldsymbol{\sigma}_1 \cdot \boldsymbol{\sigma}_2)\rangle$$
$$\alpha^{-1}\, d\mathbf{q}_N/dt = -(1+2\gamma)\mathbf{q}_N + \mathbf{q}_{N-1} + a\langle(\boldsymbol{\sigma}_N - \boldsymbol{\sigma}_{N-1})(\boldsymbol{\sigma}_N \cdot \boldsymbol{\sigma}_{N-1})\rangle \tag{9}$$

For $a \neq 0$, Eqs. (8) and (9) for the mean values $\mathbf{q}_i$ contain terms in higher-order correlations, a familiar and frequently frustrating type of event in statistical mechanics. However, as shown in the Appendix, when the system is not too far from equilibrium (i.e., small perturbations only) the terms $\langle\boldsymbol{\sigma}_i(\boldsymbol{\sigma}_i \cdot \boldsymbol{\sigma}_j)\rangle$ may be replaced by $\langle\boldsymbol{\sigma}_j\rangle d^{-1}$, when d is the dimensionality of the system.

Having made the above approximation we may rewrite Eqs. (8) and (9) in matrix form as

$$\alpha^{-1}\, d\mathbf{q}_/ dt = -(1 + ad^{-1})\mathbf{A}\mathbf{q} \tag{10}$$

where $\mathbf{q}$ is a $N \times 1$ column matrix whose elements comprise the N vectors

$\mathbf{q}_i$. The elements of the $N \times N$ matrix $\mathbf{A}$ have the values

$$\begin{aligned} A_{11} &= A_{NN} = 1 + 2\gamma(1 + ad^{-1})^{-1} \\ A_{ii} &= 2; \qquad i \neq 1, N \\ A_{ij} &= -1; \qquad |i-j| = 1 \\ &= 0; \qquad |i-j| > 1 \end{aligned} \tag{11}$$

The matrix $\mathbf{A}$ is of course familiar in the model of Rouse or indeed in almost all studies of linear systems. (It has been suggested that the "A" stands for "Archimedes.") According to Rutherford,[26] eigenvalues of $\mathbf{A}$ for finite N can be simply expressed for the special cases $A_{11} = 1$ and $A_{11} = 2$. If λ_p represents the pth eigenvalue ($p = 1, 2, \ldots, N$) of $\mathbf{A}$, then

$$\lambda_p = 4 \sin^2 [(p-1)\pi/2N] \qquad \text{for } A_{11} = 1; \tag{12a}$$

or

$$\lambda_p = 4 \sin^2 [(p\pi/2(N+1)] \qquad \text{for } A_{11} = 2 \tag{12b}$$

For simplicity, we limit the remainder of this paper to the case $A_{11} = 2$. (Note that when $A_{11} = 1$, then $w_0 = w_N = 0$; i.e., the ends of the chain are stationary.) For long molecules, the end-effects become insignificant and the difference between the two sets of eigenvalues obviously becomes negligible.

There exists a symmetric $N \times N$ matrix $\mathbf{Q}$ such that

$$\mathbf{QA} = \mathbf{\Lambda Q}, \tag{13}$$

with

$$\mathbf{Q}^{-1}\mathbf{Q} = \mathbf{QQ} = \mathbf{I} \tag{14}$$

where $\mathbf{\Lambda}$ is a diagonal matrix whose i, i element is λ_i and where $\mathbf{I}$ is the $N \times N$ unit matrix. The i, j element of $\mathbf{Q}$, is[27]

$$Q_{ij} = \left(\frac{2}{N+1}\right)^{1/2} \sin\left[\frac{ij\pi}{(N+1)}\right] \tag{15}$$

Premultiplication of Eq. (10) by $\mathbf{Q}$ and substitution of Eq. (13) gives

$$\alpha^{-1} d\xi/dt = -(1 + ad^{-1})\mathbf{\Lambda}\xi \tag{16}$$

where

$$\xi = \mathbf{Qq} \tag{17}$$

The elements of the column matrix ξ are the normal coordinates

$$\xi_p(t) = \sum_{j=1}^{N} Q_{pj}\mathbf{q}_j(t); \qquad p = 1, 2, \ldots, N \tag{18}$$

The solutions of the differential equations represented by Eq. (16) are

$$\xi_p(t) = \xi_p(0) \exp(-t/\tau_p)$$

where

$$\tau_p^{-1} = \alpha(1 + ad^{-1})\lambda_p \tag{19}$$

Thus the relaxation spectrum resulting from the "average coordinates" equation[11] of our model has the same form as that of Rouse, of Kargin and Slonimiskii, or of Bueche. In order to relate the parameters of the model to those of the Rouse theory, the time scale factor α must somehow be connected to the frictional coefficient ζ for a single subchain of a Rouse molecule. To achieve this comparison, we may[23] study the translational diffusion coefficients as computed for the two models.

The translational diffusion coefficient D is $(2d)^{-1}$ times the mean square displacement per unit time of the center of mass of the molecule. When bead i flips from its position $\mathbf{R}_i$, as measured with respect to an arbitrary origin, to a new position $\mathbf{R}_i'$, the square of the resulting displacement of the center of mass of the chain is $(\mathbf{R}_i' - \mathbf{R}_i)^2/(N+1)^2$. Since bead i flips with a frequency w_i, we have

$$D = \frac{\sum_i \langle (\mathbf{R}_i' - \mathbf{R}_i)^2 w_i \rangle}{2(N+1)^2 d} \tag{20}$$

the average to be computed for an equilibrium ensemble. According to the rules governing the motions of the individual beads which were presented earlier (see Fig. 1),

$$\mathbf{R}_i' - \mathbf{R}_i = b(\boldsymbol{\sigma}_{i+1} - \boldsymbol{\sigma}_i) \tag{21}$$

Substitution in Eq. (20) from Eqs. (3) and (21) leads to

$$D = \alpha b^2(1 + ad^{-1})/(N+1)d \tag{22}$$

since $\langle \boldsymbol{\sigma}_i \cdot \boldsymbol{\sigma}_{i+1} \rangle_{\text{eq}} = 0$ and $\langle (\boldsymbol{\sigma}_i \cdot \boldsymbol{\sigma}_{i+1})^2 \rangle_{\text{eq}} = d^{-1}$. Then for long chains $(N + 1 \simeq N)$ and for the slow modes of relaxation $(p \ll N;\ \lambda_p \simeq p^2\pi^2/N^2)$ our model leads to

$$\tau_p^{-1} = \lambda_p\, DN\, d/b^2 = p^2\pi^2\, Dd/Nb^2 \tag{23}$$

Now in the Rouse model the diffusion constant is given by the Einstein relation

$$D = kT/n\zeta \tag{24}$$

for a chain made of n subchains, each with a friction constant ζ. It may be recalled that no unique choice of the size of the subchains is offered;

they must be long enough to be gaussian in respect to the statistics of their end-to-end displacements h but they must be short enough to justify somehow the concentration of all their dissipative effects into a single friction constant ζ. Then the long relaxation times of the model, with $\lambda_p' = 4\sin^2(p\pi/2n)$, are given by[1]

$$\begin{aligned}\tau_p^{-1} &= \lambda_p' kTd/\zeta\langle h^2\rangle = \lambda_p' Dnd/\langle h^2\rangle \\ &= p^2\pi^2 Dd/n\langle h^2\rangle \qquad (p \ll n)\end{aligned} \tag{25}$$

A comparison of Eq. (25) with Eq. (23) shows that the two expressions for τ_p are identical, since $n\langle h^2\rangle = Nb^2$, the mean square end-to-end length of the entire chain. This is true for an arbitrary choice of the size of a Rouse subchain.

Furthermore, it may be seen that for *all* the normal modes of relaxation, including the most rapid, the freely jointed chain model and the Rouse model are identical if we set $n = N + 1$; that is, the relaxation time τ_p of the pth normal mode of a freely-jointed chain is the same as that of a Rouse marcromolecule composed of $N + 1$ subchains, each of mean square end-to-end length b^2. Moreover, for the special choice $a = 0$, Eq. (10) is true for arbitrarily large departures from equilibrium. We thus seem to have confirmed analytically the discovery of Verdier[24] that quite short chains executing a stochastic process described by Eqs. (1) and (3) on a simple cubic lattice display Rouse relaxation behavior. Of course, Verdier's Monte Carlo technique permits study of excluded volume effects, quite beyond the range of our present efforts.

However tempting it may be, further physical exploitation of the above results must be tempered by the realization that both the Rouse and the freely jointed chain models are in some sense artificial. We have nevertheless extended, somewhat beyond the ball-and-spring concept, the validity of the Rouse equations, and the prospect of developing the special case $a = 0$ for nonlinear phenomena is not without possible phenomenological interest.

III. ONE-DIMENSIONAL CHAIN WITH CORRELATIONS

We now undertake the formulation of a one-dimensional stochastic model describing a chain for which the directions of adjacent bonds are correlated. Since the chain is confined to one dimension, the bond vectors $\boldsymbol{\sigma}_i$ become scalars σ_i which can take on only values of ± 1 depending on whether bead i is to the right or left of bead $i - 1$. Therefore bead i can move only if bonds i and $i+1$ point in opposite directions, i.e., $\sigma_i = -\sigma_{i+1}$. When bead i flips, the bond directions σ_i and σ_{i+1} change to $-\sigma_i$ and $-\sigma_{i+1}$, respectively. (We obviously can take no account of excluded volume.) We consider only very long chains and do not trouble to describe the motions of beads at or near the ends.

To develop a formulation of the flip rates, we consider the chain in dynamic equilibrium. If $p^{(2)}(\sigma_i, \sigma_{i+1})$ is the probability that the directions of bonds i and $i+1$ are σ_i and σ_{i+1} and if $w_i(\sigma_i, \sigma_{i+1})$ is the probability per unit time that bead i flips, then at equilibrium the condition of microscopic reversibility requires

$$w_i(\sigma_i, \sigma_{i+1})p_{\text{eq}}^{(2)}(\sigma_i, \sigma_{i+1}) = w_i(-\sigma_i, -\sigma_{i+1})p_{\text{eq}}^{(2)}(-\sigma_i, -\sigma_{i+1}) \quad (26)$$

Correlations between adjacent bond vectors are introduced by assigning to the Hamiltonian H an energy E for each pair of consecutive bonds which point in opposite directions and an energy $-E$ for each pair of adjacent bonds which point in the same direction, i.e.,

$$H = -E \sum_{i=1}^{N-1} \sigma_i \sigma_{i+1} \quad (27)$$

Therefore at equilibrium

$$\frac{p_{\text{eq}}^{(2)}(\sigma_i, \sigma_{i+1})}{p_{\text{eq}}^{(2)}(-\sigma_i, -\sigma_{i+1})} = \frac{\exp[(\sigma_{i-1}\sigma_i + \sigma_{i+1}\sigma_{i+2})E/kT]}{\exp[-(\sigma_{i-1}\sigma_i + \sigma_{i+1}\sigma_{i+2})E/kT]} \quad (28)$$

It follows from Eqs. (26) and (28) that $w_i(\sigma_i, \sigma_{i+1})$ is proportional to $\exp[-(\sigma_{i-1}\sigma_i + \sigma_{i+1}\sigma_{1+2})E/kT]$. The transition probability is also again made proportional to $(1 - a\sigma_i\sigma_{i+1})$, where $|a| \leqq 1$. Thus we take

$$w_i(\sigma_i, \sigma_{i+1}) = \alpha(1 - a\sigma_i\sigma_{i+1}) \exp[-(\sigma_{i-1}\sigma_i + \sigma_{i+1}\sigma_{i+2})E/kT] \quad (29)$$

The transition probability is a property independent of the restrictions of equilibrium, so that Eq. (29) is equally applicable to nonequilibrium situations.

Since even powers of any σ_i equal unity, series expansion of the exponential factors allows Eq. (29) to be rewritten as

$$w_i(\sigma_i, \sigma_{i+1}) = \alpha(1 - \beta^2)^{-1}(1 - a\sigma_i\sigma_{i+1})(1 - \beta\sigma_{i-1}\sigma_i)(1 - \beta\sigma_{i+1}\sigma_{i+2}) \quad (30)$$

where

$$\beta = \tanh(E/kT).$$

For this model the master equation corresponding to Eq. (6) for the multidimensional model discussed earlier is

$$\frac{dp(\sigma_1, \ldots, \sigma_N, t)}{dt} = -p(\sigma_1, \ldots, \sigma_N, t)\sum_{j=0}^{N} w_j(\sigma_j, \sigma_{j+1})$$
$$+ \sum_{j=0}^{N} w_j(\sigma_{j+1}, \sigma_j)p(\sigma_1, \ldots, \sigma_{j+1}, \sigma_j, \ldots, \sigma_N, t) \quad (31)$$

After multiplication by σ_i and summation over all 2^N possible configurations,

$$-[\alpha(1+a)]^{-1}(1-\beta^2)dq_i/dt = 2q_i - (1+\beta)(q_{i-1}+q_{i+1}) + \beta(q_{i-2}+q_{i+2})$$
$$- \beta\langle\sigma_{i-2}\sigma_{i-1}\sigma_i - 2\sigma_{i-1}\sigma_i\sigma_{i+1} + \sigma_i\sigma_{i+1}\sigma_{i+2}\rangle \quad (32)$$
$$+ \beta^2\langle\sigma_{i-2}\sigma_{i-1}\sigma_{i+1} - \sigma_{i-2}\sigma_i\sigma_{i+1} - \sigma_{i-1}\sigma_i\sigma_{i+2}$$
$$+ \sigma_{i-1}\sigma_{i+1}\sigma_{i+2}\rangle$$

where

$$q_i(t) \equiv \sum_{\text{all }\{\sigma^N\}} \sigma_i\, p(\sigma^N, t) \quad (33)$$

As with Eq. (8), the inclusion of triple-correlations in Eq. (32) makes them intractable unless some trick of linearization can be invoked. Fortunately, for small departures from equilibrium the triple-correlation terms can be evaluated from an equilibrium model in which each bond can have a slightly preferred orientation, i.e., $\langle\sigma_k\rangle_{\text{eq}} \neq 0$. For this model, which is treated in detail in the Appendix, we find

$$\langle\sigma_i\sigma_j\sigma_k\rangle = \beta^{k-j}q_i - \beta^{k-i}q_j + \beta^{j-i}q_k \quad (34)$$

where $i < j < k$, and then Eq. (32) reduces to

$$\alpha^{-1}dq_k/dt = -(1+a)(1-\beta)$$
$$\times[2(1+\beta+\beta^2)q_k - (1+\beta)^2(q_{k-1}+q_{k+1}) + \beta(q_{k-2}+q_{k+2})] \quad (35)$$

The N equations of which Eq. (35) is an example can be represented by the matrix equation

$$\alpha^{-1}d\mathbf{q}/dt = -(1+a)\mathbf{Bq} \quad (36)$$

where $\mathbf{q}$ is a column matrix whose ith element is q_i as defined by Eq. (33).

The elements of the $N \times N$ symmetric matrix **B** are given by

$$\begin{aligned} B_{ii} &= 2(1-\beta^3) \\ B_{ij} &= -(1-\beta)(1+\beta)^2; \qquad |i-j| = 1 \\ &= +\beta(1-\beta) \qquad ; \qquad |i-j| = 2 \\ &= 0 \qquad ; \qquad |i-j| > 2 \end{aligned} \tag{37}$$

Comparing **B** with the familiar Rouse **A** matrix of Eq. (11), we see that the bond correlations have produced second-neighbor effects in the chain dynamics.

The matrix product $\mathbf{Q} \cdot \mathbf{B} \cdot \mathbf{Q}$ is a diagonal matrix **M** whose p, p element is μ_p, and the matrix **Q** is the same as that described in Eqs. (14) and (15). According to Rutherford[28]

$$\mu_p = 4(1-\beta)^3 \sin^2 u[1 + 4\beta(1-\beta)^{-2} \sin^2 u]; \; u \equiv p\pi/2(N+1) \tag{38}$$

The column matrix ξ whose N elements are the normal coordinates $\xi_p(t)$, $(p = 1, 2, \ldots, N)$, is defined by an orthogonal transformation of the matrix **q**

$$\xi = \mathbf{Qq} \tag{39}$$

From Eqs. (36) and (39) we have

$$\xi_p(t) = \xi_p(0) \exp(-t/\tau_p) \tag{40}$$

where

$$\tau_p^{-1} = \alpha(1 + a)\mu_p \tag{41}$$

As for the multidimensional freely jointed chain, it is possible to relate α to the parameters which describe a Rouse chain by evaluating the translational diffusion constant D for the center of mass. In the stochastic model, we determine the square of the displacement per unit time of a single bead averaged over an equilibrium ensemble. For bead j,

$$\begin{aligned} \langle (R_j' - R_j)^2 w_j \rangle_{eq} &= \sum_{\{\sigma\}} w_j(\sigma_j, \sigma_{j+1})(R_j' - R_j)^2 \\ &\quad \times p_{eq}^{(4)}(\sigma_{j-1}, \sigma_j, \sigma_{j+1}, \sigma_{j+2}) \\ &= 2\alpha b^2(1 + a)(1 + \beta)(1 - \beta)^2 \end{aligned} \tag{42}$$

where $(R_j' - R_j)$ is the difference between the final and the initial positions of bead j. The reduced probability function $p_{eq}^{(4)}(\sigma_{j-1}, \sigma_j, \sigma_{j+1}, \sigma_{j+2})$ is determined from the Hamiltonian for the system [Eq. (27)];

$$\begin{aligned} p_{eq}^{(4)}(\sigma_{j-1}, \sigma_j, \sigma_{j+1}, \sigma_{j+2}) &= (\tfrac{1}{2})[2 \cosh(E/kT)]^{-3} \\ &\quad \times \exp[(E/kT)(\sigma_{j-1}\sigma_j + \sigma_j\sigma_{j+1} + \sigma_{j+1}\sigma_{j+2})] \end{aligned} \tag{43}$$

When bead j moves by an amount $(R_j' - R_j)$, the square of the displacement of the center of mass is $(R_j' - R_j)^2/(N+1)^2$. The diffusion constant is then

$$D = \sum \langle (R_j' - R_j)^2 w_j \rangle_{\text{eq}} / 2(N+1)^2$$
$$= \alpha b^2(1+a)(1+\beta)(1-\beta)^2/(N+1). \qquad (44)$$

Entering Eq. (41) with Eq. (44) to eliminate $\alpha(1+a)$ and taking $p \ll N$ in Eq. (38), we find for the slow relaxation processes

$$\tau_p^{-1} = p^2\pi^2 D/\langle r^2 \rangle \qquad (45)$$

where

$$\langle r^2 \rangle = Nb^2(1+\beta)/(1-\beta) \qquad (46)$$

From the Appendix, Eq. (68), we have

$$\langle \sigma_i \cdot \sigma_{i+1} \rangle_{\text{eq}} = \beta \qquad (47)$$

and hence, the quantity $\langle r^2 \rangle$ of Eq. (46) is just the equilibrium mean square end-to-end length of a chain of N links with nearest-neighbor correlations. Therefore Eq. (45) demonstrates that the present model also conforms to a Rouse chain, as far as the *slow* relaxations are concerned. For $\beta = 0$, the equations reduce in all ways to their counterparts in Section II for $d = 1$.

The longer relaxation times $(p \ll N)$ when measured relative to τ_1 are independent of β; i.e., they are identical to the relaxation times in the Rouse model:

$$\tau_p = p^{-2}\tau_1; \qquad p \ll N \qquad (48)$$

For higher modes, the ratio τ_p/τ_1 becomes sensitive to the correlations. As β increases, τ_p/τ_1 decreases, as shown by Eq. (38). For illustration, this ratio is plotted semilogarithmically in Figure 2 as a function of p/N for a chain with 10^4 beads and for $\beta = 0$, 0.2, 0.5, and 0.9. It is seen that in this one-dimensional model the relaxation spectrum is broadened as the energetic preference for extended conformations $(\beta > 0)$ is increased. In particular, the longest and shortest relaxation times are related by

$$N^2\tau_N/\tau_1 = N^2\mu_1/\mu_N = \pi^2(1-\beta)^2/4(1+\beta)^2 \qquad (49)$$

As recalled in the Appendix, the rate of tensile relaxation is principally controlled by the slowest modes, while that for dielectric relaxation is most commonly dominated by the fastest modes. Hence, Eq. (49) may not be without interest in certain physical applications.

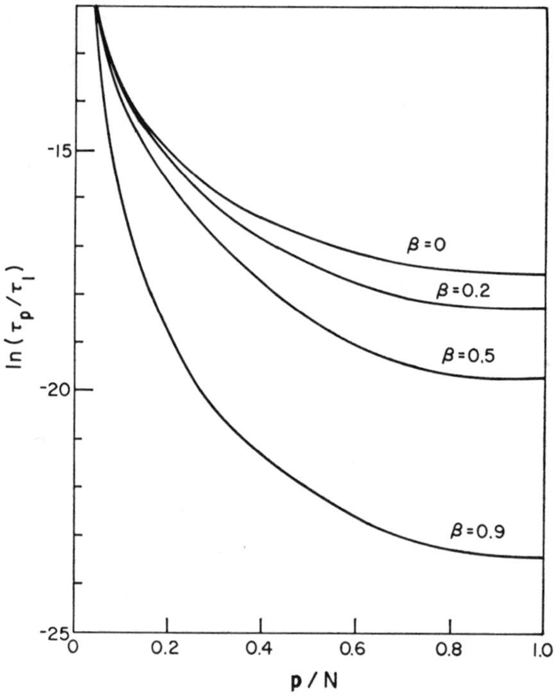

Fig. 2. Relaxation times as a function of mode number in a one-dimensional chain of 10,000 links, for different values of the bond correlation parameter β. Reduced relaxation time τ_p/τ_1 is plotted semilogarithmically against reduced mode number p/N.

The shear modulus $G(t)$ of a relaxing viscoleastic substance is a more sensitive probe of the overall distribution of relaxation times, as it does not depend so completely on either end of the relaxation spectrum. Although the present one-dimensional model cannot comprehend shear, it may be useful to study the analogous relaxation function. The relaxation function $H(\ln \tau)$, is defined[10] by

$$G(t) = \int_0^\infty H(\ln \tau) \exp(-t/\tau)\, d \ln \tau \tag{50}$$

and for Rouse models is well approximated[29] by

$$H(\ln \tau) = -ckT\, dp/d \ln \tau \tag{51}$$

where c is polymer concentration and p is treated as a continuous variable,

$1 \leqq p \leqq N$. For the present model, Eqs. (38), (41), and (51) for large N lead to

$$H(\ln \tau) = ckTN\,(\tan u)(1 + K \sin^2 u)/(1 + 2K \sin^2 u)\pi \tag{52}$$

with

$$u = p\pi/2N$$

$$K = 4\beta(1 - \beta)^{-2}$$

In Figure 3, the slope $d \ln H/d \ln \tau$ as computed from Eq. (52) has been plotted as a function of $\ln(\tau/\tau_1)$ for several values of β, again for a chain of 10^4 bonds. It is seen that the effect of local equilibrium "stiffness" as measured by the value of β, tends to counteract the well-known effect of hydrodynamic interactions among chain elements. As is well known,[10] $d \ln H/d \ln \tau = -\frac{1}{2}$ for a free-draining (Rouse) chain in the long-time region, while a nondraining (Zimm) chain gives $d \ln H/d \ln \tau = -\frac{2}{3}$ in the same region. It is interesting that Frederick, Tschoegl, and Ferry[30]

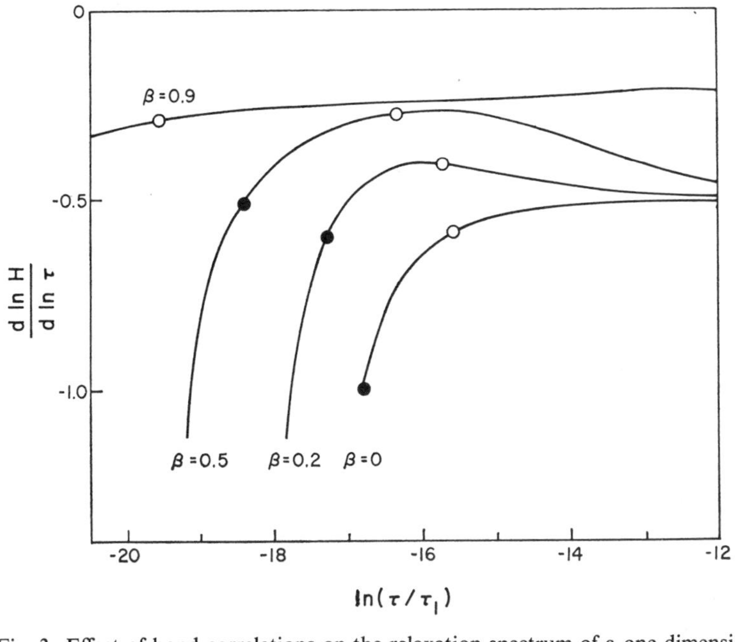

Fig. 3. Effect of bond correlations on the relaxation spectrum of a one-dimensional chain of 10,000 links. The slope $d \log H/d \log \tau$ of the relaxation function is plotted as a function of reduced relaxation time τ/τ_1. Open circles indicate points at which $p/N = 0.25$; filled circles, $p/N = 0.5$.

observed apparent deviations from Zimm theory in high-molecular weight dilute solutions of polystyrene, in a direction consistent with incomplete hydrodynamic shielding. Although the effect is considered[30] to be due principally to the difficulties of extrapolation to the region of nonoverlapping independent chains, it is conceivable that some real departures from the Zimm spectrum are involved, possibly along the lines suggested by the present one-dimensional model.

IV. DISCUSSION

Very recently Monnerie[31] has described Monte Carlo calculations for a quite realistic lattice model: nonintersecting chains confined to tetrahedral lattices performing local stochastic processes involving the simultaneous motion of three or four bonds. Without volume exclusion and with no correlations in the orientation probabilities of neighboring bonds, the model has also been treated analytically,[32] with application to the fluorescence depolarization experiment. It is easy to show that this model also leads to the long-time Rouse spectrum.

By now it may have dawned on the reader that the long-time Rouse spectrum (i.e., proportionality of τ_p to p^{-2}) is to be expected for any chain model in which the correlation lengths for both equilibrium conformations and frictional processes are small compared to the chain dimensions (and thus to the wavelength of the slow normal modes). A possible exception is that of the continuous wormlike chain of invariant contour length, which has been studied by Saito, Takahashi, and Yunoki.[33] In this latter case, the low-frequency spectrum makes τ_p proportional to p^{-4}, which resembles our special one-dimensional model in the limit $1 - \beta \ll 1$.

So far we have not been able to treat chains with bond correlations in more than one dimension. The introduction of more detailed or realistic models of local conformational processes, such as those of Reneker[34] or of Schatzki,[35] has, therefore, not been feasible. We may remark that the theory of dielectric relaxation by Work and Fujita,[36] which applies Glauber's methods[25] to *delayed* (dynamic) correlations between chain dipoles, is also in essence a one-dimensional affair.

It may be surprising that the effect of the nearest-neighbor bond correlations on the one-dimensional chain depends on the *sign* of β; i.e., that the spectrum broadens when extended conformations are favored and narrows when compact conformations are favored. No simple qualitative explanation of this result has occurred to us. The usual " internal viscosity " always produces a narrowing of the spectrum. This effect is easily introduced into a one-dimensional Rouse model; an internal viscous force is appended, proportional to the rate of extension of each subchain, and results[13] in the addition of a constant to each of the Rouse relaxation times. The treatment of internal viscosity in three dimensions is of course far more difficult, but present attempts[9,18] also appear to narrow the spectrum. It seems clear on physical grounds that in three dimensions chains for which local conformational processes (such as the bead "flips" of the present work) are strongly retarded can relax by rotational diffusion, as of a more or less rigid particle. Such a possibility is of course completely beyond the reach of any one-dimensional models; and it is also not encompassed by the Rouse-Zimm formalism. In the Rouse model, for example, there is only a single dissipative mechanism, as symbolized by the subchain frictional coefficient; furthermore, the Gaussian character of the subchains permits the separation of the unperturbed three-dimensional problem into three one-dimensional problems.

In view of the above considerations, it should be evident that our present results do not offer new formulas for practical application, but it is hoped that they will contribute to an understanding of chain dynamics.

Appendix

(*a*) In Section II, correlations $\langle \boldsymbol{\sigma}_i(\boldsymbol{\sigma}_i \cdot \boldsymbol{\sigma}_j)\rangle$ were set equal to $d^{-1}\langle \boldsymbol{\sigma}_j \rangle = d^{-1}\mathbf{q}_j$ in d-dimensional space for freely jointed chains not too far from equilibrium. To demonstrate this equality, we take a three-dimensional example, with unit cartesian vectors $\mathbf{e}_x$, $\mathbf{e}_y$, and $\mathbf{e}_z$. Then

$$\langle \boldsymbol{\sigma}_i(\boldsymbol{\sigma}_i \cdot \boldsymbol{\sigma}_j)\rangle = \langle x_i^2 x_j + x_i y_i y_j + x_i z_i z_j \rangle \mathbf{e}_x + \langle x_i y_i x_j + y_i^2 y_i + y_i z_i z_j \rangle \mathbf{e}_y + \langle x_i z_i x_j + y_i z_i y_j + z_i^2 z_j \rangle \mathbf{e}_z \tag{53}$$

Because the bond directions are not correlated, we have $\langle x_i^2 x_j \rangle = \langle x_i^2 \rangle \times \langle x_j \rangle$, $\langle x_i y_i y_j \rangle = \langle x_i y_i \rangle \langle y_j \rangle$, etc.; and at zero-field equilibrium $\langle x_i^2 \rangle = \frac{1}{3}$ while cross correlations like $\langle x_i y_i \rangle$ vanish. Hence, for small departures from equilibrium we get just

$$\langle \boldsymbol{\sigma}_i(\boldsymbol{\sigma}_i \cdot \boldsymbol{\sigma}_j)\rangle = (\tfrac{1}{3})\langle \boldsymbol{\sigma}_j \rangle \tag{54}$$

which is easily generalized to other dimensionalities.

(*b*) Multiple-spin correlations are required for a one-dimensional, N-spin Ising chain at equilibrium in a small inhomogeneous field. An obvious generalization of the treatment of Marsh[37] gives the desired results. Let the spins have unit magnitude, and let the field at the locus of spin i be $h_i kT$. Then the Hamiltonian of the spin system is

$$H = -E \sum_i \sigma_i \sigma_{i+1} - kT \sum_i h_i \sigma_i \tag{55}$$

and the partition function can be written as

$$Z = \mathrm{tr}\left[V_1 \prod_{i=2}^{N} (UV_i) X\right] \tag{56}$$

where the matrices are defined as

$$V_i = \begin{bmatrix} \exp(h_i) & 0 \\ 0 & \exp(-h_i) \end{bmatrix} \tag{57}$$

$$U = \begin{bmatrix} \exp(E/kT) & \exp(-E/kT) \\ \exp(-E/kT) & \exp(E/kT) \end{bmatrix} \tag{58}$$

and

$$X = \begin{pmatrix} 1 & 1 \\ 1 & 1 \end{pmatrix} \tag{59}$$

Equilibrium averages for products of selected sets of spin variables $\sigma_j, \sigma_k, \ldots, \sigma_s$ are given by

$$\langle \sigma_j \sigma_k \cdots \sigma_s \rangle = Z^{-1}\, \mathrm{tr}\left[V_1 \left(\prod_{i=2}^{j} UV_i\right) F \left(\prod_{i=j+1}^{k} UV_i\right) F \cdots \right.$$
$$\left. \times \cdots \left(\prod_{i=s+1}^{N} UV_i\right) X \right] \tag{60}$$

where

$$F = \begin{pmatrix} 1 & 0 \\ 0 & -1 \end{pmatrix} \tag{61}$$

The matrix U is diagonalized under the transformation

$$S^{-1}US = L; \qquad S S^{-1} = I \tag{62}$$

where

$$S = S^{-1} = 2^{-1/2} \begin{pmatrix} 1 & 1 \\ 1 & -1 \end{pmatrix} \tag{63}$$

and

$$L = \begin{bmatrix} 2\cosh(E/kT) & 0 \\ 0 & 2\sinh(E/kT) \end{bmatrix} \tag{64}$$

If we take the field to be zero except at selected spins numbered $j, k, \ldots, r, s$, the matrix V_i reduces to the idemfactor I for all other spins, and the partition function collapses to

$$Z = \mathrm{tr}\,[U^{j-1} V_j U^{k-j} V_k \cdots U^{s-r} V_s U^{N-s} X] \tag{65}$$

which after using Eq. (62) becomes

$$Z = \mathrm{tr}\,[SL^{j-1}S^{-1} V_j SL^{k-j}S^{-1} V_k \cdots SL^{s-r}S^{-1} V_s SL^{N-s}S^{-1} X] \tag{66}$$

Similar operations are performed on Eq. (60). Then for small fields, $|h_i| \ll 1$, with $l \leqq m \leqq n \leqq p$ the averages and correlations are found to be

$$\langle \sigma_l \rangle = \sum_{\alpha=j}^{s} \beta^{|l-\alpha|} h_\alpha \,;\; \beta \equiv \tanh (E/kT) \tag{67}$$

$$\langle \sigma_l \sigma_m \rangle = \beta^{m-l} \tag{68}$$

$$\langle \sigma_l \sigma_m \sigma_n \rangle = \sum_{\alpha=j}^{m-1} \beta^{n-m+|l-\alpha|} h_\alpha + \sum_{\alpha=m}^{s} \beta^{m-l+|n-\alpha|} h_\alpha \tag{69}$$

$$\langle \sigma_l \sigma_m \sigma_n \sigma_p \rangle = \beta^{m-l+p-n}, \text{ etc.} \tag{70}$$

with neglect of terms of $O(h_i^2)$. Using Eq. (67) in Eq. (69), the triple correlations can be expressed as

$$\langle \sigma_l \sigma_m \sigma_n \rangle = \beta^{n-m} \langle \sigma_l \rangle - \beta^{n-l} \langle \sigma_m \rangle + \beta^{m-l} \langle \sigma_n \rangle \tag{71}$$

Since a small arbitrary inhomogeneous field must affect the correlations like small fluctuations from zero-field equilibrium, we may use Eq. (71) in Section III to linearize Eq. (32).

(*c*) Tensile relaxation is easily calculated for the one-dimensional model of Section III. The average value of the distance separating the two ends of the chain, $\langle L(t) \rangle$, is the sum of the average values of the N-bond vectors.

$$\langle L(t) \rangle = b \sum_{i=1}^{N} q_i(t) \tag{72}$$

Since

$$q = Q\xi$$

therefore

$$\langle L(t) \rangle = \left(\frac{2}{N+1} \right)^{1/2} b \sum_{i=1}^{N} \sum_{p=1}^{N} \sin \left(\frac{ip\pi}{N+1} \right) \xi_p(0) \exp(-t/\tau_p) \tag{73}$$

Summing[38] over i, we find

$$\langle L(t) \rangle = \left(\frac{2}{N+1} \right)^{1/2} b \sum_{(p\ \text{odd})}^{N} \cot \left[\frac{p\pi}{2(N+1)} \right] \xi_p(0) \exp(-t/\tau_p) \tag{74}$$

where the remaining summation is over only odd values of the index p. If, at $t = 0$, the mean value of the end-to-end distance is

$$\langle L(0) \rangle = b \sum_i q_i(0) = Nbq(0) \tag{75}$$

then

$$\xi_p(0) = \left(\frac{2}{N+1}\right)^{1/2} \frac{\langle L(0)\rangle}{Nb} \sum_{i=1}^{N} \sin\left(\frac{ip\pi}{N+1}\right) \tag{76}$$

$$\langle L(t)\rangle = \langle L(0)\rangle \frac{2}{N(N+1)} \sum_{(p\,\text{odd})}^{N} \cot^2\left[\frac{p\pi}{2(N+1)}\right] \exp(-t/\tau_p) \tag{77}$$

For large N, the well-known[4] result is

$$\langle L(t)\rangle/\langle L(0)\rangle = 8\pi^{-2} \sum_{(p\,\text{odd})} p^{-2} \exp(-t/\tau_p) \tag{78}$$

The series converges rapidly, with the first mode making the major contribution to $\langle L(t)\rangle$. Tensile relaxation is active in only the odd modes since the motion of only these modes affects the distance between the two ends.

(*d*) Another interesting example is the dielectric relaxation of a chain on which adjacent bonds carry dipoles of magnitude m_0 and of alternating sign. The ensemble-average dipole moment in the absence of an electrical field is

$$\langle M(t)\rangle = m_0 \sum_{i=1}^{N} (-1)^i q_i(t) \tag{79}$$

Replacing $q_i(t)$ by the appropriate linear combination of normal coordinates and summing[38] over the index i, we obtain

$$\langle M(t)\rangle = -\left(\frac{2}{N+1}\right)^{1/2} m_0 \sum_{(p+N\,\text{even})}^{N} \frac{\sin \dfrac{p\pi}{N+1}}{1+\cos \dfrac{p\pi}{N+1}} \xi_p(0) \exp(-t/\tau_p) \tag{80}$$

The summation is over all odd values of p from 1 to N if N is odd, or over even values of p if N is even. If an ensemble of chains is initially at equilibrium in the presence of an electric field which is shut off at $t = 0$, then

$$q_i(0) = (-1)^i \langle M(0)\rangle / N m_0 \tag{81}$$

and

$$\xi_p(0) = -\left(\frac{2}{N+1}\right)^{1/2} \frac{\langle M(0)\rangle}{N m_0} \frac{\sin \dfrac{p\pi}{N+1}}{1+\cos \dfrac{p\pi}{N+1}} \tag{82}$$

Substitution for $\xi_p(0)$ in Eq. (80) leads to

$$\langle M(t)\rangle = \frac{2\langle M(0)\rangle}{N(N+1)} \sum_{(N+p\ \text{even})} \left[\frac{\sin\dfrac{p\pi}{N+1}}{1+\cos\dfrac{p\pi}{N+1}}\right]^2 \exp(-t/\tau_p) \qquad (83)$$

which is well approximated by

$$\langle M(t)\rangle/\langle M(0)\rangle = 8\pi^{-2} \sum_{(q\ \text{odd})} q^{-2} \exp(-t/\tau_q)$$

$$q \equiv N + 1 - p \qquad (84)$$

As has been pointed out before,[13, 14, 39] the important contributions to the dielectric relaxation are therefore from the fastest modes.

Acknowledgments

We thank the National Science Foundation for support of this work, and we thank William Gobush and Dewey K. Carpenter for valuable discussions.

References

1. P. E. Rouse, Jr., *J. Chem. Phys.*, **21**, 1272 (1953).
2. V. A. Kargin and G. L. Slonimskii, *Dokl. Akad. Nauk SSSR*, **62**, 239 (1948).
3. F. Bueche, *J. Chem. Phys.*, **22**, 603 (1954).
4. B. H. Zimm, *J. Chem. Phys.*, **24**, 269 (1956).
5. C. W. Pyun and M. Fixman, *J. Chem. Phys.*, **42**, 3838 (1965).
6. N. W. Tschoegl, *J. Chem. Phys.*, **39**, 149 (1963).
7. M. Fixman, *J. Chem. Phys.*, **45**, 785 (1966).
8. W. Kuhn and H. Kuhn, *Helv. Chim. Acta.*, **28**, 1533 (1945).
9. R. Cerf, (a) *J. Polymer Sci.*, **23**, 125 (1957); (b) *J. Phys. Radium*, **19**, 122 (1958).
10. J. D. Ferry, *Viscoelastic Properties of Polymers*, Wiley, New York, 1961.
11. A. Miyake, *Progr. Theoret. Phys. (Kyoto) Suppl.*, **10**, 56 (1959).
12. N. G. McCrum, B. E. Read, and G. Williams, *Anelastic and Dielectric Effects in Polymeric Solids*, Wiley, New York, 1967.
13. L. K. H. Van Beek and J. J. Hermans, *J. Polymer Sci.*, **23**, 211 (1957).
14. W. H. Stockmayer and M. E. Baur, *J. Am. Chem. Soc.*, **86**, 3485 (1964).
15. D. S. Thompson and S. J. Gill, *J. Chem. Phys.*, **47**, 5008 (1967).
16. R. Ullman, *J. Chem. Phys.*, **43**, 3161 (1965); *ibid.*, **44**, 1558 (1966).
17. J. D. Ferry, L. A. Holmes, J. Lamb, and A. J. Matheson, *J. Phys. Chem.*, **70**, 1685 (1966).
18. A. Peterlin and C. Reinhold, *Trans. Soc. Rheol.*, **11**, 15 (1967).
19. A. V. Tobolsky and D. B. DuPré, *J. Polymer Sci.*, Part A-2, **6**, 1177 (1968).
20. (a) J. G. Kirkwood, *Rec. Trav. Chim.*, **68**, 649 (1949); (b) J. Riseman and J. G. Kirkwood in *Rheology*, Vol. I, F. Eirich, Ed., Academic, New York, 1956, pp. 495–523.

21. (a) N. Saito, *Bull. Kobayashi Inst. Phys. Res.*, **8**, 89 (1958); (b) K. Okano, *Busseiron Kenkyu* (II), **4**, 688 (1958).
22. W. H. Stockmayer, *Pure Appl. Chem.*, **15**, 539 (1967).
23. P. H. Verdier and W. H. Stockmayer, *J. Chem. Phys.*, **36**, 227 (1962).
24. P. H. Verdier, *J. Chem. Phys.*, **45**, 2118, 2122 (1966).
25. R. J. Glauber, *J. Math. Phys.*, **4**, 294 (1963).
26. D. E. Rutherford, *Proc. Roy. Soc.* (Edinburgh) *A* **62**, 229 (1947).
27. Recall that the $\boldsymbol{\sigma}_i$ are *bond* vectors and not the loci of beads with respect to a common origin. Thus we must choose sines rather than cosines as eigenfunctions for chains with free ends. Also, since Eq. (10) is concerned only with the averages of the bond vectors, there are only N normal modes, the translational diffusion mode being automatically excluded.
28. D. E. Rutherford, *Proc. Roy. Soc.* (Edinburgh) *A* **63**, 232 (1952).
29. J. D. Ferry, in *Die Physik der Hochpolymeren*, Vol. *4*, H. A. Stuart, Ed., Springer, Berlin, 1956, p. 96.
30. J. E. Frederick, N. W. Tschoegl, and J. D. Ferry, *J. Phys. Chem.*, **68**, 1974 (1964).
31. L. Monnerie, *IUPAC Symposium on Macromolecular Chemistry, Toronto, 1968.*
32. E. Dubois-Violette, P. G. deGennes and L. Monnerie, to appear in *J. chim. phys.*
33. N. Saito, K. Takahashi and Y. Yunoki, *J. Phys. Soc., Japan*, **22**, 219 (1967).
34. D. H. Reneker, *Am. Chem. Soc.*, Div. Polymer Chem., Preprints **3** (2), 60 (1962).
35. T. F. Schatzki, *Am. Chem. Soc.*, Div. Polymer Chem., Preprints **6** (2), 646 (1965).
36. R. N. Work and S. Fujita, *J. Chem. Phys.*, **45**, 3779 (1966).
37. J. S. Marsh, *Phys. Rev.*, **145**, 251 (1966).
38. L. B. W. Jolley, *Summation of Series*, 2nd ed., Dover Press, New York, 1961, Series Nos. 417 and 455.
39. F. Bueche, *J. Polymer Sci.*, **54**, 597 (1961).

LOCAL-JUMP MODELS FOR CHAIN DYNAMICS*†

W. H. STOCKMAYER, W. GOBUSH and R. NORVICH

Department of Chemistry, Dartmouth College, Hanover, New Hampshire 03755, USA

ABSTRACT

A previously developed simple stochastic model for the rate of conformational change in freely-jointed chains is extended to a wider class of local processes, and to chains containing atoms or links of several kinds. The long-time relaxation spectrum in every case is just the same as that given by the more familiar beads-and-springs model, but the short-time behaviour depends in more detail on chain structure.

INTRODUCTION

The main purpose of this article is to illustrate and extend a treatment of chain dynamics which offers conceptual alternatives to and perhaps physical advantages over the familiar bead-and-spring models for chain diffusion which are recalled under the names of Rouse[1], Bueche[2], Kargin and Slonimsky[3], Zimm[4] and others. It is shown that the slow time-dependent behaviour of a flexible chain molecule is phenomenologically invariant to the fine details of its molecular structure; but that for short times or high frequencies the individual structural features can and must lead to differences in relaxation behaviour.

In previous papers[5,6] a simple stochastic model for chain diffusion was described. The most elementary version deals with a freely-jointed chain, there being no correlations in the directions of neighbouring links, and the local jump process was of a specially simple and restricted kind. Models were also treated which provide correlations between nearest-neighbour links, and it was further shown that a certain kind of kinetic bias could also be introduced without altering the nature of the results. In this paper, while outlining the general nature of the model and recalling some of the earlier results[5,6], the treatment is extended to a broader class of local processes and to chains whose elements need not all show equiprobable tendencies to relaxation.

It should be mentioned that somewhat related studies have been published by Monnerie and Geny[7], by Iwata and Kurata[8], by Verdier[9] and by Anderson[10].

---

* The subject matter of this article formed part of a lecture delivered by W. H. Stockmayer at the IUPAC Symposium on Macromolecules, Leiden 1970, under the somewhat misleading title of 'Statistical Mechanics of Chain Molecules'.

† Work supported by the National Science Foundation, USA.

BASIC MODEL

The close connection between the problem of random flights and the process of diffusion has long been known[11], and it is thus a natural temptation to contemplate models in which the translational, rotational and deformational motions of chain molecules result from repeated local segmental rearrangements distributed randomly along the chain backbone[12]. It appears that G. W. King [13] first suggested the possibility of studying such articulated chain motion by means of computer simulation, but it was Verdier[14] who first carried out such a programme to some incisive degree. Since one of Verdier's principal purposes was and has remained[15] the investigation of the dynamic effects of excluded volume interactions (non-intersecting restrictions for lattice chains), no simple analytical approach was feasible. The elementary results to be described here owe their simplicity to the abandonment of any attempts to treat chains with excluded volume. For further ease, we also ignore any hydrodynamic interactions between chain segments.

Let a simple chain molecule consist of $N + 1$ beads, joined by N bonds each of the same length l. Number the beads from 0 to N and the bonds from 1 to N. The direction of the bond from bead $(i - 1)$ to bead i is described by the unit vector $\boldsymbol{\sigma}_i$. If the spatial position of the zeroth bead is $\boldsymbol{r}_0$, then the location and conformation of the molecule is specified by the set of $N + 1$ vectors $\boldsymbol{r}_0, \boldsymbol{\sigma}_1, \boldsymbol{\sigma}_2, \ldots, \boldsymbol{\sigma}_N$.

To vary the chain conformation, beads are allowed to move one at a time. For interior beads ($i \neq 0$ or N) the motion of bead i consists of a jump or 'flip' whereby the vectors $\boldsymbol{\sigma}_1$ and $\boldsymbol{\sigma}_{i+1}$ are changed to new values $\boldsymbol{\sigma}'_i$ and $\boldsymbol{\sigma}'_{i+1}$. Terminal beads would require a different specification, but here we are content to deal only with long chains and thus do not trouble with end effects. We shall thus ignore translational motions. A complete treatment can be found elsewhere[5,9].

Let the probability density in the $\boldsymbol{\sigma}$-space that the chain at time t has the conformation $\{\boldsymbol{\sigma}_1, \boldsymbol{\sigma}_2, \ldots, \boldsymbol{\sigma}_N\} \equiv \{\boldsymbol{\sigma}^N\}$ be designated by $p(\boldsymbol{\sigma}^N, t)$. We must now formulate a kinetic or 'master' equation for the rate of change of this probability density. The nature of the treatment, though differing in trivial details, is inspired by and similar to that of Glauber[16] for spin relaxation on a linear Ising lattice. In the present example we shall ignore correlations between neighbouring bond vectors, so that the basic jump process for bead i depends only on the state of its two bonds to the adjacent beads. Let the conditional probability per unit time that a pair of adjacent bond vectors $\boldsymbol{\sigma}'_i$, and $\boldsymbol{\sigma}'_{i+1}$ rotate through an angle ϕ to the new conformation $\boldsymbol{\sigma}_i$, $\boldsymbol{\sigma}_{i+1}$ | be denoted by $w_i(\boldsymbol{\sigma}'_i, \boldsymbol{\sigma}'_{i+1} \mid \boldsymbol{\sigma}'_i, \boldsymbol{\sigma}_{i+1}$. The time evolution of the conformational probability density then follows the master equation

$$\partial p(\boldsymbol{\sigma}^N, t)/\partial t = - \sum_i \iint p(\boldsymbol{\sigma}^N, t) w_i(\boldsymbol{\sigma}_i, \boldsymbol{\sigma}_{i+1} \mid \boldsymbol{\sigma}'_i, \boldsymbol{\sigma}'_{i+1}) \mathrm{d}\boldsymbol{\sigma}'_i \, \mathrm{d}\boldsymbol{\sigma}'_{i+1}$$
$$+ \sum_i \iint p(\boldsymbol{\sigma}_1, \ldots, \boldsymbol{\sigma}'_i, \boldsymbol{\sigma}'_{i+1}, \ldots, \boldsymbol{\sigma}_N, t) w_i(\boldsymbol{\sigma}'_i, \boldsymbol{\sigma}'_{i+1} \mid \boldsymbol{\sigma}_i, \boldsymbol{\sigma}_{i+1}) \, \mathrm{d}\boldsymbol{\sigma}'_i \, \mathrm{d}\boldsymbol{\sigma}'_{i+1} \quad (1)$$

This is simply an expression of the fact that the rate of change of configuration of the bond vectors in time is the difference between the rates of creation and annihilation.

At equilibrium we have the condition

$$p(\boldsymbol{\sigma}_1, \ldots, \boldsymbol{\sigma}_i', \boldsymbol{\sigma}_{i+1}', \ldots, \boldsymbol{\sigma}_N, \infty) w_i(\boldsymbol{\sigma}_i', \boldsymbol{\sigma}_{i+1}' \mid \boldsymbol{\sigma}_i, \boldsymbol{\sigma}_{i+1}) = p(\boldsymbol{\sigma}^N, \infty) w_i(\boldsymbol{\sigma}_i, \boldsymbol{\sigma}_{i+1} \mid \boldsymbol{\sigma}_i', \boldsymbol{\sigma}_{i+1}') \quad (2)$$

Also, at equilibrium the probability density $p(\boldsymbol{\sigma}^N, \infty)$ for a freely-jointed chain must be a constant, having the same value for all possible sets of the bond vectors. Thus equation 2 shows that

$$w_i(\boldsymbol{\sigma}_i', \boldsymbol{\sigma}_{i+1}' \mid \boldsymbol{\sigma}_i, \boldsymbol{\sigma}_{i+1}) = w_i(\boldsymbol{\sigma}_i, \boldsymbol{\sigma}_{i+1} \mid \boldsymbol{\sigma}_i', \boldsymbol{\sigma}_{i+1}') \quad (3)$$

Now let us describe the local jump process more precisely as a rotation of bead i about an axis passing through beads $i - 1$ and $i + 1$. The old and new bond vectors are then connected by the relations

$$\boldsymbol{\sigma}_i' = \boldsymbol{\sigma}_i \cos^2 \tfrac{1}{2}\phi + \boldsymbol{\sigma}_{i+1} \sin^2 \tfrac{1}{2}\phi + (\boldsymbol{\sigma}_i \times \boldsymbol{\sigma}_{i+1}) \sin\phi / \sqrt{2(1 + \boldsymbol{\sigma}_i \cdot \boldsymbol{\sigma}_{i+1})} = \mathrm{f}(\boldsymbol{\sigma}_i, \boldsymbol{\sigma}_{i+1}, \phi) \quad (4)$$

$$\boldsymbol{\sigma}_{i+1}' = \boldsymbol{\sigma}_i \sin^2 \tfrac{1}{2}\phi + \boldsymbol{\sigma}_{i+1} \cos^2 \tfrac{1}{2}\phi - (\boldsymbol{\sigma}_i \times \boldsymbol{\sigma}_{i+1}) \sin\phi / \sqrt{2(1 + \boldsymbol{\sigma}_i \cdot \boldsymbol{\sigma}_{i+1})} = \mathrm{f}(\boldsymbol{\sigma}_{i+1}, \boldsymbol{\sigma}_i, \phi) \quad (5)$$

If the flip rate for rotations through angle ϕ to within $\mathrm{d}\phi$ is expressed as $\alpha\,\mathrm{g}(\phi)\mathrm{d}\phi$, we then have

$$w_i(\boldsymbol{\sigma}_i, \boldsymbol{\sigma}_{i+1} \mid \boldsymbol{\sigma}_i', \boldsymbol{\sigma}_{i+1}') = \alpha \int_{-\pi}^{\pi} \delta(\boldsymbol{\sigma}_i' - \mathrm{f}(\boldsymbol{\sigma}_i, \boldsymbol{\sigma}_{i+1}, \phi))\delta(\boldsymbol{\sigma}_{i+1}' - \mathrm{f}(\boldsymbol{\sigma}_{i+1}, \boldsymbol{\sigma}_i, \phi))\mathrm{g}(\phi)\,\mathrm{d}\phi \quad (6)$$

It is seen that equation 3 is consistent with equation 6 provided that $\mathrm{g}(\phi)$ is an even function, as of course is also necessary on physical grounds.

To extract simple results from the master equation it is convenient, as earlier[5,6], to work with the average values of bond vectors. Let

$$\boldsymbol{q}_j(t) \equiv \langle \boldsymbol{\sigma}_j(t) \rangle = \int \ldots \int \boldsymbol{\sigma}_j p(\boldsymbol{\sigma}^N, t)\,\mathrm{d}\boldsymbol{\sigma}_1 \ldots \mathrm{d}\boldsymbol{\sigma}_N \quad (7)$$

To evaluate the time-dependence of this quantity, multiply the master equation 1 by $\boldsymbol{\sigma}_j(t)$ and integrate over all configuration space. Since $\boldsymbol{\sigma}_j$ may only be reoriented by rotation along with either $\boldsymbol{\sigma}_{i-1}$ or $\boldsymbol{\sigma}_{i+1}$, we obtain the following general expression:

$$\begin{aligned}\mathrm{d}\boldsymbol{q}_j/\mathrm{d}t = & -\boldsymbol{q}_j \iint w_{j-1}(\boldsymbol{\sigma}_{j-1}, \boldsymbol{\sigma}_j \mid \boldsymbol{\sigma}_{j-1}', \boldsymbol{\sigma}_j')\,\mathrm{d}\boldsymbol{\sigma}_{j-1}'\,\mathrm{d}\boldsymbol{\sigma}_j'\alpha \\ & - \boldsymbol{q}_j \iint w_j(\boldsymbol{\sigma}_j, \boldsymbol{\sigma}_{j+1} \mid \boldsymbol{\sigma}_j', \boldsymbol{\sigma}_{j+1}')\mathrm{d}\boldsymbol{\sigma}_j'\,\mathrm{d}\boldsymbol{\sigma}_{j+1}' \\ & + \int \ldots \int \mathrm{d}\boldsymbol{\sigma}_1 \ldots \mathrm{d}\boldsymbol{\sigma}_N \iint \boldsymbol{\sigma}_j p(\boldsymbol{\sigma}_1, \ldots, \boldsymbol{\sigma}_{j-1}', \boldsymbol{\sigma}_j', \ldots, \boldsymbol{\sigma}_N, t) \\ & \times w_{j-1}(\boldsymbol{\sigma}_{j-1}', \boldsymbol{\sigma}_j' \mid \boldsymbol{\sigma}_{j-1}, \boldsymbol{\sigma}_j)\,\mathrm{d}\boldsymbol{\sigma}_{j-1}'\,\mathrm{d}\boldsymbol{\sigma}_j' + \int \ldots \int \mathrm{d}\boldsymbol{\sigma}_1 \ldots \mathrm{d}\boldsymbol{\sigma}_N \iint \boldsymbol{\sigma}_j \\ & \times p(\boldsymbol{\sigma}_1, \ldots, \boldsymbol{\sigma}_j', \boldsymbol{\sigma}_{j+1}', \ldots, \boldsymbol{\sigma}_N, t)\, w_j(\boldsymbol{\sigma}_j', \boldsymbol{\sigma}_{j+1}' \mid \boldsymbol{\sigma}_j, \boldsymbol{\sigma}_{j+1})\,\mathrm{d}\boldsymbol{\sigma}_j'\,\mathrm{d}\boldsymbol{\sigma}_{j+1}' \end{aligned} \quad (8)$$

Now using equation 6 for the jump probabilities, and making use of the fact that $\mathrm{g}(\phi)$ is an even function, we find

$$\mathrm{d}\boldsymbol{q}_j/\mathrm{d}t = -\alpha'(2\boldsymbol{q}_j - \boldsymbol{q}_{j-1} - \boldsymbol{q}_{j+1}) \quad (9)$$

where

$$\alpha' = \alpha \int_{-\pi}^{\pi} \mathrm{g}(\phi) \sin^2(\tfrac{1}{2}\phi)\,\mathrm{d}\phi \quad (10)$$

or in matrix form

$$\mathrm{d}\boldsymbol{q}/\mathrm{d}t = -\alpha' A\boldsymbol{q} \tag{11}$$

in which $\boldsymbol{q}$ is a column matrix with the $\boldsymbol{q}_j$ as elements and A is the familiar[1] square matrix with elements 2 on the diagonal, -1 just off diagonal, and zero otherwise. This matrix is diagonalized by the transformation $Q^{-1}AQ = \Lambda$, where

$$Q_{jp} = (2/N)^{1/2} \sin (jp\pi/N) \tag{12}$$

and the eigenvalues of A are

$$\lambda_p = 4 \sin^2 (p\pi/2N) \tag{13}$$

The normal coordinates $\xi_p(t)$, as defined by the orthogonal transformation $\xi = Q\boldsymbol{q}$, thus relax exponentially with relaxation times

$$\tau_p = 1/\alpha'\lambda_p \tag{14}$$

in exact mimicry of the bead-and-spring model results[1]. In our earlier work, the flip process had been restricted completely to 180° rotations, i.e. to $g(\phi) = \delta(\phi \pm \pi)$, which gives $\alpha' = \alpha$ in equations 10; but now we see that any mixture of rotation processes will produce the same behaviour. At another extreme, for example, we could pass to the diffusion limit by making $g(\phi)$ sharply peaked around the origin, which would lead to $\alpha' = \alpha\langle\phi^2\rangle/4$.

Although we have avoided direct consideration of the displacement of the chain as a whole, it is easy to relate the model parameters to the translational diffusion coefficient D_t of the chain. The mean square displacement per unit time of the centre of mass as a result of repeated flip processes, taken over an equilibrium ensemble, is

$$\sum_{i=0}^{N} \langle w_i(\Delta \boldsymbol{r}_i)^2\rangle_{\mathrm{eq}}/(N+1)^2 = 6D_t \tag{15}$$

in which the displacement of bead i at a flip is

$$\Delta\boldsymbol{r}_i = l(\boldsymbol{\sigma}_i' - \boldsymbol{\sigma}_i) \tag{16}$$

Performing the calculation, and neglecting the trivial difference between N and $N + 1$, we get

$$D_t = \alpha' l^2/3N \tag{17}$$

Since the equilibrium mean-square end-to-end displacement of the freely-jointed chain is just

$$\langle r^2\rangle = Nl^2 \tag{18}$$

the slower relaxation times, i.e. those for which $p \ll N$, may be expressed in the form

$$\tau_p = \langle r^2\rangle/3D_t\pi^2p^2 \tag{19}$$

which is exactly the same as the Rouse result[1], in terms of the observable quantities $\langle r^2\rangle$ and D_t. The force constants of Hookean springs and the friction constants of the beads do not appear, any more than do the flip rates of individual bond lengths of our freely-jointed chain model. The two models are both unreal in many ways; but for large scale motions, such that the precise details of chain structure are much smaller than the wavelengths of the low low-p normal coordinates, these unrealities are invisible. As an analogous case we may cite the Debye theory of crystal heat-capacity, which works well for low temperatures where the only important vibrational modes of the crystal are those with wavelengths many times the atomic separations.

A much more elegant formulation has been offered by Iwata[8], who considers more general local conformational rate processes in more realistic chains. After a 'coarse-graining' operation on his basic master equation, he obtains, for sufficiently slow motions, precisely the diffusion equation of the bead-and-spring model. Thus the physically-based assertions of the previous paragraph can apparently be substantiated quite generally.

HETEROGENEOUS CHAINS

When we abandon the long-time region of the relaxation spectrum and proceed to higher frequencies, any of the models previously described begins to display features that depend on the structural details assumed. In this region, local-jump models would appear on physical grounds to be more attractive than beads and springs. It was shown earlier[5,6] that at high frequencies the existence of correlations between neighbouring links, i.e. a departure from the strict freedom of the freely-jointed chain, causes systematic deviations from the bead-and-spring relaxation spectrum. We now briefly discuss local-jump models for heterogeneous chains, again finding agreement with Rouse behaviour at long times but systematic differences at short times.

As an elementary example, we consider a freely-jointed chain of two regularly alternating kinds of atoms, with the structure ... ABABAB .... All N bonds have the same length l, but the two kinds of atoms can have different jump rates, α and β respectively. Ignoring end effects, we can proceed as before and find for even-numbered bonds

$$\mathrm{d}\boldsymbol{q}_{2j}/\mathrm{d}t = -(\alpha+\beta)\,\boldsymbol{q}_{2j} + \beta\boldsymbol{q}_{2j-1} + \alpha\boldsymbol{q}_{2j+1} \tag{20}$$

and for odd-numbered bonds

$$\mathrm{d}\boldsymbol{q}_{2j+1}/\mathrm{d}t = -(\alpha+\beta)\,\boldsymbol{q}_{2j+1} + \alpha\boldsymbol{q}_{2j} + \beta\boldsymbol{q}_{2j+2} \tag{21}$$

To find the relaxation times, we can imitate closely the procedure for finding the vibrational frequencies of a linear diatomic lattice[17,18]. The problem differs only in the appearance of first rather than second time-derivatives. Assume solutions of the form:

$$\boldsymbol{q}_{2j} = A_2\,\mathrm{e}^{-\lambda(\alpha+\beta)t}\,\mathrm{e}^{-2jik} \tag{22}$$

$$\boldsymbol{q}_{2j+1} = A_1\,\mathrm{e}^{-\lambda(\alpha+\beta)t}\,\mathrm{e}^{-(2j+1)ik} \tag{23}$$

where $k = \pi p/N$ with integral p, and $i^2 = -1$. Substitution into the relaxation equations 20 and 21 leads to two linear equations in the amplitudes

A_1 and A_2. The condition for a non-trivial solution is that the determinant of the coefficients of A_1 and A_2 shall vanish, and this leads to the relation

$$(1 - \lambda)^2 = 1 - 4\theta(1 - \theta)\sin^2 k \tag{24}$$

where $\theta = \alpha/(\alpha + \beta)$. This equation possesses two solutions for λ, and thus the $\lambda(k)$ curve will have two branches, in exact analogy to the acoustical and optical branches of the lattice-vibration problem. It is easy to verify that for the special case $\alpha = \beta$ the two branches coalesce and the earlier results of equations 13 and 14 are recovered. The greater the disparity between α and β, the greater the gap in time scale between the slow and fast branches of the relaxation spectrum, as could be seen by generating numerical examples.

The longest relaxation times, obtained when $k \ll 1$, are specifically given by

$$\tau_p = \{(\alpha + \beta)/2\alpha\beta\}\,\{N^2/\pi^2 p^2\} \tag{25}$$

and we see that if one kind of atom is much more sluggish than the other (e.g. $\beta \ll \alpha$), the slowest chain motions are limited by the slowest backbone motions. This result is more comforting than surprising.

The method used earlier for the homogeneous chain to evaluate the translational diffusion coefficient D_t does not lend itself to the present case, and we shall not pursue this question further in this paper.

The general method of Brillouin[18] may be followed in treating chains of any backbone complexity, provided the structure be periodic. A Rouse-type spectrum always obtains for long times, and is clearly a consequence just of the linear connectivity of the chain.

CONCLUSION

The results displayed in the present paper are little more than didactic exercises. We believe, however, that models of this type can lead to more realistic results than beads and springs in the treatment of short-time relaxation processes, and we hope to confirm this belief in later work.

REFERENCES

1. P. E. Rouse, *J. Chem. Phys.* **21**, 1272 (1953).
2. F. Bueche, *J. Chem. Phys.* **22**, 603 (1954).
3. V. A. Kargin and G. L. Slonimsky, *Dokl. Akad. Nauk SSSR*, **62**, 239 (1948).
4. B. H. Zimm, *J. Chem. Phys.* **24**, 269 (1956).
5. R. A. Orwoll and W. H. Stockmayer, *Advanc. Chem. Phys.* **15**, 204 (1969).
6. W. H. Stockmayer, W. Gobush, Y. Chikahisa and D. K. Carpenter, *Disc. Faraday Soc.* **49**, 182 (1970).
7. L. Monnerie and F. Geny, *J. Polym. Sci. C*, **30**, 93 (1970).
8. K. Iwata and M. Kurata, *J. Chem. Phys.* **50**, 4008 (1969): K. Iwata, *J. Chem. Phys.* **54**, 12 (1971).
9. P. H. Verdier, *J. Chem. Phys.* **52**, 5512 (1970).
10. J. E. Anderson, *J. Chem. Phys.* **52**, 2821 (1970).
11. See, for example, S. Chandrasekhar, *Rev. Mod. Phys.* **15**, 1 (1943).
12. It is tempting to call this a 'centipede' model, but that would be inaccurate because in the stochastic model the segments move in no unique or specified order.

[13] G. W. King, private communications (*ca.* 1953).
[14] P. H. Verdier and W. H. Stockmayer, *J. Chem. Phys.* **36**, 227 (1962).
[15] P. H. Verdier, *J. Chem. Phys.* **45**, 2118 and 2122 (1966).
[16] R. J. Glauber, *J. Math. Phys.* **4**, 294 (1963).
[17] J. C. Slater, *Introduction to Chemical Physics*, Chapter XV. McGraw-Hill: New York (1939).
[18] L. Brillouin, *Wave Propagation in Periodic Structures*, Chapter IV. Dover: New York (1953).